Reproduction in Domestic Ruminants VI

Society of Reproduction and Fertility Volume 64

Proceedings of the Seventh International Symposium
on Reproduction in Domestic Ruminants

Wellington, New Zealand
August 2006

Edited by:
JL Juengel, JF Murray and MF Smith

NOTTINGHAM
University Press

Nottingham University Press
Manor Farm, Church Lane, Thrumpton
Nottingham NG11 0AX, United Kingdom
www.nup.com

NOTTINGHAM

First published 2007
© Society for Reproduction and Fertility

British Library Cataloguing in Publication Data
Reproduction in Domestic Ruminants VI
I. Juengel, J.L. II. Murray, J.F. III. Smith, M.F.

ISBN 10: 1-904761-47-X

ISBN 13: 978-1-904761-47-1

Disclaimer

Every reasonable effort has been made to ensure that the material in this book is true, correct, complete and appropriate at the time of writing. Nevertheless, the publishers and authors do not accept responsibility for any omission or error, or for any injury, damage, loss or financial consequences arising from the use of the book.

Typeset by Nottingham University Press, Nottingham
Printed and bound by The Cromwell Press, Trowbridge, Wiltshire

Contents

GAMETES AND FERTILISATION

Chairperson: C Wathes

THE ERIC LAMMING MEMORIAL SESSION

Chairperson: A McNeilly

REPRODUCTIVE MANAGEMENT

Chairperson: C Burke

DEER AND BUFFALO REPRODUCTION

Chairperson: G Martin

Preface

The Seventh International Ruminant Reproduction Symposium was held at Te Papa, Museum of New Zealand in Wellington on August 13-17 2006. There were 215 delegates from around the world including participants from Mexico and South America (20), USA and Canada (31), Europe (18), the Middle East and Asia (26), South Africa (3) and the balance from Australia and New Zealand. This volume contains the proceedings of 27 scientific presentations that were made over the 14 conference sessions. These sessions included presentations in: Male Reproduction; Ovarian Function including Follicular Development and Oocyte-Cumulus Cell Function; Neuroendocrinology; The Eric Lamming Memorial Lecture on Corpus Luteum Function; Reproductive Management; Deer, Buffalo, Yak and Camelid Reproduction; Embryo Technologies and Embryo Gene Expression; Fetal –Maternal Interactions; and Fetal Development.

A highlight of the meeting was the presentation of the Pioneer Awards for outstanding contributions to our understanding of Ruminant Reproduction to Professor Brian Setchell, University of Adelaide and Professor David Lindsay, University of Western Australia. We gratefully acknowledge the financial support of AgResearch, Dairy Insight, ICPbio Ltd, InterAg and the New Zealand Royal Society International Conference Fund. We also wish to acknowledge the New Zealand Animal Production Society, Massey University, Meat and Livestock Australia and the Society for Reproduction and Fertility for sponsoring conference sessions, speakers and/or student participation. We also acknowledge the contributions by Professor Terry Nett and Dr Vish Vishwanath who, at the last moment, agreed to present plenary lectures because two of our invited speakers and colleagues, were unexpectedly, unable to attend. The presentation of the two speakers who were unable to attend have been included in this volume. We are grateful to Nottingham University Press, UK for agreeing to include all the poster abstracts in the supplement.

The local organising committee (LOC) of Drs David Wells, Graham Barrell, John Smith, Tim Parkinson, Chris Burke, Jim Peterson, Vish Vishwanath, Jenny Juengel and Professor Ken McNatty wish to thank their international colleagues who selected the invited plenary speakers and who assisted with reviews of all the manuscripts.

Ken McNatty
Chairman, LOC.

Pioneer Awards

At the previous Ruminant Reproduction Symposia held at Colorado Springs, 1998 and Crieff, 2002, the organising committees of those meetings made special awards to acknowledge the outstanding contributions made to Ruminant Reproduction Science by individuals in the USA and UK respectively. The local organising committee of this symposium wished to continue this emerging tradition by acknowledging the contributions made by our distinguished colleagues in Australasia. It was evident during our deliberations that there were two outstanding scientists in our region that we wanted to acknowledge namely Professors David Lindsay, University of Western Australia and Brian Setchell of University of Adelaide. Accordingly, we were delighted that both David and Brian were able to attend the Symposium and to receive their Pioneer Awards by their former students Dr John Smith and Professor Simon Maddocks at the conference dinner.

Professor David Lindsay

David Roy Lindsay graduated with honours in Agricultural Science from The University of Sydney in 1957. He was awarded a PhD at The University of Sydney in 1963, followed by a post doctoral year at Washington State University. He then returned to Sydney as a Research Fellow and in 1967 went to Perth to the University of Western Australia as lecturer in Animal Science. He became Professor of Animal Science in the Faculty of Agriculture in 1984 and in 2000 was made an Honorary Research Fellow and Professor Emeritus at the University of Western Australia.

Dave's service to science and agriculture has been recognised over the years with him being elected a Fellow of the Australian Academy of Technological Sciences and Engineering in 1984, awarded the Institute Medal by the Australian Institute of Agricultural Science in 1991, elected Fellow of the Australian Society of Animal Production in 1992, Officer of the Order of Australia in 1997 and the recipient of the Australian Centenary Medal in 2000. He also has been instrumental in establishing the strong collaborative relationship, of research on ruminant reproduction, between France and Perth.

One of Dave's major contributions to Ruminant Reproduction is his role in mentoring others. He has officially supervised 21 PhD and 7 MSc students and served as an unofficial mentor to numerous other researchers at various stages of their careers. This mentoring continues today through organizing and delivering courses and workshops on "communication in science" and mentoring others in publishing of scientific papers in a wide range of disciplines.

Dave is the author of 2 books, "A guide to scientific writing" and "Breeding the flock", and author/co-author on some 20 book chapters, 92 referred manuscripts, 59 refereed conference papers and 11 non-refereed papers and reports.

The major topics of Dave's research career include the factors affecting mating efficiency of sheep; social, hormonal and environmental factors affecting the reproductive behaviour of sheep; clover disease and the metabolism of phyto-oestrogens and role of coumestans in ewe infertility; ewe-lamb behaviours and lamb survival; the mechanisms behind the ram effect and how best to utilise this phenomena; effect of high temperature and other environmental factors on ram fertility; and nutrition-ovulation relationships and the role of lupins. One particular area of endeavour of note was the development of the concept and techniques for on-farm evaluation of reproductive wastage in commercial sheep flocks. This work, which was initiated by Dave in the late 1960s, established a benchmark and the basis for subsequent research to try and answer, with practical and economical solutions, the problems identified- two in particular being low ovulation rate and high lamb mortality. This technology and approach has been adopted by other states in Australia and New Zealand.

Professor Brian Setchell

Brian Peter Setchell graduated with Honours in Veterinary Science from The University of Sydney and joined the NSW Department of Agriculture before moving to the University of Cambridge in 1955 as the Walter & Eliza Hall Scholar in Veterinary Science to undertake a PhD. Brian's thesis investigated Glucose Metabolism by the Brain and he was supervised by Alan Wilson. He completed this in 1957 and returned to the NSW Department of Agriculture before joining the CSIRO Division of Animal Physiology in 1962. He made a return visit to Cambridge in 1966 for a sabbatical year at the ARC Institute of Animal Physiology, Babraham, and then moved to Cambridge to join the Department of Biochemistry at Babraham in 1969 where he remained until 1982. He then accepted appointment as Professor of Animal Sciences at the University of Adelaide, Waite Agricultural Research Institute in South Australia and was also Head of Department for 10 years. He "retired" in 1996, and currently holds an Honorary Visiting Research Fellowship appointment in the Department of Anatomical Sciences at Adelaide where he still spends most working days.

He was awarded a DSc by the University of Cambridge in 1977 and has received numerous accolades for his contributions including being the Goding Lecturer for the ASRB in 1985, Honorary Life Membership of the ASRB in 1994, The Parkes Lecturer for the SSF (UK) in 1997, and the Distinguished Andrologist Award from the American Society for Andrology in 1997. He has supervised over 20 PhDs in Australia and the UK, many of whom are now well established researchers in the field of reproductive biology.

Brian has made numerous contributions to our understanding of reproductive physiology, with particular reference to the male, and the testis more specifically although by no means exclusively. In collaboration with Geoff Waites, Brian developed an approach to cannulate the rete testis to enable collection of the fluids emptying from the seminiferous tubules, prior to their transfer to the epididymis. It was the analysis of this fluid and the defining of the unique composition of tubule fluid that led Brian to propose and then define the nature of the blood-testis barrier as an anatomical & physiological barrier essential to normal spermatogenesis. This technique also allowed spermatozoa to be collected both before, and after transit of the epididymis, and led to our understanding of the importance of epididymal maturation of spermatozoa for their attainment of motility and full fertilising ability. His studies of the blood-testis barrier led to further investigations of the vasculature of the testis, and to numerous studies and papers addressing the importance of blood flow, and the regulation of vascular permeability for normal testis function. The significance of this for temperature regulation in turn has led to extensive studies of the effects of temperature on testis function and sperm viability.

Regarding his work on the testis, Brian has published 2 books, 74 invited reviews or chapters, 146 original scientific papers and 207 conference abstracts. In addition, he has 47 papers and 9 abstracts on blood flow and metabolism in other organs and on metabolic diseases in sheep, and 6 papers on analytical methods. Of interest to many have been his annotated translations from Latin of Regnier de Graaf's 2 books on "The Organs of Reproduction" and a translation from Italian of Sertoli's original description of the Sertoli cell, amongst other historical accounts and articles.

As a physiologist and biochemist, Brian has always had a propensity for addressing systems biology. For those of us trained in the traditions of reductionist science, putting the pieces of the jigsaw back together to understand the system is often the unmet challenge. Brian has established collaborations across the world, including France, Italy, Finland, Sweden, Thailand and the USA to name but a few. Collaboration with Brian inevitably means the establishment of a life-long friendship. His zest for life, science, and an amazing knowledge of history, invariably lead to many good dinners and fun times.

Nuclear organization of the protamine locus

RP Martins[1] and SA Krawetz[1-3]

[1]Center for Molecular Medicine and Genetics; [2]Department of Obstetrics and Gynecology, Institute for Scientific Computing, Wayne State University School of Medicine; 253 C.S. Mott Center, 275 East Hancock Detroit, MI 48201, USA

The human protamine gene cluster consists of three tightly regulated genes, protamine 1 (PRM1), protamine 2 (PRM2) and transition protein 2 (TNP2). Their products are required to repackage the paternal genome during spermiogenesis into a functional gamete. They reside within a single DNase I-sensitive domain associated with the sperm nuclear matrix, bounded by two haploid-specific Matrix Attachment Regions. The nuclear matrix is a dynamic proteinaceous network that is associated with both transcription and replication. While substantial effort has been directed toward pre- and post-transcriptional regulation, the role of the nuclear matrix in regulating haploid expressed genes has received comparatively little attention. In this regard, the functional organization of the human PRM1→PRM2→TNP2 cluster and where appropriate, comparisons to other model systems will be considered.

Introduction

We are continuing the exploration of the nucleus that commenced in 1871 when Miescher began exploiting the gonad of the spawning male salmon in the Rhine River as an alternative source of cells from which to investigate nuclear composition. A mixture of protamine encased nucleic acids termed "nuclein" was isolated (Miescher 1874). Once again attention is being focused toward understanding the carefully ordered structure of the cell nucleus. It is becoming apparent that the manner in which chromatin is organized within the nucleus provides a door to understanding gene regulation and cellular reprogramming.

Individual chromosomes occupy distinct territories within the cell nucleus. Transcriptionally active segments tend to localize to the periphery of the territories whereas transcriptionally inert regions localize to their centers (Kurz et al. 1996). In proliferating cells, their central position within the nucleus positively correlates with gene density (Boyle et al. 2001). Changes in the relative position can reflect the cell entering different stages of differentiation (Foster et al. 2005), undergoing malignant transformation (Cremer et al. 2003) or senescence (Bridger et al. 2000). Key to this organization (Ma et al. 1999) is the attachment or association of the genome to a network of proteins that lies just interior to the nuclear envelope, termed the nuclear matrix (Ma et al. 1999).

Until recently, the nuclear matrix had been regarded as a static structure, arranging chromatin into domains varying in length from ~ 100 - 200 kb in somatic cells and 20 – 50 kb in sperm (Barone et al. 1994). The points of attachment, i.e., the MARs, Matrix Attachment Regions, vary in size from 100 - 1000 base pairs. Several "families" of MAR motifs that anchor chroma-

Corresponding author E-mail: steve@compbio.med.wayne.edu

tin to the nuclear matrix or recruit specific *trans*-factors have been described (reviewed in Platts *et al*. 2006). It is slowly becoming accepted that nuclear matrix association is at the heart of a series of nuclear events (Kramer *et al*. 1997; Ostermeier *et al*. 2003; Vassetzky *et al*. 2000; Yasui *et al*. 2002; Zaidi *et al*. 2005). In some cases they may shield domains from neighbouring enhancers, the silencing effects of heterochromatin (Martins *et al*. 2004, Namciu *et al*. 1998) and may even be required for early embryonic development (Ward *et al*. 1999). Structural elements of the nuclear matrix have been shown to recruit a number of chromatin modifiers (Yasui *et al*. 2002). Their detailed molecular interacations of association and mechanisms of action are only beginning to be characterized (Heng *et al*. 2004).

Factors associated with many of the underlying nuclear processes are often co-purified with nuclear matrix proteins. One well characterized family is that of the Special AT-rich sequence Binding proteins SatB1 and SatB2. These MAR-binding proteins appear as cage-like networks (Cai *et al*. 2003) providing a MAR associating platform for specific loci throughout the nuclear interior. They modulate differentiation of several cell lineages acting as both activators and silencers likely reflecting their phosphorylation status (Pavan Kumar *et al*. 2006).

SatB1 coordinates the interaction of several factors critical for T-cell differentiation (Alvarez *et al*. 2000). For example, silencing is achieved through site-specific recruitment of histone deacetylases and ATP-dependent nucleosome remodeling complexes (Cai *et al*. 2003; Yasui *et al*. 2002) to the *IL-2Rα* locus MAR. SatB1 can also promote transcription of epsilon-globin in erythroid progenetors through its interaction with CBP (Wen *et al*. 2005). Similarly, cranio-facial patterning and bone formation are modulated through the interaction of SatB2 with several targets. For example, interaction of SatB2 with an enhancer downstream of *Hoxa2* modulates skeletal patterning whereas interaction with transcription factors Runx2 and AFT4, synergistically acts to drive osteoblast differentiation (Dobreva *et al*. 2006). Like other MAR-binding proteins, members of the Runx family are characterized by their nuclear matrix targeting signal. Visualization at interphase reveals punctuate scaffolds. Perhaps these scaffolds are preserved during mitosis to promote the next active phase of transcription (reviewed in Zaidi *et al*. 2005). Epigenetic regulation likely reflects nuclear matrix association.

Developmental systems including spermatogenesis have been used as models to study epigenetic regulation (Kramer *et al*. 1998). Encased within the walls of the seminiferous tubules, each phase of this continuous process is highly ordered. The pathway is marked by a series of morphologically distinct transitions that ultimately yield a highly compacted haploid genome encased in a unique motile delivery vehicle. For example, the paternal genome is repackaged and compacted to approximately 1/13th the volume of the oocyte nucleus. Yet, it contains a complete set of epigenetic instructions that provide a key to the correct usage of the paternal genome. An overview of the utility of this model towards understanding how nuclear organization modulates the molecular mechanisms in differentiation of the *PRM1→PRM2→TNP2* domain is presented.

Spermatogenesis

Spermatogenesis is characterized by a wave-like continuum of cellular and sub-cellular morphological, biochemical and physiological changes culminating in the repackaging of the male haploid genome (Sassone-Corsi 2002). The process initiates by the Bone Morphogenetic Protein-signaling of the proximal epiblast (reviewed in Raz 2005) giving rise to the primordial germ cells. The mitotically arrested germ cells then migrate to the genital ridge to form the presumptive gonad. During their migration they undergo epigenetic germline reprogramming, whereby the majority of the epigenetic marks, including imprints, are erased. The marks are re-established during spermatogenesis prior to the paternal genome assuming a hypermethylated state. Interestingly, even though the *PRM1→PRM2→TNP2* domain is hypermethylated, it is one of the few loci that remains in a potentiated chromatin conformation (Schmid *et al.* 2001). Following fertilization, the paternal genome is demethylated, save for the paternal imprints (Olek & Walter 1997). The underlying principle governing this state of epigenetic flux is uncertain. Together, demethylation along with the preservation of the paternal imprint likely imparts totipotency.

As summarized in Fig. 1, spermatogenesis can be divided into three phases 1) mitotic re-newal of the spermatogonial stem cells. 2) Commitment to a meiotic phase reducing tetrap-loid spermatocytes to haploid round spermatids. 3). The final phase, spermiogenesis, the mor-phogenic differentiation of round spermatids to spermatozoa. Spermiation then follows during which spermatozoa become motile. Most stages are easily identified reflecting the coordi-nated and gradual transition from a nucleo-histone organization to one that is dominated by nucleo-protamines. The morphologically distinct cell types represented within the cross sec-tion of the tubule shown in Fig. 1 can routinely be recovered by various cell-separation tech-niques. Spermatogonia are localized to the basal compartment, whereas the meiotic stages localize to the adluminal compartment. The differentiative process culminates in spermatozoa being released into the lumen where they collect and mature in the epididymus. This summa-rizes the continuous spermatogenic wave that in humans takes approximately 60 days to com-plete (Clermont 1972). Cells from the majority of stages can be enriched by their selection during the first wave and easily isolated in relatively pure form (Bellve 1993). This provides an ideal system to study differentiation.

Nuclear Organization during Spermatogenesis

Throughout the mitotic and meiotic programs, the germ cell nucleus is structured in a manner similar to that of somatic cells. The re-establishment of paternal imprints by targeted genomic methylation and histone acetylation continues through mitotic and meiotic divisions (reviewed in Rousseaux *et al.* 2005).

As shown in Fig. 2, nucleo-histone organization persists throughout the spermatocyte to just prior to spermiogenesis when the transition from histone to protamine packaging begins in earnest. During this early stage of spermatogenesis the nuclear DNA is organized as nucleo-somes. These are comprised of 146 bp supercoiled repeated segments of DNA, that wrap 1.75 times around a histone octamer composed of two heterodimers of each of H2A-H2B plus H3-H4. Each is joined by approximately 15 bp of "linker DNA," stabilized by histone H1. This repeated unit then coils to form a chromatin loop that associates with the nuclear matrix.

Spermatogonia harbor several germ cell-specific histone variants, i.e., TH2A, H2A.X, TH3, and H3.3A in addition to the somatic H2A and H3 counterparts. Variants of H2B including TH2B, H2B-RP and ssH2B, H3, including H3.3B and H3F3/B and H1 including H1t, H1t2 and HILS1 systematically begin to replace their somatic counterparts from the primary spermato-cyte stage onward (Kimmins & Sassone-Corsi 2005). Several post-translational modifications are required to ensure histone/protamine exchange and to maintain the fidelity of the genetic material. These include ubiquitination, H4 hyperacetylation, phosphorylation and ADP-ribosylation. Ubiquitination and phosphorylation of histones have been associated with the initial histone to transition protein exchange. H2A/H2B ubiquitination has been implicated in sperm chromatin reorganization (Roest *et al.* 1996). It is likely that ubiquitination targets histones for degradation, signaling their replacement by the transition proteins *TNP1, TNP2, TNP3, TNP4,* (Wouters-Tyrou *et al.* 1998) as the cell differentiates until the latter phase of the

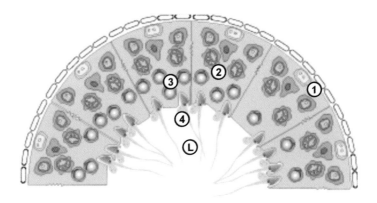

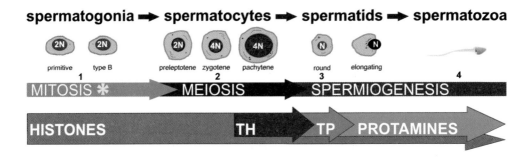

Fig. 1: Spermatogenesis. Spermatogenesis is a continual process throughout the seminiferous tubule. Cells that can be isolated in relatively pure-form by various simple differential sedimentation techniques are illustrated. Just inside basement membrane (BM), spermatogonia (1) differentiate into spermatocytes (2), then into round spermatids (3) and finally into spermatozoa (4), that traverse into the lumen (L), for transport to the epididymis where they mature further and acquire motility. Spermatogenesis consists of three parts: a mitotic amplification of primitive (Type A) spermatogonia; spermatocytic progression through meiosis and genetic recombination to culminate in haploid spermatids; and spermiogenesis, the morphological differentiation of the haploid male germ cell. Spermatogonia that differentiate to type B spermatogonia are committed (yellow asterisk) to the spermatogenic differentiative pathway. In the progression from zygotene to pachytene spermatocytes, the reorganization of the male genome commences. Initially, most somatic histones are replaced with testis-specific histone variants (TH). During spermiogenesis these are gradually displaced by the transition proteins (TP) and eventually by the protamines. Repackaging of the genome in this manner leads to a higher degree of compaction in sperm nucleus that finally yields the mature spermatozoa. This figure was adapated and is reproduced with permission from Nature Reviews Genetics Krawetz, S.A. (2005) Paternal Contribution: new insights and future challenges. Nature Reviews Genetics **6**:633-642 copyright 2005 Macmillan Magazines Ltd.

round spermatid stage (Fig. 1). Although acetylation of select residues in histone amino termini is associated with actively transcribing chromatin, hyperacetylation of H4 is thought to be a critical step towards histone replacement. Foci throughout late-stage spermatogenic nuclei exhibit high levels of H4 hyperacetylation. These mark sites for localized and intermediate replacement (Lahn et al. 2002) with the phosphorylated group of transition proteins. These are subsequently replaced with protamines.

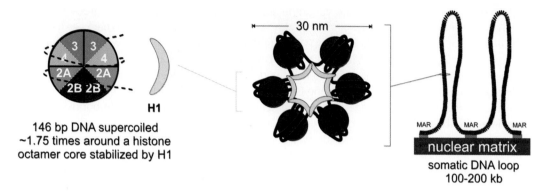

146 bp DNA supercoiled
~1.75 times around a histone
octamer core stabilized by H1

H1

30 nm

nuclear matrix

somatic DNA loop
100-200 kb

MAR   MAR   MAR

Fig. 2: Organization of somatic chromatin. The basic structural unit of eukaryotic somatic chromatin is the nucleosome. Left panel, approximately 146 bp of DNA supercoils
1.75 times around a histone octamer core composed of two H2A-H2B heterodimers and
two H3-H4 heterodimers. Successive nucleosomes are joined by approximately 15 bp of
"linker" DNA stabilized by Histone H1. The approximate diameter of the nucleosome
chain is 10 nm. This is often referred to as the 'beads on a string' conformation. Middle
panel, these structures supercoil onto themselves forming a 30 nm fiber. Right panel,
DNA is further organized into 100 - 200 kb loops by attachment to the nuclear matrix by
matrix attachment regions, MARs.

Following meiosis there is a burst of transcription, including that arising from the *PRM1* and
PRM2 genes (Kleene *et al.* 1983), most of which are then stored as inactive mRNPs (reviewed
in Kleene 2003). Spermatid transcription subsequently ceases during the elongation phase
(Kierszenbaum & Tres 1975). During this phase, the host of post-meiotically transcribed genes
are released from their quiescent inactive mRNP bound state then translated into functional
proteins (Fajardo *et al.* 1997). These include *PRM1* and *PRM2* mRNAs that are suppressed until
the late elongating spermatid stage (Balhorn *et al.* 1984).

In mice and men, *PRM1* is synthesized as a mature 51 aa protein whereas *PRM2* is synthesized as a 102 aa and 107 aa precursor protein in human and mouse, respectively. This precursor
is cleaved into a number of different peptide products to yield a 57 aa mature protein in humans
(Wouters-Tyrou *et al.* 1998). After this initial processing, both *PRM1* and *PRM2* are phosphorylated. This reduces their net positive charge, facilitating transport and *TNP* replacement (Dadoune
1995). At this juncture, the intermediary nucleo-proteins are replaced with the newly translated arginine- and cysteine-rich protamines (Dadoune 1995). The protamines successively
dephosphorylate facilitating nucleo-protein replacement with protamines then condense as
intrastrand disulfide bond formation continues during spermiogenesis (Chirat *et al.* 1993). This
completes the transition to a highly compacted structure.

With the constant upheaval of the nucleus throughout spermatogenesis, repair mechanisms are
required to ensure the fidelity of the genetic material. Although it is not yet fully understood,
ADP-ribosylation may provide at least part of the solution. Could this be part of the DNA-damage-
dependent repair mechanism or act as a signal for recombination/repair? Consistent with this
notion, members of the ADP-ribosylation pathway increase during the later stages of spermatogenesis when nuclear reorganization is prominent and DNA strand breaks accumulate (Mosgoeller *et
al.* 1996). This would include nucleo-protamine remodeling (Meyer-Ficca *et al.* 2005) and/or post-
fertilization genome decondensation and remodeling (Mosgoeller *et al.* 1996). Perhaps this reflects the management strategy to ensure the integrity of the paternal genome.

As modeled in Fig. 3, tracts of polyarginine associate directly with the major groove of DNA (Hud *et al.* 1994). The smaller size and the highly basic nature of the protamines package the DNA into a series of ~60 bp parallel sheets. One protamine dimer can span two helical turns or approximately 22 bp of DNA (Vilfan *et al.* 2004). These are stabilized as toroid ring structures through the formation of a disulfide bonded network among the adjacent protamines (Brewer *et al.* 1999; Dadoune 1995; Kimmins & Sassone-Corsi 2005). The toroid can be visualized as successive helical coils that form a hexagonal lattice, stabilized by intermolecular disulfide bonds that then associate with the sperm nuclear matrix using a unique suite of MARs. A consensus model of the protamine DNA toroid has emerged (Brewer *et al.* 1999; Vilfan *et al.* 2004) effectively resolving the comparatively dense structure of the male gamete nucleus. While this is an effective model, one must consider that approximately 15% of human DNA remains histone-associated (Tanphaichitr *et al.* 1978), while less than 2% of mouse DNA remains histone-associated (Balhorn *et al.* 1984). This begs the question how are the nucleo-histone regions organized?

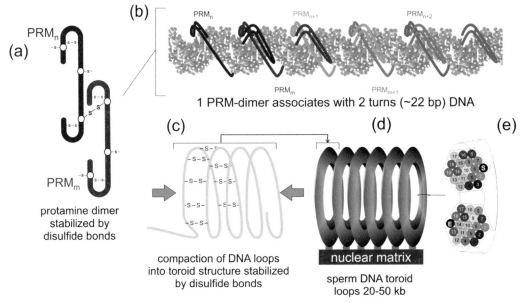

Fig. 3: **Organization of sperm chromatin by protamine.** During spermiogenesis the fundamental nucleosome structure is ostensively replaced by protamine, compacting the genome approximately six-fold. (a) As proposed (Vilfan *et al.* 2004), a single protamine dimmer is associated within the major groove (b) of 2 turns of DNA or approximately 22 bp. (c) The cysteine-rich protamines form intermolecular disulfide bonds, stabilizing the homodimer. (d) Protamine bound DNA then assembles into a series of loops that are stabilized by disulfide bridges that in turn form a toroid. This hexagonal lattice binds to the sperm nuclear matrix at approximately 20 - 50 kb intervals. (e) Entry into the toroid is indicated as S and the last loop exits the toroid through E. This figure was adapated and is reproduced with permission from *J. Biol. Chem.* Vilfan ID, Conwell CC & Hud NV 2004 Formation of Native-like Mammalian Sperm Cell Chromatin with Folded Bull Protamine. 279:20088-20095 from copyright 2004 Macmillan Magazines Ltd. The American Society for Biochemistry and Molecular Biology.

Nuclear organization of the PRM1→PRM2→TNP2 domain

After protamine deposition, specific segments of the *PRM1→PRM2→TNP2* locus remain histone enriched (Wykes & Krawetz 2003a) as shown in Fig. 4. The histone bound segments include specific MAR containing regions like the 5' MAR as well as the promoter regions of the *PRM1→PRM2→TNP2* domain. The enhancer-promoter regions are similar among the various members of both the human and mouse loci (Wykes & Krawetz 2003b). A series of DNase I hypersensitive sites have been identified flanking the ends of the *PRM1→PRM2→TNP2* domain (Wykes & Krawetz 2004). These sites colocalize to a series of predicted transcription factor binding sites including a binding site for the testes specific activator SOX-5 (Denny *et al.* 1992). This resides just downstream of the 5' MAR (Wykes & Krawetz 2004). Additional hypersensitive sites reside within the promoter regions (Wykes & Krawetz 2003b). These have been mapped to within ~ 300 nucleotides of the respective transcription initiation start sites. This includes the A box shown in Table 1, which is the most conserved cis-element among the members. It is likely indicative of the rigorous and coordinated manner in which these genes are regulated as a requisite for the survival of the species. For example, disruption at any phase can lead to a host of morphologically defective sperm and infertility phenotypes. On the one hand, ectopic or over-expression of *PRM1* prematurely condenses the nuclear DNA thus effectively halting differentiation (Lee *et al.* 1995). On the other hand, haplo-insufficient transgenic mice lacking either *PRM1* or *PRM2* produce morphologically abnormal infertile sperm (Cho *et al.* 2001). Although not as severe, disruption of *TNP2* expression by targeted mutagenesis presents as teratozoospermia and reduced fertility (Zhao *et al.* 2001).

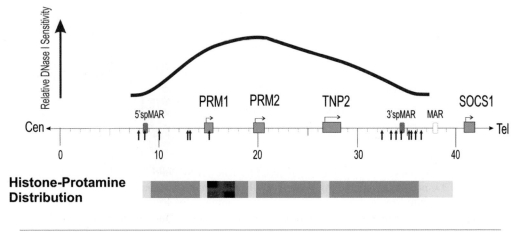

Fig. 4: The human protamine domain. The human *PRM1→PRM2→TNP2* cluster resides in a single DNase I-sensitive domain on human chromosome 16p13.13 bounded by two sperm-specific MARs and a third somatic MAR neighboring the *SOCS1* gene. Defined regions of the domain remain histone associated even after protamine deposition. These include the promoter regions of each gene and the 5' MAR. A region of complete protamine replacement localizes to the coding sequence of *PRM1*. Using a transgenic model, 5' and 3' DNase I hypersensitive sites (HSs) have been mapped to the domain boundaries encompassing the MAR regions.

The single DNase-I sensitive human *PRM1→PRM2→TNP2* domain is bounded by two sperm-specific matrix attachment regions. The mouse domain is home to a similar array of MARs and conserved regulatory regions. However, as shown in Fig. 5, the mouse *Prm1→Prm2→Tnp2* cluster is substantially compressed when compared to its human counterpart. While the mouse

Table 1. Summary of the common A box element among the promoters of the human and mouse protamine clusters.

Species	Gene	Distance from TATA (bp)	A box sequence
Human	*PRM1*	102	GGGCTGCCC
Mouse	*Prm1*	105	GCCCTGCCC
Human	*PRM2*	65	GTGCTGCCC
Mouse	*Prm2*	73	GTGCCGCCC
Human	*TNP2*	96	GAGCTGCCC
Mouse	*Tnp2*	80	AAGCTGCCC

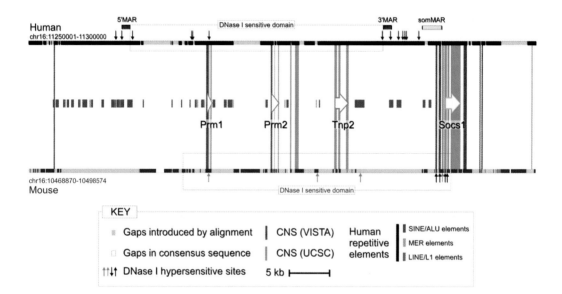

Fig. 5: Alignment of syntenic segments of human (NCBI Build 35) and mouse (NCBI Build 33) protamine containing regions on chromosomes 16. Graphical representation of VISTA (http://genome.lbl.gov/vista) genomic alignment for segments of human (black) and mouse (navy blue) chromosome 16. Gaps introduced by the alignment are shown in grey, whereas gaps in the consensus sequences are in white. Conserved non-coding regions were mapped by rVISTA. They are highlighted in red and those identified by USCS's genome browser are in cyan. Elements experimentally identified within these spans of syntenic segments of human and mouse chromosome 16 have been identified: MAR: matrix attachment regions; DNase I hypersensitive sites previously mapped to the vicinity of the human and mouse protamine domains are shown as black arrows. Those sites identified during the definition of the mouse protamine DNase I sensitive domain have been identified by green arrows are associated with the protamine domain; and orange arrows if mapping to regions previously identified in association with Socs1. Human repetitive elements were also mapped along the alignment.

protamine genes do not themselves display a large degree of sequence divergence away from their human counterparts, the expansion of the intragenic region in humans reflects the substantial accumulation of repetitive sequence elements (Nelson & Krawetz 1994). This likely occurred after the initial divergence of murines and primates, as the mouse domain is essentially devoid of repetitive elements. Interestingly, the human sperm nuclear Alu elements appear preferentially bound to protamine. This represents approximately 100,000 sequence elements in the genome. In comparison the mouse LINE/L1 elements appear histone bound (Pittoggi *et al.* 2000). Additionally, sperm histone/protamine-bound centromeres (Wykes & Krawetz 2003a) localize to the nuclear matrix (Yaron *et al.* 1998) whereas histone-bound telomeres localize to the nuclear periphery (Pittoggi *et al.* 1999; Zalenskaya *et al.* 2000). The significance of these observations remains to be delineated since the relative proportion of histones in both species varies markedly. Differential packaging may present an immediate target to assure the silencing of promoters from male haploid expressed genes like those of the *PRM1→PRM2→TNP2* domain. Alternatively it may template the genome for histone repackaging upon fertilization or enable specific zygotic transcription from the paternal contribution.

Interactions with the nuclear matrix orchestrate a host of nuclear events in somatic cells (Kramer *et al.* 1998, 2000; Ostermeier *et al.* 2003; Vassetzky *et al.* 2000; Yasui *et al.* 2002). The sperm nuclear matrix plays an important supportive and structural role. Its constituents include factors involved in spermatozoal capacitation (Calzada & Martinez 2002), factors required for early oocyte activation (Fujimoto *et al.* 2004) and factors that likely assist with the formation of the male pronucleus (Ward *et al.* 1999). The MARs of the male haploid expressed human protamine locus are situated at the boundaries of a DNase I-sensitive domain (Fig. 4; Kramer & Krawetz 1996). They have been shown to provide critical components towards assuring non-ectopic expression and minimizing position effects both in an animal models (Martins *et al.* 2004) and in humans (Kramer *et al.* 1997). Data from the transgenic mouse model of the human *PRM1→PRM2→TNP2* domain has suggested that the upstream and downstream MARs synergistically act to shield this segment from its neighboring chromatin environment. The 5' MAR may act to enhance expression whereas the 3' MAR likely provides a dominant tempering activity. Interestingly a candidate fertility associated mutation in this locus has been mapped to the 3' MAR (Kramer *et al.* 1997). In other systems, association with the nuclear matrix is transient, selective and dynamic, paralleling the differentiative state and function (Heng *et al.* 2001; Heng *et al.* 2004; Ostermeier *et al.* 2003). Whether both are representative of a haploid expressed domain remains to be determined.

Conclusions

Spermatogenesis is an elaborate differentiative program that persists throughout the adult life of the male. The relative ease that cells from virtually every stage in the program can be isolated makes it an ideal developmental model. At the heart of this process is the repackaging of the sperm genome by the systematic replacement of histones with testes-specific variants - transition proteins - then their coordinated replacement by the protamines. The protamine cluster serves as a model for a haploid-expressed gene cluster whose elaborate regulatory program involves interactions with the nuclear matrix (Kramer & Krawetz 1996; Martins *et al.* 2004) as it is regulated by a suite of transcription factors (reviewed in Lu *et al.* 2006) and then post-transcriptional associations as RNPs (Fajardo *et al.* 1997). What remains to be determined is defining what forms of epigenetic modification are utilized to initiate and complete potentiation, i.e., the opening of a chromatin domain to enable and facilitate transcription. This underscores the pivotal role of chromatin remodeling (reviewed in Cho K. S. *et al.* 2004) that has emerged as a central theme towards understanding the mechanisms of disease (reviewed in Kleinjan & van Heyningen 2005). Perhaps, the disruption of these epigenetic marks at the paternally imprinted *H19* locus are associated with oligozoospermia (Marques *et al.* 2004).

Although basal transcription can proceed from within structures of higher ordered chromatin (Georgel *et al.* 2003), there is a consensus that an open-potentiated chromatin structure is required for optimum transcription. Potentiation, that is the opening of chromatin domains, grants *trans*-acting element access to transcription promoting *cis*-regulatory sequences. The potentiative timing in mouse (Kramer *et al.* 1998) and the steadfast nature of the human domain (Kramer *et al.* 2000) have been established. What is lacking are the crucial factor(s) that set potentiation in motion. A greater understanding of the interplay of nuclear constituents, including the nuclear matrix and the factors that can recruit many of these activating and silencing terms (Yasui *et al.* 2002) is required. This understanding will shed light on the regulation of this cluster and the processes the cell has at its disposal for the discrete selection of subregions of the genome for activity.

Acknowledgements

This work was supported by NICHD grant HD36512 to SAK.

References

Alvarez JD, Yasui DH, Niida H, Joh T, Loh DY & Kohwi-Shigematsu T 2000 The MAR-binding protein SATB1 orchestrates temporal and spatial expression of multiple genes during T-cell development. *Genes and Development* **14** 521-535.

Balhorn R, Weston S, Thomas C & Wyrobek AJ 1984 DNA packaging in mouse spermatids. Synthesis of protamine variants and four transition proteins. *Experimental Cell Research* **150** 298-308.

Barone JG, De Lara J, Cummings KB & Ward WS 1994 DNA organization in human spermatozoa. *Journal of Andrology* **15** 139-144.

Bellve AR 1993 Purification, culture, and fractionation of spermatogenic cells. *Methods in Enzymology* **225** 84-113.

Boyle S, Gilchrist S, Bridger JM, Mahy NL, Ellis JA & Bickmore WA 2001 The spatial organization of human chromosomes within the nuclei of normal and emerin-mutant cells. *Human Molecular Genetics* **10** 211-219.

Brewer LR, Corzett M & Balhorn R 1999 Protamine-induced condensation and decondensation of the same DNA molecule. *Science* **286** 120-123.

Bridger JM, Boyle S, Kill IR & Bickmore WA 2000 Remodelling of nuclear architecture in quiescent and senescent human fibroblasts. *Current Biology* **10** 149-152.

Cai S, Han HJ & Kohwi-Shigematsu T 2003 Tissue-specific nuclear architecture and gene expression regulated by SATB1. *Nature Genetics* **34** 42-51.

Calzada L & Martinez JM 2002 Induction of nuclear matrix-estradiol receptor complex during capacitation process in human spermatozoa. *Archives of Andrology* **48** 221-224.

Chirat F, Arkhis A, Martinage A, Jaquinod M, Chevaillier P & Sautiere P 1993 Phosphorylation of human sperm protamines HP1 and HP2: identification of phosphorylation sites. *Biochimica et Biophysica Acta* **1203** 109-114.

Cho C, Willis WD, Goulding EH, Jung-Ha H, Choi YC, Hecht NB & Eddy EM 2001 Haploinsufficiency of protamine-1 or -2 causes infertility in mice. *Nature Genetics* **28** 82-86.

Cho KS, Elizondo LI & Boerkoel CF 2004 Advances in chromatin remodeling and human disease. *Current Opinion in Genetics & Development* **14** 308-315.

Clermont Y 1972 Kinetics of spermatogenesis in mammals: seminiferous epithelium cycle and spermatogonial renewal. *Physiological Reviews* **52** 198-236.

Cremer M, Kupper K, Wagler B, Wizelman L, von Hase J, Weiland Y, Kreja L, Diebold J, Speicher MR & Cremer T 2003 Inheritance of gene density-related higher order chromatin arrangements in normal and tumor cell nuclei. *European Journal of Cell Biology* **162** 809-820.

Dadoune JP 1995 The nuclear status of human sperm cells. *Micron* **26** 323-345.

Denny P, Swift S, Connor F & Ashworth A 1992 An SRY-related gene expressed during spermatogenesis in the mouse encodes a sequence-specific DNA-binding protein. *The EMBO Journal* **11** 3705-3712.

Dobreva G, Chahrour M, Dautzenberg M, Chirivella L, Kanzler B, Farinas I, Karsenty G & Grosschedl R 2006 SATB2 is a multifunctional determinant of craniofacial patterning and osteoblast differentiation. *Cell* **125** 971-986.

Fajardo MA, Haugen HS, Clegg CH & Braun RE 1997 Separate elements in the 3' untranslated region of the mouse protamine 1 mRNA regulate translational repression and activation during murine spermatogenesis. *Developmental Biology* **191** 42-52.

Foster HA, Abeydeera LR, Griffin DK & Bridger JM 2005 Non-random chromosome positioning in mammalian sperm nuclei, with migration of the sex chromosomes during late spermatogenesis. *Journal of Cell Science* **118** 1811-1820.

Fujimoto S, Yoshida N, Fukui T, Amanai M, Isobe T, Itagaki C, Izumi T & Perry AC 2004 Mammalian phospholipase Czeta induces oocyte activation from the sperm perinuclear matrix. *Developmental Biology* **274** 370-383.

Georgel PT, Fletcher TM, Hager GL & Hansen JC 2003 Formation of higher-order secondary and tertiary chromatin structures by genomic mouse mammary tumor virus promoters. *Genes & Development* **17** 1617-1629.

Heng HH, Krawetz SA, Lu W, Bremer S, Liu G & Ye CJ 2001 Re-defining the chromatin loop domain. *Cytogenetics and Cell Genetics* **93** 155-161.

Heng HH, Goetze S, Ye CJ, Liu G, Stevens JB, Bremer SW, Wykes SM, Bode J & Krawetz SA 2004 Chromatin loops are selectively anchored using scaffold/matrix-attachment regions. *Journal of Cell Science* **117** 999-1008.

Hud NV, Milanovich FP & Balhorn R 1994 Evidence of novel secondary structure in DNA-bound protamine is revealed by Raman spectroscopy. *Biochemistry* **33** 7528-7535.

Kierszenbaum AL & Tres LL 1975 Structural and transcriptional features of the mouse spermatid genome. *European Journal of Cell Biology* **65** 258-270.

Kimmins S & Sassone-Corsi P 2005 Chromatin remodelling and epigenetic features of germ cells. *Nature* **434** 583-589.

Kleene KC 2003 Patterns, mechanisms, and functions of translation regulation in mammalian spermatogenic cells. *Cytogenetic and Genome Research* **103** 217-224.

Kleene KC, Distel RJ & Hecht NB 1983 cDNA clones encoding cytoplasmic poly(A)+ RNAs which first appear at detectable levels in haploid phases of spermatogenesis in the mouse. *Developmental Biology* **98** 455-464.

Kleinjan DA & van Heyningen V 2005 Long-range control of gene expression: emerging mechanisms and disruption in disease. *American Journal of Human Genetics* **76** 8-32.

Kramer JA & Krawetz SA 1996 Nuclear matrix interactions within the sperm genome. *Journal of Biological Chemistry* **271** 11619-11622.

Kramer JA, McCarrey JR, Djakiew D & Krawetz SA 1998 Differentiation: the selective potentiation of chromatin domains. *Development* **125** 4749-4755.

Kramer JA, McCarrey JR, Djakiew D & Krawetz SA 2000 Human spermatogenesis as a model to examine gene potentiation. *Molecular Reproduction and Development* **56** 254-258.

Kramer JA, Zhang S, Yaron Y, Zhao Y & Krawetz SA 1997 Genetic testing for male infertility: a postulated role for mutations in sperm nuclear matrix attachment regions. *Genetic Testing* **1** 125-129.

Kurz A, Lampel S, Nickolenko JE, Bradl J, Benner A, Zirbel RM, Cremer T & Lichter P 1996 Active and inactive genes localize preferentially in the periphery of chromosome territories. *Journal of Cell Biology* **135** 1195-1205.

Lahn BT, Tang ZL, Zhou J, Barndt RJ, Parvinen M, Allis CD & Page DC 2002 Previously uncharacterized histone acetyltransferases implicated in mammalian spermatogenesis. *Proceedings of the National Academy of Sciences of the US A* **99** 8707-8712.

Lee K, Haugen HS, Clegg CH & Braun RE 1995 Premature translation of protamine 1 mRNA causes precocious nuclear condensation and arrests spermatid differentiation in mice. *Proceedings of the National Academy of Sciences of the U S A* **92** 12451-12455.

Lu Y, Platts AE, Ostermeier GC & Krawetz SA 2006 K-SPMM: a database of murine spermatogenic promoters modules & motifs. *BMC Bioinformatics* **7** 238.

Ma H, Siegel AJ & Berezney R 1999 Association of chromosome territories with the nuclear matrix. Disruption of human chromosome territories correlates with the release of a subset of nuclear matrix proteins. *Journal of Cell Biology* **146** 531-542.

Marques CJ, Carvalho F, Sousa M & Barros A 2004 Genomic imprinting in disruptive spermatogenesis. *Lancet* **363** 1700-1702.

Martins RP, Ostermeier GC & Krawetz SA 2004 Nuclear Matrix Interactions at the Human Protamine Domain: a working model of potentiation. *Journal of Biological Chemistry* **279** 51862-51868.

Meyer-Ficca ML, Scherthan H, Burkle A & Meyer RG 2005 Poly(ADP-ribosyl)ation during chromatin remodeling steps in rat spermiogenesis. *Chromosoma* **114** 67-74.

Miescher F 1874 Das Protamin—Eine neue organische Basis aus den Samenfäden des Rheinlachses. *Ber. Dtsch. Chem. Ges.* **7** 376.

Mosgoeller W, Steiner M, Hozak P, Penner E & Wesierska-Gadek J 1996 Nuclear architecture and ultrastructural distribution of poly(ADP-ribosyl)transferase, a multifunctional enzyme. *Journal of Cell Science* **109** (Pt 2) 409-418.

Namciu SJ, Blochlinger KB & Fournier RE 1998 Human matrix attachment regions insulate transgene expression from chromosomal position effects in Drosophila melanogaster. *Molecular and Cellular Biology* **18** 2382-2391.

Nelson JE & Krawetz SA 1994 Characterization of a human locus in transition. *Journal of Biological Chemistry* **269** 31067-31073.

Olek A & Walter J 1997 The pre-implantation ontogeny of the H19 methylation imprint. *Nature Genetics* **17** 275-276.

Ostermeier GC, Liu Z, Martins RP, Bharadwaj RR, Ellis J, Draghici S & Krawetz SA 2003 Nuclear matrix association of the human beta-globin locus utilizing a novel approach to quantitative real-time PCR. *Nucleic Acids Research* **31** 3257-3266.

Pavan Kumar P, Purbey PK, Sinha CK, Notani D, Limaye A, Jayani RS & Galande S 2006 Phosphorylation of SATB1, a global gene regulator, acts as a molecular switch regulating its transcriptional activity in vivo. *Molecular cell* **22** 231-243.

Pittoggi C, Zaccagnini G, Giordano R, Magnano AR, Baccetti B, Lorenzini R & Spadafora C 2000 Nucleosomal domains of mouse spermatozoa chromatin as potential sites for retroposition and foreign DNA integration. *Molecular Reproduction and Development* **56** 248-251.

Pittoggi C, Renzi L, Zaccagnini G, Cimini D, Degrassi F, Giordano R, Magnano AR, Lorenzini R, Lavia P & Spadafora C 1999 A fraction of mouse sperm chromatin is organized in nucleosomal hypersensitive domains enriched in retroposon DNA. *Journal of Cell Science* **112 (Pt 20)** 3537-3548.

Platts AE, Quayle AK & Krawetz SA 2006 In-Silico Prediction and Observations of Nuclear Matrix Attachment *Cellular and Molecular Biology Letters* **11** 191-213.

Raz E 2005 Germ cells: sex and repression in mice. *Current Biology* **15** R600-603.

Roest HP, van Klaveren J, de Wit J, van Gurp CG, Koken MH, Vermey M, van Roijen JH, Hoogerbrugge JW, Vreeburg JT, Baarends WM *et al.* 1996 Inactivation of the HR6B ubiquitin-conjugating DNA repair enzyme in mice causes male sterility associated with chromatin modification. *Cell* **86** 799-810.

Rousseaux S, Caron C, Govin J, Lestrat C, Faure AK & Khochbin S 2005 Establishment of male-specific epigenetic information. *Gene* **345** 139-153.

Sassone-Corsi P 2002 Unique chromatin remodeling and transcriptional regulation in spermatogenesis. *Science* **296** 2176-2178.

Schmid C, Heng HH, Rubin C, Ye CJ & Krawetz SA 2001 Sperm nuclear matrix association of the PRM1—>PRM2—>TNP2 domain is independent of Alu methylation. *Molecular Human Reproduction* **7** 903-911.

Tanphaichitr N, Sobhon P, Taluppeth N & Chalermisarachai P 1978 Basic nuclear proteins in testicular cells and ejaculated spermatozoa in man. *Experimental Cell Research* **117** 347-356.

Vassetzky Y, Hair A & Mechali M 2000 Rearrangement of chromatin domains during development in Xenopus. *Genes & Development* **14** 1541-1552.

Vilfan ID, Conwell CC & Hud NV 2004 Formation of native-like mammalian sperm cell chromatin with folded bull protamine. *Journal of Biological Chemistry* **279** 20088-20095.

Ward WS, Kimura Y & Yanagimachi R 1999 An intact sperm nuclear matrix may be necessary for the mouse paternal genome to participate in embryonic development. *Biology of reproduction* **60** 702-706.

Wen J, Huang S, Rogers H, Dickinson LA, Kohwi-Shigematsu T & Noguchi CT 2005 SATB1 family protein expressed during early erythroid differentiation modifies globin gene expression. *Blood* **105** 3330-3339.

Wouters-Tyrou D, Martinage A, Chevaillier P & Sautiere P 1998 Nuclear basic proteins in spermiogenesis. *Biochimie* **80** 117-128.

Wykes SM & Krawetz SA 2003a The structural organization of sperm chromatin. *Journal of Biological Chemistry* **278** 29471-29477.

Wykes SM & Krawetz SA 2003b Conservation of the PRM1 → PRM2 → TNP2 domain. *DNA Sequence* **14** 359-367.

Wykes SM & Krawetz SA 2004 A survey of the DNase I hypersensitive sites in a human transgenic PRM1→PRM2→TNP2 model system. *Trends in Comparative Biochemistry and Physiology* **10** 55-63.

Yaron Y, Kramer JA, Gyi K, Ebrahim SA, Evans MI, Johnson MP & Krawetz SA 1998 Centromere sequences localize to the nuclear halo of human spermatozoa. *International Journal of Andrology* **21** 13-18.

Yasui D, Miyano M, Cai S, Varga-Weisz P & Kohwi-Shigematsu T 2002 SATB1 targets chromatin remodelling to regulate genes over long distances. *Nature* **419** 641-645.

Zaidi SK, Young DW, Choi JY, Pratap J, Javed A, Montecino M, Stein JL, van Wijnen AJ, Lian JB & Stein GS 2005 The dynamic organization of gene-regulatory machinery in nuclear microenvironments. *EMBO Rep* **6** 128-133.

Zalenskaya IA, Bradbury EM & Zalensky AO 2000 Chromatin structure of telomere domain in human sperm. *Biochemical and Biophysical Research Communications* **279** 213-218.

Zhao M, Shirley CR, Yu YE, Mohapatra B, Zhang Y, Unni E, Deng JM, Arango NA, Terry NH, Weil MM *et al.* 2001 Targeted disruption of the transition protein 2 gene affects sperm chromatin structure and reduces fertility in mice. *Molecular and Cellular Biology* **21** 7243-7255.

Seminal plasma effects on sperm handling and female fertility

WMC Maxwell, SP de Graaf, R El-Hajj Ghaoui and G Evans

Faculty of Veterinary Science, University of Sydney, NSW 2006, Australia

The components of ruminant seminal plasma and their influence on the fertility of spermatozoa are reviewed. Seminal plasma can both inhibit and stimulate sperm function and fertility through the multifunctional actions of organic and inorganic components. These effects are now better understood because the composition of the seminal plasma, including its protein content and that of other structures, specifically membrane vesicles, has been clarified. Spermatozoa gain motility and fertilizing capacity as they transit the epididymis under the influence of factors produced by that organ. At ejaculation, inhibitory (termed "decapacitation") factors, sourced from the accessory sex glands, bind to the sperm surface. The major proteins isolated and characterised in ram seminal plasma, whose specific functions are yet to be determined, originate from the vesicular gland and comprise a spermadhesin together with proteins with fibronectin-II domains. *In vitro* handling of spermatozoa in preparation for artificial insemination (AI), involving processes such as dilution, cooling, freezing, re-warming and sperm sexing by flow cytometric sorting, can remove seminal plasma and may modify the proteins bound to the sperm surface. This destabilises the membranes and may pre-capacitate the spermatozoa, shortening their fertilizing lifespan. These changes may be reversible by seminal plasma fractions but responses differ depending on the type of sperm pre-treatment. Fertility after AI of ruminant semen may be improved if the role of seminal plasma proteins and their effect, if added individually or in combination to spermatozoa at different stages of preservation, or other manipulations such as flow cytometric sorting, can be determined.

Introduction

A large number of components, mainly proteins but also membrane vesicles, have been isolated and characterised in seminal plasma which have been associated with either positive or negative effects on sperm function and fertility. It has been hypothesised that such components, when added to spermatozoa, could either prevent or delay the natural maturation process leading to capacitation and eventual cell death (Maxwell & Watson 1996). Other components, particularly specific protein fractions, have been identified as anti-fertility or fertility-enhancing agents and have been used, by detecting their presence in the seminal plasma, as

Corresponding author E-mail: chism@vetsci.usyd.edu.au

indicators of the potential fertility of individual males (bulls: Killian et al. 1993; rams: Métayer et al. 2001). Moreover, factors in seminal plasma, mainly cytokines, have been implicated in conditioning the female reproductive tract, through the inflammatory response, to tolerate and facilitate embryo development and implantation (reviewed by Robertson 2005). These roles of seminal plasma in signalling in the female tract have been demonstrated to be important in rodents, humans and pigs, but are beyond the scope of this review and will be mentioned only where relevant to the survival or function of spermatozoa.

There has been considerable interest in the possibility of utilising seminal plasma, or its specific beneficial components, in the maintenance of sperm viability during processing for preservation or storage in preparation for artificial insemination (AI), or during other manipulations associated with controlled breeding, such as sex sorting (Maxwell & Johnson 1999). These processes usually involve the removal or dilution of seminal plasma from the semen, resulting in the loss of motility, metabolic activity and fertilising capacity of spermatozoa, termed the "dilution effect" by Mann (1954). Increasing understanding of cell physiology has led to the development of buffered extenders, containing inorganic ions and organic components which have partly alleviated the problem. Nevertheless, dilution of semen also reduces the concentration of seminal plasma and any positive effect it may be having on sperm membranes. Seminal plasma has been shown to reduce the dilution effect (Maxwell & Johnson 1999) and, when added to highly diluted semen, increases the viability of spermatozoa (rabbits: Castellini et al. 2000; cattle: Garner et al. 2001; sheep: Ashworth et al. 1994).

This review will examine the components of seminal plasma and their influence on the fertility of spermatozoa. The effects of additional seminal plasma or substitutes on the functional integrity and fertility of spermatozoa subjected to dilution, cryopreservation and sperm sexing will also be discussed. The main species of interest will be sheep and cattle, but additional insight and emphasis may be given by referring to other domesticated animals, particularly pigs and horses.

Composition of seminal plasma and its effect on fertility

The ejaculated semen of mammals comprises the spermatozoa suspended and generally swimming in a liquid medium defined as seminal plasma (Vanquelin 1791). It was initially thought that this medium was a simple filtrate of the blood that provided a nutrient-rich buffered vehicle to convey the sperm cells from the male to the female genital tract, whereupon its function ceased as the spermatozoa swam free of the plasma and commenced interaction with the fluids and cells of the female tract. Some components of the seminal plasma, such as the gel fraction found in species like the human and pig, were thought to provide a temporary physical means of retaining the spermatozoa in the female tract until they were able to establish themselves in sperm reservoirs, located further into the tract, preparatory to fertilisation (Hunter 1981).

Further functions of the seminal plasma were soon recognised. It was found that its presence, or that of a similar replacement medium, was necessary to sustain sperm viability, even for a short period of time, if the semen was held in vitro. Seminal plasma was not only a vehicle for the spermatozoa but also provided metabolic support, particularly as an energy source. The mixture of epididymal spermatozoa with seminal plasma activated their metabolic activity and motility (Mann 1964), due either to provision of special activating substances, or to dilution of inhibitory factors formerly contained in the epididymal secretion (Brooks 1990). The activation of sperm motility during epididymal transit was found to be regulated by interactions among the intracellular calcium ion concentration, cyclic AMP, adenosine and intracellular pH (Hoskins

& Vijayaraghavan 1990) as well as the phosphorylation status of specific proteins (Huang & Vijayaraghavan 2004).

As studies on the function and fertility of spermatozoa progressed, particularly with interest in the preservation of spermatozoa for AI, it became clear that the role of seminal plasma was more complex than that of a simple supportive or stimulatory medium. The contributions of the secretions of the epididymis and the interactions between the spermatozoa and the cells of the epididymis had profound effects on the maturity and function of the sperm cells (Cummins & Orgebin-Crist 1971), rendering them fully functional and able to fertilise oocytes upon their sequestration in the caudal epididymis.

At ejaculation, the accessory sex gland secretions contributed different components to the seminal plasma depending on the species. The prostate, vesicular, ampullary, bulbourethral and other minor glands located in the wall of the urethral canal showed great diversity (Mann 1964), even between closely related species (Mokkapati & Dominic 1977). The nature and effects of the secretions of these glands are still not clearly understood and are enormously variable between species. For example, the boar has large bulbourethral, prostate and vesicular glands, whilst in the ram and bull the vesicular glands are still large but bulbourethral and prostate are relatively small or disseminated. The simple explanation for these differences is that the volume and concentration of spermatozoa in the ejaculate depends on the site of semen deposition in the female tract and the length of copulation. The boar, for example, deposits a large volume of semen containing a low sperm concentration into the voluminous sow uterus during a lengthy ejaculation, whereas the bull and ram instantaneously ejaculate a small volume of highly concentrated spermatozoa into the female vagina. These differences have evolved to cope with different breeding strategies related to environmental influences.

Even within the same genus, however, the seminal plasma components can vary dramatically. In Camelids for example, the Dromedary and Alpaca, adapted to widely differing environments, produce a viscous gel ejaculate that entraps the spermatozoa until liquefaction (a process that can take several hours; Bravo *et al.* 2000) whereas the closely related Bactrian camel ejaculate contains little or no gel (Zhao 2000). The reasons for these differences are unclear but do not appear to be environmental or geographical adaptations.

Mann (1964) compiled what was known in the early 1960s from biochemical analyses about the major inorganic and organic components of seminal plasma. The vesicular glands secreted the largest volume in bulls and rams, with additional major contributions from the prostate and bulbourethral glands in boars and stallions. Nevertheless, the testes, epididymides and other accessory glands also contributed to the semen volume and, as we now know, to the important organic components of the seminal plasma and of the sperm membrane. The high water content and the presence of inorganic ions, citric acid, sugars, organic salts, prostaglandins and a number of proteins maintained the osmotic pressure of the semen and the pH of seminal plasma close to 7 in the bull and ram (Mann & Lutwak-Mann 1981). Seminal plasma also provided energy sources in the form of sugars for anaerobic and aerobic respiration. Prostaglandins were found in particularly high concentrations in ram seminal plasma compared with other species ($> 40 \mu g$ per ml) (Mann & Lutwak-Mann 1981). These were postulated as pharmacologic agents to aid motility and transport of spermatozoa by stimulating muscular contractions of the female tract. Prostaglandins may also have a role in the female tract as an inflammation-inducing agent, in synergy with other seminal cytokines, to promote sperm and embryonic survival (Robertson 2005). Unfortunately, it has not been possible to stimulate either sperm transport in the female tract or fertility by adding prostaglandins to diluted ram semen before AI (Salamon & Maxwell 1995a).

Seminal plasma was found to both stimulate and inhibit the function of the spermatozoa, not surprising given our current understanding of the multifunctional nature of seminal plasma proteins. In the early work on capacitation, Chang (1951) and Austin (1952) showed that the spermatozoa needed to be separated from the seminal plasma and spend a period of time in association with the female tract in order to attain the capacity to penetrate and fertilise oocytes. This suggested that some inhibitory factors, termed "decapacitation factors" (Chang 1957), associated with the presence of seminal plasma needed to be removed as part of the process of capacitation. In subsequent investigations, it was demonstrated that the capacity of spermatozoa to participate in fertilisation was inhibited by seminal plasma (Chang 1957). Moreover, the fertilisable lifespan of the spermatozoa in the female tract could be extended by their exposure to the decapacitation factors present in seminal plasma (Dukelow et al. 1967).

Studies of the influence of seminal plasma components on fertility have recently focused on changes occurring during epididymal transit and on the proteome contributed from accessory sex gland secretions.

Epididymal components

The spermatozoa acquire the ability to fertilise homologous oocytes and display motility gradually during epididymal transit. The increased fertility gradient results in caudal epididymal spermatozoa with better fertility than ejaculated spermatozoa, which have been mixed with the decapacitating proteins from the accessory sex glands (Dacheux & Paquignon 1980). Epididymal transit requires approximately 10 days in the bull and ram. Besides structural changes, spermatozoa undergo changes in the composition of the plasma membrane surface. For instance, in ram spermatozoa the molar ratio of cholesterol to phospholipid increases in the plasma membrane (Parks & Hammerstedt 1985), as well as changes in glycoproteins of different molecular weight that bind to the sperm plasma membrane. These changes result from a direct influence of a wide range of inorganic and organic constituents of the epididymal plasma which have been described in the past (Mann & Lutwak-Mann 1981; Robaire & Hermo 1988) and vary considerably between different regions of the epididymis. In recent times it has become clear that the most important components influencing these changes are proteins.

The cartographies of secreted (secretomes) and present proteins (proteomes) in the epididymal fluid, and their interactions with the maturing mammalian spermatozoa have been the subject of much recent study (reviewed by Gatti et al. 2004). Significant regionalized variations in these fluid proteins along the epididymis are reflected in particular modifications of the sperm plasma membrane domains. This appears to be achieved by degradation or liberation of testicular components, the absorption and integration of secreted proteins and enzyme-mediated changes to particular membrane proteins, such as by glycosylation or deglycosylation (Dacheux et al. 2003). For example, 17- and 23-kDa proteins are restricted to the caudal epididymis in the ram and can be directly integrated in specific domains of the sperm plasma membrane. The immunolocalization of the 17-kDa protein on the ram sperm tail suggests that it may have a role in sperm motility (Gatti et al. 2000).

Many testicular proteins in the seminal fluid, such as clusterin and transferin, disappear in the transition from the rete testis to the caput epididymis. However, more than 100 different proteins are secreted in the epididymal duct with most activity in the caput and corpus regions. The main secreted proteins are similar in different species and enzymatic activities, capable of controlling the sperm surface changes, are present in the fluid. These proteins are contained in both soluble and particulate compartments such as exosome-like vesicles (epididymosomes) and certain specific glycolipid-protein micelles (Gatti et al. 2004). Eight and 6 proteins represent most of the total epididymal secretion in the ram and bull, respectively, but in some zones only one protein can represent more than 50% of the secretion (Gatti et al. 2004). These proteins may be inserted and integrated into the sperm membrane during transit, with or without proteolytic processing, such as clusterin and several cysteine-rich secretory proteins (CRISP), respectively (Gati *et al*. 2004). Apart from the important role in regional modifications leading to the acquisition of motility and fertilizing capacity, epididymal proteins may provide protection against reactive oxygen species and bacteria, act further down the epididymis and some may be present in pro-forms that are activated after ejaculation (Dacheux *et al*. 2003). A number of other proteins present on the spermatozoon, which are redistributed or removed during epididymal transit, such as fertilin, cyritestin and germinal angiotensin converting enzyme (gACE), may also be important for fertility (Gatti *et al*. 2004).

How these different modifications in the compartments of the epididymis interplay to modify spermatozoa into fertile gametes during their transit remains to be revealed but the surface-modifying events appear to be critically important. Having acquired the capacity to fertilise, the spermatozoa undergo further modification, under the influence of seminal plasma, to prepare them for their transport from the site of deposition in the female tract at ejaculation to the site of fertilisation in the oviduct.

Accessory sex gland components

Ejaculation results in the confluence of spermatozoa from the tail of the epididymis with various secretions from the ampullary, bulbourethral, prostate and vesicular glands. The chemical composition of the array of substances produced by these glands, and the volume of the ejaculate, are species-specific and can vary among individuals belonging to the same species. In addition, any physiological, pathological or exogenous (seasonal) conditions which change the secretory function of one or more accessory glands, can influence the amount of fluid produced and the chemical composition of the seminal plasma. For example, changes in the abundance of particular proteins in autumn compared with other times of the year (Smith *et al*. 1999; Gundogan & Elitok 2004) have been correlated with seasonal changes in sperm quality parameters (Cardozo *et al*. 2006), freezability and resistance to cold shock in ram spermatozoa (Pérez-Pé *et al*. 2001a).

The vesicular gland produces most of the semen volume and is the major source of sperm-surface modifying proteins in ruminants (Bergeron *et al*. 2005; Fernández-Juan *et al*. 2006). The prostatic secretion has low protein content and the presence of free amino acids probably results from a combined action of proteases and transaminases in the glandular tissue (Mann & Lutwak-Mann 1981). In the ram and bull, the prostate is present as disseminated glandular tissue within the wall of the pelvic urethra. It secretes fructose, citric acid as well as ergothioneine (Mann 1964). The tissues of the bulbourethral and prostate glands are major sites of local immunoglobulin production (Foster *et al*. 1988) and the cells of the urethral and prostatic epithelium also produce serotonin, somatostatin and chromogranin A, which are important for the regulation of the emission of urine and/or semen and the inhibition of local exocrine and/or endocrine secretions (Vittoria *et al*. 1990). The bulbourethral gland is distinguished by a high content of a sialoprotein which plays an important part in the process of "gelation" of the semen in humans and pigs (Mann 1964). There is no gel in ram, goat and bull semen but type A lecithinase secreted by this gland may be involved in the fertility of spermatozoa (Corteel 1980).

While their inorganic components help to buffer the seminal plasma and maintain sperm metabolism and osmolarity, the most important contributors to fertility and sperm function from the vesicular glands in mammals are the proteins (Table 1). These fall into two main categories: the spermadhesins or heparin-binding proteins (predominating in boar, stallion and ram) and those proteins that contain fibronectin type II (Fn-2) domains, usually termed BSP (bovine seminal plasma) type proteins. The latter are the main proteins in bull seminal plasma (Bergeron *et al.* 2005).

Table 1. Major proteins originating from the vesicular glands of man and farm animals.

Type of Protein	Species	Name of Protein	References
Spermadhesins	Boar	AWN, AQN-1, AQN-3 and PSPI/PSPII heterodimer	Calvete *et al.* 1995 Varela *et al.* 1997
	Stallion	HSP-7	Reinert *et al.* 1996
	Man	Human spermadhesin-like protein (HSA)	Kraus *et al.* 2005
	Ram	Ram spermadhesin	Bergeron *et al.* 2005
	Bull	aSFP Z13	Dostalova *et al.* 1994 Tedeschi *et al.* 2000
Containing fibronectin type-II domains	Bull	BSP-A1, BSP-A2, BSP-A1/A2 and BSP-30kDa BSP-A1/A2 also known as PDC-109	Manjunath & Sairam 1987 Manjunath *et al.* 1987 Esch *et al.* 1983
	Ram	BSP-A1/A2-like protein P14 and P20 (now RSVP 14 and RSVP 20) RSP-15, RSP-16, RSP-22 and RSP-24 kDa	Jobim *et al.* 2005 Barrios *et al.* 2005 (Fernández-Juan *et al.* 2006) Bergeron *et al.* 2005
	Boar	pB1	Calvete *et al.* 1997
	Stallion	HSP-1, HSP-2 and HSP-12 kDa Fn-2 type protein	Calvete *et al.* 1995 Greube *et al.* 2004 Töpfer-Petersen *et al.* 2005
	Goat	GSP-14, GSP-15, GSP-20 and GSP-22 kDa	Villemure *et al.* 2003
	Bison	BiSV-16, BiSV-17, BiSV-18 and BiSV-28 kDa	Boisvert *et al.* 2004

A number of spermadhesins have been identified in boar seminal plasma (the AWN, AQN and PSP proteins) but only one in stallions (HSP-7; Table 1). The boar spermadhesins are further subdivided, depending on their ability to either bind heparin (AQN-1, AQN-3, AWN) or not (PSP-I/PSP-II heterodimer; Calvete *et al.* 1994). Proteins homologous to spermadhesins have been also found in human, ram (comprising about 40% of the protein according to Bergeron *et al.* 2005) and bull seminal plasma (Table 1). Proteins containing Fn-2 domains were first characterised in bull seminal plasma (BSP A1, A2, A3, A1/A2 and -30 kDa proteins). Similar proteins have been identified in ram (comprising about 20% of ram seminal plasma protein according to Bergeron *et al.* 2005), boar, stallion, goat and bison seminal plasma (Table 1). There are also proteins with Fn-2 domains originating from the epididymis (Saalmann *et al.* 2001) but these may undergo modification or have only local activity related to sperm maturation. There are significant homologies between the Fn-2 proteins identified to date in ruminants; for example, between the ram RSVP14, the ram RSP15, the bovine PDC-109 and the goat GSP-14/15 (Fernández-Juan *et al.* 2006).

Another class of proteins, the CRISP proteins, have been identified in a number of species. These proteins are expressed predominantly in the male reproductive tract and are implicated in processes ranging from spermiogenesis, post-testicular sperm maturation and capacitation to oocyte-sperm fusion (Udby *et al.* 2005). It is also possible that they have a role in penetration of the zona pellucida (Udby *et al.* 2005). They have not been implicated in the modification of sperm function in ruminants.

Many of the spermadhesins and BSP proteins have only recently been isolated and characterised in ruminant seminal plasma. An understanding of their roles in capacitation will be important to the utilization of seminal plasma as a modifier of sperm function.

Seminal plasma proteins and capacitation

Seminal plasma proteins are thought to have roles that both prevent (decapacitate) and mediate capacitation. Some vesicular gland proteins stabilise sperm membranes by binding firmly to their surface at ejaculation. As *in vivo* fertilisation requires a destabilisation of the membrane, as occurs during capacitation, this may explain why cauda epididymal spermatozoa are better able to fertilise oocytes *in vitro* than their freshly ejaculated counterparts (Dacheux & Paquignon 1980; Nagai *et al.* 1984; Rath & Niemann 1997).

The stabilising proteins are thought to be of the spermadhesin family (Romão *et al.* 1997). However, some of the BSP proteins may initially play a similar role in the early stages after ejaculation. For example, the BSP proteins in ram and bull seminal plasma interact with choline phospholipids on the sperm membrane, with high and low density lipoproteins and with heparin, conferring on them multifunctional biological roles in membrane stabilisation (decapacitation) and destabilisation (capacitation; Manjunath & Thérien et al. 2001; Fernández-Juan *et al.* 2006). On mixing with the spermatozoa at ejaculation these proteins induce cholesterol efflux from the sperm membrane (Manjunath & Sairam 1987; Swamy 2004) resulting in reorganisation of the membrane components and, through their binding with choline phospholipids, stabilise the membrane.

As spermatozoa reach the oviducts they are exposed to follicular and oviductal fluids which contain high density lipoproteins (HDL) and heparin-like glycosaminoglycans (GAGs), the physiological capacitation factors (Rodriguez-Martinez *et al.* 1998). The exact signal transduction mechanism is still unclear but some BSP proteins enable spermatozoa to bind to the oviductal epithelium (Gwathmey *et al.* 2003) and they may possibly be involved in the release of spermatozoa from the oviductal sperm reservoir as has been shown in pigs (Jelinkova *et al.* 2004). Other BSPs may induce capacitation either via protein tyrosine phosphorylation (involving interaction with GAGs) or not involving protein tyrosine phosphorylation (through interaction with HDL; Thérien *et al.* 1998; 1999). BSP-A1/A2 (PDC-109) also modulates the effects of other capacitation agents (heparin, progesterone and angiotensin II) by increasing the proportion of acrosome-reacted bull spermatozoa (de Cuneo *et al.* 2004).

Pig and horse spermadhesins display carbohydrate-binding activity (Calvete *et al.* 1995) and, like the BSP proteins, interact with the sperm surface on ejaculation. Because of their known interactions with heparin and the zona pellucida, they may be involved in capacitation (Calvete *et al.* 1995) and oocyte recognition (Töpfer-Petersen *et al.* 1998) or mediate sperm binding to the oviductal epithelium (Töpfer-Petersen 1999). The 15.5 kDa spermadhesin identified in ram seminal plasma by Bergeron *et al.* (2005), which is 70% homologous with porcine AQN-1 (Töpfer-Petersen *et al.* 1998), may play a similar role in decapacitation (initially after ejaculation) and capacitation or sperm binding to the oviducal epithelium as it does in the pig, although Barrios *et al.* (2005) and Fernández-Juan *et al.* (2006) claim this "protective and restoring" role for RSVP14 and RSVP20 (Fn-2 domain proteins). This does not rule out other, yet to be elucidated, roles for porcine, bovine and ovine spermadhesins in the preparation of spermatozoa for fertilisation.

Further proteins have been identified with decapacitating activity. These either inhibit the normal signal-transduction pathways associated with the initiation of capacitation and/or mask zona pellucida ligands on the sperm surface. ESP13.2 and PSP94, epididymal secretory proteins, coat the surface of macaque spermatozoa and are released at capacitation: ESP13.2 inhibits sperm binding to the zona pellucida when added back to the sperm surface (Tollner et al. 2004). The porcine PSP-I/PSP-II spermadhesin heterodimer is able to preserve viability and acrosome integrity, and blocks oocyte penetration by frozen-thawed but not fresh boar spermatozoa (Caballero et al. 2004a), through the prevention of sperm-zona pellucida binding (Caballero et al. 2005). RSVP14 and RSVP15 in ram seminal plasma are suggested to have a decapacitating role, stabilizing sperm membranes and protecting against cold shock when added to ram spermatozoa (Pérez-Pé et al. 2002; Barrios et al. 2005).

Recent work by Fraser and co-workers, studying mouse and human spermatozoa, has revealed important roles for a number of small peptides found in seminal plasma, which act as first messengers in the regulation of in vitro capacitation (reviewed by Fraser et al. 2006). These capacitation-inducing proteins, including adenosine, angiotensin II, calcitonin and fertilisation-promoting peptide (FPP), have been shown to stimulate cAMP production and protein tyrosine phosphorylation in spermatozoa (mouse: Mededovic & Fraser 2004). Separate signal-transduction pathways, relevant to each peptide and involving specific receptors, have a common end-point of increased production of the second messenger cAMP. With the exception of angiotensin II, which stimulates cAMP throughout capacitation, these signals result initially in capacitation but this is followed by an inhibition of both cAMP production and spontaneous acrosome loss, so that the capacitated spermatozoa retain their fertilising capacity. The transition from stimulation to inhibition involves loss of decapacitation factors from sperm membrane receptors, one of which has been identified in the mouse as phosphatidylethanolamine-binding protein 1 (Gibbons et al. 2005). It has been suggested that the actions of these first messengers observed in vitro, may have significant implications for enhancing or preserving fertilising capacity in vivo after the spermatozoa have been exposed to them at ejaculation (Fraser et al. 2006). Whether these roles for small peptides in signal transduction mechanisms leading to capacitation in mouse and human spermatozoa also operate in ruminant semen is yet to be determined.

Fertility-associated proteins

There have been a number of attempts to correlate the fertility potential of spermatozoa with the proteins present in the seminal plasma. The quantity of prostaglandin D synthase (PGDS; Gerena et al. 1998) and the presence of clusterin in seminal plasma (Ibrahim et al. 2000), both originating in the epididymis, have been suggested to be correlated with bull fertility. The former was later shown to be an unreliable marker as high fertility bulls had both high and low PGDS concentrations (Fouchécourt et al. 2002). Killian et al. (1993) reported the presence of four 'fertility-associated proteins' in bull seminal plasma (osteopontin, spermadhesin Z13, phospholipase A2 and BSP-30) that may improve fertility after AI with commercial dairy bull semen. Furthermore, 'anti-fertility' factors reputed to be present in the seminal plasma may bind to the sperm plasma membrane and reduce the fertility of high fertility bull spermatozoa (Henault & Killian 1996). The 'fertility-associated proteins' identified by Killian et al. (1993) in bull seminal plasma, reported to originate from the vesicular glands, are currently under testing for commercial application by Genex Cooperative (Moura et al. 2005). Some proteins isolated from stallion seminal plasma have also been positively correlated with fertility and may have commercial application (Brandon et al. 1999).

Gatti *et al.* (2004) hypothesized that abnormal gACE levels in ram seminal plasma (released from spermatozoa during epididymal transit) could be an indicator of an interruption to sperm maturation, as the quantity of gACE is correlated with the number of spermatozoa. The correlation of gACE activity with the number of spermatozoa in the ejaculates of young rams had been confirmed (Métayer *et al.* 2001) and low gACE levels were associated with lower fertility (Gatti *et al.* 2004). Anti-sperm antibodies (ASA) have also been identified, particularly those associated with infertility in men. For example, Carlsson *et al.* (2004) reported that prostasomes adhered to human spermatozoa were major targets for ASAs and identified PIP (prolactin-inducible protein) and clusterin as dominant antigens for sperm-agglutinating autoantibodies. The usefulness or otherwise of these correlations as a basis for assessment of individual males or of their ejaculates, and their application in semen processing and preservation, is yet to be determined.

The possibility of fertility-associated proteins in seminal plasma and on the sperm surface needs to be considered in the context of new research on the relationship between Major Histocompatability Complex (MHC) haplotype and mating outcome. This work suggests a connection between the MHC and both pre- and post-copulatory female choice of the spermatozoon participating in fertilisation. This is based on the crucial importance of the MHC in immune responses, resulting in an evolutionary female preference for mates that are MHC heterozygous and unique (Ziegler *et al.* 2005). The proposed mechanism of action is through chemoreceptors, which operate in mate choice, not only through nasal odorant receptors (Buck & Axel 1991) but also through guidance cues for spermatozoa (Spehr *et al.* 2003; Robertson 2005). Osteopontin, for example, one of the fertility-associated proteins in bovine seminal plasma (Killian *et al.* 1993) is a ubiquitous cell adhesion component involved in cell migration, chemotaxis and macrophage activation (Moura 2005). PSP-I and PSP-II may not play a role in capacitation or sperm-oocyte binding but rather display proinflammatory effects which modulate immune responses in the porcine uterus (Assreuy *et al.* 2002).

The hidden female effects that impact on the success of males in fertilizing ova have been labelled 'cryptic female choice' (Eberhard 1996) and relate partly to sperm competition mediated by mutual recognition of seminal plasma components (including soluble MHC antigens and those sequestered on the sperm surface) by spermatozoa from different males and/or by the female reproductive tract (Ziegler *et al.* 2005). Thus, in humans, the sharing of MHC alleles between partners may influence the occurrence of certain forms of human sterility and recurrent spontaneous abortions (Beydoun & Saftlas 2005). There is also evidence for MHC compatibility involvement in oocyte penetration and within the penetrated oocyte by influencing the outcome of the second meiotic division (Wedekind *et al.* 1996). It is possible that some of the fertility-associated proteins may either be the result of MHC expression or be masking ligand-receptor mechanisms that might normally block either sperm function or fertilisation.

Seminal plasma vesicles

It has been known for a long time that small vesicles are present in the seminal plasma of several mammals. These are roughly spherical organelles containing dense particles that are delimited by single, double or multiple membranes, and range in diameter from 21 to nearly 1000 nm (Table 2). The vesicles generally have been named after the accessory organ from which they were thought to originate. The membranous vesicle fraction identified in human seminal plasma, for example, was initially named pellet II (Ronquist *et al.* 1978a; 1978b) and later "prostasomes" (Brody *et al.* 1983) or "epididymosomes" (Saez *et al.* 2003) because of its prostatic or epididymal origin. Prostasomes in human seminal plasma, or prostasome-like membrane vesicles in equine seminal plasma, are particles made up of lipid and protein. It has been hypothesized that these may assist the fertilizing potential of spermatozoa by adhering to and fusing with them, decreasing the fluidity of the sperm membrane, thereby delaying the acrosome reaction and improving the motility and viability of spermatozoa, especially as they enter the female genital tract (Minelli et al. 1998; Minelli et al. 1999; Kravets et al. 2000).

Table 2. Size of vesicles identified and characterised in the seminal plasma of man and farm animals.

Species	Vesicle diameter range (nm)	References
Man	21-100	Ronquist et al. 1978a; 1978b
Ram	50-300 22-986 (mean 159.7 ± 2.92)	Breitbart and Rubinstein 1982 El-Hajj Ghaoui et al. 2004
Boar	18-577 (mean 130.9 ± 3.22)	El-Hajj Ghaoui et al. 2004
Stallion	75-175 nm 15-671 (mean 164.1 ± 4.42)	Minelli et al. 1998 El-Hajj Ghaoui et al. 2004

The fusion between prostasomes and spermatozoa (Carlini et al. 1997) may stabilise the sperm plasma membrane by enriching it with cholesterol, sphingomyelin, and saturated glycerophospholipid (Arienti et al. 1997). This is postulated to prevent the occurrence of capacitation and the acrosome reaction by producing a transient increase of intracellular calcium ion concentration (Palmerini et al. 1999). Others have suggested that human prostasomes may improve the recovery of hyperactively motile spermatozoa and consequently increase the opportunities for fertilisation (Fabiani et al. 1994). Likewise, prostasome-like membrane vesicles identified in the seminal plasma of the stallion and the occurrence of a fusion-like process between these vesicles and the sperm cells suggests that they may play a physiological role in the fertilizing capacity of equine spermatozoa (Arienti et al. 1998). These roles for membrane vesicles in capacitation and/or fertilisation have yet to be supported by convincing experimental evidence.

Breitbart & Rubinstein (1982) first isolated vesicular membranes from ejaculated ram seminal plasma and demonstrated the presence of divalent-cation-dependent ATP-ase associated with them. Vesicles with a specific protein content, and in association with cellular protein isoforms, have been identified also in the cauda epididymal secretions of the ram (Ecroyd et al. 2004). These were shown to form the majority of the vesicles retrieved in the seminal plasma of the ram and were called epididymosomes (Gatti et al. 2005). By examining the seminal plasma of vasectomized and entire rams, we demonstrated that the membrane vesicles in ram seminal plasma do not originate from the accessory sex glands (El-Hajj Ghaoui et al. 2006). A comparison of the morphology of vesicles in the seminal plasma and the cytoplasmic droplets by transmission electron microscopy led us to conclude that, at least in the ram, the vesicles originate from a combination of the droplets and the epididymis (El-Hajj Ghaoui et al. 2006). It should be noted, however, that the absence of vesicles from the seminal plasma of vasectomised rams does not completely preclude their presence in the accessory sex glands. For example, "secretomes" identified by immunostaining in the vesicular glands of rams as localized sites of RSVP-14 production, may provide an apocrine mechanism for releasing the protein, whereupon the vesicle carriers break down in the lumen of the gland (Fernández-Juan et al. 2006).

The biological importance and the functions of membrane vesicles in ram seminal plasma are still obscure. Based on the data from other mammalian species, these membrane vesicles may be involved in capacitation and have the potential to influence the fertility of spermatozoa. It may be possible to incorporate these vesicles with different seminal fluid extenders in order to improve fertility after semen preservation and AI.

Addition of seminal plasma to spermatozoa

Reports are somewhat equivocal on the benefits or otherwise of seminal plasma in association with spermatozoa of various species during artificial preservation. For example, early researchers on frozen storage of boar spermatozoa recommended that removal of the seminal plasma, after initial holding in its presence, was beneficial to post-thaw sperm survival (Pursel & Johnson 1975). More recently, the addition of seminal plasma to boar spermatozoa was reported to be detrimental to sperm survival during post-thaw incubation (Erickson *et al.* 2005) but 20% seminal plasma prevented or reversed capacitation-like changes of boar spermatozoa, as detected by the chlortetracycline assay, resulting from incubation at 39°C or cooling to 5°C (Vadnais *et al.* 2005). Removal of seminal plasma and resuspension with artificial media is also reported to improve the survival and DNA integrity of stallion spermatozoa during chilled storage (Love *et al.* 2005). Seminal plasma is generally removed from equine spermatozoa before cryopreservation as it is deleterious to sperm survival if they are exposed to it for a prolonged period before freezing (Moore *et al.* 2005).

Conversely, in ruminants, Ashworth *et al.* (1994) identified beneficial proteins 5-10kDa in size that reduced the adverse effects of dilution on ram spermatozoa. Moreover, fertility of frozen-thawed ram spermatozoa after cervical insemination was improved after addition of seminal plasma (Maxwell *et al.* 1999) although it is not known whether the addition of seminal plasma components is beneficial to fertility of frozen-thawed bull spermatozoa.

One explanation for these apparent contradictions is the relative ratio of capacitating and decapacitating, or possibly beneficial and harmful, proteins in seminal plasma, and their variation between species or even individuals and within individual males over time. It should be noted that whole seminal plasma is a complex mixture of organic and inorganic components, as well as proteins with positive and negative effects. The concentrations of these proteins vary from male to male and can appear and disappear depending on environmental factors such as season of collection, temperature, nutrition and stress (Pérez-Pé *et al.* 2001b). Most research to date on the addition of seminal plasma to spermatozoa has not controlled for this within male variation.

Moreover, the beneficial effects of seminal plasma are likely to be restricted to specific proteins and these may be negated by harmful effects of other factors present in whole seminal plasma. For example, García-López *et al.* (1996) demonstrated that protein-free ram seminal plasma contains a low molecular weight factor that interferes with the viability-stimulating effect of isolated plasma proteins, possibly by preventing their adsorption to the sperm surface. The adsorption of beneficial proteins will also depend on the origin of the spermatozoa (eg epididymal, ejaculated, oviductal) and whether they have been subjected to manipulations such as dilution or cryopreservation. These treatments may act to strip proteins from spermatozoa, revealing ligands and binding sites that render them more susceptible to interactions, such as those associated with: capacitation and the (false) acrosome reaction; cell death; or more readily able to interact with and penetrate oocytes.

The influence of seminal plasma may also depend on the other components of the medium in which the spermatozoa are extended. In the presence of egg yolk, for example, Manjunath *et al.* (2002) claim that the 14, 15 and 16 kDa RSP proteins, suggested to be detrimental to sperm survival, would be bound to and rendered inactive by the yolk low density lipoprotein fraction allowing the beneficial or decapacitating protein(s) to bind to the sperm head.

When a detailed understanding is gained of the roles of the main seminal plasma proteins in the normal maturation, transport and preparation of spermatozoa for fertilisation after deposition in the female reproductive tract, it may be possible to utilise specific proteins to improve or enhance the function of spermatozoa that have been compromised through processing and storage for AI. For the present, there have been a number of studies on the addition of whole seminal plasma, or crude fractions of it, to spermatozoa that have been through such processes. The final section of this review outlines the known effects of frozen storage and sex-sorting by flow cytometry on the function and capacitation status of ram spermatozoa. It also considers current evidence for the benefit, or otherwise, of adding seminal plasma or its membrane vesicles to frozen-thawed and sex-sorted spermatozoa of several different species.

The potential benefits of seminal plasma to the fertility and function of frozen-thawed ram spermatozoa

Seminal plasma alone is not sufficient to protect ram spermatozoa against cold shock. Therefore, the extenders or diluents used for cryopreservation need to contain components that will potentially decrease sperm cryoinjury (Salamon & Maxwell 1995a). The function of spermatozoa is sustained for many years by frozen storage (Gillan *et al.* 2004), but their fertilizing ability, especially after cervical insemination, is much lower than for fresh diluted spermatozoa (Gillan & Maxwell 1999). Although recent Scandinavian studies report 25 day non-return rates above 70% after cervical insemination with frozen-thawed semen (Paulenz *et al.* 2005) the standard pregnancy rates in Australia are closer to 20% (Maxwell & Hewitt 1986).

The development of intrauterine insemination by laparoscopy has largely overcome this problem (Salamon & Maxwell 1995b) but there are ethical limitations to the continued use of this invasive procedure and it may be necessary to revert to cervical insemination. For this reason, there has been considerable interest in the effects of seminal plasma and its constituents on sperm quality after freeze-thawing. For example, when either whole ram seminal plasma or its > 10 kDa protein fraction were added to the diluent, acrosome and plasma membrane integrity, motility, and sperm heterogeneity after thawing were significantly improved (Ollero *et al.* 1997). Furthermore, improved sperm motility, a reduction in the proportion of capacitated and acrosome reacted cells, and increased ability to penetrate cervical mucus were obtained after addition of seminal plasma to frozen-thawed ram semen (Maxwell *et al.* 1999).

Capacitation-like changes in frozen-thawed spermatozoa and their possible reversal by seminal plasma

A limitation to extended viability of frozen-thawed spermatozoa in the female tract is the so-called "pre-capacitated" state in which a high proportion of the cells emerge after freeze-thawing (Cormier *et al.* 1997; Maxwell & Johnson 1997b; Gillan & Maxwell 1999; Cormier & Bailey 2003) since their membranes become permeable to calcium ions, leading to the (false) acrosome reaction and cell death. During "normal" sperm capacitation, the spermatozoon becomes highly polarized and its plasma membrane shows regional specialization with changed characteristics and composition (Guraya 2000). The changes in sperm cells as a result of freezing and thawing have been identified as "capacitation-like" (Watson 1995; Maxwell & Watson 1996) because they resemble the late steps in the signal transduction pathway that leads from capacitation to the acrosome reaction, as identified by the fluorescent membrane probe chlortetracycline (Cormier *et al.* 1997; Gillan & Maxwell 1999), and they render the sperm functionally capacitated in both their ability to fertilise oocytes in vitro (Maxwell & Watson 1996) and interact with oviduct epithelial cells (Gillan *et al.* 2000).

While these capacitation-like changes are clearly not true capacitation (Green & Watson 2001), recent work on sperm cryopreservation has focused on the role of seminal plasma in the stabilisation of sperm membranes as well as the identification and separation of its decapacitation factors (Guraya 2000). This is potentially important as, once the sperm membrane becomes fully capacitated, the spermatozoa only have a limited lifespan in which to find and fertilise the ovum. It may be that all the surface changes which occur in spermatozoa during capacitation, for example cholesterol efflux (Parks & Ehrenwald 1989; Cross 1998; 2003), alteration in the protein composition of the surface (Myles *et al.* 1990), or changes in the distribution of in-tramembranous particles (Yanagimachi 1988; Suzuki-Toyota *et al.* 2000), are influenced by changes to the proteins initially contributed by the seminal plasma and bound to the spermatozoa. It is thought that these decapacitating factors in seminal plasma are removed or modified during the transit of the spermatozoa through the female genital tract and, if added to frozen-thawed spermatozoa, would extend their longevity by binding, or re-binding, to the plasma membrane and inhibiting further structural and physiological changes (Fraser *et al.* 1990; Maxwell & Johnson 1999; Guraya 2000; Barrios *et al.* 2000; Wolfe *et al.* 2001; Barrios *et al.* 2005).

The recent published literature on spermatology poses several hypotheses on the nature of decapacitation factors, their function and their presence on the spermatozoa themselves or in their surroundings. A number of these have been tested, mainly *in vitro*, in non-ruminant species but few have been applied experimentally to ruminant spermatozoa. However, there is considerable evidence of a direct effect of whole seminal plasma and its fractions on the function of frozen-thawed and processed ruminant spermatozoa.

Function of frozen-thawed spermatozoa after addition of seminal plasma fractions and membrane vesicles

In studies on frozen-thawed ram spermatozoa, we found that sperm function and fertility were improved by the post-thaw addition of whole seminal plasma (Maxwell *et al.* 1999). We hypothesised that the beneficial components in seminal plasma would be in its vesicle-free fraction and that this fraction would be active in both entire and vasectomised rams. In other words, the beneficial components of seminal plasma were likely to be of post-epididymal origin and probably proteins from the vesicular gland. Thermal denaturation of whole seminal plasma removed its protective effect on frozen-thawed (CA McPhie, S Mortimer, WMC Maxwell & G Evans, unpublished observations) and cold-shocked ram spermatozoa (García-López *et al.* 1996), confirming that the active constituent was proteinaceous. We further postulated that the membrane vesicle fraction of seminal plasma either had no function, or had a role in sperm maturation and membrane function confined to the epididymis, and would not influence the function or fertilizing capacity of frozen-thawed spermatozoa.

We first isolated membrane vesicles from ram seminal plasma, purified them by size exclusion chromatography and defined their structure (El-Hajj Ghaoui *et al.* 2004). Next, we demonstrated that vasectomy eliminated the membrane vesicles from ram seminal plasma (El-Hajj Ghaoui *et al.* 2006), indicating that they were of testicular or epididymal origin. Whole seminal plasma collected before and after vasectomy of four rams was separated by ultracentrifugation into two fractions; supernatant and a pellet of vesicles. The protein profiles of these fractions, characterised by SDS-PAGE, were in the same molecular weight ranges as those identified by Bergeron *et al.* (2005) and Fernández-Juan *et al.* (2006) for entire rams (Fig. 1).

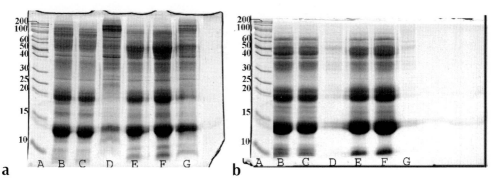

Fig. 1. SDS-PAGE pattern of whole seminal plasma and its fractions before (a) and after (b) vasectomy from the same ram. Samples were standardized (20 μg protein in 12 μl) denatured, reduced and separated on 15% polyacrylamide gel and stained with Coomassie Blue R-250. Lane A: Protein marker (PageRuler Protein Ladder; Progen Indus- tries Ltd., Darra, Qld, Australia); lanes B and E: whole seminal plasma; lanes C and F: supernatant; lanes D and G: pellet of vesicles. Whole seminal plasma was obtained by centrifugation and washing (twice: 2500xg, 4°C, 30 min) of 12 and 8 whole ejaculates, collected before and after vasectomy, respectively, and pooled within ram. Supernatant and pellet fractions were obtained by ultracentrifugation of whole seminal plasma (100,000xg, 4°C, 80 min). 1 (RSP-22 kDA and RSP-24 kDa, or RSVP-20) and 2 (RSP-15 kDa, RSP-16 kDa and 15.5 kDA spermadhesin, or RSVP-14) indicate the approximate positions of the major proteins in ram seminal plasma (Bergeron *et al.* 2005; or Fernández- Juan *et al.* 2006, respectively). (R El-Hajj Ghaoui, L Gillian, PC Thomson, G Evans and WMC Maxwell, unpublished observations).

Regardless of vasectomy, whole seminal plasma and supernatant improved motility character- istics of frozen-thawed ram spermatozoa (Fig. 2: R El-Hajj Ghaoui, L Gillian, PC Thomson, G Evans and WMC Maxwell, unpublished observations) and improved the ability of spermato- zoa to fertilise *in vitro* matured oocytes (Fig. 3: R El-Hajj Ghaoui, L Gillian, PC Thomson, G Evans and WMC Maxwell, unpublished observations). The membrane vesicle fraction from plasma collected before vasectomy had no effect on spermatozoa at its normal concentration but marginally increased fertilizing capacity of frozen-thawed spermatozoa when included at three times normal concentration in the incubation medium (Fig. 3). This may be due to protein remnants still associated with the vesicle fraction, rather than the vesicles themselves. Protein bands were identified by SDS-PAGE in the vesicle fraction of entire rams (Fig. 1a, lanes D and G) although in low abundance, particularly the 20 kDa protein. In vasectomised rams, however, only one faint protein band at 15 kDa was apparent in the vesicle fraction (Fig. 1b, lanes D and G).

These findings confirm our previous reports on the beneficial effects of whole seminal plasma on the function and fertility of frozen-thawed ram spermatozoa (Maxwell *et al.* 1999), and suggest that the beneficial components of the seminal plasma are largely contained in its non-vesicle protein fraction.

Function of sex-sorted spermatozoa after addition of whole and artificial seminal plasma

While exposure of spermatozoa to additional seminal plasma after freezing and thawing can improve their viability, the removal of seminal plasma during processing can also decrease the viable life-span of spermatozoa. An example of this is the dilution associated with the sexing of

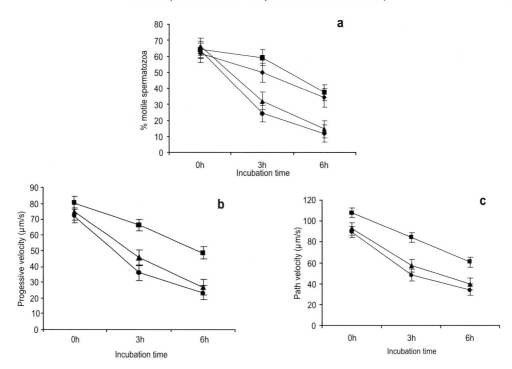

Fig. 2. The motility characteristics of frozen–thawed ram spermatozoa during 6 h incubation at 37°C with (●) control buffer, (■) whole seminal plasma, (▲) vesicles at their normal concentration or (◆) vesicles at 3X their concentration. Panel **a**: motility (%, determined subjectively), panel **b**: progressive velocity (VSL, μm.sec⁻¹ determined by CASA; HTM-IVOS; Hamilton-Thorne, USA) and panel **c**: average path velocity (VAP, μm.sec⁻¹, determined by CASA). Data are presented as mean ± s.e.m. (R El-Hajj Ghaoui, L Gillian, PC Thomson, G Evans and WMC Maxwell, unpublished observations).

mammalian spermatozoa by flow cytometric sorting. Extensive dilution of semen and insemination with a low number of spermatozoa in cattle have been standard means of applying this technology (Johnson 2000). The sperm sexing technique, based on the difference in DNA between X- and Y-chromosome-bearing spermatozoa (Moruzzi 1979), requires an extensive dilution of the spermatozoa before and after sorting but this dilution is detrimental to the motility, membrane status and fertilizing capacity of the spermatozoa (Maxwell & Johnson 1997a). In order to minimize this dilution effect, sorted spermatozoa are collected into a tube containing either an egg yolk extender (Johnson 2000) or seminal plasma may be included as a portion of the staining extender and the collection medium (Maxwell *et al.* 1997; Maxwell & Johnson 1999; Centurion *et al.* 2003).

Catt *et al.* (1997) showed that 10% ram or boar seminal plasma in the diluent increased motility and viability of ram or boar spermatozoa in semen that had been extended 400- or 20-fold, respectively. While the viability of ram spermatozoa (percentage live as assessed by non-penetration of nuclear membrane by propidium iodide) was improved by seminal plasma, this was partly due to a decrease in the proportion of agglutinated cells. Interestingly, heterologous seminal plasma is generally detrimental (boar with ram or ram with boar: Catt *et al.* 1997; CA McPhie, S Mortimer, WMC Maxwell & G Evans, unpublished observations; bull with ram: García-López *et al.* 1996), autologous plasma is usually beneficial (male with the same male)

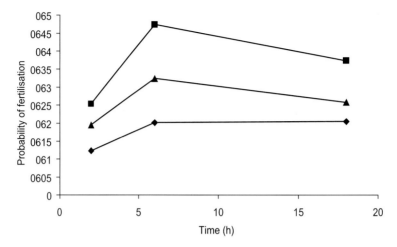

Fig. 3. The probability of *in vitro* fertilisation after 2, 6 and 18 hr co-culture of *in vitro* matured sheep oocytes with frozen-thawed ram spermatozoa, previously incubated (37°C, 3 hr) with (♦) control buffer, (■) supernatant (20%, v/v in buffer) or (▲) pellet of vesicles at 3x normal concentration (20%, v/v in buffer). Whole seminal plasma was obtained by centrifugation and washing (twice: 2500xg, 4°C, 30 min) of whole ejaculates pooled within ram. Supernatant and pellet fractions were obtained by ultracentrifugation of whole seminal plasma (100,000xg, 4°C, 80 min). Oocytes (3 replicates of 315, 341 and 198) were matured and *in vitro* fertilisation (1 × 10⁶ motile spermatozoa per ml) occurred in IVF complete medium (Cook IVF, Brisbane, Australia). After co-culture, oocytes were fixed and stained with aceto-Orcein and examined (phase contrast, 600x magnification) for evidence of fertilisation. The probability of fertilisation was determined from the binary data (fertilised or unfertilised) by fitting a generalised linear mixed model (GLMM) using GenStat 8. (R El-Hajj Ghaoui, L Gillian, PC Thomson, G Evans and WMC Maxwell, unpublished observations).

but the effects of homologous plasma vary with the donor male and can be negative (ram and bull: CA McPhie, S Mortimer, WMC Maxwell & G Evans, unpublished observations; boar: Caballero *et al.* 2004b). The beneficial components of boar seminal plasma, in the case of highly diluted boar spermatozoa, have been isolated to the PSP-II subunit of the PSP-I/PSPII spermadhesin (Garcia *et al.* 2006). In the case of ram seminal plasma the beneficial proteins may be RSVP-14/20 (Barrios *et al.* 2005) or ram spermadhesin (Bergeron *et al.* 2005) but these have been less clearly defined for ram and bull as the reported effects of seminal plasma on highly diluted and sex-sorted spermatozoa are equivocal in these species (see below).

The viability and membrane integrity of spermatozoa was improved in vitro if seminal plasma was included in the staining extenders for boar and ram spermatozoa or in the collection medium for boar or bull spermatozoa (Maxwell *et al.* 1997). It was concluded that the beneficial effects in the staining extender reflected a reduction of the 'dilution effect' and in sperm agglutination, whereas seminal plasma in the collection medium stabilised sperm membranes and prevented premature cell death independent of effects on sperm motility (Maxwell & Johnson 1999). Proportions of boar spermatozoa with capacitation-like changes were lower when seminal plasma was present in the collection medium only, than in the staining extender or absent altogether, but the former treatment substantially reduced the proportions of polyspermic, penetrated and cleaved oocytes, and the proportion of blastocysts. These findings suggested that capacitation-like changes in boar spermatozoa associated with flow-cytometric sorting could be reduced by the inclusion of seminal plasma in the collection medium but this treatment reduced the ability of spermatozoa to fertilise oocytes *in vitro* under these conditions (Maxwell *et al.* 1998). Whether similar effects would result from seminal plasma in the staining and collection media used for ram and bull spermatozoa were not determined (Maxwell & Johnson 1999).

The results of Maxwell *et al.* (1997) and Catt *et al.* (1997) were obtained using a modified Coulter Epics V flow cytometer with a 7-W argon laser (170 mW; ultra-violet) and operating at a sheath pressure of 20 psi. In the case of ram semen, the spermatozoa were diluted before sorting at 500-1000 fold (around ten times that required for boar spermatozoa) in preparation for staining with the DNA-permeant Hoechst 33342 fluorochrome that distinguishes X- from Y-bearing cells. This was followed by a second dilution by sheath fluid (phosphate-buffered saline supplemented with bovine serum albumin) and collection medium resulting in a final dilution of up to 30,000-fold. Compared with low-speed sorting with the Coulter machine, our current studies utilize a high-speed flow cytometer (SX MoFlo®) operating at a sheath pressure of 40 psi with a TRIS-based sheath fluid. The diode-pumped solid state pulse laser, operating at 125 mW, results in less exposure of individual spermatozoa to ultra-violet light than the argon laser. The spermatozoa undergo less dilution before and after sorting, because of the increased speed of sorting (5-10,000 compared with 1400 spermatozoa sec-1), resulting in a final dilution rate of about 5000-fold, and the spermatozoa are exposed to the sheath and collection medium for a shorter period of time (15-30 min) compared to low speed sorting (60 min).

Under these high speed sorting conditions, whole seminal plasma does not provide the same benefits to sorted ram spermatozoa that have been subsequently frozen and thawed, compared to non-sorted frozen-thawed spermatozoa (Maxwell *et al.* 2004). Sex-sorting of ram spermatozoa using MoFlo technology improved motility (64.7 ± 4.8 vs. $43.1 \pm 4.0\%$, $P < 0.05$), viability (53.4 ± 4.0 vs. $36.5 \pm 3.4\%$, $P < 0.05$) and mitochondrial activity (63.1 ± 4.7 vs. $41.3 \pm 3.9\%$ respiring, $P < 0.05$) of spermatozoa compared with a non-sorted population but reduced the forward progressing velocity of sex-sorted cells (measured by CASA; VAP: 83.0 ± 3.8 vs. 97.1 ± 6.9 μm.sec-1, $P < 0.05$) and their ability to penetrate cervical mucus (22.0 ± 3.5 vs. 43.3 ± 3.8 spermatozoa penetrated 1 cm, $P < 0.05$; for sorted vs. control spermatozoa, respectively) (S de Graaf, WMC Maxwell & G Evans, unpublished observations). These differences were retained when the sorted or non-sorted spermatozoa were incubated for 6 hours (Fig. 4: S de Graaf, WMC Maxwell & G Evans, unpublished observations).

The improvement in sperm motility and viability in the flow-sorted population is not surprising, as the cells are stained before sorting with a dye (originally propidium iodide but now a non-toxic food dye) that penetrates the nuclear membrane of non-viable cells, staining the nucleus red, and allowing these cells to be gated out to waste during flow sorting. The reduction in kinematics and sperm transport may explain the lower fertility obtained after AI of ewes with low (2-4 million) compared with high (20 million) doses of sex-sorted, frozen-thawed spermatozoa (Maxwell *et al.* 2004). Nevertheless, we have recently obtained indistinguishable fertility when sufficient numbers (15 million motile) of control or sex-sorted ram spermatozoa - or even frozen-thawed, sex-sorted, re-frozen and re-thawed spermatozoa - are deposited in the uterus of synchronised ewes by laparoscopic insemination (SP de Graaf, G Evans, WMC Maxwell, DG Cran & JK O'Brien, unpublished observation). The latter results demonstrate that the highly-selected population of spermatozoa resulting from sex-sorting by flow cytometry, interjacent with two rounds of freezing and thawing, are capable of normal fertility when placed in reasonable proximity to the site of fertilisation close to the time of ovulation.

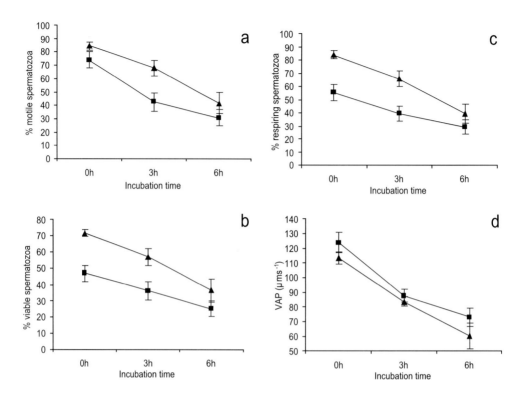

Fig. 4. The motility characteristics, viability and mitochondrial activity of [■] unsorted or [▲] sex-sorted frozen-thawed ram spermatozoa during 6 h incubation at 37°C. Panel **a**: motility (%, determined by CASA; HTM-IVOS; Hamilton-Thorne, USA), panel **b**: viability (% live spermatozoa assessed by exclusion of ethidium homodimer-1), panel **c**: mitochondrial activity (% respiring spermatozoa, assessed by accumulation of R123 in the midpiece) and panel **d**: average path velocity (VAP, μm.sec^{-1}, determined by CASA). Data are means $\pm$ s.e.m. (S de Graaf, WMC Maxwell & G Evans, unpublished observations).

The addition of whole seminal plasma, or an artificial seminal plasma (ASP) based on its inorganic composition and unlikely to influence sperm membrane status (O'Donnell 1969), to frozen-thawed ram spermatozoa improves its motility and movement characteristics assessed by CASA (Mortimer & Maxwell 2004). However, if the spermatozoa have been sex-sorted before freezing, both whole seminal plasma and ASP are detrimental to sperm motility and velocity (Fig. 5: S de Graaf, WMC Maxwell & G Evans, unpublished observations). With increasing proportions of whole seminal plasma in the post-thaw medium, the motility, viability, velocity and mitochondrial activity of sex-sorted, frozen-thawed spermatozoa decline, compared with non-sorted, frozen-thawed spermatozoa (Fig. 6: S de Graaf, WMC Maxwell & G Evans, unpublished observations). Moreover, the use of ASP as sheath fluid does not attenuate the effects of flow cytometric sorting. While initially improving post-sort kinematics, ASP markedly decreases post-thaw motility and longevity compared with a TRIS-based sheath fluid (de Graaf et al. 2004).

While these results at first seem contradictory, evidence is now emerging that the effects of seminal plasma are not simply confined to decapacitation (after addition) and capacitation (after

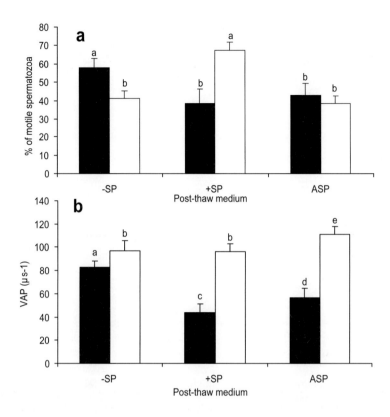

Fig. 5. The motility characteristics of [□] unsorted or [■] sex-sorted and frozen-thawed ram spermatozoa without (-SP) or with (+SP) whole or artificial seminal plasma (+ASP) in the post-thaw medium. Panel **a**: motility (%, determined by CASA; HTM-IVOS; Hamilton-Thorne, USA) and panel **b**: average path velocity (VAP, μms[-1], determined by CASA). The post-thaw medium comprised Androhep® (Minitube Australia, Smythes Creek, Australia) containing 0 (-SP) or 20 % (v/v) whole seminal plasma (+SP) pooled from 6 rams or artificial seminal plasma (+ASP) as described by Mortimer and Maxwell (2000). Data are means ± s.e.m. pooled for 0, 3 and 6 hr post-thaw incubation at 37°C. (S de Graaf, WMC Maxwell & G Evans, unpublished observations).

removal of seminal plasma) but rather consecutive actions of positive and negative regulatory factors. These factors modulate the capacitation status of spermatozoa in a manner that varies with species, males within species, ejaculates within males, stage of sperm maturity and previous sperm treatment (Ashworth *et al.* 1994; Centurion *et al.* 2003; Bergeron *et al.* 2005). The addition of whole seminal plasma to spermatozoa may initially have simple influences associated with its ionic component, as suggested by the effects of ASP (Mortimer & Maxwell 2004; and Fig. 5) but its subsequent action may be more complex. In the case of sex-sorted spermatozoa, the action of seminal plasma may depend on the proteins that remain on the sperm surface, after the dilution and mechanical agitation associated with the flow cytometric process, and their interaction with the particular factors present in the seminal plasma, or its fractions, added to them.

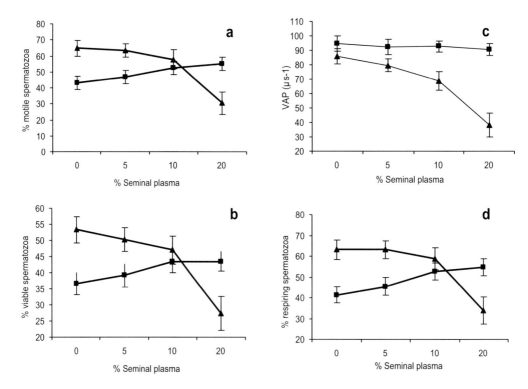

Fig. 6. The motility characteristics and mitochondrial activity of [■] unsorted or [▲] sex-sorted and frozen-thawed ram spermatozoa without (0%) or with (5, 10 or 20%) seminal plasma in the post-thaw medium. Panel **a**: motility (%, determined by CASA; HTM-IVOS; Hamilton-Thorne, USA), panel **b**: viability (% live spermatozoa assessed by exclusion of ethidium homodimer-1), panel **c**: average path velocity (VAP, μms^{-1}, determined by CASA) and panel **d**: mitochondrial activity (% respiring spermatozoa, assessed by accumulation of R123 in the midpiece). The post-thaw medium comprised Androhep® (Minitube Australia, Smythes Creek, Australia) containing 0, 5, 10 or 20 % (v/v) whole seminal plasma pooled from 6 rams. Data are means ± s.e.m. pooled for 0, 3 and 6 hr post-thaw incubation at 37°C. (S de Graaf, WMC Maxwell & G Evans, unpublished observations).

Conclusions

Our studies have demonstrated both positive and negative effects of supplementary seminal plasma on the kinematics, membrane status and fertility of ram spermatozoa that have been subjected to preservation and/or flow cytometric sex sorting. The active components of seminal plasma responsible for these changes in sperm function are proteinaceous, of post-epididymal origin (most likely from the vesicular glands) and not associated with the membrane vesicle organelles in ram seminal plasma.

The major proteins that have been isolated and characterised in ram seminal plasma originate from the vesicular gland and comprise a 15.5 kDa spermadhesin, together with four RSP proteins (15, 16, 22 and 24 kDa) with Fn-2 domains (Bergeron *et al.* 2005) or RSVP-14 (homologous with RSP 15) and RSVP-20 (Fernández-Juan *et al.* 2006). The specific functions of these proteins are yet to be determined but the spermadhesin may have a role in stabilising sperm membranes (decapacitation) at ejaculation while the 15 (or RSVP-14) and 16 kDa RSP proteins are more likely to be associated with binding and release to oviduct cells, capacitation and oocyte binding.

It is tempting to conclude that the active component of ram seminal plasma which provides protection to ram spermatozoa after freeze-thawing is the 15.5 kDa spermadhesin which has been identified by Manjunath *et al.* (2002) as comprising the majority of ram seminal plasma protein, or RSVP-14 and RSVP-20 identified as more than half the ram seminal plasma protein by Muiño-Blanco, Cebrián-Pérez and co-workers (personal communication). However, there have been no investigations to date that provide conclusive evidence on the effects of individual seminal plasma proteins in any ruminant species, or even seminal plasma protein fractions, on the fertility of spermatozoa, whether fresh or preserved. It remains to be seen whether particular proteins, present in the seminal plasma of ruminants, could be characterised, synthesized and utilized as additives to sperm preservation media to improve fertility of females after AI.

Acknowledgements

The authors are grateful to XY Inc. (Fort Collins, CO), the Australian Research Council and the Geoffrey Gardiner Foundation of Victoria for financial support. Valuable research assistance and animal care has been provided by Roslyn Bathgate, Byron Biffin, Stephen Burgun, Lindsay Gillan, Kim Heasman, Tamara Leahy, Monica Ruckholdt, Andrew Souter, Keith Tribe and Shelley Underwood. We also thank Professors Detlef Rath, Teresa Muiña-Blanco, José Cebrián-Pérez and Juan-Maria Vázquez for their constructive comments on the manuscript.

References

Arienti G, Carlini E & Palmerini CA 1997 Fusion of human sperm to prostasomes at acidic pH. *Journal of Membrane Biology* **155** 89-94.

Arienti G, Carlini E, de Cosmo AM, di Profio P & Palmerini CA 1998 Prostasome-like particles in stallion semen. *Biology of Reproduction* **59** 309-313.

Ashworth PJC, Harrison RAP, Miller NGA, Plummer JM & Watson PH 1994 Survival of ram spermatozoa at high dilution: protective effect of simple constituents of culture media as compared with seminal plasma. *Reproduction, Fertility and Development* **6** 173-180.

Assreuy AM, Calvete JJ, Alencar NM, Cavada BS, Rocha-Filho DR, Melo SC, Cunha FQ & Ribeiro RA 2002 Spermadhesin PSP-I/PSP-II heterodimer and its isolated subunits induced neutrophil migration into the peritoneal cavity of rats. *Biology of Reproduction* **67** 1796-1803.

Austin CR 1952 The capacitation of mammalian sperm. *Nature* **170** 326.

Barrios B, Pérez-Pé R, Gallego M, Tato A, Osada J, Muiño-Blanco T and Cebrián-Pérez JA 2000 Seminal plasma proteins revert the cold-shock damage on ram sperm membrane. *Biology of Reproduction* **63** 1531-1537.

Barrios B, Fernández-Juan M, Muiño-Blanco T & Cebrián-Pérez JA 2005 Immunocytochemical localization and biochemical characterization of two seminal plasma proteins that protect ram spermatozoa against cold shock. *Journal of Andrology* **26** 539-549.

Bergeron A, Villemure M, Lazure C & Manjunath P 2005 Isolation and characterization of the major proteins of ram seminal plasma. *Molecular Reproduction and Development* **71** 461-470.

Beydoun H & Saftlas AF 2005 Association of human leucocyte antigen sharing with recurrent spontaneous abortions. *Tissue Antigens* **65** 123-135.

Boisvert M, Bergeron A, Lazure C & Manjunath P 2004 Isolation and characterization of gelatin-binding bison seminal vesicle secretory proteins. *Biology of Reproduction* **70** 656-661.

Brandon CI, Heusner GL, Caudle AB & Fayrer-Hosken RA 1999 Two-dimensional polyacrylamide gel electrophoresis of equine seminal plasma proteins and their correlation with fertility. *Theriogenology* **52** 863-873.

Bravo PW, Skidmore JA & Zhao XX 2000 Reproductive aspects and storage of semen in Camelidae. *Animal Reproduction Science* **62** 173-193.

Breitbart H & Rubinstein S 1982 Characterization of Mg2+-ATPase and Ca2+-ATPase activity in membrane-vesicles from ejaculated ram seminal plasma. *Archives of Andrology* **9** 147-157.

Brody I, Ronquist G & Gottfries A 1983 Ultrastructural localization of the prostasome - an organelle in human seminal plasma. *Uppsala Journal of Medical Sciences* **88** 63-80.

Brooks DE 1990 Biochemistry of the male accessory glands. In *Marshall's Physiology of Reproduction*, vol

2, pp. 269-690. Eds GE Lamming and AS Parkes. Edinburgh, New York: Churchill Livingstone.

Buck L & Axel R 1991 A novel multigene family may encode odorant receptors: a molecular basis for odor recognition. *Cell* **65** 175-187.

Caballero I, Vazquez JM, Gil MA, Calvete JJ, Roca J, Sanz L, Parrilla I, Garcia EM, Rodriguez-Martinez H & Martinez EA 2004a Does seminal plasma PSP-I/PSP-II spermadhesin modulate the ability of boar spermatozoa to penetrate homologous oocytes in vitro? *Journal of Andrology* **25** 1004-1012.

Caballero I, Vazquez JM, Centurión F, Rodriguez-Martinez H, Parrilla I, Roca J, Cuello C & Martinez EA 2004b Comparative effects of autologous and homologous seminal plasma on the viability of largely extended boar spermatozoa. *Reproduction in Domestic Animals* **39** 370-375.

Caballero I, Vazquez JM, Rodriguez-Martinez H, Gill MA, Calvete JJ, Sanz L, Garcia EM, Roca J & Martinez EA 2005 Influence of seminal plasma PSP-I/PSP-II spermadhesin on pig gamete interaction. *Zygote* **13** 11-16.

Calvete JJ, Sanz L & Töpfer-Petersen E 1994 Spermadhesins: structure-function relationships. *Assisted Reproductive Technology/Andrology* **6** 316-330.

Calvete JJ, Sanz L, Reinert M, Dostalova Z & Töpfer-Petersen E 1995 Heparin-binding proteins on bull, boar, stallion and human spermatozoa. In *Advances in spermatozoal phylogeny and taxonomy*, pp 515-524. Eds BGM Jamieson, J Ausio & J-L Justine. Paris: Mém. Mus. Natn, Hist. nat.

Calvete JJ, Raida M, Gentzel M, Urbanke, Sanz CL & Töpfer-Petersen E 1997 Isolation and characterization of heparin- and phosphorylcholine-binding proteins of boar and stallion seminal plasma. Primary structure of porcine pB1. *FEBS Letters* **407** 201-206.

Cardozo JA, Fernández-Juan M, Forcada F, Abecia A, Muiño-Blanco T & Cebrián-Pérez JA 2006 Monthly variations in ovine seminal plasma proteins analysed by two-dimensional polyacrlamide gel electrophoresis. *Theriogenology*, **66** 841-850.

Carlini E, Palmerini CA, Cosmi EV & Arienti G 1997 Fusion of sperm with prostasomes: Effects on membrane fluidity. *Archives of Biochemistry & Biophysics* **343** 6-12.

Carlsson L, Ronquist G, Nilsson BO & Larsson A 2004 Dominant prostasome immunogens for sperm-agglutinating autoantibodies in infertile men. *Journal of Andrology* **25** 699-705.

Castellini C, Lattaioli P, Moroni M & Minelli A 2000 Effect of seminal plasma on the characteristics and fertility of rabbit spermatozoa. *Animal Reproduction Science* **63** 275-282.

Catt SL, O'Brien JK, Maxwell WMC & Evans G 1997 Assessment of ram and boar spermatozoa during cell-sorting by flow cytometry. *Reproduction in Domestic Animals* **32** 251-258.

Centurion F, Vazquez JM, Calvete JJ, Roca J, Sanz L, Parilla I, Garcia EM & Martinez E 2003 Influence of porcine spermadhesins on the susceptibility of boar spermatozoa to high dilution. *Biology of Reproduction* **69** 640-646.

Chang MC 1951 Fertilizing capacity of spermatozoa deposited into the fallopian tubes. *Nature* **168** 697-698.

Chang MC 1957 A detrimental effect of seminal plasma on the fertilizing capacity of sperm. *Nature* **179** 258-259.

Cormier N & Bailey JL 2003 A differential mechanism is involved during heparin- and cryopreservation-induced capacitation of bovine spermatozoa. *Biology of Reproduction* **69** 177-185.

Cormier N, Sirard MA & Bailey JL 1997 Premature capacitation of bovine spermatozoa is initiated by cryopreservation. *Journal of Andrology* **18** 461-468.

Corteel JM 1980 [Effects of seminal plasma on the survival and fertility of spermatozoa kept in vitro] [French]. *Reproduction, Nutrition, Development* **20** 1111-1123.

Cross NL 1998 Role of cholesterol in sperm capacitation. *Biology of Reproduction* **59** 7-11.

Cross NL 2003 Decrease in order of human sperm lipids during capacitation. *Biology of Reproduction* **69** 529-534.

Cummins JM & Orgebin-Crist MC 1971 Investigations into fertility of epididymal spermatozoa. *Biology of Reproduction* **5** 13-19.

Dacheux J-L & Paquignon M 1980 Relations between the fertilizing ability, motility and maturation process in the boar. *Annals New York Academy of Science* **438** 526-529.

Dacheux J-L, Gatti J-L & Dacheux F 2003 Contribution of epididymal secretory proteins for spermatozoa maturation. *Microscopy Research Techniques* **61** 7-17.

de Cuneo MF, Vincenti LM, Martini AC, Ponce AA & Ruiz RD 2004 Effects of PDC-109 on bovine sperm functional activity in presence or absence of heparin. *Theriogenology* **62** 207-216.

de Graaf SP, Gillan L, Evans G, Maxwell WMC & O'Brien JK 2004 The effect of sheath fluid on the quality of sex-sorted ram spermatozoa. *Reproduction, Fertility and Development* **16** 284-285.

Dostalova Z, Calvete JJ, Sanz L, Hettel C, Riedel D, Schoneck C, Einspanier R & Töpfer-Petersen E 1994 Immunolocalization and quantitation of acidic seminal fluid protein (aSFP) in ejaculated, swim-up, and capacitated bull spermatozoa. *Biol Chem Hoppe Seyler* **375** 457-461.

Dukelow WR, Cheinoff HN & Williams WL 1967 Properties of decapacitation factor and presence on various species. *Journal of Reproduction and Fertility* **14** 393-399.

Eberhard WG 1996 *Female control: sexual selection by cryptic female choice.* Princeton University Press.

Ecroyd H, Sarradin P, Dacheux JL & Gatti, JL 2004 Compartmentalization of prion isoforms within the reproductive tract of the ram. *Biology of Reproduction* **71** 993-1001.

El-Hajj Ghaoui R, Thompson PC, Evans G & Maxwell WMC 2004 Characterization and localization of membrane vesicles in ejaculate fractions from the ram, boar and stallion. *Reproduction in Domestic Animals* **39** 173-180.

El-Hajj Ghaoui R, Thompson PC, Evans G & Maxwell WMC 2006 The origin of membrane vesicles in ram seminal plasma. *Reproduction in Domestic Animals* **41** 1-8.

Ericksson BM, Bathgate R, Maxwell WMC & Evans G 2005 Effect of seminal plasma protein fractions on boar spermatozoa motility and acrosome integrity. *Theriogenology* **63** 491. (Abstract)

Esch FS, Ling NC, Bohlen P, Ying SY & Guillemin R 1983 Preliminary structure of PDC-109, a major protein constituent of bovine seminal plasma. *Biochemical and Biophysical Research Communications* **113** 861-867.

Fabiani R, Johansson L, Lundkvist O & Ronquist G 1994 Enhanced recruitment of motile spermatozoa by prostasome inclusion in swim-up medium. *Human Reproduction* **9** 1485-1489.

Fernández-Juan M, Gallego M, Barrios B, Osada J, Cebrián-Pérez JA & Muiño-Blanco T 2006 Immunohistochemical localization of sperm-preserving proteins in the ram reproductive tract. *Journal of Andrology* **27** 588-595.

Foster RA, Ladds PW, Hoffmann D & Husband AJ 1988 Immunoglobulins and immunoglobulin-containing cells in the reproductive tract of normal rams. *Australian Veterinary Journal* **65** 16-20.

Fouchécourt S, Charpigny G, Reinaud P, Dumont P & Dacheux J-L 2002 Mammalian lipocalin-type prostaglandin D-2 synthase in the fluids of the male genital tract: putative biochemical and physiological functions. *Biology of Reproduction* **66** 458-467.

Fraser LR, Harrison RAP & Herod JE 1990 Characterization of a decapacitation factor associated with epididymal mouse spermatozoa. *Journal of Reproduction & Fertility* **89** 135-148.

Fraser LR, Adeoya-Osiguwa SA, Baxendale RW & Gibbons R 2006 Regulation of mammalian sperm capacitation by endogenous molecules. *Frontiers in Bioscience* **11** 1636-1645.

García-López N, Ollero M, Cebrián-Pérez JA & Muiño-Blanco T 1996 Reversion of thermic-shock effect on ram spermatozoa by adsorption of seminal plasma proteins revealed by partition in aqueous two-phase systems. *Journal of Chromatography* B **680** 137-143.

Garcia EM, Vázquez JM, Calvete JJ, Sanz L, Caballero I, Parilla I, Gil MA, Roca J & Martinez EA 2006 Dissecting the protective effect of the seminal plasma spermadhesin PSP-I/PSP-II on boar sperm functionality. *Journal of Andrology* **27** 434-442.

Garner DL, Thomas, Gravance CG, Marshall CE, DeJarnette JM & Allen CH 2001 Seminal plasma addition attenuates the dilution effect in bovine sperm. *Theriogenology* **56** 31-40.

Gatti J-L, Druart X, Syntin P, Guerin Y, Dacheux J-L & Dacheux F 2000 Biochemical characterization of two ram cauda epididymal maturation-dependent sperm glycoproteins. *Biology of Reproduction* **62** 950-958.

Gatti J-L, Castella S, Dacheux F, Ecroyd H, Métayer S,

Thimon V & Dacheux J-L 2004 Post-testicular sperm environment and fertility. *Animal Reproduction Science* **82-83** 321-339.

Gatti J-L, Métayer S, Belghazi M, Dacheux F & Dacheux J-L 2005 Identification, proteomic profiling, and origin of ram epididymal fluid exosome-like vesicles. *Biology of Reproduction* **72** 1452-1465.

Gerena RL, Irikura D, Urade Y, Eguchi N, Chapman DA & Killian, GJ 1998 Identification of a fertility-associated protein in bull seminal plasma as lipocalin-type prostaglandin D synthase. *Biology of Reproduction* **58** 826-833.

Gibbons R, Adeoya-Osiguwa & Fraser LR 2005 A mouse sperm decapacitation factor receptor is phosphatidyethanolamine-binding protein 1. *Reproduction* **130** 497-508.

Gillan L & Maxwell WMC 1999 The functional integrity and fate of cryopreserved ram spermatozoa. *Journal of Reproduction & Fertility Supplement* **54** 271-283.

Gillan L, Evans G & Maxwell WMC 2000 The interaction of fresh and frozen-thawed ram spermatozoa with oviductal epithelial cells in vitro. *Reproduction, Fertility Development* **12** 237-244.

Gillan L, Maxwell WMC & Evans G 2004 Preservation and evaluation of semen for artificial insemination. *Reproduction, Fertility Development* **16** 447-454.

Greube A, Müller K, Töpfer-Petersen E, Herrmann A & Müller P 2004 Interaction of fibronectin type II proteins with membranes: The stallion seminal plasma protein SP-1/2. *Biochemistry* **43** 464-472.

Green CE & Watson PF 2001 Comparison of the capacitation-like state of cooled boar spermatozoa with true capacitation. *Reproduction* **122** 889-898.

Gundogan M & Elitok B 2004 Seasonal changes in reproductive parameters and seminal plasma constituents of rams in Afyon province of Turkey. *Deutsche Tierarztliche Wochenschrift* **111** 158-61.

Guraya SS 2000 Cellular and molecular biology of capacitation and acrosome reaction in spermatozoa. *International Review of Cytology* **199** 1-64.

Gwathmey T, Ignotz G & Suarez S 2003 PDC-109 (BSP-A1/A2) promotes bull sperm binding to oviductal epithelium in vitro and may be involved in forming the oviductal sperm reservoir. *Biology of Reproduction* **69** 809-815.

Henault MA & Killian GJ 1996 Effect of homologous and heterologous seminal plasma on the fertilizing ability of ejaculated bull spermatozoa assessed by penetration of zona-free bovine oocytes. *Journal of Reproduction & Fertility* **108** 199-204.

Hoskins DD & Vijayaraghavan S 1990 A new theory on the acquisition of sperm motility during epididymal transit. In *Controls of sperm motility: Biological and clinical aspects*, pp 53-62. Ed C Gagnon. Boca Raton: CRC press.

Huang Z & Vijayaraghavan S 2004 Increased phosphorylation of a distinct subcellular pool of protein phosphatase, ppl{gamma}2, during epididymal sperm maturation. *Biology of Reproduction* **70** 439-447.

Hunter RHF 1981 Sperm transport and reservoirs in the

pig oviduct in relation to the time of ovulation. *Journal of Reproduction & Fertility* **63** 109-117.

Ibrahim NM, Gilbert GR, Loseth KJ & Crabo BG 2000 Correlation between clusterin-positive spermatozoa determined by flow cytometry in bull semen and fertility. *Journal of Andrology* **21** 887-894.

Jelinkova P, Lierda J, Manaskova P, Ryslava H, Jonakova W & Ticha M 2004 Mannan-binding proteins from boar seminal plasma. *Journal of Reproductive Immunology* **62** 167-182.

Jobim MIM, Oberst ER, Salbego CG, Wald VB, Horn AP & Mattos RC 2005 BSP A1/A2-like proteins in ram seminal plasma. *Theriogenology* **63** 2053-2062.

Johnson LA 2000 Sexing mammalian sperm for production of offspring: The state-of-the-art. *Animal Reproduction Science* **61** 93-107.

Killian GJ, Chapman DA & Rogowski LA 1993 Fertility-associated proteins in Holstein bull seminal plasma. *Biology of Reproduction* **49** 1202-1207.

Kraus M, Ticha M, Zelezna B, Peknicova J & Jonakova V 2005 Characterization of human seminal plasma proteins homologous to boar AQN spermadhesins. *Journal of Reproductive Immunology* **65** 33-46.

Kravets FG, Lee J, Singh B, Trocchia A, Pentyala SN & Khan SA 2000 Prostasomes: Current concepts. *Prostate* **43** 169-174.

Love CC, Brinsko SP, Rigby SL, Thompson JA, Blanchard TL & Varner DD 2005 Relationship of seminal plasma level and extender to sperm motility and DNA integrity. *Theriogenology* **63** 1584-1591.

Manjunath P & Sairam MR 1987 Purification and biochemical characterization of three major acid proteins (BSP A1, BSP A2 and BSP A3) from bovine seminal plasma. *Biochemistry* **7** 685-692.

Manjunath P & Thérien I 2002 Role of seminal plasma phospholipids-binding proteins in sperm membrane lipid modification that occurs during capacitation. *Journal of Reproductive Immunology* **58** 109-119.

Manjunath P, Sairam MR & Uma J 1987 Purification of four gelatin-binding proteins from bovine seminal plasma by affinity chromatography. *Biosciences Reports* **7** 231-238.

Manjunath P, Nauc V, Bergeron A & Ménard M 2002 Major proteins of bovine seminal plasma bind to the low-density lipoprotein fraction of hen's egg yolk. *Biology of Reproduction* **67** 1250-1258.

Mann T 1954 *The Biochemistry of Semen*. London: Methuen and Co Ltd.

Mann T 1964 *The biochemistry of semen and of the male reproductive tract*. London: Methuan and Co. Ltd.

Mann T & Lutwak-Mann C 1981 *Male reproductive function and semen. Themes and trends in physiology, biochemistry and andrology*. Berlin: Springer-Verlag.

Maxwell WMC & Hewitt LJ 1986 A comparison of vaginal, cervical and intrauterine insemination of sheep. *Journal of Agricultural Science, Cambridge* **106** 191-193.

Maxwell WMC & Johnson LA 1997a Chlortetracycline analysis of boar spermatozoa after incubation, flow cytometric sorting, cooling, or cryopreservation. *Molecular Reproduction & Development* **46** 408-418.

Maxwell WMC & Johnson LA 1997b Membrane status of boar spermatozoa after cooling or cryopreservation. *Theriogenology* **48** 209-219.

Maxwell WMC & Johnson LA 1999 Physiology of spermatozoa at high dilution rates: The influence of seminal plasma. *Theriogenology* **52** 1353-1362.

Maxwell WMC & Watson PF 1996 Recent progress in the preservation of ram semen. *Animal Reproduction Science* **42** 55-65.

Maxwell WMC, Welch GR & Johnson LA 1997 Viability and membrane integrity of spermatozoa after dilution and flow cytometric sorting in the presence or absence of seminal plasma. *Reproduction, Fertility & Development* **8** 1165-1178.

Maxwell WMC, Long CR, Johnson LA, Dobrinsky JR & Welch GR 1998 The relationship between membrane status and fertility of boar spermatozoa after flow cytometric sorting in the presence or absence of seminal plasma. *Reproduction, Fertility & Development* **10** 433-440.

Maxwell WMC, Evans G, Mortimer ST, Gillan L, Gellatly ES & McPhie CA 1999 Normal fertility after cervical insemination with frozen-thawed spermatozoa supplemented with seminal plasma. *Reproduction, Fertility & Development* **11** 123-126.

Maxwell WMC, Evans G, Hollinshead FK, Bathgate R, de Graaf SP, Erikson B, Gillan L, Morton K & O'Brien JK 2004 Integration of sperm sexing technology into the ART toolbox. *Animal Reproduction Science* **82-83** 79-95.

Mededovic S & Fraser LR 2004 Angiotensin II stimulates cAMP production and protein tyrosine phosphorylation in mouse spermatozoa. *Reproduction* **127** 601-612.

Métayer S, Dacheux F, Guerin Y, Dacheux J-L & Gatti J-L 2001 Physiological and enzymatic properties of the ram epididymal soluble form of germinal angiotensin I-converting enzyme. *Biology of Reproduction* **65** 1332-1339.

Minelli A, Moroni M, Martinez E, Mezzasoma I & Ronquist G 1998 Occurrence of prostasome-like membrane vesicles in equine seminal plasma *Journal of Reproduction & Fertility* **114** 237-243.

Minelli A, Allegrucci C, Mezzasoma I, Ronquist G, Lluis C & Franco R 1999 Cd26 and adenosine deaminase interaction: Its role in the fusion between horse membrane vesicles and spermatozoa. *Biology of Reproduction* **61** 802-808.

Mokkapati S & Dominic CJ 1977 Morphology of the accessory reproductive glands of some male indian chiropterans. *Anatomischer Anzeiger* **141** 391-397.

Moore AI, Squires EL & Graham JK 2005 Effect of seminal plasma on the cryopreservation of equine spermatozoa. *Theriogenology* **63** 2372-2381.

Mortimer ST & Maxwell WMC 2004 Effect of medium on the kinematics of frozen-thawed ram spermatozoa. *Reproduction* **127** 285-291.

Moruzzi JF 1979 Selecting a mammalian species for the separation of X- and Y-chromosome-bearing spermatozoa. *Journal of Reproduction & Fertility* **57** 319-323.

Moura AA 2005 Seminal plasma proteins and fertility

indexes in the bull: The case for osteopontin. *Animal Reproduction* **2** 3-10.

Moura AA, Chapman DA, Killian GJ & Almquist JO 2005 Proteins in the accessory sex gland and cauda epididymis fluid as related to dairy bull fertility. *Journal of Andrology* **83** 32. (Abstract)

Myles DG, Koppel DE & Primakoff P 1990 Sperm surface domains and fertilization. In *Fertilization in mammals*, pp 169-177. Eds D Barry, JC Bavister and RSR Eduardo. Norwell Massacusetts: Serono Symposia.

Nagai T, Niwa K & Iritani A 1984 Effect of sperm concentration during preincubation in a defined medium on fertilization in vitro of pig follicular oocytes. *Journal of Reproduction & Fertility* **70** 271-276.

O'Donnell JM 1969 Intracellular levels of sodium and potassium in bull spermatozoa in relation to cell metabolism. *Journal of Reproduction & Fertility* **19** 207-209.

Ollero M, Cebrián-Pérez JA & Muiño-Blanco T 1997 Improvement of cryopreserved ram sperm heterogeneity and viability by addition of seminal plasma. *Journal of Andrology* **18** 732-739.

Palmerini CA, Carlini E, Nicolucci A & Arienti G 1999 Increase of human spermatozoa intracellular $Ca2+$ concentration after fusion with prostasomes. *Cell Calcium* **25** 291-296.

Parks JE & Ehrenwald E 1989 Cholesterol efflux from mammalian sperm and its potential role in capacitation. In *Fertilization in mammals*, pp 155-167. Eds D Barry, JC Bavister and RSR Eduardo. Norwell Massacusetts: Serono Symposia.

Parks JE & Hammerstedt RH 1985 Development changes occurring in the lipids of ram epididymal spermatozoa plasma membrane. *Biology of Reproduction* **32** 653-668.

Paulenz H, Soderquist L, Adnoy I, Nordstoga AB & Berg KA 2005 Effect of vaginal and cervical deposition of semen on the fertility of sheep inseminated with frozen-thawed semen. *Veterinary Record* **156** 372-375.

Pérez-Pé R, Cebrián-Pérez JA & Muiño-Blanco T 2001a Semen plasma proteins prevent cold-shock membrane damage to ram spermatozoa. *Theriogenology* **56** 425-434.

Pérez-Pé R, Barrios B, Muiño-Blanco T & Cebrián-Pérez JA 2001b Seasonal differences in ram seminal plasma revealed by partition in an aqueous two-phase system. *Journal of Chromatography* B **760** 113-121.

Pérez-Pé R, Grasa P, Fernández-Juan M, Peleato ML, Cebrián-Pérez JA & Muiño-Blanco T 2002 Seminal plasma proteins reduce protein tyrosine phosphorylation in the plasma membrane of cold-shocked ram spermatozoa. *Molecular Reproduction and Development* **61** 226-233.

Pursel VG & Johnson LA 1975 Freezing of boar spermatozoa: fertilizing capacity with concentrated semen and a new thawing procedure. *Journal of Animal Science* **40** 99-102.

Rath D & Niemann H 1997 In vitro fertilization of porcine oocytes with fresh and frozen-thawed ejaculated or frozen-thawed epididymal semen obtained from identical boars. *Theriogenology* **47** 785-793.

Reinert M, Calvete J, Sanz L, Mann K & Topfer-Petersen E 1996 Primary structure of stallion seminal plasma HSP-7, a zona-pellucida-binding protein of the spermadhesin family. *European Journal of Biochemistry* **242** 636-640.

Robaire B & Hermo L 1988 Efferent ducts, epididymis, and vas deferens: Structure, function and their regulation. In *The physiology of reproduction*, vol 1, pp 999-1080. Eds E Knobil and JD Neill. New York: Raven Press.

Robertson SA 2005 Seminal plasma and male factor signalling in the female reproductive tract. *Cell Tissue Research* **322** 43-52.

Rodriguez-Martinez H, Larsson B, Pertoft H & Kjellén L 1998 GAGs and spermatozoon competence in vivo and in vitro. In *Gametes: development and function*, pp 239-274. Eds A Lauria, F Gandolfi, G Enne and L Gianaroli. Italy: Serono Symposia.

Romão MJ, Kölln I, Dias JM, Carvalho AL, Romero A, Varela PF, Sanz L, Töpfer-Petersen E & Calvete JJ 1997 Crystal structure of acidic seminal fluid protein (aSFP) at 1.9 Å resolution: a bovine polypeptide of the spermadhesin family. *Journal of Molecular Biology* **274** 650-660.

Ronquist G, Brody I, Gottfries A & Stegmayr B 1978a An $Mg2+$ and $Ca2+$-stimulated adenosine triphosphatase in human prostatic fluid. Part i. *Andrologia* **10** 261-272.

Ronquist G, Brody I, Gottfries A & Stegmayr B 1978b An $Mg2+$ and $Ca2+$-stimulated adenosine triphosphatase in human prostatic fluid. Part ii. *Andrologia* **10** 427-433.

Saalmann A, Münz S, Ellerbrock K, Ivell R & Kirchhoff C 2001 Novel sperm-binding proteins of epididymal origin contain four fibronectin type II modules. *Molecular Reproduction and Development* **58** 88-100.

Saez F, Frenette G & Sullivan R 2003 Epididymosomes and prostasomes: their roles in post-testicular maturation of the sperm cells. *Journal of Andrology* **24** 149-154.

Salamon S & Maxwell WMC 1995a Frozen storage of ram semen. I. Processing, freezing, thawing and fertility after cervical insemination. *Animal Reproduction Science* **37** 185-249.

Salamon S & Maxwell WMC 1995b Frozen storage of ram semen. II. Causes of low fertility after cervical insemination and methods of improvement. *Animal Reproduction Science* **38** 1-55.

Smith JF, Parr J, Murray GR, McDonald RM & Lee RDF 1999 Seasonal changes in the protein content and composition of ram seminal plasma. *Proceedings New Zealand Society of Animal Production* **59** 223-225.

Spehr M, Gisselmann G, Poplawski A, Riffell JA, Wetzel CH, Zimmer RK & Hatt H 2003 Identification of a testicular odorant receptor mediating human sperm chemotaxis. *Science* **299** 2054-2058.

Suzuki-Toyota F, Itoh Y & Naito K 2000 Reduction of intramembranous particles in the periacrosomal plasma membrane of boar spermatozoa during in vitro capacitation: A statistical study. *Development Growth & Differentiation* **42** 265-273.

Swamy MJ 2004 Interaction of bovine seminal plasma proteins with model membranes and sperm plasma membranes. *Current Science* **87** 203-211.

Tedeschi G, Oungre E, Mortarino M, Negri A, Maffeo G & Ronchi S 2000 Purification and primary structure of a new bovine spermadhesin. *European Journal of Biochemistry* **267** 6175-6179.

Thérien I, Moreau R & Manjunath P 1998 Major proteins of bovine seminal plasma and high-density lipoprotein induce cholesterol efflux from epididymal sperm. *Biology of Reproduction* **59** 768-776.

Thérien I, Moreau R & Manjunath P 1999 Bovine seminal plasma phospholipid-binding proteins stimulate phospholipids efflux from epididymal sperm. *Biology of Reproduction* **61** 590-598.

Thérien I, Bousquet D & Manjunath P 2001 Effect of seminal phospholipid-binding proteins and follicular fluid on bovine sperm capacitation. *Biology of Reproduction* **65** 41-51.

Tollner TL, Yudin AI, Treece CA, Overstreet JW & Cherr GN 2004 Macaque sperm release ESP13.2 and PSP94 during capacitation: The absence of ESP13.2 is linked to sperm-zona recognition and binding. *Molecular Reproduction and Development* **69** 325-337.

Töpfer-Petersen E 1999 Carbohydrate-based interactions on the route of a spermatozoon to fertilization. *Human Reproduction Update* **5** 314-329.

Töpfer-Petersen E, Romero A, Varela PF, Ekhlasi-Hundrieser M, Dostalova Z, Sanz L & Calvete JJ 1998 Spermadhesins: A new protein family. Facts, hypotheses and perspectives. *Andrologia* **30** 217-224.

Töpfer-Petersen E, Ekhlasi-Hundrieser M, Kirchoff C, Leeb T & Sieme H 2005 The role of stallion seminal proteins in fertilisation. *Animal Reproduction Science* **89** 159-170.

Udby L, Lundwall A, Johnsen AH, Fernlund P, Valtonen-Andre C, Blom AM, Lilja H, Borregaard N, Kjeldsen L & Bjartell A 2005 Beta-microseminoprotein binds crisp-3 in human seminal plasma. *Biochemical & Biophysical Research Communications* **333** 555-561.

Vadnais ML, Kirkwood RN, Tempelman RJ, Sprecher DJ & Chou K 2005 Effect of cooling and seminal plasma on the capacitation status of fresh boar sperm as determined using chlortetracycline assay. *Animal Reproduction Science* **87** 121-132.

Vauquelin LN 1791 Expériences sur le sperme humain. *Ann Chim* **9**, 64.

Varela PF, Romero A, Sanz L, Romão MJ, Töpfer-Petersen E & Calvete JJ 1997 The 2.4 A resolution crystal structure of boar seminal plasma PSP-I/PSP-II: a zona pellucida-binding glycoprotein heterodimer of the spermadhesin family built by a CUB domain architecture. *Journal of Moleciuar Biology* **274** 635-649.

Villemure M, Lazure C & Manjunath P 2003 Isolation and characterization of gelatin-binding proteins from goat seminal plasma. *Reproductive Biology & Endocrinology* **1** 39.

Vittoria A, La Mura E, Cocca T & Cecio A 1990 Serotonin-, somatostatin- and chromogranin a-containing cells of the urethro-prostatic complex in the sheep. An immunocytochemical and immunofluorescent study. *Journal of Anatomy* **171** 169-178.

Watson PF 1995 Recent developments and concepts in the cryopreservation of spermatozoa and the assessment of their post-thaw function. *Reproduction, Fertility & Development* **7** 871-891.

Wedekind C, Chapuisat M, Macas E & Rulicke T 1996 Non-random fertilization in mice correlates with the MHC and something else. *Heredity* **77** 400-409.

Wolfe CA, James PS, Gunning AP, Ladha S, Christova Y & Jones R 2001 Lipid dynamics in the plasma membrane of ram and bull spermatozoa after washing and exposure to macromolecules BSA and PVP. *Molecular Reproduction & Development* **59** 306-313.

Yanagimachi R 1988 Mammalian fertilization. In *The physiology of reproduction*, pp 135-185. Eds E Knobil and JD Neill. New York: Raven Press.

Zhao XX 2000 Semen characteristics and artificial insemination in the Bactrian Camel. In *Recent advances in Camelid reproduction*. Eds JA Skidmore and GP Adams, International Veterinary Information Service (*www.ivis.org*).

Ziegler A, Kentenich H & Uchanska-Ziegler B 2005 Female choice and the MHC. *TRENDS in Immunology* **26** 496-502.

In vitro evaluation of sperm quality related to in vivo function and fertility

H Rodriguez-Martinez[1] and AD Barth[2]

[1]Division of Comparative Reproduction, Obstetrics and Udder Health, Faculty of Veterinary Medicine and Animal Science, Swedish University of Agricultural Sciences (SLU), Ullsväg 14C, Clinical Centre Box 7054, Uppsala 75007, Sweden; [2]Department of Large Animal Clinical Sciences, Western College of Veterinary Medicine, University of Saskatchewan, 52 Campus Drive, Saskatoon SK S7N 5B4, Canada

The potential fertility of a sire can not be evaluated in the field simply by assessment of mating ability and physical examination, although these procedures can expose his limitations as a breeder. Finding a laboratory test that accurately estimates the potential fertility of a semen sample or a sire is also distant, as shown by the modest correlations that present tests have with fertility. Due to the complex nature of male fertility any sought for laboratory method must include testing of most sperm attributes relevant for both fertilisation and embryo development, not only in individual spermatozoa, but within a large, heterogeneous sperm population. Although such a task has proven difficult, it is both challenging and attractive for ruminants, where methods with good estimative power are available to evaluate the many attributes required for fertilisation. Among these methods are the isolation of highly viable spermatozoa by swim-up followed by their ability to respond to capacitation or acrosome reaction challenges and their capacity to penetrate homologous or heterologous zona pellucidae (ZP). Identification of fertility markers in, for instance, seminal plasma would further aid in identifying low-fertility sires. Future efforts should concentrate on finding how many spermatozoa in the semen sample are competent for fertilisation, perhaps by screening sperm linear motion, membrane integrity and membrane stability by multi-parametric methods, linked to the ability of males to provide a stable population of spermatozoa in a repeatable manner.

Introduction

The fertility of males is best evaluated after breeding an adequate number of females either by natural mating or artificial insemination (AI). However, this is time consuming, involves major costs, good identification of bred animals and proper recording of the outcome. The potential fertility of bulls or rams can also be evaluated in the field by a clinical andrological examination, including a thorough evaluation of the ejaculate and assessment of libido and serving ability. Such screening, often referred to as a breeding soundness evaluation, has diagnostic value for assessing testicular and epididymal function, and/or the normality of the genital tract of the male (Barth 2000). Although it has proven useful to eliminate clear-cut cases of sub-

Corresponding author E-mail: heriberto.Rodriguez@kv.slu.se

fertility, breeding soundness evaluation appears insufficient to estimate the level of fertility that the males would actually achieve, whether used in natural service or for AI.

Determination of the potential level of fertility of a sire, or a particular ejaculate, by using the results of clinical and/or laboratory examinations has been the goal of veterinary andrologists for decades. However, due to the complex character of male fertility, estimating the expected level of fertility remains a challenge, particularly for sires in natural service (Parkinson 2004). Scrotal circumference (SC) is directly related to testicular size and thus indirectly to potential sperm production, but its relationship to fertility is confounded by several factors including semen quality and breeding pressure. On the other hand, SC is easily measured, it is highly heritable and relates to the age of puberty of male and female progeny; thus it is advantageous for sire selection despite its low predictive value for potential fertility. Sires can also be evaluated for their responsiveness to hormonal challenges (for example, GnRH-induced luteinising hormone [LH] or testosterone surges); however, a relationship to fertility has yet to be proven (Parkinson 2004).

Evaluation of a semen sample can determine its degree of normality before it is processed for AI or *in vitro* fertilisation (IVF). Usually, this evaluation includes recordings of volume, sperm concentration and motility and less often sperm morphology and the presence of foreign cells. In recent years, more detailed screenings of sperm function have been incorporated (reviewed by Silva & Gadella 2006), other than sperm motility which is by far the most widely used test in any species of domestic animals. Sperm numbers, sperm motility and sperm morphology are related to potential *in vivo* fertility, but only to a modest extent. Other measures of sperm function may have stronger correlations to fertility, but still provide only estimates of the outcome even when used in a battery of tests. Therefore, they are unrealistic practical alternatives and keep distant our ultimate goal to find a single *in vitro* test to reliably predict the level of fertility (Rodriguez-Martinez 2003).

Many reviews of the relationship between semen assessment and fertility in ruminants are available (Rodriguez-Martinez *et al.* 1997b; Saacke *et al.* 1998; 2000; Larsson & Rodriguez-Martinez 2000; Rodriguez-Martinez 2000; 2003). These reviews confirm that some laboratory methods can often correctly estimate the potential fertilising capacity of a semen sample and in some cases of the male, provided they are designed to test sperm attributes relevant for fertilisation and embryo development. However, such estimations depend on the type of test used to study a very heterogeneous suspension of terminal cells, the spermatozoa, whose intrinsic lability make assessments of attributes needed for fertilisation rather difficult (Holt & Van Look 2004).

The handling of semen during processes such as cooling, cryopreservation or high pressure flow cytometric sex-sorting, damages spermatozoa to various extents and further increases the complexity of fertility analyses. Moreover, these processes blur the relationship between male fertility and our (in)capacity to preserve spermatozoa. Nevertheless, the challenge of estimating level of fertility is closer to being met in ruminants, where methods are already available to evaluate an important quantity of the attributes of fertilisation in semen and relate them to the fertility outcome (most often by AI), compared to the situation in other domestic animals, where numbers of females are insufficient or proper records are not available (Amann 2005).

The present review describes the state of the art regarding *in vitro* methods available for sperm analysis as estimators of fertility, particularly those measuring sperm function and competence for physiological events such as sperm transport, storage and interaction with the female genital tract and the oocyte. It attempts to critically appraise the value and constraints of these laboratory methods as predictors of fertility when the male is used in natural mating situations and when semen is used by AI. It also considers the ability of *in vitro* methods to assess the effects of procedures such as semen extension, cool storage, cryopreservation and sperm sexing. The review hopes to avoid reiteration of the large amount of relevant descriptions of available methods and includes results obtained in the laboratories of the authors on work done on ruminant livestock; in particular, cattle and sheep.

In vitro assessment of ruminant semen: state of the art

Bull, ram or buck semen can be collected during ejaculation with an artificial vagina or by electro-ejaculation, a procedure that, albeit contested due to animal welfare issues, is routinely practiced worldwide. In bulls, spermatozoa can also be collected, primarily for diagnostic purposes, by trans-rectal massage of the ampullae (Palmer *et al.* 2005). Ejaculated semen, from either source, is immediately evaluated for volume, sperm concentration and sperm motility, usually using ocular, photometric and light microscopic aids. Sperm concentration can be evaluated manually (counting chambers) or automatically (photometers, Coulter instruments, flow cytometry). Computer assisted motility analysis (CASA) instruments are now less costly and therefore more widespread. Most of these instruments digitise microscope images of sperm trajectories providing information on proportions of motile spermatozoa, motility patterns and other kinematic variables, while other computerized sperm analysers determine the number of particles (spermatozoa) crossing fields of view, yielding a regression algorithm of sperm numbers and translation classes. While the first named can describe patterns of movement of comparatively few spermatozoa per sample, followed in most cases for a few seconds, the second one analyses thousands of spermatozoa per sample but can only provide numerical classes. In other words, these instruments are more and more sophisticated, yielding much data with a degree of objectivity. However, subjective motility evaluations are still practical for immediate assessment of sperm viability, due to easiness and low cost.

Sperm morphology is sometimes screened, but far too seldom, to determine the proportion of spermatozoa with deviating morphology or showing specific defects (Barth & Oko 1989). Morphological abnormalities may be studied in unstained wet preparations of formalin-fixed spermatozoa and/or stained spermatozoa with light microscopy. Accurate detection of some abnormalities, head defects in particular (Chacón 2001), make necessary both staining and the counting of a large number of spermatozoa in order to obtain repeatability (Kuster *et al.* 2004). Proportions of defective spermatozoa are often grouped as primary vs. secondary, or major vs. minor according to their relative importance, while in other cases, morphological deviations are grouped by origin in order to determine underlying testicular or epididymal pathology, or as artefacts caused by mishandling of the semen. Abnormalities may also be grouped as compensable or un-compensable (Saacke *et al.* 2000). Sub-fertility caused by abnormal spermatozoa which are not transported to the oviduct, or which fail to penetrate the ZP could be compensated for by increasing the dose of spermatozoa at AI in order to obtain adequate numbers of normal spermatozoa at the site of fertilisation. Sub-fertility due to spermatozoa that fertilise, but fail to participate in the development of a normal embryo, can not be compensated for by simply increasing sperm numbers in the AI-dose. For this reason, it is not only important to know the number of morphologically normal spermatozoa in a sample but also the number of un-compensable abnormalities. Ideally, as we shall see later, it is the number of spermatozoa holding most intact attributes essential for fertility within a semen sample that matters. Presence and relative amount of foreign cells other than spermatozoa are easy to account for provided discontinuous smears are prepared to form dense ridges before drying and staining. These provide valuable information in cases of inflammation, testicular degeneration or other pathologies (Rodriguez-Martinez *et al.* 1997b).

Manual assessment of sperm morphology is sometimes considered highly variable within and between technicians. This variability may be related to technician training and the assessment of too few spermatozoa per sample. In any case, automated, computer-assisted sperm head morphometry analysis instruments (ASMA, Gravance *et al.* 1998) have been designed to analyse sperm head dimensions objectively. These instruments have been found to give repeatable results when used with bull (Gravance *et al.* 1999), ram (Gravance *et al.* 1998) or goat spermatozoa (Marco-Jimenez *et al.* 2006). When combined with multivariate statistical analyses, ASMA allowed identification of sub-populations in the ejaculate (Peña *et al.* 2005a). However, ASMA is restricted to measurement of the sperm head surface area and lacks the ability to discern other morphological abnormalities.

While sperm numbers, motility and morphological features are important, sperm attributes required for interaction with the female genitalia, activation (capacitation), fertilisation and early embryo development have also been investigated. Among these attributes are the presence of intact and competent membranes, organelles (such as, the acrosome, mitochondria, proximal centriole and flagellum) and a haploid genome (sperm nucleus). While membrane abnormalities and abnormalities of many organelles would be compensable by dose, abnormalities of the proximal centriole (for example, accessory tails) and genome abnormalities (for example, pyriform sperm head shape or presence of nuclear vacuoles) would be uncompensable by dose. Several methods to assess semen (sperm) quality can be applied to ejaculated spermatozoa: including, the ability to swim through certain media (cervical mucus, artificial media); the reactivity shown by the plasma membrane to various stimuli (glycosaminoglycans or calcium-ionophores); the outcome of IVF; or the presence of fertility-related substances in seminal plasma (Rodriguez-Martinez 2000).

Ejaculated spermatozoa are handled for use in assisted reproductive technologies (ARTs); such as, AI with chilled, frozen thawed or sexed-semen, IVF *et cetera*. Spermatozoa used for ARTs may be exposed to semen extension, fluorophore loading, ultraviolet and laser illumination, high speed sorting, cooling, cryopreservation *et cetera* and each of these procedures will impose different degrees of changes in sperm function following introduced damage to sperm membranes, organelles or the DNA. Therefore, several assays have been recently developed to monitor these sperm parameters *in vitro*.

Plasma membrane integrity and stability of lipid architecture

A functionally intact plasma membrane is a pre-requisite for sperm life and function. The plasmalemma not only acts as a boundary for the sperm cytoplasm; by its semi-permeable features it maintains a chemical gradient of ions and solutes and also holds specific structural proteins that act as transporters for water, energy source substrates and signalling receptors: all relevant for sperm metabolism and the ability to interact with the surroundings. Loss of this functional integrity threatens sperm function and life to various extents, from decreased fertilizing capacity to cell death. If the plasmalemma is intact but functionally unstable, the spermatozoon is unable to interact with its environment and unable to fertilise. Plasma membrane integrity is usually assessed with membrane impermeable dyes, using the rationale that spermatozoa that can exclude these dyes are alive. Examples of impermeable dyes include eosin, many DNA-binding fluorescent probes such as Hoechst 33252, ethidium homodimer and propidium iodide (PI) (Rodriguez-Martinez *et al.* 1997b). A similar rationale lies behind the hypo-osmotic test (HOST) (Jeyendran *et al.* 1984; Correa & Zavos 1994; Zou & Yang 2000) in which sperm membranes able to react to a hypo-osmotic environment have a functional membrane. Combinations of fluorophores are now more often used. Membrane impermeable dyes are combined with penetrating acylated (AM) dyes which, taking advantage of their amphiphatic nature, can penetrate intact and functional membranes, are de-acylated in the cytoplasm, become impermeant and are retained in the living cells. Examples of these AM-probes are fluorescein diacetate (FDA) and its carboxy(methyl) derivatives, such C-FDA (Haugland 2004). There are also acylated fluorescent dyes (for example, SYBR-14) that can bind to DNA and may be combined with other non-penetrating fluorescent DNA-probes such as PI. Together, these would stain the nuclei of sperm with and without deteriorated membranes and simultaneously inform us of the percentages of membrane-intact sperm (Garner *et al.* 1994; Garner & Johnson 1995). This allows examination of ruminant spermatozoa without worrying about the medium in which they were suspended (for example, milk or egg yolk containing extenders) that usually confound measurements with other fluorophores (Nagy *et al.* 2003; 2004). The latter fluorophores can be used with fluorescent microscopy or flow cytometry, allowing us to evaluate large numbers of spermatozoa.

The plasma membrane of the spermatozoon is a highly dynamic structure with a tendency to become unstable if the glycocalix is freed from adsorbed proteins, usually of epididymal and/ or seminal plasma origin (reviewed by Flesch & Gadella 2000). Such instability starts with a loss of the order of the phospholipids present in the plasma membrane, a disorder that can be temporary (Harrison *et al.* 1996) or lead to membrane disorganisation (Harrison 1996). This membrane lipid scrambling is associated with cholesterol efflux and is considered as one of the initial steps in sperm capacitation (Harrison & Gadella 2005). Similar phenomena occur during cooling of spermatozoa and early changes in the architecture of the plasma membrane, well ahead of visible changes, can be monitored by loading spermatozoa with the lipid dye Merocyanine-540 whose significant increase in fluorescence can be determined with flow cytometry (Sostaric *et al.* 2005; Bergqvist *et al.* 2006). Recently, Annexin-V staining has been used as an alternative method to detect subtle membrane changes that involve redistribution of the phospholipid phosphatidylserine (PS) from its normal inner leaflet position to the outer leaflet of the sperm plasmalemma. (Januskauskas *et al.* 2003). Changes similar to those detected during the later part of sperm capacitation and the acrosome reaction (AR) (Tienthai *et al.* 2004) can also be indirectly visualized by the incubation of viable spermatozoa with the fluorescent antibiotic chlortetracycline (CTC) which monitors the displacement of $Ca2+$ in the sperm head plasmalemma (Fraser *et al.* 1995; Januskauskas *et al.* 2000a; Hallap *et al.* 2006).

Assessment of organelle integrity

Flagellar intactness can be easily determined by light microsopy on fixed wet preparations and during motility assessment. Specific abnormalities can be explored further using electron microscopy (Andersson *et al.* 2000). Mitochondria and their ATP production have classically been related to flagellar function, a concept that is today challenged (Silva & Gadella 2006) under the major argument that flagellar movement is related to the local ability to produce ATP anaerobically by glycolysis (Miki *et al.* 2004) while aerobically produced mitochondrial ATP is used for housekeeping metabolism at the mid-piece and head domains. This hypothesis, albeit interesting, still requires more experimental evidence. Mitochondrial integrity and functionality can be measured using specific fluorophores such as Rhodamine 123 and the family of Mitotracker dyes (Gravance *et al.* 2001) including JC-1 which differentiates whether mitochondria are functional or quiescent by detecting the changing re-polarisation of the inner mitochondrial membrane (Martinez-Pastor *et al.* 2004; Hallap *et al.* 2005). Disruption of the acrosomal outer membrane can be readily observed by phase contrast microscopy but intactness can not be assured in all cases. Acrosome integrity can be examined by microscopy or flow cytometry after using fluorescent conjugated lectins that bind to specific carbohydrate moieties of acrosomal glycoproteins. The most commonly used lectins are derived from peanuts (PNA) for assessment of the outer acrosomal membrane or from green peas (PSA) for labelling of acrosomal matrix glycoproteins. Using both PNA and PSA conjugated to fluoresceins with different wavelengths it is possible to identify intact and disrupted acrosomes (Guillan et al. 2005). Alternative methods include the use of antibodies against acrosomal membrane proteins or fluorophores such as the Lysotracker family, or Deep Red TM, that label cell lysosomes (reviewed by Guillan et al. 2005).

Simultaneous assessments of plasma membrane and organelle intactness

Several sperm traits can be simultaneously assessed using flow cytometric multi-parametric analyses after staining spermatozoa with 4 dyes to identify: (a) membrane integrity using SYBR-14/PI; (b) acrosome integrity using phycoerythrin-conjugated peanut agglutinin (PE-PNA); and (c) functional status of the mitochondria using Mitotracker deep red (Nagy et al. 2003). Such analyses can be done during several hours of sperm incubation using multi-colour flow cytometry to disclose changes occurring in spermatozoa over time (Nagy et al. 2004).

Integrity of chromatin structure and DNA integrity

The spermatozoa of ruminants are characterized by having a highly condensed chromatin where protamines tightly pack and protect the haploid DNA. Optimal sperm DNA packing seems essential for full expression of male fertility potential (Spano et al. 2000) and handling of spermatozoa, in particular cryopreservation, has been related to a deterioration of DNA integrity (Peris et al. 2004). Over the past 25 years, several methods have been designed to determine DNA damage including the Comet assay, TUNEL, the acridine orange test (AOT), the tritium-labelled 3H-actinomycin D (3H-AMD) incorporation assay, the *insitu* nick translation (ISNT) assay, DNA breakage detection-fluorescence *in situ* hybridisation (DBD-FISH) and the sperm chromatin structure assay (SCSA) (reviewed by Fraser 2004). While most assays are basically microscopy methods, the TUNEL and the SCSA assays can advantageously use flow cytometry (Evenson & Wixon 2006). Although all methods indicate DNA strand breaks, the SCSA is used most commonly. The SCSA characterizes sperm nuclear chromatin stability based on the increased susceptibility of altered DNA to *in situ* de-naturation when exposed to very low pH. De-naturation is detected by staining with the metachromatic dye acridine orange which results in green fluorescence for native DNA and red fluorescence for denatured DNA. The degree of de-naturation within each sperm nucleus is quantified by flow cytometry (Evenson et al. 1980). The SCSA provides data of spermatozoa with or without DNA fragmentation, the extent of DNA fragmentation and the proportion of immature spermatozoa (Evenson & Wixon 2006).

In vitro assessment of sperm quality and its relationship to fertility

A multitude of laboratory semen evaluation methods have been designed during the last 3 decades to estimate male fertility, aiming to avoid the costs and the time otherwise needed to measure fertility through the AI of hundreds or thousands of females (Rodriguez-Martinez 2003). Among the large variety of tests, fertility appears to be more closely related to membrane integrity than to overall sperm motility, a fact which is especially clear when a large number of spermatozoa are evaluated with fluorescence-activated cell sorting (FACS) or fluorometry (Januskauskas et al. 2001).

However, most of these statistically significant relationships between semen tests and fertility varied greatly among studies; for example, correlations between *sperm motility* and fertility ranged between 0.15 and 0.83 (Kjaestad *et al.* 1993; Stålhammar *et al.* 1994; Bailey *et al.* 1994; Amann *et al.* 2000; Januskauskas *et al.* 2003). Even analyses of semen samples from AI-bulls with CASA instruments have shown variable correlations between patterns of post-thaw motility and *in vivo* fertility (r^2 = 0.45-0.63: Zhang *et al.* 1998; Januskauskas *et al.* 2001). Statistical analyses of combinations of motility patterns yielded stronger correlations (r = 0.68-0.98: Farrell *et al.* 1998) and predictive values could be presented (r^2 = 0.83) when the outcome of motility assessments were combined with other parameters of sperm function (Januskauskas *et al.* 2001) in AI-bulls. *Mitochondrial integrity* and *functionality*, quantified with specific fluorophores, still present both variable and low relationships with AI-fertility in rams and bulls (Martinez-Pastor *et al.* 2004; Hallap *et al.* 2005).

Relationships between *sperm morphology* and fertility have also been shown to exist, but vary widely (r = 0.06-0.86: Graham *et al.* 1980), depending on the type of abnormality (Barth 1989; Barth *et al.* 1992; Ostermeier *et al.* 2001), the quality of the examined semen and the methodology standards used (Kuster *et al.* 2004). Such variability has led to frustration among scientists and clinicians alike when individual bulls often perform much better or much worse than predicted by the breeding soundness evaluation (Higdon *et al.* 2000) and has led to continuous requests for thresholds of abnormalities. Such thresholds for the tolerable levels of abnormal spermatozoa date as far back as the 1920s and 1930s (Williams & Utica 1920; Williams & Savage 1925; Lagerlöf 1934), have been confirmed by numerous studies since then (Barth & Waldner 2002) and so it is now widely accepted that morphology is a main criterion in semen quality in cattle (Fordyce *et al.* 2006). A major point, often disregarded, is the need for large numbers of confirmed pregnancies or, even better calves, before attempting to establish relationships with fertility (Amann 2005). In well-controlled field trials with beef bulls, the relationship between the proportion of morphologically normal spermatozoa and calf outcome (Fitzpatrick *et al.* 2002) or non-return rates (Phillips *et al.* 2004) has proven important. However, when the semen examined is of good quality, such relationships tend to disappear (Rodriguez-Martinez *et al.* 1997b; Saacke 1999).

Membrane integrity, when assessed in large numbers of spermatozoa by fluorometry (Alm *et al.* 2001; Januskauskas *et al.* 2001) or flow cytometry (Guillan *et al.* 2005) is related to fertility after AI but correlations are low. More subtle, early membrane changes (lipid scrambling) indicated by Merocyanine-540 (Hallap *et al.* 2006) or the outer localisation of phosphatidylserine (PS) when examined by Annexin-V staining in frozen-thawed bull spermatozoa (Januskauskas *et al.* 2003), showed low (r = -0.22 to -0.27) negative correlations to fertility.

DNA integrity evaluated by the SCSA has shown a relationship to fertility (Evenson *et al.* 1994; Evenson & Jost 2000). The proportion of sperm nuclei outside the general population [COMPat], nowadays called the DNA fragmentation index (DFI), has been related to high, moderate or very low fertility in humans, when the COMPat (DFI) was 0-15%, 16-29% and > 30%, respectively (Evenson & Jost 2000). Similarly, DFI thresholds for sub-fertility in boars and bulls have been proposed as 18% for boars and 20% for bulls (Rybar *et al.* 2004). Unfortunately, this latter report used semen from young, unproven Simmental bulls with low fertility (40-60% of 90d-NRR) and from boars without fertility data, thus leaving these thresholds of subfertility with little or no inherent value. However, in several well-controlled studies of AI-sires with proven and varied fertility it has been clear that proven bulls do not reach these high DFI values, even when there is a statistical relationship between SCSA and fertility. For instance, Januskausakas *et al.* (2001; 2003) found the COMP aat (DFI) of AI bulls with different fertility ranged from 1.2 to 8.0%; figures similar to those found in other AI-bulls in other countries

(Hallap *et al.* 2005). Therefore; significant; but largely variable relationships (r = 0.33-0.94) between this parameter and fertility have been reported for bull frozen-thawed semen used for conventional AI (Januskauskas *et al.* 2001; 2003) or heterospermic AI followed by paternity screening of fetuses using genetic markers (Ballachey *et al.* 1988).

In summary, it seems clear that only when studying handled spermatozoa, such as frozen-thawed semen for AI, are significant relationships found between modern rather than conventional semen analyses methods and the fertility achieved in the field. However, such relationships are modest and are disturbingly variable between laboratories.

How should we analyze a heterogeneous semen sample?

The ejaculate of a bull, ram or buck (or any other domestic mammal for that matter) is heterogeneous in the sense that none of the spermatozoa are equal to each other in terms of haploid genomic information and in attributes for fertilisation. Andrologists have been trying for decades to solve the question of which spermatozoa are involved in fertilization or how fertile a particular semen sample is. Sperm transport through the female genital tract imposes a series of sperm selection steps which eliminate a proportion of spermatozoa at each step, so that eventually only a small number gain access to the oocyte. Fertilisation seems, therefore, a matter of probability, based on sperm numbers and on proportions of spermatozoa with attributes that favour survival during the journey through the female genitalia and the effective interaction with the female epithelia and the oocyte. This concept suggests the presence of sub-populations of competent spermatozoa within the ejaculate, as previously discussed by Rodriguez-Martinez (2000; 2003) and Holt & Van Look (2004). Whether there is a relationship between the proportion of these competent (that is, potentially fertile) spermatozoa and the fertility of the semen sample (or the male) remains to be proven.

Considering fertilisation as a multi-factorial process, the combined assessment of a large number of parameters would lead to a higher accuracy of the test employed. To reach this goal, semen samples have been subjected to functional *in vitro* tests which are able to discern the ability of spermatozoa to undergo specific steps in the process of fertilisation and the triggering of the development of the early embryo. Among these methods we found the selection of spermatozoa via swim-up or double-phase partition in aqueous systems, the ability to bind to genital epithelia, the ability to undergo sperm capacitation or AR by exogenous stimuli, the ability to bind and penetrate the ZP and to fertilise *in vitro* (IVF). To determine whether these tests have biological value and would be able to determine the level of potential fertility of the semen before its application in the field, the results obtained have been related statistically with the semen or sire fertility (reviewed by Larsson & Rodriguez-Martinez 2000; Rodriguez-Martinez 2003).

Linearity of sperm movement - swim-up tests

Most ejaculated spermatozoa show a typical progressive and linear motility, and many display an innate ability to traverse fluids of a certain viscosity, which led to the so-called "swim-up tests" where spermatozoa are assessed for their capacity to traverse fluid barriers simulating *in vivo* conditions such as traversing the cervical barrier (Rodriguez-Martinez *et al.* 1997a). Experiments with frozen-thawed bull semen, using a simple swim-up across a column of culture medium have indicated the number of viable spermatozoa with linear motility post-swim-up reflects the innate fertilising capacity of a seminal sample (Zhang *et al.* 1998). Swim-up is, moreover, a suitable method to separate a sub-population of spermatozoa with most attributes for further testing, specially those with stable plasmalemmae (Januskauskas *et al.* 2000a; 2000b; Hallap *et al.* 2004; 2005). Another method to isolate spermatozoa with linear motion is by centrifugal counter-current distribution analysis (CCCD). This is an aqueous two-phase partition system that has proven valuable for revealing sperm heterogeneity in semen samples and results have been correlated to fertility, albeit correlations were low (r = 0.44: Pérez-Pé *et al.* 2002).

Sperm binding to oviductal epithelium

Binding of spermatozoa occurs in the sperm reservoir of the oviduct, one of the presumed barriers to sperm progression to the site of fertilisation, prolonging sperm viability. Binding to oviductal epithelial cells also prolongs sperm life *in vitro*, presumably because the binding occurs only with non-capacitated spermatozoa (Levfebre & Suarez 1996). Consequently, sperm co-culture with oviductal epithelial explants has been used to determine the capacity of a semen sample to colonise the tubal reservoir and was reported to be correlated with bull fertility, but only when high quality sperm samples were tested (De Pauw *et al.* 2002).

Measurement of sperm capacitation-like phenomena in spermatozoa

Sperm calcium levels and CTC patterns relate to *in vivo* fertility of frozen-thawed bull spermatozoa (Collin *et al.* 2000). In Swedish dairy AI-bulls of known fertility, 30-40% of frozen-thawed spermatozoa showed capacitation-like changes when screened by the CTC-assay (Thundathil *et al.* 1999). The percentage of "non-capacitated" spermatozoa showed a significant relation to fertility (r = 0.50: Thundathil *et al.* 1999; Gil *et al.* 2000), indicating that the method could be used for prognostic purposes. For instance, the proportion of CTC-assessed "uncapacitated" bull spermatozoa recovered after swim-up of frozen-thawed AI-samples correlated positively (r = 0.48, P < 0.05) with their fertility in the field (Januskauskas et al.; 2000a; 2000b). When semen from AI-boars and bulls has been examined fresh and/or following cryopreservation, increases in "capacitation-like changes" (for example, destabilisation of the plasma membrane) measured with Merocyanine-540 (Peña *et al.* 2004; Hallap *et al.* 2006) or Annexin-V (Peña *et al.* 2003; Januskauskas *et al.* 2003) were detected, and even correlated to fertility (Januskauskas *et al* 2003; Hallap *et al.* 2006). However, in the boar these capacitation-like changes appear not to be similar to *in vivo* capacitation (Guthrie & Welch 2005) and are probably a consequence, reversible or not, of the process of cryopreservation.

Capacity for in vitro acrosome reaction

The AR can be induced in vitro by exposure to homologous, solubilised ZPs (Gil *et al.* 2000). It can also be induced by exposure to particular glycosaminoglycans (GAGs) known to be present in the oviductal fluid, where spermatozoa bathe before and during fertilisation. Heparin (Parrish *et al.* 1985; Whitfield & Parkinson 1992; Januskauskas *et al.* 2000a), heparan sulphate (Bergqvist *et al.* 2006) and chondroitin sulphate (Ax & Lenz 1987; Lenz *et al.* 1988) were able to induce the AR in vitro, and the degree of AR was significantly related to fertility. The AR can also be induced by treatment with calcium ionophores such as Hoechst A23187 which promotes a massive influx of Ca^{2+} into spermatozoa similar to that occurring during binding to the ZP. Significant correlations (r = 0.60) have been reported between the degree of AR-responsiveness and fertility (Whitfield & Parkinson 1995; Januskauskas *et al.* 2000b).

Zona pellucida binding tests

Binding of spermatozoa to the ZP is a critical step in the process of fertilisation. Capacitation is a pre-requisite to binding and it is followed by the AR induced by the ZP. The combination of these progressive steps is the rationale for *in vitro* sperm-ZP binding tests using whole, or hemi-ZPs. Using ZP-binding tests, significant correlations ($r = 0.50$) have been obtained with AI-fertility in bulls (Zhang *et al.* 1998).

Accessory sperm counts and in vitro *penetration tests*

Spermatozoa that did not penetrate the ZP entirely during fertilisation due to the zona reaction (an effective block to polyspermia) are trapped in the ZP of oocytes or early embryos and named accessory spermatozoa. These spermatozoa have demonstrated the attributes needed to penetrate the ZP and should therefore be considered as potentially fertile (Saacke *et al.* 1998). Rates of in vivo fertilisation and the number of accessory spermatozoa are statistically correlated (Saacke *et al.* 2000). Fertility can be also evaluated in vitro using oocyte penetration tests (Henault & Killian 1995; Brahmkshtri *et al.* 1999), where the presence of spermatozoa or pronuclei in the ooplasm determines the success of the test. Relationships between in vitro penetration and the fertility of the sires have been reported in bulls (Henault & Killian 1996) where spermatozoa from several sires were tested simultaneously. The spermatozoa from each sire were marked with fluorophores that render different fluorescence and thus their relative penetrability was measured, substantially diminishing variation.

In vitro fertilisation (IVF)

In vitro fertilisation closely mimics gamete-to-gamete relationships occurring during fertilisation *in vivo*, allowing measurement of different end-points in the stages of early embryo development. Thus IVF has been used to determine the relative fertility of semen samples in cattle: statistically significant relations between IVF fertility and fertility *in vivo* were sometimes found, mostly in retrospective studies (Larsson & Rodriguez-Martinez 2000) but also in prospective ones (Zhang *et al.* 1999). However, IVF is largely dependent on laboratory stability, thus making comparisons among studies and controls of variation difficult (Papadopoulus *et al.* 2005). Furthermore, as with many other tests of oocyte-spermatozoa interaction, the sperm:oocyte ratio *in vitro* is largely that *in vivo* thus reducing its diagnostic value (Hunter & Rodriguez-Martinez 2002).

Indicators of male fertility in seminal plasma

Analyses of seminal plasma have shown a series of proteins and other seminal factors with indirect relationships to fertility, particularly in bulls. Some of these substances are related to sperm function; for example, heparin binding proteins (HBPs)(Bellin *et al.* 1994; Moura *et al.* 2006), platelet activating factor (PAF) (Brackett *et al.* 2004), bovine seminal plasma protein A3 (BSP-A3) (Roncoletta *et al.* 2006), acid seminal fluid protein (aSFP) (Roncoletta *et al.* 2006) or fertility associated antigen (FAA) (Bellin *et al.* 1998; Sprott *et al.* 2000). Other substances are related to oviductal attachment, such as osteopontin (Cancel *et al.* 1997; Moura 2005; Moura *et al.* 2006) and sperm membrane antigen P25b (Parent *et al.* 1999), or to prostaglandin synthesis (for example, lipocalin-type prostaglandin D2 synthase: Gerena *et al.* 1998; Moura *et al.* 2006). Although correlations with fertility have been reported, this does not mean that these factors are general predictors of fertility, but rather they indicate the complexity of factors involved in fertility. However, they may be useful to pre-screen individuals and separate extreme populations from further use.

Can the fertilizing capacity of a given semen sample be estimated *in vitro*?

A combination of results from a series of *in vitro* tests analysed by multiple regression has often been used to show significant relationships between laboratory assays and fertility in bulls (Linford *et al.* 1976; Januskauskas *et al.* 1996; Zhang *et al.* 1997; Januskauskas *et al.* 1999; 2000a; 2000b; 2001; Tanghe *et al.* 2002). These studies had a retrospective nature using bulls with wide ranges of fertility; that is, they were not predictive *per se* (Rodriguez-Martinez 2003). However, it has been argued that *in vitro* data could be predictive of *in vivo* fertility if the results of various *in vitro* methods of frozen semen evaluation could be combined to calculate the *expected fertility* of, for instance, young bulls. This approach was used with frozen-thawed semen from unproven bulls (11-13 months of age) whose potential fertility was estimated in the laboratory before their real fertility in the field was assessed by AI (Zhang *et al.* 1999). With this approach, a strong relationship (r = 0.90) was found between predicted and real fertility. This *in vitro* testing enabled identification of sub-fertile bulls, whose expected and real fertility was below the limit considered for sub-fertility (62% non-return rate), while the other young bulls predicted to have satisfactory fertility had non return rates of ≥ 65%. However, a large variation was evident among ejaculates within a young sire, both *in vitro* and *in vivo*, even when ejaculates were collected within a relatively short period (Zhang *et al.* 1998). This implies that it is necessary to analyse a large proportion of ejaculates from a large bull population in order to lower this variation to a minimum (Zhang *et al.* 1997). Moreover, semen parameters might change with age (Hallap *et al.* 2005), impairing our ability to make reliable predictions over time.

Conclusions

Male sub-fertility is a problem of complex etiology, requiring a complete andrological work-up for proper diagnosis and prognosis. At present, veterinary andrologists have access to a large battery of diagnostics aids but still lack simple comprehensive tests to estimate levels of fertility in male ruminants before they are used for natural mating or their semen is processed for breeding via AI or other ARTs. A major point derived from the present review is the lack of strategies to determine the presence of sperm subpopulations with the ability to surpass sperm selection steps present *in vivo*. Some clear examples appear in the value of swim-up procedures, in the screening of handicapping morphology, in the ability of spermatozoa to respond to pre-fertilisation stimuli (for example, bicarbonate-mediated triggering of membrane scrambling and the associated changes in metabolism), and at the same time measure their lifespan after these stimuli, and their resilience to acrosome exocytosis. Obviously, defining sperm subpopulations is difficult when handling billions of spermatozoa in an ejaculate. Common sense calls for the preparation of AI-doses with low sperm numbers that can be tested for fertility both *in vitro* and *in vivo* before the sire can be allocated into a certain level of potential fertility. This must be accompanied by a complete andrological examination before breeding and proper strategies (management as well as nutritional) to ensure good development of the male from calfhood. Measurement of LH surge and of insulin growth factor-I (IGF-I) following challenge with GnRH in bull calves are simple procedures that might monitor potential testicular development and, consequently, estimate potential fertility. Post-pubertal screening of fertility-associated factors for pre-allocation of sires is a promising strategy.

Acknowledgements

Funding has been provided by FORMAS, formerly the Swedish Council for Forestry and Agricultural Research (SJFR), and the Swedish Farmers' Foundation for Agricultural Research (SLF), Stockholm, Sweden.

References

Alm K, Taponen J, Dahlnom M, Tuunainen E, Koskinen E & Andersson M 2001 A novel automated fluorometric assay to evaluate sperm viability and fertility in dairy bulls. *Theriogenology* **56** 677-684.

Amann RP, Seidel GE Jr & Mortimer RG 2000 Fertilizing potential in vitro of semen from young beef bulls containing a high or low percentage of sperm with a proximal droplet. *Theriogenology* **54** 1499-1515.

Amann RP 2005 Weaknesses in reports of "fertility" in horses and other species. *Theriogenology* **63** 698-715.

Andersson M, Peltoniemi OAT, Mäkinen A, Sukura A & Rodriguez-Martinez H 2000 The hereditary "short tail" sperm defect - A new reproductive problem in Yorkshire boars. *Reproduction in Domestic Animals* **35** 59-63.

Ax RL & Lenz RW 1987 Glycosaminoglycans as probes to monitor differences in fertility of bulls. *Journal of Dairy Science* **70** 1477-1486.

Bailey JL, Robertson L & Buhr MM 1994 Relations among in vivo fertility, computer-analysed motility and Ca++ influx in bovine spermatozoa. *Canadian Journal of Animal Science* **74** 53-58.

Ballachey BE, Evenson DP & Saacke RG 1988 The sperm chromatin structure assay. Relationship with alternate tests of semen quality and heterospermic performance of bulls. *Journal of Andrology* **9** 109-115.

Barth AD 1989 Abaxial tail attachment of bovine spermatozoa and its effect on fertility. *Canadian Veterinary Journal* **30** 656-662.

Barth AD 2000 *Bull Breeding Soundness Evaluation Manual*, 2nd Edition. The Western Canadian Association of Bovine Practitioners, Saskatoon.

Barth AD & Oko RJ 1989 *Abnormal Morphology of Bovine Spermatozoa*. Ames: Iowa State University Press.

Barth AD & Waldner CL 2002 Factors affecting breeding soundness evaluation of beef bulls at the Western College of Veterinary Medicine. *Canadian Veterinary Journal* **43** 274-284.

Barth AD, Bowman, PA, Bo GA & Mapletoft RJ 1992 Effect of narrow sperm head shape on fertility in cattle. *Canadian Veterinary Journal* **33** 31-39.

Bellin ME, Hawkins HE, Ax RL 1994 Fertility of range bulls grouped according to presence or absence of heparin-binding proteins in sperm membranes and seminal fluid. *Journal of Animal Science* **72** 2442-2448.

Bellin ME, Oyarzo JN, Hawkins HE, Zhang H, Smith RG, Forrest DW, Sprott LR & Ax RL 1998 Fertility-associated antigen on bull sperm indicates fertility potential. *Journal of Animal Science* **76** 2032-2039.

Bergqvist A-S, Ballester J, Johannisson A, Hernández M, Lundeheim N & Rodríguez-Martínez H 2006 In vitro capacitation of bull spermatozoa by oviductal fluid and its components. *Zygote* **14** 259-273.

Brackett BG, Bosch P, McGraw RA, DeJarnette JM, Marshall CE, Massey JB & Roudebush WE 2004 Presence of platelet-activating factor (PAF) receptor in bull sperm and positive correlation of sperm PAF content with fertility. *Reproduction, Fertility and Development* **16** 265-295.

Brahmkshtri BP, Edwin MJ, John MC, Nainar AN & Krishnan AR 1999 Relative efficacy of conventional sperm parameters and sperm penetration bioassay to assess bull fertility in vitro. *Animal Reproduction Science* **54** 159-168.

Cancel A, Chapman D & Killian G 1997 Osteopontin is the 55-kilodalton fertility-associated protein in Holstein bull seminal plasma. *Biology of Reproduction* **57** 1293-1301.

Chacón J 2001 Assessment of sperm morphology in Zebu bulls, under field conditions in the tropics. *Reproduction in Domestic Animals* **36** 91-99.

Collin S, Sirard M-A, Dufour M & Bailey JL 2000 Sperm calcium levels and chlortetracycline fluorescence patterns are related to the in vivo fertility of cryopreserved bovine semen. *Journal of Andrology* **21** 938-943.

Correa JR & Zavos PM 1994 The hypoosmotic swelling test: its employment as an assay to evaluate the functional integrity of the frozen-thawed bovine sperm membrane. *Theriogenology* **42** 351-360.

De Pauw I, Van Soom A, Laevens H, Verberckmoes S & De Kruif A 2002 Sperm binding to epithelial oviduct explants in bulls with different non-return rates investigated with a new in vitro model. *Biology of Reproduction* **67** 1073-1079.

Evenson DP & Jost LK 2000 *Sperm chromatin structure assay for fertility assessment. In: Robinson P ed. Current protocols in cytometry, vol 1*, Wiley New York.

Evenson DP & Wixon R 2006 Clinical aspects of sperm DNA fragmentation detection and male infertility. *Theriogenology* **65** 979-991.

Evenson DP, Darzynkiewicz Z & Melamed MR 1980 Relation of mammalian sperm chromatin heterogeneity to fertility. *Science* **210** 1131-1133.

Evenson DP, Thompson L & Jost L 1994 Flow cytometric evaluation of boar semen by the sperm chromatin structure assay as related to cryopreservation and fertility. *Theriogenology* **41** 637-651.

Farrell PB, Presicce GA, Brockett CC & Foote RH 1998

Quantification of bull sperm characteristics measured by computer-assisted sperm analysis (CASA) and the relationship to fertility. *Theriogenology* **49** 871-879.

Fitzpatrick LA, Fordyce G, McGowan MR, Bertram JD, Doogan VJ, De Faveri J, Miller RG & Holroyd RG 2002 Bull selection and use in northern Australia. Part 2. Semen traits. *Animal Reproduction Science* **71** 39-49.

Flesch FM & Gadella BM 2000 Dynamics of the mammalian sperm plasma membrane in the process of fertilization. *Biochimica et Biophysica Acta* **1469** 197-235.

Fordyce G, Entwistle K, Norman S, Perry V, Gardiner B & Fordyce P 2006 Standardising bull breeding soundness evaluations and reporting in Australia. *Theriogenology* **66** 1140-1148.

Fraser L 2004 Structural damage to nuclear DNA in mammalian spermatozoa: its evaluation techniques and relationship with male fertility. *Polish Journal of Veterinary Sciences* **7** 311-321.

Fraser LR, Abeydeera LR & Niwa K 1995 Ca(2 +)-regulating mechanisms that modulate bull sperm capacitation and acrosomal exocytosis as determined by chlortetracycline analysis. *Molecular Reproduction and Development* **40** 233-241.

Garner DL & Johnson LA 1995 Viability assessment of mammalian sperm using SYBR-14 and propidium iodide. *Biology of Reproduction* **53** 276-284.

Garner DL, Johnson LA, Yue ST, Roth BL & Haugland RP 1994 Dual DNA staining assessment of bovine sperm viability using SYBR-14 and propidium iodide. *Journal of Andrology* **15** 620-629.

Gerena R, Irikura D, Urade Y, Eguchi N, Chapman D & Killian G 1998 Identification of a fertility-associated protein in bull seminal plasma as lipocalin-like prostaglandin D synthase. *Biology of Reproduction* **58** 826-833.

Gil J, Januskauskas A, Hååård MCH, Hååård MGM, Johannisson A, Söderquist L, Rodriguez-Martinez H 2000 Functional sperm parameters and fertility of bull semen extended in Biociphos-Plus® and Triladyl®. *Reproduction in Domestic Animals* **35** 69-77.

Graham EF, Schmehl MKL, Nelson DS 1980 Problems with laboratory assays. *Proceedings 8th National Association of Animal Breeders Technical Conference of AI & Reproduction* **1** 59-66.

Gravance CG, Champion ZJ & Casey PJ 1998 Computer-assisted sperm head morphometry analysis (ASMA) of cryopreserved ram spermatozoa. *Theriogenology* **49** 1219-1230.

Gravance CG, Garner DL, Pitt C, Vishwanath R, Sax-Gravance SK & Casey PJ 1999 Replicate and technician variation associated with computer aided bull sperm head morphometry analysis (ASMA). *International Journal of Andrology* **22** 77-82.

Gravance CG, Garner DL, Miller MG & Berger T 2001 Fluorescent probes and flow cytometry to assess rat sperm integrity and mitochondrial function. *Reproductive Toxicology* **15** 5-10.

Guillan L, Evans G & Maxwell WMC 2005 Flow cytometric evaluation of sperm parameters in relation to fertility potential. *Theriogenology* **63** 445-457.

Guthrie HD & Welch GR 2005 Impact of storage prior to cryopreservation on plasma membrane function and fertility of boar sperm. *Theriogenology* **63** 396-410.

Hallap T, Hååård MCH, Jaakma Ü, Larsson B, Rodriguez-Martinez H 2004 Does cleansing of frozen-thawed bull semen before assessment provide samples that relate better to potential fertility? *Theriogenology* **62** 702-713.

Hallap T, Nagy S, Hååård M, Jaakma Ü, Johannisson A, Rodriguez-Martinez H 2005 Sperm chromatin stability in frozen-thawed semen is maintained over age in AI bulls. *Theriogenology* **63** 1752-1763.

Hallap T, Nagy S, Jaakma Ü, Johannisson A & Rodriguez-Martinez H 2005 Mitochondrial activity of frozen-thawed spermatozoa assessed by MitoTracker Deep Red 633. *Theriogenology* **63** 2311-2322.

Hallap T, Nagy S, Jaakma Ü, Johannisson A & Rodriguez-Martinez H 2006 Usefulness of a triple fluorochrome combination Merocyanine 540/Yo-Pro 1/Hoechst 33342 in assessing membrane stability of viable frozen-thawed spermatozoa from Estonian Holstein AI bulls. *Theriogenology* **65** 1122-1136.

Harrison RAP 1996 Capacitation mechanisms, and the role of capacitation as seen in eutherian mammals *Reproduction, Fertility and Development* **8** 581–594.

Harrison RAP & Gadella BM 2005 Bicarbonate-induced membrane processing in sperm capacitation. *Theriogenology* **63** 342-351.

Harrison RAP, Ashworth PJC & Miller NGA 1996 Bicarbonate/CO2, an effector of capacitation, induces a rapid and reversible change in the lipid architecture of boar sperm plasma membranes *Molecular Reproduction and Development* **45** 378–391.

Haugland RP 2004 Asays for cell viability, proliferation and function. In: *Handbook of fluorescent probes and research products*. Molecular Probes Inc 9th ed, Eugene.

Henault MA & Killian GJ 1995 Effects of sperm preparation and bull fertility on in vitro penetration of zona-free bovine oocytes. *Theriogenology* **43** 739-749.

Henault MA & Killian GJ 1996 Effect of homologous and heterologous seminal plasma on the fertilizing ability of ejaculated bull spermatozoa assessed by penetration of zona-free bovine oocytes. *Journal of Reproduction and Fertility* **108** 199-204.

Higdon HL, Spitzer JC, Hopkins FM & Bridges WC 2000 Outcomes of breeding soundness evaluation of 2898 yearling bulls subjected to different classification systems. *Theriogenology* **53** 1321-1332.

Holt WV & Van Look KJ 2004 Concepts in sperm heterogeneity, sperm selection and sperm competition as biological foundations for laboratory tests of semen quality. *Reproduction* **127** 527-535.

Hunter RHF & Rodriguez-Martinez H 2002 Analysing mammalian fertilisation: reservations and potential pitfalls with an in vitro approach. *Zygote* **10** 11-15.

Januskauskas A, Söderquist L, Hååård MGM, Hååård MCH, Lundeheim N & Rodríguez-Martínez H 1996 Influence of sperm number per straw on the post-thaw sperm viability and fertility of Swedish Red and White AI bulls. *Acta Veterinaria Scandinavica* **37** 461-470.

Januskauskas A, Gil J, Söderquist L, Hååard MGM, Hååard MCH, Johannisson A & Rodriguez-Martinez H 1999 Effect of cooling rates on post-thaw sperm motility, membrane integrity, capacitation status and fertility of dairy bull semen used for artificial insemination in Sweden. *Theriogenology* **52** 641-658.

Januskauskas A, Gil J, Söderquist L & Rodríguez-Martínez H 2000a Relationship between sperm response to glycosaminoglycans in vitro and non return rates of Swedish dairy AI bulls. *Reproduction in Domestic Animals* **35** 207-212.

Januskauskas A, Johannisson A, Söderquist L & Rodríguez-Martínez H 2000b Assessment of the fertilizing potential of frozen-thawed bovine spermatozoa by calcium ionophore A23187-induced acrosome reaction in vitro. *Theriogenology* **53** 859-875.

Januskauskas A, Johannisson A & Rodríguez-Martínez H 2001 Assessment of sperm quality through fluorometry and sperm chromatin structure assay in relation to field fertility of frozen-thawed semen from Swedish AI bulls. *Theriogenology* **55** 947-961.

Januskauskas A, Johannisson A & Rodriguez-Martinez H 2003 Subtle membrane changes in cryopreserved bull semen in relation with sperm viability, chromatin structure, and field fertility. *Theriogenology* **60** 743-758.

Jeyendran RS, Van der Ven HH, Perez-Pelaez M, Crabo BG & Zaneveld LJD 1984 Development of an assay to assess the functional integrity of the human sperm membrane and its relationship to other semen characteristics. *Journal of Reproduction and Fertility* **70** 219–228.

Kjaestad H, Ropstad E & Andersen Berg K 1993 Evaluation of spermatological parameters used to predict the fertility of frozen bull semen. *Acta Veterinaria Scandinavica* **34** 299-303.

Kuster CE, Singer RS & Althouse GC 2004 Determining sample size for the morphological assessment of sperm. *Theriogenology* **61** 691-703.

Lagerlöf N 1934 Research concerning morphologic changes in the semen picture and in the testicles of sterile and subfertile bulls. *Acta Pathologicae Microbiologicae Scandinavica* **19** 66-77.

Larsson B & Rodriguez-Martinez H 2000 Can we use in vitro fertilization tests to predict semen fertility? In: Animal Reproduction: Research and Practice II. *Animal Reproduction Science* **60/61** 327-336.

Lefebvre R & Suarez SS 1996 Effect of capacitation on bull sperm binding to homologous oviductal epithelium. *Biology of Reproduction* **54** 575-582.

Lenz RW, Martin JL, Bellin ME & Ax RL 1988 Predicting fertility of dairy bulls by inducing acrosome reactions in sperm with chondroitin sulfates. *Journal of Animal Science* **71** 1073-1077.

Linford E, Glover FA, Bishop C & Stewart DL 1976 The relationship between semen evaluation methods and fertility in the bull. *Journal of Reproduction and Fertility* **47** 283-291.

Marco-Jimenez F, Viudes-de-Castro MP, Balasch S, Mocé E, Silvestre MA, Gomez EA & Vicente JS 2006 Morphometric changes in goat sperm heads induced by cryopreservation. *Cryobiology* **52** 295-304.

Martinez-Pastor F, Johannisson A, Gil A, Kaabi M, Anel L, Paz P & Rodriguez-Martinez H 2004 Use of chromatin stability assay, mithochondrial stain JC-1, and fluorometric assessment of plasma membrane to evaluate frozen-thawed ram semen. *Animal Reproduction Science* **84** 121-133.

Miki K, Qu W, Goulding EH, Willis WD, Bunch DO, Strader LF, Perreault SD, Eddy EM & O´Brien DA 2004 Glyceraldehyde 3-phosphate dehydrogenase-S, a sperm-specific enzyme, is required for sperm motility and male fertility. *Proceedings of the National Academy of Sciences USA* **101** 16501-16506.

Moura AA 2005 Seminal plasma proteins and fertility indexes in the bull: the case for osteopontin. *Animal Reproduction* **2** 3-10.

Moura AA, Koc H, Chapman DA & Killian GJ 2006 Identification of proteins in the accessory sex gland fluid associated with fertility indexes of dairy bulls: a proteomic approach. *Journal of Andrology* **27** 201-211.

Nagy S, Jansen J, Topper EK & Gadella BM 2003 A triple-stain flow cytometric method to assess plasma- and acrosome-membrane integrity of cryopreserved bovine sperm immediately after thawing in presence of egg-yolk particles. *Biology of Reproduction* **68** 1828-1835.

Nagy S, Hallap T, Johannisson A, Rodriguez-Martinez H 2004 Changes in plasma membrane and acrosome integrity of frozen-thawed bull spermatozoa during a 4 h incubation as measured by multicolor flow cytometry. *Animal Reproduction Science* **80** 225-235.

Ostermeier GC, Sargeant GA, Yandell BS, Evenson DP & Parrish JJ 2001 Relationship of bull fertility to sperm nuclear shape. *Journal of Andrology* **22** 595-603.

Palmer CW, Brito LFC, Arteaga AA, Söderquist L, Persson Y & Barth AD 2005 Comparison of electroejaculation and transrectal massage for semen collection in range and yearling feedlot beef bulls. *Animal Reproduction Science* **87** 25-31.

Papadopoulus S, Hanrahan JP, Donovan A, Duffy P, Boland MP & Lonergan P 2005 In vitro fertilization as a predictor of fertility from cervical insemination of sheep. *Theriogenology* **63** 150-159.

Parent S, Lefebvre L, Brindle Y & Sullivan R 1999 Bull subfertility is associated with low levels of a sperm membrane antigen. *Molecular Reproduction and Development* **52** 57-65.

Parkinson TJ 2004 Evaluation of fertility and infertility in natural service bulls. *The Veterinary Journal* **168** 215-229.

Parrish JJ, Susko-Parrish JL & First NL 1985 Effect of heparin and chondroitin sulfate on the acrosome reaction and fertility of bovine sperm in vitro. *Theriogenology* **24** 537-549.

Peña FJ, Johannisson A, Wallgren M & Rodriguez-Martinez H 2003 Assessment of fresh and frozen-thawed boar semen using an Annexin-V assay: a new method of evaluating sperm membrane integrity. *Theriogenology* **60** 677-689.

Peña FJ, Johannisson A, Wallgren M & Rodriguez Martinez H 2004 Effect of hyaluronan supplementation on boar sperm motility and membrane lipid architecture status after cryopreservation. *Theriogenology* **61** 63-70.

Peña FJ, Saravia F, García- Herreros M, Núñez-Martinez I, Tapia JA, Johannisson A, Wallgren M & Rodríguez-Martínez H 2005a Identification of sperm morphological subpopulations in two different portions of the boar ejaculate and its relation to post thaw quality. *Journal of Andrology* **26** 716-723.

Peña FJ, Saravia F, Johannisson A, Wallgren M & Rodriguez-Martinez, H 2005b A new and simple method to evaluate early membrane changes in frozen-thawed boar spermatozoa. *International Journal of Andrology* **28** 107-114.

Pérez-Pé R, Marti JI, Sevilla E, Fernandez-Sanchez M, Fantova E, Altarriba J, Cebrián-Pérez JA, Muiño-Blanco T 2002 Prediction of fertility by centrifugal countercurrent distribution (CCCD) analysis: correlation between viability and heterogeneity of ram semen and field fertility. *Reproduction* **123** 869-875.

Peris SI, Morrier A, Dufour M, Bailey JL 2004 Cryopreservation of ram semen facilitates sperm DNA damage: relationship between sperm andrological parameters and the sperm chromatin structure assay. *Journal of Andrology* **25** 224-233.

Phillips NJ, McGowan MR, Johnston SD & Mayer DG 2004 Relationship between thirty post-thaw spermatozoal characteristics and the field fertility of 11 high-use Australian dairy AI-sires. *Animal Reproduction Science* **81** 47-61.

Rodríguez-Martínez H 2000 Evaluation of frozen semen: Traditional and new approaches. In: Topics in Bull Fertility, Chenoweth PJ (ed), Recent Advances in Veterinary Medicine, International Veterinary Information Services (IVIS), *www.ivis.org* (Document No. A0502.0600, 2000a).

Rodriguez-Martinez H 2003 Laboratory semen assessment and prediction of fertility: still utopia? *Reproduction in Domestic Animals* **38** 312-318.

Rodríguez-Martínez H, Larsson B, Pertoft H 1997a Evaluation of sperm damage and techniques for sperm clean-up. *Reproduction, Fertility and Development* **9** 297-308.

Rodríguez-Martínez H, Larsson B, Zhang BR & Söderquist L 1997b In vitro assessment of viability and fertilizing capacity of bull spermatozoa. *Journal of Reproduction and Development* **43** 1-11.

Roncoletta M, Morani Eda S, Esper CR, Barnabe VH & Franceschini PH 2006 Fertility-associated proteins in Nelore bull sperm membranes. *Animal Reproduction Science* **91** 77-87.

Rybar R, Faldikova L, Faldyna M, Machatkova M & Rubes J 2004 Bull and boar sperm DNA integrity evaluated by sperm chromatin structure assay in the Czech Republic. *Veterinary Medicine (Czech)* **1** 1-8.

Saacke RG 1999 Evaluation of frozen thawed semen: Principles of different tests. *Proceedings Annual Conference Society for Theriogenology* Nashville **1** 281-292.

Saacke RG, Dalton J, Nadir S, Bame J & Nebel RL 1998 Spermatozoal characteristics important to sperm transport, fertilization and early embryonic development. In: *Gametes: development and function*, pp 320-335.Eds. A Lauria, F Gandolfi, G Enne, L Gianaroli.Italy: Serono Symposia.

Saacke RG, Dalton J, Nadir S, Nebel RL & Bame J 2000 Relationship of seminal traits and insemination time to fertilization rate and embryo quality. *Animal Reproduction Science* **60/61** 663-677.

Silva PFN & Gadella BM 2006 Detection of damage in mammalian sperm cells. *Theriogenology* **65** 958-978.

Sostaric E, van der Lest CH, Colenbrander B & Gadella BM 2005 Dynamics of carbohydrate affinities at the cell surface of capacitating bovine sperm cells. *Biology of Reproduction* **72** 346-57.

Spano M, Bonde J, Hjollund HI, Kolstadt HA, Cordelli E & Leter G 2000 Sperm chromatin damage impairs human fertility. *Fertility and Sterility* **73** 43-50.

Sprott LR, Young J, Forrest DW, Zhang HM, Oyarzo JN, Bellin ME & Ax RL 2000 Artificial insemination outcomes in beef females using bovine sperm with a detectable fertility-associated antigen. *Journal of Animal Science* **78** 795-798.

Stålhammar EM, Janson L & Philipsson J 1994 The impact of sperm motility on non-return rate in preselected dairy bulls. *Reproduction, Nutrition & Development* **34** 37-45.

Tanghe S, Van Soom A, Sterckx V, Maes D & De Kruif A 2002 Assessment of different sperm quality parameters to predict in vitro fertility of bulls. *Reproduction in Domestic Animals* **37** 127-132.

Thundathil J, Gil J, Januskauskas A, Larsson B, Söderquist L, Mapletoft R & Rodríguez-Martínez H 1999 Relationship between the proportion of capacitated spermatozoa present in frozen-thawed semen and fertility with artificial insemination. *International Journal of Andrology* **22** 366-373.

Tienthai P, Johannisson A & Rodríguez-Martínez H 2004 Sperm capacitation in the porcine oviduct. *Animal Reproduction Science* **80** 131-146.

Whitfield CH & Parkinson TJ 1992 Relationship between the fertility of bovine semen and in vitro induction of acrosome reactions by heparin. *Theriogenology* **38** 11-20.

Whitfield CH & Parkinson TJ 1995 Assessment of the fertilising potential of frozen bovine spermatozoa by in vitro induction of acrosome reactions with calcium ionophore (A23187). *Theriogenology* **44** 413-422.

Williams W & Savage A 1925 Observations on the seminal micropathology of bulls. *Cornell Veterinarian* **15** 353-375.

Williams W & Utica N 1920 Technique of collecting semen for laboratory examination with a review of several diseased bulls. *Cornell Veterinarian* **10** 87-94.

Zhang BR, Larsson B, Lundeheim N & Rodríguez-Martínez H 1997 Relation between embryo development in vitro and 56-day non-return rates of frozen-thawed semen from dairy AI bulls. *Theriogenology* **48** 221-231.

Zhang BR, Larsson B, Lundeheim N & Rodríguez-Martínez H 1998 Sperm characteristics and zona pellucida binding in relation to field fertility of frozen-thawed semen from dairy AI bulls. *International Journal of Andrology* **21** 207-216.

Zhang BR, Larsson B, Lundeheim N, Håård MG & Rodríguez-Martínez H 1999 Prediction of bull fertility by combined in vitro assessments of frozen-thawed semen from young dairy bulls entering an AI-program. *International Journal of Andrology* **22** 253-260.

Zou C & Yang Z 2000 Evaluation on sperm quality of freshly ejaculated boar semen during in vitro storage under different temperatures. *Theriogenology* **53** 1477–1488.

Control of ovarian follicular development to the gonadotrophin-dependent phase: a 2006 perspective

KP McNatty[1], K Reader[2], P Smith[2], DA Heath[2] and JL Juengel[2]

1School of Biological Sciences, Victoria University of Wellington, Wellington, New Zealand &
2AgResearch, Wallaceville Animal Research Centre, Upper Hutt, New Zealand

In sheep, as in other mammals, ovarian follicular growth is regulated mainly by intraovarian growth factors during early development with pituitary hormones increasingly important during the final phases to ovulation. Most follicles are present as primordial structures and these express many hundreds of genes that fulfil an array of housekeeping and signalling functions. Once growth has been initiated, at least two oocyte-derived growth factors, namely growth differentiation factor 9 (GDF9) and bone morphogenetic protein 15 (BMP15), are critical for ongoing development to ovulation, most likely by regulating the proliferative and differentiative functions of adjacent follicular cells. In sheep, the granulosa cell populations double some 12-14 times and a well-defined thecal layer differentiates before antrum formation and the time taken to complete this process varies between 50 -150 days with very little follicular atresia. During preantral growth, FSH and LH receptors coupled to the cyclic AMP second messenger system develop in granulosa and thecal cells respectively. From the late preantral stage, GDF9, BMP15 and perhaps other factors are thought to regulate gene expression in cumulus cells to enhance metabolic cooperativity with the oocyte and in mural granulosa cells to regulate their responses to pituitary hormones. In sheep, antral follicular development is characterized by a much faster rate of growth, additional increases in the numbers of granulosa (4-5 more doublings) and thecal cells, an increased level of steroid and inhibin secretion in response to FSH and LH, but also by most follicles undergoing atresia. The final number of follicles that go on to ovulate is dependent upon FSH as well as the intrafollicular concentrations of GDF9 and BMP15.

Introduction

In ruminants and other mammals, the initiation of ovarian follicular growth begins in late fetal or early neonatal life and continues without interruption throughout infancy, the oestrous cycle, pregnancy, lactation and regardless of ageing. Moreover, most stages of follicular growth (that is, >80% of development) occur independently of the day of the oestrous cycle, reproductive

Corresponding author E-mail: kenneth.mcnatty@vuw.ac.nz

status and age of the animal: these early stages of growth are considered to be gonadotrophin-independent and then gonadotrophin-responsive with pituitary hormones not considered to be essential (Scaramuzzi et al. 1993; McNatty et al. 2006). Indeed, it is only the final stages of development to ovulation, when follicles are critically dependent upon moment-by-moment gonadotrophic support. From gene-knockout studies in mice and naturally-occurring mutations and physiological studies in sheep, it is now evident that most stages of follicular development are regulated by factors within the ovary at a paracrine or perhaps autocrine level (Shimasaki et al. 2004; McNatty et al. 2006). In this context, it is now known that the oocyte plays a central role in orchestrating the growth of the follicle by regulating the expression of genes in the adjacent granulosa cells (Sugiura & Eppig 2005). In turn, it is thought that granulosa cells differentiate into two phenotypes namely the cumulus and mural granulosa cells and that these cell-types develop different functional roles as the follicle matures (Sugiura et al. 2005). The purpose of this review, using the ewe as the ruminant model, is to examine the evidence for local control, the role of pituitary hormones and to describe what is known about the stage specific expression of genes considered to be important during the gonadotrophin-independent and –responsive phases of follicular development.

Early follicular growth: current concepts

The primordial follicle

In the ewe, primordial follicles first form around day 75 of fetal life with the maximum numbers present between days 100 and 120 of gestation (Smith et al. 1993; 1994; Sawyer et al. 2002). With increasing age, the number of primordial follicles declines progressively. At 120 and 135 days of fetal life, 4 weeks of postnatal life, and 2, 5 and 6 years of age, the respective mean (and 95% confidence limits) numbers of primordial follicles are 100,000 (80,000, 160,000), 87,000 (67,000, 111,000), 70,000 (51,000, 86,000), 59,000 (52,000, 67,000), 20,000 (11,000, 36,000) and 16,000 (11,000, 21,000) (Smith et al. 1994; McNatty et al. 2001). While accurate numbers of primordial follicles in old age ewes are not available, there are numerous records of ewes still producing off-spring at 14 years suggesting that this species does not run out of follicles during their lifetime.

In sheep, the very-slow growing primordial follicle (Type 1; Fig. 1) contains between 3 and 52 granulosa cells (mean = 16 cells) and an oocyte between 23 and 52 μm in diameter (mean = 35 μm) (Lundy et al. 1999). Little, if any, atresia of primordial follicles occurs (Cahill, 1981). Approximately 2500 expressed sequence tags from Type 1 follicles have been matched to around 500 mRNAs linked to cytoskeletal events, DNA repair and mRNA processing, ribosomal function, protein synthesis and ubiquitination, and signalling pathways (D.A. Heath, J.L. Juengel & K.P. McNatty, unpublished data). Immunohistochemical localisation of connexins 37 and 43 suggests that there are contacts, albeit limited, between the oocyte and granulosa cells (Fig. 1: Grazul-Bilska et al. 1997). Oocytes in primordial follicles express mRNAs and proteins for c-kit, oestradiol receptor beta (ERß) and certain bone morphogenetic protein (BMP) and transforming growth factor beta (TGFß) receptors and binding proteins (betaglycan) (Fig. 2: Tisdall et al. 1999; Juengel et al. 2004; Juengel & McNatty 2005; Juengel et al. 2006a,b). In addition, oocytes express mRNAs for at least two growth factors, namely BMP6 and growth differentiating factor 9 (GDF9) with immunohistochemical evidence for the presence of GDF9 protein. Since GDF9 is also expressed in germ cells before and during follicular formation as well as in primordial and developing follicles, it seems likely that it fulfils some housekeeping roles although these have not been determined. Granulosa cells in primordial follicles synthesize stem cell factor (SCF) mRNA and protein, have the ability to metabolise progesterone and potentially respond to ligands that bind to Type 1 TGFß and/or Type II BMP receptors (Fig. 2).

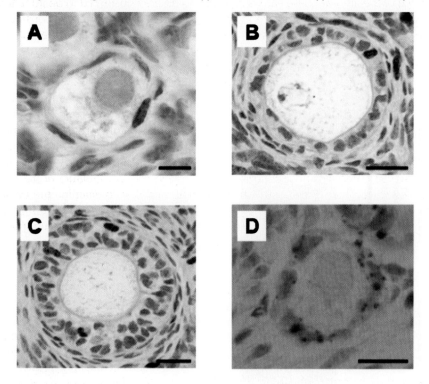

Fig. 1. Evidence for cell proliferation in early developing follicles and cell-cell contacts as determined by BrdU labelling in granulosa cells of a primordial (Type 1; A), primary (Type 2; B) and a secondary (Type 3; C) follicle and connexin 37 immunostaining in a Type 1 follicle (D). The immunohistochemical procedures used for BrdU are described by Sawyer *et al.* (2002) and for connexin 37 by Logan *et al.* (2003). The connexin 37 antibody was an affinity purified rabbit anti-mouse preparation (Alpha Diagnostics, San Antonio, Tx). The BrdU- and connexin-labelled cells are stained brown. Scale bars are 10μm (A), 20μm (B), 25μm (C) and 20μm (D).

Collectively, the evidence suggests that the primordial follicle has some awareness of its microenvironment and that it is expressing many hundreds of genes to fulfil an array of housekeeping and signalling functions.

The primordial to primary follicular growth transition

In the ewe, the primary follicle contains between 30 and 520 granulosa cells (mean = 128 cells) and an oocyte between 31 and 81μm in diameter (mean = 52μm) (Lundy et al. 1999). During this aspect of growth, the granulosa cells all become cuboidal in appearance and the population undergoes 1-2 doublings with a concomitant enlargement in the diameter of the oocyte (Figs. 1 & 2). The molecular signals which initiate the entry of a primordial or transitory (that is, Type 1 or 1a) follicle into the primary and preantral stages of development are uncertain but may involve multiple factors both stimulatory and inhibitory. What is clear is that this early growth phase does not involve pituitary hormones as there are no FSH receptors present on

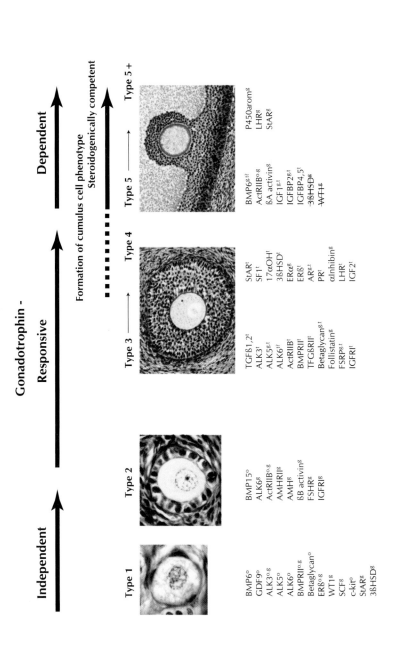

Fig. 2. Localisation of a selected number of genes expressed during ovarian follicular development in sheep. When expression is indicated at the primordial (Type 1) stage or at subsequent stages of follicular growth (Type 2, primary; Type 3–4 preantral; Type 5, small antral; Type 5 + large antral: Lundy *et al.* 1999) it is considered to continue for the remainder of follicular growth unless the gene has a line through it indicating that expression is no longer evident thereafter. No attempt has been made to indicate the level of expression at each stage of follicular growth. In some instances, such as with IGFBPs, the onset of expression has not been precisely defined. FSRP refers to follistatin related protein. The terms o, g and t indicate expression in oocytes, granulosa cells and thecal cells, respectively. Data from: Leeuenberg *et al.* 1995; Perks *et al.* 1995; Besnard *et al.* 1996; Logan *et al.* 2002, 2003; Monget *et al.* 2002; Juengel & McNatty 2005; Juengel *et al.* 2006 a,b.

granulosa cells until primary follicles are formed and LH receptors on thecal cells are only evident when this cell-type differentiates in secondary follicles (Fig. 2). Moreover, hypophysectomy does not inhibit this phase of growth (Dufour et al. 1979). Neither GDF9 nor BMP15 are likely to be involved in the early phases of this growth initiation since *in vivo* immunoneutralisation studies in the ewe show that the formation of primary follicles is not inhibited (Juengel et al. 2002) although the mean oocyte diameters are significantly larger than in the controls during the transition (Type 1a follicles) or when primary follicles are formed (Type 2). It should also be noted that BMP15 is not present in oocytes until the Type 1a or 2 stage of growth (Fig. 2). From studies in rodents, leukaemia inhibitory factor (LIF), basic fibroblast growth factor (bFGF), stem cell factor, insulin, BMP4, BMP7 and keratinocyte growth factor (KGF) have all been implicated as potential growth initiating factors (Skinner 2005). In the ewe, BMP 4 and 7 mRNA (as well as BMP 2 mRNA) were not found in non-atretic ovarian follicles although BMP4 mRNA was identified in cells of the surface epithelium and around blood vessels (Juengel et al. 2006a). Basic FGF protein has been localised to primordial follicles in the cow (van Wezel et al. 1995) but was not found to be stimulatory to bovine primordial follicles *in vitro* (Derrar et al. 2000). In the ovine ovary, the onset of expression and cellular location of bFGF, LIF or KGF are not known.

With respect to inhibitory factors, the evidence in mice is that anti-Müllerian hormone (AMH) and stromal derived factor-1 (SDF-1) have been shown to inhibit the formation of primary follicles (Durlinger et al. 2002; Holt et al. 2006). In the ewe, the onset of expression of AMH and the AMH Type II receptor is exclusive to granulosa cells at the primary (Type 2) stage (Fig. 2) but that for SDF-1 or its putative receptor, CXCR4, is not known. The molecular mechanism as to how AMH inhibits follicular growth is also unclear. Although the Type II receptor in sheep has been identified, evidence for the identity of the Type I receptor in the ovary remains to be established.

Skinner (2005) has proposed, on the basis of studies in rodents, that a key factor in the transformation of a primordial follicle to the primary stage of growth involves recruitment of thecal cells from precursors in the adjacent stroma/interstitium and that granulosa cell and oocyte signalling molecules (for example, bFGF and SCF) are involved. In turn, thecal/ stromal-derived factors such as KGF together with systemically-derived insulin may act on adjacent granulosa cells to initiate or advance development to the primary stage. This hypothesis is testable in ruminants but the experiments remain to be undertaken. Cultures of isolated or pooled populations of ovine primordial follicles using a variety of culture methodologies for up to 20 days *in vitro* in the presence or absence of isolated stroma/interstitial cells, fetal calf serum and additives such as insulin, IGF-1 and oestradiol have not been successful in stimulating the transformation of a primordial to a primary follicle (Fig. 1). By contrast, the culture of ovarian cortical explants from cow, baboon or human ovaries resulted in an unexpectedly large number of flattened granulosa cells in primordial follicles transforming into cuboidal cells and undergoing proliferation. However, growth was abnormal and there was no associated enlargement of the oocyte (reviewed by Braw-Tal 2002).

Collectively, these data are consistent with the hypothesis proposed by Skinner (2005) that the transformation of a primordial follicle into a primary follicle is the result of a communication exchange between the follicle and the adjacent stromal/interstitial and perhaps endothelial cells involving both inhibitory and stimulatory signalling pathways.

The primary to large preantral growth phase

A large preantral follicle in the ewe contains between 1090 and 3404 granulosa cells (mean =

2104 cells) and an oocyte between 71 and 120 μm in diameter (mean = 88 μm) (Lundy et al. 1999). Thus, from the primary stage, a further 6 doublings of the granulosa cell population and a 1.7-fold increase in the diameter of the oocyte takes place before a large preantral follicle is formed (Figs. 1 & 2). As follicles grow from the primary stage, the onset of expression of many more genes occurs in a stage and cell-specific manner (Fig. 2). From the primary stage, there is evidence for the continued expression in oocytes of BMP6 and GDF9 but also of BMP15, the Type IIB activin and BMPII receptors, and the Type 1 receptors ALK 3, 5 and 6 as well as c-kit (Fig. 2). Granulosa cells develop the potential to synthesize inhibin and follistatin, become responsive to steroids via receptors for androgen and oestradiol, and develop type II receptors for activin and the Type I receptors ALK 5 and 6. The theca cells develop a receptivity to progestins, androgens, oestradiol, activin, TGFβs, and BMPs and also the ability to synthesize TGFβ 1 and 2, progestins and androgens (Fig. 2). Importantly, granulosa cells develop receptors to FSH and theca cells to LH and under in vitro conditions, large preantral follicles respond to LH and FSH to synthesize cAMP and thereafter, in small antral follicles to synthesize progestins and androgens (McNatty et al. 1999). In pituitary-intact ewes, the level of follicular atresia in preantral follicles is very low (Cahill 1989). Moreover, after hypophysectomy, there is also no increase in the prevalence of atresia of preantral follicles in the short-term. However, after long-term hypophysectomy that is, > 60 days) an increase in atresia was observed (Dufour et al. 1979) but the total number of follicles up to 2-3mm in diameter remained unchanged and no difference was noted in vitro in the numbers of responsive follicles with respect to gonadotrophin-induced cAMP synthesis (McNatty et al. 1990a). From the late preantral phase through to ovulation, it is evident, at least in mono-ovulatory species such as sheep and humans, that follicles develop in a hierarchical manner and that few, if any, are at the same stage of development at any moment in time (McNatty et al. 1999). In the ewe, this pattern of hierarchical development is unaffected by hypophysectomy (McNatty et al. 1999). Moreover, ovulation can be induced in chronically hypophysectomised ewes or hypothalamic-pituitary disconnected ewes following a short-term treatment with pregnant mares serum gonadotrophin (PMSG) and human chorionic gonadotrophin (Fry et al. 1988) or with exogenous gonadotrophin-releasing hormone (GnRH) pulses, PMSG and a GnRH bolus regimen (McNatty et al. 1993b). However, it is not known whether the ovulated oocytes are capable of fertilisation and development into viable embryos.

   Collectively, the evidence suggests that the growth of follicles through the preantral stages is not critically dependent on threshold concentrations or short-term fluctuations of pituitary hormones.

Intraovarian control of preantral follicular growth

In mice lacking a functional GDF9 gene, normal follicular development was arrested shortly after primordial follicles had started to grow with evidence of continued enlargement of the oocyte but without a concomitant increase in the numbers of granulosa cells (Dong et al.1996). This study demonstrated for the first time that an oocyte-derived factor was essential for normal follicular development. Animals homozygous for the GDF9 inactivation were sterile whereas the heterozygous knockouts showed a normal phenotype. Subsequent studies with mice lacking other growth factors produced by oocytes, such as BMP6 (Solloway et al. 1998) and BMP15 (Yan et al. 2001), were fertile although a modest reduction in fertility was evident in the homozygous BMP15 knockouts. In the ewe, a number of naturally-occurring genetic mutations have been identified in either GDF9 or BMP15 which affect fertility although no mutations have been identified for BMP6 (Galloway et al. 2000; Bodin et al. 2003; Hanrahan et al. 2004).

In the ewe, as in mice, GDF9 and BMP15 are expressed exclusively within the ovary by oocytes (Fig. 2). For GDF9, one mutation has been identified in the mature region of the protein and the resulting homozygous mutant phenotype is sterile whereas the heterozygous phenotype has an increased ovulation rate around 87% above that of the wild-type (Hanrahan *et al*. 2004). For BMP15, five different point mutations have been identified, with 4 in the mature region and one in the proregion of the protein with the resulting homozygous mutant phenotypes being sterile. In 4 of the heterozygous phenotypes that have been studied, a 35-95% increase in ovulation rate was observed compared to the respective wild-types (reviewed by McNatty *et al*. 2006). Of interest is the finding that animals that were heterozygous for mutations in both GDF9 and BMP15 had ovulation rates that were at least additive to those for each mutation separately (McNatty *et al*. 2006). In women with unexplained infertility, 4 heterozygous point mutations have now been identified in the proregion of BMP15 (Di Pasquale *et al*. 2004, 2006) and at least one of these heterozygous mutant forms of BMP15 has been shown *in vitro* to be a competitive antagonist to the wild-type protein with respect to its actions on granulosa cells.

Immunisation of ewes at monthly intervals with 15-16 mer N-terminal peptides of GDF9 or BMP15 conjugated to keyhole limpet haemocyanin (KLH) generated strong immune responses. In this study, antisera raised against GDF9 did not cross-react with BMP15 and vice versa (Juengel *et al*. 2002). None of the KLH control immunised ewes had antibodies to GDF9 or BMP15 and all continued to show regular oestrous cycles. In contrast most of the GDF9 (8/9) and BMP15 (9/10) immunised animals became anovulatory with normal ovarian follicular activity being disrupted at the primary follicular stage. The ovarian phenotype was similar to that observed in the homozygous GDF9 knockout mice and homozygous GDF9 and BMP15 mutant ewes. The ovaries at recovery were significantly smaller after the GDF9 and BMP15 treatments compared to the controls but no differences between the groups were noted in the mean numbers of primordial follicles. At the transitory (Type 1a) and primary stages of growth, oocyte diameters were significantly larger after the BMP15 and GDF9 immunisations, respectively, and by the late preantral stage of growth, the numbers of follicles were significantly reduced in both of these groups compared to the controls.

In ewes, a naturally-occurring mutation in the BMP Type I receptor, ALK6, has also been found to have a profound effect on early follicular development and ovulation rate (Mulsant *et al*. 2001; Souza *et al*. 2001; Wilson *et al*. 2001). Animals heterozygous and homozygous for this mutation have ovulation rates around 76% and 130% higher respectively than the wild-types (McNatty *et al*. 2006). The effect of the ALK6 mutation is observed from the primary stage of growth onwards when the diameter of the oocyte is noticeably larger than in the equivalent-sized wild-type follicles throughout preantral follicular growth. The consequence was that in the homozygous ALK6 mutants, the oocyte reached around 130μm in a large preantral follicle whereas in the wild-type this diameter was not reached until the early antral stage. Unlike in the BMP15 and GDF9 mutants, there was a concomitant increase in granulosa cell number with oocyte growth. However, by the early antral stage, the numbers were significantly lower in the ALK6 mutants compared to the wild-types. Moreover, there was an earlier onset of maturation of the follicle in the ALK6 mutants and they ovulated at smaller diameters than in the wild-types (Montgomery *et al*. 2001; Wilson *et al*. 2001). In essence, in the ALK6 mutants, the granulosa cells undergo 1-2 less doublings and during early antral development, the granulosa cells enter a differentiative pathway earlier than in the wild-types. Interestingly, when heterozygous ALK6 mutant animals were mated with heterozygous BMP15 mutants (FecXI), the effect on ovulation rate was greater than additive for the two mutations (Davis *et al*. 1999) suggesting a level of co-operativity between BMP15 and the Type I receptor. In these animals, it would be of interest to examine the interrelationships among oocyte diameters, granulosa cell populations and diameters of the follicles at ovulation.

Collectively, the *in vivo* evidence in ewes shows that GDF9 and BMP15 are essential regulators of preantral follicular development and that a functional relationship exists between BMP15 and ALK6 and also between BMP15 and GDF9. Whether a similar relationship exists between GDF9 and ALK6 is not known. In mice, GDF9 but not BMP15 is essential for early follicular growth; whereas in humans, BMP15 is essential but the role for GDF9 remains to be determined. The evidence suggests there are species differences with respect to the onset of gene expression for GDF9, BMP15 and other BMPs, as well as differences in how oocyte-derived growth factors influence follicular development and ovulation rate.

Development from a large preantral to a gonadotrophin-dependent follicle

In the ewe, an ovarian follicle becomes gonadotrophin-dependent upon reaching around 3mm in diameter. Thus, the population of granulosa cells is required to double around 8 times from the large preantral stage (McNatty et al. 1990a; Lundy et al. 1999). During this growth phase an increasing proportion of granulosa cells transform from a mainly proliferating state to a differentiating one (Monniaux et al. 1994). Moreover, during this period, follicles are still developing in a hierarchical manner and no two follicles share a common hormonal microenvironment at any moment in time. Also, during this interval, a significant number of follicles (that is, > 30-80%) are lost by atresia (McNatty et al.1984). The number of non-atretic follicles that develop to around 3mm in diameter in sheep is highly variable (that is, between 1 and 25) and dependent upon genotype, time of the oestrous cycle and time of year (McNatty et al. 1984; 1985, 2001). However, the number of non-atretic follicles ≥3mm in diameter can be enhanced by long-term (that is several days), exogenous FSH treatment (McNatty et al. 1993a) and this occurs without changing the total population of antral follicles (≥1mm in diameter). Thus, FSH is able to influence the proportion of ovulatory follicles by preventing atresia of small antral follicles but not by 'speeding-up' the number of preantral or very small antral follicles growing ≥1mm in diameter. The growth of a follicle from the primary to the preantral stage varies from around 25 days in fetal life to many months in adult life (Smith et al.1993; 1994; Turnbull et al. 1977). However, the growth of antral follicles from around 1mm in diameter to ovulation is more rapid and can take as little as three days (Cahill 1981; Tsonis et al. 1984). This variable rate of growth is likely to be influenced by serum-derived as well as local ovarian-derived growth factors such as the availability of IGF1, GDF9 and BMP15 (Juengel et al. 2002; Monget et al. 2002). In the ewe, under *in vitro* conditions, IGF1 stimulated proliferation of granulosa cells from follicles ≤3mm in diameter. *In vivo,* this enhanced rate of proliferation is associated with a significant decline in the local follicular fluid levels of IGFBP-2 and -4 most likely due, in part, to a decrease in mRNA expression and an increase in protease activity (Monget et al. 2002). As follicles develop to 3mm in diameter, the granulosa cells become increasingly responsive to FSH-induced cyclic AMP synthesis reflecting the increased differentiative capacity of the cells. This increase is not due to a change in the number of FSH receptors or binding affinity but to a decrease in cAMP-dependent phosphodiesterase activity, increased bio-availability of IGF1 and a marked increase in adenylate cyclase activity (McNatty et al. 1990b; Monget et al. 2002).

Intraovarian control of antral follicular growth

Recently, Sugiura & Eppig (2005) have proposed that GDF9 and perhaps other oocyte-derived factors play a role in determining the cumulus and mural granulosa phenotypes during the preantral to early antral phase of growth. These authors report that oocytes control the level of cellular metabolism in cumulus cells by regulating the expression of genes involved in events such as amino-acid transport and glycolysis. In turn, this localised upregulation of gene expression, which is not observed in the mural granulosa cells, leads to an enhanced level of metabolic co-operativity between the oocyte and cumulus cells. In the bovine ovary, there is evidence of a localised concentration gradient of BMPs from the oocyte across the cumulus cells inferring that these cells are likely to be exposed to higher concentrations than the more distant mural granulosa cells (Fig 3: Hussein *et al.* 2005). In ovine follicular fluid, both GDF9 and BMP15 have been detected as promature but not as cleaved mature proteins and the presence of BMP6 and other BMPs have yet to be confirmed (McNatty *et al.* 2006). Irrespective of the forms of GDF9 or BMP15 in follicular fluid, passive immunisation of ewes with GDF9 or BMP15 antisera at the onset of luteolysis inhibited normal follicular growth, ovulation and/or corpus luteum function (Juengel *et al.* 2002). While there are marked increases in connexins and junctional communication between the oocyte and granulosa/cumulus cells during follicular growth, more information on the importance of the somatic cells to the oocyte during the preantral to antral follicular transition as well as on the utilisation of serum-derived nutrients or other factors is required.

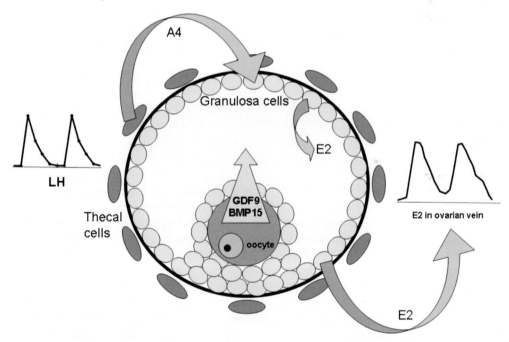

Fig. 3. Schematic outline of a gonadotrophic-responsive and steroidogenically active follicle. In response to LH pulses, androstenedione (A4) is produced by thecal cells and metabolised by granulosa cells to oestradiol (E2) with most of the E2 secreted without accumulating within the follicle. However, the steroid concentration within the follicle is not fluctuant but changes slowly compared to that entering the ovarian vein (McNatty et al. 1981). It is likely that a concentration gradient of oocyte-derived BMP15 and GDF9 exists within the follicle with the cumulus cells exposed to higher concentrations than the mural granulosa cells. It is also likely that both BMP15 and GDF9 continue to be important during the follicular phase of the oestrous cycle in ewes since passive immunisation with anti-BMP15 or –GDF9 inhibits ovulation or normal corpus luteum function (Juengel et al. 2002).

There are extensive reports of the roles of BMPs on granulosa and thecal cells from antral follicles *in vitro* (reviewed by: Matzuk *et al.* 2002; Knight & Glister 2003; Shimasaki *et al.* 2004; Juengel & McNatty 2005). It is important to note that the pattern and onset of ligand expression of the TGFß-superfamily during follicular growth is not identical across species (see Erickson & Shimasaki 2003; Knight & Glister 2003; Juengel & McNatty 2005; Juengel *et al.* 2006a). In rat granulosa cells, GDF9, BMP15 and BMP6 inhibit LHR, FSHR and StAR and p450scc mRNA, respectively. Thus the *in vitro* evidence suggests that GDF9 and BMP15 influence the responsiveness of mural granulosa cells to pituitary hormones and this is supported by *in vivo* results in the ewe (reviewed by: Shimasaki *et al.* 2004; McNatty *et al.* 2006). In the ewe, cow and rat, GDF9 and BMP15 when added together, often act co-operatively to enhance the proliferation of granulosa cells and to suppress progesterone production in vitro (McNatty *et al.* 2006). Together these ligands act via the BMP Type II receptor to affect granulosa cell function. In sheep, thecal cells in both preantral and small antral follicles have a number of Type I and II BMP and TGFß receptors (Fig. 2) and thus the potential to respond to activin, TGFß and BMP ligands. However, the issue arises as to the range of ligands that might be available locally to influence follicular growth. In the ewe, TGFß1 and 2 mRNA have been localised to thecal cells, BMP6 to oocytes at all stages of growth and possibly to the granulosa cells from the early antral development (Juengel *et al.* 2006a). Activin together with inhibin and follistatin is likely to be of granulosa cell origin (Fig. 2). Of the other possible ligands, BMP4 mRNA has been identified around blood vessels in the thecal region (Juengel *et al.* 2006a). In bovine thecal cells, BMP 4 and 6 suppressed both basal and LH induced androgen but not progesterone production (Glister *et al.* 2005).

Collectively therefore, the responsiveness of granulosa and thecal cells to gonadotrophins is likely to be modulated by locally produced members of the TGFß superfamily. In turn, this is likely to influence the level of hormonal interaction of the developing follicle with the extra-ovarian endocrine system as the follicle progresses from the preantral phase of growth.

The gonadotrophin-dependent follicle

The granulosa cells in ovarian follicles at 3mm in diameter have a capacity to synthesize inhibin, follistatin and AMH and have a FSH-responsive adenylate cyclase system, but they still require the acquisition of the P450 aromatase enzyme and a LH receptor coupled to adenylate cyclase before developing the ability to become the dominant follicle within the ovary. With respect to thecal cells, the LH-induced output of androstenedione increases during antral follicular growth but the increase is thought to be related to an increasing population of cells rather than to an increased output per cell (McNatty *et al.* 1984). During the transition to dependence upon gonadotrophins, the capacity of granulosa cells adjacent to the basement membrane to metabolise steroid substrate (that is, androstenedione) from LH-stimulated thecal cells increases. Once granulosa cells acquire aromatase activity, a very significant proportion (that is, >90%) of the oestradiol produced within 30 min of an LH pulse is exported to the ovarian vein without first accumulating in follicular fluid (Fig. 3: McNatty *et al.* 1981). In contrast, the transit time of LH and probably FSH entering follicular fluid is slow (that is, hours) as is the accumulation of steroid (McNatty *et al.*1981). Therefore, it is unlikely that the cumulus cells would be exposed to short-term fluctuations in gonadotrophins or steroids. Nevertheless, most serum-derived molecules are present in follicular fluid so that the oocyte-cumulus cell complexes are exposed to extra-ovarian as well as local follicular factors.

Conclusions

Over the past decade new insights have been gained into how ovarian follicles form and develop to ovulation. Arguably, the greatest advances have come through the identification of some of the intra-ovarian factors through the inactivation of target genes in mice (Anreu-Vieyra *et al.* 2006). Related studies with naturally-occurring mutations in sheep have also helped in providing new information on oocyte-derived growth factors and their effects on both follicular development and ovulation rate. It is clear that the oocyte is a major source of growth factors influencing follicular development, the phenotypes of the granulosa and cumulus cells and the responsiveness of granulosa and perhaps thecal cells to gonadotrophins. Several major challenges still remain including an understanding of how primordial follicles initiate their transformation to a primary follicle, the extent of the signalling and metabolic interactions between the oocyte and adjacent granulosa/cumulus cells, and the key factors that regulate the rate at which follicles may grow.

Acknowledgements

The authors thank their colleagues in the Reproductive Biology team at the Wallaceville Animal Research Centre for their technical and scientific input and especially Dr Doug Eckery for advice during preparation of the manuscript. Funding support was provided by the New Zealand Foundation for Research Science and Technology and the Marsden Fund.

References

Andreu-Vieyra C, Lin YN & Matzuk MM 2006 Mining the oocyte transcriptome. *Trends in Endocrinology and Metabolism* **17** 136-143.

Bodin L, Lecerf F, Pisselet C, San Cristhal M, Bibe M & Mulsant P 2003 How many mutations are associated with increased ovulation rate and litter size in progeny of Lacaune meat sheep. In *Proceedings of the International Workshop on Major Genes and QTL in Sheep and Goat*, pp 2-11. Publisher INRA, Toulouse, France.

Besnard N, Pisselet C, Monniaux D, Locatelli A, Benne F, Gasser F, Hatey F & Monget P 1996 Expression of messenger ribonucleic acids of insulin-like growth factor binding proteins-2, -4, and -5 in the ovine ovary: localization and changes during growth and atresia of antral follicles. *Biology of Reproduction* **55** 1356-1367.

Braw-Tal R 2002 The initiation of follicle growth: the oocyte or the somatic cells? *Molecular and Cellular Endocrinology* **187** 11-18.

Cahill LP 1981 Folliculogenesis in the sheep as influenced by breed, season and oestrous cycle. *Journal of Reproduction and Fertility Supplement* **30** 135-142

Davis GH, Dodds KG & Bruce GD 1999 Combined effect of the Inverdale and Booroola prolificacy genes on ovulation rates. *Proceedings of the Association for the Advancement of Animal Breeding Genetics* **14** 175-178.

Derrar N, Price CA & Sirard MA 2000 Effect of growth

factors and co-culture with ovarian medulla on the activation of primordial follicles in explants of bovine ovarian cortex. *Theriogenology* **54** 587-598.

Di Pasquale E, Beck-Peccoz P & Personi L 2004 Hypergonadotropic ovarian failure associated with an inherited mutation of human bone morphogenetic protein-15 (BMP15). *American Journal of Human Genetics* **75** 106-111.

Di Pasquale E, Rossetti R, Marozzi A, Bodega B, Borgato S, Cavallo L, Einaudi S, Radetti G, Russo G, Sacco M, Wasniewska M, Cole T, Beck-Peccoz P, Nelson LM & Persani L 2006 Identification of new variants of human BMP15 gene in a large cohort of women with premature ovarian failure. *Journal of Clinical Endocrinology and Metabolism* **91** 1976-1979.

Dong J, Albertini DF, Nishimori K, Rajenda Kumar T, Lu N & Matzuk MM 1996 Growth differentiation factor 9 is required for early ovarian folliculogenesis. *Nature* **383** 531-535.

Dufour J, Cahill LP & Mauleon P 1979 Short- and long-term hypophysectomy and unilateral ovariectomy on ovarian follicular populations in sheep. *Journal of Reproduction and Fertility* **57** 301-309.

Durlinger AL, Gruijters MJ, Kramer P, Karels B, Ingraham HA, Nachtigal MW, Uilenbroek JT, Grootegoed JA & Themmen APN 2002 Anti-Mullerian hormone inhibits initiation of primordial growth in the mouse ovary. *Endocrinology* **143** 1076-1084

Erickson GF & Shimasaki S 2003 The spatiotemporal

expression pattern of the bone morphogenetic protein family in rat ovary cell types during the estrous cycle. *Reproductive Biology and Endocrinology* **1** 9.

Fry RC, Clarke IJ, Cummins JT, Bindon BM, Piper LR & Cahill LP 1988 Induction of ovulation in chronically hypophysectomised Booroola ewes. *Journal of Reproduction and Fertility* **82** 711-715.

Galloway SM, McNatty KP, Cambridge LM, Laitinen MPE, Juengel JJ, Jokiranta TS, McLaren RJ, Luiro K, Dodds KG, Montgomery GW, Beattie AE, Davis GH & Ritvos O 2000 Mutations in an oocyte-derived growth factor (BMP15) cause increased ovulation rate and infertility in a dosage-sensitive manner. *Nature Genetics* **25** 279-283.

Glister C, Richards SL & Knight PG 2005 Bone morphogenetic proteins (BMP)-4, 6, and -7 potently suppress basal and luteinising hormone-induced androgen production by bovine theca interna cells in primary culture: could ovarian hyperandrogenic dysfunction be caused by a defect in thecal BMP signalling? *Endocrinology* **146** 1883-1892.

Grazul-Bilska AT, Reynolds LP & Redmer DA 1997 Gap junctions in the ovaries. *Biology of Reproduction* **57** 947-957.

Hanrahan JP, Gregan SM, Mulsant P, Mullen M, Davis GH, Powell R & Galloway SM 2004 Mutations in the genes for oocyte-derived growth factors GDF9 and BMP15 are associated with both increased ovulation rate and sterility in Cambridge and Belclare sheep (*Ovis Aries*). *Biology of Reproduction* **70** 900-909.

Holt JE, Jackson A, Roman SD, Aitken RJ, Koopman P & McLaughlin EA 2006 CXCR4/SDF1 interaction inhibits the primordial to primary follicle transition in the neonatal mouse ovary. *Developmental Biology* **293** 449-460.

Hussein TS, Froiland DA, Amato F, Thompson JG & Gilchrist RB 2005 Oocytes prevent cumulus cell apoptosis by maintaining a morphogenic paracrine gradient of bone morphogenetic proteins. *Journal of Cell Science* **118** 5257-5268.

Juengel JL & McNatty KP 2005 The role of proteins of the transforming growth factor- beta superfamily in the intraovarian regulation of follicular development. *Human Reproduction Update* **11** 143-160.

Juengel JL, Hudson NL, Heath DA, Smith P, Reader KL, Lawrence SB, O'Connell AR, Laitinen MP, Cranfield M, Groome NP, Ritvos O & McNatty KP 2002 Growth differentiation factor 9 and bone morphogenetic protein 15 are essential for ovarian follicular development in sheep. *Biology of Reproduction* **67** 1777–1789

Juengel JL, Bibby AH, Reader KL, Lun S, Quirke LJ, Haydon LJ & McNatty KP 2004 The role of transforming growth factor-beta (TGF-beta) during ovarian follicular development in sheep. *Reproductive Biology and Endocrinology* **2** 78.

Juengel JL, Reader KL, Bibby AH, Lun S, Ross I, Haydon LJ & McNatty KP 2006a The role of bone morpho-

genetic proteins 2, 4, 6 and 7 during ovarian follicular development in sheep: contrast to rat. *Reproduction* **131** 501-513.

Juengel JL, Heath DA, Quirke LD & McNatty KP 2006b Oestrogen receptor α and ß, androgen receptor and progesterone receptor mRNA and protein localisation within the developing ovary and in small growing follicles of sheep. *Reproduction* **131** 81-92.

Knight PG & Glister C 2003 Local roles of TGF-beta superfamily members in the control of ovarian follicle development. *Animal Reproduction Science* **78** 165-183.

Leeuenberg BR, Hurst PR & McNatty KP 1995 Expression of IGF-I mRNA in the ovine ovary. *Journal of Endocrinology* **15** 251-258.

Logan KA, Juengel JL & McNatty KP 2002 Onset of steroidogenic enzyme gene expression during ovarian follicular development in sheep. *Biology of Reproduction* **66** 906-916.

Logan KA, McNatty KP & Juengel JL 2003 Expression of Wilms' tumor gene and protein localization during ovarian follicular development in sheep. *Biology of Reproduction* **68** 635-643.

Lundy T, Smith P, O'Connell A., Hudson N L & McNatty KP 1999 Populations of granulosa cells in small follicles of the sheep. *Journal of Reproduction and Fertility* **115** 251–262.

McNatty KP, Dobson C, Gibb M, Kieboom L & Thurley DC 1981 Accumulation of luteinising hormone, oestradiol and androstenedione by sheep ovarian follicles in vivo. *Journal of Endocrinology* **91** 99-109.

McNatty KP, Hudson NL, Henderson KM, Lun S, Heath DA, Gibb M, Ball K, McDiarmid JM & Thurley DC 1984 Changes in gonadotrophin secretion and ovarian follicular activity in seasonally breeding sheep throughout the year. *Journal of Reproduction and Fertility* **70** 309-321.

McNatty KP, Hudson NL, Gibb M, Ball K, Henderson KM, Heath DA, Lun S & Kieboom LE 1985 FSH influences follicle viability, oestradiol biosynthesis and ovulation rate in ewes. *Journal of Reproduction and Fertility* **75** 121-131.

McNatty KP, Heath DA, Hudson NL & Clarke IJ 1990a Effect of long-term hypophysectomy on ovarian follicle populations and gonadotrophin-induced adenosine 3,5 monophosphate output by follicles from Booroola ewes with or without the F gene. *Journal of Reproduction and Fertility* **90** 515-522.

McNatty KP, Lun S, Hudson NL & Forbes S 1990b Effects of follicle stimulating hormone, cholera toxin, pertusis toxin and forskolin on adenosine 3,5-monophosphate output by granulosa cells from Booroola ewes with or without the F gene. *Journal of Reproduction and Fertility* **89** 553-563.

McNatty KP, Hudson NL, Heath DA, Shaw L, Blay L, Berry L & Lun S 1993a Effect of chronic FSH administration on ovarian follicular development, ovulation rate and corpora lutea formation in sheep. *Journal of Endocrinology* **138** 315-324

McNatty KP, Hudson NL, Lun S, Heath DA, Shaw L, Condell L, Phillips DJ & Clarke IJ 1993b Gonadotrophin-releasing hormone and the control of ovulation rate by the FecB gene in Booroola ewes. *Journal of Reproduction and Fertility* **98** 97-105.

McNatty KP, Heath DA, Lundy T, Fidler AE, Quirke L, O'Connell, A, Smith P, Groome N & Tisdall DJ 1999 Control of early ovarian follicular development. *Journal of Reproduction and Fertility Supplement* **54** 3-16.

McNatty KP, Juengel JL, Wilson T, Galloway SM & Davis GH 2001 Genetic mutations influencing ovulation rate in sheep. *Reproduction, Fertility and Development* **13** 549-555.

McNatty KP, Lawrence S, Groome NP, Meerasahib MF, Hudson NL, Whiting L, Heath DA & Juengel JL 2006 Oocyte signalling molecules and their effects on reproduction in ruminants. *Reproduction, Fertility and Development* **18** 403-412.

Matzuk MM, Burns KH, Viveiros MM & Eppig JJ 2002 Intercellular communication in the mammalian ovary: oocytes carry the conversation. *Science* **296** 2178-80.

Monget P, Fabre S, Mulsant P, Lecerf F, Elsen JM, Mazerbourg S, Pisselet C & Monniaux D 2002 Regulation of ovarian folliculogenesis by IGF and BMP system in domestic animals. *Domestic Animal Endocrinology* **23** 139-154.

Monniaux D, Pisselet C & Fontaine J 1994 Uncoupling between proliferation and differentiation of ovine granulosa cells in vitro. *Journal of Endocrinology* **142** 497-510.

Montgomery GW, Galloway SM, Davis GH & McNatty KP 2001 Genes controlling ovulation rate in sheep. *Reproduction* **121** 843-852.

Mulsant P, Lecerf F, Fabre S, Schibler L, Monget P, Lanneluc I, Pisselet C, Riquet J, Monniaux D, Callebaut I, Cribiu E, Thimonier J, Bodin L, Cognie Y, Chitour N & Elsen JM 2001 Mutation in bone morphogenetic protein receptor-1B is associated with increased ovulation in Booroola ewes. *Proceedings of the National Academy of Sciences USA* **98** 5104-5109.

Perks CM, Denning-Kendall PA, Gilmore RS & Wathes DC 1995 Localization of messenger ribonucleic acids for insulin-like growth factor 1 (IGF-1), IGF-II and the Type I IGF receptor in the ovine ovary throughout the oestrous cycle. *Endocrinology* **136** 5266-5273.

Sawyer HR, Smith P, Heath DA, Juengel JL, Wakefield SJ & McNatty KP 2002 Formation of ovarian follicles during fetal development in sheep. *Biology of Reproduction* **66** 1134-1150.

Scaramuzzi RJ, Adams NR, Baird DT, Campbell BK, Downing JA, Findlay JK, Henderson KM, Martin GB, McNatty KP, McNeilly AS & Tsonis CG 1993 A model for follicle selection and the determination of ovulation rate in the ewe. *Reproduction, Fertility and Development* **5** 459-478

Skinner MK 2005 Regulation of primordial follicle assembly and development. *Human Reproduction Update* **11** 461-471.

Shimasaki S, Moore RK, Otsuka F & Erickson GF 2004 The bone morphogenetic protein system in mammalian reproduction. *Endocrine Reviews* **25** 72–101.

Smith P, O W-S, Hudson NL, Shaw L, Heath DA, Condell L, Phillips DJ & McNatty KP 1993 Effects of the Booroola gene (FecB) on body weight, ovarian development and hormone concentrations during fetal life. *Journal of Reproduction and Fertility* **98** 41-54.

Smith P, Braw-Tal R, Corrigan K, Hudson NL, Heath DA & McNatty KP 1994 Ontogeny of ovarian follicle development in Booroola sheep fetuses that are homozygous carriers or non-carriers of the FecB gene. *Journal of Reproduction and Fertility* **100** 485-490.

Solloway MJ, Dudley AT, Bikoff EK, Lyons KM, Hogan BL & Robertson EJ 1998 Mice lacking Bmp6 function. *Developmental Genetics* **22** 321–339.

Souza CJH, MacDougall, C, Campbell BK, McNeilly AS & Baird DT 2001. The Booroola (FecB) phenotype is associated with a mutation in the bone morphogenetic receptor type IB (BMPRIB) gene. *Journal of Endocrinology* **169** R1-R6.

Sugiura K & Eppig JJ 2005 Control of metabolic cooperativity between oocytes and their companion granulosa cells by mouse oocytes. *Reproduction, Fertility and Development* **17** 667-674.

Sugiura K, Pendola FL, Eppig JJ 2005 Oocyte control of metabolic cooperativity between oocytes and companion granulosa cells: energy metabolism. *Developmental Biology* **279** 20-30.

Tisdall DJ, Fidler AE, Smith P, Quirke LD, Stent VC, Heath DA & McNatty KP 1999 Stem cell factor and c-kit gene expression and protein localization in the sheep ovary during fetal development. *Journal of Reproduction and Fertility* **116** 277-291.

Tsonis CG, Cahill LP, Carson RS & Findlay JK 1984 Identification at the onset of luteolysis of follicles capable of ovulation in the ewe. *Journal of Reproduction and Fertility* **70** 609-614.

Turnbull KE, Braden AWH & Mattner PE 1977. The pattern of follicular growth and atresia in the ovine ovary. *Australian Journal of Biological Science* **30** 229-241.

van Wezel IL, Umapathysivam K, Tilley WD & Rodgers RJ 1995 Immunohistochemical localization of basic fibroblast growth factor in bovine ovarian follicles. *Molecular and Cellular Endocrinology* **115** 133-140.

Wilson T,Wu XY, Juengel JL, Ross IK, Lumsden JM, Lord EA, Dodds KG, Walling GA, McEwan JC, O'Connell AR, McNatty KP & Montgomery GW 2001 Highly prolific Booroola sheep have a mutation in the intracellular kinase domain of bone morphogenetic protein IB receptor (ALK-6) that is expressed in both oocytes and granulosa cells. *Biology of Reproduction* **64** 1225–1235.

Yan C, Wang P, De Mayo J, De Mayo FJ, Elvin JA, Carino C, Prasad SV, Skinner SS, Dunbar BS, Dube

JL, Celeste AJ & Matzuk MM 2001. Synergistic roles of bone morphogenetic protein 15 and growth differentiation factor 9 in ovarian function. *Molecular Endocrinology* **15** 854-866.

Control of ovarian follicular and corpus luteum development for the synchronization of ovulation in cattle

WW Thatcher[1] and JEP Santos[2]

[1]Department of Animal Sciences, University of Florida, Gainesville, FL, USA and [2]Veterinary Medicine Teaching and Research Center, University of California -Davis, Tulare, CA, USA.

The objective of this review is to integrate strategies to optimize an ovulatory control program which then serves as a platform to improve the reproductive performance of lactating dairy cows. Programmed management of follicle growth, regression of the CL and induction of ovulation led to development of the Ovsynch program. Pre-synchronization of estrous cycles followed 12 to 14 days later with the Ovsynch program increased pregnancy rates to timed inseminations. Initiation of the Ovsynch program on day 3 of the estrous cycle reduced ovulation to GnRH and resulted in a smaller proportion of excellent and good quality embryos following timed insemination. The pregnancy rate to a timed insemination of Ovsynch was greater when cows ovulated to the first injection of GnRH. The Presynch-Ovsynch program provided a platform to identify factors regulating reproductive performance; such as, parity, body condition score and anovulation. Treatment with hCG at day 5 after insemination increased pregnancy rate in lactating dairy cows. Injection of bovine somatotropin at insemination increased pregnancy rate, conceptus length and interferon-τ content in uterine luminal flushings and altered endometrial gene expression at day 17 of pregnancy. During heat stress, timed embryo transfer increased pregnancy rate and using embryos cultured with IGF-I and transferred fresh resulted in a greater pregnancy rate. Induction of ovulation with estradiol cypionate, as a component of a timed insemination program, increased fertility. Manipulation of the estrous cycle to improve follicle/oocyte competence and management of the post-ovulatory dialogue between embryonic and uterine tissues should enhance embryo development and survival.

Introduction

The goal of a successful estrous synchronization program is precise control of estrus and ovulation allowing fixed-time artificial insemination (AI) with high fertility. Lactating dairy cows usually benefit the most from synchronization programs as herd pregnancy rates (PR; defined as the proportion of pregnant cows relative to all eligible cows inseminated or not in a given period of time) are often times low because of poor estrous detection, occurrence of anestrus

Corresponding author E-mail: Thatcher@Animal.ufl.edu

and low conception rates. As our knowledge regarding control of the estrous cycle has been expanded, appropriate implementation of physiological methods to control sequentially follicle turnover, CL regression and induction of ovulation have been successful (Thatcher *et al.* 2004). Prostaglandins alone do not provide acceptable synchrony because the time of ovulation depends on the stage of development of the dominant follicle at the time of the prostaglandin-induced regression of the CL. This problem has been resolved partially with the development of the Ovsynch protocol as a breeding strategy that eliminates the need for detection of estrus whilst allowing for timed AI (Pursley *et al.* 1997). The protocol is composed of an injection of GnRH at random stages of the estrous cycle to induce ovulation of the dominant follicle and synchronize the emergence of a new follicular wave. Seven days later, $PGF_{2\alpha}$ is given to regress both the original and the potential newly formed CL, followed by a second GnRH injection 48 h later to induce a synchronous ovulation approximately 28 to 32 h later. A timed AI (TAI) is done at 12 to 16 h after the second GnRH injection. Pregnancy rates are comparable to those of cows inseminated at detected estrus. This protocol has been implemented successfully in many commercial dairy farms world wide as a strategy for TAI during the first postpartum AI, as well as for re-insemination of non-pregnant cows. Although the Ovsynch protocol allows for TAI without the need for estrous detection, approximately 10 to 15% of the cows will display signs of estrus during the protocol and they should be inseminated promptly if maximum PR is to be achieved.

The objective of this review is to integrate strategies to further optimize an ovulatory control program that is applicable to lactating dairy cows and to define how it can be utilized as a platform to improve the reproductive performance of a dairy herd.

Strategies to further optimize an ovulation control program

Pre-synchronization

Vasconcelos *et al.* (1999) noted that initiation of the Ovsynch protocol between days 5 and 9 of the cycle resulted in the highest frequency of ovulation to the first GnRH injection. Fertility was decreased when the duration of dominance of the ovulatory follicle was longer than 5 days (Austin *et al.* 1999) or the Ovsynch program was initiated in the early stages of the estrous cycle (Vasconcelos *et al.* 1999). Ovulation to the first GnRH injection and initiation of a new follicular wave should improve PR because an ovulatory follicle with a reduced period of dominance is induced to ovulate. Thus the concept of pre-synchronization (Presynch; Moreira *et al.* 2001) was introduced to enhance the probability of having a dominant follicle ($\geq$ 10mm) that will be induced to ovulate to the first GnRH injection of the Ovsynch protocol and the assurance that a CL will be present throughout the synchronization period (that is, a CL will not regress prior to the injection of $PGF_{2\alpha}$). The Presynch-Ovsynch program utilizes two injections of $PGF_{2\alpha}$ 14 days apart, with the second injection given 12 days prior to the first GnRH of the TAI protocol. The Presynch-Ovsynch program increased PR 18 percentage units (that is, 25% to 43%) in lactating cyclic cows (Moreira *et al.* 2001). Likewise, El-Zarkouny *et al.* (2004) also demonstrated improvement in PR when cows were pre-synchronized prior to the Ovsynch protocol. In addition to the potential benefit of optimizing the stage of the cycle by pre-synchronization, the prior repeated injections of $PGF_{2\alpha}$ may have a therapeutic benefit on the uterine environment by stimulating re-occurring proestrous/estrous phases allowing for improved uterine defense mechanisms.

Navanukraw *et al.* (2004) demonstrated that pre-synchronizing cows with 2 injections of $PGF_{2\alpha'}$ the second given 14 days prior to initiation of the Ovsynch protocol, improved PR.

Pregnancy rate to the modified Presynch-Ovsynch program was greater than the Ovsynch program (49.6% > 37.3%; P<0.05). This slight modification makes the sequence of injections friendlier to producers as injections are given on a weekly basis.

Optimizaton of follicle turnover

Implementation of the Ovsynch protocol prior to day 10 of the estrous cycle reduced the number of cows with premature luteolysis (Vasconcelos et al. 1999) and would therefore be expected to minimize the number of cows expressing estrus prior to the second GnRH injection and ovulating prematurely prior to the time of AI. However, it does not control precisely the stage of follicle development at the time of the first injection of GnRH in the Ovsynch protocol. Regardless of pre-synchronization strategies, cows that ovulated to the first GnRH injection of the Ovsynch protocol had increased PR (Chebel et al. 2006). In cows subjected to a Presynch-Ovsynch protocol, with the first GnRH of the Ovsynch protocol given 13 days after the Presynch (which in this study utilized CIDRs [+] or no CIDRs [-]), those ovulating to the GnRH had increased PR at 31 (+ CIDR, 37% > - CIDR, 21%) and 60 (+ CIDR, 28% > - CIDR, 18%) days after AI (Chebel et al. 2006).

The importance of inducing follicle turnover is demonstrated vividly by evaluating fertilization rates and embryo quality after TAI following the induction of follicle turnover or not (Cerri et al. 2005a). Our hypothesis was that initiating Ovsynch on day 3 of the estrous cycle would lead to continued development of an ovulatory follicle that would result in poorer embryo quality compared to recruitment of a new ovulatory follicle following follicle turnover initiated by an injection of GnRH on day 6 of the cycle. Lactating cows (n = 396) were subjected to AI after one of four protocols (Fig. 1). All cows received the same pre-synchronization protocol. The first group, called "Detected estrus (DE)", received GnRH on day 6 of the estrous cycle, followed by PGF$_{2\alpha}$ 7 days later and AI upon estrus. The other three groups were all subjected to an Ovsynch protocol (that is, GnRH, 7 days later PGF$_{2\alpha}$, 2 days later GnRH followed by TAI 12h later). The difference between these three goups, OV3, OV6 and OVE, was the timing of the first injection of GnRH on days 3, 6 and 6 of the estrous cycle, respectively. Group OVE also received an injection of 0.5 mg of estradiol cypionate (ECP) 36 h before the TAI. The same technician inseminated all cows with semen from a single sire. Ovarian responses were evaluated by ultrasonography, blood was analyzed for progesterone and uteri were flushed on day 6 after AI. The incidence of ovulation in response to the first injection of GnRH was less for cows receiving GnRH on day 3 than on day 6 of the estrous cycle (OV3 = 7.1% compared to DE = 79.2%, OV6 = 87.3%, OVE = 85.2%; P<0.001) because of smaller dominant follicles (OV3 = 9.5 < DE = 14.2, OV6 = 15.4 and OVE = 15.0 mm in diameter; P<0.001). A new follicular wave was observed in less cows when the first injection of GnRH was administered on day 3 compared with day 6 (OV3 = 7.1% compared to DE = 81.2%, OV6 = 88.6% and OVE = 88.9%; P<0.001). The diameter of the ovulatory follicle at AI differed (20.7, 19.7, 18.1 and 19.7mm for OV3, DE, OV6 and OVE, respectively; P<0.001). Growth of the dominant follicle that did not turnover (that is, OV3) and the fresh dominant follicle that was allowed to ovulate spontaneously (that is, DE) grew to a larger size and reflected the differences in the duration of follicular dominance (8.1, 7.3, 5.8 and 5.7 days for OV3, DE, OV6 and OVE, respectively; P<0.001). The proportion of cows synchronized at AI (luteolysis and ovulation) was greater for DE (96.9%) than the TAI treatments (84.1%).

Non-synchronized cows in TAI protocols were not flushed on day 6 after AI. Those cows not observed in estrus in the DE group were not inseminated and, therefore, not flushed on day 6 after AI. Fertilization rates were similar (P=0.96) and averaged 86.3% across treatments. Of

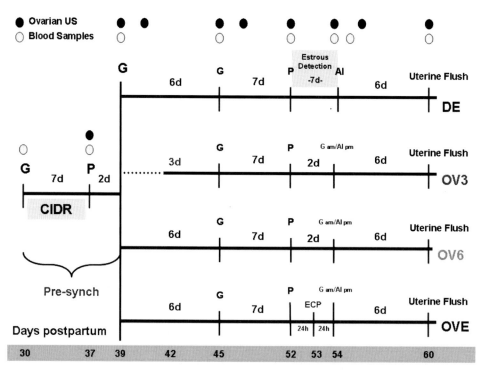

Fig. 1. Protocol to examine the effects of follicle dynamics on fertilization and embryo quality (Cerri et al. 2005a). AI = artificial insemination; DE = insemination upon detection of estrus; G = injection of GnRH; P = injection of PGF$_{2\alpha}$; OV3 = Ovsynch protocol initiated on day 3 of the estrous cycle; OV6 = Ovsynch protocol initiated on day 6 of the estrous cycle; OVE = Ovsynch protocol initiated on day 6 of the estrous cycle and with supplemental estradiol; ECP = supplemental estrogen, estradiol cypionate; Pre-synch = pre-synchronization of the estrous cycle; US = ultrasonography

the total structures recovered, the proportion of embryos graded as excellent and good was reduced in cows that began the Ovsynch protocol on day 3 (OV3) of the cycle (Table 1). Similarly, there was a decrease in the proportion of excellent and good embryos as a proportion of the fertilized ova collected for cows in the OV3 group. Both the degree of development (the number of blastomeres per embryo) and the viability (proportion of blastomeres that were alive) of the embryos were reduced in the OV3 group (Table 1). Insemination at DE did not improve fertilization rates or embryo quality. The TAI compromised embryo quality when the protocol reduced the number of cows ovulating in response to the first injection of GnRH (OV3). Consequently, the OV3 protocol resulted in prolonged dominance of the ovulatory follicle at AI.

These findings substantiate the importance of regulating follicle dynamics to optimize the period of follicle dominance to obtain high quality embryos. Increasing concentrations of estradiol via an injection of ECP prior to the induction of ovulation with GnRH at 48 h after the injection of PGF$_{2\alpha}$, did not appear to improve fertilization rates or embryo quality. Perhaps supplemental estradiol would be more beneficial if given in the proestrus period when a follicle is allowed to develop to mature ovulatory potential.

Table 1. Effect of four artificial insemination protocols on embryos recovered from uterine flushings at day 6 after AI (Cerri *et al.* 2005a).

	Treatment[a]			
	DE	OV3	OV6	OVE
Percent grades 1 and 2 of recovered structures[c]	61.0	40.0	72.0	65.9
Percent grades 1 and 2 of recovered embryos[b]	71.4	47.0	83.7	74.3
Total number of blastomeres[b]	42.3±3.39	29.7±3.44	42.5±3.29	44.1±3.16
Proportion of live blastomeres[c]	94.3±2.20	90.1±2.35	97.7±2.14	97.6±2.02

[a]Detected estrus (DE)= GnRH on day 6 of the estrous cycle, followed by $PGF_{2\alpha}$ 7 days later and AI at estrus; Ovsynch (GnRH, followed by $PGF_{2\alpha}$ 7 days later, then 2 days later GnRH followed by a timed AI 12 h later) as OV3, OV6 and OVE, which correspond to the injection of the first GnRH on day 3, 6 and 6 of the estrous cycle, respectively, but OVE also received an injection of 0.5 mg of estradiol cypionate 36 h before the timed AI.
[b]$P<0.01$, [c]$P<0.05$

Follicle size and estradiol concentrations during the periovulatory period

The injection of GnRH at 48 h after $PGF_{2\alpha}$ is at a time that reduces the occurrence of a spontaneous estrus but induces a synchronized ovulation permitting TAI. However, injection of GnRH at 48 h after $PGF_{2\alpha}$ causes a reduction in plasma estradiol concentrations and a slight rise in plasma progesterone concentrations (Thatcher & Chenault 1976); the increase in plasma progesterone concentrations was not evident when GnRH was injected at 60 h. When groups of lactating dairy cows were synchronized for estrus following injections of GnRH and $PGF_{2\alpha}$ given 7 days apart, the mean occurrence of estrus was at 72 h after $PGF_{2\alpha}$; a time indicative of a fully functional follicle with a mean diameter of 15.7 mm (Badinga *et al.* 1994). Injections of GnRH prior to 72 h would induce ovulation of a smaller size follicle that might not be fully developed and may have prematurely decreased concentrations of estradiol as a consequence of luteinization of the granulosa cells. It is typical for cows inseminated to an Ovsynch protocol to have reduced estrogenic tone at the time of AI. Injection of GnRH to induce ovulation initiates a preovulatory LH surge before some cows appear to have developed dominant follicles that are physiologically mature (Perry *et al.* 2005). In beef cows, GnRH-induced ovulation of these smaller sized immature follicles caused a decrease in PR because of increased late embryonic/fetal losses (Perry *et al.* 2005).

An additional means to enhance the programmed proestrus period is to further delay the injection of the second GnRH injection of the Presynch-Ovsynch protocol until 72 h after the injection of $PGF_{2\alpha}$ (Portaluppi & Stevenson 2005). In this approach, cows were inseminated at the time of the GnRH injection. At 72 h many cows are in estrus and inseminating at the time of the GnRH injection is a compromise in that those cows not yet expressing estrus and induced to ovulate will be inseminated early relative to ovulation. However, this strategy allows for an extended proestrus period in lactating dairy cows that results in a more mature functional follicle. Delaying the GnRH injection and AI until 72 h after $PGF_{2\alpha}$ increased PR compared to the Presynch/Ovsynch protocol with GnRH injection and AI at 48 h after $PGF_{2\alpha}$ (31.4% > 22.8%) or GnRH injection at 48 h after $PGF_{2\alpha}$ and AI 24 h later (31.4% > 23.5%). Thus extending the proestrus period improved PR (Portaluppi & Stevenson 2005). In a recent study, however, when cows were given both GnRH and TAI at either 48 or 72 h after a $PGF_{2\alpha}$-induced luteolysis, PR did not differ even when 1 mg of ECP was given 24 h after $PGF_{2\alpha}$ (Table 2; Hillegass *et al.* 2006). Brusveen *et al.* (2006) also failed to detect any difference in PR when cows were given both GnRH and TAI at either 48 (G48) or 72 h (G72) after $PGF_{2\alpha}$ induced luteolysis as part of a Presynch-Ovsynch program. However, PR was improved when the GnRH injection was delayed to 56 h (G56) followed by TAI 16 h later (that is, at 72 h). The G56, PR of 36.2% at day 31-33 post-AI was greater than 26.7% and 27.2% for G48 and G72, respectively. The beneficial effect of GnRH at 56 h prior to TAI at 72 h is most likely due to a more optimal stage of follicle development at the time of the GnRH injection (56 h compared to 48 h) combined with the proper timing of insemination that allows capacitated sperm to be at the site of fertilization when the cows ovulate.

Table 2. Effect of timed AI at 48 or 72 h after $PGF_{2\alpha}$ and treatment with estradiol cypionate (ECP) on reproductive responses of dairy cows (Hillegass et al. 2006).

	Treatment[a]						
	CoSynch 48		CoSynch 72		P value[b]		
	No ECP	ECP	No ECP	ECP	CoS	ECP	CoS*ECP
Cows, n	240	244	246	240			
Estrus at AI, %	33.3	59.8	49.8	79.6	0.001	0.001	NS
Pregnant 38 d, %	42.1	46.5	46.4	43.7	0.77	0.83	NS

[a] CoSynch 48 = GnRH and timed AI 48 h after $PGF_{2\alpha}$-induced luteolysis; CoSynch 72 = GnRH and timed AI 72 h after $PGF_{2\alpha}$-induced luteolysis; ECP (1 mg) injected 24 h after $PGF_{2\alpha}$
[b] CoS = effect of CoSynch; ECP = effect of estradiol cypionate; CoS*ECP = interaction between CoS and ECP.

Supplemental estradiol in the periovulatory period

In a low progesterone environment, during late diestrus and proestrus, exogenous estradiol can stimulate hypothalamic secretion of GnRH thereby inducing a LH surge and ovulation. The estradiol-induced LH surge mimics a spontaneous surge in that it lasts for approximately 10 h: twice the duration of a GnRH-induced surge of LH. Estradiol cypionate (ECP), an esterified form of estradiol-17ß, has been used to induce ovulation as a modification of the Presynch-Ovsynch program (Pancarci et al. 2002; Cerri et al. 2004). Cows were pre-synchronized with two injections of $PGF_{2\alpha}$ given 14 days apart with the TAI program beginning 14 days after the second injection of $PGF_{2\alpha}$ and using ECP to induce ovulation (Pancarci et al. 2002). Cows were injected with GnRH followed by $PGF_{2\alpha}$ 7 days later. The ECP (1 mg, i.m.) was injected 24 h after $PGF_{2\alpha}$ and cows received TAI 48 h later or at detected estrus if it occurred prior to 48 h. In lactating dairy cows, the frequencies of detected estrus and ovulation after ECP were 75.7% and 86.5%, respectively (Pancarci et al. 2002). Estrus occurred at 29.0 ± 1.8 h (n = 28) after ECP. Mean intervals to ovulation were 55.4 ± 2.7 h after ECP and 27.5 ± 1.1 h after onset of estrus. The interval from ECP injection to ovulation had a rather large coefficient of variation of 27.4%. This lead to the recommendation to inseminate cows at all detected estruses occurring prior to 48 h and TAI the remaining cows at 48 h after the ECP injection.

Lactating dairy cows have reduced concentrations of plasma estradiol in the preovulatory period and a reduced intensity of estrus (De la Sota et al. 1993; Sartori et al. 2002a) therefore the elevation of plasma estradiol concentrations following ECP injection will supplement for a lactational induced deficiency in estradiol. If cows are anovulatory (for example, anestrus or have not developed positive estradiol feedback) then the ECP modified Ovsynch program may not be as effective as the GnRH based Ovsynch program in which GnRH causes the direct secretion of LH. The greater occurrence of estrus, enhanced uterine tone and ease of insemination are practical advantages with the use of the Presynch/ECP modified Ovsynch program.

Cerri *et al.* (2004) clearly demonstrated that lactating dairy cows subjected to a presynchronized/ECP modified Ovsynch with a TAI had increased proportions of cows in estrus, as well as greater conception and pregnancy rates compared with cows inseminated following detected estrus induced by a presynchronized/GnRH-PGF$_{2\alpha}$ (that is, 7 day interval) protocol. Fertility responses to the presynchronized/ECP modified Ovsynch protocol were improved when lactating cows were observed in estrus after ECP treatment. Occurrence of estrus in both groups was associated with cyclic status and the size of the preovulatory follicle determined at 48 h after ECP. As cows inseminated following the presynchronized/ECP modified Ovsynch protocol had greater conception rates, it is possible that supplemental estradiol given during proestrus in the form of ECP enhanced the fertility of these high-producing dairy cows.

The dynamics of estradiol exposure during the proestrus period prior to either spontaneous or induced preovulatory surges of LH need to be evaluated precisely to determine whether estradiol supplementation enhances conception and pregnancy rates. The variables to be examined in the proestrus period following an injection of PGF$_{2\alpha}$ include the timing of the estradiol treatment relative to both follicle maturity (size, estrogenic status *et cetera*) and the preovulatory surge of LH. The surge of LH initiates oocyte maturation and luteinization of the follicular cellular components leading to a decrease in estradiol secretion and ovulation. Clearly the findings of Cerri *et al.* (2004) of enhanced conception and pregnancy rates with the use of ECP was in association with enhanced plasma estradiol concentrations during a programmed proestrus period of 53 h (that is, mean interval from the injection of PGF$_{2\alpha}$ to the onset of estrus or the LH surge). The mean diameter of preovulatory follicles at 48 h after ECP was 18 mm in diameter. Proestrus follicles that matured earlier, based on detected estruses prior to 48 h, were accommodated by prompt AI.

In subsequent studies, ECP was used in conjunction with a GnRH injection to control more precisely the timing of estradiol injections relative to the LH surge; however, reproductive responses were not improved (Cerri *et al.* 2005a; Sellars *et al.* 2006; Hillegass *et al.* 2006). An injection of 0.5 mg ECP 24 h after PGF$_{2\alpha}$ injection, which was 24 h prior to an injection of GnRH, failed to increase either fertilization rates or embryo quality at day 6 after TAI (Table 1; Cerri *et al.* 2005a). Pregnancy rates were not increased when 0.5 mg ECP were injected concurrently with GnRH at 48h after PGF$_{2\alpha}$ injection when compared to a standard Ovsynch protocol (Sellars *et al.* 2006). In contrast to the experiment of Cerri *et al.* (2004), the dose of ECP was less (0.5 mg *versus* 1.0 mg) and plasma estradiol concetrations were elevated at the time the follicle was induced to undergo luteinization in response to the LH surge induced by the concurrent injection of GnRH. Although sizes of the preovulatory follicles at 48 h were the same among the experiments (Cerri *et al.* 2004; 2005a; Sellars *et al.* 2006), the approximate proestrus periods before the induced surge-like release of LH differed (that is, 53, 48 and 48 h, respectively) as did the period of elevated plasma estradiol concentrations prior to the LH surge (~ 29, 24 and 0 h, respectively). When Souza *et al.* (2005) supplemented lactating dairy cows with 1 mg of estradiol-17ß at 8 h prior to the induction of ovulation in the Ovsynch program, treatment with estradiol-17ß increased circulating estradiol concentrations (18.4$\pm$4.6 units, n = 8 *vs.* 2.94$\pm$0.6 units, n = 8; P < 0.01) and expression of estrus (78.5%, n = 302 *vs.* 42.4%, n = 290; P < 0.01). Thus induction of a sharp rise and decline in plasma estradiol concentrations improved PR to TAI in cows during the cool season (< 21°C) (50.9%, n = 116 *vs.* 34.6%, n = 108; P < 0.02). Cows classified as anovular had greater PR when treated with estradiol-17ß. In addition, cows with intermediate-sized ovulatory follicles (15-20 mm in diameter with single ovulation, n = 416) tended (P = 0.06) to have greater PR when supplemented with estradiol-17ß prior to GnRH-induced ovulation. These data support the findings of Cerri *et al.* (2004) and suggest that the response to supplemental estradiol depends upon cyclic status, season and size of the ovulatory follicle.

Use of ovulatory control programs as a platform to improve pregnancy rates

Ovulatory control systems have provided a platform for investigators to quantitatively elucidate the factors that are limiting herd fertility, and such a platform provides a reference base to test strategies that may improve PR. The ability to synchronize follicle development and CL regression coupled with the precise timing of ovulation allows for synchronous inseminations or transfer of embryos. Consequently, the management and biological factors regulating reproductive performance can be examined and systems developed to improve PR in what is currently considered a sub-fertile population of lactating dairy cows.

The impact of anovulatory cows and body condition scores on pregnancy rates and embryo losses

The interrelationships among parity, body condition score (BCS; 1 to 5 scale), milk yield, AI protocol (inseminated at estrus or TAI) and cyclicity (cyclic or anovular) were evaluated with respect to their impacts on PR and embryonic survival following the first postpartum AI (n = 5,767) in nine studies on five dairy farms (Rutigliano & Santos 2005). All farms and cows had records on BCS at calving and at AI. For all farms, cows were inseminated at first service either upon detection of a synchronized estrus or at TAI in association with a Presynch-Selectsynch or Presynch-Ovsynch or Presynch-Heatsynch program (Thatcher *et al.* 2004). Pre-synchronization entailed two PGF$_{2\alpha}$ injections given 14 days apart with the TAI or estrous synchronization protocol initiated 12 to 14 days after the pre-synchronization. Effects of parity (that is, primiparous and multiparous) were evaluated in all of the studies and farms. Milk yield was determined once (eight studies) or twice monthly (one study). On two of the five farms used in the nine studies, cows were milked either twice or thrice daily (that is, on one farm, cows were milked thrice daily in three of the five studies whilst on another farm they were milked thrice daily in two of four studies).

Occurrence of cyclicity was greater for multiparous than primiparous cows at 65 days postpartum (81.8% *vs.* 69.5%; P < 0.001). In addition to parity, cyclicity was also influenced by milking frequency (twice = 82.7% vs. thrice = 68.7%; P < 0.01), BCS at calving and at AI, BCS change and milk yield. However, milk yield, BCS at calving and the AI protocol had no effect (P > 0.10) on PR at 30 and 58 days after AI or pregnancy loss. More (P < 0.001) cyclic than anovular cows were pregnant at 30 (40.0% *vs.* 28.3%) and 58 (34.2% *vs.* 23.0%) days after AI and anovulation tended (P = 0.09) to increase pregnancy loss (14.5% *vs.* 18.6%) between 30 and 58 days of gestation. Pregnancy loss was highest (22.5% *vs.* 16.8% *vs.* 12.2%; P < 0.01) and conception rates at day 58 lowest (21.7% *vs.* 30.4% *vs.* 35.6%; P < 0.01) in cows that lost ≥ 1 unit of BCS than those that lost < 1 or experienced no change in BCS from calving to AI. Likewise, a greater BCS at AI (≥3.75 *vs.* 3.0 to 3.5 *vs.* ≤2.75) increased conception rates at 30 (46.4% *vs.* 40.1% *vs.* 33.9%, respectively; P < 0.01) and 58 (41.8% *vs.* 34.6% *vs.* 27.9%; P < 0.01) days after AI. Consequently, minimizing loss of BCS after calving and improving cyclicity early postpartum are expected to increase PR and enhance embryonic survival. The AI protocol and milk yield did not affect pregnancy and embryonic survival after the first postpartum AI. This summary of studies with repeated measurements of BCS dynamics and anovulatory status in association with PR is in agreement with other studies indicating that pregnancy rates to first timed insemination are less for anovulatory cows (Moreira *et al.* 2001; Cordoba & Fricke 2001; Gumen *et al.* 2003).

Supplemental progesterone for anovular cows

Methods to improve fertility of anovular cows, such as the use of supplemental progesterone, have proven inconsistent in high-producing dairy cows. Use of an intravaginal progesterone insert increased the induction of cyclicity in anovular cows between 49 and 62 days postpartum (30.8 vs 46.2%; P < 0.01: Chebel *et al.* 2006) but failed to improve PR in cyclic and anovular cows. Galvao *et al.* (2004) incorporated an intravaginal progesterone insert, containing 1.38 g of progesterone, in a TAI protocol but the progesterone insert did not improve PR or embryo survival in anovular cows. Previously, incorporation of an intravaginal progesterone insert into a TAI protocol improved PR in one of two experiments (El-Zarkouny *et al.* 2004) but in none of the experiments did the progesterone insert improve PR of anovular cows. The inconsistency in reproductive performance after progesterone supplementation is quite intriguing and possibly related to the low concentrations of progesterone induced by the CIDR in high-producing lactating cows (Cerri *et al.* 2005b).

Embryotrophic actions of bovine somatotropin at first AI

Exogenous bovine somatotropin (bST) increased PR (for example, 57.0% > 42.6%: Moreira *et al.* 2001) when administered as part of a Presynch-Ovsynch TAI protocol (that is, bST was given either at the first injection of GnRH or at TAI) in cyclic lactating dairy cows for the first AI postpartum or in protocols involving cows being inseminated at detected estrus (Morales-Roura *et al.* 2001; Santos *et al.* 2004; Starbuck *et al.* 2006). Furthermore, pregnancy losses were reduced in cows that received bST at the time of the first GnRH injection of either a Presynch-Ovsynch TAI protocol or synchronization using GnRH followed 7 days later by PGF$_{2\alpha}$ with cows inseminated at detected estrus (Santos *et al.* 2004). The study of Morales-Roura *et al.* (2001) involved cows identified as having three or more prior inseminations. Pregnancy rates were stimulated when bST was given at estrus and again 10 days later. As bST was effective at AI, it is likely that bST stimulated fertilization, embryonic development and survival following AI in lactating dairy cows. Subsequent *in vitro* and *in vivo* studies indicated that both GH and IGF-I stimulated fertilization and blastocyst development (Moreira *et al.* 2002a;b) as well as embryo cell number (Moreira *et al.* 2002a).

An additional study utilized a pre-synchronization protocol followed by Ovsynch and TAI as a platform to examine the effect of bST (injected at TAI and 11 days later) on conceptus development and endometrial gene expression at day 17 after TAI, as well as daily hormonal responses between TAI (= day 0) and day 17 (Bilby *et al.* 2006a;b; Thatcher *et al.* 2006). There were increases in conceptus length (45 > 34 cm), interferon-τ concentrations in uterine luminal flushings (9.4 > 5.3 μg) and enhanced embryo survival based on a greater PR at day 17 (83% > 40%) in animals treated with bST. When the contents of interferon-τ in the uterine flushings were adjusted for the length of the extra-embryonic membranes as a covariate, the difference due to bST was no longer significant. Thus a greater uterine pool size of interferon-τ in bST-treated cows appears to be due to a greater size of the conceptuses at day 17. Treatment with bST enhanced the abundancies of both IGF-II and PGES mRNAs whilst it decreased PGFS mRNA in the endometrial tissue of day 17 pregnant cows compared to pregnant cows not treated with bST. Collectively, these uterine and conceptus responses likely contribute to enhancing embryo survival in lactating cows at first AI when administered at the proper time in a programmed and synchronous ovulation with insemination.

Timed embryo transfer

An additional use of the Ovsynch protocol is to control the time of ovulation as a platform to carry out a timed embryo transfer in recipient lactating dairy cows. Since the Ovsynch program can be used successfully to synchronize ovulation rate in a significant number of cows (that is,

~ 85%), it can be used for a synchronized transfer of either fresh or frozen embryos produced from superovulated donors or by *in vitro* production (IVP). The practicality and efficiency of recipient management is further improved by identifying, at the time of embryo transfer, whether the synchronized recipient has a CL and a dominant follicle confirmatory of a synchronized ovulation to the second injection of GnRH of the Ovsynch protocol. Use of a Presynch-Ovsynch protocol or some appropriate modification for pre-synchronization would further improve the percentage of recipients with a synchronized ovulation.

The percentage of viable embryos (that is, excellent and good quality) recovered at day 6 after AI, following a pre-synchronization program that preceded an Ovsynch protocol implemented at day 6 of the estrous cycle, was 66.5% (Table 1, DE = 61% and OV6 = 72%; Cerri *et al.* 2005a). Hence there is a considerable number of potential pregnancy losses by day 6 and this is even greater in lactating dairy cows exposed to heat stress (22% with grades 1, 2 and 3 embryos; Sartori *et al.* 2002b). Thus timed embryo transfer of excellent to good grade embryos would be a logical means to by-pass these early embryo losses prior to day 6 or 7 in lactating dairy cows.

Two studies completed in Florida during the summer heat stress season evaluated timed embryo transfer following the second injection of GnRH in the Ovsynch protocol; embryos were transferred to recipients either at day 7.5 (Ambrose *et al.* 1999) or day 8 (Al-Katanani *et al.* 2002). Fresh IVP embryos increased PR compared to TAI (16.6% > 5.6%). Frozen-thawed IVP embryos resulted in PR comparable to the TAI control groups (that is, 5.6%). Consequently, processes associated with the freezing and thawing of IVP embryos are not optimal for transfer of embryos during conditions of heat stress in lactating dairy cows. However, it is clear that under heat stress conditions in Florida, the well characterized detrimental effects of high temperature on early embryo development can be by-passed partially with the use of timed embryo transfer.

Based on the observations that IGF-I stimulates blastocyst development during *in vitro* culture (Moreira *et al.* 2002a), Block *et al.* (2003) documented clearly that embryos cultured with IGF-I subsequently transferred fresh into recipients (timed embryo transfer on day 8 after the second GnRH of an Presynch-Ovsynch protocol) resulted in greater PR during summer heat stress. This beneficial effect of IGF-I exposure in culture, that results in increased PR following embryo transfer of fresh embryos, occurred in the summer heat stress season (day 41-49 PR: +IGF-I, 28/67 = 41.8% *vs.* –IGF-I, 13/71 = 18.3%) but not during the cooler season (day 41-49 PR: +IGF-I, 16/73 = 21.9% *vs.* –IGF-I, 21/74 = 28.4%) of the year (Block & Hansen 2006).

Since early embryonic viability is reduced in lactating dairy cows (Sartori *et al.* 2002b), Sartori and coworkers (2006) hypothesized that embryo transfer would improve PR in lactating dairy cows. Ovulation was synchronized with a modified Ovsynch program (GnRH followed at 7 days with PGF$_{2\alpha}$ and then 3 days later with GnRH) and then the cows were either inseminated at the time of the second GnRH injection or they received fresh or frozen embryos (classified as excellent or good quality from superovulated cows or heifers) at day 7 after the second GnRH injection. Pregnancy rates at days 60-66 for cows that had a synchronized ovulation did not differ between AI (31.2%) and embryo transfer (30.7%). The authors suggested that lactating dairy cows might have other reproductive problems that may not be solved by embryo transfer. These could include subclinical endometritis, carry over effects of negative energy balance or hormonal imbalances that collectively may compromise the uterine environment for optimal embryo development after embryo transfer. In fact, 30 to 50% of lactating dairy cows have subclinical endometritis, characterized by increased presence of polymorphonuclear neutrophils in the uterine lumen after 40 days postpartum, which are associated with reduced PR (Sheldon *et al.* 2006).

The comparison between animals that were inseminated versus a transferred embryo is insightful. For all cows that had a synchronized ovulation of a single ovulatory follicle, those with a smaller size ovulatory follicle (10-15 mm in diameter) had lower PR at days 25-32 when inseminated compared to those receiving embryo transfer (23.7% < 42.3%; P<0.05). Similar results were observed when pregnancy was diagnosed on days 60-66 (18.4% < 38.5%; P<0.05). In contrast, no differences between groups were detected for medium- and large-size ovulatory follicles (Sartori *et al.* 2006). For all single-ovulatory cows, those classified as having smaller size ovulatory follicles had lower concentrations of progesterone at day 7 than those cows ovulating medium- or larger-size ovulatory follicles (1.59 ng/ml < 2.00 ng/ml and 2.13 ng/ml, respectively). It is plausible to suggest that the lower concentrations of progesterone associated with the synchronized ovulation of a small size ovulatory follicle results in a sub-optimal rise in progesterone that alters early embryo development resulting in a reduced PR. This early period of developmental sensitivity appears to be by-passed with the transfer of excellent or good grade embryos to single ovulation recipient cows that are lactating.

A possible method to overcome the formation of sub-functional luteal tissue is the induction of accessory CL early in diestrus. Santos *et al.* (2001) injected 3,300 IU of hCG in lactating cows 5 days after AI following detection of a synchronized estrus and cows receiving hCG had an increased number of CL and increased plasma progesterone concentrations. Pregnancy rates on days 28, 42, and 90 were improved by hCG treatment but late embryonic and fetal losses remained unaltered. Benefits of hCG treatment were clearly demonstrated in cows that were losing BCS between AI and pregnancy diagnosis. Nishigai *et al.* (2002), utilizing embryo recipient cows, demonstrated that hCG induction of an accessory CL and a subsequent increase in plasma progesterone concentration increased PR. Pregnancy rate in cows receiving hCG on day 6 was higher (67.5%) than in control cows (45.0%) or cows receiving hCG on day 1 (42.5%) after AI.

Conclusions

The management of reproductive processes in the high producing dairy cow to optimize PR can be achieved partially through implementation of a timed AI program characterized by sequentially controlling follicle turnover in concert with induced regression of the corpus luteum and timely induction of ovulation. Such systems partially optimize the proestrous period and supplemental estradiol further increases PR. Components of the postpartum period, such as parity, dynamic changes in body condition score, anovulation and uterine health, influence PR and embryo losses. Management of the post-ovulatory dialogue between embryonic and uterine tissues should enhance embryo development in lactating dairy cows, as demonstrated by the timely injection of hCG to increase circulating progesterone concentrations and the enhancement of conceptus development and PR in response to bST. Timed embryo transfer is a means to by-pass inefficiencies in fertilization and early embryo development associated with heat stress and has provided a platform to demonstrate the *in vivo* beneficial effects of culturing *in vitro* produced embryos with IGF-I. Embryos produced *in vitro* that are cultured in the presence of IGF-I result in greater PR post-transfer during seasonal periods of heat stress. Conversely, the lack of an improvement in PR of timed embryo transfers in the cool season compared to timed insemination indicates that there may be potential arrays of factors compromising fertility in lactating dairy cows beyond the optimization of the periovulatory period.

Acknowledgements

Research conducted by the authors was supported by grants from the National Research Initiative of the USDA Cooperative State Research, Education, and Extension Service (Competitive Grant no. 2004-35203-14137; Grant no. 98-35203-6367); Florida-Georgia Milk Checkoff Program; Center for Food Animal Health of the University of California Davis, USDA Formula Funds (Project no. CALV-AH-214); National Association of Animal Breeders; Pfizer Animal Health; and Select Sires.

References

Al-Katanani YM, Drost M, Monson RL, Rutledge JJ, Krininger CE III, Block J, Thatcher WW & Hansen PJ 2002 Pregnancy rates following timed embryo transfer with fresh or vitrified in vitro produced embryos in lactating dairy cows under heat stress conditions. *Theriogenology* **58** 171-182.

Ambrose JD, Drost M, Monson RL, Rutledge JJ, Leibfried-Rutledge ML, Thatcher MJ, Kassa T, Binelli M, Hansen PJ, Chenoweth PJ & Thatcher WW 1999 Efficacy of timed embryo transfer with fresh and frozen in vitro produced embryos to increase pregnancy rates in heat stressed dairy cattle. *Journal of Dairy Science* **82** 2369-2376.

Austin EJ, Mihm M, Ryan MP, Williams DH, Roche JF 1999 Effect of duration of dominance of the ovulatory follicle on onset of estrus and fertility in heifers. *Journal Animal Science* **77** 2219-2226.

Badinga L, Thatcher WW, Wilcox CJ, Morris G, Entwistle K & Wolfenson D 1994 Effect of season on follicular dynamics and plasma concentrations of estradiol-17ß, progesterone and luteinizing hormone in lactating Holstein cows. *Theriogenology* **42** 1263-1274.

Bilby TR, Sozzi A, Lopez MM, Silvestre F, Ealy AD, Staples CR & Thatcher WW 2006a Pregnancy, bST and omega-3 fatty acids in lactating dairy cows: I. ovarian, conceptus and growth hormone – IGF system response. *Journal of Dairy Science* **89** 3360–3374.

Bilby TR, Guzeloglu A, MacLaren LA, Staples CR & Thatcher WW 2006b Pregnancy, bST and omega-3 fatty acids in lactating dairy cows: II. Endometrial gene expression related to maintenance of pregnancy. *Journal of Dairy Science* **89** 3375–3385.

Block J & Hansen PJ 2006 Effect of the addition of insulin-like growth factor-1 to embryo culture medium on pregnancy rates following timed embryo transfer in lactating dairy cows. *Journal of Dairy Science* **89** (Suppl. 1) 287.

Block J, Drost M, Monson RL, Rutledge JJ, Rivera RM, Paula-Lopes FF, Ocon OM, Krininger CE III, Liu J & Hansen PJ 2003 Use of insulin-like growth factor-1 during embryo culture and treatment of recipients with GnRH to increase pregnancy rates following the transfer of in vitro produced embryos to heat-stressed, lactating cows. *Journal Animal Science* **81** 1590-1602.

Brusveen DJ, Cunha AP, Silva CD, Cunha PM, Sterry RA, Silva EPB, Guenther JN & Wiltbank MC 2006 Effects on conception rates of lactating dairy cows by altering the time of the second GnRH and AI during Ovsynch. *Journal Dairy Science* **89** (Suppl. 1) 150.

Cerri RLA, Santos JEP, Juchem SO, Galvão KN & Chebel RC 2004 Timed artificial insemination with estradiol cypionate or insemination at estrus in high-producing dairy cows. *Journal of Dairy Science* **87** 3704-3715.

Cerri RLA, Rutigliano HM, Bruno RGS, Chebel RC & Santos JEP 2005a Effect of artificial insemination (AI) protocol on fertilization and embryo quality in high-producing dairy cows. *Journal of Dairy Science* **88** (Suppl. 1) 86.

Cerri RLA, Rutigliano HM, Bruno RGS & Santos JEP 2005b Progesterone (P4) concentrations and ovarian response after insertion of a new or a 7 d used intravaginal P4 insert (IPI) in proestrus lactating cows. *Journal of Dairy Science* **88** (Suppl. 1) 37.

Chebel RC, Santos JEP, Cerri RLA, Rutigliano HM & Bruno RGS 2006 Reproduction in dairy cows following progesterone insert presynchronization and resynchronization protocols. *Journal of Dairy Science* **89** 4205-4219.

Cordoba MC & Fricke PM 2001 Evaluation of two hormonal protocols for synchronization of ovulation and timed artificial insemination in dairy cows managed in grazing-based dairies. *Journal Dairy Science* **84** 2700-2708.

De La Sota RL, Lucy MC, Staples CR & Thatcher WW 1993 Effects of recombinant bovine somatotrophin (Sometribove) on ovarian function in lactating and nonlactating dairy cows. *Journal of Dairy Science* **76** 1002-1013.

El-Zarkouny SZ, Cartmill JA, Hensley BA & Stevenson JS 2004 Pregnancy in dairy cows after synchronized ovulation regimens with or without presynchronization and progesterone. *Journal of Dairy Science* **87** 1024-1037.

Galvao KN, Santos JE, Juchem SO, Cerri RL, Coscioni AC & Villasenor M 2004 Effect of addition of a progesterone intravaginal insert to a timed insemination protocol using estradiol cypionate on ovulation rate, pregnancy rate, and late embryonic loss in lactating dairy cows. *Journal Animal Science* **82** 3508-3517.

Gumen A, Guenther JN & Wiltbank MC 2003 Follicular size and response to Ovsynch versus detection of estrus in anovular and ovular lactating dairy cows. *Journal Dairy Science* **86** 3184-3194.

Hillegass J, Santos JEP, Lima FS & Sa Filho M 2006 Effect of interval from luteolysis to timed AI and supplemental estradiol on reproductive performance of dairy cows. *In preparation*

Morales-Roura JS, Zarco L, Hernandez-Ceron J & Rodriguez G 2001 Effect of short-term treatment with bovine Somatotropin at estrus on conception rate and luteal function of repeat-breeding dairy cows. *Theriogenology* **55** 1831-1841.

Moreira FCO, Risco CA, Mattos R, Lopes F & Thatcher WW 2001 Effects of presynchronization and bovine somatotropin on pregnancy rates to a timed artificial insemination protocol in lactating dairy cows. *Journal of Dairy Science* **84** 1646-1659.

Moreira F, Paula-Lopes F, Hansen PJ, Badinga L & Thatcher WW 2002a Effects of growth hormone and insulin-like growth factor-I on development of in vitro derived bovine embryos. *Theriogenology* **57** 895-907.

Moreira F, Badinga L, Burnley C & Thatcher WW 2002b Bovine somatotropin increases embryonic development in superovulated cows and improves post-transfer pregnancy rates when given to lactating recipient cows. *Theriogenology* **57** 1371-1387.

Navanukraw C, Redmer DA, Reynolds LP, Kirsch JD, Grazul-Bilska AT & Fricke PM 2004 A modified presynchronization protocol improves fertility to timed artificial insemination in lactating dairy cows. *Journal of Dairy Science* **87** 1551-1557.

Nishigai M, Kamomae H, Tanaka T & Kaneda Y 2002 Improvement of pregnancy rate in Japanese Black cows by administration of hCG to recipients of transferred frozen-thawed embryos. *Theriogenology* **58** 1597-1606.

Pancarci SM, Jordan ER, Risco CA, Schouten MJ, Lopes FL, Moreira F & Thatcher WW 2002 Use of estradiol cypionate in a pre-synchronized timed artificial insemination program for lactating dairy cattle. *Journal of Dairy Science* **85** 122-131.

Perry GA, Smith MF, Lucy MC, Green JA, Parks TE, MacNeil MD, Roberts AJ & Geary TW 2005 Relationship between follicle size at insemination and pregnancy success. *PNAS* **102** 5268-5273.

Portaluppi MA & Stevenson JS 2005 Pregnancy rates in lactating dairy cows after presynchronization of estrous cycles and variations of the ovsynch protocol. *Journal of Dairy Science* **88** 914-921.

Pursley JR, Kosorok MR & Wiltbank MC 1997 Reproductive management of lactating dairy cows using synchronization of ovulation. *Journal of Dairy Science* **80** 301-306.

Rutigliano HM & Santos JEP 2005 Interrelationships among parity, body condition score (BCS), milk yield, AI protocol, and cyclicity with embryonic survival in lactating dairy cows. *Journal of Dairy Science* **88** (Suppl 1)178.

Sartori R, Rosa GJM & Wiltbank MC 2002a Ovarian structures and circulating steroids in heifers and lactating cows in summer and lactating and dry cows in winter. *Journal of Dairy Science* **85** 2813-2822.

Sartori R, Sartor-Bergfelt R, Mertens SA, Guenther JN, Parrish JJ & Wiltbank MC 2002b Fertilization and early embryonic development in heifers and lactating cows in summer and lactating and dry cows in winter. *Journal of Dairy Science* **85** 2803-2812.

Sartori R, Gumen A, Guenther JN, Souza AH, Caraviello DZ & Wiltbank MC 2006 Comparison of artificial insemination versus embryo transfer in lactating dairy cows. *Theriogenology* **65** 1311-1321.

Santos JEP, Thatcher WW, Pool L & Overton MW 2001 Effect of human chorionic gonadotropin on luteal function and reproductive performance of high-producing lactating Holstein dairy cows. *Journal Animal Science* **79** 2881-2894.

Santos JEP, Juchem SO, Cerri RLA, Galvão KN, Chebel RC, Thatcher WW, Dei CS & Bilby CR 2004 Effect of bST and reproductive management on reproductive performance of Holstein dairy cows. *Journal of Dairy Science* **87** 868-881.

Sellars CB, Dalton JC, Manzo R, Day J & Ahmadzadeh A 2006 Time and incidence of ovulation and conception rates after incorporating estradiol cypionate into a timed artificial insemination protocol. *Journal of Dairy Science* **89** 620–626.

Sheldon MI, Lewis GS, LeBlanc S & Gilbert RO 2006 Defining postpartum uterine disease in cattle. *Theriogenology* **65** 1516-1530.

Souza AH, Gümen A, Silva EPB, Cunha AP, Guenther JN, Peto CM, Caraviello DZ & Wiltbank MC 2005 Effect of estradiol-17ß supplementation before the last GnRH of the Ovsynch protocol in high producing dairy cows. *Journal of Dairy Science* **88** (Suppl. 1) 170.

Starbuck MJ, Inskeep EK & Dailey RA 2006 Effect of a single growth hormone (rbST) treatment at breeding on conception rates and pregnancy retention in dairy and beef cattle. *Animal Reproduction Science* **93** 349–359.

Thatcher WW & Chenault JR 1976 Reproductive physiological responses of cattle to exogenous prostaglandin $PGF_{2\alpha}$. *Journal of Dairy Science* **59** 1366-1375.

Thatcher WW, Bartolome JA, Sozzi A, Silvestre F & Santos JEP 2004 Manipulation of ovarian function for the reproductive management of dairy cows. *Veterinary Research Communications* **28** 111-119.

Thatcher WW, Bilby TR, Bartolome JA, Silvestre F, Staples CR & Santos JEP 2006 Strategies for improving fertility in the modern dairy cow. *Theriogenology* **65** 30-44.

Vasconcelos JLM, Silcox RW, Pursley JR & Wiltbank MC 1999 Synchronization rate, size of the ovulatory follicle, and pregnancy rate after synchronization of ovulation beginning on different days of the estrous cycle in lactating dairy cows. *Theriogenology* **52** 1067-1078.

Novel concepts about normal sexual differentiation of reproductive neuroendocrine function and the developmental origins of female reproductive dysfunction: the sheep model

DL Foster[1], LM Jackson[1] and V Padmanabhan[1,2]

Reproductive Sciences Program, [1]Department of Obstetrics and Gynecology, [2]Department of Pediatrics, University of Michigan, Ann Arbor, Michigan USA, 48109

The neuroendocrine regulation of GnRH secretion plays a central role in timing gamete release in both sexes. This regulation is more complex in the female because the discontinuous release of ova is more complex than the continuous release of spermatozoa. This review provides an evolving understanding of the sex differences in reproductive neuroendocrine controls and how these differences arise. The rules for sexual differentiation of steroid feedback control of GnRH secretion conceptually parallel the well-established principles that underlie the sexual differentiation of the internal and external genitalia. In the context of the neuroendocrine regulation of the ovarian cycle, and using the sheep as a model, four steroid feedback controls for GnRH secretion are inherent (default). They require no ovarian developmental input to function appropriately during adulthood. Two steroid feedback controls regulate the preovulatory surge mode of GnRH secretion, and two regulate the pulsatile mode. If the individual is a male, three steroid feedback controls of GnRH secretion become unnecessary or irrelevant, and these are abolished or become functionally inoperative through programmed reductions in hypothalamic sensitivity. This central programming occurs through exposure of presynaptic GnRH neurons in the developing male brain to the androgenic and estrogenic actions of testicular steroids. In precocial species such as ruminants, this programming begins well before birth. Understanding how GnRH secretion normally becomes sexually differentiated is of practical importance to determining how inappropriate hormonal environments during development can variously malprogram the neuroendocrine system to produce a variety of reproductive dysfunctions relating to patterning of gonadotropin secretion.

Introduction

Why should male and female neuroendocrine controls be different?

As issues of sex differences assume an expanding role in society, science is looked to for

Corresponding author: E-mail: dlfoster@med.umich.edu

explanations regarding their basis. Of recent concern is the issue of ever-increasing synthetic compounds in the environment that have the potential to produce sex-specific fertility problems. Differences in neuroendocrine controls of male and female mammals contribute to sex differences in susceptibility to reproductive disruption by exogenous substances. In this review, we note that indeed, there are several important differences between the sexes in the control of the secretion of gonadotropin releasing hormone (GnRH), the hypophysiotropic hormone central to reproduction. These differences are organized during development and become manifest to affect the timing of reproductive activity beginning as early as puberty. However, signs of sex differences in brain function are evident well before this time as evidenced by differences in patterns of hormone secretion and behavior during early postnatal life. Sex-specific reproductive neuroendocrine controls between males and females are linked to differences in the pattern of gamete production and release. Males have a relatively simple system in which sperm are made continuously in great numbers. Females, however, are born with a fixed number of eggs and these are parceled out over a prolonged period. Given a sufficiently long life, the ovary will run out of eggs, as is the case for the modern human. To return to the question posed at the beginning, the suggestion emerges that simple reproductive systems require simple controls and complex systems require more complicated ones. Added to these differences in how sperm and eggs are made and released are differences in sexual and social behaviors. How an individual responds towards another with respect to aggressive actions, territoriality, rank et cetera is also sex-specific.

In this review, we provide emerging concepts that account for the marked differences in reproductive controls in male and female sheep. Drawing upon an area of research that has a rich history from rodent studies, we hope to show the utility of using the sheep as a model that can contribute both conceptual and practical information to understand animal and human reproductive processes. As intimated above and expanded below, an understanding of male-female differences will be increasingly important to understand the various impacts of synthetic endocrine imposters on reproductive function. These compounds have the potential to misdirect developmental processes, primarily in the female to alter the complex timing mechanisms underlying the ovulatory cycle.

Hypothesis for sexual differentiation of GnRH feedback controls

The hypothesis in perspective - sexual differentiation of genitalia

At the outset, it is instructive to review highlights of the classical hypotheses for other aspects of sexual differentiation. Rather than invent a new set of principles for sexual differentiation of the reproductive neuroendocrine system, we are guided initially by those well-known tenets for sexual differentiation of internal and external genitalia, and we are interested in assessing how they might apply to the neuroendocrine system. It is important to emphasize that ours is yet a working hypothesis whose purpose is to stimulate additional work that will either add to the evolving story or modify it with new information (Fig. 1).

The first principle learned from the sexual differentiation of internal genitalia is that one sex serves as the default phenotype and the other phenotype requires both regression of the default structures and further development of others. Regardless of genotypic sex, female internal genitalia develop in the absence of gonadal hormone action while the masculine phenotype requires the presence of testicular secretions. The testes produce a systemic hormone to stimulate the development of male internal genitalia (masculinization) and a local substance to repress development of female internal genitalia (defeminization). A second principle is that although the systemic hormone (testosterone) is secreted to differentiate the bipotential external genitalia, it must be converted locally to a more potent androgen, dihydrotestosterone (DHT), to produce the normal male phenotype.

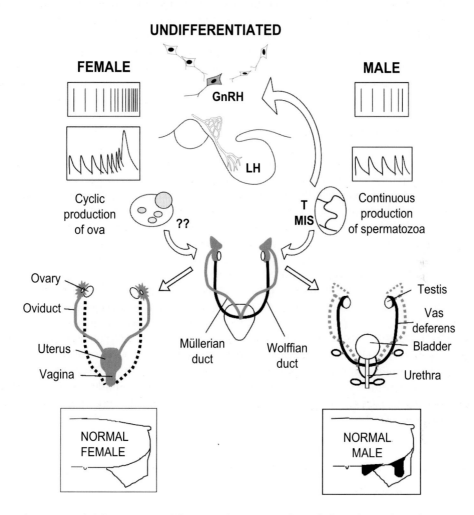

Fig. 1. Sexual differentiation of the reproductive axis (hypothalamo-hypophyseal system (top); gonads and internal genitalia (middle); external genitalia (bottom)). At all three levels of the axis, the female is the default condition. Due to the organizational effects of testosterone (T) secreted by the developing testis, feedback regulation of GnRH and LH secretion is defeminized, the undifferentiated Wolffian duct develops into male internal genitalia and the external genitalia are masculinized. A second testicular secretion, Müllerian inhibiting substance (MIS), causes regression of the Müllerian ducts which, in the female, develop into the internal genitalia. In the absence of testicular androgens the Wolffian ducts regress.

Hypothesis for sexual differentiation of reproductive neuroendocrine functions

Adapting the above well-known principles for somatic sexual structures to our system leads to the consideration that one sex has the default neuroendocrine phenotype. Logically this would be the female because of the complexity of the timing of the ovulatory cycle in comparison to spermatogenesis. Four steroid feedback mechanisms are necessary for the neuroendocrine regulation of the ovarian cycle, but only one is used for regulation in the male. Either more feedbacks are added in the female (addition hypothesis) or alternatively, unnecessary ones are eliminated in the male (subtraction hypothesis). We believe that the latter mechanism is used. We propose that all four feedback controls are inherent in the female, and they require no further developmental input to function appropriately after birth. Two feedbacks regulate the surge mode of GnRH secretion, and two feedbacks regulate the pulsatile or tonic mode. However, three of these feedback controls are not required if the individual is a male, and only a single negative feedback system is required to regulate his much simpler, noncyclic reproductive system.

In addition to accounting for the sex differences in regulation of adult reproductive function, we must also explain the sex difference in the timing of puberty that is evident in most mammals. Puberty is controlled by the single feedback control that persists in both sexes, but males and females exhibit quantitative differences in sensitivity to this feedback. We propose that for the timing of puberty, the female system is also the default, and that *when* the transition to adulthood occurs in the male depends on the rate of maturation and species-specific reproductive strategies including seasonality of reproduction and social systems. For example, in slowly maturing, non-seasonal primate species (humans) puberty generally occurs earlier in the female. In more rapidly maturing, seasonally-constrained species that reach puberty within the same year as their birth, the transition to adulthood in the male may begin well ahead of that in the female. This second strategy, used by the sheep, maximizes the likelihood that a young male will reach sexual maturity and successfully compete for females during his first breeding season. Thus, according to our working hypothesis, changing any of the feedback controls of GnRH from the female default to the male type is, by definition, a process of defeminization. Exposure to testosterone secreted by the testes and its androgenic and estrogenic metabolites alters or eliminates female GnRH control systems. As is the case for differentiation of genitalia, the organizational actions of testosterone are exerted during critical periods of development. However, multiple critical periods may exist and may even be interdependent such that the presence of organizing steroids during the early critical period(s) increases susceptibility to further organizing actions of steroids in later development.

Historical perspective

Much of our present understanding of sexual differentiation of the control of gonadotropin secretion stems from investigations in the rat. However, important observations in the sheep over the past 30 years have led to the consideration that sexual differentiation of reproductive neuroendocrine function has broad application with potential practical usefulness.

Rodent studies

In the early 1900s, Sand (as described in Pfieffer 1936) was the first to recognize that in the guinea pig, ovarian grafts functioned differently in females and males. A few years later, Pfieffer began to focus on the pituitary and the importance of the developing gonad in regulating later sex-specific differences in hypophyseal function. From a landmark study comprising 495 rats under 18 experimental conditions, he concluded that... "The hypophysis in the rat at birth is bipotential and capable of being differentiated as either male or female, depending upon whether an ovary or testis is present." With an expanded understanding of the importance of the brain's control over pituitary function, later investigators would substitute the "hypothalamus" for "hypophysis". Moreover, the ovaries would be found to be unnecessary in determining the sex of the control mechanism, this being the exclusive role of testicular testosterone. These pioneering studies inspired the elegant work in the rat by Gorski (1979, for review) to develop the concept of sexual differentiation of reproductive neuroendocrine function. He and his colleagues focused on the LH surge system for largely practical reasons, namely that it was present in the female, but absent in the male, and its assessment was relatively easy because of the massive release of LH. The eventual understanding that emerged from this large body of work is that the surge system is inherent, and becomes operational unless given information *not* to develop. The active organizing steroid that renders the surge system insensitive is an estrogen that is formed in the brain from the testosterone produced by the testes. This programming only occurs during a limited time of development, a critical period, and involves changes in a "surge center" in the anterior hypothalamus. These major findings have provided the conceptual underpinnings of all other work in this area, including the work in the sheep as discussed below. The location of controls for the GnRH surge mechanism and how they are rendered insensitive to the positive feedback action of estrogen remain under intensive study at the anatomical, electrophysiological, cellular and molecular levels in rodent models (Sullivan & Moenter 2004; Foecking et al. 2005).

Sheep studies

The earliest report of the influence of prenatal testosterone on neuroendocrine function was with a small number of sheep of various sexes and physiological states: two anestrous ewes, two ovariectomized ewes, two intact rams, two castrated rams and three freemartins (ovotestes). In addition to examining differences among the above phenotypes, the study included an experiment to probe for a critical period in females that had been treated before birth with a testosterone pellet (1g) beginning at various prenatal ages during the 147-day gestation (Days 20 n = 2, 40 n = 2, 60 n = 5, 80 n = 1) (Short 1974). From his studies, Short concluded that ... "the essential difference between the male and female hypothalamus of the sheep lies in its inability to initiate a LH discharge from the pituitary in response to oestrogen feedback. The presence of testosterone up to the 60th day of gestation is able to masculinize the hypothalamus in this respect, whereas testosterone given after day 60 is without effect." Although the number of animals in each treatment group was small, the results were variable, and the conclusion was too restrictive with respect to the timing, the observations were nevertheless very important. The findings raised the possibility that prenatal testosterone might abolish the GnRH surge and suggested the presence of an early critical period for this differentiation. Importantly, this initial report by Short led to additional detailed studies of the sexual differentiation of the GnRH surge mechanism and, of equal importance, of the differentiation of the inhibitory feedback controls of GnRH secretion that control the transitions between states of hypo-, eu- and hyper-gonadotropism, and hence, gonadal activity.

Sex differences in steroid feedback controls of GnRH

Studies of stimulatory feedback versus inhibitory feedback

Virtually all of the above work in the rat and sheep focused on the stimulatory feedback action of estrogen because the preovulatory gonadotropin surge is so clearly differentiated in those species. Moreover, systematic studies of the negative feedback control system have been few in small animals such as rodents for largely practical reasons relating to size. Evaluation of the pulsatile mode of LH secretion requires precise measurement of the frequency and amplitude of LH release, and this is difficult because of the limited blood supply in a small animal. This is made more problematic for studies of the developing rodent. Although rapid blood sampling has been conducted in the rat by several investigators in other areas of reproductive research, none have used this approach in systematic studies of sexual differentiation of the neuroendocrine system. Another factor limiting interest in negative feedback controls in the rodent model is that the corpus luteum is not functional during a luteal phase when mating does not occur, unlike several other species including the primate and sheep as noted below. In these latter species, the inhibitory feedback action of progesterone serves as a major regulator timing the ovarian cycle. The current general lack of understanding of the sex differences in the control of negative feedback is unfortunate as changes in sensitivity to estrogen feedback inhibition of GnRH secretion also time many long-term aspects of reproductive activities that are different between males and females; including puberty, transitions between non-breeding and breeding seasons, reproductive response to nutrition and inhibition of reproduction by stress. In contrast to these multiple timing responsibilities of negative feedback controls, the GnRH surge system is highly specialized. It is reflexively activated to cause ovulation in response to the presence of high circulating concentrations of estradiol, the timing of which results from a quantitative reduction in amount of negative feedback control of GnRH secretion. Thus, whereas inhibitory feedback systems are responsible for long and short-term timing, the stimulatory feedback is responsible for immediate action.

Critical periods for sexual differentiation

The timing of the critical period deserves special mention. In the rat, this has been determined to occur during the first few days before and after birth. Outside the perinatal period the organizing actions of steroids are ineffective. Birth occurs somewhat arbitrarily during development in mammals, and species are born in differing stages of somatic and neurobiological maturation. This has given rise to classifying them as altricial, those born helpless and highly dependent on parental care, and precocial, those born at an advanced stage of development and relatively independent. With this in mind, it would be predicted that the critical period for sexual differentiation of GnRH secretion analogous to that which occurs perinatally for the altricial rat should be well before birth in the precocial sheep. Perhaps the term "critical period" will need to be redefined as it may be too restrictive to further concepts about programming several GnRH feedback controls. Within a broad critical period, there are likely multiple critical periods, each defined for a given programming function that may or may not overlap with one another. Conversely, there may even be multiple, sequential critical periods programming a given function. Finally, other complex developmental relationships may emerge; for example, the presence of steroids in an early critical period may organize the development of a later critical period.

Differentiation of multiple feedback controls

According to our working hypothesis, at least four steroid feedback controls of GnRH secretion are sexually differentiated (Fig. 2). These feedback mechanisms are inherent in the female and are either abolished in the male or their sensitivities are reduced markedly. The first and most dominant feedback control is active in both sexes (*Control 1: Estrogen negative feedback*). In both females and males, high sensitivity to estrogen feedback inhibition of GnRH secretion during prepubertal development maintains hypogonadotropism to keep the gonad relatively quiescent. This sensitivity decreases during puberty to increase gonadal activity. However,

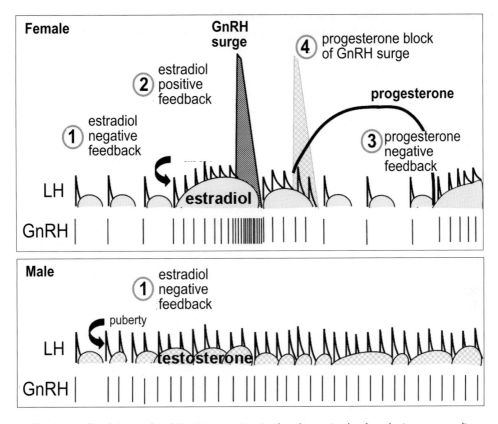

Fig. 2. Feedback controls of GnRH secretion in the sheep. In the female (upper panel), estradiol (feedback 1) and progesterone (feedback 3) have inhibitory actions on pulsatile GnRH secretion; the GnRH surge system is regulated by the stimulatory feedback action of estradiol (feedback 2) and the progesterone blockade of a GnRH surge (feedback 4). In the male (lower panel), only the estradiol inhibitory feedback mechanism is required to maintain spermatogenesis. (Redrawn from Foster *et al.* 2006).

the timing of when the sensitivity to estrogen negative feedback decreases is different between the two sexes and this accounts for the sex difference in timing of puberty in males and females of many species. Estrogen negative feedback is the only relevant feedback control of GnRH in the male. In the female beginning during and continuing after puberty, a complex interplay between the ovaries, hypothalamus, and pituitary is required to initiate and sustain repeated and regular ovarian cycles. During the pubertal transition, when the sensitivity to estradiol negative feedback is being reduced, high frequency GnRH secretion results in high frequency LH secretion which, in turn, drives the preovulatory follicle to increase its production of estradiol to high levels. The increasing concentrations of circulating estradiol from the preovulatory follicle exert a stimulatory feedback to produce a massive discharge of GnRH (*Control 2: Estrogen positive feedback*) to cause ovulation and formation of a corpus luteum. With a decrease in sensitivity to estrogen negative feedback at puberty, GnRH pulse frequency remains high, and a new negative feedback control emerges to regulate pulsatile GnRH secretion (*Control 3: Progesterone negative feedback*). After first ovulation, during the luteal phase, progesterone is secreted in high amounts for the first time to inhibit pulsatile GnRH secretion. As a consequence, GnRH pulse frequencies are too low to stimulate LH secretion to levels that promote further development of follicles to the preovulatory stage and produce high circulating concentrations of estrogen. Withdrawal of progesterone negative feedback, as the corpus luteum regresses, then permits the high frequency GnRH secretion to initiate the next follicular phase. A second action of progesterone serves as a back-up mechanism to prevent a preovulatory gonadotropin surge during the early luteal phase when concentrations of progesterone are low (*Control 4: Progesterone blockade of the GnRH surge*). Immediately after ovulation, when negative feedback is entirely absent, high frequency LH pulses persist from the follicular phase and can begin to stimulate the secretion of estradiol from developing follicles in an early luteal phase follicular wave. As the corpus luteum simultaneously forms, the resultant low circulating concentrations of progesterone prevent the GnRH surge mechanism from responding to any possible estrogen positive feedback.

An essential part of our hypothesis is that the four feedback controls of GnRH secretion are differentially organized by testosterone and its androgenic and estrogenic metabolites. We propose that the inhibitory feedback controls of GnRH secretion are developmentally organized by the androgenic actions; the stimulatory feedback controls are organized by estrogenic actions. More specifically, the negative feedback actions of both estradiol and progesterone on pulsatile GnRH secretion are developmentally programmed by androgens. By contrast, the GnRH surge mechanism consists of two parts, one of which is the stimulatory feedback action of estradiol with the other being the blockade of this action by progesterone, yet both components are developmentally programmed by estrogens. This hypothesis regarding the role of progesterone in the GnRH surge being distinct from its negative feedback effects is guided by the concept that the tonic and surge modes of GnRH release have distinctly different controls. However, it is possible that there is just one inhibitory or blocking feedback effect of progesterone, and that this is developmentally programmed by a single steroid, either an androgen or an estrogen.

Developmental programming of sex differences in feedback controls

Models to study sex differences in steroid feedback controls of GnRH

Two models have been used to study the sexual differentiation of feedback controls of GnRH secretion. The first is a neuroendocrine model in which the ovaries are removed neonatally and an estradiol implant is inserted subcutaneously to maintain chronic low concentrations of estradiol (Ovx + E). The advantage of this model is that neuroendocrine feedback controls can be studied in the presence of constant steroid conditions. In this review, we will focus on this neuroendocrine model, the one with which we have the most experience, and then provide necessary commentary about the results being obtained in the second model which is ovarian-intact. Both models are treated the same before birth, but after birth their treatments differ with respect to steroid exposure. Sexual differentiation of the feedback mechanisms will be discussed in the order in which they become relevant to an individual, beginning with the estradiol negative feedback mechanism timing the pubertal increase in GnRH (Fig. 2).

Estrogen inhibitory feedback

A sex difference in feedback inhibition of GnRH secretion is readily apparent in the developing sheep (Fig. 3). In the neuroendocrine model, the developing female remains hypersensitive to low, constant levels of circulating estradiol (implant) for several months, and LH secretion remains suppressed. By contrast, the developing male is initially hypersensitive to estradiol negative feedback but within a few weeks after birth this is no longer the case and LH rises progressively. This increase in circulating LH reflects an increase in GnRH pulse frequency (see Foster & Jackson 2006). That this decrease in estradiol negative feedback is causal to activation of the gonads is evident by the close associations of the pubertal LH rise with the increase in testicular activity or the onset of ovarian cycles. Thus, while there are obvious differences between the ovary and testis that could contribute to their differential activation during puberty, when the increase in their tropic stimulus occurs is paramount.

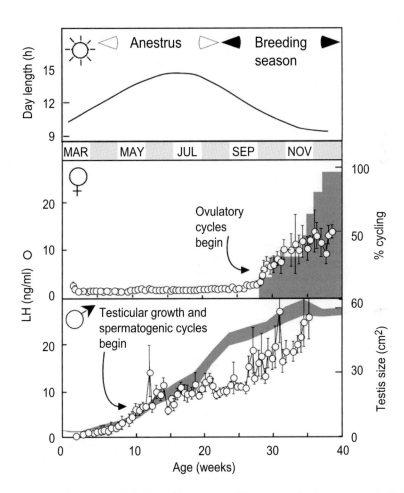

Fig. 3. Neuroendocrine model. Sex differences in the negative feedback control of GnRH secretion by estrogen in developing sheep. LH concentrations in semiweekly blood samples collected from neonatally gonadectomized and estradiol-replaced (3-5 pg/ml by sc implant) lambs. Note that the pubertal increase in LH secretion occurs earlier in the male compared to the female, and puberty in the female does not occur until the breeding season. The rise in LH concentrations coincides with the onset of progestagenic cycles in intact females and increased testicular volume in intact males (shaded areas). (Redrawn from Foster *et al.* 2006).

What times the decrease in sensitivity of the mechanisms controlling tonic/pulsatile GnRH secretion to the inhibitory feedback action of estrogen is complex (Foster & Jackson 2006). The developing individual must have achieved a minimum stage of development and a sufficiently positive energy balance and size to begin reproductive activity (Sisk & Foster 2004). Moreover, reproduction is seasonal in the sheep, so one would predict that these indices of maturity would only be expressed during the breeding season. In the adult sheep, the major environmental cue timing seasonal reproduction is photoperiod, and reproductive activity begins during decreasing day lengths (Karsch et al. 1984). In the female lamb, there is a strict photoperiod requirement for the synchronous timing of puberty, and the pubertal increase in GnRH begins during decreasing daylengths (Fig. 3 top). Reversing the photoperiod delays puberty well beyond the normal age (25-30 weeks) and prevents it from occurring synchronously among females of the same cohort (Fig. 4). Puberty cannot be delayed indefinitely and occurs both at older ages and asynchronously (Foster et al. 1986).

Interestingly, the earlier pubertal increase in GnRH begins in male lambs during increasing photoperiod which raises the possibility that either the developing male has a different photoperiod requirement or is not photoperiodic with respect to the timing of puberty. Other neuroendocrine responses are influenced by photoperiod in the developing male, including the patterns of melatonin and prolactin secretion (Foster et al. 1989), and photoresponsiveness begins before birth (Ebling et al. 1989). However, this does not seem to be the case for reproductive neuroendocrine system. Unlike the developing female, the male can initiate its pubertal increase in GnRH at the same age in increasing and decreasing photoperiods (Fig 4).

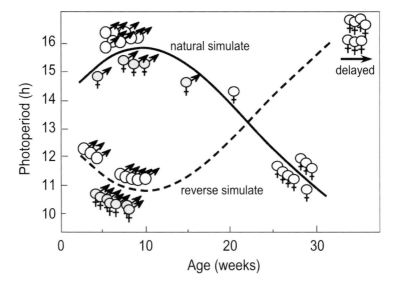

Fig. 4. Neuroendocrine model. Sex differences in the photoneuroendocrine control of the timing of the pubertal LH rise in the gonadectomized, chronically estrogen-treated sheep. The time of the increase is designated by the placement of the symbol (male, female, prenatally testosterone-treated female) in relation to age and photoperiod. Note that females are highly responsive to photoperiod and begin puberty under decreasing day lengths, but males and prenatally testosterone-treated females are not photosensitive, and they begin their pubertal rises in LH at a young age under both decreasing and increasing day lengths. (Redrawn from Foster & Jackson 2006).

The finding that the photoperiodic control of the mechanism timing the pubertal rise in GnRH is sexually differentiated leads to the hypothesis this is due to prenatal exposure to testosterone from the developing testes. This hypothesis is supported by the results in female lambs whose mothers were injected with testosterone from Days 30-90 of gestation (Wood et al. 1991). Female lambs exhibited masculinized internal and external genitalia, and in the neuroendocrine model (Ovx + E), they exhibited resistance to estradiol negative feedback beginning at an early age (Fig. 5), much like the developing male. This allows the pubertal rise to occur under an inhibitory photoperiod (long days) when untreated females typically remain hypersensitive to negative feedback and hypogonadotropic (Fig. 5). Thus, we conclude that the default control for photoperiodic regulation of the timing of puberty is defeminized (removal of a female trait) by testosterone in the developing male lamb. It is of interest that by the second year, the reproductive system of the young male also becomes regulated by photoperiodic cues, much like that for the female to result in seasonal changes in sensitivity to negative feedback of GnRH secretion. The reason for the apparent absence of photoperiodic control of reproduction in the male during the first year is speculative, but it could be necessary to increase the GnRH/LH drive to the testes early to develop fully the steroid-dependent behaviors necessary to begin competition with older males for females.

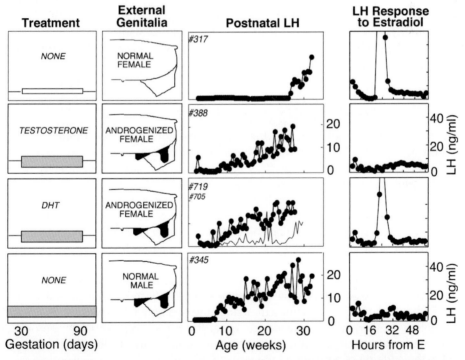

Fig. 5. Neuroendocrine model. Prenatal sexual differentiation of the inhibitory and stimulatory feedback actions of estrogen in the gonadectomized chronically estrogen treated sheep. Left column: Prenatal treatment. Second column: Schematics of postnatal external genitalia. Third column: Circulating concentrations of LH in the presence of chronic estradiol (3-5 pg/ml). Right column: Circulating concentrations of LH after administration of a late follicular phase increment in estradiol (10-15 pg/ml) designed to induce a GnRH surge. Note the advancement of the pubertal LH rise in the female with prenatal exposure to testosterone or DHT and the lack of the LH surge in testosterone-treated females (exposed to both androgen and estrogen actions) but not in females exposed to DHT (androgen only). (Redrawn from Foster & Jackson 2006).

The timing, duration, amount and type of steroid exposure all contribute to the degree of defeminization of control of pulsatile GnRH secretion. The *time of testosterone* exposure is important. A broad critical period has been identified early in development; exposure of the unborn female to testosterone between Days 30 and 90 of gestation (147 days term) advances the timing of the pubertal GnRH rise, whereas this same treatment later is without effect (Herbosa & Foster 1996). Lesser exposures (for example, Days 30-60 or Days 60-90) to testosterone are effective, but the timing of the pubertal LH rise is advanced less (Wood *et al.* 1995). The *amount of testosterone* exposure is important with the greater amounts producing the earliest decreases in response to steroid feedback, and hence, the earliest increase in LH secretion (Kosut *et al.* 1997). The *type of steroid* is important. Testosterone can be metabolized by the brain to a more potent androgen or to an estrogen, and the type of steroid action cannot be assessed by administration of only testosterone. Administration of dihydrotestosterone (DHT), the androgenic metabolite that cannot be converted to an estrogen, has been used to begin to differentiate these actions of testosterone. DHT advanced the time of the pubertal rise in LH suggesting that androgenic action is involved in this programming (Masek *et al.* 1999). However, the results were not identical to those for testosterone raising the possibility that estrogens may have some responsibility in organizing the timing of the pubertal increase in GnRH secretion. The role of estrogens is difficult to evaluate for the technical reason that administration of these steroids can act on the maternal uterus to compromise the pregnancy. Thus, other approaches must be developed to test hypotheses relating to the relative contribution of prenatal estrogenic action, if any, in the early androgen programming of the inhibitory feedback control of estrogen on pulsatile GnRH secretion.

Estrogen stimulatory feedback

Early work in the sheep indicated that the mechanism governing the stimulatory feedback action of estrogen is sexually differentiated because high amounts of estradiol were incapable of inducing a LH surge in males (Karsch & Foster 1975). A later study, directly measuring GnRH in the pituitary portal vasculature, offered the first direct proof that the absence of the LH surge was due to the failure of the male to release the massive amount of GnRH as in the female in response to follicular phase levels of estradiol (Fig. 6). This study also revealed that exposure to testosterone *in utero* renders the GnRH surge mechanism inoperative, as the large estrogen-induced surge of GnRH did not occur in female lambs treated prenatally with testosterone (Herbosa *et al.* 1996). While these observations provided the necessary direct evidence for the earlier inference of Short (1974) about prenatal testosterone and the surge mechanism, there is an important conceptual difference emerging based on the current working hypothesis. Rather than testosterone "masculinizing" the hypothalamus, we now may use different terminology in view of the consideration that the default reproductive neuroendocrine control is the female system. To distinguish removal of an innate female characteristic as opposed to the addition of a male trait the term "defeminized" is more appropriate.

The popular term "androgenization" of the LH surge mechanism is also inappropriate as this implies that the androgenic action of testosterone organizes the GnRH surge mechanism to reduce its response to the stimulatory feedback action of estrogen. This does not appear to be the case because when DHT, the potent androgenic metabolite of testosterone, was administered to the developing female lamb from Days 30-90 of gestation, the GnRH surge mechanism remained highly responsive to estrogen positive feedback when tested after birth (Masek *et al.* 1999) (Fig. 5). These same amounts of DHT were capable of advancing the pubertal increase in GnRH secretion attesting to the efficacy of the doses of androgen used to produce

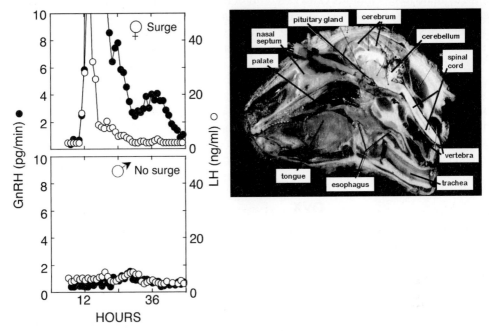

Fig. 6. Sex differences in patterns of GnRH and LH release after exposure (Hour 0) to late follicular phase concentrations of estradiol (10-15 pg/ml by implant). GnRH concentrations were assayed in blood collected from the pituitary portal vasculature. The sagittal section of the sheep's head presented in the right panel illustrates the placement of the collection device for pituitary portal sampling. In the female (top left panel), a GnRH surge occurs and is accompanied by a LH surge. The GnRH and LH surges are absent in the male (bottom left panel). (Data redrawn from Herbosa *et al*. 1996).

organizational actions on feedback control, albeit in the inhibitory feedback system (Fig. 5). Much the same as with the organizational actions of prenatal androgens on programming the sensitivity to negative feedback, the amount and timing of prenatal estrogen exposure are important in programming reduced sensitivity to positive feedback. Exposure during the early part of gestation is critical for complete defeminization of the GnRH surge system as are higher amounts of testosterone. Lesser amounts of testosterone and shorter exposure periods result in estrogen-induced LH surges that are of lower amplitude or that are delayed (Kosut *et al*. 1997; Wood *et al*. 1995). Interestingly, in most studies of the rat model (Gorski 1979), the LH surge system is rendered unresponsive by the organizational actions of estrogen. More recently, there is evidence for an androgenic action as well in the rat leading to the suggestion that either both androgens and estrogens are normally needed to inactivate the GnRH surge system or that the two actions are redundant (Foecking *et al*. 2005). To date there is no evidence for this in the sheep.

Progesterone feedback controls

Progesterone feedback regulation of GnRH secretion assumes importance only after puberty and only in the female. Its major role is to serve to inhibit feedback control of pulsatile GnRH secretion after the sensitivity to estradiol negative feedback becomes reduced during puberty, and estradiol no longer inhibits GnRH secretion. Progesterone from the newly formed corpus luteum reduces GnRH pulse frequency during the luteal phase. If no conception occurs, progesterone declines and allows GnRH pulse frequency to increase and begin the next follicular phase. In the male, this feedback hormone is not present in any appreciable amounts, and has no known physiological role in the regulation of GnRH secretion. Thus, progesterone action in the male has been generally ignored. One study has determined that the male sheep is relatively insensitive to progesterone negative feedback, and prenatal testosterone exposure of the female lamb reduces its postnatal sensitivity to progesterone feedback inhibition of LH secretion (Fig. 7) (Robinson et al. 1999). Whereas it appears that estrogen negative feedback is largely organized by the prenatal androgenic action of testosterone, whether this second inhibitory feedback control system is similarly sexually differentiated remains to be determined.

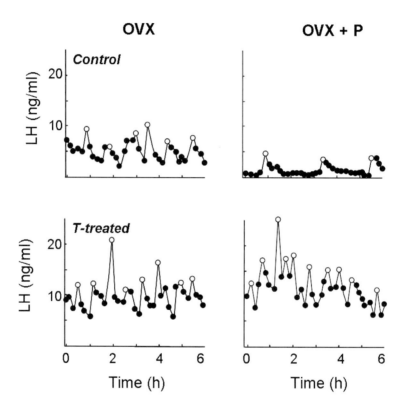

Fig. 7. Neuroendocrine model. Sex differences in the inhibitory feedback action of progesterone in the ovariectomized sheep. Whereas exogenous progesterone readily inhibits pulsatile LH secretion in control females (top), it is virtually ineffective in females exposed prenatally to testosterone (bottom), much like in the male (data not shown). (Data redrawn from Robinson et al. 1999).

Progesterone has a second important action, namely it blocks the stimulatory feedback action of estrogen on the GnRH surge mechanism during the luteal phase (Fig. 8) (Kasa-Vubu et al. 1992). We do not know if this action is also sexually differentiated by estrogen as part of the surge system. The alternative possibility, that it is differentiated by androgen, would not be expected as we propose that androgen actions primarily organize negative feedback controls.

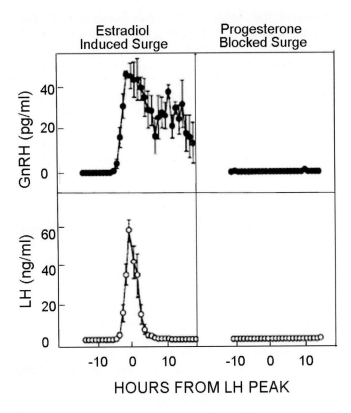

Fig. 8. Neuroendocrine model. Patterns of GnRH and LH release after exposure to high follicular phase concentrations of estradiol (10-15 pg/ml by implant). Left panels: Estradiol only. Right panels: Chronic treatment with progesterone beginning seven days before the estradiol treatment prevented both the GnRH and LH surges. (Data redrawn from Kasa-Vubu *et al.* 1992).

Postnatal programming - clues from other models

The foregoing studies in the neuroendocrine model, in which the postnatal steroid milieu remains constant (Ovx+E), provide the basis for our working hypothesis for sexual differentiation of the feedback control of GnRH. To test the working hypothesis in more physiologically relevant conditions requires a complete reproductive system, one in which the ovaries remain *in situ* and the female is not exposed chronically to exogenous steroid. A second important reason for returning to the intact model is to study the effects of prenatal programming on ovarian function. Interestingly, studies in the intact model, although identical with respect to prenatal treatments, often produce results that appear dissimilar. A conceptually interesting explanation is that such differences point to yet unrecognized facets of sexual differentiation. One such possibility that arises is that *postnatal* sexual differentiation of neuroendocrine function may be occurring, a surprising option for such a precocial species.

The differences between the two models become apparent when one predicts the expected ovarian activity in the intact model based on the results from those of the neuroendocrine model. Recall that in the neuroendocrine model prenatal testosterone programs both inhibitory and stimulatory feedback controls of GnRH to result in an early pubertal GnRH rise and a failed GnRH surge

system. In the intact model, the identical prenatal exposure should have two main consequences: 1) early hypergonadotropism (androgenic action to prematurely decrease sensitivity to negative feedback) resulting in a polyfollicular ovary; and 2) puberty does not occur (GnRH surge system rendered inoperative therefore no ovulations). Certain results for prenatally treated intact females were as predicted; females became less sensitive to estrogen negative feedback and were hypergonadotropic (Sarma *et al.* 2005). As a consequence, the ovary became polyfollicular at an early age (West *et al.* 2001) (Fig. 9). However, the most unexpected finding was that puberty did occur in most prenatally testosterone-treated lambs as evidence by repeated increases in circulating progesterone (Sharma *et al.* 2002; Birch *et al.* 2003) (Fig. 10). Although some progestagenic cycles were abnormal in this model, the fact that they even occurred was not predicted because the GnRH surge system had been demonstrated to be nonfunctional in several similar studies using the neuroendocrine model. When the positive feedback action of estradiol was tested in the ovarian-intact prenatally testosterone treated female, the results were mixed. In one study, a LH surge could be induced by exogenous estradiol but the amplitude was markedly reduced (Sharma *et al.* 2002). In another study, the surge mechanism was completely nonfunctional (Unsworth *et al.* 2005). Studying prenatally testosterone-treated females for a second year revealed that the ovarian cycle deteriorated and most females became anovulatory (Birch *et al.* 2003). Although not rigorously studied, the negative feedback of progesterone on GnRH secretion in the intact model seems to be compromised because LH pulse frequency during the luteal phase in prenatally testosterone exposed females tends to be greater compared to controls (Savabieasfahani *et al.* 2005).

That ovulatory cycles were possible in the prenatally testosterone treated female is not without precedent as this had been observed in the earliest studies of Short (1974) that preceded the development of the neuroendocrine model. Thus, the difference between the two models can no longer be ignored. An interesting possibility is that in the neuroendocrine model programming is more complete and hence is easier to demonstrate. Perhaps the chronic presence of small amounts of estradiol (2-4 pg/ml by implant) used to study activational (inhibitory and stimulatory) actions on GnRH also may have continuing *organizational* actions to program GnRH controls. According to this extended hypothesis, postnatal estradiol may be necessary to complete the organizing action of prenatal androgens and estrogens. Alternatively, prenatal steroids may increase the susceptibility of the hypothalamus to the postnatal actions of estradiol. In this context, in the absence of chronic estradiol in the intact model, incomplete defeminization or programming occurs and the prenatally testosterone-treated female is able to initiate ovarian cycles. The continued deterioration and cessation of reproductive cycles by the second year in the prenatally testosterone-treated females supports this notion as these females would be repeatedly exposed to estradiol during multiple follicular phases. Under normal physiological conditions, the male is first exposed to testosterone and its metabolites, including estrogens, from the beginning of the critical period through birth and continuing with increasing intensity during postnatal testicular development. By contrast, the postnatal ovary makes only small amounts of estradiol, and most likely, in a cyclic manner (Ryan *et al.* 1991). These multiple hypotheses are currently being tested to address the possibility of postnatal programming of GnRH feedback controls by estradiol (Malcolm *et al.* 2006). Finally, a completely different explanation is that the ovary produces one or more substances that modify the manifestation of the effects of prenatal steroids on postnatal GnRH secretion. These substances could be other steroids, peptides, or proteins that could modulate either ovarian or neuroendocrine function or both. However, a protective role of the ovary in controlling sexual differentiation of extraovarian tissues is currently without precedent.

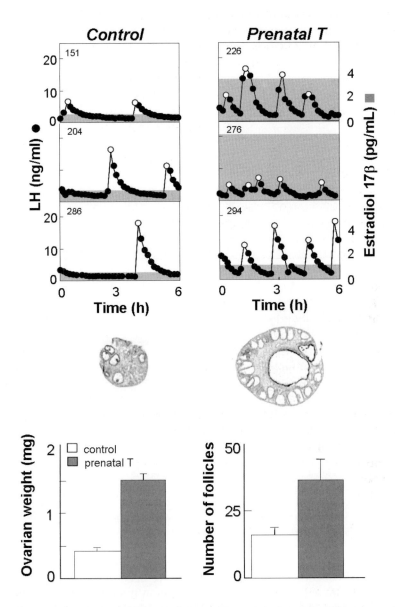

Fig. 9. Intact model. Patterns of LH release, appearance of ovaries, ovarian weight and number of preantral and antral follicles in control female lambs (left) or female lambs that had been treated prenatally with testosterone (right). Female sheep exposed to testosterone prenatally become hypergonadotropic and have large, multifollicular ovaries. (Data redrawn from Sarma *et al.* 2005 and West *et al.* 2001).

Neurobiological mechanisms underlying sexual differentiation of GnRH controls

Programming of the feedback controls of GnRH could occur either at the level of the GnRH neuron or its afferents. It is unlikely that this is within the neuron itself as GnRH neurons are generally devoid of classic estradiol-α and progesterone receptors (reviewed by Herbison 1998).

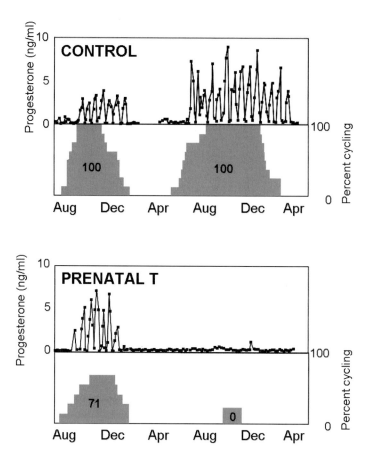

Fig. 10. Intact model. Patterns of circulating progesterone concentrations in individual female sheep during the first two breeding seasons and number of females cycling: controls (top panel) compared with female sheep prenatally treated with testosterone (bottom panel). Female sheep exposed to testosterone prenatally become anovulatory during the second year. (Data redrawn from Birch *et al.* 2003).

Furthermore, several studies found no sex differences with respect to GnRH number, morphology or location. GnRH neurons in the sheep brain have been detected as early as 43 days of gestation (Caldani 1986) and examining their anatomy in the preoptic area/hypothalamus at 85 days of gestation reveals no difference between the sexes (Wood *et al.* 1992) (Fig. 11). The total number of GnRH neurons, their relative distribution and the percentages of different morphological subtypes (unipolar, bipolar, multipolar) are the same in males and females. Moreover, because all of these characteristics were similar to those in the adult ewe (Lehman *et al.* 1986; Caldani *et al.* 1988), there appears to be no gross postnatal alteration of GnRH neurons. In a further study at the electron microscopic level, ultrastructural differences have been observed. There are more synaptic contacts on GnRH neurons of adult female sheep than male sheep, and prenatal exposure of female lambs to testosterone decreases the number of synapses in females (Kim *et al.* 1999). It is not yet known if this is due to the organizing action of androgen or estrogen on the surge or tonic mode of GnRH secretion. Finally, it is of interest that a reduction in GnRH neuronal activity in response to estradiol has been found in females exposed to testosterone *in utero* (Wood et al. 1996). This was determined by comparison of Fos co-localization with GnRH in neurons located in the preoptic area, anterior and mediobasal hypothalamus in normal and androgenized females during a surge-inducing dose of estradiol.

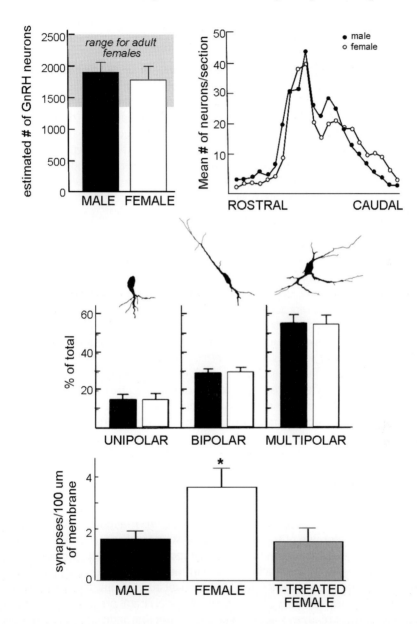

Fig. 11. Neuroendocrine model. Number of GnRH neurons in fetal male and female sheep (top left panel); the shaded area is the range for GnRH neuron counts in adults. Rostral to caudal distribution of GnRH neurons in male and female lambs (top right panel). Types of GnRH neurons in male and female lambs (middle panel). Number of synapses on GnRH neurons in male and female lambs, and in prenatally testosterone-treated female lambs (bottom panel). Although the type and location of GnRH neurons are not sexually differentiated, the number of inputs (synapses) is reduced by early testosterone exposure. (Data redrawn from Wood et al. 1992 and Kim et al. 1999).

Other considerations about sexual differentiation of the reproductive neuroendocrine control of the brain

Physiological versus pathological programming

When administering sex steroids to a developing individual one must be concerned about whether the effects are physiological or pathological. If the goal is to document the pharmacologic effects of high native sex steroids or endocrine disruptors, then higher peripheral doses of compounds that would be present in the environment can be administered to the mother. However, if the goal is to test hypotheses about normal physiology including sexual differentiation, then small doses of naturally occurring sex steroids delivered to the target within the conceptus should be the ideal approach. At present, relatively high doses of steroids are necessarily administered to the mother in an attempt to adequately expose the conceptus, given that the compounds must cross the placenta. Targeting the conceptus by a more direct route is technically difficult given its friability during known critical periods that typically occur during early development. Thus, the only present alternative is to use minimally effective doses of steroid. The general problem of embryonic/fetal accessibility is compounded for estrogens. While testosterone can cross the placenta to affect developmental functions directly or through its androgenic and estrogenic metabolites, broad administration of estrogens via the mother is problematic as they increase uterine activity during a time when it must be quiescent. Other novel strategies must be used such as administration of testosterone along with an anti-androgen so that the conceptus is functionally exposed to the estradiol metabolite. Even with the use of such strategies, experimentally administered steroids that attempt to model physiological programming of reproductive neuroendocrine functions often produce unintended consequences. These exogenous steroids not only affect reproductive tissues and their controls, but they also alter non-reproductive systems; for example, prenatal testosterone can produce metabolic alterations, such as insulin resistance and growth retardation (reviewed by Padmanabhan *et al.* 2006). Interestingly, with high doses of testosterone males show many of these non-reproductive effects suggesting that pathologies can be induced with steroid exposure that is above normal. Thus, some healthy degree of caution must be exercised in interpreting present physiological studies that broadly expose the fetus to relatively large amounts of sex steroids. Nevertheless, we believe that with our present approaches, some physiologic principals are emerging.

Prenatal programming versus sexual differentiation

Whereas most developmental programs use no external information for their execution, some do. Both prenatal programming of some metabolic systems and sexual differentiation use such external information. Arguments can be made that fetal programming and sexual differentiation are different processes conceptually with the former having the potential for some type of adaptive role and the latter being more resistant to novel adaptations. However, while we can provide a historical distinction between concepts about prenatal programming and sexual differentiation, one quickly realizes that the working distinction fades the more we learn. We now find that we must use encompassing phrases such as "prenatal programming of sex differences"; in doing so, we recognize the futility of such distinctions and relegate them to semantics.

Historically, the areas of prenatal programming and sexual differentiation have distinctly different conceptual origins. Prenatal programming, a more recent concept, considers that prevailing conditions outside the developing individual can alter its development in a graded manner to optimize one or more body functions for extrauterine survival. These conditions are transmitted through hormonal, nutritional and metabolic signals, according to the Barker Hypothesis (Barker 1994), to permanently "program" the individual's physiology. This predictive process would take advantage of the inherent plasticity in developmental processes. In theory, this flexible developmental process would work well provided the extrauterine environment does not differ from what is predicted. In a short-lived species (day, weeks, months), this could be an important adaptive mechanism. In long-lived species (years, decades), such a mechanism would likely be burdensome at best, and pathologic at worst if conditions are not totally predictable. With respect to the latter, the Barker Hypothesis was derived to account for the fetal origins of adult disease in human populations. Barker based his hypothesis on epidemiological observations of a cohort of individuals born under World War II famine conditions whose mothers had been calorically restricted during pregnancy. He linked the current epidemic of adult coronary disease to conditions in fetal life 50-60 years earlier. The idea was that the wartime caloric deprivation during pregnancy created a "thrifty phenotype" prenatally that was programmed to capture efficiently after birth the meager available calories that were being predicted. This seemingly adaptive strategy failed once calories no longer remained at a premium after birth and returned to historic norms. In postwar conditions of caloric plenty, these prenatally programmed changes led to obesity, poor glucose tolerance, Type-2 diabetes and heart disease. This most notable and dramatic example of prenatal programming that produces a mismatch between the predicted and actual environments to result in disease is now spawning a plethora of investigations about the types of environmental information that might influence the developmental trajectory of the unborn and whether such physiological changes are adaptive or maladaptive.

In contrast to prenatal programming, there are no known influences from the environment outside the mother that *normally* modulate the degree or time course of sexual differentiation in mammals. In non-mammals, such as fish, this can be a useful strategy to even alter the entire sex of the individual to respond to environmental pressures (too many or few of one sex locally). In comparison to the novel evolving concept about prenatal programming and its adaptive role for prevailing environmental conditions, sexual differentiation is generally considered to be a normal process that allows evolved sex differences to be expressed. This differentiation occurs at the level of the gonad (ovary *versus* testis), internal genitalia (Müllerian and Wolffian duct systems), external genitalia (clitoris, labia; penis, scrotum), patterns of gonadotropin secretion (cyclic *versus* noncyclic) and numerous sex behaviors. In contrast to our understanding of prenatal programming, there are presently no known adaptive effects for altering the degree of sexual differentiation at these various levels. Optimally, this differentiation should be all or none, and intermediate forms are often pathologies that interfere or prevent normal reproductive functions. An ovotestis in either sex, or retained Müllerian system and hypospadias in the male, or an enlarged clitoris and extensive Wolffian duct derivatives in the female are all deviations from established sex-specific structures. However, the distinction between concepts of sexual differentiation and fetal programming are less rigid if one takes into consideration that *within* the developing individual several substances are used to "program" "sexual differences". The physiology of roughly half of all conceptuses (males) is influenced by steroid and protein hormones from the developing gonad, and it is unlikely that all developing males are exposed to exactly the same level or pattern of these factors. Furthermore, it is crucial that the developing reproductive systems of both males and females are appropriately matched to the endocrine milieu that will be present in adult life. As is the case for prenatal growth restriction and metabolic disorders later in life, deviations from the "predicted" postnatal steroid environment could result in varying degrees of reproductive dysfunction. As detailed below, we are learning that various steroid hormones can program in a dose-dependent manner several sex differences in reproductive functions, and the boundaries between normal and abnormal function are now being defined. Some steroid "imposters" from the environment can gain access to the conceptus to partially or wholly modify its sex-specific developmental trajectory. While this seemingly calls into question the issue of external environmental influence on sexual differentiation, this type of external modulation is largely maladaptive.

Genes and steroids in perspective

A long-held belief emanating from the behavior literature is that steroids program sex differences in brain controls. The role of genes was simply to direct the indifferent gonad to develop either as a testis, or by default if a female, as an ovary. The testicular tissue then produced the differentiating steroids. More recently, our concepts have changed. We are beginning to learn that sex specific genes appear before the testes are formed (Dewing et al. 2003) and that environmental information can be inserted to modulate expression of encoded genes (Sutherland & Costa 2003; Skinner & Anway 2005). Such findings will necessitate broadening our scope of how steroids and steroid-like compounds program sex differences. Perhaps steroids serve both to reinforce inherent genetic differences and to modify genetic differences. In the case of the sheep model, genetic differences are being noted and breed differences may account for the observed severity of the effects of prenatal testosterone exposure on the initiation and maintenance of progesterone cycles; for example, Dorset females (Birch et al. 2003; Unsworth et al. 2005) appear to be more sensitive to cycle disruption than Suffolk females (Sharma et al. 2002; Manikkam et al. 2006). Finally, whether the types of neuroendocrine defects programmed by excess prenatal testosterone exposure in the sheep persist in subsequent generations remains to be determined.

Conclusions

Studies of the sexual differentiation of reproductive neuroendocrine control in the sheep lead to the hypothesis that the four major steroid feedback controls for GnRH secretion are inherent (default) (Fig. 12). They require no steroidal developmental input to function appropriately during adulthood; by contrast, the organizational actions of specific sex steroids during development selectively reduce their postnatal functions by acting on presynaptic inputs to GnRH neurons. Two steroid feedback controls regulate the preovulatory surge mode of GnRH secretion and two regulate the pulsatile mode. The inhibitory feedback action of estrogen is required in both sexes to regulate the timing of the pubertal increase in GnRH that activates the gonads; the sensitivity to this feedback is set by the organizational action of androgens produced by the developing testes. The other three feedbacks are required for the complex control of follicular development. After the first ovulation at puberty progesterone assumes the inhibitory feedback role formerly ascribed to estrogen and as an inhibitory feedback, its sensitivity is set by the organizational action of androgens beginning before birth. The early organizational actions of estrogens set the sensitivities of the stimulatory feedbacks that facilitate ovulation, the positive feedback action of estrogen and the progesterone blockade of the GnRH surge. If the individual is a male, three steroid feedback controls of GnRH secretion become unnecessary, and these irrelevant feedbacks are abolished or become functionally inoperative through the foregoing programmed reductions in hypothalamic sensitivity. While this programming begins before birth, it continues well after birth. In this respect, prenatal exposure to androgenic and estrogenic actions alters the susceptibility to the postnatal organizational actions of sex steroids.

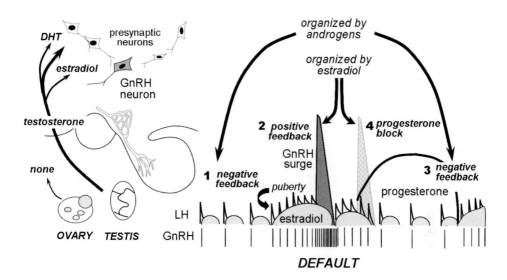

Fig. 12. Working hypothesis for sexual differentiation of steroid feedback controls of GnRH secretion in the sheep. Key elements of the hypothesis are that the four feedback controls of GnRH (right) exist in the undifferentiated (default) state and that androgens and estrogens by acting on GnRH presynaptic neurons (left) reduce, in part or totally, the sensitivity to steroid feedbacks

A major emphasis of presenting and testing hypotheses about the programming actions of testosterone and its metabolites on GnRH control systems is to develop the underlying concepts for normal sexual differentiation of reproductive functions. Once this is accomplished, understanding how programming mistakes originate should be simpler. However, in the types of studies being described, malprogramming is by necessity used to test hypotheses, and a variety of neuroendocrine phenotypes are being created with reproductive defects that become instructive. With a combination of at least four feedback controls for GnRH that can be malprogrammed, these phenotypes provide useful models to unravel the etiologies of many aspects of infertility such as premature or delayed puberty, primary anovulation, premature anovulation and a host of reproductive cycle defects. This understanding of underlying mechanisms assumes even greater urgency as it is likely that the prevalence of these defects will increase. There is a clear and increasing danger in modern societies that inadvertently produce and release into the environment numerous synthetic compounds that may act like steroids (endocrine disruptors, endocrine mimics, endocrine imposters). Compounds that can enter the conceptus during critical periods of development may program reproductive defects that will only become apparent months or years later, depending upon species. These compounds acting like estrogens or androgens or steroid antagonists may be the tragic mediators of a megafeedback loop between human populations and the environment that could ultimately impair fertility in both human and animal populations.

Acknowledgements

The authors are grateful to the several undergraduate and graduate students, postdoctoral fellows, and visiting scientists in our laboratory and to the numerous other investigators who have generated the recent data described in this chapter that contributed to the overarching hypothesis described: principally Dr. Jane Robinson, University of Glasgow and to the many colleagues of the Reproductive Sciences Program, particularly Dr. Theresa Lee. The authors also gratefully acknowledge our technical staff for assistance with animal experimentation, especially Mr. Douglas Doop; and the National Institutes of Health (currently P01-HD44232, T32-HD-07048), the National Science Foundation, and the United States Department of Agriculture for research support.

References

Barker DJP 1994 *Mothers, Babies and Health in Later Life.* Edinburgh: Churchill Livingstone.

Birch RA, Padmanabhan V, Foster DL & Robinson JE 2003 Prenatal programming of reproductive neuroendocrine function: fetal androgen exposure produces progressive disruption of reproductive cycles in sheep. *Endocrinology* **144** 1426-1434.

Caldani M 1986 Mise en évidence immunohistochimique des neurones a LHRH du mouton (Ovis aries): systématisation et ontogenese. (Thesis), Université de Paris-VI, Paris.

Caldani M, Batailler M, Thiery JC & Dubois MP 1988 LHRH immunoreactive structures in the sheep brain. *Histochemistry* **89** 129–139.

Dewing P, Shi T, Horvath S & Vilain E 2003 Sexually dimorphic gene expression in mouse brain precedes gonadal differentiation. *Molecular Brain Research* **118** 82 – 90.

Ebling FJP, Wood RI, Suttie JM, Adel TE & Foster DL 1989 Prenatal photoperiod influences neonatal prolactin secretion in the sheep. *Endocrinology* **125** 384–391.

Foecking EM, Szabo M, Schwartz NB & Levine JE 2005 Neuroendocrine consequences of prenatal androgen exposure in the female rat: absence of luteinizing hormone surges, suppression of progesterone receptor gene expression, and acceleration of the gonadotropin-releasing hormone pulse generator. *Biology of Reproduction* **72** 1475-1483.

Foster DL & Jackson LM 2006 Puberty in the sheep. In *Physiology of Reproduction,* edn 3, pp 2127-2176. Ed JD Neill. San Diego: Elsevier.

Foster DL, Karsch FJ, Olster DH, Ryan KD, Yellon SM 1986 Determinants of puberty in a seasonal breeder. *Recent Progress in Hormone Research* **42** 331–384.

Foster DL, Ebling FJP, Claypool LE & Wood RI 1989 Photoperiodic timing of puberty in sheep. In *Research in Perinatal Medicine, Volume IX: Development of Circadian Rhythmicity and Photoperiodism in Mammals,* pp 103-153. Ed SM Reppert. Ithaca: Perinatology Press.

Foster DL, Jackson LM & Padmanabhan V 2006 Programming of GnRH feedback controls timing pu-

berty and adult reproductive activity. *Molecular and Cellular Endocrinology* **254-255** 109-119.

Gorski RA 1979 The neuroendocrinology of reproduction: an overview. *Biology of Reproduction* **20** 111-127.

Herbison AE 1998. Multimodal influence of estrogen upon gonadotropin releasing hormone neurons. *Endocrine Reviews* **19** 302-330.

Herbosa CG & Foster DL 1996 Defeminization of the reproductive response to photoperiod occurs early in prenatal development in the sheep. *Biology of Reproduction* **54** 420-428.

Herbosa CG, Dahl GE, Evans NP, Pelt J, Wood RI & Foster DL 1996 Sexual differentiation of the surge mode of gonadotropin secretion: prenatal androgens abolish the gonadotropin-releasing hormone surge in the sheep. *Journal of Neuroendocrinology* **8** 627-633.

Karsch FJ & Foster DL 1975 Sexual differentiation of the mechanism controlling the preovulatory discharge of luteinizing hormone in sheep. *Endocrinology* **97** 373-379.

Karsch FJ, Bittman EL, Foster DL, Goodman RL, Legan SJ & Robinson JE 1984 Neuroendocrine basis of seasonal reproduction. *Recent Progress in Hormone Research* **40** 185-232.

Kasa-Vubu JZ, Dahl GE, Evans NP, Thrun LA, Moenter SM, Padmanabhan V & Karsch FJ 1992 Progesterone blocks the estradiol-induced gonadotropin discharge in the ewe by inhibiting the surge of gonadotropin-releasing hormone. *Endocrinology* **131** 208-12.

Kim S-J, Foster DL & Wood RI 1999 Prenatal testosterone masculinizes synaptic input to GnRH neurons in sheep. *Biology of Reproduction* **61** 599-605.

Kosut SS, Wood RI, Herbosa-Encarnacion CG & Foster DL 1997 Prenatal androgens time neuroendocrine puberty in sheep: effects of dose. *Endocrinology* **38** 1072-1077.

Lehman MN, Robinson JE, Karsch FJ & Silverman A-J 1986 Immunocytochemical localization of luteinizing hormone-releasing hormone (LHRH) pathways in the sheep brain during anestrous and the mid-luteal phase. *Journal of Comparative Neurology* **244** 19–35.

Malcolm KD, Jackson LM, Bergeon C, Lee TM, Padmanabhan V & Foster DL 2006 Long-term exposure of female sheep to physiologic concentrations of estradiol: effects on the onset and maintenance of reproductive function, pregnancy, and social development in female offspring. *Biology of Reproduction* **75** 844-852.

Manikkam M, Steckler TL, Welch KB, Inskeep EK & Padmanabhan V 2006 Fetal programming: prenatal testosterone treatment leads to follicular persistence/luteal defects; partial restoration of ovarian function by cyclic progesterone treatment. *Endocrinology* **147** 1997-2007.

Masek KS, Wood RI & Foster DL 1999 Prenatal dihydrotestosterone differentially masculinizes tonic and surge modes of luteinizing hormone secretion in sheep. *Endocrinology* **140** 3459-3466.

Padmanabhan V, Manikkam M, Recabarren S & Foster DL 2006 Prenatal testosterone excess programs reproductive and metabolic dysfunction in the female. *Molecular and Cellular Endocrinology* **246** 165-174.

Pfeiffer CA 1936 Sexual differences of the hypophyses and their determination by the gonads. *American Journal of Anatomy* **58** 195-223.

Robinson JE, Forsdike RA & Taylor JA 1999 In utero exposure of female lambs to testosterone reduces the sensitivity of the gonadotropin-releasing hormone neuronal network to inhibition by progesterone. *Endocrinology* **140** 5797-5805.

Ryan KD, Goodman RL, Karsch FJ, Legan SJ & Foster DL 1991 Patterns of circulating gonadotropins and ovarian steroids during the first periovulatory period in the developing sheep. *Biology of Reproduction* **45** 471–477.

Sarma HN, Manikkam M, Herkimer C, Dell'orco J, Welch KB, Foster DL & Padmanabhan V 2005 Fetal programming: excess prenatal testosterone reduces postnatal luteinizing hormone, but not follicle-stimulating hormone responsiveness, to estradiol negative feedback in the female. *Endocrinology* **146** 4281-4291.

Savabieasfahani M, Lee JS, Herkimer C, Sharma TP, Foster DL & Padmanabhan V 2005 Fetal programming: testosterone exposure of the female sheep during mid-gestation disrupts the dynamics of its adult gonadotropin secretion during the periovulatory period. *Biology of Reproduction* **72** 221-229.

Sharma TP, Herkimer C, West C, Ye W, Birch R, Robinson JE, Foster DL & Padmanabhan V 2002 Fetal programming: prenatal androgen disrupts positive feedback actions of estradiol but does not affect timing of puberty in female sheep. *Biology of Reproduction* **66** 924-933.

Short RV 1974 Sexual differentiation of the brain of the sheep. *INSERM* **32** 121–142.

Sisk CL, Foster DL 2004 The neural basis of puberty and adolescence. *Nature Neuroscience* **7** 1040-1047.

Skinner MK & Anway MD 2005 Seminiferous cord formation and germ-cell programming: epigenetic transgenerational actions of endocrine disruptors. *Annals of the New York Academy of Sciences* **1061** 18-32.

Sullivan SD & Moenter SM 2004 Prenatal androgens alter GABAergic drive to gonadotropin-releasing hormone neurons: implications for a common fertility disorder. *Proceedings of the National Academy of Sciences USA* **101** 7129-7134.

Sutherland JE & Costa M 2003 Epigenetics and the environment. *Annals of the New York Academy of Sciences* **983** 151-160.

Unsworth WP, Taylor JA & Robinson JE 2005 Prenatal programming of reproductive neuroendocrine function: the effect of prenatal androgens on the development of estrogen positive feedback and ovarian cycles in the ewe. *Biology of Reproduction* **72** 619-627.

West C, Foster DL, Evans NP, Robinson J & Padmanabhan V 2001 Intrafollicular activin availability is altered in prenatally-androgenized lambs. *Molecular and Cellular Endocrinology* **185** 51-59.

Wood RI, Ebling FJP, I'Anson H, Yellon SM & Foster DL 1991 Prenatal androgens time neuroendocrine sexual maturation. *Endocrinology* **128** 2457–2468.

Wood RI, Newman SW, Lehman MN & Foster DL 1992 GnRH neurons in the fetal lamb hypothalamus are similar in males and females. *Neuroendocrinology* **55** 427–433.

Wood RI, Mehta V, Herbosa CG & Foster DL 1995 Prenatal testosterone differentially masculinizes tonic and surge modes of LH secretion in the developing sheep. *Neuroendocrinology* **56,** 822-830.

Wood RI, Kim S-J & Foster DL 1996 Prenatal androgens defeminize activation of GnRH neurons in response to estradiol stimulation. *Journal of Neuroendocrinology* **8** 617-625.

Immunological influences on reproductive neuroendocrinology

GK Barrell

Agriculture and Life Sciences Division, P O Box 84, Lincoln University, Lincoln 7647, New Zealand

One of the consequences of activation of the immune system, with its associated inflammatory responses and operation of the stress axes, is a generalised inhibition of reproductive function. This can be considered as part of the all-encompassing effects of an activated immune system, included in which is the 'immunological cost' arising from the nutritional demand required to maintain a competent, responsive immune system, and the pathological effects produced by severe immune responses. Elucidation of specific immune-neuroendocrine linkages has largely involved examination of corticosteroid-based mechanisms or use of bacterial endotoxin as a model stimulus and examination of effects on GnRH and LH pulsatility, on GnRH and LH surge processes and on pituitary responsiveness to GnRH, using various sheep models. Although there is good evidence for prostaglandins as common mediators for endotoxin-induced and stress axis-induced impairment of neuroendocrine reproductive processes, both mechanisms appear to have prostaglandin-independent pathways as well. At the anterior pituitary gland level, the type II glucocorticoid receptor appears to mediate corticosteroid effects. Otherwise, the identity of specific cytokines, their sites of action and the cell level mechanisms underlying the inhibition of the reproductive axis at hypothalamic and anterior pituitary levels, especially in sheep, remains largely unresolved.

Introduction

This review is intended to provide initially a brief perspective of the immune system and its impacts on body systems in general, and on neuroendocrine pathways in particular. Thereafter, the primary objectives are to review recent research activity that has focused on mechanisms of immune stress-associated effects, largely involving corticosteroids, and cytokine-mediated effects on neuroendocrinology of reproduction with particular emphasis on studies with sheep.

The immune system

What is the immune system?

The immune system can be regarded as a defence mechanism, the impairment of which has tragic consequences, as demonstrated by genetic or acquired (e.g. AIDS) severe immunodefi-

Corresponding author E-mail: barrell@lincoln.ac.nz

ciencies. Although it influences the function of most bodily systems, enables tolerance of 'self' and can reject tumours, its genesis is largely based on the need of vertebrates to cope with challenges from pathogenic organisms.

Innate immunity includes the physical epithelial barriers encountered by pathogenic organisms attempting to invade an animal and the antimicrobial factors produced by epithelia, plus macrophages and other cells that have receptors for cell-surface molecules on the organisms. This provides for interactions that may lead to phagocytosis of the invading organisms and/or activation of macrophages causing induction of an inflammatory response, and recruitment of other cell types to the site. Such pathogens may also be recognised by the complement system causing enhancement of their phagocytosis as well as activation of inflammatory responses. Antigen-specific responses are recruited from other cells attracted to the site of infection and through uptake of antigen by dendritic cells that transport antigen to lymphoid tissue where primary immune responses are initiated.

Adaptive immunity includes primary immune responses that occur when antigenic substances interact with antigen-specific lymphocytes, most commonly causing production of specific antibody molecules and proliferation of antigen-specific helper and effector T-lymphocytes. The latter cells can produce killer T cells capable of lysing infected cells. Secondary immune responses involve the formation of immunologic memory during the primary event that produces an enhanced immune response at subsequent exposure to antigen – i.e. a state of immunity and the basis of vaccination. The specificity of the immune response is determined by lymphocytes which makes them central players in the system, but a whole raft of other leukocytes plus specialised epithelial and stromal cells are involved in the immune process by providing the anatomical environment for conducting it as well as secreting chemical factors that regulate migration, growth, and/or gene activation of cells in the immune system. Antibodies are large immunoglobulin molecules that can be expressed on the cell surface of lymphocytes and secreted into the surrounding environment. Another group of chemicals that mediate the functions of cells in the immune system are small proteins referred to as cytokines. Their definition has been discussed by Turnbull and Rivier (1999) who grouped them usefully into eight families, namely: interleukins (e.g. IL-2, IL-3, etc.), tumour necrosis factors (e.g. TNF-α, etc.), interferons, chemokines, haematopoietins, colony stimulating factors, neurotrophins, and growth factors. A complication is that many peptides with mediatory functions are common to the immune, nervous and neuroendocrine systems. For example, GnRH, a well established hypothalamic neuropeptide with a central role in reproduction, is also produced by T-lymphocytes and affects laminin receptor expression in lymphocytes (Chen et al. 2002) thus having a potential role as an immunological mediator. This illustrates the difficulty, highlighted in a recent review (Shepherd et al. 2005), of classifying peptides using nomenclature such as 'cytokines' or 'neuropeptides' as well as confounding the linkages between immune and neuroendocrine function. Interestingly, the various signalling chemicals that have evolved within one cytokine molecular family, the class-I helical cytokines, include some members generally regarded as classical peptide hormones, e.g. leptin, growth hormone, prolactin, erythropoietin (Huising et al. 2006).

Immune-nervous interactions

Involvement of the central nervous system (CNS) in regulation of immune processes at many levels is well established (e.g. Sternberg & Webster 2003; Shepherd et al. 2005) and the, now classical, arms of the neuroendocrine and neuronal responses to stress, i.e. the hypothalamic-pituitary-adrenal (HPA) axis and sympathetic adrenal medulla (SAM) system, are both important in regulation of immune responses. Both divisions of the autonomic nervous system, i.e. sympathetic and parasympathetic, modulate immune function indirectly by their various actions on blood circulation, smooth muscles, secretory glands, etc., as well as directly via receptors on virtually all immune cells, including macrophages and lymphocytes. There are other examples of neural effects on immunological processes (e.g. Shepherd *et al.* 2005), however, the focus of this review is on the converse interaction; the effects of the immune system on neuroendocrine function, specifically the reproductive neuroendocrine axis.

Modulation of nervous function by immune factors

Within the CNS immune chemicals such as cytokines act like growth factors affecting neuronal cell death and survival. These cytokines and/or their receptors are synthesised directly by cells of the CNS which include resident glia or immune cells that have invaded the CNS (Sternberg & Webster 2003). Such immunoregulation is a constitutive function of these cells that is likely to be enhanced under pathological conditions or when the immune system is activated. In contrast, cytokines that reach the CNS from outside this system are likely to act, by definition, in an endocrine manner as classical hormones. For instance some cytokines act like neuropeptides and cause such centrally mediated events as fever, sleep, changes in behaviour, or activation of hormonal stress responses (Sternberg & Webster 2003). However, the occurrence of such an endocrine role of cytokines within the CNS had remained largely unproven until sufficient quantities of pure recombinant cytokines became available to test these theories directly.

Another issue related to an endocrine role of cytokines in the CNS is that of their access to brain cells. One way in which compounds such as cytokines can surmount the blood-brain barrier is by leaving the circulation at regions in the CNS where the blood-brain barrier does not exist, or the vessels are leaky. Such regions are in the circumventricular organs which include: median eminence, area postrema, subfornical organ, organum vasculosum of the lamina terminalis, and choroid plexus (Sternberg & Webster 2003). Many of these organs are located close to areas of the CNS that are linked to neuroendocrine regulatory centres, such as the hypothalamus. Access to CNS tissue can be gained also by means of active transport processes across the endothelium of CNS blood vessels (Banks 2001) or, alternatively, cytokines can signal their presence through activation of secondary messengers located at the endothelial cell membranes and achieve their effects on CNS function indirectly (Scammell *et al.* 1996; Elmquist *et al.* 1997). Another means of indirect signalling of the CNS by cytokines involves activation of sensory nervous input. For example, paraganglia cells in lymphoid tissue can signal the brain via vagal afferents when there is some change in activity of the lymphoid organs (Goehler *et al.* 2000).

Impacts of the immune system on body function in general – 'immunological cost'

Whenever the immune system of an animal is activated there is the possibility that this can be accompanied by major impacts on body function. It is now widely recognised that these impacts largely result from the immune responses themselves. A dramatic example is toxic shock syndrome following infection by streptococcal bacteria (Stevens 1995). A recent example involving an intestinal nematode parasite was provided by Greer *et al.* (2005). Lambs in the study were treated with an immunosuppressive dose of corticosteroid that was sufficient to allow establishment of a high worm burden. In spite of the greater infection of the immunosuppressed sheep, their live-weight gain and feed conversion efficiency was not affected, showing the benefits in terms of nutrients spared from the need to maintain immune responsiveness and also indicating little actual harm caused to the sheep by the parasites *per se*. As in the case of toxic shock syndrome, most of the detrimental effects of the intestinal parasitism arose from the immune responses to the infective organism, not the organism itself. Collectively, these and similar findings lend support to the concept of an 'immunological cost', that is the metabolic cost of maintaining immunological competence or mounting an immune response (Colditz 2004) with or without the additional burden of pathology arising from tissue damage caused by severe immune responses.

Any change in nutritional status arising from activation of the immune system is likely to alter neuroendocrine function directly via changes in supply of nutrients and by actions of signalling molecules. Such molecules include leptin, which is involved in communicating nutritional status information to the reproductive neuroendocrine axis in sheep (Adam *et al.* 2003) and cows (Barb & Kraeling 2004), and grhelin, which is associated with meal-related neuroendocrine signalling in ruminants (Sugino *et al.* 2004).

Immunostimulation as a stressor

It is not surprising from evidence of its major impacts on body function that activation of the immune response cascade causes major disturbances of neuroendocrine function. Whether it is by default or by design, as it makes sense to curtail the expense of reproductive function if survival of the individual is threatened, the consequences for the reproductive axis when the immune system is activated are not unexpected. What is not so clear are the details of the mechanisms by which activation of the immune response cascade imposes its effects on the reproductive axis. Immunostimulation is one of the responses to an adverse environmental stress, such as exposure to antigens, and can be considered as being akin to responses such as those caused by thermal stress, malnutrition, psychological stress, physical injury, hypoxia or any of the stressors that activate stress responses in animals. In the current view, many or maybe all of these stressors appear to produce co-ordinated suites of responses mediated by the brain, particularly via the amygdala – the two organs of which form part of the sub cortical limbic system. These organs may be the central processors of the myriad of inputs from various stressors that in turn regulate the major responses of the body to stress. For simplicity these have been allocated into four categories, namely: (1) changes in behaviour, i.e. change of skeletal muscle function; (2) activation of the SAM system; (3) activation of the HPA axis; and (4) activation of immune responses, i.e. immunostimulation. Two of these response categories, 2 - the SAM system and 3 - HPA axis, are well-established as being regulated at various levels within the hypothalamus and there is increasing evidence (Turnbull & Rivier 1999) that the fourth category – immunostimulation – also undergoes regulatory interaction within the hypothalamus. Presumably, the motor region of the cerebral cortex plays a major part in the behavioural changes – response category 1 (above) - although these are considerably influenced by structures operating below the level of consciousness, such as in the limbic system. Activation of the SAM system causes release of adrenergic hormones which are powerful immunoregulators and activation of the HPA axis causes immuosuppresion via release of glucocorticoids. This means that much of the research on effects of stress on reproduction merges with that involving the immune system.

Corticosteroid-mediated effects

Studies involving the administration of the bacterial endotoxin, lipopolysaccharide (LPS), a commonly-used model of immune stress in sheep, have demonstrated a concurrent activation of the neuroendocrine stress axis (Battaglia *et al.* 1997; Battaglia *et al.* 1998; Dadoun *et al.* 1998; Williams *et al.* 2001). This may be centrally mediated in the hypothalamus or by direct action of cytokines on the adrenal cortex (Turnbull & Rivier 1999). To test the hypothesis that cortisol mediates the inhibitory actions of endotoxin on pulsatile GnRH and LH secretion in sheep, Debus *et al.* (2002) working in Professor Fred Karsch's laboratory at The University of Michigan separated the effects of cortisol alone from those of the endotoxin-induced release of cortisol by use of metyrapone, an inhibitor of cortisol synthesis. Their work showed that although cortisol itself inhibited pulsatile GnRH and LH secretion, the marked release of cortisol provoked by administration of LPS was not required for the inhibitory effects of the bacterial endotoxin. In addition, they ruled out the possibility that their results were attributable to progesterone which was also released in response to LPS and was probably of adrenal origin because ovariectomised ewes were used. They concluded that more than one pathway must be responsible for the immune-mediated inhibition of reproductive activity, although one of these involves enhanced secretion of cortisol. Further work in Professor Karsch's laboratory with ovariectomised ewes (Breen & Karsch 2004) indicated that the site of action of cortisol in these animals is localised to the anterior pituitary gland where it lowered responsiveness to GnRH, rather than by a direct effect on hypothalamic release of GnRH. Their results showed a cortisol-induced suppression of LH pulse amplitude but no effect on pulse frequency of LH or GnRH and no effect on pulse amplitude of GnRH. However, the ovariectomised ewe model provides a very different steroid milieu from that experienced by ovarian intact ewes and eliminates potential effects of cortisol on the ovary that would alter steroid feedback at hypothalamic and pituitary gland levels. Subsequently, studies in the same laboratory have shown that stress-like elevations of plasma cortisol concentration suppressed pulsatility of LH secretion in follicular-phase ovarian intact ewes (Breen *et al.* 2005a), interfered with the preovulatory rise in plasma oestradiol concentration and blocked or delayed the preovulatory LH and FSH surges. The observation that LH pulsatility was suppressed in these intact ewes implicates the hypothalamus as a central component of this effect of cortisol. This is in clear contrast to the earlier finding with ovariectomised ewes in which there was no evidence of an effect at the hypothalamic level (Breen & Karsch 2004), suggesting a lack of sensitivity to the pulse frequency suppressing effect of cortisol in such animals. Concern that this lack of effect in ovariectomised ewes might have arisen from inadequate duration of exposure to elevated cortisol levels has been ruled out by showing that 28 hours of cortisol treatment was also without effect (Oakley *et al.* 2006). These contrasting results indicate that central responsiveness to the effects of cortisol is dependent on the presence of ovary-derived factors (e.g. ovarian steroids). However, in the case of the pre-ovulatory surge of LH, the ability of elevated cortisol concentrations to block or delay this process is not affected by the ovarian steroid milieu, as shown by Wagenmaker *et al.* (2005). Recently, Breen and Karsch (2006) demonstrated that the cortisol-induced suppression of LH pulse amplitude in ovariectomised ewes was not influenced by season of the year, which contrasts starkly with the well-known effects of season on feedback by ovarian steroids.

In support of actions of cortisol at the level of the ovine hypothalamus is the evidence for presence of the type II glucocorticoid receptor in neurons of the preoptic area and arcuate nucleus (Dufourny & Skinner 2002). This receptor is co-expressed with the progesterone receptor and the estrogen α-receptor (ER-α) (Dufourny & Skinner 2002). By use of RU486, an antagonist, and dexamethasone, a specific agonist, Professor Karsch and his co-workers (Breen *et al.* 2004b; Case *et al.* 2005) have provided good evidence that the type II glucocorticoid receptor also mediates the cortisol-induced suppression of pituitary responsiveness in the ovariectomised ewe, i.e. a pituitary-level site of action for this receptor.

It is clear from these studies that both components of the hypothalamic-pituitary axis are involved in mediation of stress effects on reproduction. In the anterior pituitary gland of castrated male sheep, cortisol interfered with oestrogen-stimulated increase in expression of GnRH receptor (Adams *et al.* 1999), probably via classical nuclear glucocorticoid receptors and the resulting activation of transcription processes, as shown by Maya-Nunez *et al.* (2003) in rat pituitocyte cultures. However, Professor Karsch and his colleagues recently reported a study of ovariectomised ewes where there was no effect of cortisol on pituitary content of GnRH receptor (Breen *et al.* 2005b). This result appears to conflict with the findings of Adams *et al.* (1999) and leaves this involvement of GnRH receptors at the pituitary level unresolved. Also, when the pituitary gland was disconnected from hypothalamic regulation by surgical transection, GnRH-stimulated output of LH secretion was suppressed by various stressors, showing that there is a pituitary-only component involved in the inhibitory effects of stress (Stackpole *et al.* 2003). However, because the HPA axis was disrupted, cortisol would not have been involved in mediation of this effect of stress (Stackpole *et al.* 2003).

As well as the moderately confusing picture provided by some of the work reviewed above, there has to be caution about interpretation of the role and actions of corticosteroids in the responses to stress, especially in relation to immunity and inflammation. The co-involvement of cortisol in the inhibition of reproductive function induced by bacterial endotoxin, as discussed above, may appear to be at odds with the recognised immunosuppressive/anti-inflammatory roles of glucocorticoids. Also, these corticosteroids must participate in the negative feedback regulation of the HPA axis, thus inhibiting their own secretion. The complexity of involvement of these steroids in the responses to stress was comprehensively reviewed by Sapolsky *et al.* (2000). Since then, the research group led by Masugi Nishihara at the University of Tokyo has shown that glucocorticoids counteracted the TNF-α-induced inhibition of LH pulsatility (Matsuwaki *et al.* 2003) and the LH surge (Matsuwaki *et al.* 2004) in rats. Also they showed that adrenalectomy actually enhanced the inhibition of LH pulsatility in ovariectomised rats exposed to a variety of stressors including infection (Matsuwaki *et al.* 2006). These workers have implicated prostaglandins as the mediators of stress-induced suppression of LH pulsatility by using combinations of adrenalectomy, corticosterone and indomethacin (a cyclo-oxygenase inhibitor) on ovariectomised rats subjected to hypoglycaemic (2-deoxy glucose) or infectious (LPS) stress (Matsuwaki *et al.* 2006). A similar role for prostaglandins in sheep had been provided earlier (Harris *et al.* 2000, see below). However, as well as the blockade by corticosterone of the stress-induced inhibition of LH pulsatility (mainly reduced pulse amplitude), this steroid counteracted the stress-induced elevation of cyclo-oxygenase immunoreactivity in the brain (Matsuwaki *et al.* 2006) showing a direct inhibitory effect of corticosteroids on the involvement of prostaglandins, by reducing their synthesis. This and related work with rats has lead to the concept that prostaglandins may be common mediators of stress-induced inhibition of the reproductive axis (Maeda & Tsukamura 2006). Also these authors contend that the output of glucocorticoids resulting from an activated HPA axis initially has a protective role by suppressing prostaglandin synthesis and, hence, helping to maintain reproductive function in spite of the underlying stressor activity (Maeda & Tsukamura 2006). This concept of a reproductive support role for glucocorticoids runs counter to their generally reproduction-inhibitory effects, such as those reported in the sheep studies described above. The roles of glucocorticoids are complex and their actions in response to stress can fall into permissive, stimulatory or suppressive categories, depending on the physiological endpoint in question (Sapolsky *et al.* 2000). The protective role of glucocorticoids during the early phase of the stress response, as described by Maeda & Tsukamura (2006, above), seems to belong to the suppressive category of response.

Cytokine-mediated effects

As mentioned above, cytokines that are released into the circulation following stimulation of cells involved in the immune response have endocrine roles in the CNS and anterior pituitary gland, some of which are directed at components of the GnRH and gonadotrophin regulatory processes. In contrast, cytokine involvement in regulation of the selection and migration of GnRH neurons during embryonic and early fetal development is most likely to be mediated by locally derived cytokines. Previously, evidence for presence of cytokines and their receptors in the brain and anterior pituitary gland has been reviewed by Sternberg (1997), Benveniste (1998) and Turnbull & Rivier (1999). The occurrence of receptors for one cytokine, fibroblast growth factor (FGF), in regions of the brain involved in regulation of reproduction has been confirmed by Gill *et al.* (2004; 2006) using immuncytochemistry. They showed the presence of FGF receptors in developing GnRH neurons of mice and the emergence of these cells to be blocked by FGF receptor antagonists. In addition they reported a stimulatory effect of FGF on neurite outgrowth and branching. Functioning of cranial neurons can also be affected by transcriptional actions of oestrogens operating via classical oestrogen receptors (ER). However, Herbison and his colleagues, using ER-knockout mice, showed that there are rapid negative feedback effects of oestrogens mediated by membrane-level, non-genomic mechanisms (Abraham *et al.* 2004). Recently it has been shown in sheep that acute suppression of LH secretion by oestradiol or an oestradiol conjugate operates via such a non-genomic mechanism (Arreguin-Arevalo & Nett 2006). In the hypothalamus, membrane-level signalling in the negative feed-back pathway to GnRH neurons involves a G protein-coupled receptor (GPR54) for which major ligands are the KiSS-1 peptins (de Roux *et al.* 2003; Seminara *et al.* 2003; Dungan *et al.* 2006). These peptins and their receptor (GPR54) appear to be involved in both inhibitory and stimulatory feedback effects of gonadal steroids in the hypothalamus (Smith *et al.* 2005) and it has been speculated that KiSS-1 neurons may provide the link between steroidal feedback signals and GnRH neurons that lack ER-α (Tena-Sempere 2005). It is possible that inflammatory peptides such as cytokines interact with ER both at classical genomic sites and with G protein-coupled receptors at the non-genomic, membrane-level sites. As an example of the former, IL-1ß activation of transcription, presumably via the transcription factor NF-κB, is antagonised by oestradiol via ER action at this promoter site (Tyree *et al.* 2002).

Downstream from the hypothalamus, gonadotropes and folliculostellate cells of the anterior pituitary gland are also implicated as targets for cytokine action (Turnbull & Rivier 1999). For instance, the pro-inflammatory cytokines, TNF and IL-1, altered the phosphorylation status of a gap junction membrane channel protein (GJA1) in cultured rat folliculostellate cells (Fortin *et al.* 2006) indicating possible modulation of cell-to-cell communication and, thus, of growth and function of these cells. This affects the growth and secretion of hormones by endocrine cells of the anterior pituitary gland and is just one example of cytokine-pituitary interactions.

As mentioned earlier, one of the components of innate immunity is activation of an inflammatory response. This is generally mediated by the so-called pro-inflammatory cytokines, primarily IL-1, IL-6, Il-8 and TNFα, and is defined by induction of vasodilation and increased vascular permeability. Probably, it is at the endothelial cell level in organs such as the brain where these cytokines initiate phospholipase activity, thus causing disruption of cell membrane integrity and release of arachidonic acid – a precursor for prostaglandin synthesis by the cyclo-oxygenase pathway, or of leukotrienes by the lipoxygenase pathway. Harris *et al.* (2000) showed that blockade of prostaglandin synthesis in sheep by use of the cyclo-oxygenase inhibitor, flurbiprofen, prevented endotoxin-induced fever and suppression of LH secretion, thereby implicating prostaglandins as the possible central mediators of the disturbances in hypothalamo-hypophyseal-ovarian function following such inflammatory challenge. The various studies using sheep to investigate the different levels within the reproductive neuroendocrine axis where endotoxin achieves such disruption has been comprehensively reviewed (Karsch & Battaglia 2002). These authors argued that the advantage of their sheep model over previously reported studies was the large size of the animal which made frequent blood sampling possible, thus enabling high resolution studies of hormonal dynamics. Because of this they were able to dissociate effects of endotoxin at hypothalamic, pituitary and ovarian levels and at various phases of the ovarian cycle (Karsch & Battaglia 2002). This work has shown that endotoxin interferes with the reproductive neuroendocrine axis at both hypothalamic and pituitary levels. Endotoxin inhibits both pulsatile GnRH secretion and pituitary responsiveness to GnRH (Battaglia et al. 1997; Williams et al. 2001). The effects of endotoxin on LH pulsatility are acute, occurring within one hour (Battaglia et al. 1997; Harris et al 2000; Williams et al. 2001), whereas its effects on the oestrogen-induced LH surge process appear to be installed some 10 to 20 hours before the surge would be expected to occur (Battaglia et al. 1999). Further work (Breen et al. 2004a) showed that the inhibitory effect of endotoxin on the oestradiol-induced surge of LH in the ewe operated at the level of the hypothalamus by causing blockade of the GnRH surge. Administration of the cyclo-oxygenase inhibitor flurbiprofen did not interfere with this blockade of the LH surge nor with the inhibition of LH secretion prior to the surge, although it did prevent fever. This means that, in spite of the dampening effect of endotoxin on pituitary responsiveness to GnRH mentioned above, endotoxin-induced inhibition of the LH surge in ovariectomised ewes is due to inhibition of the GnRH surge mechanism in the hypothalamus. Also, the finding that inhibition of prostaglandin synthesis by flurbiprofen did not overcome the effects of endotoxin on the LH surge in sheep is interesting as it differs from endotoxin inhibition of GnRH and LH pulsatility which were blocked by flurbiprofen (Harris et al. 2000). In both studies endotoxin-induced fever was blocked by flurbiprofen, showing that fever and the effect of endotoxin on pulsatility may be mediated by prostaglandins. However, the study by Breen et al. (2004a) indicates the existence of a separate prostaglandin-independent pathway that mediates the effects of endotoxin on the surge process. These various findings could be resolved simply as follows: endotoxin inhibits the GnRH pulse mechanism in the hypothalamus via a prostaglandin mediated pathway and inhibits the surge process, also at a central level, by a prostaglandin-independent pathway. However, the authors mentioned they had gathered initial evidence that the pulse inhibitory mechanism may also not require a prostaglandin-mediated pathway. Nevertheless, because of the differences in latency between endotoxin effects on pulse and surge mechanisms, the authors inferred that the neuroendocrine processes which mediate the endotoxin-induced effects on GnRH pulsing are "separable and fundamentally different" from those mediating the effects on GnRH surges (Breen et al. 2004a). These differential effects of endotoxin had been discussed earlier by Karsch & Battaglia (2002). Endotoxin-induced blockade of the GnRH surge process at a central level does not rule out existence of a similar effect on the LH surge at the pituitary level, especially as an inhibitory effect on pituitary responsiveness to GnRH has been demonstrated (Williams et al. 2001). However, evidence that the LH surge was of normal amplitude in ewes in which the endotoxin treatment unexplainably failed to block the GnRH surge (Breen et al. 2004a) strongly rules out this possibility, unless the quantity of GnRH secreted was so great that it masked any pituitary-level deficiency.

Existence of prostaglandin-independent pathways in the mediation of endotoxin effects on neuroendocrine events does not overshadow the common role of prostaglandins in stress-induced activation of the HPA axis (described above) and immunostimulation processes, but it does indicate levels of complexity in the regulatory mechanisms that await further resolution. It is not clear whether other central mediators of inhibitory actions on gonadotrophin secretion such as the endogenous opioids and corticotrophin-releasing hormone are released directly in response to endotoxin or via prostaglandin-dependent pathways, nor to what extent cytokines released by endotoxin-activated immune cells exert direct endocrine influences in the hypothalamus and anterior pituitary gland. There is evidence that endotoxin does cause release of cytokines in sheep. Intravenous administration of endotoxin to sheep elevated circulating levels of TNF (Coleman et al. 1993; Perkowski et al. 1996; Harris et al. 2000) and increased expression of IL-1ß mRNA in the choroid plexus (Vellucci et al. 1996). However, the role of cytokines at the pituitary gland level is complicated by *in vitro* studies of cultured ovine pituitary cells that have actually shown an increase in accumulation of LH in the media on administration of either endotoxin (Coleman et al. 1993) or the pro-inflammatory cytokines, IL-1α and IL-1ß (Braden et al. 1998). Also, the possible influence of ovarian steroids on production of cytokines or on central responsiveness to immunostimulation provides another level of complexity that is largely unresolved at present.

Virtually all the direct evidence for an inhibitory role of pro-inflammatory cytokines at the central level of regulation of reproduction has been obtained from rats, largely by intracerebroventricular administrations (reviewed by Kalra et al. 1998). A general analysis of cytokine signalling in the brain has been provided by Turnbull & Rivier (1999) and more recently the receptor-level mechanisms of pro-inflammatory cytokine action in the brain were reviewed by Turrin & Rivest (2004). The latter concluded that pro-inflammatory cytokines can act on endothelial cells forming the blood-brain barrier and at circumventricular organs to cause release of prostaglandin E-2 (Turrin & Rivest 2004). Another signalling pathway in the brain involves binding of cytokines to endothelial cells with activation of nitric oxide synthase and production of nitric oxide as the secondary messenger (Wong et al. 1996; Scammell et al. 1996). Both the prostaglandin-mediated and the nitric oxide-mediated mechanisms are important in induction of various brain-level responses from peripherally derived cytokines (Sternberg & Webster 2003).

Fever is one aspect of the immune response that, in the case of endotoxin administration, is mediated centrally by prostaglandins and which may have direct effects on the reproductive axis. Low dose delivery of LPS (40 ng/kg) induced fever and elevated circulating cortisol levels in sheep (Williams et al. 2001). However, this dose appeared to be near the threshold for effects on GnRH pulsatility because suppression of GnRH pulsing occurred in only a few of the treated ewes and only 10 out of 16 ewes had disrupted LH pulsing. This occurrence of fever without complete disruption of LH pulsing is consistent with an earlier finding (Schillo et al. 1990) that elevated body temperature did not alter LH pulsing in ewes. However, the latter authors reported a reduction in mean plasma LH concentration that leaves the question of direct effects of elevated body temperature on LH secretion unresolved. Further work conducted both in Professor Karsch's laboratory in Michigan and at Lincoln University, New Zealand has attempted to determine whether the endotoxin-induced inhibition of LH pulsing is attributable to the elevation of body temperature *per se* by using cyclo-oxygenase inhibitors to block endotoxin effects in ewes exposed to environmental temperatures that raised their body temperature to values equivalent of endotoxin-induced fever. The results (unpublished) give moderate support for some influence of elevated body temperature *per se* on inhibition of LH pulsatility, which provides further evidence for a prostaglandin-independent pathway for suppression of the reproductive neuroendocrine axis.

Overview

It is well accepted that immune/inflammatory stress and activation of HPA axis inhibit the reproductive neuroendocrine axis and previous reviews of this topic (e.g. Karsch & Battaglia 2002) have shown these disruptive effects acting at several regulatory processes in this axis. In the past three to four years there has been further resolution of these pathways, giving insight into separate avenues for direct cytokine actions at central levels and for activation of the HPA axis, with the central disruption mediated by cortisol. Also, there is a clearer view of the involvement of ovarian steroids in these events, as revealed by comparison between studies using ovariectomised and ovarian intact ewes. There are many aspects of brain function in relation to regulation of reproduction in mammals that remain shrouded in mystery. These include changes associated with puberty, seasonality, nutrition, immunostimulation and other forms of stress, and aging. It is almost certain that the mechanisms of some of these changes in brain function will be linked to each other and the continued resolution of regulatory mechanisms for any one of these will unravel and help to explain others. This argument should justify continuation of the stepwise approach to resolving individual pathways that has been demonstrated by the studies reviewed above.

Conclusions

It is clear that the multiplicity of pathways by which immune-neuroendocrine interactions impinge on the reproductive axis is far from resolved and there is much further investigation required to identify the specific sites, cells and cellular mechanisms that are involved. There is some uncertainty about the specific involvement of ovarian steroids in these processes. However, it seems clear that discrepancies which have arisen between results from studies using ovarian intact ewes and those involving ovariectomised ewes or steroid-replaced ovariectomised ewe models indicate an important role of ovarian steroids in the mechanisms that mediate the effects of stress-activated pathways. Ovarian steroids appear to enhance sensitivity of these pathways, which means that the presence of ovaries is required in any conceptual model of these processes. The ovary can thus be regarded as an essential component of the immune stress-reproductive neuroendocrine axis, i.e. its secretory products are involved in the central mediation of events as well as it being a target end-organ of the female reproductive system. Fig.1 provides a schema based on earlier models (e.g. see Karsch & Battaglia 2002) that incorporates linkages revealed by recent studies, primarily from those using sheep as experimental animals.

Activation of an immune response, as commonly modelled by administration of bacterial endotoxin, causes the release of pro-inflammatory cytokines as well as activation of the HPA axis with consequent release of corticosteroids. These humoral factors activate endothelial cells within CNS blood vessels or stimulate CNS neurons and anterior pituitary cells directly, leading to suppression of the GnRH pulse generating and surge mechanisms in the hypothalamus and/or inhibiting anterior pituitary responsiveness to GnRH. There is evidence for both prostaglandin mediation and prostaglandin-independent mediation of these processes and for involvement of the type II glucocorticoid receptor in the corticosteroid-mediated events. With respect to ruminant reproduction, it is noteworthy that virtually all of the recent studies are based on sheep as the animal model and much of the sheep work has emanated from a single laboratory.

Acknowledgement

I am indebted to Professor Fred Karsch of the University of Michigan for providing me with over 15 years of collaborative interaction including many enjoyable months working in his laboratory. He has been instrumental in fostering my involvement in the neuroendocrinology of reproduction in ruminant animals and he introduced me to the role of the immune system in these processes.

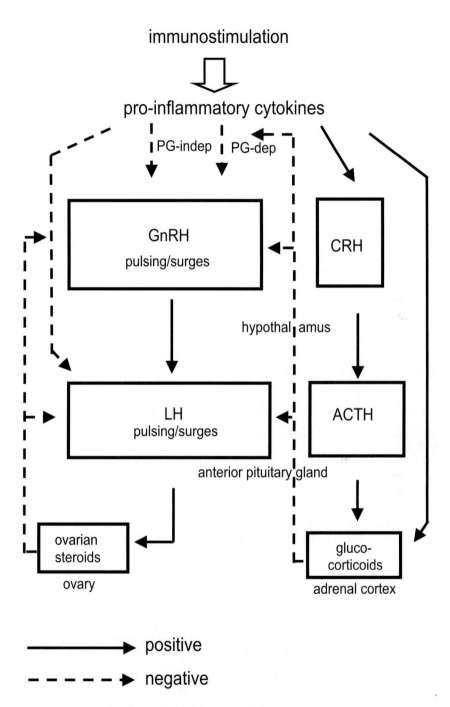

Fig 1. Pathways of linkages between pro-inflammatory cytokines released during immunostimulation and their disruptive effects on the reproductive neuroendocrine axis. Solid arrows represent stimulation, dashed arrows represent inhibition.
N.B. PG-indep = prostaglandin-independent, PG-dep = prostaglandin-dependent

References

Abraham IM, Todman MG, Korach KS & Herbison AE 2004 Critical *in vivo* roles for classical estrogen receptors in rapid estrogen actions on intracellular signalling in mouse brain. *Endocrinology* **145** 3055-3061.

Adam CL, Archer ZA & Miller DW 2003 Leptin actions on the reproductive neuroendocrine axis in sheep. *Reproduction Supplement* **61** 283-297.

Adams TE, Sakurai H & Adams BM 1999 Effect of stress-like concentrations of cortisol on estradiol-dependent expression of gonadotropin-releasing hormone receptor in orchidectomized sheep. *Biology of Reproduction* **60** 164-168.

Arreguin-Arevalo JA & Nett TM 2006 A nongenomic action of estradiol as the mechanism underlying the acute suppression of secretion of luteinizing hormone in ovariectomized ewes. *Biology of Reproduction* **74** 202-208.

Banks WA 2001 Cytokines, CVOs, and the blood-brain barrier. In *Psychoimmunology*, pp. 483-497. Eds R Ader, D Felten & N Cohen San Diego: Academic Press.

Barb CR & Kraeling RR 2004 Role of leptin in the regulation of gonadotropin secretion in farm animals. *Animal Reproduction Science* **82-83** 155-167.

Battaglia DF, Bowen JM, Krasa HB, Thrun LA, Viguié C & Karsch FJ 1997 Endotoxin inhibits the reproductive neuroendocrine axis while stimulating adrenal steroids: a simultaneous view from the hypophyseal portal and peripheral blood. *Endocrinology* **138** 4273-4281.

Battaglia DF, Brown ME, Krasa HB, Thrun LA, Viguié C & Karsch FJ 1998 Systemic challenge with endotoxin stimulates cortictropin-releasing hormone and arginine vasopressin secretion into hypophyseal portal blood: coincidence with gonadotropin-releasing hormone suppression. *Endocrinology* **139** 4175-4181.

Battaglia DF, Beaver AB, Harris TG, Tanhehco E, Viguié C & Karsch FJ 1999 Endotoxin disrupts the estradiol-induced luteinizing hormone surge: interference with estradiol signal reading, not surge release. *Endocrinology* **140** 2471-2479.

Benveniste EN 1998 Cytokine actions in the central nervous system. *Cytokine Growth Factor Reviews* **9** 259-175.

Braden ID, Fry C & Sartin JL 1998 Effects of interleukins on secretion of luteinizing hormone from ovine pituitary cells. *American Journal of Veterinary Research* **59** 1488-1493.

Breen KM & Karsch FJ 2004 Does cortisol inhibit pulsatile luteinizing hormone secretion at the hypothalamic or pituitary level? *Endocrinology* **145** 692-698.

Breen KM & Karsch FJ 2006 Does season alter responsiveness of the reproductive neuroendocrine axis to the suppressive actions of cortisol in ovariectomized ewes? *Biology of Reproduction* **47** 41-45.

Breen KM, Billings HJ, Debus N & Karsch FJ 2004a Endotoxin inhibits the surge secretion of gonadotropin-releasing hormone via a prostaglandin-independent pathway. *Endocrinology* **145** 221-227.

Breen KM, Stackpole CA, Clarke IJ, Pytiak AV, Tilbrook AJ, Wagenmaker ER, Young EA & Karsch FJ 2004b Does the type II glucocorticosteroid receptor mediate cortisol-induced suppression in pituitary responsiveness to gonadotropin-releasing hormone? *Endocrinology* **145** 2739-2746.

Breen KM, Billings HJ, Wagenmaker ER, Wessinger EW & Karsch FJ 2005a Endocrine basis for disruptive effects of cortisol on preovulatory events. *Endocrinology* **146** 2107-2115.

Breen KM, Oakley AE, Wagenmaker ER, Rispoli LA, Nett TM & Karsch FJ 2005b Rapid action of cortisol to suppress pituitary responsiveness to GnRH occurs independent of decreased GnRH receptor. *Biology of Reproduction, Special Issue for 38th Annual Meeting of the Society for the Study of Reproduction* 112 (abstract 137).

Case LC, Breen KM, Wagenmaker ER, Young EA & Karsch FJ 2005 Does a type II glucocorticoid receptor agonist inhibit pituitary responsiveness to GnRH? *Biology of Reproduction, Special Issue for 38th Annual Meeting of the Society for the Study of Reproduction* 224 (abstract W648).

Chen A, Ganor Y, Rahimipour S, Ben-Aroya N, Koch Y & Levite M 2002 The neuropeptides GnRH-II and GnRH-I are produced by human T cells and trigger laminin receptor gene expression, adhesion, chemotaxis and homing to specific organs. *Nature Medicine* **8** 1421-1426.

Colditz IG 2004 Some mechanisms regulating nutrient utilisation in livestock during immune activation: an overview. *Australian Journal of Experimental Agriculture* **44** 453-457.

Coleman ES, Elsasser TH, Kemppainen RJ, Coleman DA & Sartin JL 1993 Effect of endotoxin on pituitary hormone secretion in sheep. *Neuroendocrinology* **58** 111-122.

Dadoun F, Guillaume V, Sauze N, Farisse J, Velut JG, Orsoni JC, Gaillard R & Oliver C 1998 Effect of endotoxin on the hypothalamic-pituitary-adrenal axis in sheep. *European Journal of Endocrinology* **138** 1931-197.

Debus N, Breen KM, Barrell GK, Billings HJ, Brown M, Young EA, Karsch FJ 2002 Does cortisol mediate endotoxin-induced inhibition of pulsatile luteinizing hormone and gonadotropin-releasing hormone secretion? *Endocrinology* **143** 3748-58.

de Roux N, Genin E, Carel JC, Matsuda F, Chaussain JL & Milgrom E 2003 Hypogonadotropic hypogonadism due to loss of function of the KiSS1-derived peptide receptor GPR54. *Proceedings of the National Academy of Science, U.S.A.* **100** 10972-10976.

Dufournay L & Skinner DC 2002 Progesterone receptor, estrogen receptor a, and the type II glucocorticoid receptor are coexpressed in the same neurons of the pre optic area and arcuate nucleus: a triple immunolabeling study. *Biology of Reproduction* **67** 1605-1612.

Dungan HM, Clifton DK & Steiner RA 2006 Minireview: kisspeptin neurons as central processors in the regulation of gonadotropin-releasing hormone secretion. *Endocrinology* **147** 1154-1158.

Elmquist JK, Scammell TE & Saper CB 1997 Mechanisms of CNS response to systemic immune challenge: the febrile response. *Trends in Neuroscience* **20** 565-570.

Fortin M-E, Pelletier R-M, Meilleur M-A & Vitale ML 2006 Modulation of GJA1 turnover and intercellular communication by pro-inflammatory cytokines in the anterior pituitary folliculostellate cell line TtT/GF. *Biology of Reproduction* **74** 1-12.

Goehler LE, Gaykema RPA, Hansen MK, Anderson K, Maier SF & Watkins LR 2000 Vagal immune-to-brain communication: a visceral chemosensory pathway. *Autonomic Neuroscience* **85** 49-59.

Gill JC, Moenter SM & Tsai P-S 2004 Developmental regulation of gonadotropin-releasing hormone neurons by fibroblast growth factor signalling. *Endocrinology* **145** 3830-3839.

Gill JC & Tsai P-S 2006 Expression of a dominant negative FGF receptor in developing GnRH1 neurons disrupts axon growth and targeting in the median eminence. *Biology of Reproduction* **74** 463-472.

Greer AW, Stankiewicz M, Jay NP, McAnulty RW & Sykes AR 2005 The effect of concurrent corticosteroid induced immuno-suppression and infection with the intestinal parasite *Trichostrongylus colubriformis* on food intake and utilization in both immunologically naïve and competent sheep. *Animal Science* **80** 89-99.

Harris TG, Battaglia DF, Brown ME, Brown MB, Carlson NE, Viguié C, Williams CY & Karsch FJ 2000 Prostaglandins mediate the endotoxin-induced suppression of pulsatile gonadotropin-releasing hormone and luteinizing hormone in the ewe. *Endocrinology* **141** 1050-1058.

Huising MO, Kruiswijk CP & Klik G 2006 Phylogeny and evolution of class-I helical cytokines. *Journal of Endocrinology* **189** 1.25.

Kalra PS, Edwards TG, Xu B, Jain M & Kalra SP 1998 The anti-gonadotropic effects of cytokines: the role of neuropeptides. *Domestic Animal Endocrinology* **15** 321-332.

Karsch FJ & Battaglia DF 2002 Mechanisms for endotoxin-induced disruption of ovarian cyclicity: observations in sheep. *Reproduction Supplement* **59** 101-113.

Maeda K & Tsukamura H 2006 The impact of stress on reproduction: are glucocorticoids inhibitory or protective to gonadotropin secretion? *Endocrinology* **147** 1085-1086.

Matsuwaki T, Watanabe E, Suzuki T, Yamanouchi K & Nishikara M 2003 Glucocorticoid maintains pulsatile secretion of luteinizing hormone under infectious stress conditions. *Endocrinology* **144** 3477-3482.

Matsuwaki T, Suzuki M, Yamanouchi K & Nishikara M 2004 Glucocorticoid counteracts the suppressive effects of tumour necrosis factor-a on the surge of luteinizing hormone in rats. *Journal of Endocrinology* **181** 509-513.

Matsuwaki T, Kayasuga Y, Yamanouchi K & Nishikara M 2006 Maintenance of gonadotropin secretion by glucocorticoids under stress conditions through the inhibition of prostaglandin synthesis in the brain *Endocrinology* **147** 1087-1093.

Maya-Nunez G & Conn PM 2003 Transcriptional regulation of the GNRH receptor gene by glucocorticoids. *Molecular and Cellular Endocrinology* **200** 89-98.

Oakley AE, Breen KM, Tilbrook AJ, Wagenmaker ER & Karsch FJ 2006 Differing modes of cortisol-induced suppression of pulsatile LH secretion in ovariectomized vs. ovary-intact ewes. *Biology of Reproduction, Special Issue for 39th Annual Meeting of the Society for the Study of Reproduction* 75 (abstract 19).

Perkowski SZ, Sloane PJ, Spath Jr JA, Elsasser TH, Fisher JK & Gee MH 1996 TNF-a and the pathophysiology of endotoxin-induced acute respiratory failure in sheep. *Journal of Applied Physiology* **80** 564-573.

Sapolsky RM, Romero LM & Munck AU 2000 How do glucocorticoids influence stress responses? Integrating permissive, suppressive, stimulatory, and preparative actions. *Endocrine Reviews* **21** 55-89.

Schillo KK, Alliston CW & Malven PV 1990 Plasma concentrations of luteinizing hormone and prolactin in the ovariectomized ewe during induced hyperthermia. *Biology of Reproduction* **19** 306-313.

Scammell TE, Elmquist JK & Saper CB 1996 Inhibition of nitric oxide synthase produces hypothermia and depresses lipopolysaccharide fever. *American Journal of Physiology* **271** 333-338.

Seminara SB, Messager S, Chatzidaki EE, Thresher RR, Acierno JS, Shagoury JK, Bo-Abbas Y, Kuohung W, Schwinof KM, Hendrick AG, Zahn D, Dixon J, Kaiser UB, Slaugenhaupt SA, Gusella JF, O'Rahilly S, Carlton MB, Crowley WF, Aparicio SA & Colledge WH 2003 The GPR54 gene as a regulator of puberty. *New England Journal of Medicine* **349** 1614-1627.

Shepherd AJ, Downing JEG & Miyan JA 2005 Without nerves, immunology remains incomplete – *in verito veritas*. *Immunology* **116** 145-163.

Smith JT, Cunningham MJ, Rissman EF, Clifton DK & Steiner RA 2005 Regulation of *Kiss 1* gene expression in the brain of the female mouse. *Endocrinology* **146** 3686-3692.

Stackpole CA, Turner AI, Clarke IJ, Lambert GW & Tilbrook AJ 2003 Seasonal differences in the effect of isolation and restraint stress on the luteinizing hormone response to gonadotropin-releasing hormone in hypothalamopituitary disconnected, gonadectomized rams and ewes. *Biology of Reproduction* **69** 1158-1164.

Sternberg EM 1997 Neural-immune interactions in health and disease. *Journal of Clinical Investigation* **100** 2641-2647.

Sternberg EM & Webster JI 2003 Neural immune interactions in health and disease. In *Fundamental immunology*, edn 5, pp 1021-1042. Ed WE Paul. Philadelphia: Lippincott Williams and Wilkins.

Stevens DL 1995 Streptococcal toxic-shock syndrome: spectrum of disease, pathogenesis, and new concepts in treatment. *Emerging Infectious Diseases* **1** 69-78.

Sugino T, Hasegawa Y, Kurose Y, Kojina M, Knagawa K & Terashima Y 2004 Effects of ghrelin on food intake and neuroendocrine function in sheep. *Animal Reproduction Science* **82-83** 183-194.

Tena-Sempere M 2005 Hypothalamic KiSS-1: the missing link in gonadotropin feedback control. *Endocrinology* **146** 3683-3685.

Turnbull AV & Rivier CL 1999 Regulation of the hypothalamic-pituitary-adrenal axis by cytokines: actions and mechanism of action. *Physiological Reviews* **79** 1-71.

Turrin NP & Rivest S 2004 Unraveling the molecular details involved in the intimate link between the immune and neuroendocrine systems. *Experimental Biology and Medicine* **229** 996-1006.

Tyree CM, Zou A & Allegretto EA 2002 17ß-Estradiol inhibits cytokine induction of the human E-selectin promoter. *Journal of Steroid Biochemistry and Molecular Biology* **80** 291-297.

Vellucci SV & Parrott RF 1996 Bacterial endotoxin-induced gene expression in the choroid plexus and paraventricular and supraoptic hypothalamic nuclei of the sheep. *Molecular Brain Research* **43** 41-50.

Wagenmaker ER, Breen KM, Clarke IJ, Oakley AE, Tilbrook AJ, Turner AI & Karsch FJ 2005 Cortisol delays the estradiol-induced LH surge: is this dependent on prior ovarian steroid exposure? *Biology of Reproduction, Special Issue for 38th Annual Meeting of the Society for the Study of Reproduction* 223 (abstract W645).

Williams CY, Harris TG, Battaglia DF, Viguié C & Karsch FJ 2001 Endotoxin inhibits pituitary responsiveness to gonadotropin-releasing hormone. *Endocrinology* **142** 1915-1922.

Wong M-L. Rettori V, Al-Shekhlee A, Bongiorno PB, Canteros G, McCann SM, Gold PW & Licinio J 1996 Inducible nitric oxide synthase gene expression in the brain during systemic inflammation. *Nature Medicine* **2** 581-584.

Nutritional inputs into the reproductive neuroendocrine control system – a multidimensional perspective

D Blache[1], LM Chagas[2] and GB Martin[1]

[1]School of Animal Biology, M085, Faculty of Natural and Agricultural Sciences, The University of Western Australia, 35 Stirling Highway, Crawley, WA 6009, Australia; [2]Dexcel, Private Bag 3221, Hamilton, New Zealand

Evolution has shaped regulatory systems to improve the chance of reproductive success in a somewhat unpredictable environment. One of the more powerful regulators of reproductive function in both sexes is metabolic status, defined as the availability of nutrients and energy to the tissues. Here, we briefly review the basics of the relationship between metabolic status and the activity of the system that controls pulsatile GnRH and LH secretion. We then reflect on these relationships within the framework of a model that comprises four interdependent 'dimensions': 1) genetic, 2) structural, 3) communicational, and 4) temporal. Using two major examples, the male sheep and the post-partum dairy cow, we illustrate aspects of each dimension that seemed to have evolved to limit the risks associated with 'the decision to reproduce'. The results of recent studies have also led us to include in our model the concepts of 'metabolic memory' and 'nutrient sensing' to help explain some aspects of the temporal dimension. Throughout the review, we propose directions for future research that could shed light on pathways that have evolved to ensure that animals are able to take the least risky 'decision'.

Introduction

In this review, we are looking at how nutrition influences reproduction, an old problem that has been researched for decades and reviewed in detail many times (most recent: Butler 2005; Blache et al. 2006; Robinson et al. 2006). In this instance, we are attempting to analyse the issues from a different perspective – we will consider the multi-dimensional nature of an integrated control system that has developed over an evolutionary timescale to ensure that an animal has the best chance of reproductive success in an unpredictable environment. Of course, for farmed ruminants, the environment is nowadays far more predictable than that with which their ancestors had to cope before domestication. On the other hand, domestication has also led to intense demands, such as selection pressure for increased milk production by modern dairy cows, that greatly increase the energy requirements of the animals – in effect, this is an extreme metabolic stressor that challenges the homeostatic processes developed during evolution.

Corresponding author E-mail: dbla@animals.uwa.edu.au

Before elaborating the notion of a multi-dimensional control system, we will first define several basic concepts in the bioenergetics of reproduction, metabolic status, and their effects on reproduction. Having set the scene, we will consider each of four 'dimensions' of what we see as a 'decision-making system' that allows ruminants to optimise their reproductive success. First, we will briefly look at the genetic dimension, one that sets the limits of responsiveness to nutritional input to animals. Second, we will consider the structural dimension, a growing list of organs and tissues involved in the linkage between metabolic status and reproduction. The third dimension, the communication network of signalling pathways, will be treated in greater detail. The fourth and final dimension is temporal so concerns the role of time on the metabolism-reproduction relationship – and has led us to incorporate 'metabolic memory' and 'nutrient sensing' into our hypothetical control system (Cahill 1980; Obici & Rosetti 2003). To illustrate these concepts, we will mainly use two examples, the male sheep that has been acutely placed on a high plane of nutrition, and the high-output dairy cow that is undergoing a dramatic change in energy balance in the early postpartum period. We will conclude by discussing the four-dimensional model as a conceptual framework upon which we can base further study of the interactions between nutrition and reproduction. Most of the processes involved are autonomic by nature but, to aid clarity of expression, we will occasionally risk the use of anthropomorphic language (eg, 'decision to invest').

Basic concepts

Bioenergetics of reproduction

Energy costs are attached to all components of the reproductive process, from the expression of specific behaviours, such as sexual or maternal behaviour, to the production of morphological elements, such as gametes, fetuses and milk. In mammals, reproduction is energetically more demanding for females than for males because of the requirements for gestational development and the production of milk. The peaks in energy demand for reproduction also differ in timing between the sexes, with the male investing most energy before fertilization, while females invest most energy following fertilization, often with considerable delay (Horton & Rowsemitt 1992). For example, in grazing ewes, gestation costs only 3% of the daily energy expenditure during the first 3 months but this builds rapidly to 20% during the last 2 months (Fierro & Bryant 1990). The energy requirement for dairy cows to ovulate a follicle, form a corpus luteum, and maintain early pregnancy is minuscule compared with the requirement for lactation, which requires up to 50% of the daily energy expenditure (Graham 1964; Fierro & Bryant 1990). For the human female, it has been suggested that the reproductive axis will not be activated until the energy requirements for pregnancy are already in storage (Frisch 1994) but, in a striking contradiction of this logical concept, dairy cows have been successfully selected genetically for their ability to conceive during lactation when their energy balance is negative.

The critical time point in the reproductive cycle seems to be when the animal makes an 'investment decision', accepts the risks inherent in procreation, and 'switches on' its reproductive system. Considering the energetic requirements, the relationship between the metabolic and the reproductive regulatory systems needs to be highly tuned if the probability of success is to be reasonable. As a natural illustration of this matching of potential future resources to potential future demands, the energy regulatory systems of mammals are more strongly tied to reproductive function in females than in males because of the much higher cost for reproduction paid by females.

The concept of metabolic status

At any given time, for any given animal, the amount of energy available for reproduction depends on the difference between the amount of energy expended, including the demands for maintenance, and pool of disposable energy (Fig. 1). The pool of disposable energy includes the energy derived from feed intake plus the energy stored in body tissues, especially adipose tissue, liver and muscle. The amount of expended energy varies according to the age and the physiological status of the animal and it comprises the energy spent on functions that are responsible for the maintenance of homeostasis, as well as energy spent on extra physiological needs such as growth and reproduction. Most authors refer to 'energy balance' instead of 'metabolic status'. The terms can be seen as interchangeable but 'metabolic status' includes an integrative dimension that 'energy balance' does not.

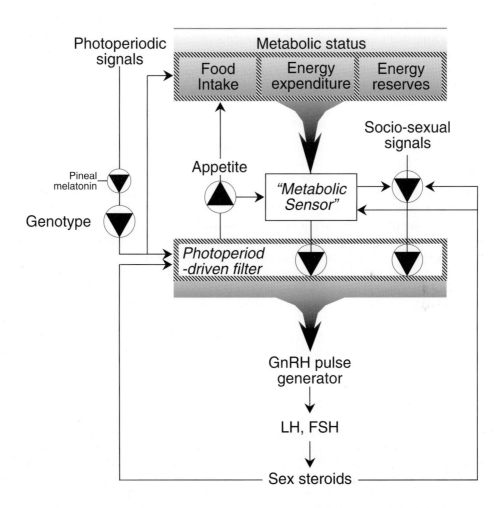

Fig. 1: A schema describing the proposed relationships between photoperiodic, nutritional and social cues and they ways that they interact with genotype and steroid feedback in the control of hypothalamo-pituitary-testicular axis in the male sheep. Nutritional input is via 'metabolic status', a reflection of the difference between energy expenditure and the sum of energy available from food intake and from energy reserves, as measured by a hypothetical 'metabolic sensor'. Redrafted after Blache et al. 2003.

The neuroendocrine system – the fundamental controller of reproduction

A full reproductive cycle starts with the emergence of sexual activity in future parents and finishes with emergence of sexual activity in the subsequent generation. The various phases of the cycle differ in their duration between the sexes but the secretion of GnRH, and consequently LH and FSH from the pituitary gland, are central to the process in both (Blache *et al.* 2002; Blache *et al.* 2003; 2006). Effectively, the 'decision' of an animal to reproduce or not is implemented by the system, as yet not described, that controls the production of pulses of GnRH – there is a threshold frequency of pulses above which males will produce sperm and females will ovulate. Thus, during postpartum anoestrous in dairy cattle, pulsatile infusion of LH decreases the time delay to first ovulation by stimulating follicular growth and steroidogenesis (Hampton *et al.* 2003). The system that controls GnRH secretion is also the final common pathway via which gonadal activity is usually influenced by external factors (Fig. 1), including socio-sexual cues, photoperiod and energy balance (review: Martin *et al.* 2004). That said, it is important to note that tissues involved in the process of reproduction, including the pituitary gland, gonads and mammary glands, can also respond independently to metabolic inputs, if not to the extent of switching the reproductive process on or off.

Metabolic status and investment in reproduction

Because reproduction is very demanding energetically, metabolic status is arguably the most powerful internal regulator of reproductive function and variation in metabolic status can profoundly affect the reproductive cycle at almost any stage. In mature male sheep, an acute increase in the intake of energy and protein induces, first, an increase in the frequency of pulses of GnRH and LH, and then an increase in the tonic secretion of FSH (Martin *et al.* 1994). An acute reduction of feed intake has the opposite effects (Martin & Walkden-Brown 1995). Similarly, in the post-partum cow, the activity of the GnRH neurons is highly positively correlated with energy balance, so pulse frequency increases when energy balance shifts from negative to positive values following an increase in intake or a decrease in expenditure (review: Butler 2000; 2005). Dairy cows have been selected heavily for high milk production and show a high level of nutrient utilisation and adipose tissue mobilization during early lactation, so the impact of milk production on reproduction can be reduced by reducing milking frequency. Reducing milking frequency to once daily increases the rate of spontaneous resumption of oestrous cycles in anoestrous cows (Rhodes *et al.* 1998) independently of any direct effect of diet (Patton *et al.* 2006), although the role played by GnRH pulse frequency in determining the length of postpartum anoestrus interval (PPAI) is yet to be proven directly.

A 4-dimensional linkage between nutrition and reproduction

Naturally, animals have evolved to increase their chances of reproducing and regulatory systems have evolved to fulfil that drive. We will now outline a framework within which we can analyse the processes and relationships involved. Following an analogy with theoretical physics, we will consider the regulatory system as having four dimensions: genetic, structural, communicational and temporal.

1) The genetic dimension

The effects of metabolic status and dietary manipulation on the reproductive axis differ between genotypes in both sheep and cattle. This genetic dimension accounts for variations in the responses to environmental inputs, whether they be natural (eg, photoperiod) or artificial (eg, genetic selection for high milk production). In our review for the previous symposium in this series, we suggested that photoperiod acts as a 'filter' of the effect of nutrition on the reproductive system and that this filtering effect depends on the genotype of the animal (Blache *et al.* 2003). This is because, in sheep, the effect of nutrition on the reproductive endocrine axis is smaller in breeds that are very responsive to photoperiod than in breeds that are less responsive (Hötzel *et al.* 2003; Martin *et al.* 1999; Martin *et al.* 2002). Similarly, a large body of work in dairy cattle has shown that genotype affects the ability of cows to resume ovulation during the postpartum period. Dairy cows in continuous calving systems that have not ovulated by 44 days postpartum are defined as having a prolonged PPAI (Lamming & Darwash 1998). It seems that the regulatory systems that control metabolism and energy balance can partition energy in different ways in cows of different merit and, in high merit animals, reproductive function has been de-prioritised in favour of milk production. This was clearly demonstrated in a study done in New Zealand that compared the performance of Friesian cows bred from semen collected in 1970s and 1990s – over that 20-year period of selection, the postpartum anovulatory interval increased from 7 to 12 days (McNaughton *et al.* 2003), despite the fact that selection for high milk production was associated with a increase in dry matter intake (Veerkamp *et al.* 2003). The genetic make-up of an animal, whether naturally or artificially acquired, is thus the first regulatory step in the influence of metabolic status on reproductive function.

2) The structural dimension

Metabolic status involves inputs from several energy pools that depend on the activity of a variety of organs so it is not surprising that recent decades of research have led to an increasing number and diversity in anatomical sites involved in the relationship between nutrition and reproduction. There has also been a parallel increase our understanding of the precise nature of the role played by each regulatory organ. The brain and the gonads have always been seen as the primary targets for nutritional input into reproductive function but we now accept major roles for the pancreas, liver and adipose tissue (Fig. 2). For adipose tissue, this transformation has been spectacular as it has been elevated from a passive storage site to a vital endocrine organ that produces a number of signals (detailed below). Similarly, the digestive system is also now implicated in the regulatory processes through which nutrition affects reproduction as its endocrine output has been detected and identified. The digestive system also produces very direct metabolic information in the form of energy metabolites (glucose, fatty acids) and amino acids, as detailed in the final section below. In summary, every organ involved in the regulation of the three compartments of metabolic status (intake, reserves, expenditure), is also involved in generating signals that influence reproductive activity (Fig. 2) – some of these have a very specific role (eg, the liver, an organ that processes first most of the circulating nutrients) whereas others have a more general input (eg, the thyroid gland).

3) The communicational dimension

The successful implementation of a 'decision to invest' in reproduction depends on the systems that regulate the reproductive axis, via changes in GnRH pulse frequency, responding accurately to changes in metabolic status. These regulatory systems can be classified as endocrine, neural and nutrient-based.

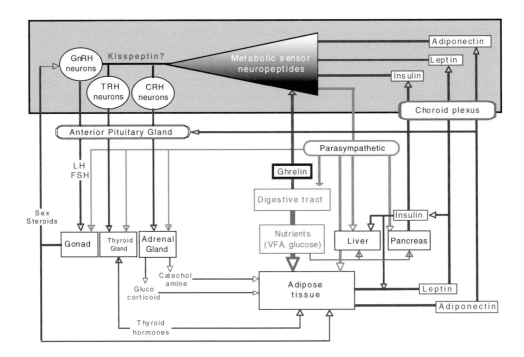

Fig. 2: Schematic summary of the potential relationships among the endocrine and neural inputs into the systems that control the reproductive system and mediate the responses to change in metabolic status. For clarity, we have omitted some hormonal systems (GH, prolactin, oxytocin) that are also involved in the control of the metabolic status but do not appear to exert any direct action on GnRH secretion. Within the brain, the neurochemical nature of the hypothetical 'metabolic sensor' that integrates the endocrine and neural inputs is still not known. Kisspeptin is proposed as a possible final link between the neuropeptide systems that respond to insulin and leptin, and perhaps adiponectin, and the GnRH neurons. It should be noted that the pituitary gland might be involved in the response of the reproductive axis to nutrition because it is responsive to many of the same hormonal signals as the hypothalamus.

Endocrine systems linking metabolic status and GnRH secretion

Insulin

Insulin is affected by energy balance and seems to be involved in the control of reproduction in male sheep because: 1) a high plane of nutrition leads to high concentrations of insulin in both plasma and CSF (Miller et al. 1998; Zhang et al. 2004; 2005); 2) with a restricted diet or with diabetes, a low dose of insulin infused into the third ventricle increases LH pulse frequency to values similar to those seen in well-fed animals (Miller et al. 1995a; Miller et al. 2002; Tanaka et al. 2000); 3) insulin receptors are present in the hypothalamus (Blache et al. 2002); 4) following an acute increase in dietary allowance, the increase in insulin secretion coincides with the increase in LH pulse frequency (Zhang et al. 2004). A central role for insulin in the regulation of GnRH secretion is also supported by studies with neuron-specific, insulin receptor knockout mice (Bruning et al. 2000).

In cattle, there are also similar associations: 1) dietary restriction and negative energy balance reduce plasma concentrations of insulin; 2) during post-partum anoestrus, the dominant follicle ovulates in response to restricted suckling only in cattle with high plasma concentrations of insulin (Sinclair *et al.* 2002); 3) dietary manipulation that increases plasma concentrations of propionate and insulin are also associated with a decrease in PPAI without any changes in either milk production or energy balance (Gong *et al.* 2002).

Thus, the accumulated evidence strongly suggests a major role for insulin in the GnRH response to variation in metabolic status, although, in dairy cattle, the evidence for this link is more mixed (review: Butler 2005). For example, during early lactation, infusion of insulin, associated with control of glycaemia, failed to stimulated LH pulse frequency, but did stimulate production of oestradiol by a direct action on the ovary (Butler *et al.* 2004). It appears that insulin does not always stimulate GnRH secretion, especially in presence of strong negative feedback.

Growth Hormone (GH)

The presence of mRNA for GH receptors in the hypothalamus and the pituitary gland (Kirby *et al.* 1996; Lucy *et al.* 1998) suggests that GH could link nutritional inputs to gonadotrophic outputs (Monget & Martin 1997). Energy balance certainly affects plasma concentrations of GH under a wide variety of conditions (Bossis *et al.* 1999) but, in male sheep at least, it is not likely to be involved in the stimulation of GnRH secretion because an increase in nutrition induces a decrease in plasma GH concentrations (Miller *et al.* 1998).

Insulin-like growth factor-I (IGF-I)

In male sheep, circulating concentrations of IGF-I are affected by diet and peripheral administration of physiological doses of IGF-I inhibits LH secretion, but this is probably through an action at pituitary level rather than by a change in GnRH pulse frequency (Adam *et al.* 1998). In addition, concentrations of IGF-I in CSF are not affected by diet (Miller *et al.* 1998) and we have not be able to demonstrate that IGF-I infusion into the third ventricle affects LH pulse frequency (Blache *et al.* 2000). It therefore seems likely that, in male sheep at least, IGF-I is not heavily involved in the effects of changes in nutritional status on the brain's reproductive centres. In the post-partum cow, plasma concentrations of IGF-I are negatively correlated with extended PPAI and positively correlated with body condition and food intake (Beam & Butler 1999). Plasma IGF-I values decrease in food-restricted heifers and, in post-partum dairy cows, IGF-I increases linearly up to the day of first ovulation (Beam & Butler 1997; Diskin *et al.* 2003). Also, IGF-I and its receptor have been found in the hypothalamus of the rat (Bohannon *et al.* 1988; Schechter *et al.* 1994) so IGF-I might play a direct role in the control of GnRH neurons and could be involved in the resumption of cyclicity in lactating cows (Beam & Butler 1997; Diskin *et al.* 2003). However, until this concept is tested directly in cattle, the information from studies with rams leads us to conclude that it is unlikely.

Thyroid hormones

Thyroid hormones play a role in the control of seasonal reproduction in sheep, probably through an input into endogenous cerebral rhythms (Karsch *et al.* 1995). However, in the mature male sheep, the concentrations of thyroid hormones in plasma and CSF are not affected by the acute elevations of nutrition that increase LH pulse frequency, suggesting that they play no role in the rapid stimulation of the GnRH neurons (Miller *et al.* 1998; Zhang *et al.* 2004; 2005). In lactating cows also, there is little or no change in thyroid hormones in relation to selection for milk production (Veerkamp *et al.* 2003). Overall, there is very little evidence for thyroid hormones being important signals linking energy balance and reproduction. Arguably, this is not surprising because the ubiquitous action of thyroid hormones is not compatible with a very specific action on the reproductive axis.

Hormones secreted by adipocytes

Our view of the role played by adipose tissue in the management of metabolic status has evolved from that of a passive energy reserve to that of a very active endocrine regulator of a multitude of bodily functions, including food intake, metabolism, immunity, thermoregulation and cardiovascular function, as well as reproduction (Ahima 2005; Cinti 2005; Kershaw & Flier 2004). Of the 20 or more endocrine products of adipose tissue, leptin appears to be the most important regulator of reproductive activity in ruminants (review: Chilliard *et al.* 2005), although leptin is also the most studied of the adipocyte hormones and future research might reveal key roles for the others. The expression and release of leptin, and the sensitivity of gonadal and brain tissues to leptin, are all altered by short- and long-term changes in metabolic status. Moreover, numerous experiments in female and male sheep have shown that leptin can affect the neuroendocrine systems that control the activity of the reproductive axis (review: Adam *et al.* 2003; Chilliard *et al.* 2005). A recent study in dairy cattle has illustrated the close link between leptin concentrations and LH pulse frequency during the post-partum period (Kadokawa *et al.* 2006), although a triggering role for leptin in the termination of anoestrus it is not always supported (Chagas *et al.* 2006). Indeed, the consensus of the large amount of leptin literature is that the role of leptin is permissive rather than triggering, and this is logical because adipose tissue is only one of the three compartments of metabolic status.

Another interesting adipose hormone, adiponectin, has yet to receive the same interest as leptin but perhaps should do so because it is exclusively produced by adipose tissue and, in contrast to leptin, it stimulates energy expenditure without any effect on feed intake when it is infused into the cerebral ventricle of the rat (Ahima 2005). In male rat pituitary cells in culture, adiponectin reduces the expression of GnRH receptor and decreases the secretion of LH (Malagon *et al.* 2006). However, an effect of adiponectin on the activity of the GnRH neurons is yet to be demonstrated.

Hormones secreted by the gastro-instestinal tract

A most interesting potential candidate in the regulation of reproduction by metabolic status is ghrelin, a gut hormone that is the endogenous ligand of the GH secretagogue (GHS) receptor. In sheep, ghrelin secretion is inhibited by an increase in feed intake and stimulated by fasting (Sugino *et al.* 2004). In rodents, the distribution of GHS receptors overlaps that of GnRH in the arcuate nucleus (St-Pierre *et al.* 2003) and ghrelin decreases the secretion of LH (Fernandez-Fernandez *et al.* 2004). Interestingly, ghrelin and GHS receptors are expressed in human and rat testis (Tena-Sempere 2005) and ghrelin is found in most reproductive tissues in both male and female sheep (Miller *et al.* 2005). In dairy cattle, Roche *et al.* (2006) measured plasma ghrelin concentration, pre- and post-feeding, during peak lactation (75 days postpartum) in animals of low and high genetic merit for milk production. High-merit cows produced more milk, consumed more pasture, and had higher plasma concentrations of ghrelin and GH, than low-merit cows. Itoh *et al.* (2005) reported a decline in ghrelin concentrations as cows progressed from early to mid- to late lactation, suggesting greater ghrelin production in cows during negative energy balance. Clearly, this should be a high priority area for further study.

Direct inputs by nutrients

Digestive processes lead to the production of critical nutrients, such as amino acids, fatty acids (volatile or non-volatile) and carbohydrates, that flow into the circulation and can act as signals in their own right. However, in ruminants there is no strong evidence that these molecules are major regulators of the secretion of GnRH pulses. In female sheep, pulsatile LH secretion is not affected by intravenous infusion of precursors of neurotransmitters of large-neutral amino acids (Downing *et al.* 1995; 1996; 1997). In contrast, studies with rams suggest that fatty acids, a major currency in energy transactions in ruminants, affect GnRH secretion (Blache *et al.* 2000). The role of glucose is less clear – in male sheep, intra-abomasal and intravenous infusions of glucose failed to stimulate LH secretion (Boukhliq & Martin 1997; Boukhliq *et al.* 1996; Miller *et al.* 1995b). In dairy cattle, however, twice-daily drenching with propylene glycol, a propionate precursor and thus a source of glucose, can profoundly reduce PPAI in primiparous cows in low body condition (Chagas 2003). Indicators of energy status (body condition score, milk production, glucose, insulin, GH, IGF-I) were not altered by this treatment, but there was an increase in LH pulse frequency and in ovarian activity and, consequently, earlier ovulation postpartum. This suggests that there are nutritional inputs, perhaps some form of 'nutrient sensing or signalling', that are not associated with whole body energy status but can nevertheless stimulate the hypothalamic-pituitary-ovary axis in the early postpartum period (discussed further below). Through such a pathway, nutrients could trigger an endocrine response that would stimulate the reproductive axis. For example, it has been proposed the effects of propylene glycol on resumption of ovulation post-partum are due to the stimulation of an insulin spike (Miyoshi *et al.* 2001) that acts as a signal to increase LH secretion and therefore evoke ovulation.

Interactions among the elements of the communicational network

Several levels of integration are needed because metabolic status at any given time depends on the status of all three key compartments – intake, storage and expenditure. The necessary interactions are managed by several hormonal systems, perhaps best exemplified by leptin secretion because it is affected by intake and expenditure as well as storage (mass of adipose tissue). In turn, leptin can stimulate the activity of three other endocrine systems involved in controlling the reproductive axis: i) pancreatic insulin in fasted cattle, albeit only at a low dose (20 μg/kg: (Zieba *et al.* 2003); ii) pituitary GH in female sheep (Henry *et al.* 2001) but apparently not normal-fed cattle (Zieba *et al.* 2005); and iii) thyroid hormones (Flier *et al.* 2000). In addition, leptin secretion is affected by other inputs, such as the products of digestion and absorption and autonomic neural activity. The autonomic effects are evident in the responses to challenges with beta-adrenergic analogues that decrease leptin sensitivity and stimulate lipolysis (Penicaud *et al.* 2000).

It is important to note that the ultimate level of integration is within the central nervous system (Fig. 2). The brain mechanisms involved in the sensing of metabolic status and in the connection of metabolic status to GnRH neuronal activity are poorly understood. Some sort of 'metabolic sensor' is thought to be localised in the arcuate nucleus and median eminence because receptors for leptin and insulin are found in these areas (Blache *et al.* 2002; Chilliard *et al.* 2005). The search for links between the GnRH cells and the neurons containing insulin and leptin receptors is continuing and there are many candidate intermediaries (see Blache *et al.* 2006). Orexins were initially among the most promising because of their sensitivity to energy balance (Taylor & Samson 2003) but intracerebral injections of orexin A or B inhibit GnRH activity in mature Merino rams (Blache et al. 2003) so their relevance is questionable. On the other hand, we need to continue investigating the roles of the orexinergic pathways because high intakes of energy and protein dramatically reduce the expression of orexin receptor 2 in the paraventricular nucleus (Blache et al. 2006). We now need to test whether this change in sensitivity to orexins is linked to a change in reproductive activity or a change in food intake.

Recently, a new neuropeptide, kisspeptin, has been shown to stimulate GnRH and FSH secretion in male rodents (Gottsch et al. 2004; Irwig et al. 2004; Navarro et al. 2005) and to stimulate GnRH and LH secretion in female sheep (Messager et al. 2005). Importantly, the kisspeptin receptor is found in over 75% of GnRH neurons in male rats (Irwig et al. 2004) and kisspeptin can be seen as major player in the processing of inputs into the systems that regulate GnRH secretion (Dungan et al. 2006). Interestingly, the ligand acts through a membrane receptor so kisspeptin pathways could mediate the rapid responses of the GnRH system to acute changes in metabolic status. Research is needed to test whether kisspeptin secretion or expression is regulated by energy balance.

The interactions among peripheral signals (insulin, leptin, ghrelin, adiponectin) and the neuroendocrine systems (kisspeptin) need to be revealed if we are to understand the integratory role of the brain in the process that leads to the 'decision to reproduce'.

4) The temporal dimension

In this section, we consider the dynamic aspects of the responses to nutritional inputs, the effects of time per se, as seen in the effects of photoperiod and fetal programming, and, finally, the concepts of 'metabolic memory' and 'nutrient sensing', as we attempt to integrate the four dimensions into a unified theoretical framework.

Dynamics of responses to change in nutrition

In the ram, the response of the GnRH neurons to an abrupt change in nutrition is initially rapid and robust, but then fades over the next few weeks (Martin et al. 1994; Zhang et al. 2004). The rapidity of these responses is also consistent with the autonomic nervous system being involved at both brain and adipose levels. This could involve input from the digestive viscera to the brain via the vagus nerve, as is the case for satiety responses following a meal (Woods et al. 2004). Some nutrients that are absorbed very quickly, such as some volatile fatty acids, could also be involved in this rapid response. In fact, plasma concentrations of insulin increase about 3 h before the start of GnRH response to the initial increase in intake (Zhang et al. 2004).

In contrast, the long-term effect of nutrition on the ram testis, measured on a scale of several weeks, seems to be independent of changes in the primary, GnRH-based, control system (Hötzel et al. 1995). The mechanism involved has not been studied further but, again, leptin might be implicated because, in the rat at least, the testis contains leptin receptor and leptin can inhibit testicular steroidogenesis (Tena-Sempere et al. 1999; Tena-Sempere et al. 2001).

On an annual timescale, the role of nutritional inputs, as well as the types of response to those inputs, can vary substantially, especially in genotypes that experience seasonal changes in appetite (Rhind et al. 2002). Some suggestions have been forwarded for control systems that might implement these strategies: i) ghrelin expression in sheep testis responds to changes in photoperiod (Miller et al. 2005); ii) leptin secretion varies among seasons (Alila-Johansson et al. 2004); iii) brain expression of relevant neuropeptides is affected by photoperiod (Adam et al. 2000); iv) adipocytes receive sympathetic and parasympathetic innervations, at least in the Siberian hamster, a seasonal animal (review: Bartness et al. 2002) – the sympathetic input controls several functions (eg, lipolysis, regulation of adipocyte number) including, most interestingly, the secretion of leptin (Bartness *et al.* 2002); v) there are autonomic nervous projections from the adipocytes to the brain (Bartness *et al.* 2005), a connection that could provide metabolic information rapidly and directly to the reproductive centres (Fig. 2).

Metabolic memory

In both sheep and cattle, previous metabolic status influences the reproductive response of the animals to an increase in energy availability. With respect to adipose stores, mature rams in low body condition, but not rams in high body condition, show a robust and repeatable increase in LH pulse frequency in response to an increase in intake. In low body condition rams, the leptin response is also blunted but the response to insulin is not (Zhang *et al.* 2005). These observations suggest that i) neither insulin nor leptin are necessary for inducing an increase in GnRH pulse frequency in response to an increase in food intake, and ii) leptin secretion does not always respond to an influx of nutrients. In contrast, in mature, grazing dairy cows calving in good body condition, pre-calving dry matter intake has no effect on PPAI or on post-calving plasma concentrations of leptin, IGF-I, or GH (Roche *et al.* 2005). In contrast, in first lactation heifers that have a low body condition score (4 on a scale of 10) 6 weeks prepartum, there are clear responses to an *ad-libitum* intake in the pre-partum period, in both metabolic hormones and LH pulse frequency (Chagas *et al.* 2006). One group of heifers in this study calved with low body condition and also received *ad libitum* feed postpartum, but they did not respond to postpartum feeding and they had low LH pulse frequency and also a longer PPAI. Together, these three studies in cattle and sheep (Chagas *et al.* 2006; Roche *et al.* 2005; Zhang *et al.* 2005) are consistent with the notion of a 'metabolic memory' that modulates the stimulatory effect of nutrient intake according to the level of either energy reserves or energy expenditure (Fig. 3).

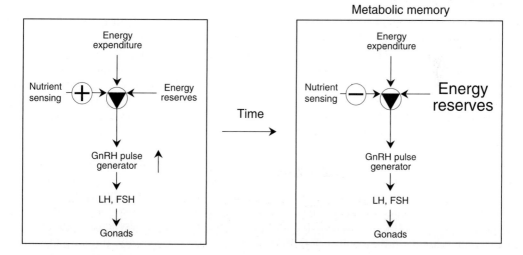

Fig. 3: A schema describing the potential influence of a 'metabolic memory' (here, an increase in body reserves) on the stimulatory effect of the nutrient influx (via 'nutrient sensing'). On the left panel, the energy reserves are low and the 'nutrient sensor' is able to stimulate the hypothetical 'metabolic sensor' in the brain and thus the GnRH pulse generator. With time, the level of energy reserves may increase, as shown in the right panel, and the input from the nutrient sensor is decreased. Note: nutrients might be still able to stimulate gonadal function by pathways that are GnRH-independent.

Recently, the processes that might underpin 'metabolic memory' have been investigated in sheep and cattle and the data suggest that leptin and insulin are central to the concept (Chilliard et al. 2005). For example, leptin stimulates LH secretion from pituitary explants of normal-fed but not fasted cows (Amstalden et al. 2003). With respect to energy expenditure, insulin increases leptin secretion in sheep maintained in their thermoneutral zone but not in sheep enduring cold stress (Asakuma et al. 2003). In addition, growth factors, such as insulin and growth hormones, also affect leptin secretion in whole animal studies and in isolated adipocytes (Chilliard et al. 2005), and could therefore also be part of 'metabolic memory'. Other endocrine factors secreted by adipose tissue or by the digestive system might also be involved: for example, several metabolic effects have been described for adiponectin in rodents, such as increased insulin resistance and fatty acid oxidation, and reduced glucose output by the liver (review: Diez & Iglesias 2003). These interactions between endocrine systems and nutrient supply may effectively act as a sort of peripheral integratory mechanism that complements the brain's integratory mechanisms, with the combination being responsible for 'metabolic memory'.

'Nutrient sensing'

Short-chain fatty acids and glucose might also contribute to the 'metabolic memory' because they are known to regulate leptin secretion in cultured ruminant adipocytes (review: Chilliard et al. 2005). However, glucose is not always stimulatory in ruminants – for example, intravenous injection of glucose does not stimulate leptin secretion (Kauter et al. 2000) but infusion of glucose modifies the post-prandial pattern of leptin secretion in sheep carrying an extra copy of the GH gene (Kadokawa et al. 2003). However, in ruminants, volatile fatty acids (acetate, propionate, butyrate) are the main nutrients absorbed from the anterior digestive tract and they are converted to heavier fats for deposition via a well-known pathway of fatty acid synthesis. In rodents, fatty acids are directly involved in the regulation of leptin gene expression and leptin secretion (Shirai et al. 2004), and studies using cell culture have led to the development of a model to explain the mechanism by which glucose regulates leptin secretion from adipocytes (Mueller et al. 1998; Shirai et al. 2004). In this model, glucose, after being metabolised into pyruvate, acts on leptin gene expression and leptin secretion via the malonyl-coenzyme A fatty acid synthesis pathway, a crucial pathway in intracellular signalling in adipocytes (Shirai et al. 2004). Should a similar pathway exist in ruminants, it would allow volatile fatty acids produced by the digestive system, such as acetate (Annison et al. 2002), to regulate leptin secretion. Interestingly, a recent paper suggested that propionate does not stimulate leptin secretion in lactating cows (Bradford et al. 2006), in contrast to the results obtained in sheep (Lee & Hossner 2002). This difference could be caused by differences in metabolic status, and differences in the responsiveness of adipocytes to short chain fatty acids. This issue needs to be resolved by further research.

Finally, nutrients might also be able to act directly on components of the reproductive axis – for example, both acetate and butyrate can modulate the responsiveness of the anterior pituitary gland to leptin, so pituitary tissue could also be a site of interaction between hormones and nutrients in the control of reproductive function. Conversely, pituitary hormones might affect the activity or responsiveness of adipose tissue because cultured adipocytes express receptors for several pituitary hormones, including LH, FSH and GH (Schaffler et al. 2005). These receptors need to be investigated but they could be part of a descending pathway which regulates the activity of the adipocytes, not only according to metabolic status, but also according to the 'decision to reproduce'

Conclusions

The 'decision to reproduce' is controlled by a range of factors and, amongst them, variation in metabolic status is one of the most important, especially for animals that have evolved in an environment where food supply is not predictable. Making a successful 'decision' depends on the integration of a large amount of information coming from a large number of organs. Over the last decade, the number and role of pathways involved in the link between metabolic status and reproductive activity has increased, and each pathway is acting in conjunction with the others. The latest developments suggest that the responses to each pathway vary with time so that, to fully understand the relationship between nutrition and reproduction, the temporal dimension (on scales of hours to years, or even a generation) should be considered because ruminants seem to be able to partition their energy, either from reserves or from nutrient intake, according to their metabolic history.

Acknowledgements

We would like to thank Ms Margaret Blackberry for her endless toil in the laboratory. These studies could not have been contemplated without the generous assistance of the students and staff of Animal Science (University of WA). Funding was supplied by the National Health & Medical Research Council, the Australian Research Council, the CSIRO Division of Animal Production, the Australian Wool Corporation, the University of WA and the New Zealand Foundation for Research Science and Technology.

References

Adam CL, Findlay PA & Moore AH 1998 Effects of insulin-like growth factor-1 on luteinizing hormone secretion in sheep. *Animal Reproduction Science* **50** 45-56.

Adam CL, Moar KM, Logie TJ, Ross AW, Barrett P, Morgan PJ & Mercer JG 2000 Photoperiod regulates growth, puberty and hypothalamic neuropeptide and receptor gene expression in female Siberian hamsters. *Endocrinology* **141** 4349-4356.

Adam CL, Archer ZA & Miller DW 2003 Leptin actions on the reproductive neuroendocrine axis in sheep. *Reproduction Supplement* **61** 283-297.

Ahima RS 2005 Central actions of adipocyte hormones. *Trends in Endocrinology and Metabolism* **16** 307-313.

Alila-Johansson A, Eriksson L, Soveri T & Laakso M-L 2004 Daily and annual variations of free fatty acid, glycerol and leptin plasma concentrations in goats (*Capra hircus*) under different photoperiods. *Comparative Biochemistry and Physiology - Part A: Molecular & Integrative Physiology* **138** 119-131.

Amstalden M, Zieba DA, Edwards JF, Harns PG, Welsh TH, Stanko RL & Williams GL 2003 Leptin acts at the bovine adenohypophysis to enhance basal and gonadotrophin-releasing hormone-mediated release of luteinizing hormones: differential effects are dependent upon nutritional history. *Biology of Reproduction* **69** 1539-1544.

Annison EF, Lindsay DB & Nolan JV 2002 Digestion and metabolism. In *Sheep Nutrition*, pp 95-118, Eds M Freer & H Dove. Wallingford: CAB International.

Asakuma S, Morishita H, Sugino T, Kurose Y, Kobayashi S & Terashima Y 2003 Circulating leptin response to feeding and exogenous infusion of insulin in sheep exposed to thermoneutral and cold environments. *Comparative Biochemistry and Physiology - Part A: Molecular & Integrative Physiology* **134** 329-335.

Bartness TJ, Demas GE & Song CK 2002 Seasonal changes in adiposity: the roles of the photoperiod, melatonin and other hormones, and sympathetic nervous system. *Experimental Biology & Medicine* **227** 363-376.

Bartness TJ, Kay Song C, Shi H, Bowers RR & Foster MT 2005 Brain-adipose tissue cross talk. *The Proceedings of the Nutrition Society* **64** 53-64.

Beam SW & Butler WR 1997 Energy balance and ovarian follicle development prior to the first ovulation postpartum in dairy cows receiving three levels of dietary fat. *Biology of Reproduction* **56** 133-142.

Beam SW & Butler WR 1999 Effects of energy balance on follicular development and first ovulation in postpartum dairy cows. *Journal of Reproduction and Fertility Supplement* **54** 411-424.

Blache D, Chagas LM, Blackberry MA, Vercoe PE & Martin GB 2000 Metabolic factors affecting the reproductive axis in male sheep. *Journal of Reproduction and Fertility* **120** 1-11.

Blache D, Adam CL & Martin GB 2002 The mature

male sheep: a model to study the effects of nutrition on the reproductive axis. *Reproduction Supplement* **59** 219-233.

Blache D, Zhang S & Martin GB 2003 Fertility in males: modulators of the acute effects of nutrition on the reproductive axis of male sheep. *Reproduction Supplement* **61** 387-402.

Blache D, Zhang S & Martin GB 2006 Dynamic and integrative aspects of the regulation of reproduction by metabolic status in male sheep. *Reproduction, Nutrition, Development* **46** 379-390.

Bohannon NJ, Corp ES, Wilcox BJ, Figlewicz DP, Dorsa DM & Baskin DG 1988 Localization of binding sites for insulin-like growth factor-I IGF-I in the rat brain by quantitative autoradiography. *Brain Research* **444** 205-213.

Bossis I, Wettemann RP, Welty SD, Vizcarra JA, Spicer LJ & Diskin MG 1999 Nutritionally induced anovulation in beef heifers: ovarian and endocrine function preceding cessation of ovulation. *Journal of Animal Science* **77** 1536-1546.

Boukhliq R & Martin GB 1997 Administration of fatty acids and gonadotrophin secretion in the mature ram. *Animal Reproduction Science* **49** 143-159.

Boukhliq R, Miller DW & Martin GB 1996 Relationships between the nutritional stimulation of gonadotrophin secretion and peripheral cerebrospinal fluid (CSF) concentrations of glucose and insulin in rams. *Animal Reproduction Science* **41** 201-204.

Bradford BJ, Oba M, Ehrhardt RA, Boisclair YR & Allen MS 2006 Propionate is not an important regulator of plasma leptin concentration in dairy cattle. *Domestic Animal Endocrinology* **30** 65-75.

Bruning J, Gautam D, Burks D, Gillette J, Schubert M, Orban P, Klein R, Krone W, Muller-Wieland D & Kahn C 2000 Role of brain insulin receptor in control of body weight and reproduction. *Science* **289** 2122-2125.

Butler WR 2000 Nutritional interactions with reproductive performance in dairy cattle. *Animal Reproduction Science* **60-61** 449-457.

Butler WR 2005 Inhibition of ovulation in the postpartum cow and the lactating sow. *Livestock Production Science* **98** 5-12.

Butler ST, Pelton SH & Butler WR 2004 Insulin increases 17ß-estradiol production by the dominant follicle of the first postpartum follicle wave in dairy cows. *Reproduction* **127** 537-545.

Cahill G 1980 Metabolic memory. *New England Journal of Medicine* **302** 396-397.

Chagas LM 2003 Propionate precursor to reduce postpartum anoestrus in heifers. *ANZ Dairy Vets Conference*, Taupo, 215-220.

Chagas LM, Rhodes FM, Blache D, Gore PJS, MacDonald KA & Verkerk GA 2006 Precalving feeding and body condition effects on metabolic responses and postpartum anestrus interval in grazing primiparous dairy cows. *Journal of Dairy Science* **89** 1981-1989.

Chilliard Y, Delavaud C & Bonnet M 2005 Leptin expression in ruminants: Nutritional and physiological regulations in relation with energy metabolism. *Domestic Animal Endocrinology* **29** 3-22.

Cinti S 2005 The adipose organ. *Prostaglandins, Leukotrienes and Essential Fatty Acids* **73** 9-15.

Diez JJ & Iglesias P 2003 The role of the novel adipocyte-derived hormone adiponectin in human disease. *European Journal of Endocrinolgy* **148** 293-300.

Diskin MG, Mackey DR, Roche JF & Sreenan JM 2003 Effects of nutrition and metabolic status on circulating hormones and ovarian follicle development in cattle. *Animal Reproduction Science* **78** 345-370.

Downing JA, Joss J & Scaramuzzi RJ 1995 A mixture of the branched chain amino acids leucine, isoleucine and valine increases ovulation rate in ewes when infused during the late luteal phase of the oestrous cycle: an effect that may be mediated by insulin. *Journal of Endocrinology* **145** 315-323.

Downing JA, Joss J & Scaramuzzi RJ 1996 The effects of N-methyl-D,L-aspartic acid and aspartic acid on the plasma concentration of gonadotrophins, GH and prolactin in the ewe. *Journal of Endocrinology* **149** 65-72.

Downing JA, Joss J & Scaramuzzi RJ 1997 Ovulation rate and the concentrations of LH, FSH, GH, prolactin and insulin in ewes infused with tryptophan, tyrosine or tyrosine plus phenylalanine during the luteal phase of the oestrous cycle. *Animal Reproduction Science* **45** 283-297.

Dungan HM, Clifton DK & Steiner RA 2006 Minireview: Kisspeptin Neurons as Central Processors in the Regulation of Gonadotropin-Releasing Hormone Secretion. *Endocrinology* **147** 1154-1158.

Fernandez-Fernandez R, Tena-Sempere M, Aguilar E & Pinilla L 2004 Ghrelin effects on gonadotropin secretion in male and female rats. *Neuroscience Letters* **362** 103-107.

Fierro LC & Bryant FC 1990 Grazing activities and bioenergetics of sheep on native range in Southern Peru. *Small Ruminant Research* **3** 135-146.

Flier JS, Harris M & Hollenberg AN 2000 Leptin, nutrition, and the thyroid: the why, the wherefore, and the wiring [comment]. *Journal of Clinical Investigation* **105** 859-861.

Frisch RE 1994 The right weight: body fat, menarche and fertility. *Proceedings of the Nutrition Society* **53** 113-129.

Gong JG, Lee WJ, Garnsworthy PC & Webb R 2002 Effect of dietary-induced increases in circulating insulin concentrations during the early postpartum period on reproductive function in dairy cows. *Reproduction* **123** 419-427.

Gottsch ML, Cunningham MJ, Smith JT, Popa SM, Acohido BV, Crowley WF, Seminara S, Clifton DK & Steiner RA 2004 A role for kisspeptins in the regulation of gonadotropin secretion in the mouse. *Endocrinology* **145** 4073-4077.

Graham NM 1964 Energy exchanges of pregnant and lactating ewes. *Australian Journal of Agricultural Research* **15** 127-141.

Hampton JH, Salfen BE, Bader JF, Keisler DH &

Garverick HA 2003 Ovarian follicular responses to high doses of pulsatile luteinizing hormone in lactating dairy cattle. *Journal of Dairy Science* **86** 1963-1969.

Henry BA, Goding JW, Tilbrook AJ, Dunshea F & Clarke IJ 2001 Intracerebroventricular infusion of leptin elevates the secretion of luteinising hormone without affecting food intake in long-term food-restricted sheep, but increases growth hormone irrespective of bodyweight. *Journal of Endocrinology* **168** 67-77.

Horton TH & Rowsemitt CN 1992 Natural selection and variation in reproductive physiology. In *Mammalian energetics : interdisciplinary views of metabolism and reproduction*, pp 160-185, Eds TE Tomasi & TH Horton. Ithaca, N.Y: Comstock Pub. Associates.

Hötzel MJ, Walkden-Brown SW, Blackberry MA & Martin GB 1995 The effect of nutrition on testicular growth in mature Merino rams involves mechanisms that are independent of changes in GnRH pulse frequency. *Journal of Endocrinology* **147** 75-85.

Hötzel MJ, Walkden-Brown SW, Fisher JA & Martin GB 2003 Determinants of the annual pattern of reproduction in mature male Merino and Suffolk sheep: response to a nutritional stimulas in the breeding and non-breeding season. *Journal of Reproduction and Fertility* **15** 1-9.

Irwig MS, Fraley GS, Smith JT, Acohido BV, Popa SM, Cunningham MJ, Gottsch ML, Clifton DK & Steiner RA 2004 Kisspeptin activation of gonadotropin releasing hormone neurons and regulation of KiSS-1 mRNA in the male rat. *Neuroendocrinology* **80** 264-272.

Itoh F, Komatsu T, Yonai M, Sugino T, Kojima M, Kangawa K, Hasegawa Y, Terashima Y & Hodate K 2005 GH secretory responses to ghrelin and GHRH in growing and lactating dairy cattle. *Domestic Animal Endocrinology* **28** 34-45.

Kadokawa H, Briegel JR, Blackberry MA, Blache D, Martin GB & Adams NR 2003 Relationships between plasma concentrations of leptin and other metabolic hormones in GH-transgenic sheep infused with glucose. *Domestic Animal Endocrinology* **24** 219-229.

Kadokawa H, Blache D & Martin GB 2006 Plasma leptin levels correlate positively with pulsatile LH secretion in early postpartum Holstein dairy cows. *Journal of Dairy Science* **89** 3020-3027.

Karsch FJ, Dahl GE, Hachigian TM & Thrun LA 1995 Involvement of thyroid hormones in seasonal reproduction. *Journal of Reproduction and Fertility Supplement* **49** 409-422.

Kauter K, Ball M, Kearney P, Tellam R & McFarlane JR 2000 The short term effect of adrenaline, insulin, and glucagon on plasma leptin levels in sheep: development and characterisation of an ovine leptin ELISA. *Journal of Endocrinology* **166** 127-135.

Kershaw EE & Flier JS 2004 Adipose tissue as an endocrine organ. *Journal of Clinical Endocrinology and Metabolism* **89** 2548-2556.

Kirby CJ, Thatcher WW, Collier RJ, Simmen FA & Lucy MC 1996 Effects of growth hormone and pregnancy on expression of growth hormone receptor, insulin-like growth factor-I, and insulin-like growth factor binding protein-2 and -3 genes in bovine uterus, ovary, and oviduct. *Biology of Reproduction* **55** 996-1002.

Lamming GE & Darwash AO 1998 The use of milk progesterone profiles to characterise components of subfertility in milked dairy cows. *Animal Reproduction Science* **52** 175-190.

Lee SH & Hossner KL 2002 Coordinate regulation of ovine adipose tissue gene expression by propianate. *Journal of Animal Science* **80** 2840-2849.

Lucy MC, Boyd CK, Koenigsfeld AT & Okamura CS 1998 Expression of somatotropin receptor messenger ribonucleic acid in bovine tissues. *Journal of Dairy Science* **81** 1889-1895.

Malagon MM, Rodriguez-Pacheco F, Martinez-Fuentes AJ, Tovar S, Pinilla L, Tena-Sempere M, Dieguez C & Castano JP 2006 Regulation of pituitary cell function by the adipokine adiponectin. *Frontiers in Neuroendocrinology* **27** 35-35.

Martin GB & Walkden-Brown SW 1995 Nutritional influences on reproduction in mature male sheep and goats. *Journal of Reproduction and Fertility Supplement* **49** 437-449.

Martin GB, Fisher JS, Blackberry MA, Boukhliq R, Hötzel MJ, Shepherd K & Walkden-Brown SW 1994 Nutritional and photoperiodic control of testicular size in Suffolk and Merino rams. *Proceedings of the Australian Society of Animal Production* **20** 427.

Martin GB, Tjondronegoro S, Boukhliq R, Blackberry MA, Briegel JR, Blache D, Fisher JA & Adams NR 1999 Determinants of the annual pattern of reproduction in mature male Merino and Suffolk sheep: modification of endogenous rhythms by photoperiod. *Reproduction, Fertility and Development* **11** 355-366.

Martin GB, Hötzel MJ, Blache D, Walkden-Brown SW, Blackberry MA, Boukliq R, Fisher JA & Miller DW 2002 Determinants of the annual pattern of reproduction in mature male Merino and Suffolk sheep: modification of response to photoperiod by annual cycle of food supply. *Reproduction, Fertility and Development* **14** 165-175.

Martin GB, Rodger J & Blache D 2004 Nutritional and environmental effects on reproduction in small ruminants. *Reproduction, Fertility and Development* **16** 491-501.

McNaughton LR, Verkerk GA, Parkinson KA, Macdonald KA & Holmes CW 2003 Postpartum anoestrous interval and reproductive performance of three genotypes of Holstein-Friesian cattle manage in a seasonal pasture-based dairy system. *Proceedings of the New Zealand Society of Animal Production* **60** 77-81.

Messager S, Chatzidaki EE, Ma, D.,, Hendrick AG, Zahn D, Dixon J, Thresher R, Malinge I, Lomet D, Carlton MBL, Colledge WH et al. 2005 Kisspeptin directly stimulates gonadotropin-releasing hormone release via G protein-coupled receptor 54. *Proceedings of the National Academy of Sciences (USA)* **102** 1761-1766.

Miller DW, Blache D & Martin GB 1995a The role of intracerebral insulin in the effect of nutrition on gonadotrophin secretion in mature male sheep. *Journal of Endocrinology* **147** 321-329.

Miller DW, Blache D & Martin GB 1995b Neuroendocrine mediation of the effects of nutrition on the reproductive system: another task for insulin? *Western Australian Endocrine Sciences Symposium*, Perth.

Miller DW, Blache D, Boukhliq R, Curlewis JD & Martin GB 1998 Central metabolic messengers and the effects of diet on gonadotrophin secretion in sheep. *Journal of Reproduction and Fertility* **112** 347-356.

Miller DW, Findlay PA, Morrison MA, Raver N & Adam CL 2002 Seasonal and dose-depedent effects of intracerebroventricular leptin on LH secretion and appetite in sheep. *Journal of Endocrinology* **175** 395-404.

Miller DW, Harrison JL, Brown YA, Doyle U, Lindsay Adam CL & Lea RG 2005 Immunohistochemical evidence for an endocrine/paracrine role for ghrelin inthe reproductive tissues of sheep. *Reproductive Biology and Endocrinology* **3** 60.

Miyoshi S, Pate JL & Palmquist DL 2001 Effects of propylene glycol drenching on energy balance, plasma glucose, plasma insulin, ovarian function and conception in dairy cows. *Animal Reproduction Science* **68** 29-43.

Monget P & Martin GB 1997 Involvement of insulin-like growth factors in the interactions between nutrition and reproduction in female mammals. *Human Reproduction* **2 Supplement 1** 33-52.

Mueller WM, Gregoire FM, Stanhope KL, Mobbs CV, Mizuno TM, Warden CH, Stern JS & Havel PJ 1998 Evidence that glucose metabolism regulates leptin secretion from cultured rat adipocytes. *Endocrinology* **139** 551-558.

Navarro VM, Castellano JM, Fernandez-Fernandez R, Tovar S, Roa J, Mayen A, Barreiro ML, Casanueva FF, Aguilar E, Dieguez C, Pinilla L, Tena-Sempere M 2005 Effects of KiSS-1 peptide, the natural ligand of GPR54, on follicle-stimulating hormone secretion in the rat. *Endocrinology* **146** 1689-1697.

Obici S & Rossetti L 2003 Minireview: nutrient sensing and the regulation of insulin action and energy balance. *Endocrinology* **144** 5172 5178.

Patton J, Kenny DA, Mee JF, O'Mara FP, Wathes DC, Cook M & Murphy JJ 2006 Effect of milking frequency and diet on milk production, energy balance, and reproduction in dairy cows. *Journal of Dairy Science* **89** 1478-1487.

Penicaud L, Cousin B, Leloup C, Lorsignol A & Casteilla L 2000 The autonomic nervous system, adipose tissue plasticity, and energy balance. *Nutrition* **16** 903-908.

Rhind SM, Archer ZA & Adam CL 2002 Seasonality of food intake in ruminants: recent developments in understanding. *Nutrition Research Reviews* **15** 43-65.

Rhodes FM, Clark BA, Macmillan KL & McDougall S 1998 Use of once daily milking compared with treatment with progesterone and oestradiol benzoate (ODB) in anoestrous cows. *Proceedings of the New Zealand Society of Animal Production* **58** 44-46.

Robinson JJ, Ashworth CJ, Rooke JA, Mitchell LM & McEvoy TG 2006 Nutrition and fertility in ruminant livestock. *Animal Feed Science and Technology* **126** 259-276.

Roche JR, Kolver ES & Kay JK 2005 Influence of precalving feed allowance on preparturient metabolic and hormonal responses and milk production in grazing dairy cows. *Journal of Dairy Science* **88** 677-689.

Roche JR, Sheahan AJ, Chagas LM & Berry DP 2006 Genetic selection for milk production increases plasma ghrelin in pasture-based dairy cows. *Journal of Dairy Science* **89** 3471-3475.

Schaffler A, Binart N, Scholmerich J & Buchler C 2005 Brain talks with fat - evidence for a hypothalamic-pituitary-adipose axis? *Neuropeptides* **39** 363-367.

Schechter R, Whitmire J, Wheet GS, Beju D, Jakson KW, Harlow R & Gavin JR 1994 Immunohistochemical and in situ hybridization of an insulin-like substance in fetal neuron cell cultures. *Brain Research* **636** 9-27.

Shirai Y, Yaku S & Suzuki M 2004 Metabolic regulation of leptin production in adipocytes: a role of fatty acid synthesis intermediates. *Journal of Nutritional Biochemistry* **15** 651-656.

Sinclair KD, Revilla R, Roche JF, Quintans G, Sanz A, Mackey DR & Diskin MG 2002 Ovulation of the first dominant follicle arising after day 21 postpartum in suckling beef cows. *Animal Science* **75** 115-126.

St-Pierre DH, Wang L & Tache Y 2003 Ghrelin: a novel player in the gut-brain regulation of growth hormone and energy balance. *News in Physiological Sciences* **18** 242-246.

Sugino T, Hasegawa Y, Kurose Y, Kojima M, Kangawa K & Terashima Y 2004 Effects of ghrelin on food intake and neuroendocrine function in sheep. *Animal Reproduction Science* **82-83** 183-194.

Tanaka T, Nagatani S, Bucholtz DC, Ohkura S, Tsukamura H, Maeda K & Foster DL 2000 Central action of insulin regulates pulsatile luteinizing hormone secretion in the diabetic sheep model. *Biology of Reproduction* **62** 1256-1261.

Taylor MM & Samson WK 2003 The other side of the orexins: endocrine and metabolic actions. *American Journal of Physiology* **284** E13-E17.

Tena-Sempere M 2005 Exploring the role of ghrelin as novel regulator of gonadal function. *Growth Hormone & IGF Research* **15** 83-88.

Tena-Sempere M, Pinilla L, González LC, Diéguez C, Casanueva FF & Aguilar E 1999 Leptin inhibits testoterone secretion from adult rat testis in vitro. *Journal of Endocrinology* **161** 211-218.

Tena-Sempere M, Pinilla L, Zhang FP, Gonzalez LC, Huhtaniemi I, Casanueva FF, Dieguez C & Aguilar E 2001 Developmental and hormonal regulation of leptin receptor (Ob-R) messenger ribonucleic acid expression in rat testis. *Biology of Reproduction* **64** 634-643.

Veerkamp RF, Beerda B & van der Lende T 2003 Effects of genetic selection for milk yield on energy balance, levels of hormones, and metabolites in lactating cattle, and possible links to reduced fertility. *Livestock Production Science* **83** 257-275.

Woods SC, Benoit SC, Clegg DJ & Seeley RJ 2004 Regulation of energy homeostasis by peripheral signals. *Best Practice & Research Clinical Endocrinology & Metabolism* **18** 497-515.

Zhang S, Blache D, Blackberry MA & Martin GB 2004 Dynamics of the responses in secretion of LH, leptin and insulin following an acute increase in nutrition in mature male sheep. *Reproduction, Fertility and Development* **16** 823-829.

Zhang S, Blache D, Blackberry MA & Martin GB 2005 Body reserves affect the reproductive endocrine responses to an acute change in nutrition in mature male sheep. *Animal Reproduction Science* **88** 257-269.

Zieba DA, Amstaldem M, Maciel MN, Keisler DH, Raver N, Gertler A & Williams GL 2003 Divergent effects of leptin on luteinizing hormone and insulin secretion are dose dependent. *Experimental Biological Medicine* **228** 325-330.

Zieba DA, Amstalden M & Williams GL 2005 Regulatory roles of leptin in reproduction and metabolism: A comparative review. *Domestic Animal Endocrinology* **29** 166-185.

Development of the dominant follicle: mechanisms of selection and maintenance of oocyte quality

R Webb[1] and BK Campbell[2]

[1]Division of Agricultural and Environmental Sciences, School of Biosciences, University of Nottingham, Loughborough, Leicestershire, LE12 5RD, UK, [2]Department of Obstetrics and Gynaecology, School of Human Development, Queens Medical Centre, University of Nottingham, Nottingham, NG7 2UH, UK

For a follicle to reach dominance, in mono-ovulatory species such as cattle, requires the integration of a number of processes involving both extra-ovarian signals and intra-follicular paracrine and autocrine regulators. Ovarian transplant studies in both cattle and sheep demonstrated that it takes approximately 4 months for primordial follicles to reach dominance. Gonadotrophins are not a prerequisite for the continued growth of pre-antral follicles, unlike antral follicles, but FSH does appear to stimulate development. Local growth factors, such as IGFs and BMPs, are expressed throughout follicle development and interact with gonadotrophins to stimulate development. As follicles become dominant, there is a transfer of dependency from FSH to LH. There are also differences in LH-responsiveness of theca and granulosa cells during follicular development, due to differential regulation and control by intricate local mechanisms altering LH receptor (LHR) mRNA expression. In addition, both the BMP and IGF systems can modulate the proliferative and differentiative responses of both granulosa and theca cells to gonadotrophins. There is a significant interaction between BMPs and the IGF system in regulating follicular development. A range of factors, including nutrition, will also determine the fate of the growing follicle and the quality of the oocyte. Nearly all follicles regress and apoptotic cell death throughout follicular development is an underlying mechanism of cell loss during follicular atresia. Several markers of follicular atresia have been identified including IGFBPs. There is a significant correlation between the presence of low molecular weight IGFBPs in bovine follicular fluid and caspase-3 activity of granulosa cells in individual follicles. In conclusion, it is the interaction between extra-ovarian and intra-ovarian factors that determine the fate of the follicle and the quality of the oocyte.

Introduction

Ovarian folliculogenesis is a lengthy and intricately regulated process marked by dramatic proliferation and precisely orchestrated differentiation of both the somatic and germ cell elements. Primordial follicles represent the source from which follicles will be recruited for growth throughout life, with paired ovaries of an individual containing around 100,000-250,000 of these follicles

Corresponding author E-mail: bob.webb@nottingham.ac.uk

at birth (human: Gougeon 1996; sheep: Turnbull *et al.* 1977). Once follicles have been initiated to grow, the granulosa cells proliferate to form multilaminar structures (pre-antral follicles) which subsequently form a fluid filled space (antrum) and a well differentiated theca layer. Follicular development in sheep and cattle takes around 4-5 months with the majority of this time (3-4 months) being spent in the pre-antral stages of development (Cahill 1981; Gougeon 1996; Turnbull *et al.* 1977). When the follicle reaches a diameter of 200-400 μm, the antrum develops (Turnbull *et al.* 1977) and this is followed by widespread atresia (50-70% for follicles over 1 mm) so that the vast majority (>99%) of follicles fail to ovulate.

It is widely accepted that control of the terminal stages of folliculogenesis lies primarily with the pituitary gonadotrophins, FSH and LH, combined with the differential expression of somatic cell-derived growth factors that modulate the action of gonadotrophins at key points during the process of follicle development (Campbell *et al.* 1995; Webb *et al.* 1999). Communication between the oocyte and the surrounding somatic cells (Carabatsos *et al.* 2000), via a range of locally produced growth factors, is essential for the coordinated development of both cell types (Eppig 2001), making folliculogenesis a highly synchronised process. These local growth factor systems include the insulin-like growth factor (IGF) system (Webb *et al.* 1999), the inhibin/activin system (Knight & Glister 2001) and the bone morphogenetic (BMP) system (Shimasaki *et al.* 2003; Knight & Glister, 2006). In addition, recent studies also suggest that the oocyte, rather than being purely a passenger within the follicle, secretes numerous factors that modulate follicle development and ovarian function. Known oocyte-secreted factors include growth differentiation factor-9 (GDF-9) (Dong *et al.* 1996) and BMP-15 (Aaltonen *et al.* 1999; Dube *et al.* 1998) as well as factors in the germline alpha (FIG-α) (Huntriss *et al.* 2002; Soyal *et al.* 2000) and c-kit receptor (Driancourt *et al.*, 2000; Reynaud *et al.* 2000; Reynaud *et al.* 2001; McNatty *et al.* 2007 this supplement). However the temporal pattern and quantity of secretion of these factors has yet to be determined.

Ovarian follicular growth is therefore a developmental process during which the follicle progressively acquires a number of properties at a specific time and sequence, each of which is an essential prerequisite for further development. The orderly expression of these somatic and oocyte-derived factors, or "intrafollicular cascade", is thought to be essential for the development of the follicle to an ovulatory size, the subsequent production of an ovulatory signal and the release of a fully developmentally competent oocyte in response to that signal. Although some components of this intrafollicular cascade have been elucidated, most have been studied in isolation and the relative importance and temporal relationships between known and novel regulatory factors at different stages of follicle development remain to be elucidated. In this review we will examine recent advances in this area with respect to follicle development in sheep and cattle, concentrating primarily on the factors regulating antral and dominant follicle development.

Pre-antral follicle development

Intra-ovarian factors

Mechanisms regulating the activation and subsequent growth of primordial follicles have been reviewed in detail, particularly in relation to sheep (McNatty *et al.* 2007 this supplement) and so will not be reviewed here. In a number of species, as well as for the initiation of primordial follicle growth, the continued growth of pre-antral follicles is dependent upon the secretion of a range of local factors including GDF-9, BMPs, activins, inhibins, basic fibroblast growth factor (bFGF) and epidermal growth factor (EGF) (Knight & Glister 2001; Smitz & Cortvindt 2002; Webb *et al.* 2003; Hunter *et al.* 2004). As will be discussed, only studies identifying genetic mutations in sheep have been carried out (see Hanrahan *et al.* 2004; McNatty *et al.* 2007). However, Armstrong *et al.* (2002a) demonstrated that bovine pre-antral follicles express mRNAs encoding both IGFBP-2, -3 and type 1 IGF receptor. EGF, also stimulates pre-antral follicle growth (Gutierrez *et al.* 2000; Saha *et al.* 2000).

Studies in our laboratory have also examined fetal ovaries for evidence of local regulatory systems operating during oogenesis. BMP-6 and BMP receptors (BMPR) IA, IB and II have been shown to be present in bovine fetal ovaries between 2-9 months of gestation (Dugan *et al.* 2004; Fouladi-Nashta *et al.* 2005). The presence of both the ligand (BMP-6) and the receptors at this early stage illustrates the presence of a fully functional BMP system early in development. The expression of BMP-6 appears to be localised both in the oocyte, with the intensity of staining increasing with fetal age, and also in granulosa cells from 6 months of gestation. These results, together with results from post-natal bovine and ovine ovaries (Dugan *et al.* 2004), suggest that expression of both the ligand and the receptors is present right through gestation, parturition and into adulthood, since BMP-6 has been shown to be abundant in both the oocyte and granulosa cells of adult bovine follicles (Glister *et al.* 2004; Campbell *et al.* 2006). However, BMP-6, but not -2, -4 or -7 mRNA expression has been detected in ovine oocytes of pre-antral follicles (Juengel *et al.* 2006). Interestingly, Fatahei *et al.* (2005) was only able to immunolocalise BMP-2 and BMP-4 in bovine oocytes and theca cells of adult antral follicles, whereas BMPRII was observed in oocytes from the primordial stage. The functional significance of the expression patterns of components of the BMPs in domestic ruminants is unknown, but in the absence of knock-out models in this species, some inferences may be drawn from naturally occurring mutations.

McNatty *et al.* (2007) has reviewed the effect of null mutations of the oocyte secreted factors GDF-9 and BMP-15 in sheep. These studies show that these factors are essential for normal pre-antral follicle development. In contrast, the FecB mutation in the BMPRIB receptor, whilst inducing precocious maturation of ovulatory follicles (see later), has not been shown to have had a major effect on pre-antral follicle development, although some changes suggesting precocious development have been detected (McNatty *et al.* 1986). At present, the available evidence suggests that the FecB mutation results in a down-regulation in BMP signalling (Fabre *et al.* 2003) and therefore it appears likely that signalling through this receptor system is not of major importance during pre-antral follicle development, although it is possible that significant redundancy exists. The abundant expression of the BMPRIA receptor tends to support this possibility.

Research into the functional role of local regulatory systems during early folliculogenesis in ruminants and other mono-ovulatory species is severely hampered by the lack of physiological culture systems that will allow normal follicle and oocyte development from the primordial/primary stages of development through to antrum formation with high efficiency. Although significant advances have been made over recent years in this regard (Gutierrez *et al.* 2000; Fortune *et al.* 2000; McCaffery *et al.* 2000, Picton *et al.* 2003), existing systems remain sub-optimal and it is a research priority to develop *in vitro* model systems for these species.

Extra-ovarian factors

It is generally agreed that gonadotrophins are not involved in the initiation of follicle growth from the primordial follicle pool. FSH receptor (FSHr) mRNA is detected in follicles with only one or two layers of granulosa cells (Bao & Garverick 1998) and both *in vivo* (Campbell *et al.* 2000) and *in vitro* (Gutierrez *et al.* 2000) studies have demonstrated that FSH can accelerate the rate of pre-antral follicle development. Expression of LH receptor (LHr) mRNA is first detected when the theca interna forms around the granulosa cells (Bao & Garverick 1998), presumably stimulating androgen precursor production. This is supported by a range of steroidogenic enzymes, these include cytochrome P450 side chain cleavage (P450scc), cytochrome P450 17α-hydroxylase (P450c17), and 3β-hydroxysteroid dehydrogenase (3β-HSD) mRNAs which are first expressed soon after formation of the theca interna (Bao & Garverick 1998), with cytochrome P450 aromatase (P450arom) being localised solely to granulosa cells. Steriod enzyme protein information is very limited, although mRNA expression patterns agree with recent results showing that both small (KJ

Dugan, M Lopez-Bejar, DG Armstrong and R Webb, unpublished observations) and larger (Thomas *et al.* 2001) pre-antral follicles are capable of producing oestradiol early in development.

Additional functional investigation into the role of gonadotrophins and their interaction with other extra-ovarian factors *in vivo* have utilised ovarian autografts treated with bovine somatotrophin (BST). In this model system, early follicle development is synchronised through the loss of growing follicles during graft re-vascularisation. In addition these studies in both sheep and cattle, in which gonadotrophin and insulin/IGF concentrations were modulated, showed that BST had a marked effect on the growing follicle population and that there was a clear interaction with gonadotrophic status (Campbell BK, Armstrong DG, Telfer EE and Webb R, unpublished observations). Thus, under normogonadotrophic conditions in hemi-ovariectomised sheep (FSH: 1.1 ± 0.1 ng/ml), BST treatment had a negative effect on the relative proportion of secondary/tertiary, pre-antral and antral follicles, whereas under hypergonadotrophic conditions (ovariectomised; FSH: 5.3 ± 0.6 ng/ml) there were less secondary and tertiary follicles, but more late pre-antral and antral follicles in BST-treated animals (Campbell BK, Armstrong D & Telfer E, unpublished observations). Whilst confirming our previous data (Campbell *et al.* 2000) that gonadotrophin levels influence the rate of pre-antral follicle development, these findings also support the hypothesis that the IGF system represents a key determinant of successful follicle and oocyte development during stages of pre-antral follicle development. Our current view is that exposure of the developing follicle and oocyte to the potent proliferative and differentiative actions of the IGFs is modulated through the abundant local expression of IGF-BP2 during the pre-antral stages of development (Webb *et al.* 2003; 2004) and findings from follicle culture studies support this hypothesis (Walters *et al.* 2006). These findings therefore demonstrate possible interactions between extra-ovarian hormones and intra-ovarian growth factors in the control of these early stages of follicle development.

Antral follicle development

Extra-ovarian factors

The formation of a fluid filled cavity or antrum occurs at a diameter of ~ 200-400μm in sheep and cattle (Turnbull *et al.* 1977). In hypophysectomised sheep antral follicle development continues to a diameter of 2-4 mm. This stage of follicle development is commonly referred to as the gonadotrophin-responsive phase in recognition of the fact that while FSH can accelerate the rate at which these follicles grow, the presence of FSH is not an essential requirement (Webb *et al.* 1999), certainly for sheep. In contrast, antral follicle growth from 2-4 mm in diameter is under gonadotrophic control (Campbell *et al.* 1995), with each wave of follicular growth being preceded by a transient increase in FSH secretion (Adams 1999; Souza *et al.* 1997). The antral follicle stage marks the transition point from a proliferative to a differentiative phenotype for the follicular somatic cells and is accompanied by a decrease in mitotic index, a marked increase in the rate of follicular atresia (Turnbull *et al.* 1977) and marked changes in gene expression (Webb *et al.* 2003). Changes in the expression patterns of mRNAs for both gonadotrophin receptors (FSHr and LHr) and steroidogenic enzymes, including P450scc, P450c17 and P450arom, and 3ß-HSD (Bao *et al.* 1997; Webb *et al.* 1999) occur at this stage of development. An increase in follicular diameter to around ~ 5mm in diameter is characterized by induction of mRNA expression for P450scc and P450arom in granulosa cells. As follicles grow further there is increased expression of mRNA for P450scc and P450arom in granulosa cells and P450c17 in theca cells.

FSH infusion in cattle, in which pituitary gonadotrophin secretion had been significantly reduced by GnRH agonist (GnRHa) or GnRH immunisation, has been shown to stimulate follicle growth up to 8.5 mm in diameter (Crowe *et al.* 2001; Garverick *et al.* 2002). Also infusion of FSH in cattle can induce an increase in mRNA expression for P450scc and P450arom in granulosa cells in small (1-4mm) follicles (Garverick *et al.* 2002). Interestingly, although there is a seven-fold variation between cattle in the maximal number of follicles >3mm in diameter during follicular waves, there appears to be high repeatability of numbers of follicles >3mm in diameter during follicular waves within individual dairy cattle (Burns *et al.* 2005). This supports further the presence of a tightly regulated compensatory mechanism that regulates follicle growth as well as ovulation rate in sheep and cattle.

The development of methods to detect and increase the number of the gonadotrophin-responsive and gonadotrophin-dependent antral follicles is of immense practical significance for the application of assisted reproduction technologies in domestic species and in humans. The number and health of the antral follicle population represents the so-called "ovarian reserve" of antral follicles that will respond to ovarian stimulation protocols employing exogenous gonadotrophins (Macklon & Fauser 2005). Conceptually, all ovarian stimulation protocols work on the principle of artificially increasing circulating FSH concentrations for a protracted period of time by administering gonadotrophin preparations that contain varying concentrations of FSH and LH. This treatment recruits and stimulates a variable proportion of the gonadotrophin-responsive follicles present in the ovaries to an ovulatory stage. The oocytes within these follicles can be retrieved by ultra-sound guided ovum pick-up (OPU) prior to ovulation or flushed from the oviducts following ovulation. An exogenous ovulatory stimulus such as hCG (in humans) or GnRH (in cattle) is usually delivered to induce the final maturational changes in the follicles and oocytes prior to OPU. Irrespective of species, aggressive and unphysiological ovarian stimulation regimes produce oocytes of variable developmental competence (Kanitz *et al.* 2002; Macklon *et al.* 2006). In human ART, indicators of ovarian reserve such as FSH concentrations, antral follicle counts, inhibin B and anti-mullerian hormone (AMH) concentrations are used increasingly to try and predict ovarian response to treatment and design suitable ovarian stimulation regimes (Macklon & Fauser 2005; Visser *et al.* 2006; Yong *et al.* 2003). In domestic ruminants, whilst such individual treatment is impractical, there have been an increasing number of reports in the literature of the use of GnRH-analogues to enhance the response to ovarian stimulation (Berlinguer *et al.* 2006; Gonzalez-Bulnes *et al.* 2004; Lopez-Alonso *et al.* 2005). These enhanced responses seem to occur due to an accumulation in the number of follicles at the gonadotrophin-responsive to gonadotrophin-dependent threshold in animals rendered hypogonadotrophic by GnRH-analogue treatment. However, whilst oocyte or embryo number may be enhanced by such GnRH-analogue treatment, some authors have reported poor embryo quality with these types of stimulation regimes (Berlinguer *et al.* 2006; Gonzalez-Anover *et al.* 2004; Lopez-Alonso *et al.* 2005). It is not known whether these effects on embryo quality reflect high rates of atresia in this pool of arrested gonadotrophin-responsive follicles or inadequacies in the gonadotrophin-stimulation regimes in these studies.

The development of follicles does not depend solely on gonadotrophins, but also the interaction with a range of local and circulating growth factors. For example, we have previously reported that treatment with BST, through increasing circulating insulin and IGF concentrations, can increase both the number of gonadotrophin-responsive follicles and their quality (Gong *et al.* 1991; 1993; 1997) and that embryo yield can be enhanced by stimulation of these follicles to the ovulatory stage using exogenous BST prior to FSH treatment (Gong *et al.* 2002).

Intra-ovarian factors

Utilising *in vitro* culture systems, a wide range of local factors, including members of the TGFß superfamily, FGFs and EGF/TGFα have been shown to be involved in the regulation of early antral follicle growth (Webb *et al.* 2003; 2004; Buratini *et al.* 2005). For example, BMP localisation studies have been carried out in adult sheep (Souza *et al.* 2002) and cattle (Glister *et al.* 2004). In line with rodent species (Shimasaki *et al.* 2004), and as discussed previously, there appears to be a functional BMP system within the ovary, with several BMPs being implicated as paracrine/ autocrine regulators of ovarian follicle development. In cattle, BMP-4 and -7 have been immunolocalised to theca cells and BMP-6 has been immunolocalised to oocytes and granulosa cells (Glister *et al.* 2004) and BMPRIB and II to granulosa cells, theca cells and denuded oocytes from bovine antral follicles (Glister *et al.* 2004; Fatahei *et al.* 2005). As discussed for pre-antral follicles, in sheep, BMPRIA, IB and II receptors have been localised to the granulosa and theca cells of growing follicles (Souza *et al.* 2002) and BMP-6 is strongly expressed in the oocyte and granulosa cells, with weak expression in the theca cell layer of antral follicles (Campbell *et al.* 2006). The action of BMPs also appear to be modulated by a range of binding proteins e.g. follistatins; (see Lin *et al.* 2006; Knight & Glister 2006), although more detailed study is required.

Functionally, BMP-2, -4, and -6 have been shown to augment FSH-stimulate oestradiol and inhibin A production by cultured granulosa cells in sheep (Souza *et al.* 2002; Campbell *et al.* 2006). Similarly BMP -4, -6 and -7 have stimulated oestradiol, inhibin –A, activin –A and follistatin, but without FSH, in bovine granulosa cells (Glister *et al.* 2004). Conversely, BMPs in both cattle (BMP-4, -6 and -7: Glister *et al.* 2005) and sheep (BMP-2, -4 and -6: Campbell *et al.* 2006) are potent inhibitors of thecal androgen production. However, in sheep these BMPs have also been shown to stimulate thecal cell proliferation *in vitro* so that at very low doses total androgen production is actually increased by BMPs. Overall, these data suggest that BMPs are acting as both autocrine and paracrine factors to enhance ovarian steroidogenesis.

Recent studies utilising *in situ* hybridisation in sheep have confirmed that oocytes express mRNA for BMP-6, but failed to show mRNA expression for BMP-2, -4 and -7 in any cells of non-atretic ovarian follicles (Juengel *et al.* 2006). These data therefore suggest that BMP-6, like BMP-15 and GDF-9, is an oocyte secreted factor that is the primary mediator of paracrine interactions with ovarian somatic cells. These results, however, conflict with those of Souza *et al.* (2002) who reported mRNA expression for BMP-2, -4, -6 and -7 in whole ovary from this species following Northern analysis. Further studies are therefore required in both sheep and cattle to confirm the tissue specific expression of potential ligands for the BMP receptors.

As an augmenter of somatic cell differentiation, the BMPs have a similar role to the IGF-system (Webb *et al.* 2003; 2004). Glister *et al.* (2004) demonstrated that BMP-4, -6 and -7 enhanced IGF-induced secretion of oestradiol, inhibin-A, activin-A and follistatin by bovine granulosa cells, but did not examine the interaction with the level of FSH stimulation. In contrast, recent data from sheep shows clearly that the BMPs are ineffective in stimulating granulosa cell differentiation in the absence of FSH, but do reveal a clear interaction between the level of IGF and BMP exposure in terms of the induction of aromatase activity (Campbell *et al.* 2006; Fig. 1). Thus, in this species both BMP and IGF act to augment FSH-stimulated cellular differentiation. Furthermore, bovine granulosa cell culture studies have recently suggested that FSH and oestradiol can down-regulate the expression of BMPRII (Jayawardana *et al.* 2006) and hence are possibly involved in the selection of bovine follicles. Moreover utilising ovarian tissue from ewes with the FecB mutation of BMPRIB showed that the mutation resulted in an increased response of both granulosa and theca cells to BMPs, gonadotrophins and IGF-I stimulation (Campbell *et al.* 2006). The increased responsiveness of ovarian somatic cells to these factors could account for the precocious maturation of antral follicles in FecB mutants, which is characterised by the development of aromatase activity and LH receptors by granulosa cells of antral follicles at markedly smaller diameters than in wild-type ewes (Driancourt *et al.* 1985; McNatty *et al.* 1985; 1987). These *in vitro* observations are therefore consistent with the profound effect of the FecB mutation in inducing precocious maturation of ovarian follicles (Webb *et al.* 1999; 2003) and hence deregulating the normal follicle selection mechanisms operating in this species.

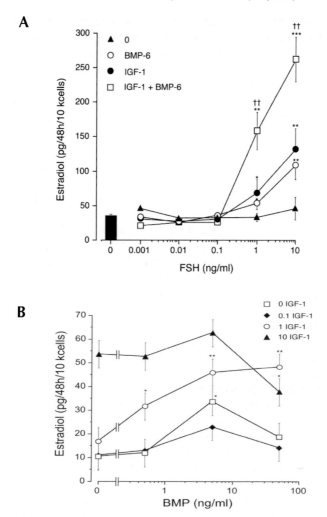

Fig. 1 A. Effect of increasing doses of oFSH in the presence of either 10 ng/ml insulin (full triangle), insulin and 5 ng/ml BMP-6 (open circle), insulin and 10 ng/ml IGF-1 LR3 (closed circle) and insulin, BMP-6 and IGF-1 LR3 combined (open squares) on ovine granulosa cell *in vitro* oestradiol production. The column represents overall mean production in the presence of all factors, but in the absence of FSH. Asterix indicate significant difference from zero dose of FSH with *P<0.05, **P<0.01 and ***P<0.001. † indicates significant difference between BMP and IGF supplementation alone compared to the combined treatment at a given dose of FSH with BMP-6 and IGF-I LR3 P<0.01.

B. *In vitro* oestradiol production by ovine granulosa with increasing doses of BMP in the presence of 10 ng/ml insulin and FSH (open square), insulin, FSH and 0.1 ng/ml IGF-1 LR3 (full diamond), insulin, FSH and 1 ng/ml IGF-1 LR3 (open circle) and insulin, FSH and 10 ng/ml IGF-1 LR3 (closed triangle). Note that the *in vitro* oestradiol production data are pooled from cultures utilising BMP-2, -4 and -6 due to similar responses to the three BMPs. A marked interaction between dose of BMP and IGF-1 is clearly evident from the data (P<0.05), with a flattening of the BMP dose response curve with increasing dose of IGF-1. Asterix indicate significant difference from zero dose of BMP within each dose of LR3 IGF-1 with *P<0.05, **P<0.01 and ***P<0.001. After Campbell *et al.* 2006.

As well as for the BMPs, the IGFs are also involved in local control mechanisms. It is around the time of antrum formation that IGF-II mRNA is first detected in thecal tissue in cattle. Type 1 IGF receptor and a range of IGFBPs (IGFBP-2, -3 and -4) have also been detected at this stage of development (Armstrong et al. 1998; 2000). Despite contradictory evidence on the production of IGF-1 by ovine and bovine granulosa cells (Webb et al. 2003; 2004), there is general agreement that IGF-II, produced by theca cells, is the major intrafollicular IGF ligand regulating the growth of bovine antral follicles (Yuan et al. 1998; Armstrong et al. 2000; Webb et al. 1999), acting through the type 1 IGF receptor (Lucy 2000). Despite the major site of IGF-II production being the thecal cells, Spicer et al. (2004) demonstrated that the stimulatory effects of IGF-II on thecal cell steroidogenesis, from large bovine follicles (> 7.9mm), is mediated via IGF type 1 receptors.

The ovarian IGF system also appears to interact directly with the oocyte. Small follicles from cattle offered high energy diets had significantly reduced expression levels of mRNA encoding IGFBP-2 and -4 (Armstrong et al. 2001), potentially regulating the bioavailability of IGF and hence influencing oocyte developmental potential. Concentrations of IGF-I that are optimal for follicle growth in vitro were seen to be detrimental to oocyte maturation (McCaffery et al. 2000). Hence, over-stimulation by IGFs, and possibly insulin, may be detrimental to oocyte quality (Armstrong et al. 2001). It appears that nutritionally induced changes in both circulating concentrations of insulin and IGF-I and the ovarian IGF system are important for follicle recruitment. However, these changes may also be detrimental to the quality of the oocyte within the growing follicle (Adamiak et al. 2005).

In conclusion, it is clear that the antral follicle stage is a transitional phase during which the follicle becomes increasingly dependent on the pituitary gonadotrophins as the rate of somatic cell proliferation declines and the cells differentiate and develop the ability to secrete increasing amounts of ovarian steroids and peptides. The control of this phase of development therefore involves complex interactions between local factors, many of which we now know are derived from the oocyte, other extra-ovarian factors circulating in the blood and the pituitary gonadotrophins. An outstanding characteristic of this follicle class is their heterogeneity and this reflects the fact that antral follicles each form their own micro-environment that will transduce gonadotrophic signals differentially. A positive response to a given level of gonadotrophic input will therefore ensure progression of a small number of individual follicles to the next large antral and/or dominant stage. The majority of follicles that reach this stage of development, however, will be unable to elicit a positive response to gonadotrophins and will be lost through the process of atresia.

Dominant follicle

Extra-ovarian factors

The precise mechanism for the selection of dominant follicles remains to be fully elucidated, but does involve the action of gonadotrophins as well as locally produced factors. As discussed, both FSH and LH exert their effects on follicular somatic cells via specific membrane bound receptors that exhibit alternate patterns of expression. It is well established that the granulosa cells of large oestrogenic (dominant) antral follicles in sheep and cattle develop LH receptors, (Webb & England 1982; Ireland & Roche 1982; Bao & Garverick 1998; Webb et al. 1999; 2003) and it is now generally accepted that this event is critical to the process of follicle selection in mono-ovulatory species. It has been suggested that after the emergence of a follicular wave it is oestradiol and inhibin, produced by the growing and selected follicles that *acts to suppress* the secretion of FSH (Webb et al. 1999; 2003) resulting in the rapid deviation in the size of the future dominant follicle and the largest subordinate follicle (Kulick *et al.* 1999). Thus the presence of LH receptors (LHR) on granulosa cells is thought to allow a follicle to switch its gonadotrophic dependence from FSH to LH and attain dominance over a follicular cohort which remains FSH-dependent. Indeed Hampton *et al.* (2004) demonstrated that while FSH can support bovine follicular growth to > 10mm, LH increases androgen production and expression of P450c17. Interestingly, changes in expression of mRNA for LHR in granulosa cells was not associated with changes in LH pulsatility and it was P450c17 mRNA expression, rather than aromatase activity, that was the most sensitive indicator of androgen production by thecal cells and oestrogen production by granulosa cells.

Study of the LHR in sheep and cattle has shown that LHR mRNA is highly polymorphic and a number of splice variants have been described. These isoforms vary due to deletion of all or part of two variable coding regions: one located between exon 3 and exon 7 and the other between exon 9 and exon 11, incorporating the first 266 bases of exon 11 (see Fig. 2A). To avoid confusion we have designated these 5´ and 3´ variable deletion sites (VDS) respectively. Several studies concentrating on the 3´VDS have described four mRNA isoforms, which are expressed in the ovary of large domestic ruminants and several other species and these have been designated 'A, B, F, and G' isoforms in sheep, (Bacich *et al.* 1999), but appear to have corresponding isoforms in a number of other species including cattle, pig, rat and humans (Kawate & Okuda 1998; Loosfelt *et al.* 1989; Reinholz *et al.* 2000). Full LHR functionality, which includes ligand specificity and nuclear signalling capacity, can only be conferred by the translation of the undeleted mRNA splice variant member of the 'A' form family. This form encodes an LH/hCG specific ectodomain (exons 1 to 10), and a membrane-bound nuclear signalling endodomain (exon 11); . Members of the 'B' and 'G' variant family mRNAs are truncated due to an early stop codon being included when part of exon 11 is spliced out resulting in an open reading frame shift. Putative proteins would therefore not incorporate any of the endodomain and would thus be soluble. The 'F' variant family loses exon 10 only and do not undergo frame shift. Therefore putative 'F' form LHR proteins should retain ligand specificity, and incorporate transmembrane and intracellular signalling regions. However to date no role has been suggested for either this soluble form or any of the putative proteins encoded by the various LHR mRNA isoforms and studies in our laboratory have shown no consistent change in the ratio between these isoforms during somatic cell differentiation in sheep either *in vivo* or *in vitro* (Marsters *et al.* 2007). This suggests that alternative splicing of the common precursor LHR primary RNA is regulated so as to produce a constant molar ratio between the mature transcripts. Thus the regulation of LHR expression in ovarian somatic cells requires further work to determine the biological role of the different splice variants.

To date several surveys of LHR mRNA in follicular somatic cells during the antral phase have been published, and provide valuable insights into the complexities of antral follicle development. The results of classic LH-binding (Carson *et al.* 1979; Webb and England 1982; Ireland & Roche 1982; Peng *et al.* 1991) and *in situ* hybridisation studies (Xu *et al.* 1995), demonstrated that granulosa cells from large pre-ovulatory follicles have LH-binding capacity. Furthermore LHR mRNA expression has been detected by RT-PCR in both granulosa and theca cells throughout the antral phase of follicle development in sheep (Abdennebi *et al.* 2002) and cattle (Robert *et al.* 2003) and is markedly upregulated in granulosa cells from large, highly oestrogenic pre-ovulatory follicles (Bao *et al.* 1997). Moreover, recent findings suggest that post- transcriptional regulation of LHR may involve LHR mRNA-binding proteins (LRBPs) that induce rapid degradation (Kash & Menon 1999). These workers have identified an LRBP binding site adjacent to the transcription start site and demonstrated that LRBP/LHR mRNA complexes are more rapidly degraded than unencumbered LHR mRNA thereby limiting time for translation. More work is therefore required to examine the mechanisms regulating the transcription and translation of LHR mRNA in the somatic cells of ovarian follicles in domestic ruminants.

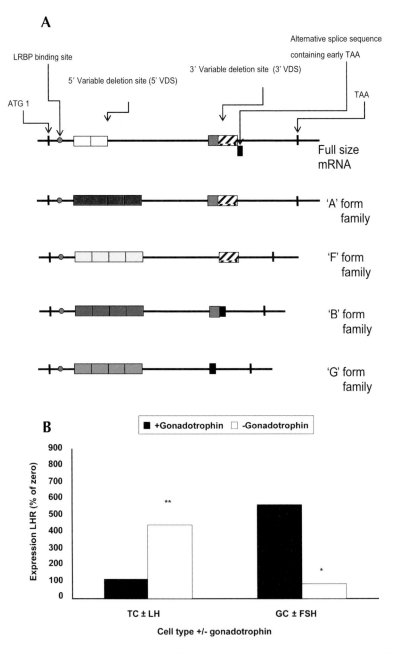

Fig. 2. A. Schematic representation of the LH receptor full size mRNA showing the positions of the 5´ and 3´ variable deletion sites (VDSs) and other important loci. Also shown the alternative splicing of the 3´ VDS which determines the variant families (A, F, B and G).

B. Semi-quantitative comparison of total LH receptor mRNA expression in theca cells (TC) and granulosa cells (GC) taken from small bovine antral follicles (3-5mm) and cultured for 96 hrs in serum-free media ± gonadotrophins (LH and FSH respectively). Expression was determined by RT-PCR. Asterisks denote individual statistical significances (*, P<0.05; **, P<0.01) between the ± gonadotrophin treatments. After Marsters *et al.* 2007.

Recent studies using somatic cells induced to differentiate *in vitro* have revealed important differences in the way that follicle stimulating hormone receptors (FSHR) and LHR expression are regulated. Granulosa cell FSHR expression appears to be constitutive from the primary stage of follicle development through to the dominant follicle. We have previously reported that FSHR expression in cultured granulosa cells declined rapidly in the period after plating and, in the presence of optimum doses of FSH and IGF-1, increase rapidly prior to induction of aromatase activity and gene expression (Marsters *et al.* 2003). Removal of FSH, but not IGF from culture media, however, results in a marked attenuation in this recovery in FSHR expression and the absence of aromatase indicating that FSH positively regulates expression of its own receptor in granulosa cells (Marsters *et al.* 2007). Further, expression of LHR in granulosa cells follows a similar pattern that parallels aromatase expression, confirming that induction of LHR in granulosa cells is an FSH dependent phenomenon that occurs concurrently with acquisition of the ability to secrete oestradiol. Conversely, LHR expression in cultured theca cells exhibits a similar profile, but is markedly up-regulated in the absence of LH in the media, indicating that thecal LHR expression is negatively regulated by LH in this species (Marsters *et al.* 2007; Fig. 2b).

At a more functional level, infusion studies in cattle have demonstrated that FSH alone, or in combination with LH, can stimulate follicles to develop to the preovulatory stage and these preovulatory follicles are capable of ovulating in response to hCG (Webb *et al.* 2003). Furthermore, adequate pulsatile LH support appears to be required to maintain the ovulatory competence of large follicles (>9 mm in diameter) when FSH concentrations are decreased. These studies also indicated that gonadotrophins are significantly involved in the control of ovulation rate. This is supported by the finding that cattle with three dominant follicles had higher FSH concentrations than cattle with two dominant follicles, with cattle with a single dominant follicle having the lowest mean FSH concentration (Lopez *et al.* 2005). These data in cattle are comparable with those generated using a similar model in sheep and agree with our current understanding of the role of declining FSH and subsequent LH support in selection of the dominant follicle. The results of recent experiments utilising a GnRH-antagonist suppression model confirm that LH is an essential requirement for normal ovulatory follicle development and subsequent luteal function, but show that a pulsatile mode of LH stimulation is not required by the ovulatory follicle for normal dominant follicle development and luteal function (Campbell *et al.* 2007; Fig. 3). These data suggest therefore that as well as an FSH threshold being necessary for the support of gonadotrophin-dependent follicles, a threshold concentration of LH is required for normal steroidogenesis and development of the dominant follicle. In addition to these direct effects of LH on dominant follicle development, LH, by modulating both oestradiol and inhibin A secretion by the ovulatory follicle, can also indirectly control the level of pituitary FSH release and hence the fate of FSH-dependent follicles.

Dominant follicles, even in the first follicular wave of the oestrous cycle, are mature enough to ovulate and corpora lutea (CL) can be generated experimentally by treatment with a number of factors including hCG (Price & Webb 1989), GnRH (Mee *et al.* 1991; 1993) and GnRH analogues (Twagiramungu *et al.* 1995). However the CL appears to be impaired with reduced progesterone production (Webb *et al.* 1992). In addition to impaired CL function, Perry *et al.* (2005) demonstrated that administration of GnRH to induce ovulation likely initiates a preovulatory gonadotrophin surge before some dominant follicles attain physiological maturity. Also GnRH-induced ovulation of follicles that are physiologically immature had a negative impact on pregnancy rates and late embryonic/fetal survival. It was suggested (Perry *et al.* 2005) that these observations in cattle may have implications for human assisted reproductive procedures. However there appears to be an optimum duration of ovulatory follicle development since an increased duration from the time of emergence (or dominance) to oestrus is associated with reduced pregnancy rates following AI in dairy cows undergoing spontaneous oestrous cycles (Bleach *et al.* 2004).

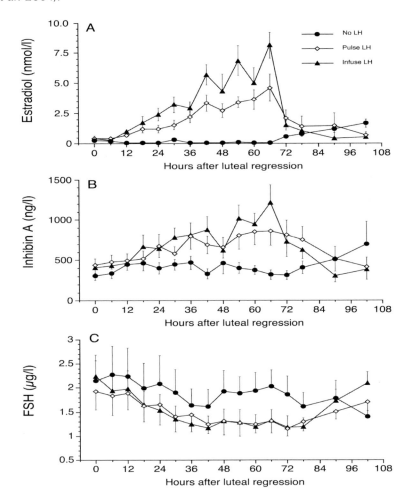

Fig. 3 Ovarian venous oestradiol (A) and inhibin A (B) and jugular venous FSH (C) concentrations in GnRH agonist suppressed ewes which received either no LH (n = 8; closed circles), pulsed LH (n = 8; open diamonds) or constant LH (n = 8; closed triangles) for 60 h after induction of luteal regression followed by an ovulatory stimulus at that time. The results demonstrate a key LH requirement for final maturation. However, these results also demonstrate that both the oestradiol and inhibin A responses occurred whether LH was given as a constant infusion or as pulses. After Campbell *et al.* 2007.

Other extra-ovarian factors are also involved in dominant follicle function. A large number of *in vitro* studies have demonstrated the direct action of metabolic factors on granulosa and theca cells (Webb *et al.* 1999; 2003; 2004; Armstrong *et al.* 2003). Bovine granulosa cells also appear to be dependent on the presence of physiological concentrations of insulin (Gutierrez *et al.* 1997) and furthermore, infusion of insulin into beef heifers increased the diameter of the dominant follicle (Simpson *et al.* 1994). Dietary-induced increases in circulating concentrations of insulin have also been correlated with increased oestradiol production in cultured granulosa cells from small antral (1-4mm) follicles (Armstrong *et al.* 2002b), demonstrating a direct action of metabolic hormones throughout the later stages of follicle development. This is supported by results of a study in post-partum dairy cows where insulin infusion increased oestradiol secretion after 30h by the dominant follicle of the first postpartum follicular wave. Interestingly these changes appeared not to be mediated through changes in pulsatile LH release (Butler *et al.* 2004), suggesting a direct effect of insulin on the follicle. However circulating free IGF-1 was also raised, which as discussed, could have increased the response of the follicle to peripheral gonadotrophins resulting in increased aromatase activity by the dominant follicle. Hence there appears to be an optimum concentration of insulin for follicle health, since it has recently been demonstrated that hyperinsulinemia can occur in cattle and that this condition is associated with impaired oocyte quality (Adamiak *et al.* 2005; 2006). Interestingly in this study, there was an interaction between body condition and level of feeding, which were cumulative, with both a positive and negative influence on oocyte quality as measured by the ability of oocytes to develop to blastocysts, with high level of feeding having a beneficial effect in animals of low body condition, but a detrimental effect in cows with a moderately high body condition.

Intra-ovarian factors

As the selected follicle reaches dominance there are also changes in the expression patterns of locally produced factors. For example, in healthy bovine follicles up to 9 mm in diameter, IGFBP-2 and -4 mRNA expression are restricted to granulosa and theca tissue, respectively (Webb *et al.* 2003; 2004). Indeed the conversion of a subordinate follicle to a future dominant follicle has been associated with a transient increase in follicular fluid activin A and oestradiol, but a decrease in IGFBP-2 (Armstrong *et al.* 1998; Ginther *et al.* 2002; Kojima *et al.* 2003). The reduction in follicular fluid IGFBP-2 and -4 concentrations has been coupled to the increase in oestradiol concentrations, in dominant follicles in cattle (Mihm *et al.* 2000). Hence, lower amounts of IGFBP-2 and increased LH receptors in granulosa cells appear to be associated with the establishment of the dominant follicle (Webb *et al.* 2003; 2004).

This reduction in IGFBPs has been associated with increased proteolytic activity (Mazerbourg *et al.* 2000). The protease that degrades IGFBP-4 and –5 has been shown to be the pregnancy-associated plasma protein –A (PAPP-A) (Monget *et al.* 2003). PAPP-A mRNA expression is also more abundant in growing dominant bovine follicles than in non-selected small follicles (Fayad *et al.* 2004). Furthermore it has been shown that PAPP-A is responsible for IGF-dependent degradation of IGFBP-2 probably leading to increased IGF bioavailability (Monget *et al.* 2003). Post-translational modification of IGFBPs are also known to occur and it has been demonstrated that at least 51 isoforms of IGFBP are present in bovine follicular fluid (Nicholas *et al.* 2002), but the physiological role of these isoforms has yet to be clarified.

Other locally produced growth factors include members of TGFß superfamily of ligands, operating through Smad signalling pathways (Knight & Glister 2006). Certainly a range of BMPs and associated factors are involved in follicular maturation as indicated by the marked increase in ovulation rate in sheep with a range of mutations (McNatty *et al.* 2007 this supplement). As with other stages of follicle development there is now evidence for a functional role of BMPs in the dominant follicle, acting in concert with other locally produced factors and gonadotrophins. However, the exact mechanisms through which all these growth factors operate and the degree of redundancy need to be elucidated.

In addition to known local factors, recent data derived from genetic array approaches (Mihm *et al.* 2006) have revealed a large number of factors that may be associated with dominance in mono-ovulatory species. Some of these effects, such as the association of dominance with the development of LH receptors on granulosa cells (Evans et al. 2004; Mihm et al. 2006) are well established (Webb & England 1982; Xu et al. 1995), whereas many others have unknown actions which may or may not be causally related to the attainment of dominance. Furthermore, other factors of unknown identity have been described through the more conventional means of the observation of a biological effect. One of the most intriguing of these activities, given the increased understanding of the importance of oocyte secreted factors in controlling follicular development, is the observation that oocytes from a number of species release a potent activity that is capable of inhibiting gonadotrophin-induced differentiation of granulosa cells in vitro without markedly affecting cellular proliferation (pig: Brankin et al. 2003; cow: Glister et al. 2003; sheep: Sfontouris 2004; Table 1). Comparison of the effect of culturing ruminant granulosa cells with a single oocyte on cell number and oestradiol production, with the known effects of candidate factors for this activity such as GDF-9, BMP-15, BMP-6 and TGFα/EGF (see Table 1), reveal that GDF-9/BMP-15 are possible candidates for this activity. This inhibitory action is consistent with the sheep models in which ovulation rates are increased in heterozygote carriers of null mutations (Hanrahan et al. 2004) and following immuno-neutralisation (Juengel et al. 2004). The role of GDF-9/BMP-15 in the control of dominance is therefore worthy of future investigation with a primary question being the mechanism whereby this putative inhibitory moiety is controlled during selection.

Table 1: Comparison of the effects of co-culture of oocytes with granulosa cells from sheep and cattle on cellular proliferation and oestradiol production under optimal conditions for induction of cellular differentiation (i.e. in the presence of IGF-1 and FSH) with other known oocyte-secreted factors. After Glister et al. 2003; Campbell et al. 2005.

Factor	Proliferation	Oestradiol production	Reference
Oocyte co-culture (+FSH/IGF)	None	Negative	Glister et al. 2003 Sfontouris 2004
BMP-6 (+FSH/IGF)	None	Positive	Glister et al. 2004 Campbell et al. 2006
TGFα/EGF (+FSH/IGF)	Positive	Negative	Glister et al. 2003 Campbell et al. 1996
GDF-9/BMP-15 (+FSH/IGF)	None	Negative	Campbell et al. 2005

Follicular atresia

As discussed, atresia occurs throughout follicular development and apoptotic cell death is an underlying mechanism of cell loss during follicular atresia (Tilly et al. 1991). Granulosa cell apoptosis may occur early in the process of atresia in rodents, before other morphological or biochemical changes are detected (Tilly et al. 1992), although in cattle a decline in intra-follicular oestradiol production has been shown to precede apoptosis of granulosa cells (Austin et al. 2001). The theca interna is also susceptible to cell death, with a reduced number of P450SCC- positive cells in atretic follicles (Clarke et al. 2004). Indeed theses authors suggested that the theca interna could be a site of initiation of atresia.

Atresia, along with follicular growth differentiation and ovulation are dependant on cyclical remodelling of the extra cellular matrix (ECM). For example plasminogen activators and inhibitors, including protease nexin-1 (PN-1) have been associated with the process of atresia in non-ovulatory dominant bovine follicles (Cao et al. 2006). In addition, gelatinolytic and caseinolytic matrix metalloproteinase's (MMPs) which degrade the proteinaceous components of the ECM are temporally and spatially regulated within the thecal and granulosa compartments of bovine follicles (Smith *et al.* 2005). Indeed, gelatinase A (MMP-2) activity was increased in response to physiological concentrations of LH. The controlled degradation of ECM proteins by MMPs and their inhibitors (TIMPs) may be essential for preserving a microenvironment conducive to follicular function.

A number of markers of atresia have been used to assess follicular status including caspases (Fenwick & Hurst 2002), Fas and Fas-L (Quirk *et al.* 2000), TUNEL staining of granulosa cells (Zeuner *et al.* 2003) and IGFBP-5 expression (Devine *et al.* 2000). Whilst several of these markers have been used to identify atretic follicles, very little evidence of their subsequent use as markers of oocyte quality exists (Nicholas *et al.* 2005). Unlike progression of apoptosis in oocytes, that relies on caspase-2, apoptosis in granulosa cells from pre-antral to preovulatory follicles is dependent upon the activity of caspases -3 and -7 (Matikainen *et al.* 2001). In cattle, it has been demonstrated that follicular atresia is accompanied by a considerable increase in the lower molecular weight IGFBPs (Nicholas *et al.* 2005). Importantly, IGFBP-5 appears to be a particularly good marker of atresia, since other IGFBPs are expressed at different stages of follicular development, whereas IGFBP-5 is exclusive to atresia (Monget *et al.* 1998). Indeed, it has recently been demonstrated that the IGFBP expression profile of follicular fluid can be used to better predict oocyte developmental competence (Nicholas *et al.* 2005). A distinctive IGFBP profile in follicular fluid of ovarian follicles has been demonstrated, with the dominant bovine follicle containing only IGFBP-3 together with a high oestradiol content, whereas the two largest subordinate follicles contain increased levels of IGFBP-2, -4 and the 29-31kDa (IGFBP-5) band, concomitant with reduced oestradiol content (Fig. 4). Furthermore, an extremely significant correlation between 29-31kDa, (IGFBP-5) expression in follicular fluid and caspase-3 activity in the granulosa cells, measured in the same follicles, has been shown (Nicholas *et al.* 2005; Fig. 4), demonstrating that IGFBP expression patterns can be used to group follicles into healthy, early atretic and late atretic follicles. Furthermore, a higher proportion of oocytes derived from follicles in early atresia progressed to the blastocyst stage after IVF (Fig. 4).

Gross morphological assessment of cumulus oocyte complexes is not an accurate method of determining whether oocytes have been derived from healthy or atretic follicles (Nicholas *et al.* 2005). Hence, selecting oocytes from follicles of defined quality using non-invasive markers, for example, IGFBP expression patterns, which can be performed rapidly enough to pre-select the oocytes prior to IVF would be a major advantage. Selection of good quality oocytes is also highly desirable for procedures that require extensive manipulations such as pronuclear microinjection, intra-cytoplasmic sperm injection (ICSI) and nuclear transplantation, which need to ensure a high developmental competence after embryo transfer. These results may enable an improvement in IVF pregnancy rates from the current success rate of ~ 30% by pre-selection of oocytes on a more biochemically sound basis than by simple morphological evaluation, which is currently used, and is a rather non-empirical method of assessment.

Conclusions

The final stages of follicle development are driven by an absolute requirement for gonadotrophic support (see Fig. 5). In addition, this requirement is developmentally regulated with dominant follicles changing their reliance from FSH to LH support. The response of these follicles to gonadotrophins is modulated by other extra-ovarian factors, such as insulin and IGF-1, which also influence follicle development, oocyte quality and a panoply of locally produced growth factors. This review has centred on those that have been studied in most detail, namely the IGF and BMP systems. Limited, but rapidly increasing, information to date has already demonstrated an interaction between these two important local regulatory systems (see Fig. 5). Fur-

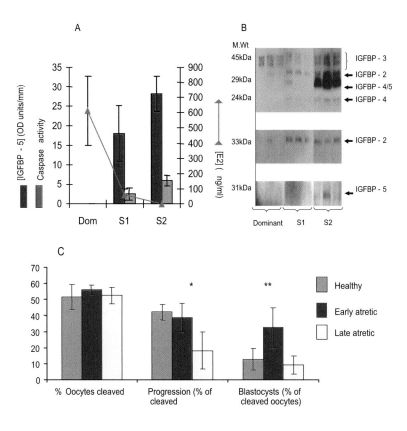

Fig. 4 Bovine granulosa cell caspase-3 activity, oestradiol 17ß follicular fluid IGFBP-S and concentrations (A), and (B) IGFBP content, visualised by Western ligand blot using biotinylated IGF-II, in follicular fluid from ovarian follicles from synchronised heifers on day 5 after oestrus and (C) the outcome of *in-vitro* fertilization on pre-selected oocytes from non-synchronised animals. Adapted from Nicholas *et al.* 2005.

thermore, gene expression profiling approaches will identify additional novel and differentially expressed genes within follicles. The challenge for the future, following identification of these factors, will be the determination of their physiological role, in particular how they interact with extra-ovarian factors. An increased understanding of these complex systems controlling follicular development will result in the *in vivo* and *in vitro* production of better quality oocytes resulting in enhanced embryo survival and pregnancy rates.

Acknowledgements

The authors would like to express their thanks to the funding agencies (BBSRC, Defra and SEERAD), who supported much of their cited work.

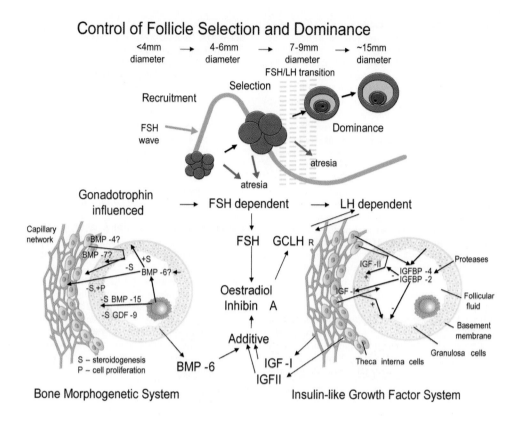

Fig. 5 Diagram showing the role of gonadotrophins in antral follicle development and the interaction with the follicular IGF and BMP systems. Note the additive effect of the IGF and BMP systems on FSH stimulated follicular development. Adapted from Webb *et al.* 2003; Campbell *et al.* 2006.

References

Aaltonen J, Laitinen MP, Vuojolainen K, Jaatinen R, Horelli-Kuitunen N, Seppa L, Louhio H, Tuuri T, Sjoberg J, Butzon R, Hovatta O, Dale L & Ritvos O 1999 Human growth differentiation factor 9 (GDF-9) and its novel homolog GDF-9B are expressed in oocytes during early folliculogenesis. *Journal Clinical Endocrinology and Metabolism* **84** 2744-2750.

Abdennebi L, Lesport AS, Remy JJ, Grebert D, Pisselet C, Monniaux D & Salesse R 2002 Differences in splicing of mRNA encoding LH receptor in theca cells according to breeding season in ewes. *Reproduction* **123** 819-826.

Adamiak SJ, Mackie K, Watt RG, Webb R & Sinclair KD 2005 Impact of nutrition on oocyte quality: cumulative effects of body composition and diet leading to hyperinsulinemia in cattle. *Biology of Reproduction* **73** 918-926.

Adamiak SJ, Powell K, Rooke JA, Webb R & Sinclair K D 2006 Body composition, dietary carbohydrates and fatty acids determine post-fertilisation development of bovine oocytes *in vitro*. *Reproduction* **131** 247-258.

Adams GP 1999 Comparative patterns of follicle development and selection in ruminants. *Journal of Reproduction and Fertility Supplement* **54** 17-32.

Armstrong DG, Baxter G, Gutierrez CG, Hogg CO, Glazyrin AL, Campbell BK, Bramley TA & Webb R 1998 Insulin-like growth factor binding protein -2 and -4 mRNA expression in bovine ovarian follicles: effect of gonadotropins and developmental status. *Endocrinology* **139** 2146-2154.

Armstrong DG, Gutierrez CG, Baxter G, Glazyrin AL, Mann GE, Woad KJ, Hogg CO & Webb R 2000 Expression of mRNA encoding IGF-I, IGF-II and type 1 IGF receptor in bovine ovarian follicles. *Journal of Endocrinology* **165** 101-113.

Armstrong DG, McEvoy TG, Baxter G, Robinson JJ, Hogg CO, Woad KJ & Webb R 2001 Effect of dietary energy and protein on bovine follicular dynamics and embryo production in vitro: associations with the ovarian insulin-like growth factor system. *Biology of Reproduction* **64** 1624-1632.

Armstrong DG, Baxter G, Hogg CO & Woad KJ 2002a Insulin-like growth factor (IGF) system in the oocyte and somatic cells of bovine pre-antral follicles. *Reproduction* **123** 789-797.

Armstrong DG, Gong JG, Gardner JO, Baxter G, Hogg CO & Webb R 2002b. Steroidogenesis in bovine granulosa cells: the effect of short-term changes in dietary intake. *Reproduction* **123** 371-378.

Armstrong DG, Gong JG &. Webb R 2003 Interactions between nutrition and ovarian activity in cattle: physiological, cellular and molecular mechanisms. *Reproduction Supplement* **61** 403-414.

Austin EJ, Mihm M, Evans ACO, Knight PG, Ireland JLH, Ireland JJ & Roche JF 2001. Alterations in intrafollicular regulatory factors and apoptosis during selection of follicles in the first follicular wave of the bovine estrous cycle. *Biology of Reproduction* **64** 839-848.

Bacich DJ, Earl CR, O'Keefe DS, Norman RJ & Rodgers RJ 1999 Characterisation of the translated products of the alternatively spliced luteinising hormone receptor in the ovine ovary throughout the oestrous cycle. *Molecular and Cellular Endocrinology* **147** 113-124.

Bao B & Garverick HA 1998 Expression of steroidogenic enzyme and gonadotropin receptor genes in bovine follicles during ovarian follicular waves: a review. *Journal of Animal Science* **76** 1903-1921.

Bao B, Garverick HA, Smith GW, Smith MF, Salfen BE & Youngquist RS 1997 Changes in messenger ribonucleic acid encoding luteinising hormone receptor, cytochrome P450-side chain cleavage, and aromatase are associated with recruitment and selection of bovine ovarian follicles *Biology of Reproduction* **56** 1158-1168.

Berlinguer F, Gonzalez-Bulnes A, Succu S, Leoni GG, Veiga-Lopez A, Mossa F, Garcia-Garcia RM, Debbere D, Galioto M, Cocero MJ & Naitana S 2006 GnRH antagonist enhance follicular growth in FSH-treated sheep but affect developmental competence of oocytes collected by ovum pick-up. *Theriogenology* **65** 1099-1109.

Bleach ECL, Glencross RG & Knight PG 2004 Association between ovarian follicle development and pregnancy rates in dairy cows undergoing spontaneous oestrous cycles. *Reproduction* **127** 621-629.

Brankin V, Mitchell MRP, Webb R & Hunter MG 2003 Paracrine effects of oocyte secreted factors and stem cell factor on porcine granulosa and theca cells *in vitro*. *Reproductive Biology and Endocrinology* **155** 1-15.

Buratini Jr J, Teixeira AB, Costa IB, Glapinski VF, Pinto MGL, Giometti IC, Barros CM, Cao M, Nicola ES & Price CA 2005 Expression of fibroblast growth factor-8 and regulation of cognate receptors, fibroblast

growthfactor receptor-3c and -4 in bovine antral follicles. *Reproduction* **130** 343-350.

Burns DS, Jimenez-Krassel F, Ireland JLH, Knight PG & Ireland JJ 2005 Numbers of antral follicles during follicular waves in cattle: evidence for high variation among animals, very high repeatability in individuals, and an inverse association with serum follicle-stimulating hormone concentrations. *Biology of Reproduction* **73** 54-62.

Butler ST, Pelton SH & Butler WR 2004 Insulin increases 17ß-estradiol production by the dominant follicle of the first postpartum follicle wave in dairy cows. *Reproduction* **127** 537-545.

Cahill L 1981 Folliculogenesis in the sheep as influenced by breed, season and oestrous cycle. *Journal of Reproduction and Fertility Supplement* **30** 135-142.

Campbell BK, Scaramuzzi RJ & Webb R 1995 Control of antral follicle development and selection in sheep and cattle. *Journal of Reproduction and Fertility Supplement* **49** 335-350.

Campbell BK, Scaramuzzi RJ & Webb R 1996 Induction and maintenance of oestradiol and immuno-reactive inhibin production with FSH by ovine granulosa cells cultured in serum free media. *Journal of Reproduction and Fertility* **106** 7-16.

Campbell BK, Telfer EE, Webb R & Baird DT 2000 Ovarian autografts in sheep as a model for studying folliculogenesis. *Molecular and Cell Endocrinology* **163** 137-139.

Campbell BK, Sharma S, Shimasaki S & Baird DT 2005 Effect of AMH, BMP-15 and GDF-9 on FSH-induced differentiation of sheep granulosa cells. *Biology of Reproduction Special Issue* Abstract 728.

Campbell BK, Souza CJH, Skinner AJ, Webb R & Baird DT 2006 Enhanced response of granulosa and theca cells from sheep carriers of the FecB mutation *in vitro* to gonadotropins and bone morphogenetic protein-2, -4, and -6. *Endocrinology* **147** 1608-1620.

Campbell BK, Kendall NR & Baird DT 2007 The effect of the presence and pattern of LH stimulation on ovulatory follicle development in sheep. *Biology of Reproduction* (in press).

Cao M, Buratini Jr J, Lussier JG, Carrière PD & Price CA 2006 Expression of protease nexin-1 and plasminogen activators during follicular growth and the periovulatory period in cattle. *Reproduction* **131** 125-137.

Carabatsos MJ, Sellitto C, Goodenough DA & Albertini DF 2000 Oocyte-granulosa cell heterologous gap junctions are required for the coordination of nuclear and cytoplasmic meiotic competence. *Developmental Biology* **226** 167-79.

Carson RS, Findlay JK, Burger HG & Trounson AO 1979 Gonadotropin receptors of the ovine ovarian follicle during follicular growth and atresia. *Biology of Reproduction* **21** 75-87.

Clarke LJ, Irving-Rodgers HF, Dharmarajan AM & Rodgers RR 2004 Theca Interna: the other side of bovine follicular atresia. *Biology of Reproduction* **71** 1071-1078.

Crowe MA, Kelly P, Draincourt MA, Boland MP & Roche JF 2001 Effects of follicle-stimulating hormone with and without luteinizing hormone on serum hormone concentrations, follicle growth, and intrafollicular estradiol and aromatase activity in gonadotropin-releasing hormone-immunised heifers. *Biology of Reproduction* **64** 368-374.

Devine PJ, Payne CM, McCuskey MK & Hoyer PB 2000 Ultrastructural evaluation of oocytes during atresia in rat ovarian follicles. *Biology of Reproduction* **63** 1245-1252.

Dong J, Albertini DF, Nishimori K, Kumar TR, Lu N & Matzuk MM 1996 Growth differentiation factor-9 is required during early ovarian folliculogenesis. *Nature* **383** 531-535.

Driancourt M, Cahill L & Bindon B 1985 Ovarian follicular populations and preovulatory enlargement in Booroola and Merino ewes. *Journal of Reproduction and Fertility* **73** 93-107.

Driancourt MA, Reynaud K, Cortvrindt R & Smitz J 2000 Roles of KIT and KIT Ligand in ovarian function. *Reviews of Reproduction* **5** 143-152.

Dube JL, Wang P, Elvin J, Lyons KM, Celeste AJ & Matzuk MM 1998 The bone morphogenetic protein 15 gene is X-linked and expressed in oocytes. *Molecular Endocrinology* **12** 1809-1817.

Dugan KJ, Campbell BK, Skinner A, Armstrong DG & Webb R 2004 Expression of bone morphogenetic protein-6 (BMP-6) in the bovine foetal ovary at different stages of gestation. *Reproduction, Abstract Series* **31** Abstract O46.

Eppig JJ 2001 Oocyte control of ovarian follicular development and function in mammals. *Reproduction* **122** 829-38.

Evans ACO, Ireland JLH, Winn ME, Lonergan P, Smith GW, Coussens PM & Ireland JJ 2004 Identification of genes involved in apoptosis and dominant follicle development during follicular waves in cattle. *Biology of Reproduction* **70** 1475-1484.

Fabre S, Pierre A, Pisselet C, Mulsant P, Lecerf F, Pohl J, Monget P & Monniaux D 2003 The Booroola mutation in sheep is associated with an alteration of the bone morphogenetic protein receptor-IB functionality. *Journal of Endocrinology* **177** 435-444.

Fatahei AN Van Den Hurk R, Colenbrander B, Daemen AJJM, Van Tol HTA, Monteiro RM, Roelen BAJ & Bevers MM 2005 Expression of bone morphogenetic protein-2 (BMP-2), BMP-4 and BMP receptors but absence of effects of BMP-2 and BMP-4 during IVM on bovine oocyte nuclear maturation and subsequent embryo development. *Theriogenology* **63** 872-889.

Fayad T, Lévesque V, Sirois J, Silversides DW & Lussier JG 2004 Gene expression profiling of differentially expressed genes in granulosa cells of bovine dominant follicles using suppression subtractive hybridization. *Biology of Reproduction* **70** 523-533.

Fenwick MA & Hurst PR 2002 Immunohistochemical localization of active caspase-3 in the mouse ovary: growth and atresia of small follicles. *Reproduction* **124** 659-665.

Fortune JE, Cushman RA, Wahl CM & Kito WS 2000 The primordial to primary follicle transition. *Molecular and Cellular Endocrinology* **163** 53-60.

Fouladi-Nashta AA, Dugan KJ & Webb R 2005 Immunohistochemical localisation of BMP receptors in bovine foetal ovaries. *Reproduction, Abstract Series* **32** Abstract 72.

Garverick HA, Baxter G, Gong J, Armstrong DG, Campbell BK, Gutierrez CG & Webb R 2002 Regulation of expression of ovarian mRNA encoding steroidogenic enzymes and gonadotrophin receptors by FSH and GH in hypogonadotrophic cattle. *Reproduction* **123** 651-661.

Ginther OJ, Beg MA, Bergfelt DR & Kot K 2002 Activin A, estradiol and free insulin-like growth factor I in follicular fluid preceding the experimental assumption of follicle dominance in cattle. *Biology of Reproduction* **67** 14-19.

Glister C, Groome NP & Knight PG 2003 Oocyte-mediated suppression of follicle-stimulating hormone and insulin-like growth factor induced secretion of steroids and inhibin-related proteins by bovine granulosa cells in vitro: possible role of transforming growth factor α. *Biology of Reproduction* **68** 758-765.

Glister C, Kemp CF & Knight PG 2004 Bone morphogenetic protein (BMP) ligands and receptors in bovine ovarian follicle cells: actions of BMP-4, -6 and -7 on granulosa cells and differential modulation of Smad-1 phosphorylation by follistatin. *Reproduction* **127** 239–254.

Glister C, Richards SL & Knight PG 2005 Bone morphogenetic proteins (BMP)-4, -6, and -7 potently suppress basal and luteinising hormone-induced androgen production by bovine theca interna cells in primary culture: could ovarian hyperandrogenic dysfunction be caused by a defect in thecal BMP signalling. *Endocrinology* **146** 1883-1892.

Gong JG, Bramley TA & Webb R 1991 The effect of recombinant bovine somatotropin on ovarian function in heifers: follicular populations and peripheral hormones. *Biology of Reproduction* **45** 941-949.

Gong JG, Bramley TA & Webb R 1993 The effect of recombinant bovine somatotrophin on ovarian follicular growth and development in heifers. *Journal of Reproduction and Fertility* **97** 247-254.

Gong JG, Baxter G, Bramley TA & Webb R 1997 Enhancement of ovarian follicle development in heifers by treatment with recombinant bovine somatotrophin: a dose response study. *Journal of Reproduction and Fertility* **110** 91-97.

Gong JG, Armstrong DG, Baxter G, Hogg CO, Garnsworthy PC & Webb R 2002. The effect of increased dietary intake on superovulatory response to FSH in heifers. *Theriogenology* **57** 1591-1602.

Gonzalez-Anover P, Encinas T, Garcia-Garcia RM, Veiga-Lopez A, Cocero MJ, McNeilly AS & Gonzalez-Bulnes A 2004 Ovarian response in sheep superovulated after pretreatment with growth hormone and GnRH antagonists is weakened by failures in oocyte maturation. *Zygote* **12** 301-304.

Gonzalez-Bulnes A, Santiago-Moreno J, Garcia-Garcia

RM, Souza CJ, Lopez-Sebastian A & McNeilly AS 2004 Effect of GnRH antagonists treatment on gonadotrophin secretion, follicular development and inhibin A secretion in goats. *Theriogenology* **61** 977-985.

Gougeon A 1996 Regulation of ovarian follicular development in primates: facts and hypotheses. *Endocrine Reviews* **17** 121-155.

Gutierrez CG, Campbell BK & Webb R 1997 Development of a long-term bovine granulosa cell culture system: induction and maintenance of estradiol production, response to follicle stimulating hormone and morphological characteristics. *Biology of Reproduction* **56** 608-616.

Gutierrez CG, Ralph JH, Telfer EE, Wilmut I & Webb R 2000 Growth and antrum formation of bovine antral follicles in long-term culture *in vitro*. *Biology of Reproduction* **62** 1322-1328.

Hampton JH, Bader JF, Lamberson WR, Smith MF, Youngquist RS & Garverick HA 2004 Gonadotrophin requirements for dominant follicle selection in GnRH agonist-treated cows. *Reproduction* **127** 695-703.

Hanrahan JP, Gregan SM, Mulsant P, Mullen M, Davis GH, Powell R & Galloway SM 2004 Mutations in the genes for oocyte-derived growth factors GDF9 and BMP15 are associated with both increased ovulation rate and sterility in Cambridge and Belclare sheep (*Ovis Aries*). *Biology of Reproduction* **70** 900-909.

Hunter MG, Robinson RS, Mann GE & Webb R 2004 Endocrine and paracrine control of follicular development and ovulation rate in farm species. *Animal Reproduction Science* **82-83** 461-477.

Huntriss J, Gosden R, Hinkins M, Oliver B, Miller D, Rutherford AJ & Picton HM 2002 Isolation, characterization and expression of the human Factor In the Germline alpha (FIGLA) gene in ovarian follicles and oocytes. *Molecular Human Reproduction* **8** 1087-1095.

Ireland JJ & Roche JF 1982 Development of antral follicles in cattle following prostaglandin-induced luteolysis: changes in serum hormones, steroids in follicular fluid and gonadotropin receptors. *Endocrinology* **111** 2077-2086.

Jayawardana BC, Shimizu T, Nishimoto H, Kaneko E, Tetsuka M & Miyamoto A 2006 Hormonal regulation of expression of growth differentiation factor-9 receptor type I and II in the bovine ovarian follicle. *Reproduction* **131** 545-553.

Juengel JL, Hudson NL, Whiting L & McNatty KP 2004 Effects of immunisation against bone morphogenetic protein 15 and growth differentiation factor 9 on ovulation rate, fertilization, and pregnancy in ewes. *Biology of Reproduction* **70** 557-561.

Juengel JL, Reader KL, Bibby AH, Lun S, Ross I, Haydon LJ & McNatty KP 2006 The role of bone morphogenetic proteins 2, 4, 6 and 7 during ovarian follicular development in sheep: contrast to rat. *Reproduction* **131** 501-513.

Kanitz W, Becker F, Schneider F, Kanitz E, Leiding C,

Nohner HP & Pohland R 2002 Superovulation in cattle: practical aspects of gonadotropin treatment and insemination. *Reproduction Nutrition Development* **42** 587-599.

Kash JC & Menon KMJ 1999 Sequence-specific binding of a hormonally regulated mRNA binding protein to cytidine-rich sequences in the lutropin receptor open reading frame. *Biochemistry* **38** 16889-16897.

Kawate N & Okuda K 1998 Coordinated expression of splice variants for luteinising hormone receptor messenger RNA during the development of bovine corpora lutea. *Molecular Reproduction and Development* **51** 66-75.

Knight PG & Glister C 2001 Potential local regulatory functions of inhibins, activins and follistatin in the ovary. *Reproduction* **121** 503-512.

Knight PG & Glister C 2006 TGF-ß superfamily members and ovarian follicle development. *Reproduction* **132** 191-206.

Kojima FN, Berfeld EGM, Wehrman ME, Cupp AS, Fike KE, Mariscal-Aguayo DV, Sanchez-Torres T, Garcia-Winder M, Clopton DT, Roberts AJ & Kinder JE 2003. Frequency of hormone pulses in cattle influences duration of persistence of dominant ovarian follicles, follicular fluid concentration of steroids, and activity of insulin-like growth factor binding proteins. *Animal Reproduction Science* **77** 187-211.

Kulick LJ, Kot K, Wiltbank MC & Ginther OJ 1999 Follicular and hormonal dynamics during the first follicular wave in heifers. *Theriogenology* **52** 913-921.

Lin JG, Lerch TF, Cook RW, Jardetzky TS & Woodruff TK 2006 The structural basis of TGF-ß, bone morphogenetic protein, and activin ligand binding. *Reproduction* **132** 179-190.

Loosfelt H, Misrahi M, Atger M, Salesse R, Thi MTVH-L, Jolivet A, Guiochon-Mantel A, Sar S, Jallal B, Garnier J & Milgrom E 1989 Cloning and sequencing of porcine LH-hCG receptor cDNA: variants lacking transmembrane domain. *Science* **245** 525-528.

Lopez-Alonso C, Encinas T, Veiga-Lopez A, Garcia-Garcia RM, Cocero MJ, Ros JM, McNeilly AS & Gonzalez-Bulnes A 2005 Follicular growth, endocrine response and embryo yields in sheep superovulated with FSH after pretreatment with a single short-acting dose of GnRH antagonist. *Theriogenology* **64** 1833-1843.

Lopez H, Sartori R & Wiltbank MC 2005 Reproductive hormones and follicular growth during the development of one or multiple dominant follicles in cattle. *Biology of Reproduction* **72** 788-795.

Lucy MC 2000 Regulation of ovarian follicular growth by somatotropins and insulin–like growth factors in cattle. *Journal of Dairy Science* **83** 1635-1647.

Macklon NS & Fauser BC 2005 Ovarian reserve. *Seminars in Reproductive Medicine* **23** 248-56.

Macklon NS, Stouffer RL, Giudice LC & Fauser BC 2006 The science behind 25 years of ovarian stimulation for in vitro fertilization. *Endocrine Reviews* **27** 170-207.

Marsters P, Kendall NR & Campbell BK 2003 Temporal relationships between FSH receptor, type 1 insulin-like growth factor receptor, and aromatase expression during FSH-induced differentiation of bovine granulosa cells maintained in serum-free culture. *Molecular and Cellular Endocrinology* **203** 117-127.

Marsters P, Kendall NR & Campbell BK 2007 Expression of luteinising hormone receptor mRNA splice variants in bovine ovarian somatic cells induced to differentiate *in vitro*. *Journal of Endocrinology* (in press).

Matikainen T, Perez GI, Zheng TS, Kluzak TR, Rueda BR, Flavell RA & Tilly JL 2001 Caspase-3 gene knockout defines cell lineage specificity for programmed cell death signaling in the ovary. *Endocrinology* **142** 2468-2480.

Mazerbourg S, Zapf J, Bar RS, Brigstock DR & Monget P 2000 Insulin-like growth factor (IGF)-binding protein-4 proteolytic degradation in bovine, equine, and porcine preovulatory follicles: regulation by IGFs and heparin-binding domain-containing peptides. *Biology of Reproduction* **63** 390-400.

McCaffery FH, Leask R, Riley SC & Telfer EE 2000 Culture of bovine pre-antral follicles in a serum-free system: markers for assessment of growth and development. *Biology of Reproduction* **63** 267-273.

McNatty K, Henderson KM, Lun S, Heath DA, Ball K, Hudson NL, Fannin J, Gibb M, Kieboom LE & Smith P 1985 Ovarian activity in Booroola x Romney ewes which have a major gene influencing their ovulation rate. *Journal of Reproduction and Fertility* **73** 109-120.

McNatty K, Hudson N, Henderson K, Gibb M, Morrison L, Ball K & Smith P 1987 Differences in gonadotrophin concentrations and pituitary responsiveness to GnRH between Booroola ewes which were homozygous (FF), heterozygous (F+) and non-carriers (++) of a major gene influencing their ovulation rate. *Journal of Reproduction and Fertility* **80** 577-588.

McNatty K, Kieboom L, McDiarmid J, Heath D & Lun S 1986 Adenosine cyclic 3',5' monophosphate and steroid production by small ovarian follicles from Booroola ewes with and without a fecundity gene. *Journal of Reproduction and Fertility* **76** 471-480.

McNatty KP, Reader K, Smith P, Heath DA & Juengel JL 2006 Control of ovarian follicular development in the gonadotrophin-dependent phase: a 2006 perspective. *Reproduction Supplement* (this issue, in press).

Mee MO, Stevenson JS, Alexander BM & Sasser RG 1991 Administration of GnRH at estrus influences pregnancy rates, serum concentrations of LH, FSH, estradiol-17 beta, pregnancy-specific protein B, and progesterone, proportion of luteal cell types, and in vitro production of progesterone in dairy cows. *Journal of Dairy Science* **74** 1573-1581.

Mee MO, Stevenson JS & Minton JE 1993 First postpartum luteal function in dairy cows after ovulation induced by progestogen and gonadotropin-releasing hormone. *Journal of Animal Science* **71** 185-198.

Mihm M, Austin EJ, Good TEM, Ireland JLH, Knight PG, Roche JF & Ireland JJ 2000 Identification of potential intrafollicular factors involved in selection of dominant follicles in heifers. *Biology of Reproduction* **63** 811-819.

Mihm M, Baker PJ, Ireland JL, Smith GW, Coussens PM, Evans AC & Ireland JJ 2006 Molecular evidence that growth of dominant follicles involves a reduction in follicle-stimulating hormone dependence and an increase in luteinizing hormone dependence in cattle. *Biology of Reproduction* **74** 1051-1059.

Monget P, Pisselet C & Monniaux D 1998 Expression of insulin-like growth factor binding protein-5 by ovine granulosa cells is regulated by cell density and programmed cell death in vitro. *Journal of Cell Physiology* **177** 13-25.

Monget P, Mazerbourg S, Delpuech T, Maurel MC, Maniere S, Zapf J, Lalmanach G, Oxvig C & Overgaard MT 2003 Pregancy-associated plasma protein A is involved in insulin-like growth factor binding protein–2 (IGFBP-2) proteolytic degradation in bovine and porcine preovulatory follicles: identification of cleavage site and characterization of IGFBP-2 degradation. *Biology of Reproduction* **68** 77-86.

Nicholas B, Scougall RK, Armstrong DG & Webb R 2002 Changes in insulin-like growth factor binding protein (IGFBP) isoforms during bovine follicular development. *Reproduction* **124** 439-446.

Nicholas B, Alberio R, Fouladi-Nashta AA & Webb R 2005 Relationship between low-molecular-weight insulin-like growth factor-binding proteins, caspase-3 activity, and oocyte quality. *Biology of Reproduction* **72** 796-804.

Peng XR, Hsueh AJW, LaPolt PS, Bjersing L & Ny T 1991 Localisation of luteinising hormone receptor messenger ribonucleic acid expression in ovarian cell types during follicle development and ovulation. *Endocrinology* **129** 3200-3207.

Perry GA, Smith MF, Lucy MC, Green JA, Parks TE, MacNeil MD, Roberts AJ & Geary TW 2005 Relationship between follicle size at insemination and pregnancy success. *Proceedings of the National Academy of Science* **102** 5268-5273

Picton HM, Danfour MA, Harris SE, Chambers EL & Huntriss J 2003 Growth and maturation of oocytes in vitro. *Reproduction Supplement* **61** 445-462.

Price CA & Webb R 1989 Ovarian response to hCG treatment during the oestrous cycle in heifers. *Journal of Reproduction and Fertility* **86** 303-308.

Quirk SM, Harman RM & Cowan RG 2000 Regulation of Fas antigen (Fas, CD95)-mediated apoptosis of bovine granulosa cells by serum and growth factors. *Biology of Reproduction* **63** 1278-1284.

Reinholz MM, Zschunke MA & Roche PC 2000 Loss of alternately spliced messenger RNA of the luteinising hormone receptor and stability of the follicle-stimulating hormone receptor messenger RNA in granulosa cell tumours of the human ovary. *Gynaecologic Oncology* **79** 264-271.

Reynaud K, Cortvrindt R, Smitz J, Bernex F, Panthier JJ & Driancourt MA 2001 Alterations in ovarian function of mice with reduced amounts of KIT receptor. *Reproduction* **121** 229-237.

Reynaud K, Cortvrindt R, Smitz J & Driancourt MA 2000 Effects of Kit Ligand and anti-Kit antibody on growth of cultured mouse preantral follicles. *Molecular Reproduction Development* **56** 483-494.

Robert C, Gagne D, Lussier JG, Bousquet D, Barnes FL & Sirard MA 2003 Presence of LH receptor mRNA in granulosa cells as a potential marker of oocyte developmental competence and characterisation of the bovine splicing isoforms. *Reproduction* **125** 437-446.

Saha S, Shimizu M, Geshi M & Izaike Y 2000 In vitro culture of bovine pre-antral follicles. *Animal Reproduction Science* **63** 27-39.

Sfontouris I 2004 Oocyte-somatic cell interactions during oogenesis and folliculogenesis in monovular species. *PhD Thesis, University Of Nottingham*

Shimasaki S, Moore RK, Erickson GF & Otsuka F 2003 The role of bone morphogenetic proteins in ovarian function. *Reproduction Supplement* **61** 323-37.

Shimasaki S, Moore RK, Otsuka F & Erickson GF 2004 The bone morphogenetic protein system in mammalian reproduction. *Endocrine Reviews* **25** 72–101.

Simpson RB, Chase Jr CC, Spicer LJ, Vernan RK, Hammond AL & Rae DO 1994 Effects of exogenous insulin on plasma and follicular insulin like growth factor I, insulin like growth factor binding activity, follicular oestradiol and progesterone and follicular growth in superovulated Angus and Brahman cows. *Journal of Reproduction and Fertility* **102** 483-492.

Smith MF, Gutierrez CG, Ricke WA, Armstrong DG & Webb R 2005 Production of matrix metalloproteinases by cultured bovine theca and granulosa cells. *Reproduction* **129** 75-87.

Smitz JE & Cortvindt RG 2002 The earliest stages of folliculogenesis in vitro. *Reproduction* **123** 185-202.

Souza CJH, Campbell BK & Baird DT 1997. Follicular dynamics and ovarian steroid secretion in sheep during the follicular and early luteal phases of the estrous cycle. *Biology of Reproduction* **56** 483-488.

Souza CJH, Campbell BK, McNeilly AS & Baird DT 2002. Effect of bone morphogenetic protein 2 (BMP2) on oestradiol and inhibin A production by sheep granulosa cells, and localization of BMP receptors in the ovary by immunohistochemistry. *Reproduction* **123** 363-369.

Soyal SM, Amleh A & Dean J 2000 FIGalpha, a germ cell-specific transcription factor required for ovarian follicle formation. *Development* **127** 4645-4654.

Spicer LJ, Voge JL & Allen DT 2004 Insulin-like growth factor-II stimulates steroidogenesis in cultured bovine thecal cells. *Molecular and Cellular Endocrinology* **227** 1-7.

Thomas FH, Leask R, Srsen V, Riley SC, Spears N & Telfer EE 2001 Effect of ascorbic acid on health and morphology of bovine pre-antral follicles during long-term cultures. *Reproduction* **122** 487-495.

Tilly JL, Kowalski KI, Johnson AL & Hsueh AJ 1991 Involvement of apoptosis in ovarian follicular atresia and postovulatory regression. *Endocrinology* **129** 2799-2801.

Tilly JL, Kowalski KI, Schomberg DW & Hsueh AJ 1992 Apoptosis in atretic ovarian follicles is associated with selective decreases in messenger ribonucleic acid transcripts for gonadotropin receptors and cytochrome P450 aromatase. *Endocrinology* **131** 1670-1676.

Turnbull K, Braden A & Mattner P 1977 The pattern of follicular growth and atresia in the ovine ovary. *Australian Journal of Biological Sciences* **30** 229-241.

Twagiramungu H, Guilbault LA, Proulx JG & Dufour JJ 1995 Buserelin alters the development of the corpora lutea in cyclic and early postpartum cows. *Journal of Animal Science* **73** 805-811.

Visser JA, de Jong FH, Laven JS & Themmen AP 2006 Anti-Mullerian hormone: a new marker for ovarian function. *Reproduction* **131** 1-9.

Walters KA, Binnie JP, Campbell BK, Armstrong DG & Telfer EE 2006 The effects of IGF-I on bovine follicle development and IGFBP-2 expression are dose and stage dependent. *Reproduction* **131** 515-523.

Webb R, Gong JG, Law AS & Rusbridge SM 1992 Control of ovarian function in cattle. *Journal of Reproduction and Fertility Supplement* **45** 141-156.

Webb R, Campbell BK, Garverick HA, Gong JG, Gutierrez CG & Armstrong DG 1999 Molecular mechanisms regulating follicular recruitment and selection. *Journal of Reproduction and Fertility Supplement* **54** 33-48.

Webb R & England BG 1982 Relationship between LH receptor concentrations in thecal and granulosa cells and in-vivo and in-vitro steroid secretion by ovine follicles during the pre-ovulatory period. *Journal of Reproduction and Fertility* **66** 169-180.

Webb R, Nicholas B, Gong JG, Campbell BK, Gutierrez CG, Gaverick HA & Armstrong DG 2003 Mechanism regulating follicular development and selection of the dominant follicle. *Reproduction* Supplement **61** 71-90.

Webb R, Garnsworthy PC, Gong J-G & Armstrong DG 2004 Control of follicular growth: local interactions and nutritional influences. *Journal of Animal Science* **82** E63–E74.

Xu Z, Garverick HA, Smith GW, Smith MF, Hamilton SA & Youngquist RS 1995 Expression of follicle stimulating hormone and luteinising hormone receptor messenger ribonucleic acids in bovine follicles during the first follicular wave. *Biology of Reproduction* **53** 951-957.

Yong PY, Baird DT, Thong KJ, McNeilly AS & Anderson RA 2003 Prospective analysis of the relationships between the ovarian follicle cohort and basal FSH concentration, the inhibin response to exogenous FSH and ovarian follicle number at different stages of the normal menstrual cycle and after pituitary down-regulation. *Human Reproduction* **18** 35-44.

Yuan W, Bao B, Garverick HA, Youngquist RS & Lucy MC 1998 Follicular dominance in cattle is associated with divergent patterns of ovarian gene expression for insulin-like growth factor (IGF) –I, IGF-II and IGF binding protein-2 in dominant and subordinate follicles. *Domestic Animal Endocrinology* **15** 55-63.

Zeuner A, Muller K, Reguszynski K & Jewgenow K 2003 Apoptosis within bovine follicular cells and its effect on oocyte development during in vitro maturation. *Theriogenology* **59** 1421-1433.

Pregnancy rates in cattle with cryopreserved sexed spermatozoa: effects of laser intensity, staining conditions and catalase

JL Schenk[1] and GE Seidel, Jr[2]

[1] XY, Inc., 2301 Research Blvd. Suite 110, Fort Collins, CO 80526-1825 USA; [2] ARBL, Colorado State University, Fort Collins, CO 80523-1683 USA

The overall aim of this research was to improve fertility of cattle inseminated with sexed spermatozoa by improving sperm sorting procedures. Six field trials were conducted in which 4,264 heifers were inseminated into the uterine body with cryopreserved sexed or unsexed control spermatozoa. Pregnancy or calving rates with doses of 2×10^6 sexed spermatozoa ranged from 32 to 51%; these averaged 69% of the pregnancy rates with 20×10^6 unsexed, control spermatozoa (range 53 to 79% of controls). Fertility of sexed spermatozoa was especially low on farms where control fertility was low. Accuracy of sexing ranged from 86 to 91%. Laser power of 150 mW for interrogating spermatozoa did not result in lower pregnancy rates (43%) than when power was decreased as much as possible for a particular sorting batch (50 to 130 mW) to still achieve sexing accuracy (38% pregnant). Addition of catalase to fluids containing spermatozoa was beneficial when thawed spermatozoa were incubated *in vitro* for 2 h but had no effect on pregnancy rates. There also was no effect on pregnancy rates between two concentrations of Hoechst 33342 for staining spermatozoa. Freezing 2×10^6 sexed spermatozoa at 20×10^6/ml resulted in a slightly higher rate of pregnancy ($P < 0.05$) than at 10×10^6/ml. The information obtained in these trials, along with other improvements, notably lowering pressure in the sorting system from 50 to 40 psi, has been used to improve procedures for sexing spermatozoa commercially.

Introduction

The only reliable method available for separation of spermatozoa capable of fertilization into relatively pure X- and Y-chromosome-bearing populations is flow cytometry/cell sorting. The 3.8% difference in DNA content between X- and Y-chromosome-bearing bovine spermatozoa is the basis of sorting, routinely done to about 90% accuracy (Tubman *et al.* 2004). This technology is now commercially available via bovine genetic companies in a number of countries. However, much remains to be done to optimize the many steps that are superimposed on routine semen processing as spermatozoa are sexed. Herein are results from six field trials from artificial insemination of heifers with sexed frozen-thawed spermatozoa that were processed by different treatments.

Corresponding author E-mail: gseidel@colostate.edu

Efforts described in this communication reflect modifications and improvements to those reported by Schenk *et al.* (1999) and Seidel *et al.* (1999) as the commercialization of sexed spermatozoa is a work in progress. Flow sorting of spermatozoa presents many challenges to overcome to assure that the resultant cells are capable of fertilization at rates approaching those of unsexed spermatozoa. These studies were done to determine the effect of Hoechst 33342 stain concentration, laser intensity, spermatozoa concentration during freezing and catalase on pregnancy rates in beef and dairy heifers inseminated artificially with sexed spermatozoa. A preliminary report on trials 3 and 4 was published earlier (Seidel and Schenk 2002).

Hoechst 33342

Hoechst 33342 (Bis-Benzimide, H-33342; #H21492, Molecular Probes, Eugene, OR), a bisbenzimide vital dye that fluoresces blue when excited by the 351 and 364 nm lines of energy from an argon laser, can be used to distinguish X- and Y-chromosome bearing populations (Johnson *et al.* 1987a). The dye is non-intercalating, as it binds to the minor groove of the DNA helix, with an affinity for adenine-thymine-rich regions (Muller & Gautier 1975; Johnson *et al.* 1987b). However, Hoechst 33342 is suspected of inducing chromosomal damage in spermatozoa under some conditions (Munné 1994). McNutt and Johnson (1996) reported an increase in fetal mortality in rabbits during the early stages of pregnancy following insemination of flow-sorted spermatozoa. Cran *et al.* (1994) reported similar cleavage but reduced blastocyst development rates and reduced pregnancy rates following embryo transfer of bovine oocytes fertilized with flow-sorted spermatozoa, in comparison with unsorted, unstained spermatozoa, but many aspects of sorting spermatozoa could have been responsible. Furthermore, Parrilla *et al.* (2004) showed that exposure of porcine spermatozoa to Hoechst 33342 was not genotoxic when evaluating resulting offspring.

Sperm sorting is accomplished by optimizing the balance between laser power, Hoechst 33342 staining concentration, spermatozoa concentration during staining, spermatozoa concentration during sorting and sort rates. Previously, we reported staining 400×10^6 spermatozoa/ml with 224 µM Hoechst 33342 (Schenk *et al.* 1999). Little was known at that time regarding the use and effects of Hoechst 33342 on long-term spermatozoa viability and function. There is a threshold for uniformly staining spermatozoa with Hoechst 33342. Furthermore, as the number of spermatozoa/ml increases so does the required amount of stain. While too much stain can be toxic to spermatozoa, inadequate staining results in poor resolution of X- and Y-chromosome bearing populations. As a result of concerns related to Hoechst 33342 toxicity, we now routinely lower the spermatozoa concentration for staining from the previously reported 400×10^6 spermatozoa/ml to 100 to 200×10^6 spermatozoa/ml, which enables lowering the Hoechst 33342 stain concentration accordingly.

Laser intensity

Hoechst 33342 usually is excited with 351 to 364 nm lines of light, wavelengths that appear not to be absorbed by nucleic acids or proteins. However, Libbus *et al.* (1987) reported that staining with Hoechst 33342 and irradiating spermatozoa during flow sorting could increase the incidence of chromosome aberrations. Meanwhile, Teng *et al.* (1988) suggested that Hoechst 33342 bound to DNA provided protection against radiation damage, at least for some somatic cells. Garner *et al.* (2001a) studied different aspects of flow-sorting bovine spermatozoa. DNA integrity of spermatozoa expressed as COMPαt (% cells outside of the main population) (Darzynkiewicz *et al.* 1975) was studied by the sperm chromatin structure assay (SCSA) (Ballachey *et al.* 1987) for spermatozoa samples processed as: (1) unsorted controls; (2) sorted spermatozoa with no stain and no laser illumination; (3) sorted spermatozoa with laser illumination (150 mW) but no stain; (4) sorted spermatozoa with stain but no illumination; or (5) sorted spermatozoa with both stain and laser excitation. They found that staining and exposure of spermatozoa to wavelengths of light used for sexing had little effect on the integrity of spermatozoa DNA. Furthermore, they concluded that the mechanical stresses of sorting and/or post-sorting centrifugation appeared to increase the proportion of spermatozoa with damaged DNA, but that neither Hoechst 33342 staining nor exposure to laser illumination during sorting increased the potential damage as determined by COMPαt.

Catalase

Shannon and Curson (1972) reported that addition of catalase eliminated peroxide, a byproduct produced by dead spermatozoa. They found that catalase increased viability of spermatozoa stored 44 h at 37°C and eliminated toxicity of dead spermatozoa. Klinc *et al.* (2005) used an extender containing catalase for sexed bovine spermatozoa that resulted in excellent pregnancy rates. To test potential benefits for sorting spermatozoa, laboratory experiments were conducted to determine the optimal concentration of catalase in different sorting media.

Materials and methods

Semen collection

Semen for all trials was collected from bulls on a routine collection schedule using an artificial vagina (Schenk 1998). Ejaculates had to contain >50% progressively motile and >75% morphologically normal spermatozoa for use in experiments. Antibiotics were added to the raw ejaculate as described by Shin (1986) within 15 min of collection, and the concentration of spermatozoa was determined using a spectrophotometer.

Procedures for evaluation of spermatozoa in vitro in the presence or absence of catalase

Catalase (#C40, Sigma Chemical Co., St. Louis, MO) was added to each of the media to give final concentrations of 0, 0.2 or 2 µg/ml after spermatozoa staining, collection and processing and therefore was added to 4% egg yolk-TALP (after spermatozoa were stained), 20% egg yolk catch buffer (during sorting) and 20% egg yolk Tris-AB extender (prior to cryopreservation). Neat ejaculates from each of 6 bulls were incubated at room temperature (22°C) for 1 and 5 h prior to staining and sorting to simulate the amount of oxidation that occurs in the neat ejaculate throughout the sorting day. Spermatozoa were "bulk sorted" (Schenk *et al.* 1999) and frozen in 0.25 ml straws at 10×10^6 total spermatozoa/ml (2×10^6 spermatozoa/dose). Straws were thawed for 30 sec at 37.5°C for evaluation. Quality of spermatozoa was assessed using visual estimates of progressive motility (2 observers, blind to treatments) after 30 and 120 min of incubation at 37.5°C.

Procedures for sexing, processing and insemination of spermatozoa

Sexing of spermatozoa at 85 to 90% accuracy was accomplished with a MoFlo® SX sperm sorter (Dako, Fort Collins, CO), operated at: 50 psi (Seidel & Garner 2002); 20,000 events per sec; and sort rates of 2,500 to 4,000 spermatozoa per sec. The sheath fluid composition was as described by Schenk *et al.* (1999). Spermatozoa for all field trials were stained for 45 min at 34.5°C at 200 x 10[6] spermatozoa/ml, unless indicated otherwise, in a modified TALP buffer (Schenk *et al.* 1999) that was adjusted to pH 7.4. After staining, spermatozoa were diluted to half the staining concentration with an equal volume of TALP with 4% egg yolk, adjusted to pH 5.5 after which 0.002% food coloring dye (FD&C #40: Johnson & Welch 1999) was added. The final pH of the stained spermatozoa sample was ~ 6.8 to 7.0.

Just prior to flow sorting, samples were filtered at unit gravity through a 50 μm CellTrics® disposable filter (#04-0042-2317, Partec GmbH, Münster, Germany). Spermatozoa were stained with 112.5 μM Hoechst 33342 and interrogated with 150 mW of laser intensity except where Hoechst 33342 staining concentration and laser intensity were used as treatments. Sexed spermatozoa were frozen in 0.25-ml polyvinylchloride straws in a 20% egg yolk-Tris extender (6% glycerol) as described previously (Schenk *et al.* 1999). Except as described for Trial 4, sexed spermatozoa were frozen at 20 x 10[6]/ml, so for 2 x 10[6] total spermatozoa, only 100 μl of the 0.25-ml straw were filled. This was accomplished by centering the 100 μl column inside the straw with air pockets to either sides of the sperm column and Tris A-B extender without spermatozoa to assure a proper seal on each end of the straw. Inseminates contained at least 30% progressively motile spermatozoa after thawing for 30 sec in 37.5°C water. The occasional batches of sorted spermatozoa that did not meet this standard were discarded. For all trials, the dose of sexed spermatozoa was 2 x 10[6] total spermatozoa inseminated conventionally into the uterine body. Data from 5 field trials (Seidel & Schenk, unpublished) in addition to previous research (Seidel *et al.* 1999) showed no consistent improvement in pregnancy rates above 2 x 10[6] frozen sexed spermatozoa per dose when inseminating heifers into the uterine body 12 to 24 h after observed standing estrus.

Each study had an unsexed control of 20 x 10[6] spermatozoa/dose to determine the intrinsic fertility of the cattle using standard procedures. All cattle were in commercial herds in agricultural settings. University personnel conducted the estrus detection, semen handling, insemination, and pregnancy diagnoses for Trials 1 and 2. Similar personnel were on site for Trial 6 to assist with semen thawing and handling, estrus detection, and record keeping. For Trials 3 to 5, farm personnel carried out all of these procedures.

Statistical procedures

Statistical treatment of data for all trials was by analysis of variance with a fixed effect model including treatments, bulls, and herds where appropriate plus first order interactions. Since subclass means were almost always in the range of 20 to 80% pregnant, data were not transformed for analysis. Numbers of animals pregnant are presented, as well as least squares means of the percentage of females pregnant. Meta-analyses were done by single degree of freedom χ^2 with the Fisher-Yates correction.

Procedures for Trial 1 – Angus heifers in Colorado

The experimental objective was to compare pregnancy rates in Angus heifers following insemination of 2 x 10[6] total sexed frozen/thawed spermatozoa isolated by flow sorting methods with different laser intensities. The high laser intensity was fixed at 150 mW. The low laser intensity varied by sorting day, bull and ejaculates within bulls and was defined as the lowest laser intensity that would result in sufficient resolution between X- and Y-chromosome bearing spermatozoa to assure a ~85+% purity; range = 55 to 130 mW (median power = 99 mW).

Both X- and Y-chromosome-bearing spermatozoa were sorted and processed. Sorted spermatozoa (2×10^6) were packaged at 20×10^6 spermatozoa/ml into 0.25 ml straws as 100 μl columns. An unsorted control treatment containing 20×10^6 total spermatozoa was included in this and subsequent field trials.

Estrus of Angus heifers (N = 412) was synchronized by feeding 500 mg melengesterol acetate with pelleted grain for 12 d followed by i.m. injection of 25 mg prostaglandin $F_2\alpha$ 18 d later (MGA-PGF$_2\alpha$). Heifers were inseminated with spermatozoa from one of three Angus bulls 12 to 24 h after visually observed standing estrus. All inseminates were deposited into the uterine body. Two-thirds of the heifers received sexed spermatozoa; heifers were randomly allocated to semen treatment from one of the three bulls and one of the three inseminators. Pregnancy was diagnosed with ultrasound ~ 60 d after insemination.

Procedures for Trial 2 – Angus heifers in Colorado

Laser intensity was again studied, using spermatozoa from the same treatments and only two of the bulls in Field Trial 1. Angus heifers (N = 169) were inseminated with 2×10^6 total X- or Y-chromosome-bearing sexed spermatozoa. Control inseminates contained 20×10^6 total unsexed spermatozoa. Heifers were synchronized for estrus using variations on the MGA-PGF$_2\alpha$ system and were mass-mated 70-74 h after the prostaglandin injection (no visual estrus detection). Pregnancy was diagnosed with ultrasound ~ 60 d after insemination.

Procedures for Trial 3 – Angus heifers in South Dakota

The experimental objective was to compare pregnancy rates for spermatozoa sorted with different laser intensities and unsexed control spermatozoa. Once again, the high laser intensity was fixed at 150 mW and the low laser intensity was reduced to 50 to 90 mW (median power = 65 mW). Sexed inseminates were cryopreserved in 0.25 ml straws at 20×10^6 spermatozoa/ml as 100 μl columns containing 2×10^6 total spermatozoa. Unsexed controls contained 20×10^6 total spermatozoa/dose. An additional control containing 30×10^6 total spermatozoa/dose was processed and frozen by a North American bull genetic center. Spermatozoa from a single Angus bull were used to inseminate 195 Angus heifers synchronized with the MGA-PGF$_2\alpha$ protocol. Heifers were inseminated 12 h after observed standing estrus by one of two inseminators. Pregnancy was determined using ultrasonography at ~ 60 d after artificial insemination.

Procedures for Trial 4 – Holstein heifers in Wisconsin

The effects of stain concentration, laser intensity and spermatozoa concentration for freezing of sex-sorted, frozen-thawed spermatozoa on pregnancy rates were determined. Sexed inseminates containing 2×10^6 total spermatozoa were cryopreserved in 0.25 ml straws at either 10×10^6 spermatozoa/ml (full straws) or 20×10^6 spermatozoa/ml (100 μl columns). Unsexed control inseminates contained 20×10^6 total spermatozoa. Spermatozoa for sorting were stained with 112.5 μM Hoechst 33342 (high stain) at 200×10^6 spermatozoa/ml or 55.8 μM Hoechst 33342 (low stain) at 100×10^6 spermatozoa/ml. The high laser intensity used to excite the dye was fixed at 150 mW. For the low intensity treatment, laser power was reduced as much as possible (50 to 130 mW) as described for Trial 1. These methodologies were combined into the following four treatments for sexed spermatozoa at 2×10^6 spermatozoa per dose: (1) 10×10^6 spermatozoa/ml, high laser, high stain; (2) 20×10^6 spermatozoa/ml, high laser, high stain; (3)

20×10^6 spermatozoa/ml, low laser, high stain; and (4) 20×10^6 spermatozoa/ml, low laser, low stain. Semen was collected and prepared from one of three Holstein bulls and used to treat approximately equal numbers of Holstein heifers. Holstein heifers (N=816) from 19 herds were inseminated with spermatozoa from one of the treatments or control. Herds in which fewer than 20 heifers were inseminated (9 herds, 101 inseminations) were not included in the statistical analyses; herd effects were included in the statistical model. Fetal sex was diagnosed in some of the herds by ultrasound at 8 to 10 weeks of gestation.

Procedures for Trial 5 – Holstein heifers in five states

The objective of this field trial was to determine pregnancy rates in Holstein heifers inseminated with sexed spermatozoa stained with either 63.0 µM or 112.5 µM Hoechst stain and treated with or without 0.5 µg/ml catalase. Additionally, these treatments were further studied as sorted spermatozoa frozen either 8 or 13 h post-collection. We also included a Non-Sort Control (NSC) inseminate containing 2×10^6 total spermatozoa treated with 0.5 µg/ml catalase that were stained with 112.5 µM Hoechst, diluted with sheath fluid (800,000 spermatozoa/ml) and processed as sorted spermatozoa, but without sorting. Unsexed control inseminates were not treated with catalase and contained 20×10^6 total spermatozoa per dose. All control spermatozoa were frozen 8 h post-collection. This field trial included 3 Holstein bulls and 10 dairy herds located in California, Nebraska, New York, Utah and Wisconsin.

Procedures for Trial 6 – Crossbred heifers in Wyoming

To test the effects of catalase further, we conducted a field trial to determine calving rates following insemination of spermatozoa that were processed with or without 0.5 µg/ml catalase. Further, we had been filling straws of sexed spermatozoa as columns at 20×10^6 spermatozoa/ml. However, automated straw filling would be a requirement to commercialize procedures. Therefore, we also compared calving rates after insemination of 2×10^6 sexed spermatozoa frozen at 10×10^6 spermatozoa/ml in automatically filled straws and 20×10^6 spermatozoa/ml in manually filled straws.

Finally, we wanted to know if 2×10^6 total spermatozoa were sufficient to maintain calving rates at commercially acceptable levels. Therefore, we included a high dose unsexed control (20×10^6 total spermatozoa) and a low dose unsexed control (2×10^6 total spermatozoa). Unsexed control spermatozoa were not treated with catalase; half of the inseminates with sexed spermatozoa were supplemented with catalase.

Non-estrus synchronized crossbred Salers heifers (N = 1,262) were inseminated by a single inseminator ~12 h after first visually detected estrus with spermatozoa from one of two Red Angus bulls. Two-thirds of the heifers received sexed inseminates, while ~16% of heifers received semen from the high dose control and ~16% of heifers were inseminated with the low dose control.

Results

Laboratory catalase experiment

There were large bull differences (P< 0.0002) throughout this experiment. Post-thaw progressive spermatozoa motility averaged over all other treatments did not differ due to ejaculate age at 1 h (30 min incubation = 47%, 120 min incubation = 40%) compared to 5 h (30 min incubation = 45%, 120 min incubation = 38%). Catalase had no effect on post-thaw spermatozoa motility after 30 min post-thaw incubation. Catalase at both concentrations resulted in significantly higher (P< 0.001) spermatozoa motility (0.2 μg/ml = 42% and 2.0 μg/ml = 41%) than for controls (34%) at 120 min of incubation post-thaw. The results of this *in vitro* experiment suggest that catalase only had a beneficial anti-oxidant effect during post-thaw spermatozoa incubation.

Trial 1 – Angus heifers in Colorado

Pregnancy rates were similar for X- and Y-chromosome bearing inseminates, 52 and 45% respectively (P> 0.2); pooled data are presented in Table 1. Furthermore, pregnancy rates for controls and the high laser treatment were not significantly different (P> 0.1) but were lower (P< 0.05) for spermatozoa interrogated with low laser intensity (43%) (Table 1). There were no effects of bulls or inseminators (P> 0.1).

Table 1. Results of Trial 1. Effects of laser power during interrogation of spermatozoa for determining DNA content on pregnancy rates.

Treatment	N	Number pregnant	Least squares means (%)
20 x 10⁶ unsexed (control)	129	83	65[a]
2.0 x 10⁶ sexed – high laser	143	79	55[a]
2.0 x 10⁶ sexed – low laser	140	60	43[b]

[a,b]Least squares means without common superscripts differ (P <0.05).

Trial 2 – Angus heifers in Colorado

Unlike the results found in Field Trial 1, laser intensity did not affect pregnancy rates for sexed spermatozoa. However, pregnancy rates were higher (P< 0.05) for unsexed controls compared to sexed spermatozoa (Table 2). There were no significant bull effects (P> 0.1).

Table 2. Results of Trial 2. Effects of laser power during interrogation of spermatozoa for determining DNA content on pregnancy rates.

Treatment	N	Number pregnant	Least squares means (%)
20 x 10⁶ unsexed (control)	59	38	65[a]
2.0 x 10⁶ sexed – high laser	49	21	48[b]
2.0 x 10⁶ sexed – low laser	61	29	46[b]

[a,b]Least squares means without common superscripts differ (P <0.05).

Trial 3 – Angus heifers in South Dakota

In this field trial, we again found no effect of laser intensity on pregnancy rates (Table 3). Fertility of sexed spermatozoa was not significantly lower than unsexed spermatozoa but the numbers of observations per treatment were low. Unsexed control pregnancy rates were nearly identical after insemination with spermatozoa from the two treatments. Pregnancy rates for sexed inseminates were 78% of controls. There were no effects due to the inseminators (P> 0.05). Sexed inseminates resulted in 91% female fetuses compared to 42% female fetuses for controls.

Table 3. Results of Trial 3. Effects of laser power during interrogation of spermatozoa for DNA content on pregnancy rates.

Treatment[a]	N	Number pregnant	Least squares means (%)
Genetic center control, 30 x 10^6 unsexed	50	22	50
Same ejaculate control, 20 x 10^6 unsexed	49	21	48
2.0 x 10^6 sexed – high laser	50	18	38
2.0 x 10^6 sexed – low laser	46	17	38

[a]No significant differences (P > 0.1).
Control: 18/43 (42% female fetuses).
Sexed spermatozoa: 32/35 (91%) female fetuses.

Trial 4 – Holstein heifers in Wisconsin

In this trial, all treatments with sexed spermatozoa (2 x 10^6 spermatozoa per inseminate) resulted in lower pregnancy rates than for unsexed control spermatozoa (20 x 10^6 spermatozoa per inseminate) (Table 4). There also appeared to be a higher pregnancy rate when freezing sexed spermatozoa at 20 x 10^6/ml (half the volume) compared to 10 x 10^6/ml (P < 0.054; the second and third treatments listed in Table 4). However, there was no effect of laser intensity (third *versus* fourth treatments listed in Table 4) nor of Hoechst 33342 stain concentration (fourth *versus* fifth treatments listed in Table 4).

Table 4. Results of Trial 4. Effects of Hoechst 33342 stain concentration and laser power during interrogation of spermatozoa for DNA content and of spermatozoa concentration during cryopreservation on pregnancy rates.

Treatment	N	Number pregnant	Least squares means (%)
20 x 10^6 unsexed spermatozoa (87 x 10^6/ml)	143	83	59[c]
2.0 x 10^6 sexed spermatozoa – high laser[a], high stain[b] (10 x 10^6/ml)	177	43	28[d]
2.0 x 10^6 sexed spermatozoa – high laser, high stain (20 x 10^6/ml)	192	67	38[e]
2.0 x 10^6 sexed spermatozoa – low laser, high stain (20 x 10^6/ml)	153	44	30[d,e]
2.0 x 10^6 sexed spermatozoa – low laser, low stain (20 x 10^6/ml)	151	42	30[d,e]

[a]High laser = 150 mW; low laser ranged from 50 – 130 mW.
[b]Low stain was 55.8 µM Hoechst 33342; high stain was 112.5 µM Hoechst 33342.
[c,d,e]Least squares means without common superscripts differ (P < 0.05).

There were large herd-to-herd differences in fertility (P < 0.01). A more in-depth analysis divided herds into low, medium, and high fertility groups based on least-squares means. Pregnancy rates of sexed spermatozoa (all treatments) as a percent of control, unsexed spermatozoa were 60, 64 and 36% for high, medium, and low fertility herds, respectively. Thus, fertility of sexed spermatozoa was particularly low in low fertility herds, which also tended to be larger herds. There were no significant bull effects (P > 0.1). The sex ratio of the 137 fetuses examined resulting from sexed spermatozoa was 86% females; the sex ratio of 60 control fetuses was 52% females.

Trial 5 – Holstein heifers in five states

Pregnancy rates (Table 5) with sexed spermatozoa were lower (P<0.05) than those for unsexed control inseminates. There was no difference in pregnancy rates when heifers were inseminated with 2 or 20 x 10^6 total unsexed spermatozoa (P>0.05), but note that the low dose also contained catalase. Overall, pregnancy rates were higher (42%) for spermatozoa frozen 8 h post-collection than for those frozen 13 h post-collection (34%) (P<0.03). There was no significant effect of stain concentration on pregnancy rates (36 vs. 41% for high and low stain, respectively: P>0.1). There were no significant bull or herd effects (P>0.1).

Table 5. Results of Trial 5. Effects of Hoechst 33342 concentration for staining spermatozoa and catalase on pregnancy rates.

| | | 1st Freeze | | | 2nd Freeze | |
Treatment/Spermatozoa	N	No. preg	LSM (%)*	N	No. preg	LSM (%)*
20 x 10^6 unsexed, without catalase	240	142	59[a]			
2 x 10^6 unsexed, with catalase (NSC)	240	125	52[a]			
2 x 10^6 sexed, high stain, without catalase	138	58	39[b]	90	37	38
2 x 10^6 sexed, high stain, with catalase	136	50	35[c]	96	32	30
2 x 10^6 sexed, low stain, without catalase	87	46	47[b,d]	153	58	35
2 x 10^6 sexed, low stain, with catalase	101	47	43[b]	129	52	37

[a,b,c,d] Means within columns without common superscripts differ (P<0.05).
* Overall, pregnancy rates for sexed, first batch inseminates were higher than second batch inseminates (42 vs 34%, P<0.03).

Trial 6 – Crossbred heifers in Wyoming

Calving rates with sexed spermatozoa were approximately 75% of those for unsexed controls (P<0.05; Table 6). Freezing sorted spermatozoa at 20 x 10^6/ml was superior to 10 x 10^6/ml, 225/422 (53%) *versus* 196/414 (47%) calving (P<0.05, 1-tail $\chi2$, data not presented in Table 6). The catalase treatment with sorted spermatozoa had no beneficial effect on calving rates. Furthermore, insemination of 2 x 10^6 unsexed control spermatozoa did not decrease calving rates significantly compared to 20 x 10^6 unsexed control spermatozoa (P>0.05). Sexed inseminates resulted in excess birth of heifers (88%) compared to that for unsexed controls (51%). There were no significant bull effects (P>0.1).

Table 6. Results of Trial 6. Effects of catalase on pregnancy rates of heifers inseminated with sexed frozen spermatozoa (pooled over spermatozoa concentrations in straws) or unsexed control spermatozoa.

Treatment	N	Number calving to AI	Least squares means (%)
20 x 10^6 unsexed (high control)	209	144	70[a]
2 x 10^6 unsexed (low control)	217	139	64[a]
2 x 10^6 sexed – without catalase	421	206	49[b]
2 x 10^6 sexed – with catalase	415	214	52[b]

[a,b]Least squares means without common superscripts differ (P<0.05).

Discussion

General characteristics of the field trials

The main objective of these experiments was to improve procedures for sexing spermatozoa with the important endpoint of pregnancy. Unfortunately, it is prohibitively expensive to obtain data on pregnancy rates in cattle under strictly controlled experimental conditions. We, therefore, worked with commercial producers. While there is considerable value in obtaining data under field conditions, there also are limitations: including varying levels of compliance with protocols; different levels of herd management; differing service sires and ages, weights, and breeds of animals; varying herd size *et cetera*. Some of these differences can be ameliorated by statistical procedures such as blocking; a particular problem is obtaining or accumulating sufficient numbers of animals to achieve reasonable statistical power for binomial responses such as percent pregnant.

We attempted to be sensitive to not reducing pregnancy rates greatly in these herds; we always included a standard AI control with normal doses of spermatozoa to established fertility with normal practices. Although this was not the best control for certain hypothesis testing, it was useful for several reasons, including showing that normal fertility was low in some herds, and having a portion of these heifers with normal pregnancy rates, even if our treatments resulted in low pregnancy rates. Ideally, we would have had a low spermatozoa dose unsexed control treatment in each experiment to compare with the low-dose sexed treatments. However, our main goals were to compare different approaches to sexing spermatozoa and maintaining their fertility, not to compare success rates of sexed spermatozoa treatments to non-sexed controls. Therefore, we usually did not use up the limited numbers of experimental animals for low-dose controls.

The pregnancy rates with doses of 2×10^6 sexed spermatozoa were lower than with 20×10^6 unsexed spermatozoa, ranging from 53.4 to 79.2% of those controls; average of 69.5% of controls (least squares mean comparisons) for the six trials. The pregnancy rates with sexed spermatozoa ranged from 31.5 to 50.5% in respective trials, averaged over all sexed treatments within each trial. Management appeared to account for most of these differences. Our university personnel did most of the estrus detection, semen handling, insemination and pregnancy diagnoses for Trials 1, 2, and 6. Pregnancy rates were clearly highest in these 3 trials, 46 to 50% with sexed spermatozoa, and were among the highest pregnancy rates with sexed spermatozoa as a percent of control 71 to 75%. The highest pregnancy rate with sexed spermatozoa as a percent of controls was Trial 3, with 79%; this trial was managed by a single person, although with two inseminators. One other telling statistic was from Trial 4, in which pregnancy rates with sexed spermatozoa as a percent of control were 60, 64, and 36% for high, medium, and low fertility herds (respective terciles). Thus use of low doses of sexed, frozen spermatozoa resulted in very poor fertility under suboptimal conditions and about 70 to 80% of controls with good management and strict compliance with protocols when inseminating heifers.

For Trials 3 through 5, we had no control over estrus detection, semen handling and insemination practices, pregnancy diagnoses or other management factors. Anecdotally, we know that standing estrus was not monitored in some cases, but rather any sign of estrus, such as mounting, mucus, rubbed off tail paint *et cetera*, was sufficient cause for insemination. Not only was overall fertility low in these trials but in most cases pregnancy rates with sexed spermatozoa were especially low.

In most trials, there were no effects due to bulls or inseminators. There are several plausible explanations for this. There were quality control standards in place for individual spermatozoa sorting batches; those not meeting standards were discarded so only good quality semen was used for all bulls. In cases where inseminators could be included in the statistical model, they were experienced, well-trained individuals. One other obvious constraint was that there were insufficient numbers of inseminations within most trials to compare bulls or inseminators with ideal statistical power. In most cases, differences would have had to exceed 12 to 15 percentage points for statistical significance.

Treatment effects

Here again, greater statistical power would have been desirable, which was compensated to some extent by studying effects in multiple trials. We studied two laser powers in four trials, the standard 150 mW compared with the lowest intensity that could be used and still resolve X- and Y-spermatozoa (50 to 130 mW, depending on the particular sort run). Surprisingly, the higher laser power resulted in a significantly higher pregnancy rate than low laser power in Trials 1 (Table 1) and 4 (third *versus* fourth treatment listed in Table 4). A similar effect was noted by Guthrie *et al.* (2002). However, in two other field trials, the opposite effect was found, and if the data of the four trials are pooled in a meta-analysis, there is no significant effect of laser intensity (means of 43 and 38% for high and low laser power with over 400 inseminations per group). It appears that 150 mW of laser power as used was not any more detrimental than lower laser power. This is important because sorting can be done faster and more accurately at the higher laser power.

The concentration of Hoechst 33342 was studied in two trials, and no effect of concentration on fertility was observed. Previous studies (unpublished) with *in vitro* endpoints indicated that it would be desirable to keep the concentration of Hoechst 33342 on the low side. *In vivo* however there appears to be no major detriment to fertility at the higher stain concentration used.

The effect of supplementing spermatozoa-containing fluids with catalase *in vitro* and in two trials with substantive numbers of inseminates was studied. While catalase was beneficial when incubating spermatozoa *in vitro* for 2 h, there was no effect on pregnancy rates of the treatments used. We postulated that hydrogen peroxide might be generated by the excessive manipulations that spermatozoa undergo in air in the process of sorting and that catalase might help. A meta-analysis combining treatments of Trials 5 and 6 indicates virtually identical fertility with and without catalase: 296/647 (46%) and 301/649 (46%), respectively. Catalase clearly is beneficial for bovine spermatozoa under some conditions (Shannon & Curson 1972). Possibly the quality of the catalase we used was suboptimal.

The effects of spermatozoa concentration/dilution during cryopreservation were also studied. We observed a strong hint in Trial 4 that more concentrated spermatozoa may be desirable, and this was confirmed in Trial 6. The published literature in this area is inconsistent for bull spermatozoa, although dilution is clearly harmful to spermatozoa of some species under some conditions (Maxwell & Johnson 1999; Garner *et al.* 2001b). Use of 0.25 ml straws instead of larger ones was chosen to minimize potential dilution effects, particularly since so few sexed spermatozoa are used per dose. However, dilution effects may not be sufficiently large to justify partial filling of 0.25 ml straws to maintain a high spermatozoa concentration; thus conventional filling and sealing may be appropriate even though the spermatozoa concentration will be about 10×10^6/ml during freezing of 2×10^6 spermatozoa/dose.

Other considerations

The art and science of sorting spermatozoa by sex chromosomes continues to evolve. Particularly noteworthy is that sorting at 40 psi results in considerably higher spermatozoa quality than

50 psi for most bulls and stallions (Suh *et al.* 2005). This also has been corroborated by fertility trials (Schenk *et al.*, unpublished). The work reported in this paper was all done at 50 psi so maybe pregnancy rates would have been higher if sorting had been done with currently recommended procedures.

Another advance is using a much simpler laser that is less burdensome to operate and does not affect pregnancy rates adversely compared to a standard laser (Schenk *et al.* 2005). Still other advances, including that no egg yolk was needed in the glycerol fraction of extender for cryopreserving sexed spermatozoa (Schenk & Crichton 2006), are making incremental improvements in the ease and speed of sorting while minimizing damage to spermatozoa.

Conclusions

The main findings from this research are that higher laser power, higher concentrations of Hoechst 33342 and lower concentrations of spermatozoa during cryopreservation were not detrimental, or only slightly detrimental, to the fertility of sexed, frozen bovine spermatozoa. Unfortunately, none of these findings is likely to have much effect on improving fertility of sexed spermatozoa. However, they will allow spermatozoa to be sexed more rapidly and efficiently, thus lowering the costs of sexing spermatozoa and making this technology available from a larger population of bulls.

Acknowledgements

We acknowledge the collaborators throughout the United States of America who allowed us to use their cattle including: Ben and Skylar Houston and staff of Aristocrat Angus, Platteville, CO; Ric Miller of Kiowa Creek Ranch, Elbert, CO (Trial 1), Robert Tointon of Tointon LLC, Walden, CO (Trial 2); staff at Cannon River Ranches, Highmore, SD (Trial 3); Roger Hanson and Lloyd Sorenson, D.V.M. of Accelerated Genetics, Baraboo, WI (Trials 3 and 4); Mel DeJarnette and Clif Marshall of Select Sires Inc., Plain City, OH (Trial 5); and Scott James of Padlock Ranches, Ranchester, WY (Trial 6). We greatly appreciate the livestock services of Pat Hemming, D.V.M. (Aristocrat Angus), Zella Brink and Rick Brandes (Colorado State University). The staff at XY, Inc.: Drs. Duane Garner and John Hasler assisted with experimental design; Lisa Massoudi, Kathy Mean, Mindy Meyers and Jodi Rasmussen provided technical assistance in sperm sorting and processing; Dr. TaeKwang Suh assisted with statistical analyses of data.

References

Ballachey BE, Hohenboken WD & Evenson DP 1987 Heterogeneity of sperm nuclear chromatin structure and its relationship to fertility of bulls. *Biology of Reproduction* **36** 915–925.

Cran DG, Cochrane DJ, Johnson LA, Wei H, Lu KH & Polge C 1994 Separation of X- and Y-chromosome bearing bovine sperm by flow cytometry for use in IVF. *Theriogenology* **41** 183 (abstr).

Darzynkiewicz Z, Tranganos F, Sharples T & Melamed M 1975 Thermal denaturization of DNA in situ as studied by acridine orange staining and automated cytofluorometry. *Experimental Cell Research* **90** 411–428.

Garner D, Schenk J & Seidel G Jr 2001a Chromatin stability in sex-sorted sperm. In *Proceedings of the VIIth International Congress of Andrology, Montreal. Andrology in the Twenty-first Century*, Canada: Short Communications, pp 3–7. Eds B Bobaire, H Chemes and CR Morales. Englewood, NJ: Medimond.

Garner DL, Thomas CA, Gravance CG, Marshall CE, DeJarnette JM & Allen CH 2001b Seminal plasma addition attenuates the dilution effect in bovine sperm. *Theriogenology* **56** 31-40.

Guthrie HD, Johnson LA, Garrett WM, Welch GR & Dobrinsky FR 2002 Flow cytometric sperm sorting: effects of varying laser power on embryo development in swine. *Molecular Reproduction and De-*

velopment **61** 87–92.

Johnson LA, Flook JP & Look MV 1987a Flow cytometry of X and Y chromosome-bearing sperm for DNA using an improved preparation method and staining with Hoechst 33342. *Gamete Research* **17** 203–212.

Johnson LA, Flook JP, Look MV & Pinkel D 1987b Flow sorting of X and Y chromosome-bearing spermatozoa into two populations. *Gamete Research* **16** 1–9.

Johnson LA & Welch GR 1999 Sex preselection: high-speed flow cytometric sorting of X and Y sperm for maximum efficiency. *Theriogenology* **52** 1323–1341.

Klinc P, Frese D, Osmers H, Sieg B, Struckmann C, Kosec M & Rath D 2005 Successful insemination in heifers with sex-sorted bull spermatozoa using a new extender concept (Sexcess®). *Reproduction in Domestic Animals* **40** 360 (abstr).

Libbus GL, Perreault SD, Johnson LA & Pinkel D 1987 Incidence of chromosome aberrations in mammalian sperm stained with Hoechst 33342 and UV-laser irradiated during flow sorting. *Mutation Research* **182** 265–274.

Maxwell WMC & Johnson LA 1999 Physiology of spermatozoa at high dilution rates: The influence of seminal plasma. *Theriogenology* **52** 1353-1362.

McNutt TL & Johnson LA 1996 Flow cytometric sorting of sperm: Influence on fertilization and embryo/fetal development in the rabbit. *Molecular Reproduction and Development* **43** 261–267.

Muller W & Gautier F 1975 Interactions of heteroaromatic compounds with nucleic acids. *European Journal of Biochemistry* **54** 358-394.

Munné S 1994 Human sperm sex selection: Flow cytometry separation of X and Y spermatozoa could be detrimental for human embryos. *Human Reproduction* **9** 758.

Parrilla I, Vázquez JM, Cuello C, Gil MA, Roca J, Di Berardino D & Martínez EA 2004 Hoechst 33342 stain and u.v. laser exposure do not induce genotoxic effects in flow-sorted boar spermatozoa. *Reproduction* **128** 615-621.

Schenk JL 1998 Bull semen collection procedures. In *Proceedings of the 17th NAAB Technical Conference on AI and Reproduction*, pp 48-58. Columbia, MO: National Association of Animal Breeders.

Schenk JL, Suh TK, Cran DG & Seidel GE Jr 1999 Cryopreservation of flow-sorted bovine sperm. *Theriogenology* **52** 1375–1391.

Schenk JL, Brink Z & Suh TK 2005 Use of competitive fertilization to evaluate a simpler laser for flow cytometric sexing of bovine sperm. *Reproduction, Fertility and Development* **17** 306 (abstr).

Schenk J & Crichton E 2006 Insemination of Holstein heifers with sexed sperm processed with or without egg yolk in the glycerol-containing freezing medium. *Reproduction, Fertility and Development* **18** 282-283 (abstr).

Seidel GE Jr & Garner DL 2002 Current status of sexing mammalian spermatozoa. *Reproduction* **124** 733–743.

Seidel GE Jr & Schenk JL 2002 Field trials with sexed, frozen bovine semen. In *Proceedings of the 19th NAAB Technical Conference on AI and Reproduction*, pp 64-69. Columbia, MO: National Association of Animal Breeders.

Seidel GE Jr., Schenk JL, Herickhoff LA, Doyle SP, Brink Z, Green RD & Cran DG 1999 Insemination of heifers with sexed sperm. *Theriogenology* **52** 1407–1420.

Shin S 1986 The control of Mycoplasmas, Campylobacter fetus, Haemophilus somnus in frozen bovine semen. In *Proceedings of the 11th NAAB Technical Conference on AI and Reproduction*, pp 33-38. Columbia, MO: National Association of Animal Breeders.

Shannon P & Curson B 1972 Toxic effect and action of dead sperm on diluted bovine semen. *Journal of Dairy Science* **55** 614–620.

Suh TK, Schenk JL & Seidel GE Jr. 2005 High pressure flow cytometric sorting damages sperm. *Theriogenology* **64** 1035–1048.

Teng M, Usman N, Frederick CA & Wang AHJ 1988 The molecular structure of the complex Hoechst 33258 and the DNA dodecamer d(CGCGAATTCGCG). *Nucleic Acids Research* **16** 2671–2690.

Tubman LM, Brink Z, Suh TK & Seidel GE Jr. 2004 Characteristics of calves produced with sperm sexed by flow cytometry/cell sorting. *Journal of Animal Science* **82** 1029–1036.

Metabolism of the bovine cumulus-oocyte complex and influence on subsequent developmental competence

JG Thompson, M Lane and RB Gilchrist

Research Centre for Reproductive Health, School of Paediatrics and Reproductive Health, The University of Adelaide, Adelaide, 5005

The two types of cells that make up the cumulus-oocyte complex (i.e. the oocyte and cumulus cells) have very different metabolic demands, with glucose occupying a central role in metabolic activity. Cumulus cells have a significant requirement for and utilise high levels of glucose, yet appear to have little need for oxidative metabolism. In contrast, oocytes have a requirement for oxidative metabolism, although limited glucose metabolism may also be an important aspect of meiotic and developmental competence. Nevertheless, because of the metabolic and communication link between the cumulus and the oocyte, glucose availability and metabolism within the cumulus can have a significant impact on oocyte meiotic and developmental competence. In particular, the role of the hexosamine biosynthesis pathway within cumulus cells appears critical for the supply of substrate from glucose for extracellular matrix production, yet if overstimulated can significantly decrease developmental competence of the oocyte. Current static systems for in vitro maturation are clearly incompatible with meeting substrate demands, especially glucose. In the future, in vitro maturation will include a more dynamic approach, which will adjust nutrient components to meet the changing functional requirements of cumulus-oocyte complexes during the final process of maturation.

Introduction

Understanding oocyte meiotic and developmental competence and unravelling the factors that mediate their fulfilment remain as great challenges in the field of reproductive biology, although recent attempts to describe the molecular basis of oocyte competence have taken significant steps forward (Pan *et al.* 2005; Reis *et al.* 2006). In contrast, the benefits that will result from a greater understanding of oocyte meiotic and developmental competence have not altered. We are still inept in our ability to exploit fully the genetic value held within every female gamete, which could be far greater by capturing the significant numbers of oocytes within an individual female, each capable of producing a viable offspring. We are also still inept in our ability to store oocytes in any developmental state so as to efficiently derive viable embryos when needed.

Corresponding author E-mail: Jeremy.Thompson@adelaide.edu.au

Oocyte developmental competence and oocyte capacitation

Oocyte developmental competence is the term that refers to the biochemical and molecular state that allows a mature oocyte to support development of the embryo and ultimately a healthy offspring at term. We still have a very poor understanding of what constitutes oocyte developmental competence, including the role that the environment surrounding an oocyte plays in its progress towards acquisition of developmental competence. The difficulty of studying the oocyte within the developing follicle remains the major barrier to rapid progress in this area. It is clear that oocytes gradually and sequentially acquire meiotic and developmental competence during the course of folliculogenesis. During the early stages of antral follicle growth, non-rodent oocytes are actively synthesizing RNA and as antral follicles progress towards ovulatory size, the synthetic activity of the oocyte is gradually reduced until the oocyte reaches a quiescent state (de Smedt *et al.* 1994; Fair *et al.* 1995). This latter phase of oocyte development has been termed "oocyte capacitation" (Hytell *et al.* 1997) as it is during this phase of oogenesis that the oocyte acquires the cytoplasmic machinery necessary to fully support preimplantation embryo development (Brevini-Gandolfi & Gandolfi 2001). The somatic cells of the follicle, and the cumulus cells in particular, certainly play a key role in the acquisition of developmental competence in vivo. However, we have only a limited understanding of the nature and diversity of compounds that transfer between the cumulus cells and the oocyte via gap junctions during this final phase of follicular development (Albertini *et al.* 2001) and we still have no idea of how dynamic this process is and whether or not dynamic changes to levels of transferring molecules impacts the process of developmental competence. We also have a poor understanding of the role that granulosa and cumulus cells have in modifying the extracellular environment that surrounds the oocyte; speculation with little data is usually the norm here (e.g. nutrients) and little consideration is taken for the influence of events such as gap junction breakdown and cumulus expansion have on nutrient availability to the oocyte.

Our laboratory has taken an interest in the metabolism of the cumulus-oocyte complex (COC) while undergoing in vitro maturation (IVM) and what role this may have on oocyte developmental competence, especially in the bovine. It is noteworthy that oocyte IVM involves the artificial removal of oocytes, generally from mid-sized antral follicles, that presumably have not completed oocyte capacitation. Incubating COCs in vitro leads effectively to a precocious spontaneous meiotic maturation, that occurs in the absence of these crucial events and environments that are required for complete cytoplasmic maturation of the oocyte. Many may argue that this is a poor model for studying the far more complex interactions that take place within the follicle during final antral follicle development. Whilst this is reasonable, the fact remains that oocyte IVM technologies, especially in domestic animals, must deal with the very mixed population of oocytes collected from follicles at varying stages of development and atresia. Our efforts are directed at trying to understand the complex metabolic interplay between oocytes and cumulus cells as they undergo maturation in vitro, in an attempt to develop IVM systems that actually confer developmental competence on oocytes that would otherwise undergo natural demise in vivo by subordination and atresia.

Here we review the progress we, and others, have made in our understanding of the metabolic conditions that influence both meiotic and developmental competence during IVM of bovine COCs in recent times. Other reviews in the area that precede this and may be of useful reference include Sutton *et al.* (2003b) and Krisher (2004).

Energy demands by the oocyte – developmental competence

Mammalian oocytes are essentially reliant on oxidative phosphorylation for the generation of ATP (Rieger and Loskutoff 1994; Steeves and Gardner 1999). Not surprisingly, the level of ATP within a mature oocyte appears now as a key indicator of developmental competence (Hashimoto et al. 2000a; Stojkovic et al. 2001). Van Blerkom (2004) has long argued the importance of mitochondria within human oocytes in the development of competence. This has only recently been examined in several other species, e.g. mouse (Thouas et al. 2004) and bovine (Stojkovic et al. 2001; Tarazona et al. 2006), which are all in agreement on the importance of mitochondria. However, there is also evidence that the activity of the glycolytic pathway within the oocyte following maturation also relates to developmental competence (Krisher and Bavister 1999; Steeves and Gardner 1999), despite the low levels of glycolytic activity within bovine oocytes (Rieger and Loskutoff 1994; Cetica et al. 2002).

Concentration of substrates in follicular fluid

Several recent publications have described levels of energy substrates available to immature, compact ruminant COCs from antral follicles prior to the LH surge (Table 1). Concentrations of glucose and pyruvate have been measured at approximately 2 - 3 mM and 0.4 mM respectively in these studies. Although studies have found lactic acid is also present within follicular fluid, care must be used in interpreting this, as most data has been derived from ovaries collected in an abattoir in which the time from slaughter to fluid collection will affect lactic acid levels.

Table 1. Glucose and carboxylic acid concentrations (Mean ± S.E.) in follicular fluid from various sizes of follicles (small = ~ 3mm, medium = ~ 5mm, large = ~ 8mm)

	Small	Medium	Large	
Glucose (mM)	2.01±0.10	2.85±0.16	3.75±0.18	Leroy et al. (2004)
	1.4±0.2	2.2±0.3	2.3±0.2	Sutton McDowall et al. (2005)
	3.5±0.5	-	3.9±0.4	Iwata et al. (2004).
Pyruvate (mM)	-	-	-	"
	0.4±0.1	0.4±0.1	0.4±0.0	"
	-	-	-	"
Lactate (mM)	14.4±0.35	9.4±0.35	5.6±0.37	"
	3.0±0.7	6.4±1.7	3.2±0.6	"
	-	-	-	"

Less is known regarding other substrates. Recently, Leroy et al. (2004) measured, amongst other parameters, levels of ß-hydroxybutyrate, urea and non-esterified fatty acids within bovine follicular fluid. Berg and colleagues (Berg et al. 2003) have reported that O_2 concentrations within dominant follicles in vivo are around 67 – 86 mmHg, (i.e. around 9 - 12% O_2), which seems higher than many other species (reviewed by Sutton et al. 2003b).

The uptake of oxygen by bovine COCs

Although oxygenation of the follicle has been linked with oocyte developmental competence, (Van Blerkom 1998; Berg et al. 2003), there is little known regarding the oxygen consumption by intact COCs from any mammalian species and what role this plays in oocyte growth and development. Sutton and colleagues (Sutton *et al.* 2003a) measured oxygen uptake by bovine COCs over the course of a 24 h in vitro maturation period (in the presence of FSH). On a per ng DNA basis, COCs increase their oxygen consumption from approximately 50 pmol/h at the start of maturation, when the COC is compact, to 90 pmol/h for fully expanded COCs, which equates to from around 1.5 to 2.5 nl/COC/h. This total uptake of oxygen is remarkably low for a complex of cells such as the COC and is comparable to about double the respiration rate of a bovine blastocyst (0.9 nl/h, Thompson *et al.* 1996). Mouse COCs consume an average of 4 nl/ h O_2 during maturation (Downs *et al.* 1997), which is higher, but within a similar range as the bovine COC. These values led us to ask what level of O_2 actually reaches the oocyte within the cumulus complex? In particular, we wanted to know if there is significant loss of oxygen within the COC as O_2 diffuses towards the oocyte, especially in immature, compact COCs. We developed a mathematical model (Clark *et al.* 2006), which assumes the COC is spherical, but accounts for a non-linear consumption of oxygen dependent on the external concentration. Our model also accounts for the two layers of cells, the cumulus layer and the oocyte and differs from other such models (e.g. Byatt-Smith *et al.* 1991). The model revealed that, surprisingly, the loss of oxygen across the compact bovine COC is negligible (less than 4% of the external concentration), even if hypoxic levels of oxygen are utilised. Thus the level that reaches the oocyte is effectively that which surrounds the COC in the follicular antrum. This has led us to postulate that the structure and metabolic activity of antral follicles is such that follicular oxygen is preserved for the oocyte, as oocytes are dependent on oxidative phosphorylation for ATP production (Rieger and Loskutoff 1994) and have little capacity for glycolysis (Cetica *et al.* 2002).

The central role and importance of glucose to oocyte meiotic and developmental competence

Uptake of glucose by the COC

The bovine COC from antral follicles consumes substantial quantities of glucose - 23 times more uptake on a per volume basis compared to oocytes (Thompson *et al.* 2005), pointing to the importance of glucose to COC metabolism (Fig. 1). Indeed, under standard IVM conditions (TCM199 medium containing 5.6 mM glucose), where the COC density is 1:5 μl medium, we have shown that over 70% of glucose will be metabolised and this may well impact glucose uptake and metabolism kinetics (Sutton-McDowall *et al.* 2004). Over the course of FSH-stimulated in vitro maturation, the uptake rate in TCM199-based medium is initially 23 pmol/ ngDNA/h, increasing to 40 pmol/ngDNA/h. This is entirely due to uptake by the cumulus cells, as the bovine oocyte itself has a very low capacity for glucose uptake (Rieger and Loskutoff 1994; Zuelke and Brackett 1992). There is evidence that ruminant granulosa cells contain both GLUT1 and the insulin-sensitive GLUT4, which may explain the high uptake rates (Williams *et al.* 2001).

   The induction of cumulus expansion increases glucose uptake by approximately 25% and is associated with an increase in hyaluronic acid synthesis via the hexosamine biosynthesis pathway (Sutton-McDowall *et al.* 2004). Using a non-metabolised fluorescent glucose analogue, we have observed the temporal uptake of glucose within the immature compact and fully expanded COC (Thompson *et al.* 2005). Although preliminary at this stage, in immature COCs, the non-metabolised glucose forms a gradient of fluorescence through the cumulus cell layers towards the oocyte. Within 40-60 min of exposure, significant accumulation occurs within the corona radiata. From approximately 40 min, fluorescence is observed within the oocyte. As this is a non-metabolised glucose, the question that remains to be answered is: Does this reflect the transfer of unmetabolised glucose or some metabolite of glucose? How much actual glu-

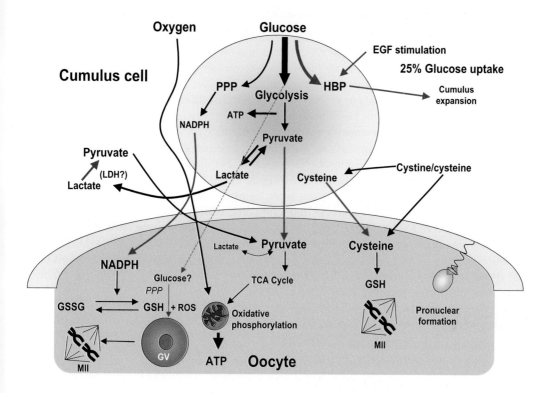

Fig. 1. Schematic representation of the metabolism of glucose and carboxylic acids, plus the amino acid cysteine in the cumulus oocyte complex during maturation. Direction of arrows represents the direction of substrate transfer and density of arrows suggests possible activity of pathways. Colours of arrows (other than black) represent stage of maturation, with blue specific for substrate transfer occurring prior to gap junction breakdown and red specific for activity following cumulus expansion. Abbreviations: ATP = Adenosine triphosphate; EGF = Epidermal Growth Factor; GSH = glutathione (reduced); GSSG glutathione (oxidised); GV = Germinal vesicle; HBP = Hexosamine biosynthesis pathway; LDH = Lactate dehydrogenase; MII = Metaphase II; PPP = Pentose Phosphate Pathway; NADPH = Nicotinamide adenine dinucleotide phosphate; ROS = reactive oxygen species.

cose reaches the oocyte within the immature compact cumulus complex? To answer this, we are developing a much more sophisticated mathematical model to understand the dynamics of facilitated transport, diffusion and metabolism across entities such as the COC. Nevertheless, Saito and others (Saito *et al.* 1994) have shown that in the mouse both glucose and glucose-6-phosphate levels are high in the immature oocyte derived from preovulatory follicles, falling dramatically with (presumably) the loss of cumulus cell gap-junction connection, strongly suggesting that prior to gap junction breakdown, both glucose and glucose metabolites are transferred from cumulus cells into the oocyte. In expanded COCs, over the same time course, it is clear that cumulus cells play little role in glucose transport into the oocyte as no fluorescence appears to enter the oocyte, similarly to that described in the mouse (Saito *et al.* 1994). This further supports the work of others that glucose uptake is very low in mature bovine oocytes (Rieger and Loskutoff 1994).

The effect of glucose on meiotic and developmental competence of the bovine oocyte

Several studies have demonstrated that the presence of glucose during spontaneous in vitro maturation is necessary for complete meiotic maturation to occur (Hashimoto *et al.* 2000b). However, recent studies have revealed that glucose availability (i.e. concentration vs. depletion as a result of uptake over the course of maturation) can have significant affects on the progression of maturation (Iwata *et al.* 2004; Sutton-McDowall *et al.* 2005). Indeed, the use of sequential IVM systems, where fresh medium is used to "replenish" glucose levels during the maturation, can alleviate poor meiotic responses in media with low levels of glucose. Nevertheless, delayed maturation appears to have little consequence in terms of developmental competence (Iwata *et al.* 2004; Sutton-McDowall *et al.* 2006). In contrast, also using a sequential maturation media, Atef and colleagues (Atef *et al.* 2005) found exposure to FSH for 6 hours during IVM was beneficial to development when using a low (1.5 mM) glucose concentration maturation medium (Synthetic Oviduct Fluid) compared with continuous exposure (no change in medium) or 2 h of FSH (and exchange at that time). We suggest these latter two treatments also induce significant glucose depletion at the volume per COC that was utilized (5 μl/COC). Depletion of glucose may prevent oocytes from maintaining glucose-6-phosphate levels (Saito *et al.* 1994), which is known to have a stimulatory effect on meiotic progression via activity of the pentose phosphate pathway (Downs *et al.* 1998).

The hexosamine biosynthesis pathway and oocyte competence

In most somatic cells the hexosamine biosynthesis pathway (HBP, Fig. 2) is a minor glucose metabolic pathway, accounting for 1-3% of glucose metabolism (Marshall *et al.* 1991). A major role of the HBP is the generation of N-linked acetyl sugars for the production of glycosaminoglycans. For example, the end product of the HBP is UDP-N-acetylglucosamine (UDP-N-GlcNAc), which is then synthesized into hyaluronic acid by the action of the enzyme, hyaluronic acid synthase (Fig. 2). The HBP is considered a self-regulating pathway; the rate-limiting enzyme is also the first step of the pathway, glutamine-fructose-6-phosphate transaminase (GFPT). Inhibition of GFPT activity is regulated by the pathway's end product, UDP-N-GlcNAc (McKnight *et al.* 1992). Cumulus cells have a great capacity for hyaluronic acid synthesis as a result of initiation of expansion by epidermal growth factor like factors in vivo (Park *et al.* 2004) or by FSH/EGF in vitro. As previously mentioned, glucose uptake increases 25% with the initiation of cumulus expansion (Sutton-McDowall *et al.* 2004), which appears to be entirely due to up regulation of the HBP pathway. How such a dramatic increase in activity is initiated remains to be determined, but it appears that factors regulating cumulus expansion do so by regulating GFPT activity.

    An alternate fate for UDP-N-GlcNAc is O-linked glycosylation, via O-linked glycosyltransferase, an X-linked enzyme (O'Donnell *et al.* 2004). The importance of O-linked glycosylation within protein signalling systems has only recently come to light, with many major signalling proteins now recognized as being regulated by O-linked glycosylation (e.g. Sp1, Sp3, CREB, p53, in addition to many others). In particular, serines and threonines that are normally phosphorylated can also be O-linked glycosylated, thereby changing the activity (either up or down-regulated) of phosphorylated signalling proteins (Wells *et al.* 2003; Zachara and Hart 2004). One of the more widely studied effects of HBP up-regulation leading to increased O-linked glycosylation is insulin signalling through the insulin receptor substrate-1 (Andreozzi *et al.* 2004), where perturbed phosphorylation of the insulin receptor substrate significantly down-regulates the protein kinase B pathway.

    Glucosamine is a sugar related to glucose and can be transported into cells by members of the facilitated glucose transporter family, similarly to glucose (albeit with different kinetics). However, unlike glucose, there is only one metabolic fate for glucosamine, phosphorylation to

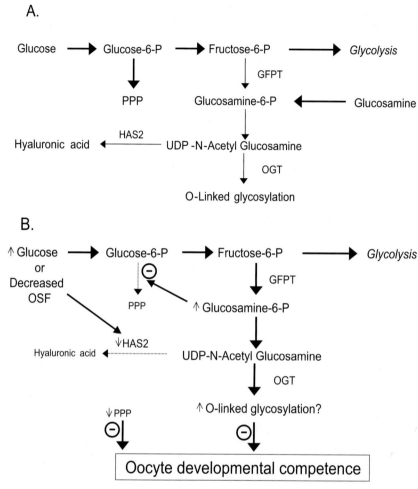

Fig. 2. A. Glucose metabolism, with particular focus on the hexosamine biosynthesis pathway. B. The effect of reduction of oocyte secreted factors or hyperglycaemia on developmental competence. An increase in the hexosamine biosynthesis pathway activity or diversion of UDP-N-acetylglucosamine towards O-linked glycosylation (either by hyperglycaemia or by inhibition of cumulus expansion) causes a reduction in developmental competence of the oocyte. Abbreviations: GFPT = glutamine:fructose-6-phosphate transaminase; HAS2 = hyaluronic acid synthase; OSF = oocyte secreted factors; PPP = Pentose phosphate pathway.

glucosamine-6-phosphate (Fig. 2) and thus bypasses the rate-limiting enzyme of the hexosamine pathway. The consequence is that UDP-N-GlcNAc levels increase significantly, which can lead to increased O-linked glycosylation. Glucosamine is therefore regarded as a hyperglycaemic mimetic.

Glucosamine addition to a bovine IVM medium significantly reduced the level of glucose incorporated within the expanding cumulus mass (Sutton-McDowall *et al.* 2004), signifying the role the hexosamine pathway plays in extracellular matrix formation. Surprisingly, although not significantly affecting meiotic maturation or early cleavage rates following fertilization, glucosamine addition during IVM also substantially reduced subsequent developmental competence (Sutton-McDowall *et al.* 2006), with almost complete inhibition at levels of 2.5 mM glucosamine (when incubated with 5.6 mM glucose). This loss of developmental competence was somewhat reversed when inhibition of O-linked glycosyltransferase in the presence of glucosamine was performed (Sutton-McDowall *et al.* 2006). However, we have not yet fully determined if the glucosamine effect is mediated through cumulus cell signalling, or if it has a direct affect on the oocyte, although immunohistochemical staining of O-linked glycosylation was largely isolated to within cumulus cells, indicating the former rather than the latter.

Hashimoto and others (Hashimoto *et al.* 2000b) have also found that under hyperglycaemic conditions, developmental competence was compromised. This fits well with our model (Fig. 2B), where either hyperglycaemic conditions or compromised cumulus expansion mechanisms leads to an inhibition of developmental competence. However, this inhibition was overcome and, furthermore, developmental competence improved when oxygen concentration was reduced from atmospheric levels (20.9%) to 5% O_2 (Hashimoto *et al.* 2000a). Indeed, only when hyperglycaemic conditions were utilized, was low oxygen found to benefit developmental competence (Hashimoto *et al.* 2000a). This interaction between oxygen and glucose may possibly be explained by the involvement of the transcription factor, hypoxia inducible factor (HIF). Lowering pO_2 may activate HIF to increase glycolytic enzyme activities, which are regulated by HIF (Iyer *et al.* 1998), increasing the capacity to metabolise glucose via glycolysis, thus reducing the option for glucose to be utilised for the HBP and O-linked glycosylation (see Fig. 2B). HIF is also believed to mediate selected FSH-induced responses (Alam *et al.* 2004), so that low oxygen levels may increase the effectiveness of FSH within the maturation system.

Role of carboxylic acids

Most of the glucose taken up by an immature bovine COC is accounted for by the production of lactic acid by cumulus cells and efflux into the surrounding environment (Sutton *et al.* 2003a). Initially in IVM, a positive correlation exists between lactic acid production and glucose uptake, which then weakens as maturation proceeds (Sutton *et al.* 2003a). This, as well as the unchanging lactic acid production over the course of FSH-stimulated cumulus expansion and glucose uptake, provides further evidence that the increasing glucose uptake that occurs during maturation is directed into pathways other than glycolysis (Sutton *et al.* 2003a). Exogenous pyruvate uptake by the FSH stimulated bovine COC occurs, even in the presence of glucose (Sutton *et al.* 2003a). A role for exogenous pyruvate, however, is largely unknown, especially in the presence of adequate glucose. Glycolysis is particularly active in cumulus cells and they have significant levels of several isoforms of lactate dehydrogenase (LDH, the close to equilibrium enzyme for pyruvate conversion to lactate) (Cetica *et al.* 1999; 2002). Therefore it remains to be shown that there is a need for exogenous pyruvate, at least for much of the first nine hours of bovine IVM, where gap-junction communication is patent (Thomas *et al.* 2004). On the other hand, one can imagine that following initiation of cumulus expansion (and hence rapid loss of the oocyte's source of glycolytic metabolites), direct exposure of the oocyte to adequate levels of pyruvate for the latter period of maturation could be necessary for the oocyte. Interestingly, exogenous pyruvate appeared to have no influence on developmental competence, whereas exogenous lactate did (Rose-Hellekant *et al.* 1998). One possible explanation may be the high levels of lactate present in the medium as maturation proceeds and the release of LDH from apoptotic cumulus cells, especially those on the outer edge of the COC (Hussein *et al.* 2005), providing a source of pyruvate for the oocyte towards the end of maturation.

Amino acids

The role of cysteine (and related sulphated amino acids, such as cysteamine), along with glutamine and proline as substrates for glutathione, is now well characterised as promoting developmental competence in bovine oocytes, and oocytes from other species (de Matos et al. 1995; de Matos and Furnus 2000). Inclusion of these sulphated amino acids during IVM and the subsequent increase in oocyte glutathione levels is thought to reduce the level of reactive oxygen species production within oocytes, which has beneficial effects on subsequent developmental competence (de Matos and Furnus 2000). Consequently, cysteamine is now a standard additive in most IVM culture systems. In addition, glutamine is known to be stimulatory for cumulus expansion (Rose-Hellekant et al. 1998), most likely through its involvement in hyaluronic acid synthesis through the hexosamine biosynthesis pathway (see above). Zuelke and Brackett (1993) also found that under the influence of luteinizing hormone, glutamine oxidation increased within the COC, although the relevance of this observation remains obscure.

Lipids

Little new work has emerged on the role of lipids during oocyte IVM, with the exception of the efforts of Leroy and co-workers (Leroy et al. 2005), who have an interest in the hypothesis that non-esterified fatty acids within follicular fluid play a role in oocyte competence during periods of negative energy balance in high performing dairy cows. Their recent work has revealed that addition of stearic and palmitic acids, but not oleic acid, had negative affects on oocyte developmental competence when included during IVM (Leroy et al. 2005).

Conclusions

The bovine cumulus oocyte complex contains two cell types that appear to have antagonistic nutritional demands. Cumulus cells have a great need and appetite for glucose. Too little glucose during maturation appears to have significant adverse effects on oocyte meiotic competence, whilst too much appears to affect oocyte developmental competence. The former may involve glucose and glucose-6-phosphate within the oocyte itself, the latter appears to be mediated through the hexosamine biosynthesis pathway and links the process of supplying substrate for cumulus expansion with oocyte developmental competence. However, achieving consistent glucose concentrations under static IVM conditions is problematic due to the high uptake rate. In contrast, the demand for oxygen and oxidative substrates appear low.

Oocytes, on the other hand, require oxygen and oxidative phosphorylation, and although some glycolytic activity occurs, we question its significance prior to cumulus expansion and loss of gap-junction communication.

The anatomical design of the cumulus oocyte complex fits these antagonistic demands, as oxygen appears to readily diffuse through the cumulus layer to the oocyte, with little being consumed by the cumulus cells themselves. In contrast, glycolytic activity within cumulus cells is such that some glucose appears to reach the immature oocyte whilst the cumulus layer remains compact. This changes once cumulus expansion occurs, so that little glucose enters the oocyte.

Achieving such a delicate balance in substrate supply during oocyte in vitro maturation will remain difficult whilst the current static culture systems employed continue to be utilized. Novel culture systems, utilizing principles of perfusion, microfluidics and small culture chambers with large media reservoirs - the WOW system (Vajta et al. 2000), will facilitate more purpose-built conditions that satisfy the complex demands required for successful in vitro maturation of oocytes.

Acknowledgements

Many thanks to Melanie Sutton-McDowell, Karen Kind and members of the Early Development Group, RCRH, The University of Adelaide for their assistance with this work. Research support for the authors was provided by the Australian Research Council and Cook Australia Pty Ltd, in addition to National Health and Medical Research Council (Australia) and the National Institute of Health (USA).

References

Alam H, Maizels E T, Park Y, Ghaey S, Feigert Z J, Chandel, N S and Hunzicker-Dunn M 2004 Follicle-stimulating hormone activation of Hypoxia-inducible factor-1 by the phosphatidylinositol 3-Kinase/AKT/Ras homolog enriched in brain (Rheb)/mammalian target of rapamycin (mTOR) pathway is necessary for induction of select protein markers of follicular differentiation. *Journal of Biological Chemistry* **279** 19431-19440.

Albertini D F, Combelles C M H, Benecchi E and Carabastsos M J 2001 Cellular basis for paracrine regulation of ovarian follicle development. *Reproduction* **121** 647-653.

Andreozzi F, D'Alessandris C, Federici M, Laratta E, Del Guerra S, Del Prato S, Marchetti P, Lauro R, Perticone, F and Sesti G 2004 Activation of the hexosamine pathway leads to phosphorylation of IRS-1 on Ser307 and Ser612 and impairs the phosphatidylinositol 3-kinase/Akt/mTOR insulin biosynthetic pathway in RIN pancreatic {beta}-cells. *Endocrinology* **145** 2845-57.

Atef A, Francois P, Christian V and Sirard M-A 2005 The potential role of gap junction communication between cumulus cells and bovine oocytes during in vitro maturation. *Molecular Reproduction and Development* **71** 358-367.

Berg D K, Beaumont S E, Berg M C, Hull C D and Tervit H R 2003 Oxygen and carbon dioxide tension in days 14-15 dominant bovine follicles measured in vivo or 4 hours post-mortem. *Theriogenelogy* **59** 406.

Brevini-Gandolfi T A L and Gandolfi F 2001 The materna legacy to the embryo: cytoplasmic components and their effects on early development. *Theriogenology* **55** 1255-1276.

Byatt-Smith J G, Leese H J and Gosden R G 1991 An investigation by mathematical modelling of whether mouse and human preimplantation embryos in static culture can satisfy their demands for oxygen by diffusion. *Human Reproduction* **6** 52-57.

Cetica P, Pintos L, Dalvit G and Beconi M 2002 Activity of key enzymes involved in glucose and triglyceride catabolism during bovine oocytes maturation *in vitro Reproduction* **124** 675-681.

Cetica P D, Pintos L N, Dalvit G C and Beconi M T 1999 Effect of lactate dehydrogenase activity and isoenzyme localisation in bovine oocytes and utilisation of oxidative phosphorylation on in vitro maturation. *Theriogenology* **51** 541-550.

Clark A R, Stokes Y M, Lane M and Thompson J G 2006 Mathematical modelling of oxygen concentration in bovine and murine cumulus-oocyte complexes. *Reproduction* **131** 999-1006.

de Matos D G and Furnus C C 2000 The importance of having high glutathione (GSH) level after bovine in vitro maturation on embryo development effect of B-mercaptoethanol, cysteine and cystine. *Theriogenology* **53** 761-771.

de Mato, D G, Furnus C C, Moses D F and Baldasarre H 1995 Effect of cysteamine on glutathione level and developmental capacity of bovine oocyte matured in vitro. *Molecular Reproduction and Development* **42** 432-436.

de Smedt V, Crozet N and Gall L 1994 Morphological and functional changes accompanying the acquisition of meiotic competence in ovarian goat oocyte. *Journal of Experimental Zoology* **269** 128-139.

Downs S, Humpherson P and Leese H 1998 Meiotic induction in cumulus cell-enclosed mouse oocytes: Involvement of the Pentose Phosphate Pathway. *Biology of Reproduction* **58** 1084-1094.

Downs S M, Houghton F D, Humpherson P G and Leese H J 1997 Substrate utilization and maturation of cumulus cell-enclosed mouse oocytes: evidence that pyruvate oxidation does not mediate meiotic induction. *Journal of Reproduction and Fertility* **110** 1-10.

Fair T, Hyttel P and Greve T 1995 Bovine oocyte diameter in relation to maturational competence and transcriptional activity. *Molecular Reproduction and Development* **42** 437-442.

Hashimoto S, Minami N, Takakura R, Yamada M, Imai H and Kashima N 2000a Low oxygen tension during in vitro maturation is beneficial for supporting the subsequent development of bovine cumulus-oocyte complexes. *Molecular Reproduction and Development* **57** 353-360.

Hashimoto S, Minami N, Yamada M and Imai H 2000b Excessive concentration of glucose during in vitro maturation impairs the developmental competence of bovine oocytes after in vitro fertilisation: Relevance to intracellular reactive oxygen species and glutathione contents. *Molecular Reproduction and Development* **56** 520-526.

Hussein T S, Froiland D A, Amato F, Thompson J G and Gilchrist R B 2005 Oocytes prevent cumulus cell apoptosis by maintaining a morphogenic paracrine gradient of bone morphogenetic proteins. *Journal of Cell Science* **15** 5257-5268.

Hytell P, Fair T, Callesen H and Greve T 1997 Oocyte growth, capacitiation and final maturation in cattle. *Theriogenology* **47** 23-32.

Iwata H, Hashimoto S, Ohota M, Kimura K, Shibano K and Miyake M 2004 Effects of follicle size and electrolytes and glucose in maturation medium on nuclear maturation and developmental competence of bovine oocytes. *Reproduction* **127** 159-164.

Iyer N V, Kotch L E, Agani F, Leung S W, Laughner E, Wenger R H, Gassmann M, Gearhar, J D, Lawler A M, Yu Q Y and Semenza G L 1998 Cellular and developmental control of O$_2$ homeostasis by hypoxia-inducible factor 1a. *Genes and Development* **12** 149-162.

Krisher R L 2004 The effect of oocyte quality on development. *Journal of Animal Science* **82** 14-23.

Krisher R L and Bavister B D 1999 Enhanced glycolysis after maturation of bovine oocytes in vitro is associated with increased developmental competence. *Molecular Reproduction and Development* **53** 19-26.

Leroy J L M R, Vanholder T, Mateusen B, Christophe A, Opsomer G, de Kruif A, Genicot G and Van Soom A 2005 Non-esterified fatty acids in follicular fluid of dairy cows and their effect on developmental capacity of bovine oocytes *in vitro*. *Reproduction* **130** 485-495.

Leroy J L M R, Vanholder T R D J, Opsomer G, Van Soom A J B P E and De Kruif A 2004 Metabolite and ionic composition of follicular fluid from different-sized follicles and their relationship to serum concentrations in dairy cows. *Animal Reproduction Science* **80** 201-211.

Marshall D L, Bacote V and Traxinger R R 1991 Discovery of a metabolic pathway mediating glucose-induced desenitition of the glucose transport system. *Journal of Biological Chemistry* **266**, 4706-4712.

McKnight G L, Mudri S L, Mathewes S L, Traxinger R R, Marshall S, Sheppard P O and O'Hara,P 1992 Molecular cloning, cDNA sequence, and bacterial expression of human glutamine:fructose-6-phosphate amidotransferase. *Journal of Biological Chemistry* **267** 25208-25212.

O'Donnell N, Zachara N E, Hart G W and Marth JD 2004 Ogt-Dependent X-Chromosome-Linked Protein Glycosylation is a Requisite Modification in Somatic Cell Function and Embryo Viability. *Molecular and Celluarl Biology* **24** 1680-1690.

Pan H, O'Brien M J, Wigglesworth K, Eppig J J and

Shultz R M 2005 Transcript profiling during mouse oocyte development and the effect of gonadotropin priming and development in vitro. *Developmental Biology* **286** 493-506.

Park J Y, Su Y Q, Ariga M, Law E, Jin S L and Conti M 2004 EGF-like growth factors as mediators of LH action in the ovulatory follicle. *Science* **303** 682-684.

Reis A., Chang HY and Levasseur M J K 2006 APCcdh1 activity in mouse oocytes prevents entry into the first meiotic division. *Nature Cell Biology* **8** 539-540.

Rieger D and Loskutoff N M 1994 Changes in the metabolism of glucose, pyruvate, glutamine and glycine during maturation of cattle oocytes in vitro. *Journal of Reproduction and Fertility* **100** 257-262.

Rose-Hellekant T A, Libersky-Williamson E A and Bavister B D 1998 Energy substrates and amino acids provided during in vitro maturation of bovine oocytes alter acquisition of developmental competence. *Zygote* **6** 285-294.

Saito T, Hiroi M and Kato T 1994 Development of glucose utilization studied in single oocytes and preimplantation embryos from mice. *Biology of Reproduction* **50** 266-270.

Steeves T E and Gardner D K 1999 Metabolism of Glucose, pyruvate, and glutamine during the maturation of oocytes derived from pre-pubertal and adult cows. *Molecular Reproduction and Development* **54** 92-101.

Stojkovic M, Machado S A, Stojkovic P, Zakhartchenko V, Hutzler P, Goncalves P B and Wolf E 2001 Mitochondrial distribution and adenosine triphosphate content of bovine oocytes before and after in vitro maturation: correlation with morphological criteria and developmental capacity after in vitro fertilization and culture. *Biology of Reproduction* **64** 904-909.

Sutton M L, Cetica P D, Beconi M T, Kind K L, Gilchrist R B and Thompson J G 2003a Influence of oocyte-secreted factors and culture duration on the metabolic activity of bovine cumulus cell complexes. *Reproduction* **126** 27-34.

Sutton M L, Gilchrist R B and Thompson J G 2003b Effects of in-vivo and in-vitro environments on the metabolism of the cumulus-oocyte complex and its influence on oocyte developmental capacity. *Human Reproduction* **9** 35-48.

Sutton-McDowall M L, Gilchrist R B and Thompson J G 2004 Cumulus expansion and glucose utilisation by bovine cumulus-oocyte complexes during *in vitro* maturation: the influence of glucosamine and follicle-stimulating hormone. *Reproduction* **128** 314-319.

Sutton-McDowall M L, Gilchrist R B and Thompson J G 2005 Effect of hexoses and gonadotrophin supplementation on bovine oocyte nuclear maturation during in vitro maturation in a synthetic follicle fluid medium. *Reproduction, Fertility, and Development* **17** 407-415.

Sutton-McDowall M L, Mitchell M, Cetica P, Dalvit G,

Pantaleon M, Lane M, Gilchrist R and Thompson J G 2006 Glucosamine supplementation during in vitro maturation inhibits subsequent embryo development: Possible role of the hexosamine pathway as a regulator of developmental competence. *Biology of Reproduction* **74** 881-888.

Tarazona A M, Rodriguez J I, Restrepo L F and Olivera-Angel M 2006 Mitochondrial activity, distribution and segregation in bovine oocytes and in embryos produced in vitro. *Reproduction in Domestic Animals* **41** 5-11.

Thomas R E, Thompson J G, Armstrong D T and Gilchrist R B 2004 Effect of specific phosphodiesterase isoenzyme inhibitors during in vitro maturation of bovine oocytes on meiotic and developmental capacity. *Biology of Reproduction* **71** 1141-1149.

Thompson J, Clark A, Froiland D, Stokes Y and Lane M 2005 A "nutrient" sparing hypothesis to explain the nutrition of the immature bovine cumulus oocyte complex. *Biology of Reproduction* **Special Issue** 417.

Thompson J G, Partridge R J, Houghton F D, Cox C I and Leese H J 1996 Oxygen uptake and carbohydrate metabolism by in vitro derived bovine embryos. *Journal of Reproduction and Fertility* **106** 299-306.

Thouas G A, Trounson A, Wolvetang E J and Jones G M 2004 Mitochondrial dysfunction in mouse oocytes results in preimplantation embryo arrest in vitro. *Biology of Reproduction* **71** 1936-1942.

Vajta G, Peura T T, Holm P, Paldi A, Greve T, Trounson A O and Callesen H 2000 New method for culture of zona-included or zona-free embryos: the well of the well (WOW) system. *Molecular Reproduction and Development* **55** 256-264.

Van Blerkom J 1998 Epigenetic influences in oocyte developmental competence: perifollicular vascularity and intrafollicular oxygen. *Journal of Assisted Reproduction and Genetics* **15** 226-234.

Van Blerkom J 2004 Mitochondria in human oogenesis and preimplantatioin embryogenesis: engines of metabolism, ionic regulation and developmental competence. *Reproduction* **128** 269-280.

Wells L, Whelan S A and Hart G W 2003 O-GlcNAc: a regulatory post-translational modification. *Biochemical and Biophysical Research Communications.* **302** 435-441.

Williams S A, Blache D, Martin G B, Foot R, Blackberry M A and Scaramuzzi R J 2001 Effect of nutritional supplementation on quantities of glucose transporters 1 and 4 in sheep granulosa and theca cells. *Reproduction* **122** 947-956.

Zachara N E and Hart G W 2004 O-GlAc a sensor of cellular state: the role of nucleocytoplasmic glycosylation in modulating cellular function in response to nutrition and stress. *Biochimica et Biophysica Acta* **1673** 13-28.

Zuelke K A and Brackett B G 1992 Effects of luteinizing hormone on glucose metabolism in cumulus-enclosed bovine oocytes matured in vitro. *Endocrinology* **131** 2690-2696.

Zuelke K A and Brackett B G 1993 Increased glutamine metabolism in bovine cumulus cell-enclosed and denuded oocytes after in vitro maturation with lutenising hormone. *Biology of Reproduction* **48** 815-820.

Judge, jury and executioner: the auto-regulation of luteal function

GD Niswender, TL Davis, RJ Griffith, RL Bogan, K Monser, RC Bott, JE Bruemmer and TM Nett

Animal Reproduction and Biotechnology Laboratory, Colorado State University, Fort Collins, CO 80523-1683, USA

Experiments were conducted to further our understanding of the cellular and molecular mechanisms that regulate luteal function in ewes. Inhibition of protein kinase A (PKA) reduced ($P < 0.05$) secretion of progesterone from both small and large steroidogenic luteal cells. In addition, the relative phosphorylation state of steriodogenic acute regulatory protein (StAR) was more than twice as high ($P < 0.05$) in large vs small luteal cells. Large steroidogenic luteal cells appear to contain constitutively active PKA and increased concentrations of phosphorylated StAR which play a role in the increased basal rate of secretion of progesterone.

To determine if intraluteal secretion of prostaglandin (PG) $F_2\alpha$ was required for luteolysis, ewes on day 10 of the estrous cycle received intraluteal implants of a biodegradable polymer containing 0, 1 or 10 mg of indomethacin, to prevent intraluteal synthesis of $PGF_2\alpha$. On day 18, luteal weights in ewes receiving 1 mg of indomethacin were greater ($P < 0.05$) than controls and those receiving 10 mg were greater ($P < 0.05$) than either of the other two groups. Concentrations of progesterone in serum were also increased ($P < 0.05$) from days 13 to 16 of the estrous cycle in ewes receiving 10 mg of indomethacin. Although not required for decreased production of progesterone at the end of the cycle, intraluteal secretion of $PGF_2\alpha$ appears to be required for normal luteolysis.

To ascertain if oxytocin mediates the indirect effects of $PGF_2\alpha$ on small luteal cells, the effects of 0, 0.1, 1 or 10 mM oxytocin on intracellular concentrations of calcium were quantified. There was a dose-dependent increase ($P < 0.05$) in the number of small luteal cells responding to oxytocin. Thus, oxytocin induces increased calcium levels and perhaps apoptotic cell death in small luteal cells. Concentrations of progesterone, similar to those present in corpora lutea (~ 30 $\mu g/g$), prevented the increased intracellular concentrations of calcium ($P < 0.05$) stimulated by oxytocin in small cells. In large luteal cells the response to progesterone

Corresponding author E-mail: gordon.niswender@colostate.edu

was variable. There was no consistent effect of high quantities of estradiol, testosterone or cortisol in either cell type. It was concluded that normal luteal concentrations of progesterone prevent the oxytocin and perhaps the $PGF_2\alpha$-induced increase in the number of small and large luteal cells which respond to these hormones with increased intracellular concentrations of calcium.

In summary, large ovine luteal cells produce high basal levels of progesterone, at least in part, due to a constituitively active form of PKA and an enhanced phosphorylation state of StAR. During luteolysis $PGF_2\alpha$ of uterine origin reduces the secretion of progesterone from the corpus luteum, but intraluteal production of $PGF_2\alpha$ is required for normal luteolysis. Binding of $PGF_2\alpha$ to receptors on large luteal cells stimulates the secretion of oxytocin which appears to activate PKC and may also inhibit steroidogenesis in small luteal cells. $PGF_2\alpha$ also activates COX-2 in large luteal cells which leads to secretion of $PGF_2\alpha$. Once intraluteal concentrations of progesterone have decreased, oxytocin binding to its receptors on small luteal cells also results in increased levels of intracellular calcium and presumably apoptosis. Increased secretion of $PGF_2\alpha$ from large luteal cells activates calcium channels which likely results in apoptotic death of this cell type.

Introduction

It has been known for decades that the parenchyma of the corpus luteum in many mammalian species includes two distinct types of steroidogenic cells (Corner 1919; Warbritton 1934; Mossman & Duke 1973), both of which secrete progesterone (Lemon & Loir 1977; Ursely & Leymarie 1979; Koos & Hansel 1981; Fitz et al. 1982; Rodgers & O'Shea 1982). The two types of steroidogenic luteal cells can be differentiated based on morphological and biochemical characteristics, as well as the follicular source of origin (Niswender et al. 1985).

Small steroidogenic luteal cells range from 12 to 22 mm in diameter (Fitz et al. 1982; Rodgers & O'Shea 1982). They are more numerous than large luteal cells and can be identified at the light microscopic level by their size, characteristic elongated shape, irregularly shaped nucleus and the presence of lipid droplets (O'Shea et al. 1979; Farin et al. 1986). The diameter of large steroidogenic luteal cells ranges from 22 to 35 μm with occasional cells having diameters greater than 35 μm (Donaldson & Hansel 1965; Deane et al. 1966; Fitz et al. 1982). At the light microscopic level large luteal cells usually appear spherical or polyhedral and the nucleus appears rounded. Small protein-containing secretory granules are usually detectable in the cytoplasm and large luteal cells in ewes, but not in cows, are essentially devoid of lipid droplets.

At the ultrastructural level both small and large luteal cells have numerous mitochondria and an abundance of smooth endoplasmic reticulum characteristic of other steroid-secreting cells (Christensen & Gillim 1969). However, large luteal cells also possess numerous Golgi complexes, an abundance of rough endoplasmic reticulum and numerous membrane-bound secretory granules characteristic of protein-secreting cells (Deane et al. 1966; Enders 1973; O'Shea et al. 1979). For further details regarding the morphological characteristics of the two steroidogenic luteal cells types see Rodgers & O'Shea (1982) or Niswender et al. (1985).

In addition to their morphological differences, small and large luteal cells have many biochemical differences in receptor content, second messenger function, and steroidogenic characteristics. Small cells have receptors for luteinizing hormone (LH) coupled to the protein kinase A (PKA) second messenger pathway which stimulates many components of the synthetic pathway for progesterone. Secretion of progesterone can be increased as much as 20-fold following stimulation of small cells with LH or analogs of cAMP (Fitz *et al.* 1982; Alila *et al.* 1988). Activation of the PKC pathway in small luteal cells inhibits LH-stimulated secretion of progesterone (Wiltbank *et al.* 1991). Large luteal cells have receptors for prostaglandin (PG) $F_2\alpha$ and PGE2 but are devoid of receptors for LH except under some conditions (Fitz *et al.* 1982; Harrison *et al.* 1987). Large cells have little, if any, response to LH in terms of increased secretion of progesterone (Hoyer & Niswender 1986). Basal secretion rates of progesterone by large luteal cells are 10- to 20-fold higher on a per cell basis than for small luteal cells (Fitz *et al.* 1982). The mechanisms responsible for this increased, but apparently unregulated, rate of progesterone secretion are not clear. Diaz *et al.* (2002) suggested that constitutive activity of the PKA pathway was likely involved since specific inhibitors of this pathway reduced secretion of progesterone by large luteal cells. Stimulation of large cells by $PGF_2\alpha$ results in activation of the PKC second messenger pathway which inhibits synthesis of progesterone by acting at multiple sites in the steroidogenic pathway (Wiltbank *et al.* 1991; Juengel & Niswender 1999). Large luteal cells also respond to $PGF_2\alpha$ with increased concentrations of intracellular calcium (Wiltbank *et al.* 1989; Wegner *et al.* 1990) which appears to be the key event in initiating the processes involved in apoptotic cell death (Vinatier *et al.* 1996).

The purpose of the research presented in the remainder of this publication was to address four specific questions regarding the cellular regulation of luteal function.

Regulation of progesterone biosynthesis in the two cell types

One of the more intriguing questions remaining to be answered regarding regulation of progesterone biosynthesis: What mechanisms are responsible for the high basal rate of secretion of progesterone from large luteal cells? To determine if PKA was tonically active in large cells resulting in continual stimulation of steroidogenesis we utilized a specific PKA inhibitor (PKI) and the adenylate cyclase activator forskolin as a positive control. The effects on progesterone and cAMP production were quantified following treatment of partially purified preparations of small and large ovine luteal cells.

Treatment of small cells with forskolin (10 μM) caused an approximately 8-fold increase in progesterone secretion (P < 0.05) while co-treatment with the PKI (50 μM) prevented the forskolin-induced increase (Fig. 1A). These data once again document the important role of PKA in regulating the steroidogenic pathway in small luteal cells. Concentrations of cAMP in media were not different between forskolin-treated small cells in the presence or absence of the PKI, which suggested that the effect of PKI on progesterone secretion was not due to treatment-induced toxicity. In large luteal cells forskolin treatment increased progesterone production (P < 0.05) and both basal secretion and forskolin-stimulated secretion of progesterone was decreased more than 5-fold (P < 0.05) by treatment with PKI (Fig. 1B). The modest but significant increase in progesterone production following forskolin treatment may have been due to contamination of the preparation of large luteal cells with a small number of clumps of small luteal cells. Forskolin treatment increased cAMP concentrations almost 6-fold (P < 0.05) in large luteal cells while progesterone production only doubled. The increase in cAMP was prevented by treatment with PKI. Thus, the effects of PKI were dramatic on reducing progesterone production by large luteal cells but the effects on cAMP concentrations were non-existent (basal) or less dramatic (forskolin treated). These data are consistent with the hypothesis that large luteal cells contain a constitutively active form of PKA

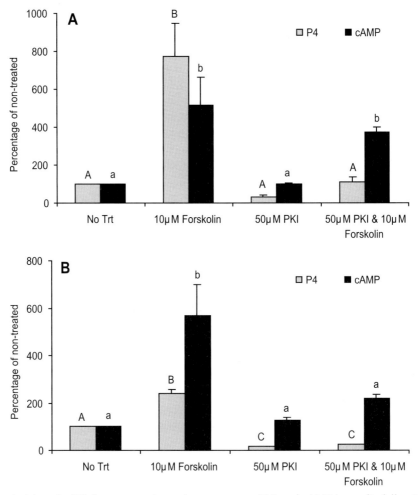

Fig. 1. Mean ($\pm$ SEM) concentrations of progesterone (P4) and cAMP in media following culture of small (A) and large (B) luteal cells. Columns with different letters are different (P < 0.05) for mean concentrations of progesterone or cAMP.

Steroidogenic acute regulatory protein (StAR) is an important component for the transport of cholesterol to the mitochondrial membrane, the rate-limiting step in progesterone biosynthesis (Niswender 2002). Since phosphorylation is thought to be key for regulating the biological activity of StAR (Arakane *et al.* 1997) we next quantified the relative concentrations and phosphorylation state of this protein. A modified sandwich enzyme linked immunoassay (ELISA) was developed to quantify the concentrations and relative phosphorylation states of StAR in the two cell types as a biochemical marker of PKA activity (Bogan 2006). There were no effects of treatment on concentrations of StAR protein in either cell type, and no correlation between concentrations of StAR protein and progesterone in media (data not shown). However, StAR was more extensively phosphorylated under basal conditions in large compared to small luteal cells (Fig. 2; P < 0.05) and there was a significant decrease (P < 0.05) in the phosphorylation state of StAR in control or forskolin-treated large luteal cells following PKI treatment. Treatment of small luteal cells with PKI did not significantly reduce the phosphorylation state of

StAR in control cells but did reduce this parameter in forskolin-treated small cells. The relative phosphorylation state of StAR and concentrations of progesterone in media were significantly correlated when all treatments in both cell types were evaluated ($r = .71$; $P < 0.05$). Thus, the high basal rate of secretion of progesterone by large luteal cells is associated with tonically active PKA and increased phosphorylation of StAR. Degree of phosphorylation, but not changes in the quantity of StAR, appears to be a primary effect of the PKA pathway on activity of StAR protein in ovine large luteal cells.

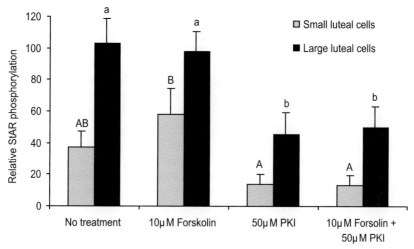

Fig. 2. Relative StAR phosphorylation state in small and large luteal cells, expressed as percentage of StAR phosphorylation compared to that in standards. Forskolin/PMA-treated mixed luteal cells were used as the 100% phosphorylated standard. Error bars indicate one SEM; columns with different letters are different ($P < 0.05$) for each cell type.

Regulation of luteolysis

The generally accepted model for regression of the corpus luteum in ruminants is that oxytocin secretion from the neuro-hypophysis or the corpus luteum stimulates the secretion of $PGF_2\alpha$ from the uterus (Bazer *et al.* 1991; Silvia *et al.* 1991). The $PGF_2\alpha$ then stimulates oxytocin secretion from large luteal cells in the corpus luteum which in turn stimulates more secretion of $PGF_2\alpha$ from the uterus establishing a positive feedback loop which ultimately causes decreased secretion of progesterone and death of the luteal cells. There are three unanswered questions regarding the cellular and molecular mechanisms involved in luteolysis that are not addressed by the existing model. First, does normal luteolysis require intraluteal as well as uterine production of $PGF_2\alpha$? Second, since small luteal cells do not have receptors for $PGF_2\alpha$, how does this hormone cause these cells to undergo apoptosis? Third, does progesterone act within the corpus luteum to influence the process of luteolysis? We have conducted experiments designed to directly address these questions.

Is intraluteal production of $PGF_2\alpha$ required for luteolysis?

Prostaglandin $F_2\alpha$ can be synthesized by corpora lutea of women (Shutt *et al.* 1976; Swanston *et al.* 1977; Patwardhan & Lanthier 1980), sows (Guthrie *et al.* 1978), ewes (Rexroad & Guthrie 1979; Tsai & Wiltbank 1997), cows (Pate 1988), and rodents (Olofsson *et al.* 1992). Therefore, we investigated whether intraluteal secretion of PGF$_2\alpha$ played a role in normal luteolysis (Griffith 2002). Ewes were administered an implant of Atrigel (100 μl) that contained 0 (n = 6), 1 (n = 8) or 10 (n = 5) mg of indomethacin to inhibit intraluteal production of PGF$_2\alpha$ into the corpus luteum on day 10 of the estrous cycle. Atrigel is a biodegradable polymer that is in a liquid state but rapidly solidifies when exposed to an aqueous environment. Indomethacin would be expected to prevent synthesis of all prostaglandins and is not specific for PGF$_2\alpha$. Samples of jugular blood were collected daily from the time of implantation until day 18 when corpora lutea were collected and weighed. Progesterone concentrations in serum were quantified by radioimmunoassay (Niswender 1973).

Although concentrations of progesterone in serum tended to be greater in ewes receiving an implant of 1 mg indomethacin compared to controls, the differences were not significant (Fig. 3). Concentrations of progesterone were greater (P < 0.02) from days 13-16 in ewes which received an implant containing 10 mg of indomethacin than in controls. It seems likely that there is continuous negative regulation of luteal production of progesterone by PGF$_2\alpha$ during the mid-luteal phase of the estrous cycle. By days 17-18 progesterone concentrations had reached follicular phase levels in all treatment groups indicating that the decreased secretion of progesterone associated with luteolysis had occurred. However, when luteal weights were analyzed (Fig. 4), corpora lutea which received implants containing 1 mg of indomethacin were heavier (P < 0.05) than controls and corpora lutea which received implants containing 10 mg indomethacin were heavier than either of the other groups with weights comparable to those observed on day 10 of the estrous cycle in similar sheep (Niswender et al. 1985). Thus, the normal processes responsible for reduced luteal weights did not occur in corpora lutea that contained an indomethacin implant presumably due to the lack of luteal production of PGF$_2\alpha$.

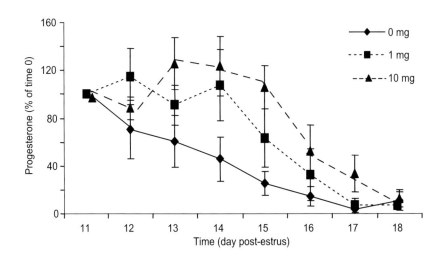

Fig. 3. Mean ($\pm$ SEM) concentrations of progesterone in sera collected from ewes administered Atrigel alone (0 mg) or Atrigel containing 1 or 10 mg of indomethacin. Progesterone concentrations were greater (P < 0.02) on days 13-16 in ewes administered 10 mg indomethacin compared to controls.

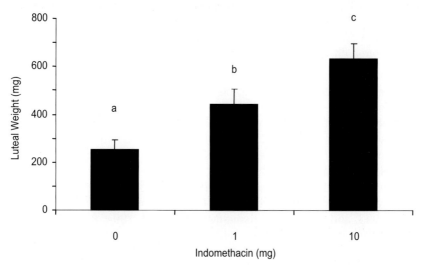

Fig. 4. Mean ($\pm$ SEM) luteal weights following administration of intraluteal Atrigel implants containing 0, 1 or 10 mg indomethacin. Means with different letters were different ($P < 0.05$).

To determine if the effect of indomethacin was local, within the corpus luteum, or via a systemic mechanism an additional experiment was performed in ewes which had a single corpus luteum on each ovary. On day 11 of the estrous cycle ewes ($n = 8$) were administered an Atrigel implant in one corpus luteum and the corpus luteum on the opposite ovary received an implant containing 1 mg indomethacin. On day 18 of the estrous cycle the corpora lutea which received an implant containing 1 mg of indomethacin weighed more (412 ± 83 mg) than corpora lutea treated with Atrigel alone (275 ± 24 mg; $p < 0.05$). Thus, the effects of the intraluteal indomethacin implants appeared to be local and not systemic.

It was concluded that intraluteal production of $PGF_2\alpha$ is an important component of the normal luteolytic process. These data also support the previous suggestions that the anti-steroidogenic actions of $PGF_2\alpha$ can be separated from its luteolytic actions. McGuire *et al.* (1994) first demonstrated that intraluteal administration of PMA to activate the PKC second messenger pathway decreased serum concentrations of progesterone to follicular phase levels but had no effect on luteal weights, even though the ewes had re-ovulated. Juengel *et al.* (2000) suggested that low doses of $PGF_2\alpha$ decreased progesterone synthesis but that higher doses were required for structural regression of the corpus luteum. The current data support this suggestion and further indicate that the lower concentrations of $PGF_2\alpha$ associated with de-creased progesterone production are produced by the uterus. Thus, the uterus provides the $PGF_2\alpha$ that initiates the process of luteolysis and specifically provides the signal for reduced secretion of progesterone. However, intraluteal production of $PGF_2\alpha$ is required for loss of luteal weight and cell death.

It is likely that attenuation of intraluteal secretion of $PGF_2\alpha$ may play an essential role in maintenance of the corpus luteum of pregnancy. Silva *et al.* (2000) clearly demonstrated that luteal mRNA encoding 15-hydroxyprostaglandin dehydrogenase (PGDH) and the activity of this enzyme, a cytosolic enzyme responsible for catabolism of $PGF_2\alpha$, is increased in ovine corpora lutea during the time of maternal recognition of pregnancy. The signal for increasing mRNA for PDGH activity in the corpus luteum may be interferon-tau since Ott *et al.* (personal communication) have suggested that this hormone has direct effects at the level of the corpus luteum.

Does oxytocin control intracellular calcium concentrations in small luteal cells?

Since small luteal cells do not have receptors for $PGF_2\alpha$ (Fitz et al. 1982) the luteolytic effects of this hormone on this cell type have to be indirect. Oxytocin inhibited the LH-stimulated secretion of progesterone from small luteal cells which suggests that small luteal cells have receptors for oxytocin (Mayan & Niswender, unpublished data). Since large luteal cells release oxytocin when stimulated by $PGF_2\alpha$, this hormone was a likely candidate to mediate the effects of $PGF_2\alpha$ on small luteal cells. A series of experiments was performed to ascertain if oxytocin treatment of small luteal cells influenced intracellular concentrations of calcium. Increases in intracellular calcium can perturb cell function by activating endonucleases which cause DNA fragmentation (Vinatier et al. 1996), which occurs during regression of bovine (Juengel et al. 1993) and ovine luteal tissue (McGuire et al. 1994; Rueda et al. 1995; Juengel et al. 2000). Further, increases in intracellular calcium concentrations also induce nitric oxide production (Yi et al. 2004), which has also been implicated as a mediator of luteal regression (Jaroszewski & Hansel 2000; Korzekwa et al. 2006).

Corpora lutea were collected from 1-3 ewes for each replicate (n = 8) on days 9 or 10 of the estrous cycle, dissociated into single cell suspensions and partially purified into preparations of small and large steroidogenic luteal cells by centrifugal elutriation. Small luteal cells (50,000 per plate) were plated overnight on Matrigel-coated (BD Biosciences) 35 mm glass bottom dishes (MatTek) in DMEM containing 10% fetal bovine serum, 50 units/ml penicillin, and 50 μg/ml streptomycin. Media were removed and cells were washed twice with fluorescence buffer (145 mM NaCl, 5 mM KCl, 1 mM Na_2HPO_4, 0.5 mM $MgCl_2$ $6H_2O$, 1 mM $CaCl_2$ $2H_2O$, 10 mM HEPES, and 5 mM glucose). Cells were loaded with 3 μM Fura-2, acetoxymethlyester (Fura-2, AM) containing 0.1% Pluronic F-127 (Molecular Probes) for 30 minutes. Cells were then washed in fluorescence buffer and allowed to recover for 30 minutes before a final wash. Calcium responses were measured using an InCyt2 imaging system (Intracellular Imaging Inc.) as previously described (Shlykov & Sanborn 2004). Changes in intracellular concentrations of calcium in individual cells were measured at 340 nm and 380 nm excitation and 510 nm emission wave lengths. After measuring basal intracellular calcium concentrations for 30s, cells were stimulated with 0.1 μM- 10 μM oxytocin (Sigma Chemical Co., St. Louis, MO) and intracellular calcium concentrations measured for 5 minutes. The oxytocin receptor in ovine uterus has a Kd of approximately 10 nM (Dunlap & Stormshak 2004) so addition of 100 nM to 10 μM allowed rapid occupancy of receptors. It also is likely that intraluteal levels of oxytocin become quite high since large luteal cells produce this hormone. Data collected included percentage of cells responding to oxytocin with increased concentrations of intracellular calcium as well as the average peak height.

There was a dose-dependent increase in both the percentage of cells exhibiting increased calcium concentrations and peak height of intracellular concentrations of calcium in cells that responded following treatment with increasing quantities of oxytocin (Fig. 5). It seems likely that the increased intracellular concentrations of calcium initiate the mechanisms leading to apoptosis in this cell type. Oxytocin had no effect on intracellular concentrations of calcium in large luteal cells when studied under the same conditions as those used for small luteal cells.

Does progesterone influence luteolysis?

It is generally accepted that the corpus luteum undergoes luteolysis by apoptotic mechanisms. Apoptotic processes have been well characterized. Chromosomal condensation along the

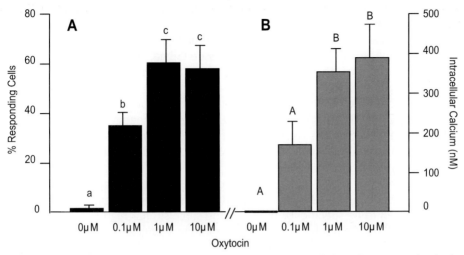

Fig. 5. Effect of various doses of oxytocin to increase intracellular calcium in individual small luteal cells. Percent of cells responsive to oxytocin is shown in panel A while the peak height of intracellular calcium is shown in panel B. Doses of oxytocin with different lower case superscripts are statistically different (P < 0.05). Data are from 8 replicates for percent responsive cells and 4 replicates for maximum intracellular calcium concentrations.

periphery of the nucleus and shrinkage of the cell body are early signs of apoptosis. Later the nucleus and the cytoplasm fragment forming apoptotic bodies which are phagocytosed by macrophages (Lodish *et al.* 2000). Another distinguishing characteristic of apoptosis is the formation of 185 bp fragments of genomic DNA (oligonucleosomes) as a result of activation of calcium-dependent endonucleases (Arends *et al.* 1990). Oligonucleosome formation has been used as an indicator of apoptosis in ovine (McGuire *et al.* 1994; Rueda *et al.* 1995; Juengel *et al.* 2000) and bovine (Juengel *et al.* 1993) corpora lutea following treatment with $PGF_2\alpha$.

Progesterone appears to inhibit apoptosis in a number of reproductive tissues in a variety of species. In bovine luteal cells, treatments that reduced progesterone secretion increased oligonucleosome formation (Rueda *et al.* 2000) while treatment with cAMP analogs increased progesterone production and attenuated apoptosis (Tatsukawa *et al.* 2006). Progesterone rescues luteal structure and function in the rat (Goyeneche *et al.*, 2003) by inhibiting DNA fragmentation and apoptosis. Friedman *et al.* (2000) demonstrated that progesterone protected luteal endothelial cells from TNFα-induced apoptosis. In the endometrium, decreased progesterone concentration is followed by apoptosis in epithelial cells (Rotello *et al.* 1992). Pecci *et al.* (1997) demonstrated that addition of anti-progestins increased apoptosis in rat endometrial cell lines. Thus, there is considerable evidence that progesterone may prevent apoptosis in the corpus luteum and the endometrium.

Since intracellular concentrations of calcium are thought to be a primary regulator of endonuclease activity and apoptosis, the next series of experiments was performed to evaluate the role of progesterone on the ability of oxytocin and $PGF_2\alpha$ to increase intracellular concentrations of calcium in small and large luteal cells, respectively. The first step in this study was to determine what "physiological" concentrations of progesterone were within the corpus luteum. We collected blood samples from the ovarian veins of nine ewes on day 10 of the estrous cycle and serum concentrations of progesterone were 0.9 ± 0.1 μg/ml. We also collected six blood samples following puncture of the surface of the corpus luteum with a needle and serum concentrations of progesterone were $0.9 \pm 0.3 \mu g/ml$. It was clear that intraluteal concentrations of progesterone were at least 1 $\mu g/ml$ and likely much higher than that in the extracellular fluid surrounding the luteal cells which produce this hormone. Luteal content of progesterone during the mid-luteal phase ranges from 10-30 $\mu g/g$ tissue (Silvia, 1985) and there is no evidence for storage of this hormone. Therefore, concentrations 1 to 30 $\mu g/ml$ of progesterone were used for the following experiments.

In the first series of experiments 50,000 small luteal cells per plate were incubated overnight with media containing 0.1% DMSO or 1, 10 or 30 μg of progesterone, estradiol, testosterone or cortisol per ml. Steroids were dissolved in DMSO and 0.1% was added in all treatments. Intracellular concentrations of calcium were quantified as described above after addition of 10 μM oxytocin.

Both 1 and 10 $\mu g/ml$ progesterone tended to decrease the number of small luteal cells responding to oxytocin while 30 $\mu g/ml$ essentially abolished the response to oxytocin (Fig. 6). The effect appeared to be specific for progesterone since similar high quantities of estradiol, testosterone or cortisol did not have a consistent inhibitory effect.

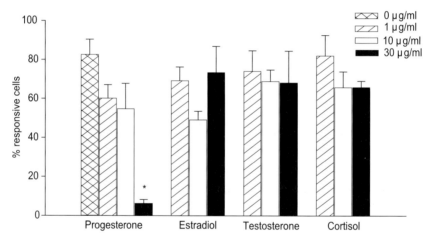

Fig. 6. Effect of steroid hormones (0, 1, 10 or 30 $\mu g/ml$) on the percent of cells responding with an increase in intracellular concentrations of calcium following oxytocin treatment (10 μM) in individual small luteal cells. Thirty $\mu g/ml$ progesterone suppressed the response to oxytocin ($P < 0.0001$). Data are from 3 replicates.

The suppressive effect of progesterone on the ability of oxytocin to increase intracellular calcium concentrations is likely a direct, non-genomic action of progesterone on the oxytocin receptor. Recently it has been reported that progesterone competes with oxytocin for oxytocin receptors in the uterine endometrial plasma membrane (Bogacki et al. 2002; Dunlap & Stormshak 2004; Duras et al. 2005; Bishop & Stormshak 2006) and in CHO cells transfected with oxytocin receptor (Grazzini et al. 1998). Binding of oxytocin to its receptor in endometrial cells stimulates hydrolysis of phospholipase C to phosphoinositides (Mirando et al. 1990). Oxytocin stimulated IP3 production is attenuated in ovine uterine explants exposed to progesterone (Bishop & Stormshak 2006). Progesterone is able to suppress oxytocin-stimulated $PGF_2\alpha$ secretion from both ovine (Bishop & Stormshak 2006) and bovine endometrium (Duras et al. 2005). Burger et al. (1999) reported that large concentrations of progesterone similar to those used in the present studies abolished calcium signaling stimulated by ligand activated G-protein coupled receptors (GPCR). In contrast to the suggestion that progesterone directly interacts with the oxytocin receptor, these authors suggest that microgram concentrations of progesterone alter the cholesterol content of caveolae where many GPCR reside and regulate their ability to increase calcium concentrations. Progesterone also attenuates oxytocin-induced increases in concentrations of intracellular calcium in bovine endometrial cells (Bogacki *et al.* 2002). Burger *et al.* (1999) suggested that progesterone may act at plasma membrane calcium channels or the membrane of the endoplasmic reticulum to alter calcium signaling. The recent publication by Ashley *et al.* (2006) who identified a membrane receptor for progesterone in sheep corpora lutea, that appears to be localized to the membranes of the endoplasmic reticulum, makes the role of progesterone in luteal function even more confusing.

In large luteal cells the ability of progesterone to inhibit $PGF_2\alpha$ induced increases in intraluteal concentrations of calcium was not so apparent (Fig. 7). There was no significant decrease in the number of cells responding to $PGF_2\alpha$ compared to controls. However, in six replicates of this experiment it was clear that the response was highly variable. In three replicates there was no response to progesterone while in the other three progesterone at 30 μg/ml reduced (P<0.05) the number of large luteal cells which responded to $PGF_2\alpha$ treatment by over 50% (Fig. 8).

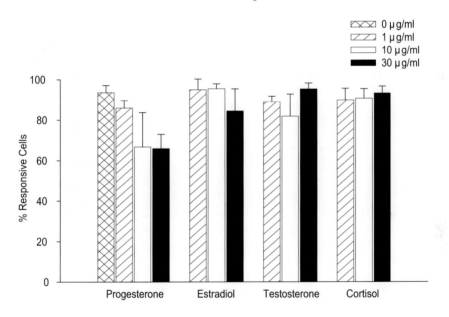

Fig. 7. Effect of steroid hormones (0, 1, 10 or 30 μg/ml) on the number of large luteal cells responding with increased intracellular concentrations of calcium following $PGF_2\alpha$ treatment (1 μM). There was no significant effect of any steroid hormone. Data are from 6 replicates.

It may not be surprising that large luteal cells were not as responsive to progesterone as small luteal cells. Large cells secrete high levels of progesterone at all times and thus, physiological levels at the cell membrane may always be quite high. In fact, it would not be surprising if the levels of progesterone in extracellular fluid surrounding large luteal cells was much higher than the maximum dose (30 μg/ml) used for these studies. We attempted to address this issue in two ways. First, we removed the media from large cells to reduce progesterone content and then incubated them for only 4 hr in the presence of the differing levels of progesterone. Results following these procedures were not consistent. We also attempted to use RU486, a

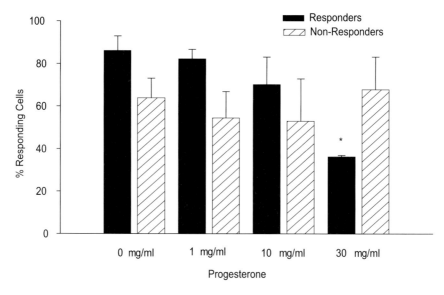

Fig. 8.   Effects of 1 μM PGF$_2\alpha$ on the percent of large cells responding with an increase in intracellular calcium concentrations in the presence of various concentrations of progesterone. Data are shown as replicates in which cells responded (n = 3) or did not respond (n = 3) to progesterone (30 μg/ml) treatment. *Percent responding cells with 30 μg/ml progesterone is statistically different (P < 0.05) than % responding with 0 μg/ml progesterone.

known inhibitor of progesterone's genomic actions but this analog had no effect. This was not surprising since the actions of progesterone on the response to oxytocin and PGF$_2\alpha$ are likely mediated by a membrane receptor for this steroid hormone. It would seem to be unusual that any steroid hormone receptor would require such high concentrations of ligand to act through any currently accepted model. However, the data described above clearly suggest that progesterone secretion and therefore concentrations within the corpus luteum must decrease before oxytocin and likely PGF$_2\alpha$ can induce the increases in intracellular calcium concentrations that are associated with apoptotic cell death.

It is clear that cytokines and the immune system also play a role in the structural demise of the corpus luteum (Pate & Keyes 2001; Davis & Rueda 2002). However, it was outside the scope of this review to address the ever increasing information regarding the role of the immune system in regulating the functions of the corpus luteum.

Conclusions

Our current model of the mechanisms controlling luteolysis in ewes (Fig. 9) is that pulses of PGF$_2\alpha$ of uterine origin activate PKC in steroidogenic luteal cells (Wiltbank et al. 1991) which reduces the activity of many of the components of the progesterone synthetic pathway (Juengel & Niswender 1999). PGF$_2\alpha$ also induces secretion of oxytocin from large luteal cells which, once intra-luteal levels of progesterone are reduced, allows increased concentrations of calcium and ultimately death of small luteal cells. This is supported by the findings of Braden et al. (1988) who demonstrated that numbers of small luteal cell decreased sooner than numbers of large luteal cells following treatment with PGF$_2\alpha$ in ewes. Uterine PGF$_2\alpha$a also increases the activity of COX-2 and synthesis of PGF$_2\alpha$ from large luteal cells. Once the secretion of progesterone has been decreased, PGF$_2\alpha$ also acts to increase intracellular concentrations of calcium in large luteal cells and induces apoptotic death of this cell type.

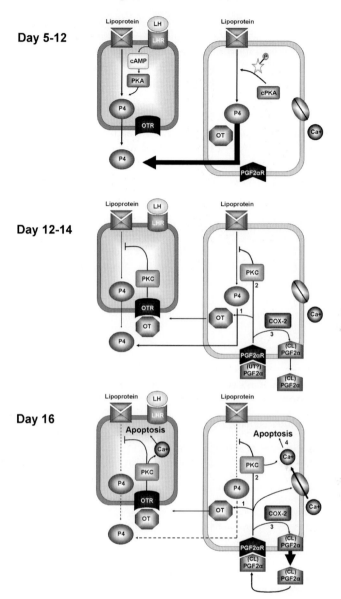

Fig. 9. Current hypothetical model of the intraluteal mechanisms involved in regulation of luteal function in ruminants. During the mid-luteal phase of the estrous cycle (days 5-12) large luteal cells produce the majority of the progesterone, at least in part, due to constitutively active PKA (cPKA) and the increased phosphorylation of steroidogenic acute regulatory protein (StAR). Small luteal cells respond to LH with increased secretion of progesterone. During the early luteolytic phase (days 12-14) large luteal cells respond to small doses of PGF$_2\alpha$ from the uterus with: 1) secretion of oxytocin; 2) activation of PKC which reduces steroidogenesis; and 3) increased activity of cycloxygenase (COX)-2. During this time the high intraluteal levels of progesterone likely prevent the actions of oxytocin on small luteal cells and the ability of PGF$_2\alpha$ to increase calcium in large luteal cells. During the late luteolytic phase of the estrous cycle (day 16), when progesterone secretion has diminished by 80%, oxytocin binds to its receptor in small luteal cells reducing further the secretion of progesterone and increasing intracellular levels of calcium and apoptosis. In large luteal cells secretion of PGF$_2\alpha$ is further increased and this hormone continues to stimulate oxytocin release, activate PKC, and stimulate COX-2 activity. However, the high levels of PGF$_2\alpha$ secreted by large luteal cells also induces increases in intracellular calcium which likely leads to apoptosis and death of this cell type.

We propose that the previous models for luteolysis are incomplete. The data presented herein strongly suggest that the decision regarding ultimate fate of luteal cells is orchestrated by the corpus luteum itself. The initiation of the events is controlled by the uterus but an intricate system of checks and balances allows the corpus luteum to control its own destiny. The large luteal cells preside over this activity responding initially to $PGF_2\alpha$ with an attenuation of progesterone production and an increase in secretion of oxytocin. Together the small and large luteal cells, now no longer protected by production of progesterone, can respond to oxytocin or $PGF_2\alpha$ with increased intracellular concentrations of calcium and apoptosis. During early pregnancy, $PGF_2\alpha$ production by large luteal cell is decreased due to the increased activity of PGDH which is likely important for pregnancy to proceed. Thus, the corpus luteum plays the role of being its own judge, jury, and executioner.

Acknowledgements

Funding for these studies was provided by the Colorado Agricultural Experiment Station and USDA grant 98-35203-6376. Dr. T.L. Davis and R.L. Bogan were supported by NIH Training Grant HD07031. The authors would like to thank Dr. Barbara Sanborn, Dr. Yoon-Sun Kim, Dr. M. Hans Mayan, and Ryan Ashley for their assistance with this research. They also thank Dr. E.K. Inskeep for scientific discussion and input into many of these experiments.

References

Alila HW, Dowd JP, Corradino RA, Harris WV & Hansel W 1988 Control of progesterone production in small and large bovine luteal cells separated by flow cytometry. *Journal of Reproduction and Fertility* **82** 645-655.

Arakane F, King SR, Du Y, Kallen CB, Walsh LP, Watari H, Stocco DM & Strauss JF III 1997 Phosphorylation of steroidogenic acute regulatory protein (StAR) modulates its steroidogenic activity. *Journal of Biological Chemistry* **272** 32656-32662.

Arends MJ, Morris RG & Wyllie AH 1990 Apoptosis – the role of endonuclease. *American Journal of Pathology* **136** 59231-608.

Ashley RL, Clay CM, Farmerie TA, Niswender GD & Nett TM 2006 Cloning and characterization of an intracellular seven transmembrane progesterone receptor that mediates calcium mobilization. *Endocrinology* **147** 4151-4159.

Bazer FW, Thatcher WW, Hansen PJ, Mirando MA, Ott TL & Plante C 1991 Physiological mechanisms of pregnancy recognition in ruminants. *Journal of Reproduction and Fertility Supplement* **43** 39-47.

Bishop CV & Stormshak F 2006 Nongenomic action of progesterone inhibits oxytocin-induced phosphoinositide hydrolysis and prostaglandin $F_{2\alpha}$ secretion in the ovine endometrium. *Endocrinology* **147** 937-942.

Bogacki M, Silvia WJ, Rekawiecki R & Kotwica J 2002 Direct inhibitory effect of progesterone on oxytocin-induced secretion of prostaglandin $F_{2\alpha}$ from bovine endometrial tissue. *Biology of Reproduction* **67** 184-188.

Bogan RL 2006 Regulation of steroidogenic acute regulatory (StAR) protein and evidence that constitutive steroidogenesis in ovine large luteal cells is mediated by tonically active protein kinase A. PhD Dissertation, Colorado State University, Fort Collins, CO.

Braden TD, Gamboni F & Niswender GD 1988 Effects of prostaglandin F2a-induced luteolysis on the populations of cells in the ovine corpus luteum. *Biology of Reproduction* **39** 245-253.

Burger K, Fahrenholz F & Gimpl G 1999 Non-genomic effects of progesterone on the signaling function of G protein-coupled receptors. *FEBS Letters* **24** 25-29.

Christensen AK & Gillim SW 1969 The correlation of fine structure and function in steroid-secreting cells, with emphasis on those in the gonads, pp 415-488. Ed KW McKerns. New York: Appleton-Century-Crofts.

Corner GW 1919 On the origin of the corpus luteum of the sow from both granulose and theca interna. *American Journal of Anatomy* **26** 118-183.

Davis JS & Rueda BR 2002 The corpus luteum: an ovarian structure with maternal instincts and suicidal tendencies. *Frontiers in Bioscience* **7** 1949-1978.

Deane HW, Hay MF, Moor RM, Rowson LEA & Short RV 1966 The corpus luteum of the sheep: Relationships between morphology and function during the oestrous cycle. *Acta Endocrinologica* **51** 245-263.

Diaz FJ, Anderson LE, Wu YL, Rabot A, Tsai SJ & Wiltbank MC 2002 Regulation of progesterone and prostaglanding $F_{2\alpha}$ production in the CL. *Molecular and Cellular Endocrinology* **191** 65-80.

Donaldson L & Hansel W 1965 Histological study of bovine corpora lutea. *Journal of Dairy Science* **48** 905-909.

Dunlap KA & Stormshak F 2004 Nongenomic inhibition of oxytocin binding by progesterone in the ovine uterus. *Biology of Reproduction* **70** 65-69.

Duras M, Mlynarczuk J & Kotwica J 2005 Non-genomic effect of steroids on oxytocin-stimulated intracellular mobilization of calcium and on prostaglandin F2a and E2 secretion from bovine endometrial cells. *Prostaglandins and Other Lipid Mediators* **76** 105-116.

Enders AC 1973 Cytology of the corpus luteum. *Biology of Reproduction* **8** 158-182.

Farin CE, Moeller CL, Sawyer HR, Gamboni F & Niswender GD 1986 Morphometric analysis of cell types in the ovine corpus luteum throughout the estrous cycle. *Biology of Reproduction* **35** 1299-1308.

Fitz TA, Mayan MH, Sawyer HR & Niswender GD 1982 Characterization of two steroidogenic cell types in the ovine corpus luteum. *Biology of Reproduction* **27** 703-711.

Friedman A, Weiss S, Levy N & Meidan R 2000 Role of tumor necrosis factor-a and its type 1 receptor in luteal regression: induction of programmed cell death in bovine corpus luteum-derived endothelial cells. *Biology of Reproduction* **63** 1905-1912.

Goyeneche AA, Deis RP, Gibori G & Telleria CM 2003 Progesterone promotes survival of the rat corpus luteum in the absence of cognate receptors. *Biology of Reproduction* **68** 151-158.

Grazzini E, Guillon G, Mouillac B & Zingg HH 1998 Inhibition of oxytocin receptor function by direct binding of progesterone. *Nature* **392** 509-512.

Griffith RJ 2002 Intraluteal prostaglandin $F_2\alpha$ production during normal and prostaglandin $F_2\alpha$- induced luteolysis in sheep. MS Thesis, Colorado State University, Fort Collins, CO.

Guthrie HD, Rexroad CE Jr, & Bolt DJ 1978 In vitro synthesis of progesterone and prostaglandin F by luteal tissue and prostaglandin F by endometrial tissue from the pig. *Prostaglandins* **16** 433-440.

Harrison LM, Kenny N & Niswender GD 1987 Progesterone production, LH receptors, and progesterone secretion by ovine luteal cell types on days 6, 10 and 15 of the oestrous cycle and day 25 of pregnancy. *Journal of Reproduction and Fertility* **79** 539-548.

Hoyer PB & Niswender GD 1986 Adenosine 3',5'-cyclic monophosphate-binding capacity in small and large ovine luteal cells. *Endocrinology* **119** 1822-1829.

Jaroszewski JJ & Hansel W 2000 Intraluteal administration of a nitric oxide synthase blocker stimulates progesterone and oxytocin secretion and prolongs the life span of the bovine corpus luteum. *Proceedings of the Society of Experimental Biology and Medicine* **224** 50-55.

Juengel JL, Garverick HA, Johnson AL, Youngquist RS & Smith MF 1993 Apoptosis during luteal regression in cattle. *Endocrinology* **132** 249-254.

Juengel JL & Niswender GD 1999 Molecular regulation of luteal progesterone synthesis in domestic ruminants. In *Reproduction in Domestic Ruminants IV*, suppl. 54, pp.193-205. Eds: WW Thatcher, EK Inskeep, GD Niswender & C Doberska. Cambridge, UK: Journal of Reproduction and Fertility Ltd.

Juengel JL, Haworth JD, Rollyson MK, Silva PJ, Sawyer HR & Niswender GD 2000 Effect of dose of prostaglandin $F_{2\alpha}$ on steroidogenic components and oligonucleosomes in ovine luteal tissue. *Biology of Reproduction* **62** 1047-1051.

Koos RD & Hansel W 1981 The large and small cells of the bovine corpus luteum: Ultrastructural and functional differences. In *Dynamics of Ovarian Function*, pp 197-203. Eds NB Schwartz and M. Hunzicker-Dunn. New York: Raven Press.

Korzekwa AJ, Okuda K, Woclawek-Potocka I, Murakami S & Skarzynski DJ 2006 Nitric oxide induces apoptosis in bovine luteal cells. *The Journal of Reproduction and Development* **52** 353-61.

Lemon M & Loir M 1977 Steroid release in vitro by two luteal cell types in the corpus luteum of the pregnant sow. *Journal of Endocrinology* **72** 351-359.

Lodish H, Berk A, Zipursky SL, Matsdaira P, Baltimore D & Darnell J 2000 Cell death and its regulation, pp. 1044-1050 In *Molecular Cell Biology*. New York: WH Freeman & Co.

McGuire WJ, Juengel JL & Niswender GD 1994 Protein kinase C second messenger system mediates the antisteroidogenic effects of progesterone $F_{2\alpha}$ in the ovine corpus luteum in vivo. *Biology of Reproduction* **51** 800-806.

Mossman HW & Duke KL 1973 Some comparative aspects of the mammalian ovary. In *Handbook of Physiology*, section 7, volume 2, part 1, pp 389-402. Ed SR Geiger. Bethesda, MD: American Physiological Society.

Mirando MA, Ott TL, Vallet JL, Davis M & Bazer FW 1990 Oxytocin-stimulated inositol phosphate turnover in endometrium of ewes is influenced by stage of the estrous cycle, pregnancy, and intrauterine infusion of ovine conceptus secretory proteins. *Biology of Reproduction* **42** 98-105.

Niswender GD 1973 Influence of the site of conjugation on the specificity of antibodies to progesterone. *Steroids* **22** 413-424.

Niswender GD 2002 Molecular control of luteal secretion of progesterone. *Reproduction* **123** 333-339.

Niswender GD, Schwall RH, Fitz TA, Farin C & Sawyer HR 1985 Regulation of luteal function in domestic ruminants: new concepts. In *Recent Progress in Hormone Research*, vol 4, pp 101-142. Ed RO Greep. New York: Academic Press.

Olofsson J, Norjavaara E & Selstam G 1992. Synthesis of prostaglandin F2, E2 and prostacyclin in isolated corpora lutea of adult pseudopregnant rats through the luteal lifespan. *Prostaglandins, Leukotrienes, and Essential Fatty Acids* **46** 151.

O'Shea JD, Cran DG & Hay MF 1979 The small luteal cell of the sheep. *Journal of Anatomy* **128** 239-251.

Pate JL 1988 Regulation of prostaglandin synthesis by progesterone in the bovine corpus luteum. *Prostaglandins* **36** 303-315.

Pate JL & Keyes PL 2001 Immune cells in the corpus luteum: friends or foes? *Reproduction* **122** 665-676.

Patwardhan VV & Lanthier A 1980 Concentration of prostaglandins PGE and PGF, estrone, estradiol, and progesterone in the human corpora lutea. *Prostaglandins* **20** 963-969.

Pecci A, Scholz A, Pelster D & Beato M 1997 Progestins prevent apoptosis in a rat endometrial cell line and increase the ratio of bcl-XL to bcl-XS. *Journal of Biological Chemistry* **272** 11791-11798.

Rexroad CE Jr & Guthrie HD 1979 Prostaglandin F2 and progesterone release in vitro by ovine luteal tissue during induced luteolysis. *Advances in Experimental Medicine and Biology* **112** 639-644.

Rodgers RJ & O'Shea JD 1982 Purification, morphology, and progesterone production and content of three cell types isolated from the corpus luteum. *Australian Journal of Biological Sciences* **35** 441-455.

Rotello RJ, Lieberman RC, Lepoff RB & Gerschenson LE 1992 Characterization of uterine epithelium apoptotic cell death kinetics and regulation by progesterone and RU-486. *American Journal of Pathology* **140** 449-456.

Rueda BR, Wegner JA, Marion SL, Wahlen DD & Hoyer PB 1995 Internucleosomal DNA fragmentation in ovine luteal tissue associated with luteolysis: in vivo and in vitro analyses. *Biology of Reproduction* **52** 305-312.

Rueda BR, Hendry IR, Hendry WJ III, Stormshak F, Slayden OD & Davis JS 2000 Decreased progesterone levels and progesterone receptor antagonists promote apoptotic cell death in bovine luteal cells. *Biology of Reproduction* **62** 269-276.

Shlykov SG & Sanborn BM 2004 Stimulation of intracellular Ca2 + oscillations by diacylglycerol in human myometrial cells. *Cell Calcium* **36** 157-164.

Shutt DA, Clarke AH, Fraser IS, Goh P, McMahon GR, Saunders DM & Shearman RP 1976 Changes of concentration of PGF2 and steroids in human corpora lutea in relation to growth of the corpus luteum and luteolysis. *Journal of Endocrinology* **73** 453-454.

Silva PJ, Juengel JL, Rollyson MK & Niswender GD 2000 Prostaglandin metabolism in the ovine corpus luteum: catabolism of prostaglandin F2α (PGF2α) coincides with resistance of the corpus luteum to PGF2α. *Biology of Reproduction* **63** 1229-1236.

Silvia WJ 1985 Maintenance of the corpus luteum of early pregnancy in the ewe. PhD Dissertation, Colorado State University, Fort Collins, CO.

Silvia WJ, Lewis GS, McCracken JA, Thatcher WW & Wilson L Jr, 1991 Hormonal regulation of secretion of prostaglandin F2 during luteolysis in ruminants. *Biology of Reproduction* **45** 655-663.

Swanston IA, McNatty KP & Baird DT 1977 Concentration of prostaglandin F2alpha and steroids in the human corpus luteum. *Journal of Endocrinology* **73** 115-122.

Tatsukawa Y, Bowolaksono A, Nishimura R, Komiyama J, Acosta TJ & Okuda K 2006 Possible Roles of Intracellular Cyclic AMP, Protein Kinase C and Calcium Ion in the Apoptotic Signaling Pathway in Bovine Luteal Cells. *The Journal of Reproduction and Development* **52** 517-522.

Tsai SJ & Wiltbank MC 1997 Prostaglandin F2 induces expression of prostaglandin F/H synthase-2 in the ovine corpus luteum: a potential feedback loop during luteolysis. *Biology of Reproduction* **57** 1016-1022.

Ursely J & Leymarie P 1979 Varying response to luteinizing hormone of two luteal cell types isolated from bovine corpus luteum. *Endocrinology* **83** 303-310.

Vinatier D, Dufour P & Subtil D 1996 Apoptosis: a programmed cell death involved in ovarian and uterine physiology. *European Journal of Obstetrics, Gynecology and Reproductive Biology* **67** 85-102.

Warbritton V 1934 The cytology of the corpora lutea of the ewe. *Journal of Morphology* **56** 181-202.

Wegner JA, Martinez-Zaguilan R, Wise ME, Gillies RJ & Hoyer PB 1990 Prostaglandin F2a-induced calcium transient in ovine large luteal cells: I. Alterations in cytosolic-free calcium levels and calcium flux. *Endocrinology* **127** 3029-3037.

Wiltbank MC, Guthrie PB, Mattson, MP, Kater SB & Niswender GD 1989 Hormonal regulation of free intracellular calcium concentrations in small and large ovine luteal cells. Biology of Reproduction **41** 771-778

Wiltbank MC, Diskin MG & Niswender GD 1991. Differential actions of second messenger systems in the corpus luteum. In *Reproduction in Domestic Ruminants*, Vol. 3, pp 65-75. Ed BJ Weir. Cambridge, UK: Journal of Reproduction and Fertility Ltd.

Yi FX, Magness RR & Bird IM 2004 Simultaneous imaging of [Ca2 +]i and intracellular NO production in freshly isolated uterine artery endothelial cells: effects of ovarian cycle and pregnancy. *American Journal of Physiology. Regulative, Integrative and Comparative Physiology* **288** R140-148.

Socio-sexual signalling and gonadal function: Opportunities for reproductive management in domestic ruminants

R Ungerfeld

Departamento de Fisiología, Facultad de Veterinaria, Lasplaces 1550, Montevideo 11600, Uruguay

The aims of this review are to summarize the common biological basis of the responses to social stimulus in domestic ruminants and to consider the research still required in order to put this knowledge to practical use on the farm. The mechanisms involved in the stimulation of sheep and goat females, including both the expected ovarian and behavioural responses, are described. In most breeds, the male effect may be used effectively to induce ovulation during seasonal anoestrus. Although good responses have been obtained in most sheep trials, in some experiments using more seasonal breeds of sheep, poor responses were observed. In goats, it seems that this can be partially overcome if teaser bucks are adequately stimulated (by light treatment and melatonin administration). The strategic use of these stimuli to induce fertile ovulations during the postpartum period is also discussed. In cattle, less is known about the physiological mechanisms by which cows respond to stimulation from the bull. Most trials have focused in trying to advance postpartum rebreeding, with very diverse outcomes. The wide variety of interacting factors and the paucity of data make it difficult to draw conclusions regarding the use of social stimuli in postpartum management. The challenge for researchers is to develop social management techniques that will induce oestrus and ovulation whenever farmers require them. Although more research is necessary to improve efficacy in some sheep breeds and in postpartum animals, social stimulation emerges as an inexpensive and hormone-free strategy that may be useful for farmers.

Introduction

Reproduction is a consequence of endogenous neuroendocrine regulating mechanisms and external factors that interact with them. Environmental conditions interact with the endogenous system by either stimulating or inhibiting physiological mechanisms and many of these mechanisms are related to the reproductive axis. In ruminants, social cues may act to either stimulate or inhibit reproductive activity. Social hierarchies or suckling are examples of inhibitory cues, which may be considered in management practice. On the other hand, there is considerable evidence that males may stimulate oestrus and ovulation in anovulatory females: the so called "male effect". Research on the use of the male effect has been increasing recently (Ungerfeld

---

Corresponding author E-mail: piub@internet.com.uy

2005), probably as a consequence of being described as a "clean, green, and ethical" practice (Martin *et al.* 2004).

The effects of social stimuli on reproductive physiology have been widely demonstrated in rodents and several farm species. In ruminants, several reproductive responses may be obtained in females following male stimulation (as in the male effect: Table 1). Although there is only preliminary information, male stimuli on female reproductive physiology may also exist in caribou, musk deer and camels (Green 1987; Claus *et al.* 1999; Adams *et al.* 2001).

Table 1. Ruminant species in which the male effect has been demonstrated and the main effects on the females.

Species	Effect	Reference
Sheep	Induction of oestrus in seasonal anoestrus	Underwood *et al.* 1944
	Shortening of postpartum in ewes	Geytenbeek *et al.* 1984
	Advancement of puberty in lambs	O'Riordan & Hanrahan 1989
Goat	Induction of oestrus in seasonal anoestrus	Chemineau 1985
	Advancement of puberty in goats	Mellado *et al.* 2000
	Synchronization of puberty onset	Amoah & Bryant 1984
Cattle	Advancement of postpartum rebreeding	Zalesky *et al.* 1984
	Advancement of puberty	Rekwot et al. 2000
Red deer	Advancement of the breeding season	Moore & Cowie 1986
	Advancement of puberty	Fisher *et al.* 1995
Eld's deer	Advancement of oestrus and the LH peak	Hosack *et al.* 1999
Fallow deer	Advancement of the breeding season	Komers *et al.* 1999
Reindeer	Advancement of onset of breeding season	Shipka *et al.* 2002
Moose	Synchronization of the breeding season	Whittle *et al.* 2000
	Induction of ovulation	Miquelle 1991
Antelope	Modification of oestrous cycle duration	Skinner *et al.* 2002
Oryx	Advancement of puberty	Blanvillain *et al.* 1997
Impala	Advancement of the breeding season	Skinner *et al.* 1992
Blesbok	Increase in length of breeding season	Skinner *et al.* 1992

The female reproductive responses to the male effect have been reviewed before (Walkden-Brown *et al.* 1999; Delgadillo *et al.* 2004; Ungerfeld *et al.* 2004a; Delgadillo *et al.* 2006). In the present review, information regarding the female reproductive responses to male stimulation of sheep and goats during seasonal anoestrus, and sheep, goats and cattle during the postpartum period are compared and summarized and areas in which further research is needed are proposed. In addition new previously unpublished information on sheep and cattle is included.

Male effect: evolutionary advantages

What is the evolutionary origin of the male effect? Seasonal patterns of reproduction should be a consequence of selective processes related to best environmental conditions for parturition and nursing. To measure the appropriate time for conception, animals would then be forced to use different environmental cues such as photoperiod, temperature, rainfall and nutrition. However, social cues may also trigger the onset of the breeding season in wild and feral ruminants. Domestic animals display some differences in their reproductive physiology compared with their wild ancestors. Wild or feral cattle and swine have short breeding seasons compared to their domesticated counterparts, which have been selected to breed throughout the year (Mauget 1981; Rowlands & Weir 1984; Reinhardt *et al.* 1986). The reproductive pattern of sheep is a bit different: wild sheep have a short breeding season and even the most developed breeds retain a seasonal reproductive pattern.

The high degree of breeding synchrony observed in wild and feral female ruminants may be at least partially a consequence of male introduction and other social interactions. The social structure throughout the year is similar in wild and feral sheep breeds (such as, Soay, Rocky Mountain Bighorn [*Ovis canadensis canadensis*], Punjab Urial [*Ovis orientalis* ssp. *punjabiensis*] and Mouflon), goats and in farmed breeds (Geist 1965; 1971; Grubb & Jewell 1973; Schaller & Mirza 1974; Knight *et al.* 1998; McClelland 1991). Outside the rutting period, social groups are comprised of several females with their offspring while males live in small same-sex groups (Stricklin & Mench 1987). When male offspring become mature they disperse from the female group (Shackleton & Schank 1984). Wethers (castrated male sheep) remain together or with females (Jewell 1997), suggesting that the testis (probably through androgens) is involved in segregation. As the time of breeding approaches, males join female groups. Nudging, block-ing, rubbing and aggressive behaviours (Jewell 1976; Lincoln & Davidson 1977) by males begins before females come into oestrus, probably as a consequence of the earlier activation of the male reproductive system (including, increased pulsatile LH secretion and increased circu-lating concentrations of both FSH and testosterone: Lincoln & Short 1980). In sheep, natural joining may trigger, through the ram effect, an earlier onset of the breeding season.

The rut period causes an increase in the energy expenditure of males (Jewell 1997). So what is the significance of beginning the reproductive season before females are cyclic? Moreover, what is the sense of having a mechanism where males trigger the female reproductive system? The mechanism may be especially important in breeds that display a very short breeding sea-son (for example, Soay have 1 to 3 oestrous cycles: Grubb & Jewell 1973). Moreover, late conception in Rocky Mountain Bighorn sheep during the breeding season increases lamb and ewe mortality (Hogg *et al.* 1992). The male stimuli may also promote an advancement of puberty in females, which may increase their reproductive success throughout their lifetime (Bérubé *et al.* 1999). The period from joining of males and females until the peak of oestrus may also be useful for males to sort out hierarchical ranks. Oestrous synchronization allows different males to mate different females thereby decreasing the risks of inbreeding and the reduction of genotype variation that would result. Komers *et al.* (1999) working with fallow deer postulated that the onset of females' cyclic activity in response to males is related to the "quality" of males. If so, females may be able to detect the better quality stimulus emitted by individual males and, if advantages may be obtained, advance their cyclic activity. Overall, the available information suggests that there is an evolutionary mechanism underlying the reproductive response of domestic ewes to the introduction of rams. Understanding this mecha-nism may allow us to manipulate the breeding activity of farmed breeds.

Considering all this information together, we speculate that most ruminant females would respond to the introduction of males at least 45 to 60 days before the onset of the breeding season. However, the challenge for researchers is to develop social managements to induce ovulation and oestrus whenever it is desired, and not only close to the breeding season onset, as happens in wild ruminants.

Sheep and goats: the basis of anoestrus and endocrine changes induced by males

Seasonal anoestrus is associated with reduced LH pulsatility compared to the breeding season and with an absence of preovulatory surges of FSH and LH. Low LH pulsatility is due to two inhibitory mechanisms: (1) an increased negative feedback effect of oestradiol in the hypo-thalamus; and (2) a direct effect of photoperiod on the hypothalamo-hypophyseal system con-trolling LH secretion.

The introduction of males induces an increase in the pulsatile secretion of LH in ewes and does (reviewed by Walkden-Brown et al. 1999), which may end with a LH surge followed by ovulation. In ewes, it has been demonstrated that after the introduction of the rams the negative oestradiol-LH feedback switches to a positive, stimulatory feedback but there is also a stimulus independent from oestradiol, acting directly at the hypothalamus-pituitary level (Martin et al. 1983). This response ends with a LH surge, similar to that observed during a follicular phase in cyclic ewes. A silent ovulation occurs, without signs of oestrus. When the corpus luteum (or corpora lutea) regresses after the first ovulation, another ovulation accompanied by oestrus occurs 17 to 19 days after the introduction of the rams. However, in some ewes, the corpus luteum regresses after 4 to 5 days and another silent ovulation takes place. After this ovulation, a corpus luteum of normal life span is formed, followed by oestrus 21 to 25 days after joining ewes and rams. Recently, we observed ultrasonographically other ovarian responses, like delayed ovulations occurring 5 to 7 days after the introduction of rams followed by normal or short luteal phases, or luteinized follicles (Ungerfeld et al. 2002; 2004b).

In goats, the ovarian and behavioural responses differ. An initial ovulation associated with oestrus is observed 2 to 3 days after the introduction of bucks, which is followed in most goats by a short ovarian cycle of 5 to 7 days. Afterwards a second ovulation which is also associated with oestrus occurs and is followed by a normal luteal phase (Delgadillo et al. 2004; 2006). Fig. 1A and 1B illustrates the more common patterns of response in sheep and goats, respectively. In Fig. 1C, a synthesis of the expected periods in which oestrus may be expected to occur in sheep flocks and goat herds is shown.

If ewes or does are primed with progestogens and then stimulated by teasers, all the first corpora lutea will have normal function. Although some attempts have been made to determine the physiological mechanisms producing each response pattern (normal versus short luteal phase) little has been elucidated. Studies determining the relationship between the ovarian response pattern with the follicle status present before the introduction of rams (Ungerfeld et al. 2004b), the follicular development after the introduction of rams (Pearce et al. 1987), the uterine effect on early luteolysis (Lassoued et al. 1997) and the possible existence of spontaneous short luteal phases produced by luteinised follicles or short-lived corpora lutea during the anoestrous season (Ferreria et al. 2005) have all failed to provide a definitive answer. A recently published review provides a working hypothesis of the physiological mechanisms involved (Chemineau et al. 2006).

Apparently in Merino sheep, in response to the introduction of the male a normal oestrous cycle will be induced in approximately 50% of the females while the other 50% will have first cycles with short luteal phases (Martin et al. 1986). However, in several experiments with Corriedale ewes we observed very few ewes in oestrus during the first expected period (days 17-19: Ungerfeld et al. 2003; Ungerfeld & Rubianes 2004). Some factors related to the percentage of ewes that initially respond with normal or short luteal phases are summarized in Table 2. Oldham et al. (1985: see Table 2) in sheep and Luna-Orozco et al. (2005) in goats, observed a higher incidence of short luteal phases in adult compared to young females. However, in sheep this has not been confirmed by other researchers (Thimonier et al. 2000). In agreement with Khaldi (1984) and Thimonier et al. (2000), we observed that Corriedale ewes in poor body condition as a consequence of postpartum status, showed delayed oestrous behaviour compared to control animals (25.8 ± 0.3 versus 24.6 ± 0.4 days after the introduction of the rams, P=0.03; R Ungerfeld, unpublished data), suggesting that first ovulations were delayed or that a higher percentage of ewes responded initially with short luteal phases.

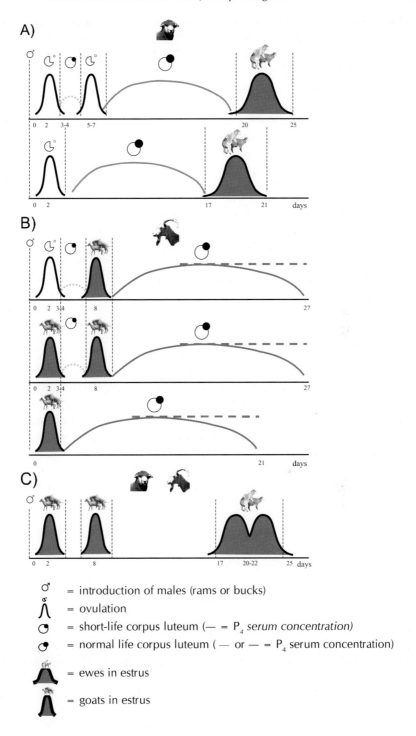

Fig. 1. Ovarian and behavioural response of A) ewes and B) does to the introduction of males. C) Synthetic scheme of the distribution of oestrus after the introduction of males in doe herds and ewe flocks.

Table 2. Factors that influence the length of the first luteal phase after the introduction of rams. NLP = normal luteal phase; SLP = short lutel phase.

Factor	Effect	Reference
Related to the stimuli		
High ram's sexual behaviour	Decreases the incidence of SLP	Perkins & Fitzgerald, 1994
Related to female status		
Percentage of cyclic ewes	Decreases the incidence of SLP	Lassoued, 1998
Underfeed ewes	Increases the incidence of SLP	Khaldi & Lassoued, 1991
Increase of nutritional level	Decreases the incidence of SLP	Khaldi 1984; Thimonier et al. 2000
Postpartum interval	Longer postpartum interval decreases the incidence of SLP	Thimonier et al. 2000
Parity	Maiden have a higher percentage of NLP than adult	Oldham et al. 1985
Breed	More Dorset than Hampshire ewes seem to have SLP	Nugent III et al. 1988

Response to the male effect during the postpartum period in sheep and goats

The physiology of postpartum anoestrus is influenced by different factors; such as suckling, season, nutritional status and age. Although this period has not been studied as much as seasonal anoestrus, the male effect has been used to induce breeding during the postpartum period in both sheep and goats.

The postpartum interval may be reduced if rams are introduced to ewes that have lambed in autumn or spring (Wright et al. 1989; Silva and Ungerfeld 2006). In autumn-lambing Merino ewes, it was observed that the percentage of ewes ovulating during the first 4 days after the introduction of the rams increases progressively from 21 to 45 days postpartum (Geytenbeek et al. 1984). In contrast to what happens in cattle, in autumn-lambing ewes, parity does not seem to influence the response to the ram effect. We have recently compared the response to the introduction of rams of multiparous and primiparous non-cyclic ewes during the breeding season (30 to 60 days postpartum) primed with medroxy-progesterone acetate for 12 days. Although the percentage of multiparous ewes that ovulated, which was determined by ultrasound scanning, tended to be higher than that of primiparous ewes (92.5 versus 79.3 %, P = 0.08), the percentage that came into oestrus (88.8 versus 75.8 %) was not significantly different (SP González-Pensado, MA Ramos, T de Castro & R Ungerfeld, unpublished results).

In one study performed during the non-breeding season, a similar percentage of postpartum ewes and ewes that had lambed several months before exhibited oestrus after the introduction of rams (Ungerfeld et al. 2004a). However, in agreement with Wright et al. (1990), probably as a consequence of suckling and low body condition, the conception rate was lower in the postpartum ewes. In ewes that had lambed during the non-breeding season Khaldi (1984) observed that the percentage of ewes that ovulated after the introduction of rams was higher at 75 days than at 15, 30, 45 or 60 days after parturition. It is interesting that in several experiments we observed a widespread distribution of oestrus, with ewes coming into oestrus on almost any day until 30 days after the introduction of rams. It remains to be elucidated as to how the range of physiological status between individuals present in a sheep flock affects the pattern of endocrine and ovarian responses to the introduction of rams.

Determinants of the response of sheep and goats to the male effect

In feral or wild populations, when males stimulate anoestrous females early before the onset of the breeding season, males have begun their reproductive recovery at least 40 to 60 days before. Therefore, females are stimulated by active males, which display their own maximum reproductive capacity. However, at the time of the year that may be of productive interest for farmers to induce ovulation and obtain pregnancies in their female flocks/herds, the males may not have recovered their maximum reproductive capacity from the preceding low capacity period. Thus, both the male and the female, and the stimuli derived from interactions between them, may determine the final response in a sheep flock or a goat herd.

Male condition and stimulating capacity

Stimulating signals

The stimulation of anoestrous females by males involves different cues; such as, odour, sound, visual or behavioural signals (reviewed by Walden-Brown *et al.* 1999; Delgadillo *et al.* 2006). The relative importance of each signal has varied in different experiments but this may be a consequence of using females in different states of responsiveness. In both sheep and goats, it has been reported that castrated males treated with androgens may be effective teasers suggesting that the involved cues are at least partially androgen-dependent (Fulkerson *et al.* 1981; Mellado & Hernández 1996).

The scent of wool and wax from intact rams can be enough to trigger a response, in terms of ovulation, in ewes (Knight & Lynch 1980). Similarly, odours from bucks' hair have been proven to stimulate LH secretion in does and ewes, and ovulation in anoestrous does (Over *et al.* 1990; Walkden-Brown *et al.* 1993). Interestingly, odours produced by the goat buck can also stimulate LH pulse frequency and ovulation in anoestrous ewes (Knight *et al.* 1983; Over *et al.* 1990). While Morgan *et al.* (1972) observed that ewes with impaired smell did not respond to rams, a normal LH response was observed in ewes reported without vomeronasal activity (Cohen-Tannoudji *et al.* 1989). In agreement, Gelez *et al.* (2004a) demonstrated that inactivation of the neural projections of the main olfactory bulb (which includes projections from the olfactory epithelium) blocks the endocrine response to male odour, whereas inactivation of the projections from the accessory olfactory bulb (which come from the vomeronasal organ) has no effect. The destruction of the neuroreceptors of the olfactory epithelium blocked the endocrine response to the ram odour (Gelez & Fabre-Nys 2004). A general overview of the neural pathways involved in the ewe's response to male odours has been recently provided by Gelez & Fabre-Nys (2006). They observed that although male odours activate pathways involving the accessory olfactory bulb, its role seems to be limited compared to that of the main olfactory bulb.

The maximum response of anoestrous ewes to the male was obtained after full ram contact compared with either contact through an open or an opaque fence, or the application of masks containing ram's wool (odour) (Pearce & Oldham 1988). Similarly, a higher percentage of does ovulated after full contact with bucks than after odour stimulation alone (Walkden-Brown *et al.* 1993). Moreover, in sheep other authors observed that other sensory signals may completely replace the pheromone stimulus (Cohen-Tannoudji *et al.* 1986; 1989). Perkins & Fitzgerald (1994) demonstrated the importance of the sexual behaviour of rams: a higher number of ewes ovulated after mixing them with high- than with low-serving capacity rams.

Overall, although the main stimulating cues are known, it seems that we are still far from determining the relative importance of each cue and how to manage them, as happens in swine, in which synthetic pheromones are utilized in production management.

Male characteristics and management

Little is known regarding those characteristics of the ram that may improve the percentage of anoestrous ewes that respond to teasing. Some authors observed that Dorset rams are better teasers than Suffolk, Romney, Romney X Finn or Coopworth rams (see Table 1 in review: Ungerfeld et al. 2004a).

At least under subtropical conditions, the reproductive activity of bucks is strongly depressed during the non-breeding season: plasma testosterone concentrations, sexual behaviour assessed by ano-genital sniffing, nudging and mounts, sexual odour and vocalizations are all at a low level (reviewed by Delgadillo et al. 2006). However, treatments with artificial long days and melatonin implants improve the biostimulatory effect of males. Therefore, these authors consider the reproductive condition of the buck as the limiting factor determining the response of anoestrous does to the male effect.

Another strategy to improve the stimulus of the male effect is to include a group of oestrous females when teasers are joined with anoestrous females (Rodríguez-Iglesias et al. 1991). Ewes in oestrus also influence the reproductive activity of rams, mainly by inducing an increase in LH pulse frequency and testosterone concentrations (Yarney & Sanford 1983). The increases may continue for at least 4 days if ewes in oestrous are still present (Ungerfeld & Silva 2004). While the increase in testosterone concentrations is similar in adult or yearling teaser rams, as happens during the developmental period (Sanford et al. 1982), LH concentrations reach higher values in yearling rams (Ungerfeld 2003). A possible explanation is that although concentrations are similar, testosterone may be less effective in yearling than in adult rams, as is reflected by a weak effect on the inhibition of pituitary activity. This may be attributed to an increase of feedback inhibition by androgens on the hypothalamo-pituitary axis during the developmental period (Foster et al. 1978) and/or to the maturation of the negative feedback mechanism (Courot et al. 1975). Therefore, as stimulating signals are related to androgen concentration, we compared the response of ewes to adult or yearling teasers. Significantly more ewes came into oestrus and ovulated after teasing with adult rather than with yearling teasers (MA Ramos & R Ungerfeld, unpublished data: Fig. 2). Moreover, significantly more ewes ovulated after being stimulated with masks containing adult than yearling ram's wool (53.3 versus 29.8%: V Miller, M Cuadro & R Ungerfeld, unpublished data). Therefore, the explanation of the greater effectiveness of adult than of yearling teasers is at least partially explained by quantitative and/or qualitative differences in odour secretion.

The effect of the presence of oestrous ewes is similar if teaser rams had been in contact with oestrous ewes for a short period before being joined with anoestrous ewes (Knight 1985). Rams that have been isolated from ewes and then placed with ewes in oestrus are more effective in stimulating ovulation in anoestrous ewes than are rams that have been in contact with ewes long before the procedure takes place (Knight et al. 1998).

Depth of anoestrus

Several parameters have been used to describe the "depth of anoestrus": such as, length of time until the ewe would spontaneously begin to cycle; arbitrary percentages of how many animals are cyclic; responsiveness to the introduction of rams; ovulation rate; LH pulse frequency; and basal circulating LH concentrations (Ungerfeld 2003). Thomas et al. (1988) proposed that differences between breeds in the depth of anoestrus could be related to differences in the sensitivity of the hypothalamus to both negative feedback by oestradiol and the direct effects of photoperiod. Overall, LH pulsatility has been consistently proposed as one parameter to assess the depth of anoestrus. We also observed that FSH concentrations before the introduction of rams were significantly higher in ewes that subsequently had a luteal phase than in those that do not (Ungerfeld et al. 2004a).

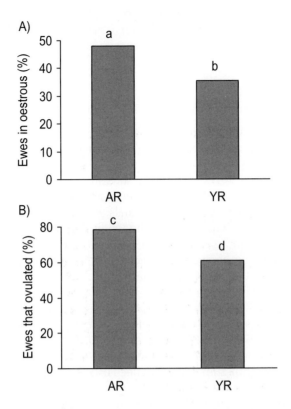

Fig. 2. Percentage of ewes that A) came into oestrus and B) ovulated after a 6-days medroxy-progesterone priming and the introduction of adult rams (AR) or yearling rams (YR) (MA Ramos & R Ungerfeld, unpublished data). The experiment was performed during October (Southern Hemisphere, mid non-breeding season) with Merilin sheep.

The percentage of females that respond to the introduction of males is related to factors such as period of the anoestrous season, parity, previous experience of contact with males and breed (Oldham *et al.* 1985; Nugent III *et al.* 1988; Rodríguez-Iglesias *et al.* 1991; Restall 1992; Gelez *et al.* 2004b; Gelez & Fabre-Nys, 2006). More ewes from a less seasonal breed (for example, Dorset) than from a more seasonal breed (for example, Hampshire) ovulate and conceive after stimulation with rams (Nugent III *et al.* 1988). Ewes from more seasonal breeds may not necessarily respond to the ram effect by ovulating even if they display an increase in LH pulsatility (Minton *et al.* 1991).

Management practices that may improve the effectiveness of stimulation in other breeds may be ineffective in strong seasonal breeds. Three hundred Lincoln ewes, strong seasonal breeders, were stimulated in two consecutive years during January (Southern Hemisphere) joining them with 10 rams from the same breed and twenty ewes in estrus. Forty-eight and 46 ewes were examined laparoscopically 5 days after joining in each year: 2/48 and 0/46 had ovulated (H Irazoqui & RM Rodríguez-Iglesias, personal communication). In another recent experiment using another strong seasonal breed, 46 and 47 Texel ewes were joined during November (Southern Hemisphere) with 6 Texel rams that had been either isolated (n = 3) or kept near oestrous ewes for 3 days before (n = 3), respectively. Fifteen and nine ewes respectively were marked by the rams over the 25 days they were together although none of them became pregnant (MA Ramos, SP González-Pensado & R Ungerfeld, unpublished data). As we could not discriminate the effects of photoperiod on males and females, it remains to be tested in these breeds whether strategies that are effective with bucks may overcome seasonal inhibition (reviewed by Delgadillo *et al.* 2006).

Response to the male effect during the postpartum period in cattle

In cattle, less is known about the physiological mechanisms by which cows respond to stimulation by the bull. Most trials have focused in trying to advance postpartum rebreeding, with very diverse outcomes. The diversity of outcomes and paucity of data makes it difficult to draw conclusions about the possibilities of using social stimuli in postpartum management.

As primiparous beef cattle have an especially long postpartum anoestrous period, it is interesting to determine the potential use of social-signals as stimulators of early rebreeding. The postpartum anoestrous interval in beef cows is shortened after bull introduction (Zalesky *et al.* 1984; Custer *et al.* 1990; Fernández *et al.* 1993; 1996; Rekwot *et al.* 2000; Landaeta-Hernández *et al.* 2004). Similar results have been also obtained with androgen-treated cows, suggesting (as in rams and bucks) that the bull stimulatory effect is also mediated by androgen stimulation (Burns & Spitzer 1992). A female-female effect, has also been described which may result in ovulation (Wright *et al.* 1994). In both the bull and cow effect, it has been demonstrated that odours (probably pheromones) are the main stimulating cues (Izard & Vandenbergh 1982; Anderson *et al.* 2002; Berardinelli & Joshi 2005a). Moreover, it has been recently reported that social interactions between bulls and cows seem not to be necessary to obtain the maximum response (Berardinelli & Joshi 2005a). Bull stimulation has been used to stimulate postpartum rebreeding in zebu and buffalo cattle as well as to advance puberty in beef cattle (Izard & Vandenbergh 1982; Roberson *et al.* 1987; Bolaños *et al.* 1998; Roberson *et al.* 1991; Ingawale & Dhoble 2004).

However, in some experiments cows failed to respond to bull stimulation (Naasz & Miller 1987; Gifford *et al.* 1989). In the few experiments performed with dairy cattle, no positive response has been observed (Shipka & Ellis 1998; 1999). However, in dual purpose cattle, bull teasing advanced rebreeding (Pérez-Hernández *et al.* 2002). There is also some contradictory information on the physiological mechanisms by which cows respond to bull stimulation: while Fernàndez *et al.* (1996) observed higher LH serum concentrations and pulsatility, this was not consistent with the results of Custer *et al.* (1990). If the increase of LH pulsatility is slower than that observed in small ruminants, the increase in LH may have not been detected due to the fact that individual animals may show an increase at different times after bull exposure.

The influence of many factors, such as body condition, nutritional status, season and postpartum interval, on the response to bull stimulation has been reported (Alberio *et al.* 1987; Monje *et al.* 1992; Stumpf *et al.* 1992; Larson *et al.* 1994; Berardinelli & Joshi 2005b). Other factors, like bull familiarity to cows and the bulls' age (in contrast to what we observed with rams), seem not to affect the response (Cupp *et al.* 1993; Berardinelli *et al.* 2005). Considering the previous information, it is possible that when bulls are introduced, the diversity of physiological status present in a herd of postpartum cows, at least under extensive conditions, determines that many of them do not respond to the stimulation. If this happens with a significant percentage of cows, no response may be observed in the herd. Therefore, when cows gradually enter a responsive condition, the acute stimuli of bull introduction may have disappeared. Thus, the greater variation between individuals, compared with ewes or goats during seasonal anoestrus, may explain the lack of consistent and predictable results.

Therefore, one strategy for synchronizing parturition in cattle may be to obtain more homogeneous conditions within the herd. Another strategy may be to prolong the stimuli, allowing cows to respond individually as they develop better physiological conditions. We recently tested this strategy by comparing the time of postpartum rebreeding in two groups of multiparous cows which were extensively managed. At 30-60 days postpartum bulls were introduced to suckling Hereford and Angus cows. In one group, two bulls were changed every week to prolong the stimuli time, while the other group remained continuously with the same two bulls. After 7 weeks 91.1% and 47.8% of the cows from the groups with or without bull rotation, respectively, were cycling (P< 0.001: V Miller, M Rodríguez-Irazoqui & R Ungerfeld, unpulished results).

Overall view

Overall, insights into the mechanisms involved in the natural socio-sexual stimulation of wild or feral ruminants will promote the inclusion of these management techniques in domestic ruminant production. Better comprehension of the involved mechanisms, strategies to strengthen the stimuli according to female condition and the use of different stimulating techniques may open interesting perspectives. The challenge for researchers is to develop standardized treatments with predictable results in strong seasonal breeds of sheep and goats and during the postpartum period in the three species. We are still far from developments such as those used in swine, in which the application of synthetic stimulating odours to induce ovulation in females is commonly used in reproductive management.

Acknowledgements

I gratefully acknowledge: Dr. Edgardo Rubianes and Dr. Mats Forsberg, for their participation in many of the experiments and continuous support; Dr. Andrea Pinczak, Bettina Carbajal, Leticia Silva and María Alejandra Ramos, my collaborators for some of the experiments; and both Milton Pintos and Alejandro Salina, for caring for the animals in many of the experiments. I am especially grateful to Jennifer Brown for her helpful work with language revision. Financial support for the experiments was provided by CSIC (Universidad de la República, Uruguay), CIDEC (Facultad de Veterinaria, Uruguay), Fondo Clemente Estable (Ministerio de Educación y Cultura, Uruguay) and the Department of Clinical Chemistry (Swedish University of Agricultural Sciences, Sweden).

References

Adams CA, Bowyer RT, Rowell JE, Hauer WE & Jenks JA 2001 Scent marking by male caribou: an experimental test of rubbing behavior. *Rangifer* **21** 21-27.

Alberio H, Schiersmann G, Carou N & Mestre J 1987 Effect of a teaser bull on ovarian and behavioural activity of suckling beef cows. *Animal Reproduction Science* **14** 263-272.

Amoah EA & Bryant M 1984 A note on the effect of contact with male goats on occurrence of puberty in female goat kids. *Animal Production* **38** 141-144.

Anderson KA, Berardinelli JG, Joshi PS & Robinson B 2002 Effects of exposure to bull or bull excretory products of bulls on the breeding performance of first-calf restricted suckled beef cows using a modified CO-Synch protocol. In *Realistic solutions for maintaining the sustainability of Montana's livestock industry.* Proceedings of the Montana Nutrition Conference, May 7-8, GranTree Inn, Bozeman, Montana, USA.

Berardinelli JG & Joshi SA 2005a Initiation of postpartum luteal function in primiparous restricted-suckled beef cows exposed to a bull or excretory products of bulls or cows. *Journal of Animal Science* **83** 2495-2500.

Berardinelli JG & Joshi SA 2005b Introduction of bulls at different days postpartum on resumption of ovarian cycling activity in primiparous beef cows. *Journal of Animal Science* **83** 2106-2110.

Berardinelli JG, Joshi SA & Tauck SA 2005 Postpartum resumption of ovarian cycling activity in first-calf suckled beef cows exposed to familiar or unfamiliar bulls. *Animal Reproduction Science* **90** 201-209.

Bérubé CH, Festa-Bianchet M & Jorgenson JT 1999 Individual differences, longevity, and reproductive senescence in Bighorn ewes. *Ecology* **80** 2555-2565.

Blanvillain C, Ancrenaz M, Delhomme A, Greth A & Sempére A 1997 The presence of the male stimulates puberty in captive female Arabian oryx (*Oryx leucoryx* Pallas, 1777). *Journal of Arid Environments* **36** 359-366.

Bolaños JM, Forsberg M, Kindahl H & Rodríguez-Martínez H 1998 Biostimulatory effects of estrous cows and bulls on resumption of ovarian activity in postpartum anestrous zebu (*Bos indicus*) cows in the humid tropics. *Theriogenology* **49** 629-636.

Burns PD & Spitzer JC 1992 Influence of biostimulation on reproduction in postpartum beef cows. *Journal of Animal Science* **70** 358-362.

Chemineau P 1985 Effects of a progestagen on buck-induced short ovarian cycles in the Creole meat goat. *Animal Reproduction Science* **9** 87-94.

Chemineau P, Pellicer-Rubio M-T, Lassoued N, Khaldi G & Monniaux D 2006 Male-induced short oestrous and ovarian cycles in sheep and goats: a working hypothesis. *Reproduction Nutrition Development* **46** 417–429.

Claus R, Kaufmann B, Dehnhard M & Spitzer V 1999 Demonstration of 16-unsaturated C-19 steroids ("boar pheromones") in tissues of the male camel (*Camelus dromedarius*). *Reproduction in Domestic Animals* **34** 455-458.

Cohen-Tannoudji J, Lavenet C, Locatelli A, Tillet Y & Signoret JP 1989 Non-involvement of the accessory olfactory system in the LH response of anoestrous ewes to male odour. *Journal of Reproduction and Fertility* **86** 135-144.

Cohen-Tannoudji J, Locatelli A & Signoret JP 1986 Non-pheromonal stimulation by the male of LH release in the anoestrous ewe. *Physiology & Behavior* **36** 921-924.

Courot M, De Reviers M-M & Pelletier J 1975 Variations in pituitary and blood LH during puberty in the male lamb. Relation to time of birth. *Annales de Biologie Animale Biochimie Biophysique* **15** 509-516.

Cupp AS, Roberson MS, Stumpf TT, Wolfe MW, Werth LA, Kojima N, Kittok RJ & Kinder JE 1993 Yearling bulls shorten the duration of postpartum anestrus in beef cows to the same extent as do mature bulls. *Journal of Animal Science* **71** 306-309.

Custer EE, Berardinelli JB, Short RE, Wehrman M & Adair R 1990 Pospartum interval to estrus and patterns of LH and progesterone in first-calf suckled beef cows exposed to mature bulls. *Journal of Animal Science* **68** 1370-1377.

Delgadillo JA, Fitz-Rodríguez G, Duarte G, Veliz FG, Carrillo E, Flores JA, Vielma J, Hernández H & Malpaux B 2004 Management of photoperiod to control caprine reproduction in the subtropics. *Reproduction Fertility and Development* **16** 471-478.

Delgadillo JA, Flores JA, Veliz FG, Duarte G, Vielma J, Hernández H & Fernández IG. 2006. Importance of the signals provided by the buck for the success of the male effect in goats. *Reproduction Nutrition Development* **46** 391–400.

Fernández DL, Berardinelli JG, Short RE & Adair R 1993 The time required for the presence of bulls to alter the interval from parturition to resumption of ovarian activity and reproductive performance in first-calf suckled beef cows. *Theriogenology* **39** 411-419.

Fernández DL, Berardinelli JB, Short RE & Adair R 1996 Acute and chronic changes in luteinizing hormone secretion and postpartum interval to estrus in first-calf suckled beef cows exposed continuously or intermittently to mature bulls. *Journal of Animal Science* **74** 1098-1103.

Ferrería J, Rodríguez-Iglesias RM, Ciccioli NH, Pevsner DA & Rosas CA 2005 Características del cuerpo lúteo de ovejas Corriedale en anestro tratadas con acetato de medroxiprogestrona (MAP) en diferentes momentos previo a una bioestimulación. In 28 Congreso Argentino de Producción Animal. Bahía Blanca, Buenos Aires, Argentina. *Revista Argentina de Producción Animal* **25**, Supplement 1 117.

Fisher MW, Meikle LM & Johnstone PD 1995 The influence of the stag on pubertal development in the red deer hind. *Animal Science* **60** 503-508.

Foster DL, Mickelson IH, Ryan KD, Coon GA, Drongowski RA & Holt JA 1978 Ontogeny of pulsatile luteinizing hormone and testosterone secretion in male lambs. *Endocrinology* **102** 1137-1146.

Fulkerson WJ, Adams NR & Gherardi PB 1981 Ability of castrated male sheep treated with oestrogen or testosterone to induce and detect oestrus in ewes. *Applied Animal Ethology* **7** 57-66.

Geist V 1965 On the rutting behavior of the mountain goat. *Journal of Mammalogy* **45** 551-568.

Geist V 1971 *Mountain sheep. A study in behaviour and evolution.* University of Chicago Press, Chicago, IL, USA.

Gelez H & Fabre-Nys C 2004 The "male effect" in sheep and goats: a review of the respective roles of the two olfactory systems. *Hormones and Behavior* **46** 257– 271.

Gelez H & Fabre-Nys C 2006 Role of the olfactory systems and importance of learning in the ewes' response to rams or their odours. *Reproduction Nutrition Development* **46** 1-10.

Gelez H, Archer E, Chesneau D, Magallon T & Fabre-Nys C 2004a Inactivation of the olfactory amygdala prevents the endocrine response to male odour in anoestrus ewes. *European Journal of Neuroscience* **19** 1581-1590.

Gelez H, Archer E, Chesneau D, Campan R & Fabre-Nys C 2004b Importance of learning in the response of ewes to male odor. *Chemical Senses* **29** 555–563.

Geytenbeek PE, Oldham CM & Gray SJ 1984. The induction of ovulation in the post-partum ewe. *Proceedings of the Australian Society of Animal Production* **15** 353–356.

Gifford DR, D'Occhio MJ, Sharpe PH, Weatherly T & Pittar RY 1989 Return to cyclic ovarian activity following parturition in mature cows and first-calf heifers exposed to bulls. *Animal Reproduction Science* **19** 209-212.

Green MJB 1987 Scent-marking in the Himalayan musk deer (*Moschus chrysogaster*). *Journal of Zoology, London* (B) **1** 721-737.

Grubb P & Jewell PA 1973 The rut and the occurrence of oestrus in the Soay sheep on St Kilda. *Journal of Reproduction and Fertility* (Supplement) **19** 491-502.

Hogg JT, Hass CC & Jenni DA 1992 Sex-biased maternal expenditure in Rocky Mountain bighorn sheep. *Behavioral Ecology and Sociobiology* **31** 242-251.

Hosack DA, Miller KV, Ware LH, Mashburn KL, Morrow CJ, Williamson LR, Marchinton RL & Monfort SL 1999 Stag exposure advances the LH surge and behavioral estrus in Eld's deer hinds after CIDR device synchronization of estrus. *Theriogenology* **51** 1333-1342.

Ingawale MV & Dhoble RL 2004 Buffalo reproduction in India: an overview. *Buffalo Bulletin* **23** 4-9.

Izard MK & Vandenbergh JG 1982 The effects of bull urine on puberty and calving date in crossbred beef heifers. *Journal of Animal Science* **55** 1160-1168.

Jewell PA 1976 Selection for reproductive success. In *The evolution of reproduction*, pp 71-109. Eds CR Austin and RV Short. Cambridge: Cambridge University Press.

Jewell PA 1997 Survival and behaviour of castrated Soay sheep (*Ovis aries*) in a feral island population on Hirta, St. Kilda, Scotland. *Journal of Zoology* **243** 623-636.

Khaldi G 1984 Variations saisonnières de l'activité ovarienne, du comportement d'oestrus et de la durée de l'anoestrus post-partum des femelles ovines de race Barbarine: influence du niveau alimentaire et de la présence du mâle. PhD Thesis, UST Languedoc, France.

Khaldi G & Lassoued N 1991 Interactions nutrition-reproduction chez les petits ruminants en milieu méditerranéen. In Proceedings of the International Symposium on Nuclear and Related Techniques in Animal Production and Health. AIEA/FAO, 15–19 April, Vienna pp 379–390.

Knight TW 1985 Are rams necessary for the stimulation of anoestrous ewes with oestrous ewes? *Proceedings of the New Zealand Society of Animal Production* **45** 49–50.

Knight TW & Lynch PR 1980 Source of ram pheromones that stimulate ovulation in the ewe. *Animal Reproduction Science* **3** 133–136.

Knight TW, Tervit HR & Lynch PR 1983 Effects of boar pheromones, ram's wool and presence of bucks on ovarian activity in anovular ewes early in the breeding season. *Animal Reproduction Science* **6** 129-134.

Knight TW, Ridland M & Litherland AJ 1998 Effect of prior ram-ewe contact on the ability of rams to stimulate early oestrus. *Proceedings of the New Zealand Society of Animal Production* **58** 178-180.

Komers PE, Birgersson B & Ekvall K 1999 Timing of estrus in fallow deer is adjusted to the age of available mates. *The American Naturalist* **153** 431-436.

Landaeta-Hernández AJ, Giangreco M, Meléndez P, Bartolomé J, Bennet F, Rae DO, Hernández J & Archbald LF 2004 Effect of biostimulation on uterine involution, early ovarian activity and first postpartum estrous cycle in beef cows. *Theriogenology* **61** 1521-1532.

Larson CL, Miller HL & Goehring TB 1994 Effect of postpartum bull exposure on calving intrval of first-calf heifers bred by natural service. *Canadian Journal of Animal Science* **74** 153-154.

Lassoued N 1998 Induction de l'ovulation par "effet bélier" chez les brebis de race Barbarine en anoestrus saisonnier. Mécanismes impliqués dans l'existence du cycle ovulatoire de courte durée. PhD Thesis, Univ Tunis II.

Lassoued N, Khaldi N, Chemineau P, Cognie Y & Thimonier J 1997 Role of the uterus in early regression of corpora lutea induced by the ram effect in seasonally anoestrous Barbarine ewes. *Reproduction Nutrition Development* **37** 559-571.

Lincoln GA & Davidson W 1977 The relationship between sexual and aggressive behaviour, and pituitary and testicular activity during the seasonal sexual cycles of rams, and the influence of the photoperiod. *Journal of Reproduction and Fertility* **49** 267-276.

Lincoln GA & Short RV 1980 Seasonal breeding: nature's contraceptive. *Recent Progress in Hormone Research* **36** 1-43.

Luna-Orozco JR, Flores JA, Hernández H, Delgadillo JA & Fernández IG 2005 La respuesta de la actividad estral de las cabras nulíparas no difiere de las multíparas al someterlas al efecto macho. In *Proceedings of the Memorias de la XX Reunión Nacional sobre Caprinocultura*, October 5-7, Universidad Autónoma de Sinaloa, Asociación Mexicana de Producción Caprina, Culiacán Sinaloa, México, pp 397-401.

Martin GB, Scaramuzzi RJ & Lindsay DR 1983 Effect of the introduction of rams during the anoestrous season on the pulsatile secretion of LH in ovariectomized ewes. *Journal of Reproduction and Fertilility* **67** 47-55.

Martin GB, Oldham CM, Cognie Y & Pearce DT 1986 The physiological responses of anovulatory ewes to the introduction of rams - a review. *Livestock Production Science* **15** 219-247.

Martin GB, Milton JTB, Davidson RH, Banchero Hunzicker GE, Lindsay DR & Blache 2004 Natural methods for increasing reproductive efficiency in small ruminants. *Animal Reproduction Science* **82/83** 231-246.

Mauget R 1981 Behavioural and reproductive strategies in wild forms of Sus scrofa (European wild and feral pigs). In *The welfare of pigs*. Ed W Sybesma. *Current Topics in Veterinary Medicine and Animal Science* **11** 3-13.

McClelland BE 1991 Courtship and agonistic behavior in mouflon sheep. *Applied Animal Behaviour Science* **29** 67-85.

Mellado M & Hernández JR 1996 Ability of androgenized goat wethers and does to induce estrus in goats under extensive conditions during anestrus and breeding seasons. *Small Ruminant Research* **23** 37-42.

Mellado M, Olivas R & Ruiz F 2000 Effect of buck stimulus on mature and pre-pubertal norgestomet-treated goats. *Small Ruminant Research* **36** 269-274.

Minton JE, Coppinger TR, Spaeth CW & Martin LC 1991 Poor reproductive response of anestrus Suffolk ewes to ram exposure is not due to failure to secrete luteinizing hormone acutely. *Journal of Animal Science* **69** 3314–3320.

Miquelle DG 1991 Are moose mice? The function of scent urination in moose. *American Naturalist* **138** 460-477.

Monje AR, Alberio R, Schiersmann G, Chedrese J, Carou N & Callejas S 1992 Male effect on the post-partum sexual activity of cows maintained on two nutritional levels. *Animal Reproduction Science* **29** 145-156.

Moore GH & Cowie GM 1986 Advancement of breeding in non-lactating adult red deer hinds. *Proceedings of the New Zealand Society of Animal Production* **46** 175-178.

Morgan PD, Arnold EW & Lindsay DR 1972 A note on the mating behaviour of ewes with various senses impaired. *Journal of Reproduction and Fertility* **30** 151-152.

Naasz CD & Miller HL 1987 Effects of bull exposure on postpartum interval and preproductive performance in beef cows. *Journal of Animal Science* **65** (Supplement) 426.

Nugent III RA, Notter DR & Beal WE 1988 Effects of ewe breed and ram exposure on estrous behavior in May and June. *Journal of Animal Science* **66** 1363-1370.

O'Riordan EG & Hanrahan JP 1989 Advancing first estrus in ewe lambs. *Farm and Food Research* **20** 25-27.

Oldham CM, Pearce DT & Gray SJ 1985 Progesterone priming and age of ewe affect the life-span of corpora lutea induced in the seasonally anovulatory Merino ewe by the "ram effect". *Journal of Reproduction and Fertility* **75** 29-33.

Over R, Cohen-Tannoudji J, Dehnhard M, Claus R & Signoret JP 1990 Effect of pheromones from male goats on LH-secretion in anoestrous ewes. *Physiology & Behavior* **48** 665-8.

Pearce GP & Oldham CM 1988 Importance of non-olfactory ram stimuli in mediating ram-induced ovulation in the ewe. *Journal of Reproduction and Fertility* **84** 333-339.

Pearce DT, Oldham CM, Haresign W & Gray SJ 1987 Effects of duration and timing of progesterone prim-

ing on the incidence of corpora lutea with a normal life-span in Merino ewes induced to ovulate by the introduction of rams. *Animal Reproduction Science* **13** 81-89.

Pérez-Hernández P, García-Winder M & Gallegos-Sánchez J 2002 Bull exposure and an increased within-day milking to suckling interval reduced postpartum anoestrus in dual purpose cows. *Animal Reproduction Science* **74** 111-119.

Perkins A & Fitzgerald JA 1994 The behavioral component of the ram effect: the influence of ram sexual behavior on the induction of estrus in anovulatory ewes. *Journal of Animal Science* **72** 51–55.

Reinhardt C, Reinhardt A & Reinhardt V 1986 Social behaviour and reproductive performance in semi-wild Scottish highland cattle. *Applied Animal Behavior Science* **15** 125-136.

Rekwot PI, Ogwu D, Oyedipe EO & Sekoni VO 2000 Effects of bull exposure and body growth on onset of puberty in Bunaji and Friesian X Bunaji heifers. *Reproduction Nutrition Development* **40** 359-367.

Restall BJ 1992 Seasonal variation in reproductive activity in Australian goats. *Animal Reproduction Science* **27** 305-318.

Roberson MS, Ansotegui RP, Berardinelli JG, Whitman RW & McInerney MJ 1987 Influence of biostimulation by mature bulls on occurrence of puberty in beef heifers. *Journal of Animal Science* **64** 1601-1605.

Roberson MS, Wolfe MW, Stumpf TT, Werth LA, Cupp AS, Kojima ND, Wolfe PL, Kittok RJ & Kinder JE 1991 Influence of growth rate and exposure to bulls on age at puberty in beef heifers. *Journal of Animal Science* **69** 2092-2098.

Rodriguez-Iglesias RM, Ciccioli N, Irazoqui H & Rodriguez BT 1991 Importance of behavioural stimuli in ram-induced ovulation in seasonally anovular Corriedale ewes. *Applied Animal Behavior Science* **30** 323-332.

Rowlands IW & Weir BJ 1984 Mammals: non-primate eutherians. In *Marshall´s Physiology of Reproduction. Vol.1. Reproductive cycles of vertebrates*, edn 4, pp 455-658. Ed GE Lamming. Churchill Livingstone, Edinburgh, London, Melbourne, and New York.

Sanford LM, Palmer WM & Howland BE 1982 Influence of age and breed on circulating LH, FSH and testosterone levels in the ram. *Canadian Journal of Animal Science* **62** 767-776.

Schaller G & Mirza ZB 1974 On the behaviour of Punjal Urial Ovis orientalis punjabensis. In *The behaviour of ungulates and its relation to management. Vol. 1.* Eds V Geist & FR Walther. IUCN Publications, Morges, Switzerland.

Shackleton DM & Shank CC 1984 A review of the social behavior of feral and wild sheep and goats. *Journal of Animal Science* **58** 500-509.

Shipka MP & Ellis LC 1998 No effects of bull exposure on expression of estrous behavior in high-producing dairy cows. *Applied Animal Behaviour Science* **57** 1-7.

Shipka MP & Ellis LC 1999 Effects of bull exposure on

postpartum ovarian activity of dairy cows. *Animal Reproduction Science* **54** 237-244.

Shipka MP, Rowell JE & Ford SP 2002 Reindeer bull introduction affects the onset of the breeding season. *Animal Reproduction Science* **72** 27-35.

Silva L & Ungerfeld R 2006 Reproductive response in suckling Corriedale ewes to the ram effect during the non-breeding season: effect of postpartum condition and of the use of medroxyprogesterone primings. *Tropical Animal Health and Production* **38** 365 - 369

Skinner DC, Cilliers SD & Skinner JD 2002 Effect of ram introduction on the oestrous cycle of springbok ewes (*Antidorcas marsupialis*). *Reproduction* **124** 509-513.

Skinner JD, Jackson TP & Marais AL 1992 The "ram effect" in three species of African Ungulates. *Ungulates* **91** 565-568.

Stricklin WR & Mench JA 1987 Social Organization. In *Farm Animal Behavior*. Ed EO Price, WB Saunders Company. *The Veterinary Clinics of North America, Food Animal Practice* **3** 307-322.

Stumpf TT, Wolfe MW, Wolfe PL, Day ML, Kittok RJ & Kinder JE 1992 Weight changes prepartum and presence of bulls pospartum interact to affect duration of postpartum anestrus in cows. *Journal of Animal Science* **70** 3133-3137.

Thimonier J, Cognie Y, Lassoued N & Khaldi G 2000 L'effet mâle chez les ovins: une technique actuelle de maîtrise de la reproduction. *INRA Productions Animales* **13** 223-231.

Thomas GB, Pearce DT, Oldham CM, Martin GB & Lindsay DR 1988 Effects of breed, ovarian steroids and season on the pulsatile secretion of LH in ovariectomized ewes. *Journal of Reproduction and Fertility* **84** 313-324.

Underwood EJ, Shier FL & Davenport N 1944 Studies in sheep husbandry in W.A.V. The breeding season of Merino, crossbreed and British Breeds ewes in the agricultural districts. *Journal of Agriculture W. A.* **11, Series 2** 135-143.

Ungerfeld R 2003. The reproductive response of anestrous ewes to the introduction of rams. Ph.D. Thesis, Swedish University of Agricultural Sciences, Uppsala, Sweden. *Acta Universitatis Agriculturae Sueciae. Veterinaria* **163**, pp. 66.

Ungerfeld R 2005 Sixty years of the ram effect (1944-2004): how we have learned what we know about it? *Journal of Animal and Veterinary Advances* **4** 698-701.

Ungerfeld R & Rubianes E 2004 Estrous distribution in anestrous ewes primed with estradiol-17beta 3 or 5 days before the introduction of the rams (the ram effect). In *Proceedings of the International Congress of Animal Reproduction*, August 8-12, Porto Seguro, Brasil.

Ungerfeld R & Silva L 2004 Ewe effect: endocrine and testicular changes in adult and young Corriedale rams used for the ram effect. *Animal Reproduction Science* **80** 251-259.

Ungerfeld R, Pinczak A, Forsberg M & Rubianes E 2002 Ovarian and endocrine responses of Corriedale ewes to "ram effect" in the non-breeding season. *Canadian Journal of Animal Science* **82** 599–602.

Ungerfeld R, Suárez G, Carbajal B, Silva L, Laca M, Forsberg M & Rubianes E 2003 Medroxyprogesterone priming and response to the ram effect in Corriedale ewes during the nonbreeding season. *Theriogenology* **60** 35–45.

Ungerfeld R, Forsberg M & Rubianes E 2004a Overview of the response of anoestrous ewes to the ram effect. *Reproduction Fertility and Development* **16** 479-490.

Ungerfeld R, Dago AL, Rubianes E & Forsberg M 2004b Response of anestrous ewes to the ram effect after follicular wave synchronization with a single dose of estradiol-17beta. *Reproduction Nutrition Development* **44** 1-10.

Walkden-Brown SW, Restall BJ & Henniawati. 1993 The male effect in Australian cashmere goats 2. Role of olfactory cues from the male. *Animal Reproduction Science* **32** 55–67.

Walkden-Brown SW, Martin GB & Restall BJ. 1999. Role of male-female interaction in regulating reproduction in sheep and goats. *Journal of Reproduction and Fertility* (Suppl.) **52** 243-257.

Whittle CL, Bowyer RT, Clausen TP & Duffy LK 2000 Putative pheromones in urine of rutting male moose (*Alces alces*): evolution of honest advertisement? *Journal of Chemical Ecology* **26** 2747-2762.

Wright PJ, Geytenbeek PE, Clarke IJ & Hoskinson RM 1989 The efficacy of ram introduction, GnRH administration, and immunisation against androstenedione and oestrone for the induction of oestrus and ovulation in anoestrous post-partum ewes. *Animal Reproduction Science* **21** 237-247.

Wright PJ, Geytenbeek PE & Clarke IJ 1990 The influence of nutrient status of post-partum ewes on ovarian cyclicity and on the oestrous and ovulatory responses to ram introduction. *Animal Reproduction Science* **23** 293-303.

Wright IA, Rhind SM, Smith AJ & Whyte TK 1994 Female-female influences on the duration of the postpartum anoestrous period in beef cows. *Animal Production* **59** 49-53.

Yarney TA & Sanford LM 1983 The reproductive-endocrine response of adult rams to sexual encounters with estrual ewes is season dependent. *Hormones and Behavior* **17** 169-182.

Zalesky DD, Day ML, García-Winder M, Imakawa K, Kittok RJ, D'Occhio MJ & Kinder JE 1984 Influence of exposure to bulls on resumption of estrous cycles following parturition in beef cows. *Journal of Animal Science* **59** 1135-1139.

Technologies for fixed-time artificial insemination and their influence on reproductive performance of *Bos indicus* cattle

GA Bó[1,2], L Cutaia[1,3], LC Peres[1,3], D Pincinato[1,3], D Maraña[1,3] and PS Baruselli[4]

[1]Instituto de Reproducción Animal Córdoba (IRAC), J.L. de Cabrera 106, X5000GVD, Córdoba, Argentina, [2]Universidad Católica de Córdoba, [3]Universidad Nacional de Córdoba, [4]Departamento de Reprodução Animal, FMVZ-USP, Brazil.

The adaptation of *Bos indicus* cattle to tropical and subtropical environments has led to their widespread distribution around the world. Although artificial insemination (AI) is one of the best alternatives to introduce new genetics into *Bos indicus* herds, the peculiarity of their temperament and the tendency to show short oestrus (many of them during the night) greatly affects the effectiveness of genetic improvement programs. Therefore, the most useful alternative to increase the number of females that are inseminated is the use of protocols that allow for AI without the need for oestrus detection, usually called fixed-time AI (FTAI). Besides, the development of protocols to advance the resumption of cyclicity during the early postpartum period has a great impact on beef production and will allow for the inclusion of a significantly larger population of animals into genetic improvement programs. Fixed-time AI protocols using progestin devices, oestradiol and eCG have resulted in consistent pregnancy rates in suckled *Bos indicus* and *Bos indicus x Bos taurus* cows. Furthermore, fertility in the successive cycles and the overall pregnancy rates at the end of the breeding season, have been shown to be improved by the use of progestin devices at the beginning of the breeding season. In summary, exogenous control of luteal and follicular development has facilitated the application of assisted reproductive technologies in *Bos indicus*-influenced cattle, by offering the possibility of planning programs without the necessity of oestrus detection and may provide the opportunity to improve reproductive performance of beef cattle in tropical climates.

Introduction

Most beef herds are located in tropical regions where *Bos indicus* breeds predominate. Data on reproductive performance, such as calving rate, calf survival and weaning rate have indicated both inferior and superior results for *Bos indicus* cattle (Chenoweth 1994). However, there is little doubt that *Bos indicus* breeds, and their crosses, are superior to *Bos taurus* cattle when they are both kept in tropical or subtropical environments, where stressors such as high tem-

Corresponding author E-mail: gabrielbo@iracbiogen.com.ar

peratures and humidity, ectoparasites and low quality forages predominate. Artificial insemina-tion is one of the best alternatives to introduce new genetics into *Bos indicus* herds (especially from *Bos taurus* breeds); however, only a small percentage of beef animals are subjected to AI. In Argentina for example, only 4.5% of the beef breeding females are artificially inseminated and 80% of those are heifers (Marcantonio 2003). Among the main factors that affect the exten-sive use of AI in the beef herd are those related to nutrition, management and inefficient oestrus detection. The most useful alternative to significantly increase the number of animals involved in AI programs is the use of protocols that allow for AI without the need for oestrus detection, usually called fixed-time AI (FTAI) protocols. Also, the development of protocols for suckled cows will allow for the inclusion of a significantly larger population of animals, and not just limit the application of these technologies to heifers. The intention of this manuscript is to present data from studies in which current methods of manipulation of follicular waves and ovulation for FTAI have been successfully applied in *Bos indicus* and *Bos indicus* x *Bos taurus* crossbred herds, and discuss how these protocols may impact the overall fertility of these herds, paying particular attention to those currently applied in extensively managed *Bos indicus* or *Bos indicus* x *Bos taurus* crossbred herds in South America.

Oestrous behaviour and reproductive physiology in *Bos indicus* cattle

The characteristics of the oestrous cycle and follicular dynamics in *Bos indicus* cattle have been recently reviewed (Bó *et al.* 2003). *Bos indicus* cattle usually have a very particular tempera-ment that makes oestrus detection a very difficult task. "Silent" or "missed" heats have been reported, after a regular oestrus detection (Galina and Arthur 1990; Galina *et al.* 1996). Further-more, duration of oestrus has been reported to be shorter in *Bos indicus* than in *Bos taurus* cattle (Galina and Arthur 1990). The average duration of standing oestrus in *Bos indicus* cattle has been shown to be about 10 h, with variations between 1.3 to 20 h (Galina and Arthur 1990; Barros *et al.* 1995; Pinheiro *et al.* 1998). Other studies utilizing radiotelemetry have confirmed that crossbred *Bos indicus* x *Bos taurus* females have a shorter duration of oestrus (approxi-mately 10 h; Bertam Membrive 2000; Rocha 2000), and found more mounting activity during the night (56.6%). These findings are in agreement with the results obtained by Pinheiro *et al.* (1998), who reported 53.8% of the oestrous expression at night, with 30.7% of these beginning and ending during the night. Mizuta (2003), using radiotelemetry, found that the mean dura-tion of standing oestrus was 3.4 h shorter in Nelore (12.9 h) and Nelore x Angus crossbred (12.4 h) than in Angus (16.3 h) cows. However, the interval from the onset of oestrus to ovulation was 27.1 ± 3.3 h and 26.1 ± 6.3 h in Nelore and Angus cows, respectively (Mizuta 2003). Thus, the interval from the onset of oestrus to ovulation in *Bos taurus* and *Bos indicus* cows would not appear to differ.

Several studies have also characterized follicular-wave dynamics in *Bos indicus* cattle (re-viewed in Bó *et al.* 2003). *Bos indicus* cattle have two, three or four waves of follicular growth during their oestrous cycle and have a smaller diameter of the dominant follicle and corpus luteum (CL; Bó *et al.* 2003) and lower serum progesterone concentrations (Segerson *et al.* 1984) relative to those of *Bos taurus* cattle.

In more recent studies, the diameter of the dominant follicle at the time of deviation has been reported to be smaller in Nelore (6.0 to 6.3 mm, Sartorelli *et al.* 2005; Gimenes *et al.* 2005b) than in Holstein (8.5 mm; Ginther *et al.* 1996) cattle. Futhermore, the diameter at which the dominant follicle acquired the capacity to ovulate in response to a treatment with pLH (Lutropin-V, Bioniche Animal Health, Canada) in Nelore heifers was found to be between 7 and 8.4 mm (Gimenes *et al.* 2005a); whereas, it was 10 mm in Holstein cows (Sartori *et al.* 2001).

Seasonality has also been shown to affect cyclicity in *Bos indicus* cattle. Randel (1984) reported that *Bos indicus* cows had a decreased incidence of preovulatory LH-surges and their luteal cells *in vitro* were less responsive to LH during the winter. Furthermore, conception rates of Brahman cattle were higher in the summer (61%) than in the fall (36%; Randel 1994). Stahringer *et al.* (1990) and McGowan (1999) also reported an increased occurrence of anoestrus and anovulatory oestrus in Brahman females during the winter.

Physiology of the postpartum period

Following parturition, there is a dramatic increase in FSH that is followed by the emergence of the first follicular wave (2 to 7 d postpartum; reviewed in Wiltbank *et al.* 2002). However, ovulatory capacity of the dominant follicle only occurs when it is exposed to adequate LH-pulse frequency (approximately 1 pulse/hour) to grow and increase oestradiol production, which will result in an LH surge and ovulation (reviewed in Wiltbank *et al.* 2002). Gonadotrophin secretion patterns in the postpartum period have been shown to differ between *Bos taurus* and *Bos indicus* cattle. Thirty days after calving, Hereford x Shorthorn suckled cows had higher plasma LH concentrations (0.7 ± 0.1 ng/ml) than suckled Brahman cows (0.6 ± 0.1 ng/ml) and this difference appeared to increase over time (D'Occhio *et al.* 1990). In addition, a higher proportion of *Bos taurus* cows had greater pulsatile LH secretion than *Bos indicus* cows (D'Occhio *et al.* 1990). Futhermore, a greater proportion of *Bos taurus* cows became pregnant during mating between 50 to 120 d after calving compared to Brahman cows. In this study, circulating concentrations of LH were also affected by body condition and postpartum management (D'Occhio *et al.* 1990), confirming the notion that nutrition is one of the major factors affecting postpartum ovarian activity in cattle. In that regard, Ruiz-Cortez and Olivera-Angel (1999) observed that *Bos indicus* suckled cows kept on natural pasture in Colombia re-established their cyclicity from 217 to 278 d after calving. During the first 6 months after parturition, many of these cows had only small follicles (< 6 mm in diameter, exceptionally 8 mm). From 7 to 12 months postpartum, follicular waves were more regular and when cyclicity re-commenced at 217 to 278 d postpartum, oestrus preceded ovulation in 43% of the cases and cows had normal (21.0 ± 3.0 d), short (10.0 ± 2.0 d) or long (50.0 ± 4.0 d) first oestrous cycles. This condition may not be uncommon in *Bos indicus* cattle and has to be taken into consideration when deciding to begin an AI program. Cows in low body condition would rarely respond to oestrus synchronization treatments (Wiltbank *et al.* 2002; Bó *et al.* 2002a; 2002b).

Synchronization of oestrus and ovulation

Prostaglandin F2α

Prostaglandin F2α (PGF) has been the most commonly used treatment for synchronization of oestrus in cattle (Odde 1990). However, the variable interval from PGF treatment to expression of oestrus and ovulation (Kastelic & Ginther 1991) makes oestrus detection essential to attain high pregnancy rates in AI programs. In *Bos indicus* cattle, oestrus response was about 30% less than that reported for *Bos taurus* cattle under the same conditions (reviewed in Galina and Arthur 1990). In two other studies, although 80 to 100% of the cows treated with PGF had luteal regression, only 29 to 60% were detected in oestrus (Moreno *et al.* 1986; Alonso *et al.* 1995) and 51% (29/57) ovulated (Alonso *et al.* 1995) within 5 d of treatment. The combination of low and variable oestrus response and the high incidence of anoestrus common in animals grazing tropical grasses explain the wide variability in oestrus response and pregnancy rates after PGF treatments (Galina and Arthur 1990; Moreno et al. 1986; Kerr et al. 1991; Alonso et al. 1995; Pinheiro et al. 1998). These studies emphasize the need for treatments that control follicular and luteal development to obtain high pregnancy rates to FTAI without the necessity of oestrous detection. Furthermore, treatment protocols should be capable of inducing oestrus and ovulation in anoestrus animals.

GnRH-based protocols

GnRH-based treatment protocols have been used extensively in recent years for FTAI in beef and dairy cattle (Pursley et al. 1995; 1997; Geary et al. 2001). These treatment protocols consist of an injection of GnRH followed by PGF 7 d later and a second injection of GnRH 48 h after PGF treatment. In Co-Synch protocols cows are FTAI at the time of the second GnRH (Geary et al. 2001), whereas in Ovsynch protocols, cows are FTAI 16 h after GnRH (Pursley et al. 1995).

The Ovsynch protocols have also been used in FTAI programs in Bos Indicus cattle (Barros et al. 2000; Lemaster et al. 2001; Williams et al. 2002; Baruselli et al. 2004). However, overall pregnancy rates have often been lower than those rates reported in Bos taurus cattle (Baruselli et al. 2004; Saldarriaga et al. 2005), with low conception rates in anoestrus cows (Fernandes et al. 2001; Baruselli et al. 2004). The addition of a progestin-releasing device increased pregnancy rates in anoestrus Bos taurus cows (Lamb et al. 2002); however, this approach has not resulted in increased pregnancy rates in Bos indicus and Bos indicus x Bos taurus crossbred cattle (Saldarriaga et al. 2005; Pincinato et al. 2006) and is probably related to a low ovulation rate following the first GnRH treatment (Saldarriaga et al. 2005).

Treatments using progestins and oestradiol

Oestradiol and progestin treatments have been increasingly used over the past several years in oestrus synchronization programs in cattle (Macmillan & Burke 1996; Bó & Baruselli 2002; Yelich 2002; Bo et al. 2003). Treatments consist of insertion of a progestin-releasing device and the administration of oestradiol on Day 0 (to synchronize follicular wave emergence), PGF at the time of device removal on Days 7, 8 or 9 (to ensure luteolysis), and the subsequent application of a lower dose of oestradiol 24 h later or GnRH/LH 48 to 54 h later to synchronize ovulation (Bo et al. 2002a; 2002b; Martinez et al. 2002). The most commonly used treatment for FTAI using progesterone-releasing devices in beef cattle in South America consists of the administration of 2 mg of oestradiol benzoate (EB) im upon insertion of the device (Day 0); on Day 7 or 8 the device is removed and PGF is administered im, and 24 h later, 1 mg of EB im is given (Bó et al. 2002b); FTAI is done between 52 and 56 h after device removal. Data from 13,510 inseminations in Bos taurus and Bos indicus x Bos taurus crossbred cattle, performed between December 2000 and December 2004, resulted in a mean pregnancy rate of 52.7%, ranging from 27.8% to 75.0%. The factors that most affected pregnancy rates were body condition score (BCS) and cyclicity of the cows (Bó et al. 2005).

Progestin based treatments for FTAI in suckled cows

Under favourable conditions, a cow has the potential to produce one calf per year, with an interval of 12 months between calvings. However, suckled beef cattle under grazing conditions often have a high incidence of postpartum anoestrus, which extends the calving to conception interval and, consequently, negatively affects their reproductive performance. The insertion of subcutaneous norgestomet ear implants or intravaginal progesterone devices, combined with the application of eCG at the time of device removal, has been extensively used in Bos indicus herds with high incidence of postpartum anoestrus (reviewed in Baruselli et al.

2004). The use of 400 IU of eCG at the time of progestin device removal resulted in increased pregnancy rates in cows without a CL at the time of insertion of the progestin device (Baruselli *et al.* 2003; Cutaia *et al.* 2003a). In another study (Baruselli *et al.* 2004), eCG treatment increased plasma progesterone concentrations and pregnancy rates in suckled cows treated during postpartum anoestrus. Therefore, eCG treatment may be an important tool for increasing pregnancy rates at FTAI, to reduce the postpartum period, and to improve reproductive efficiency in postpartum *Bos indicus* and *Bos indicus* x *Bos taurus* beef cows. Analysis of data collected from 9,668 FTAI done from December 2000 through December 2003 has shown that *Bos taurus* and *Bos indicus* x *Bos taurus* crossbred animals treated with progestin-devices must have a BCS higher than 2.5 (scale 1 to 5) and ideally ≥3 to achieve pregnancy rates of 50% or higher (Bó *et al.* 2005). Conversely, the addition of eCG allowed for pregnancy rates close to 50% in cows with a BCS of 2 (Bó *et al.* 2005). It is very important to note that these results have been achieved only when cows were gaining body condition during the breeding season. If drought conditions or lack of feed prevent cattle from improving BCS during the breeding season, pregnancy rates will most probably be 35% or less, even after the administration of eCG (Cutaia *et al.* 2003a; Bó *et al.* 2005; Maraña *et al.* 2006). Another analysis performed with 1,987 FTAI in Nelore cows confirmed that BCS is critical to achieve pregnancy rates and that the beneficial effect of eCG treatments was significant in cows with a BCS ≤3 (Fig. 1; Baruselli *et al.* 2005). Since BCS is usually associated with cyclicity (D'occhio *et al.* 1990), it is conceivable that most cows in the lower BCS were anoestrus at the time that treatments were initiated. When 485 *Bos indicus* x *Bos taurus* suckled cows were examined by real time ultrasonography at the time of device insertion, pregnancy rates in cows that had a CL when treatments were initiated did not differ between cows treated (56.3%) or not treated with eCG (56.5%; Bó *et al.* 2005). However, eCG treatments increased pregnancy rates (eCG: 49.5% vs no eCG: 40%; P<0.05) in cows that only had follicles at the time of progestin device insertion. In yet another retrospective analysis of 2,489 FTAI in suckling Nelore cows from two commercial farms in Brazil, pregnancy rates were not different between cows that were 40 to >80 d postpartum at the time of FTAI (40-49 d: 57/142, 52.8%; 50-59 d: 419/759, 55.2%; 60-69 d: 137/263, 52.1%, 70-79 d: 361/684, 56.3% and >80 d: 334/641, 52.1%; Marques *et al.* 2006).

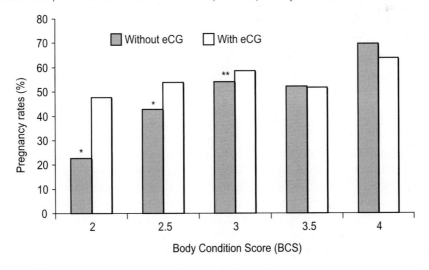

Fig. 1. Effect of body condition scores (1 to 5 scale) on pregnancy rates in Nelore cows (n = 1,984) treated with progestin-releasing devices with or without 400 IU eCG at device removal (* P < 0.05; ** P<0.1). Adapted from Baruselli *et al. 2005.*

Restricted suckling or calf removal associated with progestin devices has also been used for the induction of cyclicity in *Bos indicus* cows (Williams 1990; Soto Belloso *et al*. 2002). We have recently conducted two experiments to compare the effects of eCG treatment and temporary weaning (TW) on ovulation and pregnancy rates in postpartum cows. In the first experiment, 39 lactating multiparous *Bos indicus x Bos taurus* crossbred cows, 60 to 80 d postpartum with a BCS between 2.0 to 2.5 (scale 1 to 5) were randomly allocated to 1 of 4 treatment groups, in a 2 by 2 factorial design (Maraña *et al*. 2006). On Day 0, all cows received a DIB device (intravaginal progesterone-releasing device with 1 g of progesterone, Syntex SA, Buenos Aires, Argentina) and 2 mg EB im (Benzoato de Estradiol, Syntex SA). On Day 8, DIB devices were removed and all cows received PGF and were randomly divided to receive 400 IU eCG im (Novormon, Syntex SA) at the same time or no further treatment. In addition, half of the cows in each treatment group had their calves weaned for 56 h from the time of DIB removal; the other half remained with their calves. All cows received 1 mg EB im on Day 9 and were examined every 8 h by ultrasonography, from the time of DIB removal until ovulation. The interval to ovulation (eCG, 72.0 ± 1.4 h vs no eCG, 75.6 ± 2.0 h and TW, 73.8 ± 1.6 h vs no TW, 73.0 ± 1.8 h) did not differ among groups ($P > 0.05$). However, TW increased (7/10, 70.0%; $P < 0.05$) and eCG treatment tended to increase (12/20, 60.0%; $P < 0.09$) the proportion of cows ovulating compared to control cows (no TW or eCG treatment: 2/9, 22.2%). Although there was no effect of eCG treatment on the size of the preovulatory follicle (eCG, 11.1 ± 0.4 mm vs no eCG, 10.1 ± 0.6 mm), the growth rate of the ovulatory follicle was greater ($P < 0.02$) in cows treated with eCG (1.1 ± 0.1 mm/d) than in those not treated with eCG (0.6 ± 0.1 mm/d). Conversely, the ovulatory follicle was smaller in TW cows (9.9 ± 0.4 mm), compared to those not TW (11.8 ± 0.3 mm; $P < 0.05$).

The second experiment was conducted over 2 years; 769 *Bos indicus x Bos taurus* crossbred suckled cows (year 2004, n = 393 and year 2005, n = 376) with a BCS of 2 to 2.5 were used (Maraña *et al*. 2006). All animals were examined by palpation per rectum at the time of initiating the treatment to determine ovarian status. Cows were randomly assigned to 4 treatment groups in a 2 by 2 factorial design (Control, eCG, TW and TW + eCG), so that cows with a CL (22.5%), follicles > 8 mm (30.0%) or ovaries with small follicles (<8 mm; 47.5%) were equally represented in each group. Temporarily weaned calves were separated from their dams by approximately 1000 m, to prevent any kind of contact between cows and calves. All cows were FTAI between 52 and 56 h after DIB removal. Data were analyzed by logistic regression. There was no interactions between years and treatments or between treatments ($P > 0.7$). The overall pregnancy rate was lower in 2005 (109/376, 29.0%; $P < 0.01$) than in 2004 (173/393, 44.0%), due to a drought during that breeding season; but in both years eCG treatment increased pregnancy rates (eCG, 154/377, 40.8% vs no eCG, 128/392, 32.6%; $P < 0.01$). Conversely, no differences were found between cows that were TW (141/379, 37.2%) and those that were not (141/390, 36.1%; $P > 0.7$). It was concluded that the use of eCG but not TW improved pregnancy rates following FTAI in postpartum *Bos indicus x Bos taurus* crossbred cows in moderate to low body condition. Results also suggest that the eCG-related increase in pregnancy rates may be due to the final growth rate of the ovulatory follicle. On the other hand, the absence or little effect of TW on pregnancy rates contrasts with data from another study done with Nelore cows (Penteado *et al*. 2004), and those from other studies (reviewed in Baruselli *et al*. 2004). In the experiment with Nelore cows (Penteado *et al*. 2004), 459 suckled cows were treated with Crestar (Intervet, Sao Paulo, Brazil) for 9 d and were divided into 1 of 4 treatment groups to receive or not receive 400 IU eCG im (Folligon, Intervet), and have calves TW for 56 h or not. In this case, both eCG and TW significantly increased ($P < 0.05$) pregnancy rates (eCG, 126/227, 55.5% vs no eCG, 98/232 42.2%; TW, 121/229, 52.8% vs no TW, 103/230, 44.8%). Therefore, the beneficial effects of temporary weaning may differ, depending on the management and body condition of the cows. Moreover, to set up a temporary weaning program creates logistical problems in several farms and is probably the most resisted management technique by beef producers, at least in Argentina and Brazil. Nevertheless, the results from both studies confirmed those reported previously that eCG increased pregnancy rates in suckled cows enrolled in a FTAI program utilizing progestin devices and oestradiol (Cutaia *et al.* 2003a; Baruselli *et al.* 2004).

Impact of Fixed-time AI programs on the overall fertility of beef herds

One of the main advantages of implementing FTAI programs in a beef herd is that more cows can be impregnated earlier in the breeding season to genetically improved bulls, resulting in heavier weaning weights (Cutaia *et al.* 2003b). Fifty percent of the cows could potentially become pregnant on the first day of the breeding season and result in a higher number of cows calving at the beginning of the calving season. Therefore, their calves will be older and heavier at weaning. Besides, the use of genetically superior bulls will also result in heavier calves at weaning (Cutaia *et al.* 2003b). The impact of FTAI has proven to be equally efficient for different beef operations in Argentina and Brazil (Bó & Baruselli. 2002; Baruselli *et al.* 2005; Bó *et al.* 2005) and examples will be shown in the following.

In 2002, the "Estancia El Mangrullo" (Lavalle, Santiago del Estero, Argentina) started implementing FTAI programs. This operation is located in the semiarid region of Argentina, with seasonal rainfalls of 600 mm per year from November-December to May-June (Summer and Fall). Animals are all zebu-derived and a cross-breeding program with Bonsmara (*Bos taurus* adapted breed) has been implemented with the use of semen and embryos. Table 1 shows the evolution of the number of animals involved in FTAI programs and the pregnancy rates obtained.

As shown in Table 1, a FTAI program was implemented in heifers and suckled cows which resulted in pregnancy rates between 40 and 50%. It is important to highlight that the summer of 2005 (i.e. the breeding season) was especially dry, with no rains between December and March which, undoubtedly affected the pregnancy rates. However, it is apparent that an aggressive FTAI program may still result in acceptable pregnancy rates, even in the presence of the drought experienced that year. Probably, the main aspect of applying this system was its effect on calving distribution as shown in Fig. 2. The progression of calvings throughout the calving season was compared between years using Kaplan Meier's Method for comparison of survival curves. Survival curves across years were significantly different (P< 0.01). In 2002/3 (no FTAI), calvings were distributed over 6 months with a high number of cows calving from December to March (late calvers). This was changed with the limited use of FTAI in 2003/04 (Table 1). However, with a more aggressive FTAI program, calvings began earlier, with a high proportion of heifers calving in September (i.e. 30 d prior to the cows) and a higher percentage of mature cows calving earlier in the breeding season (October onwards) in 2004/05.

We also evaluated the impact of FTAI on weaning weights of the calves obtained through natural service compared to that of calves obtained through FTAI in 2004 (Bó *et al.* 2005). Only one group of animals in which all calving data could be collected was used. The cows from the Natural Service Group were bred with Bonsmara bulls at a rate of 3% (i.e. 3 bulls per 100 cows) for 90 d. Cows in the FTAI Group were inseminated at the beginning of the breeding season and exposed to clean-up bulls at a rate of 1.5%. All cows were monitored during the calving season and calves born were identified with ear tags and weighed. Table 2 shows the weaning weights of calves produced through FTAI or natural service. Weight of the calves was adjusted to 205 d to determine the proportion of the weight difference between groups that was due to the age of the calves and the proportion that was due to a genetic improvement introduced with

Table 1. Pregnancy rates following FTAI programs implemented in "Estancia El Mangrullo" in Lavalle, northeast of the Province of Santiago del Estero, Argentina. Adapted from Bó *et al.* 2005.

Category	Year 2002/03	Year 2003/04	Year 2004/05	Total
Heifers	148/292	341/619	564/1233	1053/2144
	(50.7%)	(55.1%)	(45.7%)	(49.1%)
Dry cows	—	189/394	—	189/394
		(47.9%)		(47.9%)
Suckled cows	156/289	345/790	450/1199	951/2278
	(54.0%)	(43.7%)	(37.5%)	(41.7%)

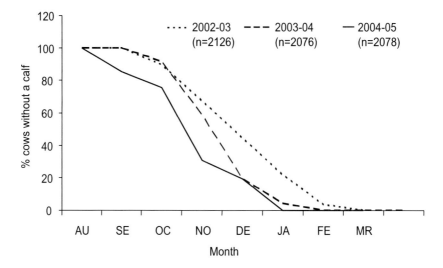

Fig. 2. Survival curves for calving distribution at "Estancia El Mangrullo" Santiago del Estero, Argentina in three consecutive years. Curves differ significantly among the three years (P < 0.01). Adapted from Bó *et al.* 2005.

Table 2. Weaning weights (means ±SEM) of Zebu x Bonsmara calves produced through FTAI or Natural Service. "Estancia El Mangrullo", Santiago del Estero, Argentina, 2004. Adapted from Bó *et al.* 2005.

	N	Weaning weight (Kg)	Adjusted 205 d-weight (Kg)
FTAI	138	178.1 ± 1.9[a]	184.2 ± 1.6[a]
Natural service	181	149.4 ± 1.5[b]	173.8 ± 1.4[b]
Difference		28.7	10.4

Means in the same column with different superscripts differ ([ab] P < 0.01)

the bulls through FTAI. Calves from the FTAI Group were heavier at weaning than calves in the Natural Service Group. Part of this difference (18.3 Kg) was attributed to age, because the calves from the FTAI Group were born earlier than those in the Natural Service Group. There was also a 10.4 Kg weight advantage for the FTAI calves due to genetic improvement. These data confirm previous results in Angus cattle (Cutaia *et al.* 2003b) where differences in weaning weights were 34.6 Kg for calves produced through FTAI compared to those produced through natural service, and showed that it was possible to improve production in a beef herd with a FTAI program at the beginning of the breeding season.

Another study was performed in Brazil using Nelore cows (Baruselli *et al.* 2005). In this study, 594 suckled Nelore cows (55 to 70 d postpartum) were randomly allocated to 1 of 4 treatment groups. Cows in Group 1 were FTAI on Day 0 of the breeding season and were exposed to bulls for a further 90 d. Cows in Group 2 were FTAI on Day 0, then AI based on oestrus detection for 45 d and then exposed to bulls for the last 45 d of the breeding season. Cows in Group 3 were AI based on twice daily oestrus detection for 45 d and then exposed to clean-up bulls for another 45 d. Cows in Group 4 simply were exposed to bulls for 90 d. In order to determine the progression of the pregnancies during the breeding season, cows were examined by ultrasonography on Days 30, 70 and 120 after the beginning of the breeding season. Results are shown in Table 3 and survival curves for days open are shown in Fig. 3. The use of FTAI improved fertility by having more cows pregnant at the beginning of the breeding season. Survival curves in the FTAI + Bulls and FTAI + OED & AI + Bulls breeding schemes differed from those in the OED & AI + Bulls breeding schemes and the cows that were bred by bulls for the entire breeding season (P< 0.01; Figure 3). Compared to the cows bred by bulls only, the insertion of FTAI hastened the mean day of conception by about 17 d and increased the pregnancy rates after the first 45 d by 30% and at the end of the breeding season by about 9% (Table 3). Conversely, the application of a traditional scheme of oestrus detection and AI for 45 d was the least efficient program; a reflection of the difficulty of oestrus detection in suckled *Bos indicus* cows.

Table 3. Reproductive parameters in suckled Nelore cows managed under four different breeding programs during a 90 d breeding season, Camapua, MS, Brazil. Adapted from Baruselli *et al.* 2005.

Breeding strategy	FTAI	First 45 d of the breeding season			Overall (90-d breeding season)	
	Pregnancy rate	Oestrus detection rate	Conception rate	Pregnancy rate	Pregnancy rate	Mean interval to conception (d)
FTAI[1] + Bulls[2]	76/150 (50.7%)	-	-	113/150 [a] (75.3%)	139/150 [a] (92.7%)	29.3±2.0 [a]
FTAI[1] +OED & AI[3] +Bulls[2]	81/148 (54.3%)	17/67 (25.4%)	13/17 (76.5%)	94/148 [b] (63.5%)	136/148 [a] (91.9%)	31.1±2.2 [a]
OED & AI[3] + Bulls[2]	-	59/150 (39.3%)	35/66 (53.0%)	35/150 [d] (23.3%)	125/147 [b] (85.0%)	57.3±2.3 [b]
Bulls[2]	-	-	-	66/149 [c] (44.3%)	124/149 [b] (83.2%)	46.5±1.9 [c]

Means and percentages differ significantly ([abcd], P <0.05).
[1]FTAI: fixed-time artificial insemination on Day 10 of the breeding season.
[2]Bulls: bulls until Day 90 of the breeding season.
[3]OED & AI: oestrus detection and AI until Day 45 of the breeding season.

Another example is the "Hofig Ramos Agricultura e Pecuaria", located in Brasilandia, Brazil. In this farm, an aggressive FTAI program was implemented in 5,579 suckled Nelore cows in 2005. Cows were enrolled in a FTAI program early in the postpartum period (i.e. 35 to 45 d postpartum) using Crestar ear implants and 400 IU eCG at the time of implant removal and 10 d later were exposed to clean-up bulls for the remainder of the breeding season. Pregnancy rate to the FTAI was 50.5% (2817/5579) and the overall pregnancy rate after two cycles of re-breeding with bulls was 80.7% (4390/5579). As in the previous examples, comparison of survival curves

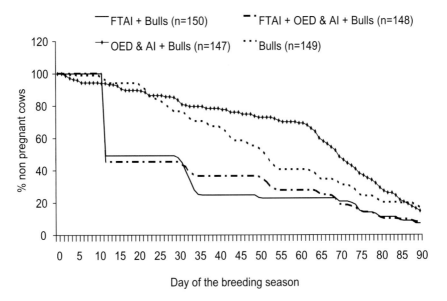

Fig. 3. Survival curves for days open in suckled Nelore cows managed under four different breeding programs during a 90 d breeding season, Camapua, MS, Brazil. Survival curves in the FTAI+Bulls and FTAI+OED&AI+Bulls breeding schemes differ from the OED&AI+Bulls and the Bulls breeding schemes (P<0.01). Adapted from Baruselli *et al.* 2005.

for calving distributions in 2005 and the projected calvings for 2006 confirmed the notion that the use of a progestin-based FTAI program at the beginning of the breeding season increased the number of calvings early in the calving season (Fig. 4).

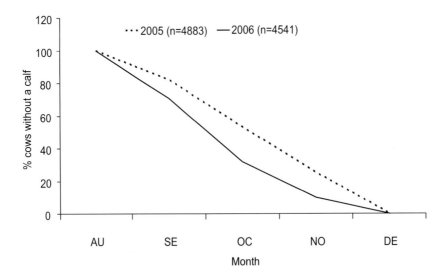

Fig. 4. Survival curves for calving distribution at "Hofig Ramos Agricultura e Pecuaria", located in Brasilandia, Brazil. Curves differ significantly between years (P<0.01). Marques *et al.* 2006.

Another program worth mentioning is that applied at "Cabaña Ministaló", in Río Ceballos, Córdoba, Argentina. This operation is located in an area that is more temperate than the others, with about 800 to 1000 mm of rain per year, in a seasonal fashion from October to June. This is a mixed operation (soybean and corn crops and beef cattle) with purebred Brangus and Braford cattle (3/8 *Bos indicus* and 5/8 *Bos taurus*). Fixed-time AI has been done in November and December of the last 5 years in 22 to 26 month old heifers and suckled cows that were 45 to 70 d postpartum. In this herd, animals have always been in good BCS (2.5 to 3.5) at the beginning of the breeding season; the FTAI treatment consisted of a progesterone-releasing device (Triu-B, 1 g of progesterone, Biogénesis, Argentina; DIB, 1 g of progesterone, Syntex SA, Argentina or CIDR-B, 1.9 g of progesterone, Pfizer Animal Health, Argentina), with 2 mg EB on Day 0, device removal and PGF on Days 7 or 8, 1 mg EB 24 h later and FTAI from 52 to 56 h after device removal. As the goal was to increase the number of offspring produced by AI, animals were re-synchronized by re-insertion of the progesterone-releasing device on Day 13 after FTAI. EB (1 mg im) was also given to cows (but not to heifers) on Day 13. In this case, oestrus was detected for 5 d after device removal (Days 20 to 25 after the first FTAI) and all animals were inseminated 8 to 12 h after the onset of oestrus. As is shown in Table 4, pregnancy rates with FTAI were similar over the 5 years (P > 0.88). Overall pregnancy rates with AI decreased during 2004 (P < 0.05), compared to the two previous years, due to failures in oestrus detection after the re-synchronization protocol, which demonstrates the sensitivity of systems that depend on oestrus detection in beef cattle. This was corrected in 2005 by the use of a re-synchronization protocol that included a second FTAI. Briefly, the re-synchronization treatment consisted of re-insertion of a once-used progesterone-releasing device from Day 16 to Day 21 and GnRH was given on Day 21. On Day 28, all cows and heifers were examined by ultrasonography and those that were open received PGF at that time, followed by 1 mg EB im 24 h later and FTAI 30 h after EB. As is shown in Table 4, avoiding oestrus detection for the second AI overcame the problem and approximately 70% of the cows were pregnant after two inseminations.

Table 4. Pregnancy rates with FTAI, oestrus detection rate, conception and pregnancy rates following re-synchronization and overall pregnancy rates on a purebred Brangus and Braford farm, "Cabaña Ministaló", Córdoba, Argentina. Adapted from Bó *et al.* 2005.

Year	Pregnancy rate FTAI	Re-synchronization			Cumulative pregnancy rate
		Oestrus detection rate	Conception rate	Pregnancy rate	
2001	107/189	44/82 [ab]	24/44	24/82 [ab]	131/189 [ab]
	(56.6%)	(53.7%)	(54.5%)	(29.3%)	(69.3%)
2002	104/192	49/88 [ab]	35/49	35/88 [b]	139/192 [b]
	(51.2%)	(55.7%)	(71.4%)	(39.7%)	(72.4%)
2003	128/228	71/100 [b]	36/71	36/100 [b]	164/228 [b]
	(56.1%)	(71.0%)	(50.7%)	(36.0%)	(71.9%)
2004	149/279	50/130 [a]	25/50	25/130 [a]	174/279 [a]
	(53.4%)	(38.4%)	(50.0%)	(19.2%)	(62.4%)
2005[1]	164/309	—	—	65/145 [b]	229/309 [b]
	(53.1%)			(44.8%)	(74.1%)
Total	652/1197	214/400	120/214	185/544	837/1197
	(54.5%)	(53.5%)	(56.1%)	(34.0%)	(69.9%)

[ab] Proportions in the same column with different superscripts differ (P < 0.05).
[1] In 2005 cows and heifers were not observed for oestrus and were re-inseminated based on a FTAI protocol.

Conclusions

Currently, the world's economic situation requires efficient management practices to increase the profitability of beef cattle operations. Optimal reproductive efficiency is crucial to increase net returns. The use of animal breeding technologies has become of great importance, particularly in tropical and subtropical areas where AI is the only alternative to introduce *Bos taurus* genetics into Zebu-based herds. However, variability in response to the traditional PGF-based hormonal treatments and the time and effort required to perform oestrus detection, particularly in *Bos indicus* influenced cattle, have limited the widespread application and success of these technologies. The incorporation of techniques designed to control follicular wave dynamics and ovulation in recent years has reduced problems associated with oestrus detection. Furthermore treatments with progestin-releasing devices, oestradiol and eCG have provided possibilities for the application of FTAI in suckled cows and to advance the resumption of cyclicity in cows that were in anoestrus. Furthermore, fertility in successive cycles and overall pregnancy rates at the end of the breeding season have been shown to improve with the use of progestin-releasing devices at the beginning of the breeding season in cows that were well managed and on an increasing plane of nutrition. However, it is very important to recognize that the success of the program will also depend on many management factors such as the nutritional and health management, availability of qualified personnel, facilities and the objectives of the breeding program.

Aknowledgments

Research was supported by FAPESP, Brazil (08363-0/2002), the Instituto de Reproducción Animal Córdoba (IRAC) and Estancia "El Mangrullo S.A.", Argentina. We also thank Syntex S.A., Argentina; Biogenesis, Argentina; Pfizer Animal Health, Argentina and Brazil; Intervet, Brazil and Tecnopec, Brazil for the hormones used in the studies. Special thanks to our colleagues of IRAC, U. of São Paulo, GeraEmbryo and FIRMASA IATF for technical assistance. e-mail: gabrielbo@iracbiogen.com.ar

References

Alonso A, Mapletoft RJ, Bó GA, Tribulo HE, Carcedo J, Tribulo R & Menajovsky JR 1995 Niveles de hormona luteinizante y de estrógeno en hembras *bos indicus* tratadas con prostaglandina F2a. Resultados preliminares. XIV Reunión Latinoamericana de Producción Animal, Mar del Plata, Argentina. *Revista Argentina de Producción Animal* **15** 961-963.

Barros CM, Figueiredo RA & Pinheiro OL 1995 Estro, ovulação e dinâmica folicular em zebuínos. *Revista Brasileira de Reprodução Animal* **19** 9-22.

Barros CM, Moreira, MBB, Figueiredo RA, Teixeira AB & Trinca LA 2000 Synchronization of ovulation in beef cows (Bos indicus), using GnRH, PGF2a and estradiol benzoate. *Theriogenology* **53** 1121-1134.

Baruselli PS, Marques MO, Nasser LF, Reis EL & Bó GA 2003 Effect of eCG on pregnancy rates of lactating zebu beef cows treated with CIDR-B devices for timed artificial insemination. *Theriogenology* **59** 214.

Baruselli PS, Reis EL, Marques MO, Nasser LF & Bó GA 2004 The use of treatments to improve reproductive performance of anestrus beef cattle in tropical climates. *Animal Reproduction Science* **82-83** 479-486.

Baruselli PS, Bó GA, Reis EL, Marques MO & Sá Filho MF 2005 Introdução da IATF no manejo reproductivo de rebanhos bovinos de corte no Brasil. *Proc VI Simposio Internacional de Reproducción Animal, June 24-26 2005*, Córdoba, Argentina, pp 151-176.

Bertram Membrive CM 2000 Estudo da sincronização das ondas foliculares e das características de estros, por radio telemetría, em novilhas cruzadas (*Bos indicus x Bos taurus*) tratadas com acetato de melengestrol e prostaglandina associados a hCG, GnRH ou 17â estradiol + progesterona. Tesis de Maestría. Universidade de São Paulo, Faculdade de Medicina Veterinária e Zootecnia, São Paulo, SP, Brazil.

Bó GA & Baruselli PS 2002 Programas de Inseminación Artificial a Tiempo Fijo en el Ganado Bovino en Regiones Subtropicales y Tropicales. Capítulo XXXI. In *Avances en la Gandería doble propósito*, pp 499-

514. Eds C Gonzalez-Stagnaro, E Soto Belloso & L Ramírez Iglesia. Maracaibo, Venezuela: Fundación Girarz.

Bó GA, Cutaia L & Tribulo R 2002a Tratamientos hormonales para inseminación artificial a tiempo fijo en bovinos para carne: algunas experiencias realizadas en Argentina. Primera Parte. *Taurus* **14** 10-21.

Bó GA, Cutaia L & Tribulo R. 2002b. Tratamientos hormonales para inseminación artificial a tiempo fijo en bovinos para carne: algunas experiencias realizadas en Argentina. Segunda Parte. *Taurus* **15** 17-32.

Bó GA, Baruselli PS & Martinez MF 2003 Pattern and manipulation of follicular development in *Bos indicus* cattle. *Animal Reproduction Science* **78** 307-326.

Bó GA, Cutaia L, Chesta P, Balla E, Pincinato D, Peres L, Maraña D, Avilés M, Menchaca A, Veneranda G & Baruselli PS 2005 Implementacion de programas de inseminación artificial en rodeos de cría de argentina. *Proc VI Simposio Internacional de Reproducción Animal, June 24-26 2005, Córdoba, Argentina,* pp 97-128.

Chenoweth PJ 1994 Aspects of reproduction in female *Bos indicus* cattle: a review. *Australian Veterinary Journal* **71** 422-426.

Cutaia L, Tríbulo R, Moreno D & Bó GA 2003a Pregnancy rates in lactating beef cows treated with progesterone releasing devices, estradiol benzoate and equine chorionic gonadotropin (eCG). *Theriogenology* **59** 216.

Cutaia L, Veneranda G, Tribulo R, Baruselli PS & Bó GA 2003b Programas de Inseminación Artificial a Tiempo Fijo en Rodeos de Cría: Factores que lo Afectan y Resultados Productivos. *Proc V Simposio Internacional de Reproducción Animal. June 25-27 2003, Huerta Grande, Córdoba, Argentina,* pp 119-132.

D'Occhio MJ, Neish A & Broadhurst L 1990 Differences in gonadotrophin secretion postpartum between Zebu and European breed cattle. *Animal Reproduction Science* **22** 311-317.

Fernandes P, Teixeria AB, Crocci AJ & Barros CM 2001 Timed artificial insemination in beef cattle using GnRH agonist, PGF2a and estradiol benzoate (EB). *Theriogenology* **55** 1521-1532.

Galina CS & Arthur GH 1990 Review on cattle reproduction in the tropics. Part 4. Oestrus cycles. *Animal Breeding Abstracts* **58** 697-707.

Galina CS, Orihuela A & Rubio I 1996 Behavioural trends affecting oestrus detection in Zebu cattle. *Animal Reproduction Science* **42** 465-470.

Gimenes LU, Carvalho NAT, Sá Filho MF, Santiago LL, Carvalho JBP, Mapletoft RJ, Barros CM & Baruselli PS 2005a Capacidade ovulatória em novilhas bos indicus. *Acta Scientiae Veterinariae* **33** (Supplement) 209.

Gimenes LU, Sá Filho MF, Madure EH, Trinca LA, Barros CM, Baruselli PS 2005b Estudo ultrasonográfico da divergência folicular em novilhas Bos indicus. *Acta Scientiae Veterinariae* **33** (Supplement) 210.

Ginther OJ, Wiltbank MC, Fricke PM, Gibbons JR & Kot K 1996 Selection of the dominant follicle in cattle. *Biology of Reproduction* **55** 1187-1194.

Geary TW, Whittier JC, Hallford DM & MacNeil MD 2001 Calf removal improves conception rates to the Ovsynch and Co-synch protocols. *Journal of Animal Science* **79** 1-4.

Kastelic JP & Ginther OJ 1991 Factors affecting the origin of the ovulatory follicle in heifers with induced luteolysis. *Animal Reproduction Science* **26** 13-24.

Kerr DR, McGowan MR, Carroll CL, Baldock FC 1991 Evaluation of three estrus synchronization regimens for use in extensively managed *bos-indicus* and *bos indicus/taurus* heifers is northern Australia. *Theriogenology* **36** 129-138.

Lamb GC, Stevenson JS, Kesler DJ, Garverick HA, Brown DR, Salfen BE 2002 Inclusion of an intravaginal progesterone insert plus GnRH and prostaglandin F2α ovulation control in postpartum suckled beef cows. *Journal of Animal Science* **79** 2253-2259.

Lemaster JW, Yelich JV, Kempfer JR, Fullenwider JK, Barnett CL, Fanning MD, Selph, JF 2001 Effectiveness of GnRH plus prostaglandin F2a for estrus synchronization in cattle of Bos indicus breeding. *Journal of Animal Science* **79** 309-316.

Macmillan KL & Burke CR 1996 Effects of oestrus cycle control on reproductive efficiency. *Animal Reproduction Science* **42** 307-320.

Maraña D, Cutaia L, Peres L, Pincinato D, Borges LFK & Bó GA 2006 Ovulation and pregnancy rates in postpartum bos indicus cows treated with progesterone vaginal devices and estradiol benzoate, with or without eCG and temporary weaning. *Reproduction Fertility and Development* **18** 116-117.

Marcantonio SA 2003 El mercado del semen bovino en Argentina. *Taurus* **19** 11-17.

Martinez MF, Kastelic JP, Adams GP & Mapletoft RJ 2002 The use of a progesterone-releasing device (CIDR) or melengestrol acetate with GnRH, LH or estradiol benzoate for fixed-time AI in beef heifers. *Journal of Animal Science* **80** 1746-1751.

McGowan MR 1999 Sincronización de celos y programas de inseminación artificial a tiempo fijo en ganado *bos-indicus* y cruza *bos indicus*. *Proc III Simposio Internacional de Reproducción Animal, June 19-21 1999, Carlos Paz, Córdoba, Argentina,* pp 71-82.

Mizuta K 2003 Estudo comparativo dos aspectos comportamentais do estro e dos teores plasmáticos de LH, FSH, progesterona e estradiol que precedem a ovulação em fêmeas bovinas Nelore (*Bos taurus indicus*), Angus (*Bos taurus taurus*) e Nelore × Angus (*Bos taurus indicus* × *Bos taurus taurus*), PhD Thesis. Universidade de São Paulo, Faculdade de Medicina Veterinária e Zootecnia, São Paulo, SP, Brazil.

Moreno I, Galina CS, Escobar FJ, Ramirez B & Navarro-Fierro R 1986 Evaluation of the lytic response of PGF2α in Zebu cattle based on serum progesterone. *Theriogenology* **25** 413-421.

Odde KG 1990 A review of synchronization of estrus in postpartum cattle. *Journal of Animal Science* **68** 817-830.

Penteado L, Ayres H, Madureira E.H. & Baruselli PS 2004 Efeito do desmame temporário na taxa de prenhez de vacas Nelore lactantes inseminadas em tempo fixo. *Acta Scientiae Veterinariae* **32** (Supplement) 223.

Pincinato D, Coelho Peres L, Miranda G, Cutaia L & Bó GA 2006 Follicular dymanics and fertility in beef suckled cows synchronized with progesterone releasing devices and GnRH. *Acta Scientiae Veterinariae* **33** (Supplement) Submitted.

Pinheiro OL, Barros CM, Figueredo RA, Valle ER, Encarnação RO & Padovani CR 1998 Estrous behaviour and the estrus-to-ovulation interval in Nelore cattle (*Bos indicus*) with natural estrus or estrus induced with prostaglandin F2α or norgestomet and estradiol valerate. *Theriogenology* **49** 667-681.

Pursley JR, Mee MO & Wiltbank MC 1995 Synchronization of ovulation in dairy cows using PGF2á and GnRH. *Theriogenology* **44** 915-923.

Pursley JR, Wiltbank MC, Stevenson JS, Ottobre JS, Garverick HA & Anderson LL 1997 Pregnancy rates per artificial insemination for cows and heifers inseminated at a synchronized ovulation or synchronized estrus. *Journal of Dairy Science* **80** 295-300.

Randel RD 1984 Seasonal effects on female reproductive functions in the bovine (Indian breeds). *Theriogenology* **21** 170-185.

Randel RD 1994 Unique reproductive traits of Brahman and Brahman based cows. *In: Factors affecting calf crop*, pp 23-43. Eds MJ Field & RS Sand. Boca Ratón, FL, USA: CRC Press.

Rocha JL 2000 Sincronização hormonal da onda folicular e do estro em novilhas de corte mestiças monitoradas por radio telemetría. Thesis Doctoral. Universidade de São Paulo, Faculdade de Medicina Veterinária e Zootecnia, São Paulo, SP, Brazil.

Ruiz-Cortes ZT & Olivera-Angel M 1999 Ovarian follicular dynamics in suckled zebu (*Bos indicus*) cows monitored by real time ultrasonography. *Animal Reproduction Science* **54** 211-220.

Saldarriaga JP, Cooper DA, Cartmill JA, Stanko RL & Williams GL 2005 Sincronización de la ovulación e inseminación artificial a tiempo fijo en ganado cruza *bos indicus* en Estados Unidos. *VI Simposio Internacional de Reproducción Animal, June 24-26, Córdoba, Argentina*, pp 97-128.

Sartorelli ES, Carvalho LM, Bergfelt DR, Ginther OJ & Barros CM 2005 Morphological characterization of follicle deviation in Nelore (Bos indicus) heifers and cows. *Theriogenology* 2005 **63** 2382-2394.

Sartori R, Fricke PM, Ferreira JCP, Ginther OJ & Wiltbank MC 2001 Follicular deviation and acquisition of ovulatory capacity in bovine follicles. *Biology of Reproduction* **65** 1403-1409.

Segerson EC, Hansen TR, Libby DW, Randel RD & Getz WR 1984 Ovarian and uterine morphology and function in Angus and Brahman cows. *Journal of Animal Science* **59** 1026-1046.

Soto Belloso E, Portillo Martinez G, De Ondíz A, Rojas N, Soto Castillo G, Ramírez Iglesia L, Perea Ganchou F 2002 Improvement of reproductive performance in crossbred zebu anestrus suckled primiparous cows by treatment with norgestomet implants or 96 h calf removal. *Theriogenology* **57** 1503-1510.

Stahringer RC, Neuendorff DA & Randel RD 1990 Seasonal variations in characteristics of estrous cycles in pubertal Brahman heifers. *Theriogenology* **34** 407-415.

Williams GL 1990 Suckling as a regulator of pospartum rebreeding in cattle. A review. *Journal of Animal Science* **68** 831-852.

Williams SW, Stanko RL, Amstalden M, Williams GL 2002 Comparison of three approaches for synchronization of ovulation for timed artificial insemination in *Bos indicus*-influenced cattle managed on the Texas gulf coast. *Journal of Animal Science* **80** 1173-1178.

Wiltbank MC, Gumen A, Sartori R 2002 Physiological classification of anovulatory conditions in cattle. *Theriogenology* **57** 21-52.

Yelich JV 2002 A vaginal insert (CIDR) to synchronize estrus and timed-AI. *In: Factors affecting calf crop: biotechnology of reproduction*, pp 87-100. Eds MJ Fields, RS Sands & JV Yelich. Boca Raton, FL USA: CRC Press.

Fertility in high-producing dairy cows: Reasons for decline and corrective strategies for sustainable improvement

MC Lucy

158 Animal Sciences Research Center, University of Missouri, Columbia MO USA 65211

The fertility of dairy cows has declined worldwide and this change is surprising given the importance of good fertility to the dairy industry. The decline in fertility can be explained by management changes within the dairy industry and also negative genetic correlations between milk production and reproduction. Four primary mechanisms that depress fertility in lactating cows are anovulatory and behavioral anestrus (failure to cycle and display estrus), suboptimal and irregular estrous cyclicity (this category includes ovarian disease and subnormal luteal function after breeding), abnormal preimplantation embryo development (may be secondary to poor oocyte quality), and uterine/placental incompetence. The solution for improving fertility in high-producing dairy cows will include both short-term and long-terms components. For the immediate short-term, using high fertility sires and implementing controlled breeding programs will help. Controlled breeding programs improve reproductive efficiency in confinement-style dairy herds and can be combined with post-insemination treatments to enhance fertility. An additional immediate short-term solution involves changing the diet so that dietary ingredients invoke hormonal responses that benefit the reproduction of the cow. The short-term solutions described above do not address the fundamental need for correcting the underlying genetics for reproduction in high-producing dairy cows. Crossbreeding will improve reproductive performance perhaps because it alleviates inbreeding and also lowers production in cows with an extreme high milk production phenotype. The current crisis in dairy reproduction will be permanently solved, however, when the genetics for dairy reproduction are improved through a balanced genetic selection strategy.

Introduction

Reproduction is important for sustainable dairying worldwide but reproductive efficiency has declined for dairy cows. Reproductive traits have low heritability so a major component of reproductive decline can be attributed to changes in the dairy industry (larger farms, less-skilled labor, etc.) that make reproductive management more difficult (Lucy 2001). Nonetheless, fertility breeding values for dairy cows have shown evidence of decline since 1957 (Lucy

Corresponding author E-mail: LucyM@missouri.edu

2005; data from United States dairy cows). The genetic decline in dairy fertility can be explained in part by negative genetic correlations between milk production and reproduction (Hansen 2000). One widely held theory is that the strategic use of adipose tissue for energy and milk substrates in early lactation leads to low postpartum body condition that in turn leads to poor reproductive performance (Lucy 2003; Pryce & Harris 2006). The dissemination of high-milk producing genetics (predominately Holstein) from relatively few sire families led to a global problem in dairy reproduction (Lucy 2001). Reproductive decline is perhaps most acutely felt in seasonal pasture-based systems where non-pregnant cows are not carried over until the next calving season (Harris and Kolver 2001). Confinement-style dairy farmers that practice continuous calving also view reproduction as an area of concern (Lucy et al. 2004; Moore & Thatcher 2006).

Reproductive rates are declining in lactating dairy cows but reproductive rates in dairy heifers (non-lactating) remain relatively high. A conception rate of 64% was found when over 330,000 inseminations to over 220,000 United States Holstein heifers were examined (January 2003 to October 2004; Kuhn & Hutchinson 2005). The conception rate is considerably higher than the 20 to 40% conception rate typically reported for lactating cows in the United States. One conclusion from the Holstein heifer data is that the reproductive system of modern dairy cattle is essentially normal when lactation demands are not imposed. Perhaps unexpectedly, however, the same study found a positive association between heifer fertility and daughter pregnancy rate (DPR; the cow fertility trait used in the United States). Subsequent analyses have reported negative genetic correlations between heifer fertility and breeding values for lactation traits (milk, fat, and protein; VanRaden 2006). Failing to address the antagonistic genetic relationship between milk production traits and reproductive traits, therefore, will erode the fertility of both lactating cows and heifers.

Four primary components of infertility in dairy cows

Infertility in dairy cattle is multi-faceted and will require a holistic approach that addresses the problem. The scientific literature on dairy cow infertility is extensive. Indeed, the key words "dairy cow infertility" returned at least twice as many citations when compared to equivalent citations in other farm animals (Fig. 1). Dairy cattle, like any species, have a theoretical optimum for conception rate that is probably above 70% (a conception rate that can be achieved in dairy heifers selected for fertility; Andersen-Ranberg et al. 2005). Factors that impinge upon the lactating cow act collectively to decrease conception rate from the optimum (Fig. 2). Four primary mechanisms that depress fertility in lactating cows will be discussed herein. They are anovulatory and behavioral anestrus (failure to cycle and display estrus), suboptimal and irregular estrous cyclicity (this category includes ovarian disease and subnormal luteal function after breeding), abnormal preimplantation embryo development (may be secondary to poor oocyte quality), and uterine/placental incompetence. It is not surprising that reproduction is in decline when one considers that each of these primary components acting alone can cause infertility.

Anovulatory and behavioral anestrus

A period of anovulatory anestrus (ovarian follicular development without ovulation; also termed the anovulatory period) is normal for postpartum cows. In beef cows (considered highly fertile relative to dairy), suckling inhibits LH pulsatility and the lack of LH pulsatility leads to anovulatory anestrus (Williams & Griffith 1995). As long as nutrition is adequate, the anestrous beef

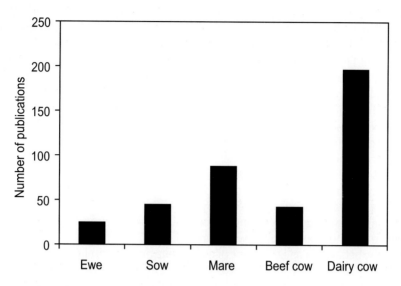

Fig. 1. Number of scientific publications for "infertility" for different farm animal species returned in a June 2006 search of the PubMed scientific citation database (National Center for Biotechnology Information; National Library of Medicine, Bethesda, Maryland, USA).

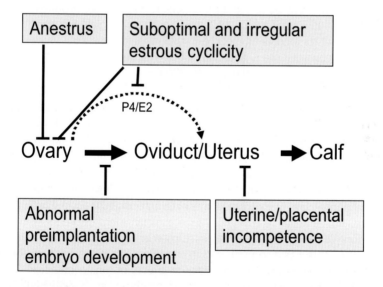

Fig 2. Four primary components of infertility in dairy cows (light gray boxes) that act at different levels of the reproductive process. The primary components act collectively to inhibit (blunt end lines) ovarian and oviductal/uterine function (see text for details). P4/E2 = progesterone/estradiol.

cow is fertile once she starts cycling (Lucy *et al.* 2001). Likewise, New Zealand dairy cattle (grazing management system) are anestrus for longer than North American dairy cattle (confinement management system) but have better fertility (Meyer *et al.* 2004). Thus, the length of postpartum anovulatory anestrus *per se* may not be a major factor contributing to infertility of cattle unless the anestrus period extends into the breeding period.

High-producing dairy cows have extended periods of anovulatory anestrus (Roche *et al.* 2000; Gong 2002; Lucy 2003). The anestrus is caused by negative energy balance that contributes to a hormonal milieu (low blood LH, insulin, and IGF-I) that is inadequate for preovulatory follicular development, the LH surge and ovulation (Lucy 2003). In dairy cows, anovulatory anestrus is symptomatic of a catabolic state (Beam & Butler 1999). A benign state of anovulatory anestrus in beef cows and low-producing dairy cows is not the same biologically for high-producing dairy cows because the underlying causes of anestrus are different. Although a beef cow can be treated and recover from anestrus with normal fertility (Lucy *et al.* 2001), an anestrous dairy cow that is treated has lower fertility (Gumen *et al.* 2003) perhaps because her anestrum is a consequence of negative energy balance. Anovulatory anestrus can be treated with progestogen supplementation (Wiltbank *et al.* 2002; Rhodes *et al.* 2003). Routine progesterone treatment of anestrous dairy cows has been questioned, however, particularly in New Zealand. McDougall & Compton (2006) found that cows treated for anestrus calved early in the next calving season but failed to retain any advantage during the subsequent breeding period. Treating an anestrous cow and keeping her may only perpetuate the anestrous problem within a herd.

Until recently, behavioral anestrus (lack of estrus behavior for a cyclic cow) was viewed as only occurring at first postpartum ovulation (Inskeep 1995). Otherwise, cows classified as behaviorally anestrus were thought to have had estrus activity but this activity was not observed by humans (failed detection of estrus). Examination of data from electronic mount detectors demonstrated that high milk-producing cows had shorter estrous periods, fewer standing events, and less standing time when compared to low-milk producing cows (Lopez *et al.* 2004). The differences in estrous expression were linked to low blood estradiol concentrations in high-producing dairy cows. Low rates of estrous expression caused by low blood estradiol may explain the popularity of injectable formulations of estradiol for increasing blood estradiol, enhancing estrous expression and promoting ovulation in dairy cows (Moore & Thatcher 2006). The use of estradiol for this purpose is only approved in some dairying countries (Lucy *et al.* 2004).

Suboptimal and irregular estrous cyclicity

Cows with estrous cycle abnormalities have poorer reproductive performance than their normal-cycling herdmates (Hommeida *et al.* 2005; Mann *et al.* 2005; Petersson *et al.* 2006). The prevalence of estrous cycle abnormalities can approach 50% in some herds. Estrous cycle abnormalities are classified into three primary types: 1) extended period of anovulatory anestrus, 2) temporary cessation of luteal phases, and 3) long luteal phases (greater than 20 days). Factors known to affect postpartum cows such as negative energy balance, periparturient disorders, and postpartum diseases are risk factors for abnormal estrous cycles (Opsomer *et al.* 2000). The incidence of twinning and cystic ovarian disease has also increased in modern dairy cattle because there are positive genetic correlations between these abnormalities and level of milk production (Vanholder *et al.* 2006; Wiltbank *et al.* 2006). Lucy (2003) proposed that a common physiological mechanism (low LH pulsatility, low blood growth factor concentrations, and enhanced steroid metabolism) may underlie the increased incidence of anovulatory anestrus, abnormal estrous cycles, and twinning. This common mechanism appears to be a consequence of the hormonal and metabolic state that supports a high level of milk production. Cystic ovarian disease apparently arises through a completely different physiological mechanism because cystic cows have high blood LH activity (Vanholder *et al.* 2006).

There is accumulating evidence that one component of infertility in dairy cows is caused by low blood progesterone concentrations after insemination (Stronge *et al.* 2005; McNeill *et al.* 2006a; Starbuck *et al.* 2006). A slow rise in progesterone delays embryonic development because early embryonic growth is partially dependent on progesterone perhaps acting at the level of the oviduct or endometrium (Green *et al.* 2005; Mann *et al.* 2006; McNeill *et al.* 2006b). Cows with low progesterone had equivalent blood LH concentrations and in vitro steriodogenic capacity when compared to normal cows (Robinson *et al.* 2006). Thus, the corpus luteum (CL) may be normal but its capacity to elevate blood progesterone may be less in lactating cows. The relatively large body size of dairy cows may create a large tissue pool size and steroid metabolism may be greater (Wiltbank *et al.* 2006). The combined effects of pool size and turnover rate may lead to low blood progesterone. The same mechanism may lead to low blood estradiol in high-producing dairy cows (see above).

Abnormal preimplantation embryo development

The metabolic state of high-producing dairy cows may have a direct effect on the oocyte. Snijders *et al.* (2000) found that in vitro fertilized oocytes from dairy cows in low body condition had a lower cleavage rate and a lower developmental rate when compared with oocytes from dairy cows in better body condition. Oocyte quality could be improved, however, when low body condition heifers were fed at a high level (Adamiak *et al.* 2005). Sartori *et al.* (2002) flushed the reproductive tract of lactating cows and found fewer cleavage stage embryos when compared to similar flushes in nonlactating cows. Nonesterified fatty acids (NEFA) are released from adipose tissue in early lactation and their concentrations are increased in follicular fluid (approximately 40% of serum concentrations; Leroy *et al.* 2005). The increase in NEFA within follicular fluid may decrease the proliferation of granulosa cells (Vanholder *et al.* 2005) and may also affect the oocyte directly. Both Burkhart *et al.* (2005) (confinement system dairy cows) and McDougall *et al.* (2005a) (pasture system dairy cows) found that high NEFA concentrations were predictive of low fertility postpartum. The addition of NEFA to in vitro maturation medium decreased maturation rate, fertilization rate, cleavage rate, and blastocyst yield for in vitro cultured embryos (Leroy *et al.* 2005). Collectively the data suggest that early embryonic development is compromised by lactation perhaps through elevated NEFA that enters the follicular fluid and damages the oocyte. Fewer embryos reach the cleavage stage because oocyte quality is low.

If the primary mechanism leading to infertility involves the oocyte or early embryo then embryo transfer should improve conception rates in postpartum dairy cows. Heat stress is known to negatively affect the early embryo, for example, and heat-stressed dairy cows subjected to embryo transfer have a higher conception rate than those inseminated by conventional AI (Hansen *et al.* 2001). Embryo transfer pregnancy rates for dairy cows were increased by greater than 10% over control dairy cows inseminated at estrus in two studies (Demetrio *et al.* 2006; Vasconcelos *et al.* 2006) but a third study failed to demonstrate an effect (Sartori *et al.* 2006). Thus, there is some evidence to support the concept that fertility can be recovered in dairy cows by circumventing the period of oocyte and early embryonic development. The condition of the recipient cow, however, is one factor that can potentially affect the outcome because Mapletoft *et al.* (1986) reported that body condition has a large effect on embryo transfer success (higher body condition score cows have higher pregnancy rates after transfer). Thus uterine environment as affected by body condition plays some role in the fertility of dairy cows.

Uterine/placental incompetence

The uterus of postpartum cows may appear grossly normal but nonetheless fail to support pregnancy (Gilbert et al. 2005). Failure to support the pregnancy typically manifests itself during the early embryonic period (Santos et al. 2004). Embryonic loss was traditionally viewed as occurring during the period of maternal recognition of pregnancy (days 17 to 21 of pregnancy; Thatcher et al. 2001). Losses during this early period extended the estrous cycle in inseminated cows but were manageable because cows returned to estrus during the fourth week after AI. Ultrasonography revolutionized pregnancy detection in beef and dairy cattle because pregnancy could be detected as early as day 25 after insemination (about 1 to 2 weeks earlier than manual palpation; Miyamoto et al. 2006). Examination of ultrasound data revealed that appreciable numbers of embryos died after the initial pregnancy examination (done between days 25 and 28 after AI; Santos et al. 2004). This later period of embryonic loss leads to the "phantom cow syndrome" where inseminated cows fail to return to estrus and are difficult to resynchronize using conventional methods (Cavalieri et al. 2005). A likely period for embryonic death may be during placentation (fourth to sixth week of pregnancy) because placentation involves intricate communication between maternal and fetal tissues (King et al. 1982). Whether or not the incidence of embryonic loss is greater now than in the past is debated because the capacity to routinely detect early pregnancy in cattle evolved with the use of ultrasound in cattle (after 1984; Pierson & Ginther 1984).

Embryonic loss in modern dairy cattle probably arises from predisposing factors that are common in dairy systems. Dairymen may inseminate cows early postpartum because they are fearful that they will not observe a subsequent estrus. Cows inseminated early postpartum are more likely to have embryonic loss whether they are in confinement or pasture-based systems (McDougall et al. 2005b; Meyer et al. 2006). Disease is a predisposing factor as well. Gilbert et al. (2005) found that 53% of cows had evidence of uterine inflammation (endometritis) at 40 to 60 days postpartum. Cows with endometritis had lower first service conception rates, required more services per conception, and had pregnancy rates at 300 days postpartum that were 26 percentage points lower than cows with a healthy endometrium. Similar effects were observed in pasture-based dairy cows (McDougall et al. 2006). A link between mastitis and early embryonic loss has also been established (Chebel et al. 2004). The mastitic mammary gland activates immune cells whose inflammatory cytokines adversely affect the ovary and uterus (Hansen et al. 2004).

The final predisposing factor for embryonic loss arises from the relatively low body condition of postpartum dairy cows. Several studies have tied high postpartum milk production or low postpartum body condition to early embryonic loss (Grimard et al. 2006; Vasconcelos et al. 2006). Silke et al. (2002) found that embryonic loss after day 28 of pregnancy was highest in cows losing the greatest amount of body condition.

Short-term strategies for increasing fertility in dairy cows

There are both short and long-term solutions for solving dairy infertility. Some short-term solutions have no conceivable drawbacks and should be enacted immediately. Other short-term solutions can be implemented immediately but they are not necessarily sustainable solutions for dairy cow infertility because they may be too expensive, too difficult to enact, or unacceptable in the eye of the public or the dairy farmer. Individual solutions have more or less merit depending on the laws governing dairy production and the economics of the dairy production system.

Using high fertility sires

The importance of semen handling and AI technique to successful reproduction cannot be understated. Assuming that semen is handled properly and placed appropriately within the female reproductive tract then the next obvious step is to use highly fertile dairy sires. Dairy sire fertility is calculated and published in the United States by the Animal Improvement Programs Laboratory (AIPL; Beltsville, Maryland, USA) as Estimated Relative Conception Rate (ERCR). Dairy sires have inherent differences in fertility that are related to capacitation time and sperm survival in the female reproductive tract (Saacke *et al.* 2000). The ERCR is the estimated deviation from herd conception rate that can be expected with the use of a specific sire (i.e., +4 is four percentage points above herd average, etc.). Given inherent variation in reproductive data, ERCR should only be used for sires with a large number of services. Cornwell *et al.* (2006) used high versus low fertility sires in a timed AI program and demonstrated a tendency for an increase in conception rate when high fertility sires were used (6 percentage point increase; P = 0.12). There is little difference in Net Merit (NM$) for sires that are stratified across a wide range of ERCR (Fig. 3). Thus, it is possible to achieve genetic gain while using sires with superior fertility in an AI system.

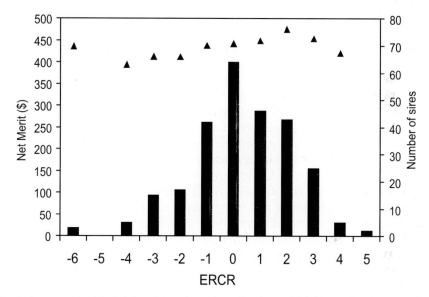

Fig. 3. Average Net Merit (triangle) and number of active artificial insemination sires (bar) for different estimated relative conception rates (ERCR; statistics are for the May 2006 evaluation; Animal Improvement Programs Laboratory; Beltsville, Maryland, USA).

Intensive reproductive management programs (synchronization and resynchronization)

An immediate solution to combat infertility in dairy herds involves intensive management of the estrous cycle and ovulation (estrous synchronization and timed AI). Numerous review articles have been published on the methods that can be used to do this (Diskin *et al.* 2002; Rhodes *et al.* 2003; Lucy *et al.* 2004; Moore & Thatcher 2006). Most approaches employ a method for controlling follicular wave development, promoting ovulation in anestrous cows, regressing the corpus luteum in cyclic cows, and synchronizing estrus and (or) ovulation at the end of treatment. In many dairy herds, cows are inseminated after spontaneous estrus for a predetermined period and then cows that have not been inseminated are managed intensively (timed AI). More intensive approaches to reproductive management involve programmed breeding for all inseminations without any type of estrous detection.

There is variation between countries in availability and regulatory requirements for hormonal treatments used in estrous synchronization. For example, estradiol benzoate is actively used in New Zealand and Australia but is not registered in the European Union or the United States. The only approved estradiol for United States dairy cows (estradiol cypionate) was voluntarily removed from the market by its manufacturer (Pfizer Animal Health). New Zealand and Australia have had intravaginal progesterone releasing devices for use in lactating cows for over 15 years but the devices were only recently approved for United States dairy cattle. In the United Kingdom, $PGF_{2\alpha}$ must be administered by a veterinary surgeon and this requirement makes its routine use too expensive. There is global public concern about the blanket application of hormones to food-producing animals.

Timed AI is popular in large confinement dairies because the benefits of a timed AI system increase under conditions of poor estrous detection rate (Lucy *et al.* 2004). A popular method for timed AI practiced in North American herds is "Ovsynch" (GnRH; wait seven days; $PGF_{2\alpha}$; wait two days; GnRH; Lucy *et al.* 2004). Cows are either inseminated at the same time as the last GnRH (Cosynch) or 16 to 24 hours after the final GnRH treatment (Ovsynch). In a meta analysis of 53 research papers, Rabiee *et al.* (2005) concluded that conception and pregnancy rates after synchronization programs with estrous detection and after Ovsynch were similar. The Ovsynch program has distinct advantages over estrous synchronization procedures because estrous detection is not required and every cow is inseminated at the end of treatment (100% submission rate). A progesterone-containing device (CIDR) can be added into the Ovsynch program (inserted after the first GnRH and removed after the first $PGF_{2\alpha}$) and this will improve conception rate in some herds (Stevenson *et al.* 2006). Follicular wave synchronization followed by timed AI is more efficacious when cows are between days 5 and 12 of the estrous cycle. Thus, a pre-synchronization strategy can be employed in which cows are treated with a series of $PGF_{2\alpha}$ injections before the Ovsynch protocol (Thatcher *et al.* 2002). Pre-synchronization improves conception rate after Ovsynch by 5 to 10%.

Cows that are not pregnant after first insemination can be resynchronized for second AI. Progesterone-alone can be used for the purpose of grouping estruses in cows that are not pregnant after first insemination (McDougall 2003). For resynchronization timed AI, the first GnRH injection of Ovsynch can be given to all cows approximately one week before pregnancy diagnosis (Chebel *et al.* 2003). Cows that are subsequently diagnosed non-pregnant can be injected with $PGF_{2\alpha}$ and 48 hours later injected with GnRH before timed AI. An alternative method is to simply start cows back on Ovsynch once they are diagnosed non-pregnant (Sterry *et al.* 2006).

An obvious detraction for pre-synchronization timed AI methods (for example, Presynch-Ovsynch) is that a series of five injections is required and the injections occur over a 45-day period. If a post-insemination treatment is applied (see below) and cows are placed back on a re-synchronization program then a cow that is not pregnant to first insemination (the most-probable outcome) will receive nine injections before her second insemination and a total of ten injections if she is again treated post-insemination (Fig. 4). Assuming a 35% timed AI conception rate, a group of 100 cows would receive 860 injections and achieve 58 pregnancies after two inseminations (about 15 injections per pregnancy). Many managers of large herds feel that scheduling reproductive treatments and inseminations is simpler and more effective than multiple daily sessions of estrous detection. Their approach has merit, given the difficult nature of estrous detection in large herds (Lucy 2001). The example stated above assumes no estrous detection. In reality, North American herds that use Presynch-Ovsynch and re-synchronization typically inseminate any cow that is seen in estrus (essentially terminating the program until pregnancy examination; Stevenson & Phatak 2005).

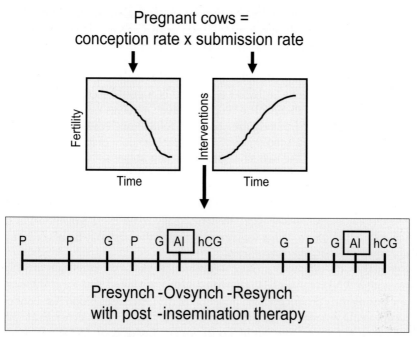

Fig. 4. The number of pregnant cows is a function of conception rate and submission rate. As conception rates (fertility) have declined over time, farmers have increased submission rates by intensive reproductive intervention such as timed AI. An example of a timed AI program is the Presynch-Ovsynch-Resynch program that includes a post-insemination treatment (hCG) to increase fertility. In this system a cow receives six injections before or shortly after first AI (P = $PGF_{2\alpha}$; G = GnRH). If she is not pregnant for first AI then she will receive four additional injections for second AI. The success rate of the program is about 35% for first AI and somewhat less for second AI so that only 50 to 60% of cows are pregnant after two AI. Repeated hormonal injections that return relatively low pregnancy rates may not be a sustainable option for maintaining fertility on dairy farms.

An average of fifteen injections per dairy cow pregnancy is a troublesome number. A question raised by Macmillan *et al.* (2003), and restated here, concerns the effectiveness of our current programs for estrous synchronization. When is the point of diminishing returns reached? Estrous synchronization and timed AI programs have become the primary method to combat the declining trend in fertility within North American dairy herds. What was once a method to control the estrous cycle and to group cows in estrus is now the only possible means of achieving acceptable submission rates in large confinement dairies. The situation for dairy cows contrasts greatly from that of beef cows where timed AI programs with fewer injections can achieve conception rates above 60% (Schafer *et al.* 2005). Improving the underlying fertility of dairy cattle (see below) may increase pregnancy rates for timed AI in dairy cows and simplify timed AI programs because fewer injections will be needed.

Treating cows after insemination

Treatments can be applied to dairy cows after insemination in an effort to improve fertility. The reader is referred to several recent reviews of the subject (Macmillan *et al.* 2003; Thatcher et *al.* 2006). There are three primary strategies. The first strategy addresses low progesterone during the first week after breeding. An injection of GnRH or hCG given between days 5 to 8 of the estrous cycle will cause the ovulation of accessory CL in some cows and may also improve CL function (hCG through its LH-like activity). A progesterone-containing device may also be inserted during this period and left in place for approximately one week. The aforementioned treatments may increase progesterone in blood and there is a positive correlation between blood progesterone and fertility (see above). The second strategy involves GnRH treatment later in the estrous cycle. Turning over or ovulating the dominant follicle decreases estradiol and blocks the luteolytic mechanism. Delaying luteolysis will increase the amount of time that the embryo has to signal the mother. In practice, the two strategies described above suffer from herd by treatment interactions where there is a positive response in some herds but not others. The underlying cause of the herd effect is unknown. The treatments may be most-successful when they are applied to lactating cows in low body condition (i.e., targeting cows with the greatest risk of infertility; Thatcher *et al.* 2006). It may be necessary to periodically retest these treatments as dairy cattle continue to evolve in the future. What was not effective in the past may be effective in the future because dairy cows have changed genetically.

The third strategy is the administration of recombinant bovine somatotropin (rbST) around the time of insemination. Recombinant bST is inexpensive in the United States and is approved for the purpose of increasing milk production. An injection of rbST increases blood IGF-I concentrations. Embryonic and uterine tissues typically respond positively to IGF-I (Thatcher *et al.* 2006). Cows with elevated blood IGF-I are more fertile. For example, Taylor *et al.* (2004) reported that dairy cows with blood IGF-I concentrations greater than 50 ng/ml had a five-fold increase in pregnancy rate. Application of rbST at AI has been shown to be efficacious for increasing pregnancy rate for cows inseminated by timed AI and at estrus (Thatcher *et al.* 2006).

Feeding diets that are designed to improve fertility

Developing diets that increase the fertility of dairy cows has always been an attractive option to scientists and farmers. In North American confinement-style herds and in pasture-based systems, farmers have some flexibility in terms of the diets and supplements that they feed. The diet is mixed and fed along a fence-line or in the milking parlor so there is no need to handle individual cows when feeding a specially-designed diet.

Negative energy balance, weight loss, and decreased body condition score occur during early lactation when nutrient requirements for maintenance and lactation exceed the ability of the cow to consume energy in the feed. Cows in negative energy balance have lower blood concentrations of insulin and IGF-I (Lucy 2004). Low blood IGF-I causes reduced negative feedback on growth hormone (GH) and an increase in blood GH concentrations (Lucy 2000). Greater blood GH increases liver gluconeogenesis and promotes lipolysis (NEFA release) from adipose tissue. High blood GH and NEFA concentrations antagonize insulin action and create a state of insulin resistance in postpartum cows. The insulin resistance blunts glucose utilization by non-mammary tissues and conserves glucose for milk synthesis. The cycle described above (low IGF-I, high GH, low glucose, low insulin, and insulin resistance) is gradually reversed during the first 4 to 8 weeks of lactation.

The aforementioned endocrine hormones (insulin and IGF-I) that are metabolically controlled can influence GnRH and LH secretion (Lucy 2003). Insulin and IGF-I can also act directly on the ovary to increase the sensitivity of the ovary to LH and FSH. Postpartum dairy cows are thought to be less sensitive to LH and FSH because their insulin and IGF-I concentrations are low. Although typically thought to affect ovarian function, the insulin/IGF system is clearly resident within the uterus and embryo (Watson *et al.* 1999). Therefore, insulin and IGF-I may be tied to an effect of body condition on the uterus and embryo.

Roche *et al.* (2006) found that cows with North American pedigrees in a New Zealand system increased milk production (and not body condition) when fed additional energy. Likewise, North American dairy cows consume feed ad libitum but nevertheless use all available nutrients for milk production at the expense of body condition (Bauman & Currie 1980). Nutrient partitioning prevents the transition out of the catabolic state because all of the available glucose is consumed, primarily for milk synthesis (Lucy 2004). Feeding more energy may not solve reproductive problems in dairy cows selected for milk production because the cows will partition additional nutrients toward milk production and not toward adipose or reproductive tissues. The metabolic state of low insulin, low IGF-I and elevated GH is maintained despite the higher level of feeding.

Perhaps a more realistic approach to feeding dairy cows for fertility is to provide specific nutrients that are designed to impinge upon the endocrine system of the cow (nutraceutical-type approach). Examples of this include feeding hyperinsulinemic diets (Gong *et al.* 2002) and supplementing with propylene glycol (Miyoshi *et al.* 2001; Butler *et al.* 2006). In each case, blood glucose and insulin concentrations are strategically increased and fertility may be improved because the cow is "tricked" into thinking that she is anabolic. It is also possible to tailor the fatty acid composition of the diet. Feeding polyunsaturated fatty acids may improve reproduction in dairy cows because the $PGF_{2\alpha}$-synthesizing luteolytic mechanism is attenuated (Mattos *et al.* 2000).

Long-term strategies for increasing fertility in dairy cows

The problems facing reproduction in dairy cattle are not simple. A reversal in the current trends can be achieved only through a variety of approaches. The short-term solutions provided above should be pursued. In the long-term, the current trend in inbreeding needs to be attenuated and cows should be actively selected for improved reproductive efficiency.

Inbreeding in dairy cows

Inbreeding in dairy cattle breeds has increased dramatically since 1980 and may play a role in reproductive decline (Funk 2006). Present levels of inbreeding for United States cows are greater than 5% and continue to increase in most breeds. Inbreeding negatively affects reproductive and longevity traits in dairy cows (Sewalem *et al.* 2006; VanRaden & Miller 2006). One way to correct inbreeding is through crossbreeding. Dairy farmers in New Zealand routinely crossbreed their cows; so much so that crossbred cows may soon outnumber purebred Holstein-Friesian (Harris 2005). Crossbreeding in the United States is practiced by a small number of farmers. The major limitation is that the Holstein breed is superior to all others in terms of milk production. Thus, although there is heterosis for milk production, the crossbred cow produces less milk than the Holstein (Heins *et al.* 2006a). Holstein-Jersey crossbred cows had better fertility than Holstein cows when studied within a university research herd (Heins *et al.* 2006b). In the long-term, it may be necessary to develop multiple lines of dairy cattle with equivalent capacity for milk production so that crossbreeding can be used to maintain genetic diversity and capitalize on heterosis.

Improving dairy cattle genetics for reproduction traits

Genetic selection programs for dairy cattle have capitalized on partitioning nutrients away from adipose tissue. This was not a preplanned strategy of genetic selection (i.e., cows were not purposely selected for low body condition in early lactation) but instead was a consequence of the genetic selection for milk production (the highest producing cows had genetics that supported the low body condition phenotype). The homeorhetic mechanisms that supported the low body condition phenotype were viewed as positive and highly desirable, particularly in the North American system. Low body condition during lactation, however, antagonizes reproduction (Pryce & Harris 2006). Dairy fertility has economic value but how much value does it have relative to the value of milk? Reproduction will not improve if it is undervalued relative to other traits in the selection index but reproduction should not be over-valued relative to other traits simply to correct a perceived problem.

There has clearly been a change in the way we select dairy cattle. A historical examination of the primary selection indices in the United States clearly shows a shift toward longevity and functional traits since the mid-1990's (Fig. 5). The worldwide decline in dairy fertility is being addressed by including fertility traits in selection indices (Lucy 2005). The Scandinavians were the first to do this, and other countries followed in the past decade (Lindhe & Philipsson 1998). It is impossible to capture each of the individual fertility components listed above. Instead, time to pregnancy, i.e. the most meaningful outcome, is measured. The United States has adopted daughter pregnancy rate (DPR) for fertility weightings (VanRaden *et al.* 2004). The DPR is based on days open, i.e. the number of days from calving to conception. A 1% increase in DPR is equivalent to a 4-day reduction in days open. In untreated cattle, the DPR captures cyclicity, expression of estrus and fertility (conception rate), in a single measure. The DPR breeding value for North American Holstein and Jersey cows has declined since 1957 but appears to have stabilized (Lucy 2005). The correlation between DPR and NM$ for United States dairy sires is nearly zero; meaning that sires at the top of the selection index are neutral for DPR (Fig. 6). There is clearly a negative correlation between DPR and milk traits such as protein yield. Thus, the selection index (that theoretically reflects profitability) may be the best method for selecting future sires because balanced selection does not place reproductive traits at a disadvantage.

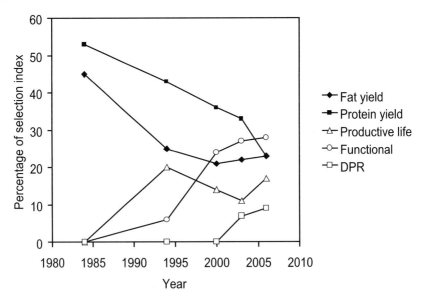

Fig. 5. Relative emphasis of different traits in United States dairy selection indexes (Animal Improvement Programs Laboratory, Beltsville, Maryland, USA). Since 1994, the weightings for fat and protein yield have decreased whereas the weightings for productive life, functional traits (somatic cell score, udder, feet and legs, etc.) and daughter pregnancy rate (DPR) have increased.

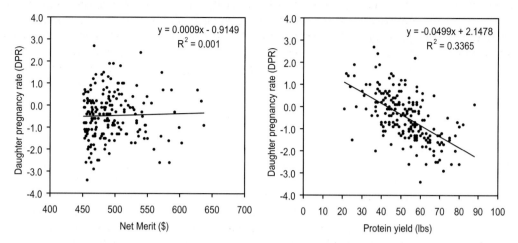

Fig. 6. Regression of daughter pregnancy rate (DPR) on Net Merit (left panel) and protein yield (right panel) for Holstein sires (Animal Improvement Programs Laboratory, Beltsville, Maryland, USA). There is no relationship between DPR and Net Merit. This indicates that DPR does not change across sire rankings within the Net Merit index. Using a single trait such as protein yield, however, clearly leads to lower DPR when the top sires are selected.

Most of the available literature suggests that dairy cattle have a genetically-determined set point for body condition during lactation (Stockdale 2001). Once dairy cows begin lactation, they will migrate toward their body condition set point through the coordinated depletion of adipose tissue. The magnitude of adipose tissue loss does not depend on nutrient demands *per se* but instead depends on the available adipose tissue mass and the genetically-determined set point for the individual cow. There is wide-spread consensus that the genetically-determined set point for body condition during lactation affects the reproductive performance of dairy cows (Berry *et al.* 2003; Pryce & Harris 2006). The lower body condition of modern dairy cows reflects the genetic predisposition to direct nutrients away from body fat during lactation (homeorhetic mechanism that supports milk production).

Since there are strong positive genetic trends between body condition and reproductive performance then selection programs based on postpartum body condition score should alleviate some reproductive loss (Pryce *et al.* 2002). Milk that is made in early lactation could be made in later lactation if selection indices emphasized a persistent lactation instead of a high peak milk yield. This change in emphasis would improve the body condition of cows during the breeding period and could improve reproductive success simply because cows are in better body condition when they are inseminated. There has been little attention paid to residual feed intake (RFI; the difference between an animal's actual feed intake and its expected feed intake based on nutrient requirements; Crews 2005). Such an energetic efficiency measure may have utility particularly when feed has limited availability (pasture systems) or represents a high percentage of costs (confinement systems). If RFI were applied to lactating dairy cows then the RFI calculation would have to account for the milk energy gained from adipose tissue mobilization. Otherwise, cows that lose excessive adipose tissue would have a low RFI but also a low body condition.

Reproductive traits have low heritabilities but the coefficient of variation for reproductive traits is large. Therefore, genetic selection for good fertility is possible in dairy cattle (Weigel

& Rekaya 2000). The current situation with dairy reproduction genetics is not a blind alley. The problem can be corrected without retrenching in terms of milk production. It is, of course, scientifically appealing to predict how dairy cows will change if they are simultaneously selected for both milk production and reproduction traits. The above discussion implies that high fertility cows will have lactation curves that are flatter so that the cow maintains better body condition throughout lactation. An alternative and equally reasonable possibility is that high-producing cows with good fertility will have reproductive systems that function efficiently under conditions of high milk production. There is no biological reason that this cannot occur in as much as some species reproduce naturally under extreme metabolic conditions. For example, Bauman & Currie (1980) described the work of Miescher who in 1880 noted that the reproductive organs of salmon developed extensively during their migration up the Rhine River when 55% of muscle mass was lost. Selection for milk production was done without any preconceived notion as to how it would change the cow. Likewise, genetic selection for reproductive traits should be practiced in the same manner with the sole objective of achieving better fertility through a balanced approach. Simply maintaining current fertility levels may be unacceptable because pregnancy rates are low and the level of reproductive intervention is high on modern dairy farms.

Conclusions

Single-pronged approaches will probably not reverse the current decline in dairy fertility because the underlying causes are multi-faceted and appear to affect the reproductive process at nearly every level. In the short-term, aggressive reproductive management (treatment of anestrus, use of high fertility sires, estrous synchronization and re-synchronization, post-insemination treatments, etc.) should maintain current reproductive rates. Routine treatment of food animals with hormones will likely become a concern to the public so these approaches may become unavailable to farmers. In the United States, for example, we have recently witnessed the removal of a popular estrous synchronization product from the market (estradiol cypionate). Formulating diets to improve reproduction is perhaps a more sustainable option because farmers are used to changing diets to suit their management objectives. A long-term solution is to improve the reproductive genetics of the dairy cow. This includes addressing the potential impact of inbreeding and also reversing the genetic trends that underlie the current pattern of reproductive decline. Although progress toward greater milk production may be less, the cow will be healthier and easier to manage because she will become pregnant more easily.

References

Andersen-Ranberg IM, Klemetsdal G, Heringstad B & Steine T 2005 Heritabilities, genetic correlations, and genetic change for female fertility and protein yield in Norwegian Dairy Cattle. *Journal of Dairy Science* **88** 348-355.

Adamiak SJ, Mackie K, Watt RG, Webb R & Sinclair KD 2005 Impact of nutrition on oocyte quality: cumulative effects of body composition and diet leading to hyperinsulinemia in cattle. *Biology of Reproduction* **73** 918-926.

Bauman DE & Currie WB 1980 Partitioning of nutrients during pregnancy and lactation: a review of mechanisms involving homeostasis and homeorhesis. *Journal of Dairy Science* **63** 1514-1529.

Beam SW & Butler WR 1999 Effects of energy balance on follicular development and first ovulation in postpartum dairy cows. *Journal of Reproduction and Fertility Supplement* **54** 411-424.

Berry DP, Buckley F, Dillon P, Evans RD, Rath M & Veerkamp RF 2003 Genetic relationships among body condition score, body weight, milk yield, and fertility in dairy cows. *Journal of Dairy Science* **86** 2193-204.

Burkhart M, Youngquist R, Spain J, Sampson J, Bader J, Vogel R, Lamberson W & Garverick HA 2005 NEFA and glucose levels in serum of periparturient dairy cows are indicative of pregnancy success at first service. *Journal of Animal Science* **83** (Supplement 1) 299 (abstract).

Butler ST, Pelton SH & Butler WR 2006 Energy balance, metabolic status, and the first postpartum ovarian follicle wave in cows administered propylene glycol. *Journal of Dairy Science* **89** 2938-2951.

Cavalieri J, Rabiee AR, Hepworth G & Macmillan KL 2005 Effect of artificial insemination on submission rates of lactating dairy cows synchronised and resynchronised with intravaginal progesterone releasing devices and oestradiol benzoate. *Animal Reproduction Science* **90** 39-55.

Chebel RC, Santos JE, Cerri RL, Galvao KN, Juchem SO & Thatcher WW 2003 Effect of resynchronization with GnRH on day 21 after artificial insemination on pregnancy rate and pregnancy loss in lactating dairy cows. *Theriogenology* **60** 1389-1399.

Chebel RC, Santos JE, Reynolds JP, Cerri RL, Juchem SO & Overton M 2004 Factors affecting conception rate after artificial insemination and pregnancy loss in lactating dairy cows. *Animal Reproduction Science* **84** 239-255.

Cornwell JM, McGilliard ML, Kasimanickam R & Nebel RL 2006 Effect of sire fertility and timing of artificial insemination in a Presynch + Ovsynch protocol on first-service pregnancy rates. *Journal of Dairy Science* **89** 2473-2478.

Crews DH Jr 2005 Genetics of efficient feed utilization and national cattle evaluation: a review. *Genetics and Molecular Research* **4** 152-165.

Demetrio DGB, Santos RM, Demetrio CGB, Rodrigues CA, & Vasconcelos JLM 2006 Factors affecting conception of AI or ET in lactating cows. *Journal of Animal Science* **84** (Supplement 1) 207-208 (abstract).

Diskin MG, Austin EJ & Roche JF 2002 Exogenous hormonal manipulation of ovarian activity in cattle. *Domestic Animal Endocrinology* **23** 211-228.

Funk DA 2006 Major advances in globalization and consolidation of the artificial insemination industry. *Journal of Dairy Science* **89** 1362-1368.

Gilbert RO, Shin ST, Guard CL, Erb HN & Frajblat M 2005 Prevalence of endometritis and its effects on reproductive performance of dairy cows. *Theriogenology* **64** 1879-1888.

Gong JG 2002 Influence of metabolic hormones and nutrition on ovarian follicle development in cattle: practical implications. *Domestic Animal Endocrinology* **23** 229-241.

Gong JG, Lee WJ, Garnsworthy PC & Webb R 2002 Effect of dietary-induced increases in circulating insulin concentrations during the early postpartum period on reproductive function in dairy cows. *Reproduction* **123** 419-427.

Green MP, Hunter MG & Mann GE 2005 Relationships between maternal hormone secretion and embryo development on day 5 of pregnancy in dairy cows. *Animal Reproduction Science* **88** 179-189.

Gumen A, Guenther JN & Wiltbank MC 2003 Follicular size and response to Ovsynch versus detection of estrus in anovular and ovular lactating dairy cows. *Journal of Dairy Science* **86** 3184-3194.

Grimard B, Freret S, Chevallier A, Pinto A, Ponsart C & Humblot P 2006 Genetic and environmental factors influencing first service conception rate and late embryonic/foetal mortality in low fertility dairy herds. *Animal Reproduction Science* **91** 31-44.

Hansen LB 2000 Consequences of selection for milk yield from a geneticist's viewpoint. *Journal of Dairy Science* **83** 1145-1150.

Hansen PJ, Drost M, Rivera RM, Paula-Lopes FF, al-Katanani YM, Krininger CE & Chase CC Jr. 2001 Adverse impact of heat stress on embryo production: causes and strategies for mitigation. *Theriogenology* **55** 91-103.

Hansen PJ, Soto P & Natzke RP 2004 Mastitis and fertility in cattle - possible involvement of inflammation or immune activation in embryonic mortality. *American Journal of Reproductive Immunology* **51** 294-301.

Harris BL 2005 Breeding dairy cows for the future of New Zealand. *New Zealand Veterinary Journal* **53** 384-389.

Harris BL & Kolver ES 2001 Review of Holsteinization on intensive pastoral dairy farming in New Zealand. *Journal of Dairy Science* **84** E56-E61.

Heins BJ, Hansen LB & Seykora AJ 2006a Production of pure Holsteins versus crossbreds of Holstein with Normande, Montbeliarde, and Scandinavian Red. *Journal of Dairy Science* **89** 2799-2804.

Heins BJ, Hansen LB, Seykora AJ, Hazel AR, Linn JG, Johnson DG, & Hansen WP 2006b Crossbreds of Jersey/Holstein compared to pure Holsteins for production, calving difficulty, stillbirths, and fertility. *Journal of Dairy Science* **89** (Supplement 1) 245-246 (abstract).

Hommeida A, Nakao T & Kubota H 2005 Onset and duration of luteal activity postpartum and their effect on first insemination conception rate in lactating dairy cows. *Journal of Veterinary Medical Science* **67** 1031-1035.

Inskeep EK 1995 Factors that affect fertility during oestrous cycles with short or normal luteal phases in postpartum cows. *Journal of Reproduction and Fertility Supplement* **49** 493-503.

King GJ, Atkinson BA & Robertson HA 1982 Implantation and early placentation in domestic ungulates. *Journal of Reproduction and Fertility Supplement* **31** 17-30.

Kuhn M & Hutchison J 2005 Factors affecting heifer fertility in US Holsteins. *Journal of Dairy Science* **88** (Supplement 1) 11 (abstract).

Leroy JL, Vanholder T, Mateusen B, Christophe A, Opsomer G, de Kruif A, Genicot G & Van Soom A 2005 Non-esterified fatty acids in follicular fluid of dairy cows and their effect on developmental capacity of bovine oocytes in vitro. *Reproduction* **130** 485-495.

Lindhe B & Philipsson J 1998 Conventional breeding programmes and genetic resistance to animal diseases. *Reviews in Science and Technology* **17** 291-301.

Lopez H, Satter LD & Wiltbank MC 2004 Relationship between level of milk production and estrous behavior of lactating dairy cows. *Animal Reproduction Science* 81 209-223.

Lucy MC 2000 Regulation of ovarian follicular growth by somatotropin and insulin-like growth factors in cattle. *Journal of Dairy Science* 83 1635-1647.

Lucy MC 2001 Reproductive loss in high-producing dairy cattle: Where will it end? *Journal of Dairy Science* 84 1277-1293.

Lucy MC, Billings HJ, Butler WR, Ehnis LR, Fields MJ, Kesler DJ, Kinder JE, Mattos RC, Short RE, Thatcher WW, Wettemann RP, Yelich JV & Hafs HD 2001 Efficacy of an intravaginal progesterone insert and an injection of PGF2a for synchronizing estrus and shortening the interval to pregnancy in postpartum beef cows, peripubertal beef heifers, and dairy heifers. *Journal of Animal Science* 79 982-995.

Lucy MC 2003 Mechanisms linking nutrition and reproduction in postpartum cows. *Reproduction Supplement* 61 415-427.

Lucy MC 2004 Mechanisms linking the somatotropic axis with insulin: Lessons from the postpartum dairy cow. *Proceedings of the New Zealand Society of Animal Production* 64 19-23.

Lucy MC, McDougall S & Nation DP 2004 The use of hormonal treatments to improve the reproductive performance of lactating dairy cows in feedlot or pasture-based management systems. *Animal Reproduction Science* 82-83 495-512.

Lucy MC 2005 Non-lactational traits of importance in dairy cows and applications for emerging biotechnologies. *New Zealand Veterinary Journal* 53 406-415.

Macmillan KL, Segwagwe BV & Pino CS 2003 Associations between the manipulation of patterns of follicular development and fertility in cattle. *Animal Reproduction Science* 78 327-344.

Mann GE, Mann SJ, Blache D & Webb R 2005 Metabolic variables and plasma leptin concentrations in dairy cows exhibiting reproductive cycle abnormalities identified through milk progesterone monitoring during the post partum period. *Animal Reproduction Science* 88 191-202.

Mann GE, Fray MD & Lamming GE 2006 Effects of time of progesterone supplementation on embryo development and interferon-tau production in the cow. *Veterinary Journal* 171 500-503.

Mapletoft RJ, Lindsell CE & Pawlshyn V 1986 Effects of clenbuterol, body condition, and nonsurgical embryo transfer equipment on pregnancy rates in bovine recipients. *Theriogenology* 25 172 (abstract).

Mattos R, Staples CR, Thatcher WW 2000 Effects of dietary fatty acids on reproduction in ruminants. *Reviews in Reproduction* 5 38-45.

McDougall S 2003 Resynchrony of previously anoestrous cows and treatment of cows not detected in oestrus that had a palpable corpus luteum with prostaglandin F2 alpha. *New Zealand Veterinary Journal* 51 117-124.

McDougall S, Blache D, Rhodes FM 2005a Factors affecting conception and expression of oestrus in anoestrous cows treated with progesterone and oestradiol benzoate. *Animal Reproduction Science* 88 203-214.

McDougall S, Rhodes FM & Verkerk G 2005b Pregnancy loss in dairy cattle in the Waikato region of New Zealand. *New Zealand Veterinary Journal* 53 279-287.

McDougall S & Compton C 2006 Reproductive performance in the subsequent lactation of dairy cows previously treated for failure to be detected in oestrus. *New Zealand Veterinary Journal* 54 132-140.

McDougall S, Macaulay R & Compton C 2006 Association between endometritis diagnosis using a novel intravaginal device and reproductive performance in dairy cattle. *Animal Reproduction Science* (in press).

McNeill RE, Diskin MG, Sreenan JM & Morris DG 2006a Associations between milk progesterone concentration on different days and with embryo survival during the early luteal phase in dairy cows. *Theriogenology* 65 1435-1441.

McNeill RE, Sreenan JM, Diskin MG, Cairns MT, Fitzpatrick R, Smith TJ & Morris DG 2006b Effect of systemic progesterone concentration on the expression of progesterone-responsive genes in the bovine endometrium during the early luteal phase. *Reproduction Fertility and Development* 18 573-583.

Meyer JP, Verkerk GA, Gore PJ, Macdonald KA, Holmes CW & Lucy MC 2004 Effect of genetic strain, feed allowance, and parity on interval to first ovulation and the first estrous cycle in pasture-managed dairy cows. *Journal of Animal Science* 82 (Supplement 1) 66 (abstract).

Meyer JP, Radcliff RP, Rhoads ML, Bader JF, Murphy CN & Lucy MC 2006 Timed Artificial Insemination of Two Consecutive Services in Dairy Cows Using Prostaglandin F2α and Gonadotropin-releasing Hormone *Journal of Dairy Science (in press)*.

Miyamoto A, Shirasuna K, Hayashi KG, Kamada D, Awashima C, Kaneko E, Acosta TJ & Matsui M 2006 A potential use of color ultrasound as a tool for reproductive management: New observations using color ultrasound scanning that were not possible with imaging only in black and white. *Journal of Reproduction and Development* 52 153-160.

Miyoshi S, Pate JL & Palmquist DL 2001 Effects of propylene glycol drenching on energy balance, plasma glucose, plasma insulin, ovarian function and conception in dairy cows. *Animal Reproduction Science* 68 29-43.

Moore K & Thatcher WW 2006 Major advances associated with reproduction in dairy cattle. *Journal of Dairy Science* 89 1254-1266.

Opsomer G, Gröhn YT, Hertl J, Coryn M, Deluyker H & de Kruif A 2000 Risk factors for post partum ovarian dysfunction in high producing dairy cows in Belgium: a field study. *Theriogenology* 53 841-857.

Petersson KJ, Gustafsson H, Strandberg E & Berglund B 2006 Atypical progesterone profiles and fertility in Swedish dairy cows. *Journal of Dairy Science* 89 2529-2538.

Pierson RA & Ginther OJ 1984 Ultrasonography of the bovine ovary. *Theriogenology* **21** 495-504.

Pryce JE, Coffey MP, Brotherstone SH & Woolliams JA 2002 Genetic relationships between calving interval and body condition score conditional on milk yield. *Journal of Dairy Science* **85** 1590-1595.

Pryce JE & Harris BL 2006 Genetics of body condition score in New Zealand dairy cows. *Journal of Dairy Science* **89** 4424-4432.

Rabiee AR, Lean IJ & Stevenson MA 2005 Efficacy of Ovsynch program on reproductive performance in dairy cattle: a meta-analysis. *Journal of Dairy Science* **88** 2754-2770.

Rhodes FM, McDougall S, Burke CR, Verkerk GA & Macmillan KL 2003 Invited review: Treatment of cows with an extended postpartum anestrous interval. *Journal of Dairy Science* **86** 1876-1894.

Robinson RS, Hammond AJ, Nicklin LT, Schams D, Mann GE & Hunter MG 2006 Endocrine and cellular characteristics of corpora lutea from cows with a delayed post-ovulatory progesterone rise. *Domestic Animal Endocrinology* **31** 154-172.

Roche JF, Mackey D & Diskin MD 2000 Reproductive management of postpartum cows. *Animal Reproduction Science* **60-61** 703-712.

Roche JR, Berry DP & Kolver ES 2006 Holstein-Friesian strain and feed effects on milk production, bodyweight and body condition score profiles in grazing dairy cows. *Journal of Dairy Science* **89** 3532-3543.

Saacke RG, Dalton JC, Nadir S, Nebel RL & Bame JH 2000 Relationship of seminal traits and insemination time to fertilization rate and embryo quality. *Animal Reproduction Science* **60-61** 663-677.

Santos JEP, Thatcher WW, Chebel RC, Cerri RLA, Galvao KN 2004 The effect of embryo death rates in cattle on the efficiency of estrus synchronization programs. *Animal Reproduction Science* **82-83** 513-535.

Sartori R, Sartor-Bergfelt R, Mertens SA, Guenther JN, Parrish JJ & Wiltbank MC 2002 Fertilization and early embryonic development in heifers and lactating cows in summer and lactating and dry cows in winter. *Journal of Dairy Science* **85** 2803-2812.

Sartori R, Gumen A, Guenther JN, Souza AH, Caraviello DZ & Wiltbank MC 2006 Comparison of artificial insemination versus embryo transfer in lactating dairy cows. *Theriogenology* **65** 1311-1321.

Schafer DJ, Bader JF, Meyer JP, Haden JK, Ellersieck MR, Smith MR, & Patterson DJ 2005 A comparison of progestin-based protocols to synchronize ovulation prior to fixed-time artificial insemination in postpartum beef cows. *Journal of Animal Science* **83** (Supplement 1) 85.

Sewalem A, Kistemaker GJ, Miglior F & Van Doormaal BJ 2006 Analysis of inbreeding and its relationship with functional longevity in Canadian dairy cattle. *Journal of Dairy Science* **89** 2210-2216.

Silke V, Diskin MG, Kenny DA, Boland MP, Dillon P, Mee JF & Sreenan JM 2002 Extent, pattern and factors associated with late embryonic loss in dairy cows. *Animal Reproduction Science* **71** 1-12.

Snijders SE, Dillon P, O'Callaghan D & Boland MP 2000 Effect of genetic merit, milk yield, body condition and lactation number on in vitro oocyte development in dairy cows. *Theriogenology* **53** 981-989.

Starbuck GR, Gutierrez CG, Peters AR & Mann GE 2006 Timing of follicular phase events and the postovulatory progesterone rise following synchronisation of oestrus in cows. *Veterinary Journal* **172** 103-108.

Sterry RA, Welle ML & Fricke PM 2006 Effect of interval from timed artificial insemination to initiation of resynchronization of ovulation on fertility of lactating dairy cows. *Journal of Dairy Science* **89** 2099-2109.

Stevenson JS & Phatak AP 2005 Inseminations at estrus induced by presynchronization before application of synchronized estrus and ovulation. *Journal of Dairy Science* **88** 399-405.

Stevenson JS, Pursley JR, Garverick HA, Fricke PM, Kesler DJ, Ottobre JS & Wiltbank MC 2006 Treatment of cycling and noncycling lactating dairy cows with progesterone during Ovsynch. *Journal of Dairy Science* **89** 2567-2578.

Stockdale CR 2001 Body condition at calving and the performance of dairy cows in early lactation under Australian conditions: a review. *Australian Journal of Experimental Agriculture* **41** 823-839.

Stronge AJ, Sreenan JM, Diskin MG, Mee JF, Kenny DA & Morris DG 2005 Post-insemination milk progesterone concentration and embryo survival in dairy cows. *Theriogenology* **64** 1212-1224.

Taylor VJ, Cheng Z, Pushpakumara PG, Beever DE & Wathes DC 2004 Relationships between the plasma concentrations of insulin-like growth factor-I in dairy cows and their fertility and milk yield. *Veterinary Record* **155** 583-588.

Thatcher WW, Guzeloglu A, Mattos R, Binelli M, Hansen TR & Pru JK 2001 Uterine-conceptus interactions and reproductive failure in cattle. *Theriogenology* **56** 1435-1450.

Thatcher WW, Moreira F, Pancarci SM, Bartolome JA & Santos JE 2002 Strategies to optimize reproductive efficiency by regulation of ovarian function. *Domestic Animal Endocrinology* **23** 243-254.

Thatcher WW, Bilby TR, Bartolome JA, Silvestre F, Staples CR & Santos JE 2006 Strategies for improving fertility in the modern dairy cow. *Theriogenology* **65** 30-44.

Vanholder T, Leroy JL, Soom AV, Opsomer G, Maes D, Coryn M & de Kruif A 2005 Effect of non-esterified fatty acids on bovine granulosa cell steroidogenesis and proliferation in vitro. *Animal Reproduction Science* **87** 33-44.

Vanholder T, Opsomer G & de Kruif A 2006 Aetiology and pathogenesis of cystic ovarian follicles in dairy cattle: a review. *Reproduction Nutrition and Development* **46** 105-119.

VanRaden PM, Sanders AH, Tooker ME, Miller RH, Norman HD, Kuhn MT & Wiggans GR 2004 Development of a national genetic evaluation for cow fertility. *Journal of Dairy Science* **87** 2285-2292.

VanRaden PM 2006 Fertility trait economics and correlations with other traits. Interbull Annual Meeting Proceedings. *Interbull Bulletin* **34** 53-56.

VanRaden PM & Miller RH 2006 Effects of nonadditive genetic interactions, inbreeding, and recessive defects on embryo and fetal loss by seventy days. *Journal of Dairy Science* **89** 2716-2721.

Vasconcelos JL, Demetrio DG, Santos RM, Chiari JR, Rodrigues CA & Sa Filho OG 2006 Factors potentially affecting fertility of lactating dairy cow recipients. *Theriogenology* **65** 192-200.

Watson AJ, Westhusin ME & Winger QA 1999 IGF paracrine and autocrine interactions between conceptus and oviduct. *Journal of Reproduction and Fertility Supplement* **54** 303-315.

Weigel KA & Rekaya R 2000 Genetic parameters for reproductive traits of Holstein cattle in California and Minnesota. *Journal of Dairy Science* **83** 1072-1080.

Williams GL & Griffith MK 1995 Sensory and behavioral control of gonadotrophin secretion during suckling-mediated anovulation in cows. *Journal of Reproduction and Fertility Supplement* **49** 463-475.

Wiltbank MC, Gumen A & Sartori R 2002 Physiological classification of anovulatory conditions in cattle. *Theriogenology* **57** 21-52.

Wiltbank M, Lopez H, Sartori R, Sangsritavong S & Gumen A 2006 Changes in reproductive physiology of lactating dairy cows due to elevated steroid metabolism. *Theriogenology* **65** 17-29.

Gestation length in red deer: genetically determined or environmentally controlled?

AgResearch, Ltd, Invermay Agricultural Centre, Private Bag 50034, Mosgiel, New Zealand

The red deer *(Cervus elaphus)* of European origin (e.g. subspecies *scoticus, hispanicus, hippelaphus)* is a medium sized (100-150kg mature hind weight) ruminant that exhibits highly seasonally patterns of autumn conceptions and summer births. Historic data indicate average ($\pm$ s.d.) gestation length of 233-234 ($\pm$ 2-4) days. Recently, however, there has been growing awareness that there is considerably greater variation in gestation length than earlier indicated and that there is a significant element of environmental, and possibly even social, control over the duration of pregnancy in this species. Imposition of variable levels of nutrition over late pregnancy of red deer hinds has been observed to influence fetal growth trajectory and gestation length, with no apparent effect on birth weight. This supports a hypothesis that under conditions of modest feed imbalance, variation in gestation length compensates for variation in fetal growth trajectory to ensure optimisation of birth weight. More recent studies on primiparous (24 month old) red deer hinds have identified surprisingly large variation in gestation length (193-263 days) compared with adult hinds (228-243 days), with earlier conceiving individuals within the primiparous cohort expressing significantly longer gestation than the later conceiving hinds, resulting in a higher level of calving synchrony than expected from known conception dates. This introduces an intriguing hypothesis of social indicative effects on parturition timing to promote within-cohort birth synchrony. Collectively, these data debunk the commonly held notion that gestation length of red deer is genetically fixed within strict limits. A review of the literature points to this as possibly a common phenomenon across a range of non-domesticated ruminant species but this conclusion is not supported by numerous conflicting studies on domestic sheep and cattle.

Introduction

A species-specific gestation period (that is, interval from fertilisation to birth) is a profound feature of mammalian reproductive cycles. The trophoblast has a finite lifespan, serving as a fundamental clock in the timing of events which terminate pregnancy (Holm 1966). The metabolism of the conceptus however determines the absolute timing of gestation length, with the most important fetal contribution to the initiation of parturition (in ruminants at least) being

E-mail: geoff.asher@agresearch.co.nz

the secretion of glucocorticoids from the fetal adrenal glands (Liggins 1979). As a general rule across species, gestation length is positively correlated with adult body size although seasonal imperatives for relatively longer mating-to-birth intervals for some species may extend apparent gestation length though delayed development processes such as "delayed implantation/embryonic diapause" (for example, roe deer, *Capreolus capreolus*: Aitken *et al.* 1973).

There is a general underlying acceptance that gestation length for any given species is genetically programmed and relatively robust in the face of variations in environmental conditions. Recognition is given to the influence of fetal genotype as being the single most important determinant of gestation length; as well as determining interspecific differences in gestation length it also accounts for subspecies and breed differences (Kenneth & Ritchie 1953; Racey 1981). Indeed, this has been demonstrated for the red deer species (*Cervus elaphus*) in which there is considerable subspecies variation in observed mating-birth intervals ranging from 233 days for *C. e. scoticus* (Scottish red deer) (Guinness *et al.* 1971) to 247 days for *C. e. nelsoni, roosevelti, manitobensis* (North American Wapiti subspecies) (Haigh 2001). Red deer hinds gestating F1 crossbred red x Wapiti fetuses show a mating-calving interval of ~ 239 days (Asher *et al.* 2005a). Similarly, although Père David's deer (*Elaphurus davidianus*), with a reported gestation length of 280 + days, are classified within a different genus, hybridisation with red deer hinds can result in fertile offspring. Red deer hinds gestating F1 hybrid red x Père David fetuses exhibit a wide range in mating-calving intervals from 262 to 274 days (Asher *et al.* 1988).

Environmental modifiers of gestation length have been documented and include photoperiod, stress, nutrition and temperature, to name a few (Racey 1981). However, such modifiers have exerted relatively minor effects (1-2% change in established species-specific gestation length). In the case of medium-sized ruminants (that is, 50-350kg mature female weight), species-specific gestation lengths generally range between 150-300 days, with a co-efficient of variation of 1-2%.

Perhaps the most notable departure from rigorous genetic control of gestation length has been described for members of the Camelidae family. Alpacas (*Lama pacos*), for example, not only control the parturition process to ensure births occur during daylight hours and not during inclement weather (Reiner & Bryant 1983), but also exhibit considerable seasonal variation in gestation length (Knight *et al.* 1995; Davis *et al.* 1997). A comparison of pregnancies from spring and autumn matings showed that gestation lengths were 10-12 days longer with spring conceptions. For each day later in spring that a female was mated there was an increase in gestation length of 0.11 days (Davis *et al.* 1997). A similar phenomenon has been noted in the dromedary camel (*Camelus dromedaries*) (Elias *et al.* 1991). Such variations in gestation length are greater than the natural variance observed within cohorts and clearly indicates unspecified environmental modifiers.

Gestation Length in Red Deer

The red deer, of temperate European origin (for example, subspecies *scoticus, hispanicus, hippelaphus*), is a medium sized (100-150kg mature hind weight) ruminant that exhibits highly seasonal patterns of autumn conceptions and summer births (Lincoln and Short 1980). In most natural environments in which red deer exist, there are harsh penalties on calf survival if births occur late in the proscribed season due to the inability of late-born calves to attain sufficient body mass to cope with their first winter (Albon *et al.* 1983). Thus, selection pressure for optimum calving seasonality has been a major feature of red deer evolution. While numerous studies have investigated seasonality drivers around conception, little attention has been given to further effects around the precise timing of parturition due to the assumption that gestation length is genetically-fixed within strict limits. Numerous data sources indicate an average ($\pm$ s.d.) gestation length of 233-234 ($\pm$ 2-4) days (for example: Guinness *et al.* 1971; Krzywinski & Jaczewski 1978; Kelly *et al.* 1982; Asher *et al.* 1988). Recently, however, there has been growing awareness that there is considerably greater variation in gestation length than earlier indicated and that there is a significant element of environmental, and possibly even social, control over the duration of pregnancy. The first indications of a greater variation in the gestation lengths for red deer than expected were from anecdotal accounts related to artificial insemination (AI) to artificially synchronised oestrus. Given a red deer oestrous cycle length of 18-19 days (Guinness *et al.* 1971; Asher *et al.* 2000), it is in theory a simple process to determine which hinds conceived to AI (to red deer sires) and which hinds returned to oestrus 18-19 days later based on calving date relative to AI data; those hinds with apparent gestation lengths > 240 days, were likely to have had return-service conceptions. However, there were several apparently ambiguous situations in which DNA genotyping clearly identified calves from such late parturitions as being the result of conceptions to the AI semen. This indicated a wider variance for red deer gestation length than previously thought. However, the mechanisms behind the variation were obscure.

More recently, Asher *et al.* (2005b) observed a marked effect of level of dam nutrition between days 150 and 220 of pregnancy on fetal growth and gestation length. In the study, conducted on individually penned red deer hinds, (*C.e. scoticus* x *hippelaphus*) mated to similar genotype stags, a range of nutritional allowances from *ad-libitum* intake to a 50% restriction was offered. There were no discernable effects on calf (sex corrected) birth weights but dams expressed unexpectedly wide variation (range = 27 days) in gestation length, with low-nutrition hinds exhibiting longer gestations (> 240 days) than high nutrition hinds (< 230 days). Furthermore gestation length was negatively correlated with change in hind live weight (reflecting nutritional level) between days 150 and 220 of pregnancy. It was concluded from this study that variation in nutrition to hinds during the last trimester may strongly influence fetal development but, under conditions of feed imbalance, variation in gestation length compensates to ensure optimisation of birth weight (and maximise calf survival) (Asher *et al.* 2005b). Such a compensatory mechanism would seem to be driven by a model of fetal induction of parturition dependent on attainment of specific size or stage of development being influenced by fetal growth trajectory (Liggins 1979).

More recently, Garcia *et al.* (2006) observed that Spanish red deer hinds (*C.e. hispanicus*) induced to conceive 6-8 weeks before the natural breeding season exhibited extended gestation lengths by 8-10 days compared with control hinds, but there were no discernable effects on birth weight. While they postulated that optimisation of birth date was the principal driver of the effect, their study was confounded by lower live weight gains over pregnancy in the early-conceiving hinds. Their data, therefore, show some support for a nutritional-fetal growth influence on gestation length in red deer.

There are, however, some apparent contradictions to this hypothesis, not the least of which being the consistency in average gestation length and its variance in previous studies of red deer. In these cases it could be argued that such observations are based on cohorts in which individuals are subjected to similar levels of nutritional management and unlikely to include groups subjected to severe nutritional constraints (Asher *et al.* 2005b). More pertinently though, considerable variation in birth weight has been recorded previously for red deer populations subjected to severe variations in their nutritional environment. Most notably, Albon *et al.* (1983) demonstrated a significant positive correlation between mean daily temperature during late pregnancy and subsequent mean birth weight across a number of years for red deer on the

Isle of Rhum, Scotland. From this observation they contended that differential nutritional environments across years influenced mean birth weight via fetal growth. Similarly, Thorne et al. (1976) demonstrated for North American Wapiti (C.e. nelsoni) that birth weights of calves was correlated with hind nutrition during the later half of pregnancy.

The major difference between these studies and that of Asher et al. (2005b) and Garcia et al. (2006) is the apparent severity of nutritional constraint. It is possible, based on live weight and body condition data presented, that the earlier studies of Albon et al. (1983) and Thorne et al. (1976) imposed or compared nutritional levels considerably more extreme than the late studies. Thus the compensatory hypothesis presented by Asher et al. (2005b) may apply only under conditions of moderate nutritional imbalance.

Interestingly, such compensatory mechanisms have not been demonstrated for fallow deer (Dama dama) subjected to modest levels of feed deprivation. Mulley (1989) described fetal growth retardation and low birth weight associated with a 20-30% reduction in metabolisable energy intake in the second and third trimesters of pregnancy, with no significant effects on gestation length. However, some support for the hypothesis of compensatory control in gestation length in red deer lies with a few studies on other non-domesticated ruminants.

Verme (1965) showed that the gestation length of white-tailed deer (Odocoileus virginianus) increased by 4-6 days when feed intake was restricted during pregnancy. Similarly, analysis of the birth seasons of Alaskan moose (Alces alces gigas) indicates that the timing of births occurred earlier for hinds with the thickest rump fat during pregnancy (Keech et al. 2000). In a non-cervid example, Skinner and van Zyle (1969) found that African eland (Taurotragus oryx) in poor nutritional habitats had longer pregnancies than those in richer habitats. These examples hint at a more widespread phenomenon of environmental control of gestation length in non-domesticated ruminants. It should be noted, however, that similar nutritional studies on domesticated sheep and cattle show contradictory and smaller effects (Alexander 1956; Bewg et al. 1969; Tudor 1972; Wallace et al. 1999) demonstrating that such compensatory mechanisms are not universal.

Can the social environment influence gestation length?

A recent study on a single cohort of 78 primiparous (24 month old) red deer hinds, in which conception date was established by direct observation of mating and fetal age measured by ultrasonography, revealed an unexpected and surprisingly large variation in yearling gestation length (193-263 days) compared with a contemporary cohort of 80 adult hinds (228-243 days) (IC Scott & GW Asher, unpublished data). Interestingly, the earlier conceiving individuals within the primiparous cohort expressed significantly longer gestations than the later conceiving hinds, resulting in a higher level of within-herd calving synchrony than expected from the known conception dates. There was no evidence of any live weight or body condition effects.

In the absence of any other explanatory variable, it was considered that a putative "social facilitation" effect may be operating to promote within-herd synchrony of births. This hypothesis is presently being tested under more controlled conditions that take seasonality variables into account. However, while there may be strong evolutionary drivers to promote birth synchrony within herds (for example, prey saturation), these have always been seen previously to act via selection pressure on conception synchronisation (often including a social component) and uniformity of gestation length.

Conclusions

Collectively, these studies debunk the commonly held notion that gestation length of red deer is genetically fixed within restricted limits, and indicates environmental (and social?) modifiers acting as buffers to enhance neonate survival. Such phenomena may not be limited to red deer and have, to various degrees, been observed in other cervid and non-cervid ruminants.

References

Aitken RJ, Burton J, Hawkins J, Kerr-Wilson R, Short RV & Steven DH 1973 Histological and ultrastructural changes in the blastocyst and reproductive tract of the roe deer *(Capreolus capreolus)* during delayed implantation. *Journal of Reproduction and Fertility* **34** 481-493.

Albon SD, Guinness FE & Clutton-Brock TH 1983 The influence of climatic variation on the birth weights of red deer *(Cervus elaphus)*. *Journal of Zoology (London)* **200** 295-297.

Alexander G 1956 Influence of nutrition upon duration of gestation in sheep. *Nature (London)* **178** 1058-1059.

Asher GW, Adam JL, Otway W, Bowmar P, van Reenan G, Mackintosh CG & Dratch P 1988 Hybridisation of Père David's deer *(Elaphurus davidianus) and red deer (Cervus elaphus)* by artificial insemination. *Journal of Zoology (London)* **215** 197-203.

Asher GW, O'Neill KT, Scott IC, Mockett BG & Fisher MW 2000 Genetic influences on reproduction of female red deer (Cervus elaphus). (1) Seasonal luteal cyclicity. *Animal Reproduction Science* **59** 43-59.

Asher GW, Scott IC, O'Neill KT & Littlejohn RP 2005a Influence of level of nutrition during late pregnancy on reproductive productivity of red deer (2) Adult hinds gestating wapiti x red deer crossbred calves. *Animal Reproduction Science* **86** 285-296.

Asher GW, Mulley RC, O'Neill KT, Scott IC, Jopson NB & Littlejohn RP 2005b Influence of level of nutrition during late pregnancy on reproductive productivity of red deer (1) Adult and primiparous hinds gestating red deer calves. *Animal Reproduction Science* **86** 261-283.

Bewg WP, Plasto AW & Daly KK 1969 Studies on reproductive performance of beef cattle in a sub-tropical environment. 1. Conception rate, length of oestrous cycle and length of gestation. *Queensland Journal of Agricultural Animal Science* **26** 629-637.

Davis GH, Dodds KG, Moore GH & Bruce GD 1997 Seasonal effects on gestation length and birth weight in alpacas. *Animal Reproduction Science* **46** 297-303.

Elias E, Degen AA & Kam M 1991 Effect of conception date on length of gestation in the dromedary camel *(Camelus dromedaricus)* in the Negev Desert. *Animal Reproduction Science* **25** 173-177.

Garcia AJ, Landete-Castillejos T, Carrion D, Gaspar-Lopez E & Gallego L 2006 Compensatory extension of gestation length with advance of conception in red deer *(Cervus elaphus)*. *Journal of Experimental Zoology* **305A** 55-61.

Guinness FE, Lincoln GA & Short RV 1971 The reproductive cycle of the female red deer, *Cervus elaphus*. *Journal of Reproduction and Fertility* **27** 427-438.

Haigh JC 2001 The gestation length of wapiti *(Cervus elaphus)* revisited. *Animal Reproduction Science* **65** 89-93.

Holm LW 1966 Prolonged pregnancy. *Advanced Veterinary Science* **11** 159-205.

Keech MA, Bowyer RT, ver Hoef JM, Boertje RD, Dale BW & Stephenson TR 2000 Life-history consequences of maternal condition in Alaskan moose. *Journal of Wildlife Management* **64**(2) 450-462.

Kelly RW, McNatty KP, Moore GH, Ross D & Gibb M 1982 Plasma concentrations of LH, prolactin, oestradiol and progesterone in female red deer *(Cervus elaphus)* during pregnancy. *Journal of Reproduction and Fertility* **64** 475-483.

Kenneth JH & Ritchie GR 1953 Gestation periods, a table and bibliography. *Technical Communications of the Commonwealth Bureau of Animal Breeding and Genetics* No 5.

Knight TW, Ridland M, Scott I, Death AF & Wyeth TK 1995 Foetal mortality at different stages of gestation in alpacas *(Lama pacos)* and the associated changes in progesterone concentrations. *Animal Reproduction Science* **40** 89-97.

Krzywinski A & Jaczewski Z 1978 Observations on the artificial breeding of red deer. *Symposium of Zoology Society (London)* **43** 271-287.

Liggins GC 1979 Initiation of parturition. *Breeding Medical Bulletin* **45** 145-150.

Lincoln GA & Short RV 1980 Seasonal breeding: nature's contraceptive. *Recent Progress in Hormone Research* **36** 1-52.

Mulley RC 1989 Reproduction and performance of farmed fallow deer *(Dama dama)*. Ph.D Thesis. The University of Sydney, Australia.

Racey PA 1981 Environmental factors affecting the length of gestation in mammals. In *Environmental Factors in Mammal Reproduction*, pp 199-213. Eds B Cook and DF Gilmore. London: Macmillan.

Reiner R & Bryant F 1983 A different sort of sheep. *Rangelands* **5**(3) 106-108.

Skinner JD & van Zyle JMH 1969 Reproductive performance of the common eland, *Taurotragus oryx*, in two environments. *Journal of Reproduction and Fertility Supplement* **6** 319-322.

Thorne ET, Dean RE & Hepworth WG 1976 Nutrition during gestation in relation to successful reproduction in elk. *Journal of Wildlife Management* **40**(2) 330-335.

Tudor GD 1972 The effect or pre- and post-natal nutrition on the growth of beef cattle. 1. The effect of nutrition and parity of the dam on calf birth weight. *Australian Journal of Agriculture and Research* **23** 389-395.

Verme LJ 1965 Reproduction studies on penned white-tailed deer. *Journal of Wildlife Management* **29** 74-79.

Wallace JM, Bourke DA & Aitken RP 1999 Nutrition and foetal growth: paradoxical effects in the overnourished adolescent sheep. *Journal of Reproduction and Fertility Supplement* **54** 385-399.

Reproductive endocrinology and biotechnology applications among buffaloes

ML Madan[1] and BS Prakash[2]

[1]Livestock Production Systems, 842, Sector –6, Urban Estate, Karnal -132001 India; [2]Division of Dairy Cattle Physiology, National Dairy Research Institute, Karnal – 132001, India

Buffalo, as the major livestock species for milk and meat production, contribute significantly to the economy of many countries in south & south-east Asia, South America, Africa and the Mediterranean. Improved buffalo production could significantly enhance the economy and the living standards of farmers in countries where buffaloes predominate; particularly, in countries with a tropical climate. The major factors limiting the efficient utilization of buffaloes in countries with a tropical climate are: late maturity; poor estrus expressivities, particularly in summer months; long postpartum calving intervals; low reproductive efficiencies and fertility rates which are closely linked with environmental stress; as well as managerial problems. As good reproductive performance is essential for efficient livestock production, the female buffalo calves must grow rapidly to attain sexual maturity, initiate estrous cycles, ovulate and be mated by fertile males or inseminated with quality semen to optimize conception and production. In the last two decades, considerable attention has been focused on understanding some of the causes for the inherent limitations in reproduction among buffaloes by studying their reproductive endocrinology as well as developing biotechniques for augmenting their reproductive efficiency. This review provides an overview of buffalo reproductive endocrinology and also of the research done to date towards the enhancement of buffalo reproductive efficiency through endocrine and embryo biotechniques.

Introduction

There are about 158 million buffaloes in the world and roughly 153 million (97%) of these are water buffaloes essentially found in the Asian Region. The overall buffalo numbers are increasing by about 1.3% annually (FAO 2000). Apart from being the mainstay of the milk production system in many south & south-east Asian countries, buffalo also contribute to the rural economy in terms of meat production and draught. The riverine buffalo is better adapted than cattle to tropical climates especially with respect to utilization of poor quality roughages and resistance to some of the tropical diseases (Chauhan 1995). This makes buffaloes easy to maintain using the locally available roughage and crop residue. In recent years the buffalo has gained more attention around the world compared to the cow, not just due to its reasonable growth rate on

Corresponding author E-mail: mlmadan@hotmail.com

roughage feeding, but also due to its high milk yield with high fat percentage, tolerance to hot and humid climates, lean meat and draught ability.

The buffalo, however, is a sluggish breeder and is beset with various constraints which adversely influence its fertility; such as, problems of silent heat coupled with late maturity, poor expression of estrus, irregular estrous cycles, seasonality in breeding, anestrus, low conception rates, long postpartum calving intervals and repeat breeding (Madan 1990). Considerable attention has been focused on the reproductive endocrinology of the buffalo as a means to identify specific problems and devise means to augment reproductive performance. An understanding of hormonal interplay is required for alleviating reproductive problems of an endocrine origin. This knowledge is paramount for biotechnological applications for enhancing the reproductive efficiency in this animal. This review presents a) the state of knowledge on the reproductive endocrinology of the riverine buffalo (Murrah breed) and b) research studies carried out for improving fertility in these species using endocrine biotechniques.

Estrous behaviour

Reproductive efficiency among large ruminants is greatly dependent upon the detection of estrus. This is even more important with reference to small herds managed under tropical or subtropical environments because high air temperatures shorten the duration of estrus and lower its intensity (Madan & Johnson 1973; 1975) as demonstrated under controlled environments in cattle. The intensity of estrous behavior in tropical buffaloes has been found to be much less than cows. The usual weak symptoms of estrus in the normal breeding season (September to February) become even weaker during the hot months of summer. Among Murrah buffaloes diurnal patterns of estrous behavior have been observed with 59% of estruses recorded between 10pm and 6am (Prakash 2002). The maximum occurrences of various heat symptoms were seen in the winter months of November to February while the lowest occurrences were during March to August in a selected group of buffaloes observed throughout the year (Fig. 1) (Prakash 2002). Out of the 8 major symptoms of estrus, 5 symptoms (that is, vulval engorgement, frequent urination, bellowing, bull mounting and restlessness) contributed to 85 percent of the total observations (Fig. 2 and 3). Mucus discharge, licking of the female by the bull and chin resting by the bull were minor symptoms. During the summer months frequent urination was the most prominent heat symptom recorded (Fig. 2).

In another study, the incidence of silent heat occurrences throughout the year was determined in buffaloes by milk progesterone monitoring with the objective of studying the influence of changing environmental temperatures on heat occurrences (Prakash et al. 2005). Out of a total of 292 estruses detected by milk progesterone monitoring 108 estruses (37%) went unobserved. The incidence of silent heat was lowest in December (10.5%) while the peak was seen in April (70%). There was a gradual decline in incidence of silent heat occurrence from May onwards (Fig. 4). Due to the high incidence of silent heat, large numbers of buffaloes are left unbred and contribute substantially to a high service period in this animal (139 days among 89 buffaloes in this study). Season of calving had a profound influence on the service period. The mean service period of animals calving from December till June was more than 140 days and was significantly higher than mean service period of animals calving in the months of July to November (< 110 days). The high service period of buffaloes in the former group of animals was attributed to the high incidence of silent estrus, which the animals would exhibit in the summer months once they commence cycling postpartum (Fig. 5). The effect of different seasons on both the resumption of ovarian activity and embryo survival may be a function of temperature and/or photoperiod: further elucidation by conducting systematic studies uncoupling temperature from photoperiod influences is required.

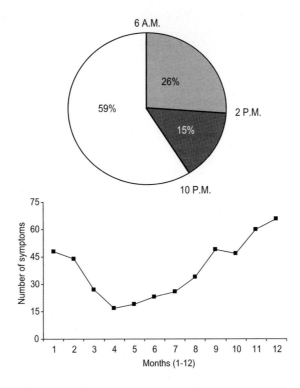

Fig. 1. Variation in number of heat symptoms observed in Murrah buffaloes (n = 13) during different months

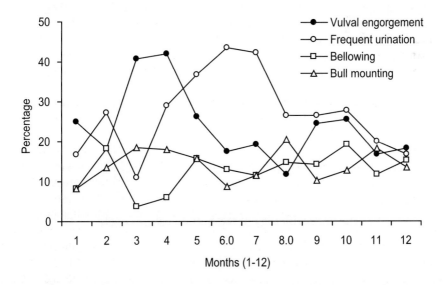

Fig. 2. Incidence of various heat symptoms in Murrah buffaloes

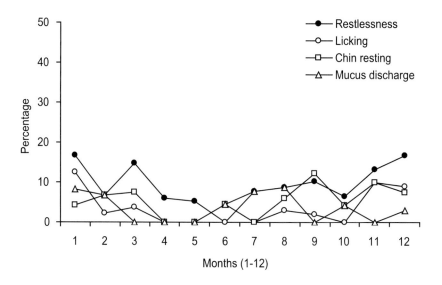

Fig.3. Incidence of various heat symptoms in Murrah buffaloes

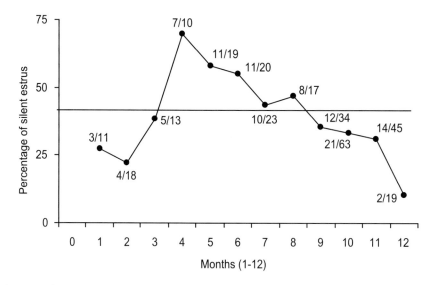

Fig.4. Incidence of silent estrus in buffaloes during different months

The endocrinology of the buffalo estrous cycle

Progesterone

Progesterone at appropriate cyclic concentrations is essential for expression of estrus, preparing the uterus for implantation and the maintenance of pregnancy. Essentially, the concentrations of progesterone in peripheral plasma of cycling buffaloes rise and fall in coincidence with the growth and regression of the corpus luteum (CL) (Bachlaus *et al.* 1979; Kamboj & Prakash 1993). The reports indicate that concentrations of progesterone in peripheral blood plasma are

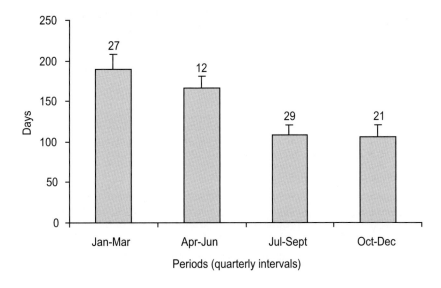

Fig.5. Influence of month of delivery on service period (n = 89)

minimal on the day of estrus (< 0.4 ng/ml) and rise to a peak concentrations (1.0 ng/ml to 4 ng/ml) on days 12-16 of the cycle before declining to a basal level at the onset of next estrus. Circulating progesterone concentrations are lower in hotter months and may therefore be responsible for poor expression of estrus and low conception rates (Rao & Pandey 1982). Progesterone concentrations have been found to vary with nutritional status in buffaloes. Kaur and Arora (1984) studied circulating progesterone concentrations in a group of buffaloes fed normally (7 kg straw, 2 kg concentrates and 0.5 kg dry matter) and in an underfed group (65% less digestible crude protein, 23.8% less total digestible nutrients than normally fed group). Their results showed that progesterone concentrations were lower in summer than in winter in normally fed buffaloes. The optimum fed buffaloes exhibited weak estruses of short duration during night hours in summer whereas underfed buffaloes did not exhibit any estruses, which lead to persistent infertility. Under-nutrition coupled with high environmental temperature stress was responsible for a long anestrus in buffaloes. The pattern of progesterone in milk is similar to that in plasma but concentrations in milk are higher than those observed in blood plasma due to lipid solubility of steroid (Batra *et al.* 1979; Kamboj & Prakash 1993).

Estradiol

Estrogen induces behavioral symptoms of estrus by its action on the central nervous symptom. Some reports on circulating estradiol concentrations during the estrous cycles of Murrah buffaloes are available (Bachalaus *et al.* 1979; Kanai & Shimizu 1984; Avenell *et al.* 1985). Circulating concentrations of estradiol-17ß increase following luteolysis and reach a peak value either a day before or on the day of estrus. After attaining the peak concentration, estradiol concentrations in circulation decline to basal levels 2 days following estrus. Minor peaks are detected during the midluteal phase, suggesting the occurrence of additional follicular waves in this species. Buffaloes have been reported to have two or three waves of follicular growth during an estrous cycle with the second wave occurring during days 10-11 of the cycle (Baruselli *et al.* 1997; Manik *et al.* 1998). Palta *et al.* (1996a) reported estradiol-17ß concentrations (ng/ml) of

2.40 ± 0.85, 6.27 ± 2.74 and 30.29 ± 11.69 in small (3-5 mm in diameter), medium (6-9 mm in diameter) and large (10-12 mm in diameter) buffalo follicles, respectively. The concentrations of estradiol in whole milk are higher and positively correlated with that in plasma during the estrous cycle in buffaloes (Batra *et al.* 1980). Seasons have been shown to affect estradiol concentrations in blood plasma. Rao and Pandey (1983) reported that whilst estradiol concentrations increase significantly on the day before and on the day of estrus in hot-dry (April to June), hot-humid (July to September), warm (October to December) and cold (January to March) weather, the peak concentrations are lower in summer compared to cooler months. A decrease in the peak concentrations of estradiol around estrus, at the time of decreased progesterone concentrations, may be responsible for the higher incidence of silent estrus during summer (Rao & Pandey 1982).

Inhibin

Inhibin, a glycoprotein hormone produced by the granulosa cells of ovarian follicles, suppresses pituitary production and/or secretion of gonadotropins (preferentially FSH) through negative feedback on the pituitary. There are several studies that report the peripheral plasma inhibin concentrations in relation to cyclicity in buffaloes (Palta *et al.* 1996b; 1997; Mondal *et al.* 2003a;b). The studies indicated that peripheral inhibin concentrations increase to a maximum concentration 2-4 days preceding estrus and decline subsequently to basal concentrations during the early luteal phase. The significant increase in inhibin concentrations through the late luteal to periestrus phase is suggestive of a vital role of inhibin in folliculogenesis. Inhibin concentrations were significantly higher during the winter season compared to summer (Palta *et al.* 1997). The lower inhibin concentrations during the summer season could be due to a decline in ovarian activity of buffalo ovaries (Roy *et al.* 1972). Changing photoperiods do not influence circulating concentrations of either gonadotrophins (LH and FSH) or gonadal hormones (progesterone and estradiol-17ß) which suggests that environmental stress due to increasing temperature may be influencing follicular turnover and hence inhibin production (Singh 1990). Inhibin concentrations in buffalo follicular fluid are directly correlated with the size of the antral follicle (Palta *et al.* 1996b;c; 1998). As the population of the follicles of all size categories contributes to the circulating inhibin levels, the pattern of peripheral inhibin concentrations observed in the Singh (1990) study is probably a reflection of the overall follicular population at different stages of estrous cycle.

Follicle Stimulating Hormone

Gonadotrophin (both LH and FSH) estimations in buffalo plasma have been done using heterologous radioimmunoassays with bovine standards since purified buffalo gonadotrophins are not available. Circulating FSH plays a vital role in the initiation and regulation of the buffalo estrous cycle (Razdan *et al.* 1982). Several reports are available on peripheral plasma FSH concentrations during the oestrous cycles of buffaloes (Heranjal *et al.* 1979; Janakiraman *et al.* 1980; Razdan *et al.* 1982). Peak concentrations of FSH were detected on the day of estrus in Murrah buffaloes and FSH concentrations declined gradually over the next 3-6 days. Kaker *et al.* (1980) observed the occurrence of preovulatory surges of FSH. Seasons have been shown to influence peripheral concentrations of FSH. Janakiraman *et al.* (1980) reported significantly higher FSH concentrations at estrus and during the luteal phase during the peak breeding seasons (November to December; 61.6 ng/ml) in comparison to the corresponding times of the cycle during both the medium (July to October; 46.9 ng/ml) and low (March to June; 49.1 ng/ml) breeding seasons in Surti buffaloes. Lower FSH concentrations during the low breeding seasons were associated with lower follicular activity and the highest incidence of anovulatory cycles.

Luteinizing Hormone

During most of the buffalo oestrous cycle the circulating concentrations of LH are low (0.72–2.0 ng/ml) and peak concentrations (20–40 ng/ml) are measured on the day of estrus (Heranjal *et al.* 1979; Kaker *et al.* 1980; Arora & Pandey 1982; Kanai & Shimizu 1984; Avenell *et al.* 1985). Higher concentrations of LH on the day of estrus were observed in cooler months compared to hotter months (Rao & Pandey 1983). The decrease in peak LH concentrations in hotter months has been attributed, among other causes, to the direct effects of heat stress (Madan & Johnson 1975). As estrous behavior is controlled by estrogen and progesterone, the decrease in peak concentrations of LH around estrus would contribute to the higher incidence of silent estrus during the summer.

The accurate detection of preovulatory events facilitates efficient reproductive management in artificial insemination (AI) and embryo transfer programs; for example, knowledge of the timing of ovulation permits the precise timing of AI. In order to assess the exact time of ovulation after onset of estrus, Murrah buffaloes were observed for the onset of estrus by visual observations of the signs of estrus and the females response to a vasectomized bull (teaser) at 0600, 1200, 1800 and 2400 h every day. The time of ovulation was determined by changes in the follicular surface, from turgid at the onset of estrus to flaccid after ovulation, felt by rectal palpation at 2h intervals from the onset of estrus till ovulation (Prakash *et al.* 2005). Blood samples were collected at 2h intervals from the onset of estrus till 2h after ovulation and the concentration of LH was assayed in the corresponding plasma samples. Ovulation occurred 42.2 ± 2.8h (range = 28 to 60h) after the onset of spontaneous estrus and 23.3 ± 3.5h (range = 18 to 40h) post onset of the peak LH concentration (Fig. 6). However, initially LH profile concentrations gradually declined from 2.71 ± 1.22 ng/ml at the onset of estrus (0h) to basal levels of ≤ 0.31 ng/ml 16h later (Prakash *et al.* 2005).

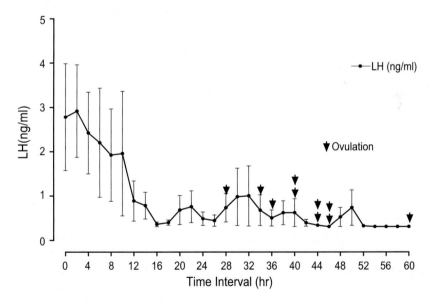

Fig.6. Plasma LH (Mean±SEM) profile and timing of ovulation after onset of spontaneous estrus in Murrah buffaloes (n=10)

Prolactin

The concentrations of prolactin in peripheral plasma of buffaloes have been reported by several workers (Heranjal *et al.* 1979; Razdan & Kaker 1980; Pahwa & Pandey 1984; Galhotra *et al.* 1988;

Singh & Madan 1993). Circulating prolactin concentrations are 2-6 fold higher throughout the estrous cycle in the low breeding season (summer) compared to those in either the medium (monsoon) or peak breeding seasons. High concentrations of prolactin during the low breeding season (summer) may be due to an influence of photoperiod on the pineal gland activities of prolactin controlling factors. Plasma prolactin concentrations in cycling buffalo heifers were 248.50 ± 16.03 to 369.63 ± 25.13 ng/ml in the summer months in comparison to several fold lower concentrations of those observed in the winter months (3.10 ± 0.48 to 9.14 ± 1.39 ng/ml) without exhibiting a definite trend during cyclicity (Roy & Prakash 2006). Singh and Madan (1993) reported that prolactin concentrations during summer were higher in non-lactating females than in lactating buffaloes whereas during winter the concentrations were lower in non-lactating females than in lactating buffaloes. The circadian variation in prolactin concentrations was more pronounced in summer than in winter because of greater variation in ambient temperature between morning, noon, evening and night during summer than winter. Very high prolactin concentrations in the summer months were associated with low estrogen concentrations and summer anestrus (Singh & Madan 1993).

Prostaglandins

$PGF_{2\alpha}$ plays an important role in ovarian function and is involved in both the induction and synchronization of estrus in buffaloes (Bachalaus et al. 1979; Kamonpatana et al. 1976). Mishra et al. (2003) observed a pulsatile pattern of plasma PGFM release prior to estrus when PGFM was determined in blood samples collected at hourly intervals of time. PGFM pulsatility was not observed when a blood sampling frequency of either 4 or 12 hours was used (Batra & Pandey 1983; Mishra et al. 2003). The increase in PGFM concentrations before estrus coincided with the decline of progesterone concentrations (Batra et al. 1979; Mishra et al. 2003).

Testosterone, cortisol, T3 and T4

Singh and Madan (1985) reported that mean testosterone concentrations were low from the day of estrus up to day 5 of the cycle with peak concentrations, of approximately 0.10 ng/ml, measured on days 8 and 9 of the cycle. Peripheral plasma cortisol concentrations have been reported to be higher on the day before and day of estrus in Murrah buffalo heifers (Kumar et al. 1991; Madan et al. 1993). During the estrous cycle, mean plasma T3 and T4 concentrations vary between 0.84 ± 0.16 - 1.88 ± 0.10 ng/ml and 33.28 ± 5.08 - 50.76 ± 8.23 ng/ml, respectively (Khurana & Madan 1985). On the day of estrus the highest concentrations of T3 and T4 are detected: the high concentrations of T3, T4 and cortisol may be due to stress of estrus. Sarvaiya and Pathak (1992) reported that circulating concentrations of T3 and T4 were significantly higher in cycling buffalo heifers compared to anestrus heifers. Higher concentrations of thyroid hormones in cycling animals have also been attributed to the influence of ovarian steroids (D'Angelo & Fisher 1969) as well as pituitary gonadotrophins (Maqsood 1954).

Endocrine changes in the peripubertal buffalo

The buffalo suffers from a slow growth rate (Arora 1979) and hence delayed puberty (Pandey 1979): in particular, Indian Murrah, an important dairy breed of buffalo, has been recorded as attaining puberty as late as 33 months of age (National Dairy Research Institute, Karnal, India 1996). Buffaloes have been administered with GnRH at 12, 24, 30 and 36 months to test the responsiveness of the hypophysis in terms of both FSH and LH release (Singh & Madan 1998a;b; 2000a). The mean basal circulating concentrations of plasma FSH among the heifers of 12 months of age were lower than at 24 and 30 months. Whereas the basal concentrations of plasma LH among the heifers at 12 months were higher than plasma LH concentrations measured among buffaloes of 24 and 30 months of age. Peri-pubertal animals displayed a steady increase in basal LH production and a decline in LH peak concentrations with advancing age indicating that the process of sexual maturation of buffaloes was associated with an increase in basal circulating LH. Though the gonadotrophin release pattern after administration of GnRH (single injection) was nearly similar in these animals, animals of different age groups and reproductive cycle status differed in terms of magnitude of release of gonadotrophin and time taken to reach the peak response post GnRH injection (Singh & Madan 1998a;b; 2000b; 2002a).

Buffalo heifers at 36 months of age have lower circulating concentrations of both estrogen and progesterone than heifers of 24 and 30 months of age. These results are suggestive of the possible lack of intrinsic hypothalmo-hypophyseal-gonadal interplay as seen among the adults. Administration of a non-hypophyseal gonadotrophin (pregnant mare's serum gonadotrophin: PMSG) to animals of 36 months of age resulted in a gradual increase in estrogen concentrations associated with follicular development (Singh & Madan 1999a;b) and phasic production of progesterone. In terms of follicular and ovulatory responses to exogenous PMSG adminstration, heifers of 12 months responded better with an increased number of follicles ovulating (7.1 ± 0.7) than the heifers of 24 months (5.3 ± 1.0) and 36 months (4.0 ± 1.0) of age (Singh & Madan 1999a;b). A GnRH injection at PMSG-induced estrus in prepubertal buffaloes caused synchronized ovulations of large sized follicles and reduced the intervals from onset of estrus to ovulation and also between the first and last ovulation. The plasma prolactin concentrations at 12 months of age were significantly higher than at 24, 30 and 36 months and administration of GnRH did not affect prolactin release (Singh & Madan 2000c; 2002b).

In buffalo heifers the beginning of ovarian activity depends on live weight; in fact, it has been determined that buffaloes have normal estrous cycles when they attain two thirds of their adult body weight (Esposito *et al.* 1992). Nutrition therefore plays an important role in the timing of the beginning of reproductive activity. In the Mediterranean buffalo (Esposito *et al.* 1992) there is a delay in the onset of puberty when in the months preceding puberty the diet has insufficient energy. In a study that evaluated the effect of long term and short term nutritional management on conception patterns at first mating in Mediterranean buffalo heifers, it was found that long term nutritional management had a major influence on the age and live weight at first conception (Campanile *et al.* 2001). Heifers that were kept on a constant and relatively high plane of nutrition conceived at a younger age (540 days of age) compared with heifers that were fed on a low plane of nutrition (840 days of age). Similar results have also been observed by Barkawi *et al.* (1989) and Barnabe *et al.* (1997) in like studies.

The interplay of endocrine mechanisms leading to the transition from sexual quiescence to sexual function and the interaction with growth rate, live weight, metabolic changes and age in buffaloes were investigated (Haldar & Prakash 2005). A group of Murrah buffalo heifers (21.92 ± 1.09 months of age, 269.67 ± 7.97 kg body weight) were assigned to a diet to provide weight gain of 0.4 kg/day. Heifers attained puberty at an average age of 31.53 ± 0.88 months with 380.67 ± 6.42 kg body weight. Circulating progesterone concentrations were very low (0.20 - 0.30 ng/ml) during the pre-pubertal period. Plasma LH and GH concentrations increased ($P < 0.05$) over the months preceding puberty and were highest during the month before puberty: plasma GH and LH concentrations were positively correlated prior to ($r = +0.59$; $P < 0.05$) as well as after puberty ($r = +0.42$; $P < 0.05$). There was a positive correlation between plasma LH concentrations and body weight during the pre-pubertal period ($r = +0.61$; $P < 0.05$) and thereafter a negative correlation during the post-pubertal period ($r = -0.64$; $P < 0.05$). Plasma

GH concentrations and body weight were positively correlated both before puberty (r = +0.92: P< 0.01) and after puberty (r = +0.32; P< 0.05). These results suggest that both GH and LH are equally important and vital cues in inducing the onset of ovarian function in buffalo heifers.

Endocrinology of gestation and parturition in buffaloes

Palta and Madan (1996) measured a significant reduction in plasma FSH concentrations in Murrah buffaloes from days 60 to 240 of gestation. However, they did not observe any significant changes in the basal plasma LH concentrations in the same period. During the last stages of pregnancy, LH concentrations fluctuated narrowly between 0.4 and 0.9 ng/ml in riverine buffaloes (Galhotra et al. 1981; Barkawi et al. 1986). Kamonpatana (1984) also reported low circulating concentrations of LH in Swamp buffaloes during the 10 days before and after parturition.

In peripartum Murrah buffaloes mean plasma estradiol-17ß concentrations increased gradually from 30 days preterm to 5 days preterm followed by a steep increase to peak concentrations on 1 day prepartum (Prakash & Madan 1984a; 1986). The mean plasma progesterone concentrations declined gradually from 1.82 ng/ml 30 days preterm to 1.21 pg/ml 2 days prepartum, then fell sharply to low levels at calving (Prakash & Madan 1985a; 1986). Mean PGF concentrations remained low (< 0.75 ng/ml) up to day 2 prepartum, increasing substantially to 1.86 ng/ml on day 1 prepartum, followed by a steep rise to a peak value of 4.16 ng/ml at partum (Prakash & Madan 1985a). The mean plasma cortisol concentrations remained more or less constant during the pre- and postpartum periods rising sharply to a peak at calving (Prakash & Madan 1984b; 1986). The pattern of change in the concentrations of the hormones was essentially similar to those recorded in peripartum cows (Thorburn et al. 1977).

Endocrine Changes in the buffalo postpartum

Postpartum anestrus remains a major reproductive limitation in buffalo. The calving to calving interval in 48-66% of buffalo is > 14 months as a result of the environment and unpredictable management (Perera 1999). Buffaloes are susceptible to many stressors that seriously affect reproductive efficiency; especially, day length, high temperatures and nutritional deficiencies. Suckling is encouraged in buffalo to enhance calf survival rate and facilitate milk let down but unfortunately this practice also attenuates the neuroendocrine signals required for resumption of ovarian activity. Studies have been attempted for reducing post-partum anestrus in buffaloes but protocols still require refinement (Tiwari & Pathak 1995).

Progesterone

Routine determinations of milk progesterone concentrations indicated that the interval from calving to first postpartum commencement of cyclicity is highly variable with a mean of around 68 days in Murrah buffaloes (Prakash et al. 2005). This also contributes to the large service period in this animal. Postpartum anestrus has also been classified in terms of a) non detection of animals in estrus due to poor estrus expressivity and behavioral estrus signs or b) as "true anestrus" when the ovaries are non-cyclic with no follicular development or CL formation (Madan et al. 1984). Such animals do not respond to luteolytic compounds for estrus induction. However, animals showing phasic cycles with progesterone undulations in diestrous phase, respond to such compounds.

FSH and LH

During the postpartum period baseline plasma FSH concentrations were significantly higher ($P < 0.01$) on day 20, compared to the concentrations obtained on day 2, but did not differ significantly from the concentrations obtained on day 35 postpartum (Palta & Madan 1996). However when milked buffaloes were compared with suckled buffaloes there were no significant differences in plasma FSH concentrations between days 3 and 90 postpartum or between milked and suckled buffaloes (Arya & Madan 2001a). In two separate studies, no significant changes in basal plasma LH concentrations were detected between days 3 to 90 postpartum or the first 4 months postpartum during which period the buffaloes did not exhibit any estrus (Arya & Madan 2001a; Galhotra et al. 1981). There were also no significant changes in plasma LH concentrations between milked and suckled Murrah anestrous buffaloes (Arya & Madan 2001a). However, in another study there was a progressive increase in basal plasma LH concentrations from days 2 to 35 postpartum (Palta & Madan 1995). The responsiveness of the pituitary gland to GnRH administration in terms of LH and FSH release was drastically increased ($P < 0.01$) between days 2 and 20 postpartum, although the responsiveness to GnRH administration during advancing gestation in buffaloes showed a declining trend in gonadotrophin release (Palta & Madan 1996). In conclusion, there appears to be insufficient gonadotrophic support for follicular growth and development during the early postpartum period.

Prolactin and GH

Prolactin is absolutely crucial for galactopoesis and whilst the suckling stimulus is of a certain quality, and that quality is thought to be read by the hypothalamus by the pattern and concentration of prolactin release, the hypothalamic-pituitary-ovarian axis is suppressed although it is not clear by what mechanism(s). Growth hormone is a known anabolic hormone and has been seen to play a role in enhancing growth and early commencement of puberty in buffaloes (Mondal & Prakash 2003; 2004; Haldar & Prakash 2006). Hence the principle hormones, which could be playing a major role in determining the length of postpartum period for cyclicity commencement, are prolactin and GH. While no significant correlation was found between GH concentrations and days to commencement of cyclicity, the correlation of plasma prolactin concentrations with commencement of cyclicity was highly significant ($r = 0.90$, $P < 0.01$: Prakash et al. 2005): the high plasma prolactin concentrations are probably associated with low release of gonadotrophins and hence the inhibitory effect on cyclicity as has been recorded in cattle in earlier investigations.

Estrogens

After parturition the plasma estradiol–17ß concentrations decline steeply during the first 24-72 h (Arora & Pandey 1982; Pahwa & Pandey 1983; Prakash & Madan 1984a). Circulating estradiol–17ß concentrations were basal between days 2 and 7 after calving (Prakash & Madan 1986; Arya & Madan 2001b) and continued to stay low until 45 days postpartum (Madan et al. 1984). These observations are also indicative of delayed follicular development postpartum in this species.

Reproductive Technologies for augmenting buffalo reproduction

Over the years, arrays of options have been studied for the reproductive management of buffaloes. Such management systems are now being fine-tuned with the aim of increasing pregnancy rates and thereby increasing the overall reproductive efficiency among buffaloes. Endocrine techniques, like estrus detection, estrus synchronization, pregnancy diagnosis, induction of parturition, corpus luteum control *et cetera*, have been found useful for augmenting fertility. Second generation strategies aimed at maximizing genetic improvement are also being introduced; for example, assessment of spermatozoa quality through fertilization capacity, sexing spermatozoa, synchronization and fixed time insemination, superovulation (SO), embryo transfer (ET) and *in vitro* embryo production (IVEP). Reproductive technologies are also credited with controlling the incidence of reproductive diseases when the procedures and protocols are accurately followed.

Augmentation of growth for early puberty in buffaloes

Growth hormone (GH) has been long recognized as a potent regulator of an animal's growth however repeated direct administration of exogenous GH for increasing growth has met with limited success due to negative feed back mechanism mediated by IGF-I (Berelowitz *et al.* 1981). As a better alternative, growth hormone releasing factor (GRF), also called Somatoliberin, which controls the expression of GH in more physiological context, has been reported as a better potential growth promoter in cattle (Lapierre *et al.* 1992; Moseley *et al.* 1987; Sejrsen *et al.* 1996). In recent studies the effectiveness of repeated exogenous GRF administrations was proved with respect to a sustained release of GH along with the growth promotion in growing buffalo calves during a long course of treatment and led to the hypothesis that repeated GRF administration had the potential for inducing early onset of puberty in buffalo species (Mondal & Prakash 2005). Subsequently, the successful application of GRF administration for advancing puberty in buffaloes has been demonstrated (Haldar & Prakash 2006).

Augmentation of fertility through progesterone monitoring

Monitoring changing concentrations of progesterone in blood plasma or milk provides an objective evaluation of ovarian activity. Concentration changes in samples collected twice weekly provide a very convenient indicator of buffalo luteal activity (Gupta & Prakash 1990; Kaul *et al.* 1993). The sequential monitoring of progesterone can be used to detect ovulation, estimate the proportion of cycling females in a group, determine whether inseminations have been correctly performed and diagnose and treat reproductive disorders (Kamboj & Prakash 1993; Kaul *et al.* 1993; Kaul & Prakash 1994a). Another important use of progesterone determination in blood plasma or milk is in terms of early (around days 22-24 post insemination in buffaloes) pregnancy and non-pregnancy diagnosis for which the test is about 75% and 100% accurate, respectively (Gupta & Prakash 1990; Kaul & Prakash 1994b).

The tremendous utility of progesterone assays has evolved considerable interest in simplifying the assaying procedure. A simple radioimmunoassay (RIA) method for progesterone estimation, which is completed in a few hours, has been developed and requires no prior extraction of the samples as was previously required (Prakash & Madan 1986; Kamboj & Prakash 1993; Gupta & Prakash 1993). This new assay has been further refined by the use of a highly sensitive and specific antiserum (Prakash & Madan 2001). Quantitative enzyme-immunoassays (EIA) for estimating the hormone in blood plasma and milk of buffaloes have also been developed (Prakash *et al.* 1990; 1992a). The EIA technique is advantageous in situations where facilities do not exist for radioisotope handling.

Pregnancy confirmation through estrone sulphate determination

Estrone sulphate has been found to be quantitatively one of the major estrogens in the blood plasma and milk of pregnant and lactating cows and buffaloes. During the first half of pregnancy its concentrations increase gradually so that after 100 days of pregnancy it is present in all milk samples taken from pregnant cows and buffaloes, whereas it is low or undetectable in non-pregnant animals (Hung & Prakash 1990; Prakash & Madan 1993). Measurement of estrone sulphate in milk on a routine basis could serve as a viable test for pregnancy confirmation and hence also the detection of the presence of mummified fetuses (Prakash & Madan 1994).

Estrus synchronization

Various methods to induce the regression of the CL and the induction of estrus in buffaloes have been developed but are beset with their own advantages and disadvantages. The effectiveness of all these protocols is dependent upon the precision of estrus detection and knowing the time of ovulation after synchronization (Singh et al. 2000). Several methods of estrus synchronization using progestogens, $PGF_{2\alpha}$ and the combination of both are available (Macmillan & Burke 1996). The use of two $PGF_{2\alpha}$ injections at an interval of 11 to 14 days is the most popular technique for estrus synchronization in buffaloes. However, the behavioural signs of estrus expressed after a synchronized estrus are much weaker than those expressed after a spontaneous estrus making estrus detection more difficult. A new estrus synchronization protocol in cattle, called Ovsynch, has been developed: it makes use of a combination of $GnRH$-$PGF_{2\alpha}$-$GnRH$ injections (Pursley et al. 1995). The efficacy of using this estrus synchronization protocol in Murrah buffaloes in tropical conditions has been examined (Paul & Prakash 2005). The conception rates obtained after fixed time AI using the Ovsynch protocol were similar to those obtained from buffaloes inseminated at observed heat. The authors have also compared their results using the Ovsynch protocol with the two injections of $PGF_{2\alpha}$ protocol for synchronization (V Paul & BS Prakash, unpublished data: Tables 1 and 2).

Table 1. Conception rates of buffaloes inseminated after spontaneous estrus and induced estrus using two different protocols for the synchronization of estrus (V Paul & BS Prakash, unpublished data)

Treatment	Number of buffaloes		Conception rate (%)
	Inseminated	Pregnant	
Spontaneous estrus (Control group)	75	23	30.7
Ovsynch (GnRH- $PGF_{2\alpha}$-GnRH)	15	5	33.3
Two injections of $PGF_{2\alpha}$	15	4	26.7

Table 2. Analysis of efficacy of estrus synchronization and timed artificial insemination (V Paul & BS Prakash, unpublished data)

Treatment	Number of estrusesobserved	Number of estruses unobserved	Total number of estruses	Number of animals conceived	% Conception
Control	75	25*	100	23	23.0
Ovsynch	15	-	15	5	33.3
Two injections of $PGF_{2\alpha}$	15	-	15	4	26.7

*During this period of study the incidence of silent estrus in the buffalo herd of the National Dairy Research Institute (NDRI) was reported to be 25% (Prakash 2002)
** Timed artificial insemination was performed 24 h after the second GnRH treatment for Ovsynch group; and 72 and 96 h after second $PGF_{2\alpha}$ injection for the two injections of $PGF_{2\alpha}$ group.

Prolactin inhibition in repeat breeding buffalo heifers in the summer

A study was undertaken to investigate the efficacy of the Ovsynch protocol for estrus synchronization with or without anti-prolactin (Norprolac) administration in repeat breeding Murrah buffalo heifers following timed artificial insemination during the summer season (KS Roy & BS Prakash, unpublished data). A dose of 10 mg Norprolac administered intramuscularly resulted in prolactin suppression. Plasma progesterone profiles indicated that a high percentage (36 to 45%) of repeat breeding buffalo heifers became acyclic during the peak summer months. The Ovsynch protocol without prolactin inhibition was beneficial in terms of reducing the incidence of anestrus from 45% before treatment to only 18% after treatment. Norprolac induced prolactin suppression improved the efficiency of the Ovsynch treatment as there was no incidence of acyclicity post-treatment compared to 36% acyclicity before treatment. The Ovsynch protocol plus Norprolac treatment induced more estrus symptoms per animal than the Ovsynch protocol alone (3.7 vs. 2.5).

Parturition induction

Synthetic glucocorticoid, namely dexamethasone, has been successfully used to induce parturition in buffaloes (Prakash & Madan 1985b). Inducing parturition in buffaloes is useful in clinical cases of uterine prolapse, which is often seen in this species, and other cases of maternal ill health (such as, limb fractures, pericarditis, reticulitis, hydroamnios and hydroallantois) as well as a potential management tool for synchronizing parturition. It can also be useful in terminating prolonged gestations. Parturition induction in buffaloes may however result in placental retention (Prakash & Madan 1986).

Recombinant cytokines and their potential application in the maintenance of pregnancy

In a recent study, recombinant interferon alpha administration on days 14 to 16 of the buffalo estrous cycle not only prolonged the length of the estrous cycle but also significantly reduced the oxytocin-stimulated increase in plasma PGFM concentrations normally detected on day 17 of the cycle (DP Mishra & BS Prakash, unpublished data). These results suggested that interferon alpha may indeed be a signal for the maternal recognition of pregnancy in buffaloes as has already been demonstrated in cattle by earlier workers. The authors went on to demonstrate an improvement in fertility (60% conception rates) in buffaloes treated with recombinant interferon (16 mg/day on days 14 to 16: the dose was divided between morning and evening adminstrations) post AI (the normal conception rate in the herd around the same time was 35%).

Assessment of semen quality and artificial insemination

The most important and early technology that has helped in the improvement of buffalo reproductive efficiency has been the processing and evaluation of semen for national AI programs. Unfortunately the desired impact in animal improvement schemes has remained wanting because both the fertility and conception rates in field programs are very low. Most of the semen banks use only spermatozoa motility as a criterion for semen evaluation despite considerable advances in semen evaluation processes. There is also considerable information available regarding the processing, storage and thawing of buffalo semen (Sansone *et al.* 2000) however the processing and handling procedures of many semen processing laboratories and banks leave much to be desired. The ultimate fertility/conception rate results obtained in AI programs often suffer from this acute inadequacy.

Embryo transfer

Multiple ovulation and embryo transfer (MOET) is one of the major reproductive technologies that facilitates the genetic improvement of the bovine. Unfortunately, high variability in the ovarian follicular response to gonadotrophin stimulation continues to be the major problem in commercial MOET programs. MOET, that takes AI one-step further, both in terms of genetic gains possible and the level of technical capacity and organization required, is one of the basic technologies required for the application of more advanced reproductive biotechnologies; such as, cloning and the generation of transgenics.

The response to superovulation treatments in zebu cattle and buffaloes is considerably inconsistent compared to results obtained in *Bos taurus* (Barros & Nogueira 2001). However, it is important to note that the average number of transferable embryos produced by zebu donors has improved in the last 10-15 years: ranging from 2.4 to 5.8 embryos per flush in the late 1980s to 5.6 to 9.9 transferable embryos in 2000 (Barros & Nogueira 2001). The application of ET technology in buffalo has had limited success (Singla *et al.* 1996; Misra *et al.* 1999). The low embryo production is thought to be due to several reasons: the inherently low reproductive efficiency of buffalo (Singh *et al.* 2000); their poor superovulatory response (Madan *et al.* 1996); and a very low primordial follicle population combined with a higher incidence of atresia (Manik *et al.* 2002). Poor superovulatory responses in buffaloes were also due to the failure to respond optimally to the Lutalyse treatment used for estrus induction (Prakash *et al.* 1992a;b). It has also been hypothesized that ova trapping by the fimbriae of the fallopian tubes may not be as efficient in this species, especially following superovulation. Over the years numerous trials in buffalo using various types of hormonal treatments for induction of superovulation have resulted in highly variable responses with a mean recovery of around 2 transferable embryos per flush (Madan *et al.* 1996). However, in subsequent studies Misra *et al.* (1999) found better superovulatory responses using Folltropin with a viable embryo recovery of 2.8. Following the transfer of embryos to recipients, the conception rates are very low (16%: Singla *et al.* 1996). Since the introduction of embryo transfer technology in buffaloes in India, more than 186 calves have been produced (Misra *et al.* 2005).

In vitro *production of embryos*

Since the birth of the first buffalo calf from an IVF oocyte (Madan *et al.* 1991), there have been a number of publications on *in vitro* embryo production systems describing the effects of different protocols and medium conditions on buffalo oocyte and embryo development. Two extensive reviews have been published recently (Gasparrini 2002; Nandi *et al.* 2002). However, practical use of IVF is limited due to the high production costs and lower overall efficiency under field conditions. The high rates of oocyte maturation (70 to 90%), fertilization (60 to 70%) and cleavage (40 to 50%) in contrast to the moderate to low rates of blastocyst formation (15 to 30%) and calf production (10.5%) have been reviewed (Nandi *et al.* 2002). Viable buffalo blastocysts can be produced from abattoir ovaries (Madan *et al.* 1994a;b) but the rate of transferable embryos remains low (15 to 39%: Chauhan *et al.* 1997; 1998; 1999; Nandi *et al.* 1998; 2002). *In vitro* produced embryos have been successfully used for producing pregnancies and live calves in buffalo (Madan *et al.* 1994b; Chauhan *et al.* 1997) but the success rates in terms of yield of transferable embryos and number of calves born has been low.

Ovum pickup (OPU) and *in vitro* embryo production (IVEP) has been successfully applied to buffaloes even though the efficiency is low in terms of the number of punctured follicles (Galli *et al.* 2001; Manik *et al.* 2002). The use of OPU and IVEP may represent a valid approach to speed up genetic improvement by decreasing the generation interval. Repeated OPU techniques yield a large number of meiotically competent oocytes from individual donors, which can then be used for IVEP programs. It has been demonstrated that OPU is competitive with superovulation because it can yield more transferable embryos per donor on a monthly basis (2.0 *vs.* 0.6: Gasparrini 2002). It has been estimated that a selection scheme based on OPU and IVEP if applied in a closed nucleus of farms will decrease the generation interval from 6.28 to 3.25 yrs compared when progeny testing is used. In fact, the use of OPU combined with IVEP, allows the selection of young bulls on the basis of half and full sibling's milk production rather than their daughters.

The application of nuclear transfer procedures (Singla *et al.* 1997) also holds the potential for producing large numbers of identical offspring (cloning): a technology which can be used for the multiplication of genotypes of superior economic value. The technology of embryo sexing and the production of buffalo calves of pre-determined sex has also been achieved (Appa Rao *et al.* 1993).

On account of the low milk production and considerable variability in the genetic potential of buffaloes, the technique of embryo transfer has been exploited for practical application by utilizing the female population as surrogate mothers for the production of high yielding calves. Embryo cryopreservation and the easy transport of cryopreserved embryos have already proved useful under field conditions to obtain large numbers of progeny from young high yielding donors. This procedure has greater relevance in buffaloes, since high potential progeny tested bulls are not available in large numbers as they are in cattle. The technique of *in vitro* fertilization also holds great promise in harvesting the ova from high producing buffaloes which are transported to metropolitan cities to supply milk and then subsequently lost as they are slaughtered after lactation. Realizing the full potential of the practical application of these technologies, particularly the techniques of nuclear transfer and embryo sexing, will be difficult because of the high costs involved.

Conclusions

This review provides a comprehensive account of the information available on buffalo endocrinology during different physiological states associated with reproduction, along with the endocrine techniques and potential technologies for augmentation of reproduction. An important research objective is to produce recombinant buffalo gonadotrophins: these may improve the consistency of the responses to superovulatory treatment, an extremely important requirement for the greater success and practical application of embryo transfer in this species. Research efforts also need to be intensified in understanding the mechanism(s) that regulate the expression of estrus in buffaloes; including further investigations into the molecular endocrinology of the ovarian follicle, follicular recruitment, atresia and follicular dominance. Understanding all these aspects are essential for unraveling the causes for silent heat in this species. Understanding the causes of early embryonic mortality, particularly the specific genes involved at the level of the ovum/embryo-uterine interaction, are potentially critical approaches for fertility augmentation. The positive correlation obtained between plasma prolactin concentrations and the delay in the postpartum commencement of cyclicity in buffalo needs further investigation. The control of prolactin secretion could hold the key for reducing the postpartum interval to first estrus in the species. Although, a number of biotechniques have been developed for the improvement in reproductive efficiency of the buffalo, these have to be adopted at the field level in a big way if they are to make an impact towards increasing milk production from this species.

References

Appa Rao KBC, Pawshe CH & Totey SM 1993 Sex determination of in vitro developed buffalo (Bubalus bubalis) embryos by DNA amplification. *Molecular Reproduction and Development* **36** 291-296.

Arora SP 1979 Management and feeding of calves from birth for early maturity. In *Proceedings of FAO / SIDA / Govt. of India Symposium on Buffalo Reproduction and Artificial Insemination*, pp 319-326, Dec 4-15, National Dairy Research Institute, Karnal, India.

Arora RC & Pandey RS 1982 Plasma concentrations of progesterone, oestradiol-17b and luteinizing hormone in relation to repeat breeding in buffalo (Bubalus bubalis). *Animal Production* **34** 139-144.

Arya JS & Madan ML 2001a Postpartum gonadotrophins in suckled and weaned buffaloes. *Indian Veterinary Journal* **78** 406-409.

Arya JS & Madan ML 2001b Postpartum reproductive cyclicity based on ovarian steroids in suckled and weaned buffaloes. *Buffalo Journal* **17** 361-369.

Avenell JA, Saepudin Y & Fletcher IC 1985 Concentrations of LH and progesterone in the peripheral plasma of Swamp buffalo cow (*Bubalus bubalis*) around the time of estrus. *Journal of Reproduction and Fertility* **74** 419-424.

Bachalaus NK, Arora RC, Prasad A & Pandey RS 1979 Plasma levels of gonadal hormones in cycling buffalo heifers. *Indian Journal of Experimental Biology* **17** 823-825.

Barkawi AH, Shafie MM & Abul-Ela MB 1986 Prepartum hormonal profile in Egyptian buffaloes. *Buffalo Journal* **2** 117-124.

Barkawi AH, Mokhless EM & Bedeir LH 1989 Environmental factors affecting age at puberty in Egyptian buffaloes. *Buffalo Journal* **5** 71-78.

Barnabe RC, Barnabe VH, Guido MC, Baruselli PS & Zogno MA 1997 Onset of ovarian activity in Murrah buffalo heifers. In *Proceedings of V World Buffalo Congress*, pp 702-705, Oct 13-16.

Barros CM & Nogueira MF 2001 Embryo transfer in Bos indicus cattle. *Theriogenology* **56** 1483-1496.

Baruselli PS, Mucciolo RG, Vistin JA, Viana WG, Arruda RP, Maduriera EH, Oliveira CA & Molero-Filho JR 1997 Ovarian follicular dynamics during the oestrous cycle in buffalo (*Bubalus bubalis*). *Theriogenology* **47** 1531-1547.

Batra SK & Pandey RS 1983 Relative concentrations of 13, 14 dihydro–15 keto-prostaglandin F2a in blood and milk of buffaloes during oestrous cycle and early pregnancy. *Journal of Reproduction and Fertility* **67** 191-195.

Batra SK, Arora RC, Bachalaus NK & Pandey RS 1979 Blood and milk progesterone in pregnant and non-pregnant buffalo. *Journal of Dairy Science* **62** 1390-1393.

Batra SK, Arora RC, Bachalaus NK, Pahwa GS & Pandey RS 1980 Quantitative relationships between oestradiol-17b in the milk and blood of lactating buffaloes. *Journal of Endocrinology* **84** 205-209.

Berelowitz M, Szabo M, Frohman LA, Firestone S & Hintz RL 1981 Somatomedin-C mediates growth hormone negative feedback by effects on both the hypothalamus and the pituitary. *Science* **212** 1279-1281.

Campanile G, Di Palo R, Gasparrini B, D'Occhio MJ & Zicarelli L 2001 Effects of early management system and subsequent diet on growth and conception in maiden buffalo heifers. *Livestock Production Science* **71** 183-191.

Chauhan TR 1995 Buffalo Nutrition in India: a review. *Buffalo Journal* **2** 131-148.

Chauhan MS, Katiyar PK, Singla SK, Manik RS & Madan ML 1997 Production of buffalo calves through in-vitro fertilization. *Indian Journal of Animal Science* **67** 306-308.

Chauhan MS, Singla SK, Palta P, Manik RS & Madan ML 1998 In vitro maturation and fertilization, and subsequent development of buffalo *(Bubalus bubalis)* embryos: effects of oocyte quality and type of serum. *Reproduction, Fertility and Development* **10** 173-177.

Chauhan MS, Singla SK, Palta P, Manik RS & Madan ML 1999 Effect of epidermal growth factor on the cumulus expansion, meiotic maturation and development of buffalo oocytes in vitro. *Veterinary Record* **144** 266-267.

D'Angelo SA & Fisher JS 1969 Influence of oestrogen on the pituitary-thyroid system of the female rat. Mechanism of Lactation. *Endocrinology* **84** 117-122.

Esposito L, Di Palo R, Campanile G, Boni R & Montemurro N 1992 Onset of ovarian activity in Italian buffalo heifers. In *Prospect of Buffalo Production in the Mediterranean and Middle East*, pp374-377, 9-12 Nov, Cairo, Egypt.

FAO 2000 Food and Agriculture Organisation of United Nations. *Livestock Production Statistics*, Rome, Italy

Galhotra MM, Kaker ML & Razdan MN 1981 Serum LH levels during pre- and post-puberty, pregnancy and lactation in Murrah buffaloes. *Theriogenology* **16** 477-481.

Galhotra MM, Kaker ML, Lohan IS, Singal SP & Razdan MN 1988 Thyroid and prolactin status in relation to production in lactating Murrah. In *Proceedings of 2nd World Buffalo Congress*, Vol I, 32-34,

New Delhi, India.

Galli C, Crotti G, Notari C, Turini P, Duchi R & Lazarri G 2001 Embryo production by ovum pick up from live donors. *Theriogenology* **55** 1341-1357.

Gasparrini B 2002 In vitro embryo production in buffalo species: state of art. *Theriogenology* **57** 237-256.

Gupta M & Prakash BS 1990 Milk progesterone determination in buffaloes post-insemination. *British Veterinary Journal* **146** 563-570.

Gupta M & Prakash BS 1993 Development of direct rapid radioimmuno-assays for progesterone in milk and fat-free milk of buffaloes. *Indian Journal of Animal Science* **63** 1206-1211.

Haldar A & Prakash BS 2005 Peripheral patterns of growth hormone, luteinizing hormone and progesterone before, at and after puberty in buffalo heifer. *Endocrine Research* **31** 295-306.

Haldar A & Prakash BS 2006 Growth hormone releasing factor (GRF) induced growth hormone advances puberty in female buffaloes. *Animal Reproduction Science* **92** 254-267.

Heranjal DD, Sheth AR, Desai R & Rao SS 1979 Serum gonadotropins and prolactin levels during oestrous cycle in Murrah buffaloes. *Indian Journal of Dairy Science* **32** 247-249.

Hung NN & Prakash BS 1990 Changes in estrone and estrone sulphate concentrations in blood plasma of Karan Swiss cows and Murrah buffaloes throughout gestation. *British Veterinary Journal* **146** 449-456.

Janakiraman K, Desai MC, Amin DR, Sheth AR, Moodbidr SB & Wadadekar KB 1980 Serum gonadotropin levels in buffaloes in relation to phases of oestrous cycle and breeding periods. *Indian Journal Animal Science* **50** 601-606.

Kaker ML, Razdan MN & Galhotra MM 1980 Serum LH concentrations in cyclic buffalo (*Bubalus bubalis*). *Journal of Reproduction and Fertility* **60** 419-424.

Kamboj M & Prakash BS 1993 Relationship of progesterone in plasma and whole milk of buffaloes during cyclicity and early pregnancy. *Tropical Animal Health and Production* **25** 185-192.

Kamonpatana M, Luvira Y, Bodipaksha P & Kunawongkrit A 1976 Serum progesterone, 17-hydroxy progesterone and 17b-oestradiol during oestrous cycle in Swamp buffalo in Thailand. In *Nuclear Techniques in Animal Production and Health*, pp. 569-578, International Atomic Energy Agency, Vienna, Austria.

Kamonpatana M 1984 Application of hormone assay and endocrine pattern in buffalo. In *Proceedings of the 10th International Congress on Animal Reproduction and Artificial Insemination*, Vol IV, pp. 1-9, Urbana, Illinois, USA.

Kanai Y & Shimizu H 1984 Plasma concentrations of LH, progesterone and oestradiol during oestrous cycle in Swamp buffaloes (*Bubalus bubalis*). *Journal of Reproduction and Fertility* **70** 507-510.

Kaul V & Prakash BS 1994a Application of milk progesterone estimation for determining the incidence of false oestrus detection and ovulation failures in zebu

and crossbred cattle and Murrah buffaloes. *Indian Journal of Animal Science* **64** 1054-1057.

Kaul V & Prakash BS 1994b Accuracy of pregnancy/non-pregnancy diagnosis in zebu and crossbred cattle and Murrah buffaloes by milk progesterone determination post insemination. *Tropical Animal Health and Production* **26** 187-192.

Kaul V, Rao SVN & Prakash BS 1993 Reproductive status monitoring by milk progesterone determination in rural anestrus and repeat breeding bovines. *Indian Journal of Animal Science* **63** 1132-1135.

Kaur H & Arora SP 1984 Annual pattern of plasma progesterone in normal cycling buffaloes (*Bubalus bubalis*) fed two different levels of nutrition. *Animal Reproduction Science* **7** 323-332.

Khurana ML & Madan ML 1985 Thyroidal hormone during oestrous cycle in cattle and buffaloes. *Indian Journal of Dairy Science* **38** 119-123.

Kumar R, Jindal R & Rattan PJS 1991 Plasma hormonal profiles during oestrous cycle of Murrah buffalo heifers. *Indian Journal of Animal Science* **61** 382-385.

Lapierre H, Tyrrell HF, Reynolds CK, Elsasser TH, Gaudreau P & Brazeau P 1992 Effects of growth hormone–releasing factor and feed intake on energy metabolism in growing beef steers: whole body energy and nitrogen metabolism. *Journal Animal Science* **70(3)** 764-772.

Macmillan KL & Burke CR 1996 Effect of oestrous cycle control on reproductive efficiency. *Animal Reproduction Science* **42** 307-320.

Madan ML 1990 Factors limiting superovulation responses in embryo transfer programs among buffaloes. *Theriogenology* **33** 280.

Madan ML & Johnson HD 1973 Environmental heat effects upon bovine LH. *Journal of Dairy Science* **56** 1420-25.

Madan ML & Johnson HD 1975 Heat stress effects upon bovine reproductive cycle. In *Selected Topics in Environmental Biology*, p 207. Eds B Bhatia, G Chinna & Baldev Singh. New Delhi, India: Interprint publications.

Madan ML, Singh Mahendra, Prakash BS, Naqvi SMK & Roy AK 1984 Postpartum endocrinology of buffalo. In *Proceedings of the 10th International Congress on Animal Reproduction and Artificial Insemination*, Vol. III, pp 402-404, Urbana, Illinois, USA.

Madan ML, Singla SK, Jailkhani S & Ambrose D 1991 In-vitro fertilization and birth of first ever IVF buffalo calf. In *Proceedings of the Third World Buffalo Congress*, Vol 2, pp. 11-17. Varna, Bulgaria.

Madan ML, Prakash BS, Jailkhani S, Singla SK, Palta P & Manik RS 1993 Buffalo endocrinology with special reference to embryo transfer. Technical report publication No.265, p 32, National Dairy Research Institute, Karnal, India.

Madan ML, Chauhan MS, Singla SK & Manik RS 1994a Pregnancies established from water buffalo (Bubalus bubalis) blastocysts derived from in vitro fertilized oocytes and co-cultured with and oviductal cells. *Theriogenology* **42** 591-600.

Madan ML, Chauhan MS, Singla SK & Manik RS 1994b

In vitro production and transfer of embryos in buffaloes. *Theriogenology* **41** 139-143.

Madan ML, Das SK & Palta P 1996 Application of reproductive technology to buffaloes. *Animal Reproduction Science* **42** 299-306.

Manik RS, Singla SK, Palta P & Madan ML 1998 Real time ultrasound evaluation of change in follicular populations during oestrus cycle in Buffalo. *Indian Journal of Animal Science* **68** 1157-1159.

Manik RS, Palta P, Singla SK, Sharma V & Madan ML 2002 Folliculogenesis in buffalo (Bubalus bubalis): A review. *Reproduction and Fertility Development* **14** 315-325.

Maqsood M 1954 Role of thyroid hormone in reproduction. *Indian Veterinary Journal* **31** 23-30.

Mishra DP, Meyer HHD & Prakash BS 2003 Validation of a sensitive enzyme immunoassay for 13, 14-dihydro-15-keto-PGF2a in buffalo plasma and its application for reproductive health status monitoring. *Animal Reproduction Science* **78** 33-46.

Misra AK, Muthu Rao M, Kasiraj R, Ranga Reddy NS & Pant HC 1999 Factors affecting pregnancy rate following non-surgical embryo transfer in buffalo (Bubalus bubalis) – a retrospective study. *Theriogenology* **52** 701-707.

Misra AK, Pragsad S & Taneja VK 2005 Embryo transfer technology (ETT) in cattle and buffalo in India: A review. *Indian Journal of Animal Science* **75** 842-857.

Mondal M & Prakash BS 2003 Changes in plasma GH, LH and progesterone and blood metabolites following long term exogenous somatoliberin administration in growing buffaloes (Bubalus bubalis). *Journal of Animal and Veterinary Advances* **2(4)** 259-270.

Mondal M & Prakash BS 2004 Effects of long-term growth hormone-releasing factor (GRF) administration on pattern of GH and LH secretion in growing female buffaloes (Bubalus bubalis). *Reproduction* **127** 45-55.

Mondal M & Prakash BS 2005 Effects of long–term growth hormone–releasing factor treatment on growth, feed conversion efficiency and dry matter intake in growing female buffaloes (Bubalus bubalis). *Journal of Animal Physiology and Animal Nutrition* **89** 260-267.

Mondal S, Prakash BS & Palta P 2003a Relationship between peripheral plasma inhibin and progesterone concentrations in Sahiwal cattle (Bos indicus) and Murrah buffaloes (Bubalus bubalis). *Asian-Australian Journal of Animal Science* **16 (1)** 6-10.

Mondal S, Prakash BS & Palta P 2003b Peripheral plasma inhibin concentrations in relation to expression of oestrus in Murrah buffaloes (Bubalus bubalis). *Indian Journal of Animal Science* **73(4)** 405-407.

Moseley WM, Huisman J & Van Weerden EJ 1987 Serum growth hormone and nitrogen metabolism responses in young bull calves infused with growth hormone-releasing factor for 20 days. *Domestic Animal Endocrinology* **4** 51.

Nandi S, Chauhan MS & Palta P 1998 Influence of cumulus cells and sperm concentration on cleavage rate and subsequent embryonic development of buffalo (Bubalus bubalis) oocytes matured and fertilized in vitro. *Theriogenology* **50** 1251-1262.

Nandi S, Raghu HM, Ravindranatha BM & Chauhan MS 2002 Production of buffalo (Bubalus bubalis) embryos in vitro: premises and promises. *Reproduction in Domestic Animals* **37** 65-74.

Pahwa GS & Pandey RS 1983 Gonadal steroid hormone concentrations in blood plasma and milk of primiparous and multiparous pregnant and non pregnant buffaloes. *Theriogenology* **19** 491-505.

Pahwa GS & Pandey RS 1984 Prolactin in blood plasma and milk of buffalo during oestrous cycle and early pregnancy. *Journal of Dairy Science* **67** 2001-2005.

Palta P & Madan ML 1995 Alterations in hypophyseal responsiveness to synthetic GnRH at different postpartum intervals in Murrah buffaloes (Bubalus bubalis). *Theriogenology* **44** 403-411.

Palta P & Madan ML 1996 Effect of gestation on GnRH induced LH and FSH release of buffalo (Bubalus Bubalis). *Theriogenology* **46** 993-998.

Palta P, Jailkhani S, Prakash BS, Manik RS & Madan ML 1996a Development of direct radioimmunoassay for oestradiol-17b determination in follicular fluid from individual buffalo ovarian follicles. *Indian Journal of Animal Science* **66** 126-130.

Palta P, Prakash BS & Madan ML 1996b Peripheral inhibin levels during oestrous cycle in Murrah buffalo (Bubalus Bubalis). *Theriogenology* **45** 655-664.

Palta P, Prakash BS, Manik RS & Madan ML 1996c Inhibin in individual buffalo ovarian follicles in relation to size. *Indian Journal of Experimental Biology* **34** 606-608.

Palta P, Mondal S, Prakash BS & Madan ML 1997 Peripheral plasma inhibin levels in relation to climatic variation and stage of oestrous cycle in buffalo (Bubalus bubalis). *Theriogenology* **47** 989-995.

Palta P, Bansal N, Prakash BS, Manik RS & Madan ML 1998 Follicular fluid inhibin in relation to follicular diameter and oestradiol-17b, progesterone and testosterone concentrations in individual buffalo ovarian follicles. *Indian Journal of Experimental Biology* **36** 768-774.

Pandey RS 1979 Hormonal status of female and induced breeding in Murrah buffaloes. In *Proceedings of FAO / SIDA / Govt. of India Symposium on Buffalo Reproduction and Artificial Insemination*, pp 185-197, Dec 4-15, National Dairy Research Institute, Karnal, India.

Paul V & Prakash BS 2005 Efficacy of the Ovsynch protocol for synchronization of ovulation and fixed-time artificial insemination in Murrah buffaloes (Bubalus bubalis). *Theriogenology* **64** 1049-1060.

Perera BMAO 1999 Reproduction in water buffalo: comparative aspects and implications for management. *Journal of Reproduction and Fertility* **45** 157-168.

Prakash BS 2002 Influence of environment on animal reproduction. In *National Workshop on Animal Climate Interaction*, pp 33-47, held at Izatnagar, India.

Prakash BS & Madan ML 1984a Radioimmunoassay of

cortisol in peripheral blood plasma of buffaloes peripartum. *Theriogenology* **22** 241-245

Prakash BS & Madan ML 1984b Radioimmunoassay of estradiol 17ß in buffaloes peripartum. *Indian Journal of Experimental Biology* **22** 104-105.

Prakash BS & Madan ML 1985a Periparturient plasma progesterone and prostaglandin F (PGF) levels in buffaloes (*Bubalus bubalis*). *Indian Journal of Animal Science* **55** 642-646.

Prakash BS & Madan ML 1985b Induction of parturition in water buffaloes (*Bubalus bubalis*). *Theriogenology* **23** 325-331.

Prakash BS & Madan ML 1986 Peripheral plasma oestradiol-17B, progesterone and cortisol in buffaloes induced to calve with dexamethasone and vetoestrol. *Animal Reproduction Science* **11** 111-122.

Prakash BS & Madan ML 1993 Influence of gestation on oestrone sulphate concentration in milk of zebu and crossbred cows and Murrah buffaloes. *Tropical Animal Health and Production* **25** 94-100.

Prakash BS & Madan ML 1994 Oestrone sulphate determination in milk for detection of foetal death in bovines. *Indian Journal of Animal Science* **64** 56-58.

Prakash BS & Madan ML 2001 Production and characterization of a sensitive antiserum against progesterone. *Indian Journal of Animal Science* **71** 251-253.

Prakash BS, Madan ML, Jailkhani S & Singla SK 1990 Development of a simple direct, microtiterplate Enzymeimmunoassay (EIA) for progesterone determination in whole milk of buffaloes. *British Veterinary Journal* **146** 571-576.

Prakash BS, Singla SK, Ambrose JD, Jailkhani S & Madan ML 1992a Determination of milk progesterone profiles by a sensitive, direct, enzymeimmunoassay in superovulated buffaloes. *Theriogenology* **37** 897-905.

Prakash BS, Jailkhani S, Singla SK & Madan ML 1992b Application of a sensitive, heterologous enzyme-immunoassay for progesterone determination in unextracted buffalo plasma samples collected in an embryo transfer experiment. *Animal Reproduction Science* **27** 67-74.

Prakash BS, Sarkar M, Paul Vijay, Mishra DP, Mishra A & Meyer HHD 2005 Postpartum endocrinology and prospects for fertility improvement in the lactating riverine buffalo (*Bubalus bubalis*) and yak (*Poephagus graunnies* L.) *Livestock Production Science* **98** 13-23.

Pursley JR, Mee MO & Wiltbank MC 1995 Synchronization of ovulation in dairy cows using PGF2a and GnRH. *Theriogenology* **44** 915-923.

Rao LV & Pandey RS 1982 Seasonal changes in the plasma progesterone concentration in buffalo cow (*Bubalus bubalis*). *Journal of Reproduction and Fertility* **66** 57-61.

Rao LV & Pandey RS 1983 Seasonal variations in oestradiol-17b and luteinizing hormone in the blood of buffalo cows (*Bubalus bubalis*). *Journal of Endocrinology* **98** 251-255.

Razdan MN & Kakar ML 1980 Summer sterility and endocrine profiles of buffaloes. *Indian Dairyman* **32** 459-464.

Razdan MN, Kaker ML & Galhotra MM 1982 Serum FSH levels during oestrus and a 4-week period following mating in Murrah buffaloes (*Bubalus bubalis*). *Theriogenology* **17** 175-181.

Roy KS & Prakash BS 2006 Development and validation of a simple, sensitive enzymeimmunoassay (EIA) for quantification of prolactin in buffalo plasma. *Theriogenology* **67** 572-579.

Roy DT, Bhattacharya AR & Luktuke SN 1972 Estrus and ovarian activity of buffaloes in different months. *Indian Veterinary Journal* **49** 54-60.

Sansone G, Nastri MJE & Fabbrocini A 2000 Storage of buffalo (*Bubalus bubalis*) semen. *Animal Reproduction Science* **62** 55-76.

Sarvaiya NP & Pathak MM 1992 Profiles of progesterone, 17ß-oestradiol, triiodothyronine and blood biochemical parameters in Surti buffalo. *Buffalo Journal* **8** 23-30.

Sejrsen K, Oksbjerg N, Vestergaard M & Sorensen MT 1996 Growth hormone and related peptides as growth promoters. In *Proceedings of Scientific Conference on Growth Promotion in Meat Production*, pp 87-119, Brussels, Belgium.

Singh J 1990 Prolactin and hypophysial-gonadal functions in buffaloes. *PhD thesis*. National Dairy Research Institute, Deemed University, Karnal, India.

Singh M & Madan ML 1985 Plasma testosterone during oestrous cycle and post partum period among buffaloes. *Indian Journal of Animal Reproduction* **6** 13-16.

Singh J & Madan ML 1993 RIA of prolactin as related to circadian changes in buffaloes. *Buffalo Journal* **9** 159-164.

Singh C & Madan ML 1998a Pituitary and gonadal response to GnRH to prepubertal buffaloes (*Bubalus bubalis*). *Asian-Australasian Journal of Animal Science* **11** 78-83.

Singh C & Madan ML 1998b Hypophyseal and gonadal response to GnRH in buffalo heifers (*Bubalus bubalis*) *Asian-Australasian Journal of Animal Science* **11** 416-421.

Singh C & Madan ML 1999a The ovarian response of prepubertal buffaloes (*Bubalus bubalis*) to superovulation in equine chorionic gonadotrophin with and without treatment with GnRH. *Veterinary Journal* **158** 155-158.

Singh C & Madan ML 1999b Ovarian response to equine chorionic gonadotrophin (eCG) and GnRH in buffalo heifers (*Bubalus bubalis*). *Indian Journal of Animal Science* **69** 492-493.

Singh C & Madan ML 2000a circulatory level of follicle stimulating hormone (FSH) in Murrah buffalo heifers (*Bubalus bubalis*). *Indian Journal of Animal Science* **70** 403-404.

Singh C & Madan ML 2000b Plasma prolactin concentration in Murrah buffalo heifers (*Bubalus bubalis*) with and without GnRH administration. *Indian Jour-*

nal of Animal Science **70** 933-935.

Singh C & Madan ML 2000c Effect of PGF2α and GnRH during different ovarian status at onset of puberty in Murrah buffalo heifers (*Bubalus bubalis*). *Asian-Australasian Journal of Animal Science* **13** 1059-1062.

Singh C & Madan ML 2002a Hypophyseal and gonadal response to PGF2α and GnRH at onset of puberty in Murrah buffalo heifers. *Indian Journal of Animal Science* **72** 431-533.

Singh C & Madan ML 2002b Plasma prolactin concentration in non-cycling Murrah buffalo heifers. *Indian Journal of Animal Science* **72** 841-843.

Singh J, Nanda AS & Adams GP 2000 The reproductive pattern and efficiency of female buffaloes. *Animal Reproduction Science* **60-61** 593-604.

Singla SK, Manik RS & Madan ML 1996 Embryo biotechnology in buffaloes: A review. *Bubalus bubalis* **1** 53-63.

Singla SK, Manik RS & Madan ML 1997 Micromanipulation and cloning studies on buffalo oocytes and embryos using nucleus transfer. *Indian Journal of Experimental Biology* **35** 1273-1283.

Thorburn GD, Challis Jr RG & Currie WB 1977 Control of parturition in domestic animals. *Biology of Reproduction* **16,** 18-29.

Tiwari SR & Pathak MM 1995 Influence of suckling on postpartum reproduction performance of Surti buffaloes. *Buffalo Journal* **2** 213-217.

The challenges and progress in the management of reproduction in yaks

S J Yu

Faculty of Veterinary Medicine, Gansu Agricultural University, Lanzhou, Gansu, 730070, P.R.China

This paper deals with the progress in the management of reproduction in domestic yaks, including female reproductive biology, male reproductive biology and the main reproductive technologies in both female and male yaks. Further studies and actions with some immediate measures are also recommended.

Introduction

The yak (*Poephagus grunniens*) belongs to the bovine subfamily *Bovidae* and is the only species of the genus *Poephagus*. While the camel is regarded as "a ship of the desert", the yak is reasonably termed as "a ship of plateau" or more precisely, "a sole vehicle of the Asia Highland". The yak is indeed the only large domestic animal which populates the central Asian Highlands up to altitudes of 6000m. The population of yak worldwide is estimated to be about 14 million, distributed in China, Mongolia, Southern Russia, Tajikistan, Kirgies, Nepal, Bhutan, Myanmar, Pakistan and Afghanistan, of which 90% is located in China. It is therefore of extreme economic importance, providing the indigenous nomads and farmers with the essentials for their livelihood. Milk, dairy products and meat are the main foodstuffs. Skin, wool and dung are utilized to meet requirements for shelter, housing and fuel. The animals are also used in farming and transportation of goods.

Currently, most of the literature on the yak is represented by a wealth of articles in scientific journals, reports and proceedings of technical meetings which are mainly in Chinese or Russian languages. These languages are not generally accessible to international communities, and there are only a few books and papers dealing with the yak written in English, especially about its reproduction. The present paper therefore tries to give an account of the current knowledge, new findings and techniques on all aspects of yak reproduction.

Female reproductive physiology

Reproductive morphology and follicular systems

The structure of the reproductive organs of the female yak differs in some aspects from those of dairy cows. The cervix in yaks has three transverse circles consisting of many tight folds, with an average length of 5.0 ± 0.9 cm and an average external diameter of 3.2 ± 0.7 cm. The body of the uterus is rather short, being 2.1 ± 0.8 cm in length. A long and distinct septum (approx 6 cm in length) extends downward from the bifurcation of the uterine horns towards the uterine body (Fig.1). Because of the short uterine body and the long septum, especially the relatively free cervix within the pelvic cavity, the uterus can be readily held for palpation. The length of the uterine tube is 18– 24 cm (Cui & Yu 1999a).

E-mail: sjyu@163.com

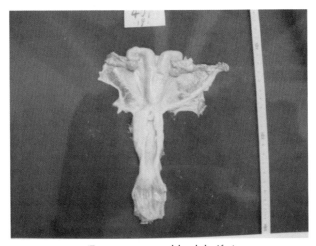

(From one year old yak heifer)

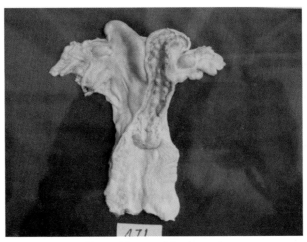

(From 12 year old yak cow)

Fig. 1 The inner reproductive organs of female yaks.

Cui and Yu (1999b) carried out a study on forty-five female yaks of different ages with know reproductive histories to determine the morphology and follicular systems of ovaries. The histological structure of the ovaries resembled that of cattle and buffalo and was similar in animals of different ages. Atresia was recognized at all stages of follicular development: atresia of primordial follicles took one form (i.e. atresia of the oocyte); that of growing follicles could be divided into two stages (early and late); and that of Graafian follicular were classified as early, definite- and late-stage. Details of follicle populations in this study are given in Table 1.

Table 1. Follicular characteristics of female yaks

Total per ovary pair,	1-month-old calves	1-year-old heifers,	2-year-old heifers	7-10-year-old cow
Primordial follicles (n)	53,500 ± 6300	32,870 ± 4500	22,850 ± 2800	9500 ±1200
Atretic primordial follicles (%)	51.6	55.5	56.7	47.4
Growing follicles (n)	210 ± 76	815 ± 95	895 ± 142	445 ± 88
Atretic growing follicles (n)	119.5 ± 21.5	605.5 ± 74.3	721.6 ± 78.5	275.8 ± 66.3
Graafian follicles (n)	36.5 ± 14.2	41.7± 12.3	37.8 ± 9.8	42.5 ±14.5
Atretic Graafian follicles (%)	22.1 ± 5.6,	21.2 ± 7.6	21.5 ± 4.7	25.3 ± 6.7

Puberty

First oestrus in yaks generally occurs in the second or third summer and autumn following birth, i.e. at between 13 and 30 months of age. In Mongolia, around 10% of yaks have their first oestrus in the second summer of their life, but most females do not show estrus until their third summer when they are more than 2 years old. Onset of first oestrus is determined more by the body condition at the beginning of the breeding season than by age (Magash 1991). Very similar results were also reported by Katzina & Maturova (1989) based on observations on yak in Tuva Autonomous Republic.

Yu & Li (2001) found, for observation of 60 yak heifers, that cyclic activity began at a mean age of 33 months. In China, the majority of yaks are mated for the first time at the age of 3 years (i.e., in the 4th warm season following birth) but, under favorable conditions, some yaks may be mated a year earlier. Such conditions prevailed among 197 primaparous Jiulong yaks in the Sichuan province studied by Cai (1980). In that study, 32.5% calved at 3 years, 59.9% at 4, 6.1% at 5, and the remaining 1.5% at 6 years of age. At these ages, yaks had reached between 75% and 100% of their mature weight. In this context, yaks in Tuva reached fertile estrus at approximately 90% of mature body weight, compared with 60% for *Bos taurus* cattle in that region (Katzina & Maturova 1989). First mating at an age of 2 years occurs to a limited extent in China but is more common amongst yaks in other countries.

Breeding season

Yaks are considered to be seasonal breeders. However, information about the breeding season is rather conflicting. The onset and the end of the breeding season are affected by ecological factors such as climate, grass growth, latitude and altitude. When temperature and humidity start to rise and grass starts to grow, the body condition of yaks improves following their long period of deprivation and weight loss over the winter and, at that time, females come into the breeding season. On the northwestern grasslands of Sichuan the season begins around June (Hu et al. 1960) whereas at the higher elevation of Laqu in Tibet, the breeding season does not start until July. Similar observations have been reported from Kirgizia, where the annual onset of the breeding season started on May 25th at an elevation of 1400m and occurred progressively later until at the altitude of 2700m oestrus started after June 22nd (Denisov 1958). The breeding season reaches its peak in July and August when temperature is at its highest and grass growth is at its best. Thereafter, the frequency of oestrus decreases and stops altogether around November.

Oestrus

The average length of oestrous cycle in yaks has been reported to be around 18 to 22 days. However, the great variation in the length of the oestrous cycle is one of the problems in yak reproduction. The main reasons for the variation could be silent or non-detected oestrus, delayed ovulation, implantation failure, or embryonic death. Oestrus in yaks is greatly affected by the environment, and when the weather is unfavorable the onset of oestrus is delayed, and when in favorable circumstances the onset of oestrus in female yaks is advanced.

The duration of the oestrus is not easily determined in the yak, since the signs of oestrus are not always clear. Estimates from northwestern Sichun suggest that oestrus lasts 12 to16 hours. In a study of 41 female yaks (Yu *et al*. 1993a), oestrus lasted 24 hours or less in 26 of them, whilst 4 yaks showed oestrus for up to 72 hours. More than 80% of these animals ovulated within 24h after the end of oestrus. There is a tendency for the proportion of yaks with heat periods of 1 to 2 days to increase later in the breeding season when ambient temperature begins to decline. Purevzav & Beshlebnov (1967) recorded that among 54 Mongolian yaks, 26 were recorded in heat for only 0.5 to 6.5h. A further 17 females showed oestrus for between 6.5 and 12.5h, 7 females showed oestrus for 12.5 to 18.5h, and only 4 females showed a longer duration of oestrus. Most of these yaks ovulated 12 to 24h after the end of oestrus (Yu *et al*. 1993a).

Changes in the appearance of the reproductive organs at the time of oestrus are more obvious than behavioral changes (Yu *et al*. 1993a). The vulva becomes swollen and the vagina reddens. Mucus is discharged from the vulva in a majority of oestrous females, although a substantial minority shows no such discharge. The vagina and cervix dilate, the female tends to raise her tail and urinates frequently. As in other cattle, yak females on heat search out and ride other females and like to be approached by male yaks. However, these signs are less pronounced than in *Bos taurus* cattle. The most pronounced signs of oestrus are being followed and mounted by mature bulls, swelling of the vulva, reddening of the vaginal mucosa and the discharge of mucous.

The average duration of post-partum anoestrous at Xiandong farm in Sichuan Province was reported to be 125 days (Cai & Weiner 1995); that figure, however, was subject to much variation. However, the duration of the postpartum aneostrous period was much shorter (70.5 ± 18.5 days) for yaks in good body condition than for those in poor body condition (122.3 ± 11.8 days; Liu & Liu 1982). The duration of postpartum anoestrous period has been reported to be related to month of calving: for females calving in March, April, May and June, the periods were 131, 124, 90 and 75 days respectively, whilst for animals calving between March and August, the periods for each month were 134, 130, 105, 89 and 37 days (Lei 1964).

Likewise, Magasch (1990) provided results on the interval between calving and first postpartum oestrus for yak females in Mongolia, showing clearly a relationship with month of calving, the earlier the calving the longer the interval. However, there was considerable variation in these results around the average interval periods. Magasch (1990) reported that the service period following calving showed a very similar seasonal pattern for the interval between calving and first postpartum oestrus.

Gestation and parturition

Conception rates after mating at the first oestrus of the breeding season are generally high. In a trial with 265 yaks that had previously calved, 72.4% became pregnant following the first oestrus, a further 23.4% following the second, and 3.4% and 0.8% following the third and fourth cycles respectively (Liu 1981). In a similar investigation of 342 yaks in Mongolia, it was found that 70.5% were pregnant after the first service, 19.3% conceived to a second service and 4.6% to a third service; giving an overall pregnancy rate of 94.4% (Magasch 1990). It also appears from the above results that Mongolian yaks that fail to conceive at their first oestrus are capable of returning to oestrus up to three times in the same breeding season. In one particularly well-maintained herd of yaks on grassland in Gansu, in which animals received some supplementary feed in late winter and early spring, a conception rate of 93.4% was achieved (Yu *et al.* 1993a).

Almost all births take place during the day and only very few at night when yak cows are normally at the herders' campsite. Typical behavior of the yak during labour includes lying on her side and standing up again for delivery. Dystocia is a rare occurrence in yak carrying pure-breed calves. The umbilical cord is severed by mechanical stretching as the cow gets up or as the calf falls down after delivery. Yak cows with hybrid calves, however, require help for delivery. For example, there were 28 cases of dystocia among 681 such calvings (4.1%) over a period of 10 years in one study in Sichuan (Xu 1964). Twins are rare and represent only about 0.5% of all births, but higher rates have exceptionally been recorded (Cai & Weiner 1995). The dam generally licks the newborn calf for about 10 min, after which the calf attempts to stand up and suckle.

The placenta is normally passed within half an hour of parturition, although this may not occur until up to 6 hours later. The dam is intensely protective of her calf in the period shortly after birth and will attack any person coming close. Bonding between dam and calf depends mainly on smelling and licking. Longer parturition times and dystocia militate against such bonding and thus place Pian Niu (hybrid) calves at a disadvantage versus purebred yak calves.

Yu & Chen (1994a) studied a total of 1953 yak cows (3 to 11 years of age) in two populations differing in calving rates. The calving rate was 78% in the high calving rate group and 50% in the other. It appears that the main differences were due to milking frequency and feed supplementation between the two populations. The high calving rate group was milked once daily and received supplementary feed from late winter to early spring, whilst the low calving rate group was milked twice daily and did not received any supplementary feed.

Calving interval is an important economic index in the yak industry. Wang *et al.* (1997) observed 439 calvings from 161 yaks over a 7-year period. The average calving interval was 459 ± 131 days and showed a tendency for gradual decrease with increase in parity.

Reproductive endocrinology

Since 1985 we have carried out a series of studies to examine the hormonal control of puberty, the oestrous cycle, pregnancy, parturition and the postpartum period in the yak (Yu *et al.* 1993b; 1993c; Yu & Chen 1993; 1994b; 2000; Yu & Li 2001).

Fifteen yak heifers were used to analyze plasma progesterone before and at the attainment of puberty. They were divided into three age groups: group1 (10 ~ 14 months, n = 5), group 2 (20 ~ 24 months, n = 5) and group 3 (30 ~ 36 months, n = 5). Group 1 yak heifers were found to have two progesterone profiles: IO (inactive ovary) profile and LPSC (low progesterone short cycle) profile; group 2 had three profiles: IO profile, LPSC profile and LPNC (low progesterone normal cycle) profile; group 3 had three profiles: LPSC profile, LPNC profile and a normal estrous cycle profile. It would be concluded that one or two, or even more brief rise(s) in circulating progesterone were present in yak heifers. These rises, however, were not followed by a normal luteal phase except in two yak heifers that came into oestrus.

In yak cows, the LH surge appeared 12-15 h after the onset of oestrus. Concentrations of oestradiol-17ß were high (28.07 ± 7.24 pg/mL in plasma) at the onset of oestrus, but continued to increase thereafter, reaching peak values (49.88 ± 10.58 pg/mL in plasma) about 2 h after the

LH surge; then it fell. Progesterone concentrations were low at the onset of oestrus, rose markedly within 16 h and reached a small peak 18 h after the onset of oestrus (i.e. about 4 h after the LH surge). However, concentrations promptly dropped thereafter, with similar values at 20 h as had been present at the onset of oestrus (Fig. 2).

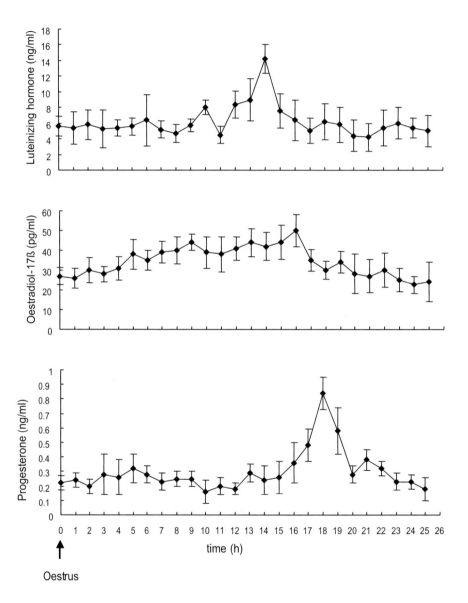

Fig. 2 Mean concentration (+SD) of luteininizing hormone (n = 6), oestradiol-17ß (n = 5) and progesterone (n = 5) in peripheral plamsa taken at 1hr intervals around the time of oestrus.

Concentrations of oestradiol-17ß and progesterone in plasma and milk of the six yaks during normal cycles were measured. There were three peaks of oestradiol-17ß in plasma and milk on the day of oestrus, and on days 5 and 14 of the cycle. Progesterone levels in plasma and milk were low during oestrus but peaks were seen on day 15. The levels of oestradiol-17ß and progesterone in milk were about 4 or 5 times higher than those in plasma during the normal cycle. Three yak cows were used to determine the oestradiol-17ß and progesterone profiles in plasma during the short cycle (Fig. 3). It was found that the pattern of both oestradiol-17ß and progesterone during the short cycle were similar to those during the normal cycle, but the values were lower. The result indicated that some follicles must have been developing; this was confirmed in all 3 cows examined rectally. We also found that all yak cows showed some oestrus-like symptoms and some of them presented a short cycle at the beginning of the breeding season before a normal cycle. It might be concluded that a transition to the normal cyclic activity occurs at the beginning of the breeding season in all yak cows.

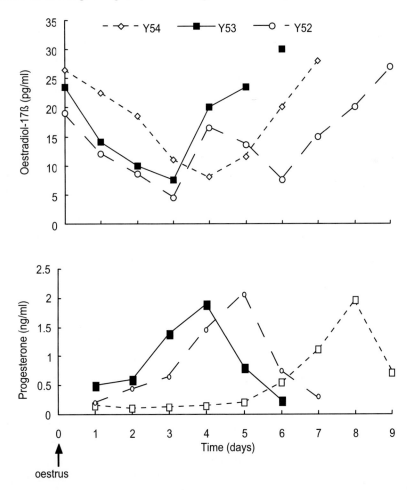

Fig. 3 Concentrations of plasma progesterone and oestradiol-17ß during the short oestrous cycle in three yaks.

The profiles of progesterone were similar in pregnant and non-pregnant yaks over the first 14 days following oestrus, but concentrations were significantly higher in pregnant than non-pregnant yaks on Day 19, and tended to progressively increase thereafter. Plasma progesterone concentrations declined around Day 120, then increased to reach its maximum on Day 210; declined again 20 days before parturition and were basal by the time of parturition. Oestradiol-

17ß concentrations in plasma and milk increased gradually from Day 23 post-conception, decreased abruptly on Day 60, then increased to reach maximum values at parturition. After parturition, oestradiol concentrations returned to baseline.

After calving, four patterns of milk progesterone profiles in suckled yak cows were noted. Type I: Cows had cyclic changes of milk progesterone concentrations within 40 days postpartum. Type II: Cows had a short progesterone rise within 20 days postpartum, then progesterone concentrations remained low (<0.5 ng/ml) until 90 days postpartum. Type III: Cows had milk progesterone concentrations of below 0.5 ng/ml until 90 days postpartum. Type IV: Cows had milk progesterone concentrations of above 1.0 ng/ml from 20 days until 90 days postpartum.

Male reproductive physiology

Testis and sperm morphology

The yak scrotum is small, with abundant hair, which is apparently an adaptation to the cold environment (Xu 1964). Yak testes are similar anatomically and histologically to those of other bovine bulls, except that they are smaller that those of B. taurus bulls (approximately 300 versus 550 - 650 g). In general, the testicular weight in Charolais and Holstein bulls is ~ 2x that of yak, while that of Zebu was ~0.65x (Yan et al. 1997). The height of the seminiferous epithelium and volume density of seminiferous tubules and seminiferous epithelium increase with age; but the volume density of the lumen of the seminiferous tubule and interstitial tissue decrease gradually with age. At 24 months of age, the volume density of seminiferous tubule of yak is 0.786 $\mu m^3/ \mu m^3$, the volume density of seminiferous epithelium is 0.677 $\mu m^3/ \mu m^3$, the height of seminiferous epithelium is 85.66 μm and the volume percentage of seminiferous tubules is 78.6%. The height of seminiferous epithelium and the capacity rate of seminiferous tubule in yak bulls at 24 months of age is 85.66 μm and 78.58%, respectively. These figures are similar to those in mature males of common cattle breeds (79.4%; Yan et al. 1997).

Spermatogenesis

The yak sperm consists of a head, neck and tail that are very similar to those of other cattle breeds. Compared to sperm from B. taurus bulls, the dimensions for yak sperm are 9.50 versus 8.32 μm, the length of the midpiece is 14.20 versus 14.40 μm, and the length of the principal piece is 51.83 versus. 47.5 μm (Xia et al. 1990a).

Spermatogenesis in the yak is similar to that in other species inasmuch as it is divided into three major processes: spermatocytogenesis, meiosis and spermiogenesis. Xia et al. (1990c) reported that the spermatogenetic cycle and the seminiferous epithelium phase of yak bulls are very similar to that of cattle. It appears that Phases VII and XI are very long, whilst Phases IX, X, and XII are rather shorter than in cattle.

Xia et al. (1990b) concluded that the numbers of both Sertoli cells and all kinds of germ cells in yak testes are approximately 80% of those in cattle: a difference which was attributed to both intrinsic differences between the species and the effects of nutrition. The rate of seminiferous cell production in yaks is similar to that in cattle during spermatogenesis; however, the proliferation rate in yak is about 10% lower than in cattle. Hence, daily sperm production in yak is maintained at a point somewhere between that of B. taurus and B. indicus cattle

Sexual activity

Males start to show mounting behaviour around 6 months of age. This behaviour continues and intensifies to include searching and mounting females in the following year. No sperm, however, are found in epididymal fluid of yak bulls before the age of 2 years. Puberty thus occurs in the third warm season following birth, when the males are over 2 years old. In practice, bulls start to mate from the age of 3, reaching their peak ability at around 6 to 7 years old after establishing their dominant position in the mating hierarchy of the herd by fighting. After 8 years of age, yak bulls start to lose to younger bulls in the competition for females (Zhang 1994).

A sexually productive life expectancy of not more than 10 years for the yak bull was suggested by results from an artificial insemination (AI) stud farm of 38 yak bulls in Tibet. The ejaculate volume, concentration and motility of sperm rose steadily from the age of 3 to 9 years old and then declined. Semen quality and quantity also change with season. In Dangxi, Tibet, the highest values for volume, sperm concentration and motility were obtained from June to September, with the peak occurring in September (Zhang 1994).

In Mongolian yaks, the older the bull the more females it is able to serve; which is consistent with the courtship behavior and dominance hierarchy of bulls. However, it was also found that the younger bulls, with fewer females at their disposal, mounted their mates more frequently (Magash 1991). Fertilization seemed to be more dependent on the number of services than on the age of the bull, which shows that overall pregnancy rate of females increased with the number of services.

Reproductive biotechnologies

Pregnancy confirmation

Various methods for the determination of pregnancy have been developed in the yak, but only method which has been investigated sufficiently to be recommended for use in practice is the clinical method of rectal palpation. This method is the same as for cattle and buffalo. Early pregnancy diagnosis can be made (Yu *et al.* 1993a) by examining the changes in the uterus (the uterine horns are of equal size, but became flaccid), the furrow between the bases of the horns (become less distinct) and especially the size of the ovaries (the ovary on the gravid horn side is significantly enlarged to about twice the size of the other one).

Assaying for progesterone is another method for pregnancy diagnosis, based on the presence of continuously elevated progesterone concentrations during pregnancy. Blood should be sampled $20 \sim 24$ days after mating and progesterone be assayed by radioimmunoassay. Females with blood progesterone concentrations above 1.0 ng/ml are considered to be pregnant (Yu *et al.* 1993b).

Oestrus synchronization

Oestrus synchronization in the yak has been effectively accomplished with products commonly used in cattle, including $PGF_{2\alpha}$ and progesterone. Synchronization of oestrus is most effective during the breeding season, but it is also possible to induce oestrus outside the breeding season. Regardless of the method used, considerable effort must be made to minimize stress, or fertility will be adversely affected.

When $PGF_{2\alpha}$ is used, the animal must be cycling and have a responsive corpus luteum. Following a single injection of $PGF_{2\alpha}$ (during the ovulatory season), approximately 50% of yaks will be detected in oestrus, but the interval from treatment to the onset of oestrus is usually slightly longer than in cattle. A second injection of $PGF_{2\alpha}$ 11 days after the first will effectively synchronize the group. In one study (Magash 1997), 135 yaks were administered twice daily in 2.0 ml and 5.0 ml doses with two $PGF_{2\alpha}$ preparations (Oestrophan & Enzaprost; the former was made in Czech and later in Hungary). Most animals (55 to 70 %) were detected in oestrus 3 days after treatment. However, there was a seasonal effect on the proportion in oestrus, with 50.0 % and 46.2 % detected in oestrus following administration of the two products, respectively, in July and a better response rate (61.5 % and 58.3 %) in August. Following two treatments at 10-day intervals, the oestrus synchronization rates were 86.2% and 90.0 %, and the conception rates following insemination with frozen-thawed Holstein semen were 77.9 and 78.9% for the two products respectively.

Inducing oestrus

Yu et al. (1996) used various exogenous hormones (LRH-A3, FSH, eCG and hCG) to induce oestrus in 180 female yaks of different ages. The animals included 80 milking yak-cows that had calved and had their calves with them throughout the experiment year, 80 non-milking cows that had not calved during the experiment year, and 20 heifers. After a single intramuscular injection of LRH-A3, the non-milking yak-cows showed the best response, with an oestrus rate of 95%. Yak heifers responded with an oestrus rate of 80%, which was better than occurred in the milking yak cows. The oestrus rates were 82% and 88% respectively in the non-milking yak-cows after eCG or hCG injection. The oestrus rate in the milking yak-cows reached 70% only after a combined treatment of an intramuscular injection of FSH on day 0 and LRH-A3 on day 2. Oestrus in treated yaks occurred within 1 to 10 days after normal treatment, with the non-milking yak-cows and heifers showing oestrus earlier than milking cows. While there is considerable variation in fertility after oestrus induction in female yaks in this study, overall pregnancy rates were 86% in non-milking yak-cows and 73% in milking yak-cows.

Superovulation

Superovulation has been attempted in yak females using a variety of treatments and under varying conditions. However, satisfactory responses have not been consistently achieved. In a small trial (Zagdsuren et al. 1997), 3 barren females (4- 8 years old) were given an intravaginal insert containing 1.2 g progesterone (CIDR) for 12 days, injected with $PGF_{2\alpha}$ (Lutalyse) on the third day after the first injection of FSH and inseminated 24 - 36 hours after the last injection of FSH. The average number of corpora lutea and non-ovulated follicles were 5.0 ± 0.6 and 1.3 ± 0.9, respectively, but embryo recovery was not reported.

Davaa et al. (2000) used FSH and PMSG to induce superovulation in yak cows. Oestrus was detected 34.1 ± 0.52 h after the prostaglandin treatment. The average number of ovarian follicles was 5.4 ± 0.65 and 4.5 ± 0.43, of which 2.6 ± 0.30 and 1.9 ± 0.20 ovulated from the right and left ovaries, respectively.

In vitro embryo production (IVP)

Despite success in other species, there are no reports of yak calves being produced following in vitro maturation and fertilization of oocytes collected from nonstimulated females. However, it is noteworthy that these embryos are capable of developing to the blastula phase in vitro.

Using standard procedures, Chen et al. (1997) studied the efficiency of IVP methods in yak and local breeds of cattle. Results showed that the cleavage and blastocyst rates for yak IVF were 35/56 (62.5%) and 9/56 (16.1%), respectively, indicating that these procedures, derived from those used in cattle, produced acceptable results which were at least as good as those obtained with IVP using oocytes and semen from local cattle.

Luo & Yang (1994) obtained ovaries from 4 yaks after slaughter. In this study, 25 oocytes were aspirated from 8 ovaries and, after incubation for 20 h, 20 oocytes developed to meiosis II. Fertilization and cleavage rates were 12/20 (60%) and 8/20 (40%), respectively. Although two normal morrulae were transferred into the uterus of a Holstein-Friesian cow, she was subsequently diagnosed nonpregnant.

In our laboratory, He & Cui (2005) carried out a comparative study to improve the techniques of IVP in yak. They established a procedure that achieved a cleavage rate of 62.2%, and the developmental rates of 4-cell and 8-cell were 46.7% and 28.9%, respectively.

Embryo transfer

Although further studies are still needed to optimize the procedures of superovulation, oestrus synchronization, embryo recovery and transfer in yaks, we have made a successful embryo transfer in Tianzhu white yaks (Ju & Yu 2006). In our experiment, a comparative test was taken and a standard protocol was determined as follows. The donors (Tianzhu white yak) received a CIDR-B program containing 1.9g progesterone at an undetermined stage of the oestrous cycle (Day 0). Superovulation was induced with 10 mg of p-FSH, divided into eight decreasing *doses* at intervals of 12 h from Day 5. The donors once again received two injections of $PGF_{2\alpha}$ on Days 7 and 8 and CIDRs were removed in the afternoon of Day 8. AI was performed 12 and 22 h after the onset of behavioural oestrus and embryos were collected 7 days after AI. The mean numbers of total corpora lutea, non-ovulated follicles, viable (transferable) and degenerated embryos were 4.8 ± 2.2, 1.1 ± 0.8, 2.5 ± 1.3 and 1.4 ± 0.9, respectively. Twelve viable embryos were finally transplanted to 10 oestrus-synchronized black yak females (recipients). As a result, 6 of the 12 black yak females became pregnant. All 6 animals delivered full-term calves normally. The results of our study are the first confirmation that the embryo transfer is feasible in the yak.

Interspecies hybridization and AI

Hybridization between yak and cattle of other species is recorded in ancient historical records. The cattle originally used were local breeds, generally referred to as 'yellow' cattle (*B. taurus*) in China, and both *B. taurus* and *B. indicus* (Zebu) cattle elsewhere. Although this practice continues, more recently improved breeds of cattle have also been used (facilitated by AI of frozen-thawed semen). However, due to the reproductive isolation of different animals and the low conception rate of hybrids produced from yak and other cattle breeds, crossbreeding between yaks and cattle is limited to only the F1 and F2 generations. The F3 generation (containing about 12.5% of yak genetics) have difficulty surviving in the high-altitude areas (above 3000 m) in the Qinghai-Tibet Plateau and therefore are seldom produced. Since AI is inevitably restricted to more accessible areas and maintenance of improved breeds is difficult and expensive, most crossbreeding programs use locally available cattle.

Although crossbreeding and the infertility of male hybrids have been extensively studied, there have been no clear conclusions. The ability of hybrids to produce sperm increases gradually with successive generations. Crossbreeding is generally restricted to the F1 and F2 generations, but sperm production does not resume until at least the third backcross (15/16 yak or cattle), and often not until the fourth backcross.

Semen from improved breeds of cattle, such as Holstein-Friesian and Simmental, have been used to breed yaks but are not popular due to the lower survival of the hybrids. Recently, semen from Hereford, Angus, Simmental, Limousin and Charolais cattle has been used extensively to breed yak females. In Nepal, Zebu cattle are regularly crossed with yaks. It has been suggested that crossing yaks with Highland cattle (from the United Kingdom) results in hybrids with superior productivity.

Experiments were conducted in the 1920's and 1930's (at Buffalo Park, Wainwright, Canada) to develop a meat animal for their cold northern conditions. In these experiments, a small number of crosses were successfully made between yak males and female American bison and half-bison (bison x cattle cross) (Deakin et al. 1935).

Conclusions and recommendations

Although it is a credit to those involved that much scientific investigation is done with the reproduction of yak, in spite of the difficulties, it is also important to acknowledge that large gaps in knowledge still exist. Therefore, the following general policy guidelines are recommended for the further studies:

Reproductive performance: yaks under the existing management conditions breed once in two years though some of the yaks are annual breeder. Delayed sexual maturity, the long postpartum period, and seasonality of breeding are some of the reproductive problems confronted in the yak husbandry. The age of sexual maturity has been markedly reduced in the other domestic cattle by selective breeding. Since it is clear that, besides genetic make up, inadequate nutrition including deficiency and imbalance of minerals contribute to delayed maturity, there are a number of possible strategies to use to achieve a reduction in the age of sexual maturity. Indeed, it is likely that once the nutrition is addressed properly, many of the existing problems of reproductive performance in yak would be automatically overcome. The long winter, with scanty feeding resources, keep yaks in a maximum of physiological stress which also impacts on other systems.

Reproductive physiology: Basic information of embryology in yak is scanty. Information on factors including hormones and other related materials which may play a major role in various reproductive periods is still limited. However, studies in this area should mainly be concentrated on physiology of reproductive adaptability to the extreme environment.

Breeding: Crossbreeding of yaks should only be encouraged up to F1 and should preferably utilize a breed of bull with high production potential but smaller in size to avoid parturition complications. The F1 male will be made available to the different pockets on exchange with the non-description in bred bulls. Meanwhile, breeding should be restricted within the breeds/types to conserve and preserve genetic identity of each breed.

Reproduction and nutrition: There is a paucity of information on whether natural yak diets have deficiencies, excesses or imbalances of, for example, trace minerals. So more detailed study about the relationship of yak reproductive physiology and nutrition are needed to determine whether specific nutrients may be limiting factors and whether specific supplements at critical times would be cost-effective.

Reproductive technologies: Although the study of reproductive events in yak has generally lagged far behind that of domestic livestock, recent interest and investigations have in part corrected this deficit. Nonetheless, there is still much to be learned before these technologies are commonplace in yak. In that regard, short-term preservation of yak ovaries and/or oocytes, and cryopreservation of oocytes and embryos would be of considerable benefit. Remarkably, many techniques developed and proven in cattle have been adapted to yak with only a few, minor, modifications. The success of IVP and embryo transfer techniques in yak suggests that there are many opportunities for similar approaches in other animals. In that regard, IVM and IVF utilizing epididymal spermatozoa and IVC with oviduct cell co-culture are likely to be of use in many different species in the near future. It is expected that this will promote assisted reproduction in exotic and endangered species and facilitate opportunities for the international movement of gametes.

Overall, it is now widely recognized that many of the questions at a fundamental and a practical level, which it is hoped would lead to an improvement in yak productivity, are concerned not with single issues, or single characteristics of the yak, but with the understanding of complex interactions. For example, it may no longer be necessary to ask what hormones are involved in affecting reproductive rate, but how hormonal responses to nutritional deprivation affect calving intervals and how specific nutrient inputs at critical times may improve matters. It is possibly no longer necessary to ask about the occurrence of heterosis from interspecific hybrids with yak, even though the actual levels of heterosis are still largely unknown, but it is more relevant to current practice to ask how such heterosis is expressed at different levels of nutritional and environmental stress and how it fits into a comprehensive breeding strategy.

Cooperative and coordinated design, execution and analysis of experiments and development trials with yak may well provide a total benefit which is greater than the sum of its parts. It deserves serious further consideration.

Acknowledgements

Grateful acknowledgments to International Foundation for Science (B/1424-1&2; B/1696-1&2), National Natural Science Foundation of China (39100096, 39970546 & 30371069), Gansu Science Foundation of Application of Biological Techniques in Agriculture (GNSW-2004-07) and Science Foundation of Gansu Province (2GS042-A41-001-05) for funding our studies.

References

Cai L 1980 Studies on the reproductive organs of female yak. *Journal of Chinese Yak* **3** 10-16.

Cai L & Winer G 1995 *The Yak*. Regional Office for Asia and the Pacific of the Food and Agriculture Organization of the United Nations, Bangkok, Thailand.

Chen JB, Hai LQ & Yi H 1997 Comparison on effect on in vitro fertilization between the yak and ordinary cattle. *Proceedings of the Second International Congress on Yak* China 195-197.

Cui Y & Yu SJ 1999a An Anatomical study of the internal genital organs of the yak at different age. *The Veterinary Journal* **157** 192-196.

Cui Y & Yu SJ 1999b Ovarian morphology and follicular systems in yaks of different ages. *The Veterinary Journal* **157** 197-205.

Davaa M, Altankhuag N & Zagdsuren Y 2000 The preliminary experiment to induce superovulation in female yaks. *International Yak Newsletter* **5** 78 (Abstract).

Deakin A, Muir GW & Smith AG 1935 Hybridization of domestic cattle, bison and yak. *Report of Warinwright experiment*. Publication 479, Technical Bulletin 2, Dominion of Canada, Department of Agriculture, Ottawa.

Denisov VF 1958 Domestic yak and their hybrids. *Selhozgiz* P116.

He JF & Cui Y 2005 Effect of fertilization medium and time on in vitro fertilization of follicular oocytes of yaks. *Chinese Journal of Veterinary Science and Technology* **11** 900-903.

Hu AG., Cai L & Du SD 1960 An investigation on yak in Ganzi County. *Journal of Southwest Nationality College* **4** 46-50.

Ju XH & Yu SJ 2006 Successful embryo transfer in Tianzhu white yaks by standard protocol. *Animal Reproduction Science* (In Press)

Katzina EV & Maturova ET 1989 The reproductive function of yak cows. *Animal Breeding Abstracts* **352** 58.

Lei HZ 1964 Observation on the genital, physiological and reproductive features of yak. *Journal of Chinese Animal Science* **7** 1-3.

Liu WL & Liu SY 1982 The observation and analysis for female yaks' characters of Qinghai yak. *Journal of Chinese Yak* **4** 5-7.

Liu ZQ 1981 The reproductive characters of Qinghai yak. *Journal of Chinese Yak* **4** 5-7.

Luo XL & Yang YZ 1994 Research on in vitro maturation and development of oocyte in slaughtered yak. *Proceedings of the First International Congress on Yak* China 311-313.

Magasch A 1990 Statische Massmalen diagnostischer Zuchthygienemerkmale bei Yakkuhen in der Mongolei. *Wiss Zeitschrift der Humboldt - Universität zu Berlin R. Agrarwiss.* **39** 359-366.

Magash A 1991 Ergebnisse von Untersuchungen uber die Physiologie des Sexualzyklus beim weiblichen Yak. *Monatschefte fur Veterinärmedicin* **46** 257-258.

Magash A 1997 The use of biotechniques in yak reproduction. *Proceedings of the Second International Congress on Yak* China 175-178.

Purevzav Z & Beshlebnov AV 1967 Some data on the physiology of reproduction in the yak. *Zhivotnovdstvo Mask* **29** 92-94.

Wang MQ, Li PL & Bo JL 1997 Yak calving interval and calving efficiency. *Proceedings of the Second International Congress on Yak* China 29-31.

Xia LJ, Ben ZhK & Liu H 1990a Ultrastructure of yak spermatozoa. *Journal of Chinese Yak* **3** 4-7.

Xia LJ, Liu H & Lai MR 1990b Analysis on the spermatogenesis in yaks and comparison with common cattle. *Journal of Chinese Yak* **3** 12-17.

Xia LJ, Liu H & Tian HP 1990c Seminiferous cycles in yak bulls. *Journal of Chinese Yak* **3** 8-11.

Xu KZ 1964 The study on the histology and anatomy of reproductive organs and the reproductive functions of yak and its hybrid. *Journal of Chinese Animal Science* **3** 18-21.

Yan P, Liu ZL, Pan HP & Zi DJ 1997 Quantitative histology studies on the testes in hybrid bull of wild and domestic yak. *Proceedings of the Second International Congress on Yak* China 167-169.

Yu SJ & Chen BX 1993 Traditional systems of management of yak cows in China and observations on reproductive characteristics using milk and plasma progesterone. *Strengthening Animal Reproduction Research in Asia through the Application of Immunoassay Techniques* Vienna 127-135.

Yu SJ, Huang YM & Chen BX 1993a Reproductive patterns of the yak. I. Reproductive phenomena of female yak. *British Veterinary Journal* **149** 579-583.

Yu SJ, Huang YM & Chen BX 1993b Reproductive patterns of the yak. II. Progesterone and oestradiol-17ß levels in plasma and milk just before the breeding season; also during normal and short oestrous cycles. *British Veterinary Journal* **149** 583-593.

Yu SJ, Huang YM &Chen BX 1993c Reproductive patterns of the yak. III. Levels of progesterone and oestradiol-17ß during pregnancy and the periparturient period. *British Veterinary Journal* **149** 595-601.

Yu SJ & Chen BX 1994a Traditional systems of management of yak cows in China and observations on reproductive characteristics using milk and plasma progesterone. *Proceedings of the Final Research Coordination Meeting of an FAO/IAEA Coordinated Research Programmer* Thailand 127-136.

Yu SJ & Chen BX 1994b Postpartum ovarian function in yak cows as revealed by concentrations of progesterone in defatted milk. *Recent Advances in Animal Production* **3** 93.

Yu SJ & Chen BX 2000 Peripheral concentrations of luteinizing hormone, oestradiol-17ß and progesterone around estrus in six yaks. *The Veterinary Journal* **160** 157-161.

Yu SJ & Li FD 2001 Profiles of plasma progesterone before and at the onset of puberty in yak heifers. *Animal Reproduction Science* **65** 67-73

Yu SJ, Liu ZP & Chen BX 1996 Use of exogenous hormones to induce oestrus in yak. *International Yak Newsletter* **2** 31-34.

Zagdsuren Yo, Davaa M, Magash A & Altankhung N 1997 Superovulation in female yaks. *Proceedings of the Second International Congress on Yak* China 193-194.

Zhang Y 1994 The relationship between season and age of stud yak bull in Dangxin. *Proceedings of the First International Congress on Yak* China 303-307.

Current knowledge and future challenges in camelid reproduction

A Tibary[1], A Anouassi[2,3], A Sghiri[2] and H Khatir[3]

[1]Department of Veterinary Clinical Sciences and Center for Reproductive Biology, Washington State University, Pullman, WA, USA 99164-6610, [2]Department of Reproduction and Artificial Insemination, Institut Agronomique et Vétérinaire Hassan II, B.P. 6202, Rabat-Institut, Moroccco, [3]Veterinary research Center, Abu Dhabi, United Arab Emirates, PO Box. 44479

Reproductive biology research on camelids offers some interesting peculiarities and challenges to scientists and animal production specialists. The objective of this paper is to review camelid reproduction, advances in reproductive physiology and reproductive biotechnologies in camelids and discuss some areas for further research. In the female, the focus has been on understanding follicular dynamics. This has allowed development of synchronization and superovulation strategies to support embryo transfer technologies which are now commonly used in camels. Some advances have been achieved in preservation of embryos by vitrification. Fertilization, early embryo development and embryo signaling for maternal recognition of pregnancy are still not fully understood. New information on the interaction of the developing embryo and the endometrium may shed some light on this signaling as well as the mechanism of prevention of luteolysis. The presence of a seminal ovulation-inducing factor (OIF) was confirmed in llamas and alpacas. Chronology of oocytes maturation has been described. In vitro production of embryos has been achieved resulting in successful pregnancies and births in the dromedary. These techniques offer a new tool for the production and study of interspecies/cross-species embryos and their effect on pregnancy. Male reproductive function remains poorly studied. Semen preservation and artificial insemination still present many challenges and are not used in production at the moment. The involvement of climatic and nutritional conditions as well as the role of leptin in the regulation of reproductive function need to be evaluated.

Introduction

Camelidae are economically important in many countries. There are six species, two old world camelids (OWC); *Camelus dromedarius* and *Camelus bactrianus* and four new world camelids (NWC); *Lama glama, Lama pacos* (renamed *Vicugna pacos*), *Lama guanacoe* and *Vicugna vicugna*. Domestic camelids show a variety of "breeds" with specific production characteristics. Camelid reproductive biology research has seen tremendous development in recent years due to renewal of interest for these species. The application of hormone assays and ultrasonog-

Corresponding author E-mail: tibary@vetmed.wsu.edu

raphy have permitted great advances in this area (Anouassi *et al.* 1988; Tibary & Anouassi 1996). The objective of the present paper is to review camelid reproduction and to point out areas for future research, focusing on reproductive physiology and applied biotechnologies. Recent reviews are available for discussion of reproductive disorders in camelids (Tibary *et al.* 2005b; Tibary & Vaughan 2006; Vaughan & Tibary 2006).

Reproduction in the male camelid

Reproductive parameters of the male are critical for evaluation of the prospective herd sire and are summarized in Table 1.

Table 1: Reproductive parameters in male camelids (adapted from Tibary & Anouassi 1997; Bravo 2002; Tibary & Vaughan 2006)

Parameter	C. dromedaries	C. bactrianus	L. pacos	L. glama
Testicular descent	Present at birth	Present at birth	Present at birth	Present at birth
Puberty	3 to 5 years	3 to 5 years	15 to 24 months	15 to 24 months
Sexual maturity (years)	5 to 6	5	4	4
Seasonality	Yes	Yes	sometimes	sometimes
Testicular size at maturity	36.4 cm*	?	3.7 x 2.5**	5.4 x 3.3**
Duration of Spermatogenesis	?	?	?	?
Sperm production per gram of testis	20 to 61 x 10^6	?	?	?
Daily sperm production	0.751 x 10^9	?	?	?
Gonadal sperm reserve reserves	1.7 – 3.4 x 10^9	?	?	?
Epididymal transit	4.3. days	?	?	?
Epididymal sperm reserve	2.3-6.1 x 10^9	?	?	?
Mating duration (minutes)	7 – 20	3 – 20	5 -50	5-65
Ejaculate characteristics				
Volume (ml)	2-13	2.5-12.5	0.4-4.3	0.2-7.9
Concentration (x 10^6/ml)	140-1,300	450-5,590	0.06-170	15-640
Total spermatozoa (x 10^6)	600- 5,690	?	0.06-167	2.9 -246
Motility (%)	20-80	70-90	85	50-95
Normal Morphology (%)	40-70	50-90	40-75	40-85
Sperm cell dimensions (μm)				
Total length	51.1 ± 0.9	42 ± 1.9	?	49.5
Head length	6.6 ± 0.5	6.0 ± 0.6	?	5.3
Head width	3.8 ± 0.1	3.9 ± 0.1	?	3.8
Midpiece	6.8 ± 0.5	6.2 ± 0.7	?	5.3
Tail	37.6 ± 0.9	30.8 ± 1.9	?	36.6

* scrotal circumference, **Length by width

Sexual development and puberty

Anatomy of genital organs has been reviewed extensively (Tibary & Anouassi 1997; Tibary & Vaughan 2006). The most notable feature is the absence of seminal vesicles. The testicles are already descended at birth but may not be easily palpable until 6 months of age in OWC.

Male NWC can display sexual activity as early as 11 months of age. Preputial detachment and complete erection and intromission are possible at 15 months but may take up to 3 years. Presence of spermatozoa in the seminiferous tubules was reported at 15-18 months of age in the alpaca. Testicular growth is slow and does not plateau until 3 years of age (Tibary & Vaughan

2006). In the vicuna, spermatogenesis is present at 16 months of age. Guanaco and vicuna males live in a bachelor band until they reach maturity between 4 and 6 years of age, at which point they start forming their own harem (Tibary & Vaughan 2006).

In the OWC, rutting behavior may be seen as early as 2 years of age, but spermatogenesis and fertilization are not possible until 3 years at the earliest. Sexual maturity (5 to 6 years) is signaled by the plateau in spermatogeneic activity (Al-Qarawi *et al.* 2001). Reduction of spermatogenic activity, libido and expression of male behavior are observed after 20 years (Tibary & Anouassi 1997).

Spermatogenesis, sperm production and sperm maturation

The cycle of the seminiferous tubule in the male llama and dromedary presents 8 stages (Delhon & Von Lawzewitsch 1987; Bustos-Obregon *et al.* 1997). Differences in the frequency of each stage exist between the two species. Epididymal maturation and transit has been estimated to be 4.3 days in the dromedary. Histologically and histochemically different segments were identified in the epididymis in llamas (Delhon & Von Lawzewitsch 1994) and camels (Tingari & Moniem 1979).

Testicular weight is correlated with spermatogenesis and epididymal sperm reserves. However, sperm production and gonadal sperm reserves are lower than those of ruminants. Sperm production is affected by age and season (Tibary & Anouassi 1997; Tibary & Vaughan 2006).

Seasonality of reproduction in the male OWC and vicuna has been established through endocrinological, behavioral and histological studies (Tibary & Vaughan 2006). Testicular size decreases in winter in vicunas (Urquieta *et al.* 1994) and in summer in camels (Tibary & Anouassi 1997). Sperm production and quality is lower in summer in llamas (Tibary & Vaughan 2006). The mechanism regulating seasonality in male camelids remains poorly studied.

Mating behavior and ejaculation

Mating behavior has been extensively described (Tibary & Anouassi 1997; Tibary & Vaughan 2006). In OWC, rutting season is evidenced by increase poll gland secretions, exteriorization of the soft palate and increased urine marking behavior. All camelids mate with the female in a sternal recumbency and mating lasts an average of 20 minutes. Duration of copulation in alpacas is affected by the age of females (longer for multiparous) and by the presence of other males (shorter) (Tibary & Vaughan 2006).

Semen is delivered continuously throughout mating, in non-fractioned small quantities, via a combination of pelvic thrusts and rhythmic urethral pulses throughout copulation (Lichtenwalner *et al.* 1996; Lichtenwalner *et al.* 1997; Tibary & Anouassi 1997). Copulation length affects ovulation induction rate but does not affect conception rates of ovulating females (Table 2). Sperm cells are present in the ejaculate within 2 to 5 minutes of copulation. Semen is deposited partly deep in both uterine horns and partly in the cervix and vagina (Tibary & Anouassi 1997; Tibary & Vaughan 2006).

Semen characteristics

Specially designed artificial vaginas allow reliable collection of ejaculates (Lichtenwalner *et al.* 1996; Tibary & Vaughan 2006), unfortunately a large number of ejaculates do not contain spermatozoa (Flores *et al.* 2002; Deen *et al.* 2003).

Table 2: Effect of copulation length on ovulation and embryo recovery in dromedary camels
(Tibary & Anouassi 1997)

Mating duration (minutes)	Number of females	% non-ovulating females	% females with at least one embryo
Less than 1.5 minutes	45	35.6	44.9
1.5 to 3	232	13.8	53.5
4	102	13.9	56.8
4.5-5	102	16.7	65.9
5.5-6	57	14	77.6
>6	94	8.5	55.3

Camelid semen has high viscosity, low sperm concentration and poor motility (Garnica *et al.*
1993; Tibary & Anouassi 1997). Viscosity is attributed to mucopolysaccharides secreted from
the bulbourethral glands or the prostate (Garnica *et al.* 1993). Liquefaction of camelid semen
may take several hours and may be enhanced by exposure to hydrolytic and proteolytic enzymes (trypsin, collagenase, fibrolysin, hyalurodinase). A trypsin solution of 1:250 seems to
be effective with minimal negative effects on spermatozoa (Bravo *et al.* 2000a).

The effects of the various sperm abnormalities on fertility have not yet been determined.
Incidence of sperm abnormalities in the ejaculates is highly variable (Lichtenwalner *et al.*
1996; Flores *et al.* 2002; Urquieta *et al.* 2005). The incidence of cytoplasmic droplets is
relatively high. The incidence of abnormalities is not affected by rank of ejaculate (Bravo *et
al.* 1997). Concentration of the ejaculate remains stable if a period of at least 12 hours is
allowed between successive ejaculations (Bravo *et al.* 1997; Urquieta *et al.* 2005). Male
camelids may differ in their ability to induce ovulation (Tibary & Anouassi 1997).

Reproduction in the female camelid

Reproductive parameters in the female (Table 3) present striking differences compared to
ruminants. There are several distinct anatomical differences in the genital organs compared to
the ruminant (i.e. large ovarian bursa, absence of intercornual ligaments, a well-developed
papillae of the utero-tubal junction, arrangement of the cervical rings) (Fig. 1) (Tibary & Anouassi
1997; Vaughan & Tibary 2006).

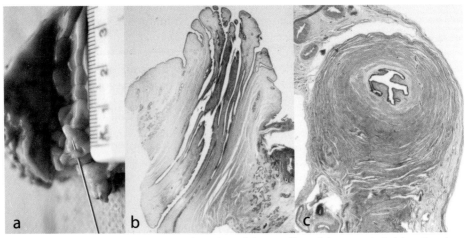

Fig. 1 Gross (a) and microanatomy (b and c) of the papillae of the utero-tubal junction in
camelidae. Note the projection into the uterine lumen (2 to 5 mm) and the strong smooth
muscle sphincter. The utero-tubal junction serves as a sperm reservoir and selective pas-
sage of healthy embryos into the uterus

Table 3: Reproductive parameters in female camelids (adapted from Tibary & Anouassi 1997; Bravo 2002; Vaughan & Tibary 2006,)

Parameter	C. dromedarius	C. bactrianus	L.pacos	L. glama
Puberty (months)	24-48	18-24	4-6	6-8
Age at first breeding (months)	36-48	24-36	12-18	15-18
Follicular wave phases duration (days)				
Growth (days)	10.5 ± 0.5	10.9 ± 3	3-9	3-9
Maturation	7.6 ± 0.8	7 ± 4.2	2-8	2-8
Regression	11.9 ± 0.8	11.9 ± 4.2	3-8	3-8
Ovulatory follicle characteristic				
Minimum size (mm)	9	9	7	8
Growth rate/day (mm)	1.8	1.8	0.43	0.5-0.9
Average size(mm)	10-18	10-18	8-10	9-12
Maximum size(mm)	25	22	12	13
Incidence of anovulatory follicles (%)	40-50	?	5	10-40
Anovulatory follicle regression (days)	8-45	?	?	4-22
Corpus luteum characteristics				
Size (mm)	15-25	15-25	11-15	11-18
Day at CL maximum size	7.2 ± 1.7	7.3	7-8	8
Luteolysis day	10 ± 1.2	10.5	10-12	10-12
Pregnancy and Parturition				
Implantation (days)	14	14	12-14	12-14
Average and range of pregnancy length (days)	384 (330-396)	402 (374-419)	344 (330-376)	345 (333-382)
Expulsion of fetus (minutes)	10.5 ± 4.7 (5-45)	26.8 ± 12 (5-45)	10-15	10-15
Delivery of placenta (minutes)	65 ± 4.2 (30-180)	69 ± 47	88 (40-143)	77 (45-240)
Postpartum				
Uterine involution (days)	20-45	?	20	17
Postpartum ovarian activity (days)	14-300*	?	7	5

*Extreme variation in onset of postpartum ovarian follicular activity is primarily due to nutritional condition and effect of lactation anoestrus and seasonality

Puberty

Puberty is generally defined in the female as the age at which ovulation and fertilization are possible. Endocrinological and clinical studies show that recruitment, growth and ovulation of follicles are possible in female NWC as early as 4 months of age. However early breeding may result in poor fertility due to ovulation failure and increased early pregnancy loss. Females are bred when they reach 60% of the adult weight (Vaughan & Tibary 2006).

In OWC, normal follicular dynamics is present at 8 to 12 months of age in some breeds and under intensive management. However in nomadic conditions sexual activity does not start until 2 or 3 years of age. Nutritional status, season of birth, and breed of camel can affect the age at onset of puberty, age at first conception, and consequently the age at first parturition (Tibary et al. 2005b).

Seasonality of reproduction in the female

Seasonal pattern of reproduction in the female camelid is probably largely due to nutritional conditions (OWC and wild NWC) and management (NWC) (Tibary & Anouassi 1997; Sghiri &

Driancourt 1999; Tibary et al. 2005b). The breeding season coincided with warmer months when ample forage is available. However, the relationship between nutrition and reproduction has not been thoroughly studied in camelids. In the dromedary, body condition score and particularly hump fat content was shown to have a direct effect on ovarian activity (Tibary & Anouassi 1997). Also, leptinemia is positively related to hump fat content and is correlated to water and food deprivation. Recent studies show that this species has much higher hepatic lipogenic enzymes activity, an expression of leptin and leptin receptors, than in other ruminants. This suggests that leptin may be an important regulator of reproduction in this species (Bartha et al. 2005; Chillar et al. 2005a; 2005b; Sayed-Ahmed et al. 2005).

Ovarian follicular activity

In the absence of mating, there is a succession of overlapping follicular waves with highly variable rhythm showing 3 phases: growth, maturation and regression (Fig. 2) [(dromedary; (Skidmore et al. 1996a; Tibary & Anouassi 1996); llamas (Chaves et al. 2002), alpacas (Vaughan et al. 2004) and vicunas (Miragaya et al. 2004)]. Follicular wave patterns are orchestrated by endocrinological events involving both hypothaloamo-pituitary-gonadal axis and local (inhibin) regulation (Vaughan & Tibary 2006). Little is known about the mechanisms of recruitment of each follicular wave but these may involve FSH.

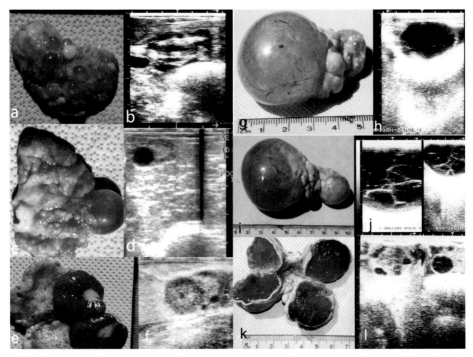

Fig. 2 Gross and ultrasonographic appearance of ovarian structures in camelids. Emergence of a new follicular wave growth (a,b), dominant follicle (c,d), mature corpora lutea (e,f), anovulatory follicle (g,h), anovulatory hemorrhagic follicle (i,j) luteinized hemorrhagic follicles (k,l)

Follicle size is correlated with plasma 17ß-oestradiol (oestradiol), oestrone sulphate and urinary oestrone sulphate in alpacas (Aba et al. 1995), llamas (Aba et al. 1995; Chaves et al. 2002) and dromedaries (Skidmore *et al.* 1996a; Tibary & Anouassi 1997). Peak plasma oestradiol concentrations coincide with maximum follicle size (Chaves *et al.* 2002). Mature follicles may develop into large anovulatory follicles (Fig. 2). These regress slowly but do not seem to affect follicular growth. Some may become luteinized (Tibary & Anouassi 1996).

Ovulation

Camelids are induced ovulators. An LH surge is observed within 15 to 40 minutes after mating and ovulation occurs in most females within 24 to 48 hours after mating (Marie & Anouassi 1987; Bravo *et al.* 1992; Aba *et al.* 1995). Incidence of ovulation following mating varies from 60 to 100%, occurring equally in the left and right ovaries (Tibary & Anouassi 1996; Vaughan & Tibary 2006).

Mechanism of induction of ovulation

Ovulation is induced by intramuscular injection of seminal plasma. The presence of an ovulation-induction factor (OIF) in semen was demonstrated in the bactrian camel (Zhao *et al.* 2001) and NWC recently (Adams *et al.* 2005). Absorption of the OIF seems to be dependant on hyperemia and endometrial irritation caused by copulation (Ratto *et al.* 2005b).

Natural multiple ovulations

Multiple ovulations (double and even triple) occur in 5 to 20 % of natural matings (Tibary & Anouassi 1996; Vaughan & Tibary 2006). However, twin pregnancies and twin births are extremely rare.

Spontaneous ovulations in camelidae

Spontaneous ovulations have been reported, particularly in the postpartum female. Their incidence range from 3.5% to 40% (Nagy *et al.* 2005; Vaughan & Tibary 2006). Some of these spontaneous ovulations could simply be luteinization of anovulatory follicles.

Corpus luteum (CL) and lutyeolysis

The luteal phase is short (Marie & Anouassi 1987; Aba *et al.* 1995). The CL reaches its maximum size by 7 days post-breeding and luteolysis is initiated 9 to 10 days post-mating through prostaglandin $F_{2\alpha}$ (PGF$_{2\alpha}$) release (Aba *et al.* 1995; Skidmore *et al.* 1998). The role of oxytocin in luteal regression in camelids is unknown. There seems to be a difference in the luteolytic effects of the right vs. left uterine horn. Luteolytic activity of the right uterine horn is local while the left uterine horn has both a local and a systemic effect (Fernandez Baca *et al.* 1979). It is postulated that the left horn may induce luteolysis of a corpus luteum in the right ovary via a local veno-arterial pathway (Ghazi 1981; Del Campo *et al.* 1996).

Pregnancy

Fertilization and early embryo development

Fertilization rate is very high and embryo development is rapid. Embryos enter the uterine cavity approximately 5-6 days after ovulation at the hatching or early hatched blastocyst stage (McKinnon et al. 1994; Tibary & Anouassi 1997; Tibary 2001b; Tibary 2001c). The embryo starts to elongate by Day 10 and the trophoblast establishes close contact with the endometrium by day 14 (Fig. 3) (Tibary 2001b).

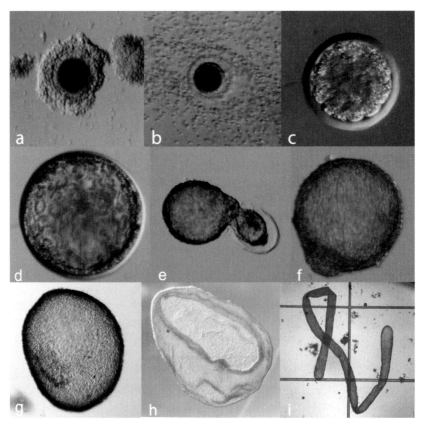

Fig. 3 Early embryo development in camelids. a) immature compact cumulus-oocyte-complex, b) expanded cumulus after maturation, c) day 5 morula (260 μm), d) day 6.5 blastocyst (380 μm), e) day 7 hatching blastocysts; f) day 7.5 hatched blastoyst, g) day 8.5 collapsing blastocyst (5 mm); h) day 9.5 embryo (9 mm), i) day 13 elongating trophoblast (5.5 cm)

Maternal recognition and embryo migration

The majority (98%) of pregnancies occur in the left uterine horn regardless of the side of ovulation. Differential luteolysis occurs in the left and right uterine horns and may explain embryo migration from the right to the left uterine horn (Tibary & Anouassi 1997; Vaughan & Tibary 2006).

Embryonic signals have not been thoroughly investigated but they must be present before Day 9 after mating in order to rescue the corpus luteum of pregnancy (Aba et al. 1997). In the dromedary, the early embryo (Day 10) exhibits high aromatizing activity and synthesizes large amounts of estrogens. This coincides with the time of luteolysis and may be the first signal for maternal recognition and prevention of luteolysis. No interferon-like protein has yet been identified (Skidmore et al. 1994). High levels of oestrone sulphate are observed between Days 21 and 27 of pregnancy in alpacas and may be involved in maternal recognition of pregnancy (Vaughan & Tibary 2006).

Placentation and fetal development

The macro and microanatomy of the uteroplacental interface was recently described. Camelid placentation is eptheliochorial, diffuse microcotyledonary with the fetus completely covered by an epidermal membrane (Skidmore et al. 1996b; Tibary & Anouassi 1997; Olivera et al. 2003a; Olivera et al. 2003b).

Implantation and placental development begins in the left horn around day 14 then extends to the right (Olivera et al. 2003a). Complete interdigitations between the membranes of uterine epithelial and trophoblastic cells are observed exclusively in the left horn in areas around the embryo. Specialized pre-contact zones, ectoplasmic pads, provide initial points of attachment and early metabolic maternal-fetal exchange (Abd-Elnaeim et al. 2003). Aerolae are present at the gland regions and may play a role in nutrient transfer (Abd-Elnaeim et al. 2003; Olivera et al. 2003b). The contact surfaces of trophoblast and uterine epithelial cells show a richly glycosylated layer and many classes of glycan that may play a role in attachment, materno-fetal interactions and embryo fixation at the implantation site (Jones et al. 2002; Olivera et al. 2003a).

Tropoblastic binucleated and multinucleated cells are present, but their function is not clear. There is no evidence of production of chorionic hormones (Jones et al. 2002; Olivera et al. 2003a).

Endocrinology of pregnancy

Ovulation, CL growth and progesterone production are similar in pregnant and non-pregnant camelids for the first 8 days after ovulation. A temporary decline in blood progesterone was reported from day 8 to 12 after mating during the period of maternal recognition of pregnancy (Aba et al. 1995). The CL is required to maintain pregnancy (Tibary & Anouassi 1997; Vaughan & Tibary 2006). Plasma progesterone concentration remains above 2 ng/mL but fluctuates throughout gestation then declines in the last 2 weeks of gestation (Tibary & Anouassi 1997; Raggi et al. 1999).

Plasma oestradiol levels in pregnant camelids are generally low in early gestation, rise in mid-gestation, are very high in the last month of pregnancy then decline rapidly when the feto-placental unit is disrupted at term (Aba et al. 1995; Tibary & Anouassi 1997).

Parturition

Gestation length is highly variable and is affected by season (Davis et al. 1997; Tibary & Anouassi 1997; Vaughan & Tibary 2006). Parity of the dam and sex of the cria do not influence gestation length in NWC (Davis et al. 1997). However, these factors, as well as nutrition, are important in OWC (Tibary & Anouassi 1997). Stages of parturition have been thoroughly described and are relatively short compared to ruminants (Table 3).

Postpartum

Uterine involution is rapid (Tibary & Anouassi 1997; Bravo 2002). Gross anatomical involution is completed by 21 days postpartum. The interval from parturition to resumption of ovarian activity is 5 to 7 days in NWC. In NWC, mating and ovulation are possible by 10 days postpartum (Bravo et al. 1994). Conception rate returns to normal by 2 to 3 weeks postpartum, allowing production of one cria per year (Vaughan & Tibary 2006). Lactational anoestrus is a problem in lactating camels particularly under desert conditions. Early weaning and nutritional supplementation allow early return to cyclicity and fertility (Tibary et al. 2005b). In the dromedary, early weaning (3 to 7 days of life) resulted in earlier onset postpartum ovarian activity (9.07 ±

7.77 days) and an average interval from calving to mating of 17.92 $\pm$ 9.11 days with a 76.8% pregnancy rate at the first postpartum mating (Khorchani *et al.* 2005). In addition, lactating females with good body condition at the time of calving were shown to have an earlier onset of postpartum ovarian activity and a shorter intercalving interval (i.e. 13 to 14 months instead of 22 to 24 months) (Sghiri 1988).

Reproductive biotechnologies

Semen preservation and artificial insemination

The use of artificial insemination (AI) has been reported in camelidae since the 60's (Tibary & Anouassi 1997; Tibary 2001a). Most of the reported work is on *Camelus bactrianus* (Zhao *et al.* 1994).

Semen preservation technology in camelids

Short-term preservation of camelid semen has been attempted at different temperatures (25°C, 30°C or 4°C) using various extenders (Tibary 2001a). Complete liquefaction of semen before adding the extender poses some technical problems (Bravo *et al.* 2000b). Normal liquefaction of camelid semen may take several hours and may be enhanced by exposure to hydrolytic and proteolytic enzymes (trypsin, collagenase, fibrolysin, hyalurodinase). A trypsin solution of 1:250 seems to be effective with minimal negative effects on spermatozoa (Bravo *et al.* 2000a).

Semen has been frozen successfully for over 3 decades (Bravo *et al.* 2000b; Tibary 2001a). However, insemination trials were carried out mainly in the bactrian camel (Zhao *et al.* 1994). Llama and alpaca semen has been frozen using a variety of common ruminant techniques after liquefaction (Bravo *et al.* 2000b). Skim milk-egg-yolk ethylene glycol seems to result in higher motility and acrosome integrity (Santiani *et al.* 2005). Freezing and thawing procedures depend on the packaging method used (Bravo *et al.* 2000b; Tibary 2001a; Deen *et al.* 2003; Miragaya *et al.* 2006).

Artificial insemination

Insemination trials using cooled or frozen-thawed semen are scarce. Pregnancy rates following AI with fresh un-extended or extended semen are variable. AI is usually carried out 24 hours after induction of ovulation with hCG or GnRH. However, the overall pregnancy rate after AI is lower compared to natural service in most species except the bactrian camel (Tibary 2001a).

The method of insemination used in large camelids is similar to that used in the bovine (i.e. rectal-vaginal manipulation and use of either a cassou gun or an infusion pipette). In alpacas, insemination is performed by guiding an insemination pipette through a vaginoscope. A dose of 1 ml of semen containing 400×10^6 spermatozoa seem to be adequate to guarantee ovulation and fertilization in the bactrian camel (Zhao *et al.* 1994). This dose can be reduced if ovulation is induced by hormonal treatment and the timing of insemination is determined with precision. In alpacas, conception was achieved following AI of 8×10^6 sperm cells (Bravo *et al.* 2000b). In our laboratory, deposition of 15×10^6 motile sperm cells, deep into the uterine horn ipsilateral to the side of ovulation 24 hours after administration of buserelin (25 ìg), resulted in a 77.8% conception rate in ovulating females (n = 18) (A. Anouassi unpublished data).

Conception rates following artificial insemination with frozen semen vary between 50 and 80% in bactrian camels. On the other hand, conception rate achieved with frozen-thawed dromedary semen exhibiting very good post-thaw motility has been disappointing (10 to 15%). There are few reports on AI with frozen-thawed NWC semen (Bravo *et al.* 2000b; Tibary 2001a; Miragaya *et al.* 2006)

<u>Interspecies artificial insemination trials</u>

Cross species insemination has long been performed in South America (Tibary 2001a). Studies using vicuna semen and paco-vicuna semen with female alpacas and llamas have yielded variable results. Trials to cross the guanaco with the dromedary camel have been reported resulting in the birth of three calves, however these pregnancies register a high rate of early loss (Skidmore *et al.* 2000). Cross-breeding between species of the same genus (i.e. *C. bactrianus* x *C. dromedaries*, *L. pacos* x *L. glama*, *L. guanacoe* x *L. glama*) are fertile. Offspring resulting from *C. dromedaries* and *L. guanacoe* do not seem to be fertile.

Multiple Ovulation and Embryo Transfer (MOET)

MOET procedures for camelids are similar to those used for ruminants (McKinnon *et al.* 1994; Tibary & Anouassi 1997; Tibary 2001c; Skidmore *et al.* 2002; Miragaya *et al.* 2006).

<u>Donor management</u>

Donors are superovulated using ovine (oFSH), porcine (pFSH), camel (cFSH), eCG or a combination of FSH and eCG. Treatment is initiated after synchronization of a follicular wave with progesterone, elimination of the dominant follicle by GnRH or hCG treatment, or at the early stage of the follicular wave (no follicle > 2mm). Optimal dose and frequencies of gonadotropin treatments vary from species to species (McKinnon *et al.* 1994; Tibary & Anouassi 1997; Tibary 2001c; Miragaya *et al.* 2006). Development of mature follicles takes 6 to 8 days after initiation of gonadotropin treatment. Dromedary females seem to become refractory to eCG and oFSH or pFSH following multiple treatments. Superovulation problems are of two types: lack of response (20%) or overstimulation (15 to 20%) (Tibary & Anouassi 1997). Embryo collection can also be performed without superovulation resulting in up to 29 pregnancies per female per season (A. Anouassi & A. Tibary unpublished data).

Preliminary experiments on immunization of dromedaries and alpacas against a synthetic peptide fragment of a-inhibin are very encouraging. An increased number of ovulations (4 to 10) was observed in 60% of the immunized females (Tibary & Anouassi 1997).

A single mating or insemination of donors is sufficient once follicles reach mature size. Donors are usually treated with hCG or GnRH agonist after mating in order to maximize ovulation rate.

<u>Embryo collection</u>

Non-surgical collection of embryos is done in the same manner described for cattle. Low epidural anesthesia is recommended in llamas and alpacas and in young dromedaries because of the smallness of the pelvis (Tibary 2001c).

Camelid embryos enter the uterus 6 to 6.5 days after ovulation. For maximum embryo recovery, flushing should be performed 7 to 8 days after ovulation. Embryo recovery rates (embryos recovered/ovulations) are highly variable and depends on many factors including superovulation treatment, fertility, management, collection date and technician experience (McKinnon *et al.* 1994; Tibary & Anouassi 1997; Tibary 2001c). Reported recovery rates in non-superovulated llamas range from 52 to 61%. Our recovery rate from the dromedary is 85% in single ovulators and 165% in double ovulators (Tibary & Anouassi 1996). Only hatched blastocysts are transferable (Tibary 2001c).

Management of recipients

Our results on over 6000 transfers show that recipients should ovulate one or two days after the donor. Synchronization of follicular development in donors and recipients has been attempted using progestagen with variable degrees of success (Tibary 2001c; Miragaya et al. 2006). We select females on the basis of follicular size from a large pool of recipients and induce ovulation with GnRH or hCG (Tibary & Anouassi 1997). Some recipients lose their primary CL, however, induction of new accessory corpora lutea by treating recipients with eCG/hCG during the first 2 months of pregnancy may be an approach for such recipients (Tibary 2001c).

Transfer of embryos into oestradiol/progesterone-treated, bilaterally ovariectomized recipient resulted in a 30% pregnancy rate. However, daily progesterone treatment is thereafter required throughout pregnancy. Spontaneous parturition has been reported in dromedary females receiving exogenous progesterone but this is not always the case. Moreover, progesterone treatment seems to increase the risk of dystocia, incomplete dilation of the cervix, premature placental separation and inadequate milk production.

Pregnancy results with fresh embryos

Embryos are generally transferred non-surgically in OWC, llamas and large alpacas. Whilst surgical transfer may be indicated in alpacas and vicunas, laparoscopy has been used in our laboratory for transfer of embryos in alpacas (Tibary 2001c).

Factors affecting pregnancy rates have been extensively studied in the dromedary (McKinnon et al. 1994; Tibary & Anouassi 1997; Skidmore et al. 2002). Pregnancy rates are affected by season, quality of recipients and synchronization between donors and recipients. However, factors such as side of transfer, method of transfer, and age of the embryo do not have any effect on pregnancy rates. In our laboratory, MOET has been practiced on a commercial basis since 1992. During the period between 1992 and 1998, a total of 2653 fresh embryos were transferred, resulting in an overall pregnancy rate of 62%. Pregnancy rates improved steadily from 30% to 70% over that period. It is not uncommon to achieve a pregnancy rate of 100% with some batches of embryos. Low pregnancy rates and high early pregnancy loss rates are observed when transfers are performed during the hottest months of the year (McKinnon et al. 1994; Tibary & Anouassi 1997).

Alpaca embryos have been successfully transferred into llama recipients, resulting in normal births. Transfer of vicuna embryos into llamas or alpacas could be a good technique for the multiplication of this wild species (Miragaya et al. 2006).

Cryopreservation of camelid embryos

Freezing camelid embryos according to the protocols that are in widespread use for cryopreservation of ruminant embryos results in poor (0 to 15%) pregnancy rates (McKinnon et al. 1994; Tibary & Anouassi 1997), probably due to the stage of development at which they are collected (i.e hatched blastocyst) and to their size (i.e. 400 to 2500 microns). More recently the use of vitrification techniques resulted in higher pregnancy rates in llamas and dromedaries (Aller et al. 2002; Nowshari et al. 2005; Skidmore et al. 2005).

In vitro production of embryo

Oocyte collection

Ovarian slicing or follicular aspiration have both been used as sources of oocytes in llamas (Del Campo et al. 1995) and camels (Khatir et al. 2004; Nili et al. 2004; Kafi et al. 2005; Nowshari

2005). In llamas, follicular aspiration results in a 62% recovery rate (i.e. number of oocytes complexes per aspirated follicle) with an average of 6 oocytes per female. Mincing produces higher yields of oocytes (27 per female) but a lower maturation rate (Del Campo *et al.* 1995). In dromedaries, recovery rates range from 31 to 49% following aspiration (Khatir *et al.* 2004) and 94% after follicle dissection (94%) (Nowshari 2005). Cumulus-oocyte-complexes (COCs) are embedded in a large sheet of granulosa cells and are difficult to retrieve without direct visualization (Nowshari 2005). In dromedaries, more oocytes are recovered during the breeding season and in the absence of a corpus luteum (Tibary *et al.* 2005a).

Oocyte recovery rates of up to 80% have been reported in alpacas and vicunas using laparotomy. Laparoscopic ovum pickup results in slightly lower recovery rate but is safer. Recovery rates range from 75 to 85% after FSH/eCG stimulation (Tibary *et al.* 2005a; Miragaya *et al.* 2006).

Transvaginal ultrasound-guided aspiration has been used in NWC and camels with collection rates ranging from 30 to 75% (Ratto *et al.* 2005a; Tibary *et al.* 2005a). Gonadotropin stimulation does not improve recovery rates but increases the number of oocytes obtained per female (Ratto *et al.* 2005a; Tibary *et al.* 2005a). Ovarian stimulation with eCG or FSH provides a uniform population of oocytes and offers a means to collect in vivo (cumulus expanded) matured COCs.

In vitro oocyte maturation (IVM)

Camelid oocytes can be matured *in vitro* in conditions similar to those described for ruminants (Tibary *et al.* 2005a). Camelid oocytes display a dark cytoplasm, due to presence of lipid particles, which makes evaluation very difficult. The perivitelline space (PVS) increases as the maturation process progresses until 24 h to a size which is maintained until 36 h of culture. A large number of microvilli are observed in the PVS in Metaphase II oocytes. Cortical granules migrate towards the peripheral areas of the ooplasm and form a lining in the suboolemmal area (Kafi *et al.* 2005).

In llamas, the optimal maturation rate (62%) of oocytes is obtained after incubation for 32 to 36 h (Del Campo *et al.* 1995). Higher maturation rate (80.6%) is obtained after 30 h incubation of oocytes aspirated after ovarian stimulation but oocyte quality and developmental ability may suffer (Ratto *et al.* 2005a). A 62% maturation rate in a medium without added hormones was reported for oocytes collected 22 h after induction of an LH surge (Miragaya *et al.* 2006). In alpacas, the oocyte maturation rates are 40 to 46% for compact COCs collected 18 to 24 h after hCG administration to stimulated females and incubation of 26 h. Similar results were obtained with vicuna oocytes (Miragaya *et al.* 2006).

In dromedaries, oocytes maturation rates vary from 50 to 83% after 30 to 36 hours of culture depending on conditions (Tibary *et al.* 2005a). Maturation rate of oocytes from non-stimulated animals are lower than those from stimulated animals (63% vs. 83%). The addition of EGF and cysteamine to the maturation medium has a beneficial effect on nuclear and cytoplasmic maturation (Khatir *et al.* 2004; 2005). In the bactrian camel, 46.7% of oocytes achieved meiotic maturation after 24 to 26 h of culture. As with ruminant oocytes, maturation rate and development ability post-fertilization are dependant on follicular size (Tibary *et al.* 2005a).

In vitro fertilization

The first successful production of embryos by IVM/IVF was reported in llamas using epididymal sperm enriched by percoll gradient in presence of heparin (2 or 5µg/ml) (Del Campo *et al.* 1995). The oocyte penetration and development to the pronuleus stage rates were 29.2% and 57.1%, respectively. Llama epididymal sperm was used to produce interspecies (*L. pacos* x *L. glama*) embryos. In vitro production of camel embryos was reported using fresh ejaculated

(Khatir *et al.* 2004; Tibary *et al.* 2005a) and epididymal sperm (Nowshari 2005). Penetration (68%) and cleavage rates (40%) are promising with fresh ejaculated semen (Khatir *et al.* 2004; Khatir *et al.* 2005). The first dromedary offspring from transfer of IVP embryos using IVM/IVF was recently reported (Khatir & Anouassi 2006).

Intracytoplasmic sperm injection (ICSI)

ICSI may be a valuable tool for production of interspecies embryos within the camelidae family. The only report of ICSI produced morulas was undertaken with llamas (Miragaya *et al.* 2006).

Cloning

Production of embryos by nuclear transfer using adult cells was reported in llamas. Fusion of the couplets was successful in 62.5% of attempts (n = 80) followed by cleavage rates of 32% to 40%. Oviductal transfer of 8- to 32-cell embryos and uterine transfer of a morula did not result in any pregnancy (Sansinena *et al.* 2003). Eight cells embryos were recently obtained by nuclear transfer in the dromedary after activation of oocytes with ionomycine followed by incubation in cyclohexidine (A. Anouassi unpublished data).

Culture of IVP embryos

In camels, we recently showed that embryos obtained by IVM/IVF and cultured to the hatched stage in semi-defined medium (mKSOMaa) have better *in vivo* development ability after transfer than those cultured with oviductal cells (Khatir *et al.* 2005). Factors affecting developmental ability of IVP embryos are currently being investigated.

Conclusion

Camelid reproductive biology offers interesting challenges to the scientist. Camelids are reputed to have poor reproductive performance, but in fact, these highly adapted species can maintain excellent reproductive performances even under harsh condition. Although our understanding of some of the reproductive phenomena has improved, there is still a need for research in areas that are critical for the management of individuals and herds. Breed differences within each species should be investigated. In males, spermatogenesis and sperm production studies are needed to elucidate the effect of season, ability to induce ovulation (OIF concentration) and interaction with some production characteristics (such as fiber, draught, etc.).

In the female, the mechanisms of luteolysis and maternal recognition of pregnancy remain unclear. This area is critical to understanding the causes of early pregnancy loss that typify IVP embryos and heat stressed animals. Studies using interspecific embryo transfer, interspecific embryo production by ICSI, or cross-species nuclear transfer would allow more genomic studies on reproductive phenomena.

Semen physiology and artificial insemination studies are limited by the lack of an easy and reliable method for semen collection, as well as the gelatinous nature of the semen. Data on use of frozen-thawed semen remains limited to the bactrian camel. Further studies on factors affecting oocyte maturation and activation, and developmental ability of IVP embryos are needed. Embryo transfer technology is well established and could be an excellent tool to study interspecies placentation. Interspecific pregnancies (alpaca in llama, bactrian in dromedary) have been obtained and resulted in births but no trials have been conducted between OWC and NWC (e.g. bactrian in camel, alpaca in camel, ect.).

Finally, to improve herd performance it is necessary to study the interaction among season, nutrition and lactation with special attention to the role of leptin in regulation of reproductive activity and effect of metabolism on hormone levels.

References

Aba MA, Forsberg M, Kindahl H, Sumar J & Edqvist LE 1995 Endocrine changes after mating in pregnant and non-pregnant llamas and alpacas. *Acta Veterinaria Scandinavica* **36** 489-498.

Aba MA, Bravo PW, Forsberg M & Kindahl H 1997 Endocrine changes during early-pregnancy in the alpaca. *Animal Reproduction Science* **47** 273-279.

Abd-Elnaeim MMM, Saber A, Hassan A, Abou-Elmagd A, Klisch K, Jones CJP & Leiser R 2003 Development of the areola in the early placenta of the one-humped camel (Camelus dromedarius): a light, scanning and transmission electron microscopical study. *Anatomia Histologia Embryologia* **32** 326-334.

Adams GP, Ratto MH, Huanca W & Singh J 2005 Ovulation-inducing factor in the seminal plasma of alpacas and llamas. *Biology of Reproduction* **73** 452-457.

Al-Qarawi AA, Abdel-Rahman HA, El-Belely MS & El-Mougy SA 2001 Intratesticular morphometric, cellular and endocrine changes around the pubertal period in dromedary camels. *Veterinary Journal* **162** 241-249.

Aller JF, Rebuffi GE, Cancino AK & Alberio RH 2002 Successful transfer of vitrified llama (Lama glama) embryos. *Animal Reproduction Science* **73** 121-127.

Anouassi A, Lahlou-Kassi A, Combarnous Y, Sghiri A & Martinat N 1988 Performance and endocrine approach for reproductive management in the camel (*Camelus dromedarius*). Proceedings of FIS/SIDA Workshop on Camels. Addis-Ababa, Ethiopia.

Bartha T, Sayed-Ahmed A & Rudas P 2005 Expression of leptin and its receptors in various tissues of ruminants. *Domestic Animal Endocrinology* **29** 193-202

Bravo PW, Stabenfeldt GH, Fowler ME & Lasley BL 1992 Pituitary response to repeated copulation and/or gonadotropin-releasing hormone administration in llamas and alpacas. *Biology of Reproduction* **47** 884-888.

Bravo PW, Fowler ME & Lasley BL 1994 The postpartum llama: fertility after parturition. *Biology of Reproduction* **51** 1084-1087.

Bravo PW, Flores D & Ordonez C 1997 Effect of repeated collection on semen characteristics of alpacas. *Biology of Reproduction* **57** 520-524.

Bravo PW, Ccallo M & Garnica J 2000a The effect of enzymes on semen viscosity in Llamas and Alpacas. *Small Ruminant Research* **38** 91-95.

Bravo PW, Skidmore JA & Zhao XX 2000b Reproductive aspects and storage of semen in Camelidae. *Animal Reproduction Science* **62** 173-193

Bravo PW (2002) The reproductive process of South American camelid. Printed by Seagull Printing, Salt Lake City, IT, USA

Bustos-Obregon E, Rodriguez A & Urquieta B 1997 Spermatogenic cycle stages in the seminiferous epithelium of the vicuna (*Vicugna vicugna*). In Recent advances in microscopy of cells, tissue and organs, Motta P.M. ed. Delfino Editor, Roma, Italy 579-584. 1997.

Chaves MG, Aba M, Aguero A, Egey J, Berestin V & Rutter B 2002 Ovarian follicular wave pattern and the effect of exogenous progesterone on follicular activity in non-mated llamas. *Animal Reproduction Science* **69** 37-46

Chillard Y, Bengoumi M, Delavaud C, Faulconnier Y & Faye B 2005a Body lipids and adaptation of camel to foot and water shortage: New data on adipocyte size and plasma leptin. In: Desertification combat and food safety; The added value of camel producers, Faye B and Esenov P (ed) NBato Science Series I, Vo 362, Pages 135-145

Chilliard Y, Delavaud C & Bonnet M 2005b. Leptin expression in ruminants: Nutritional and physiological regulations in relation with energy metabolism. *Domestic Animal Endocrinology* **29** 3-22

Davis GH, Dodds KG, Moore GH & Bruce GD 1997 Seasonal effects on gestation length and birth-weight in alpacas. *Animal Reproduction Science* **46** 297-303

Deen A, Vyas S & Sahani MS 2003 Semen collection, cryopreservation and artificial insemination in the dromedary camel. *Animal Reproduction Science* **77** 223-233

Del Campo MR, Del Campo CH, Adams GP & Mapletoft RJ 1995 The application of new reproductive technologies to South American camelids. *Theriogenology* **43** 21-30

Del Campo MR, Del Campo CH & Ginther OJ 1996 Vascular provisions for a local utero-ovarian crossover pathway in New World camelids. *Theriogenology* **46** 983-991

Delhon GA & von Lawzewitsch I 1987 Reproduction in the male llama (*Lama glama*), a South American camelid I. Spermatogenesis and organizations of the intertubular space of the mature testis. *Acta Anatomia* **129** 59-66

Delhon G & Von Lawzewitsch I 1994 Ductus epididymidis compartments and morphology of epididymal spermatozoa in llamas. *Anatomia Histologia Embryologia* **23** 217-225

Fernandez Baca S, Hansel W, Saatman R, Sumar J & Novoa C 1979 Differential luteolytic effects of right and left uterine horns in the alpaca. *Biology of Reproduction* **20** 586-595

Flores P, Garcia-Huidobro J, Munoz C, Bustos-Obregon E & Urquieta B 2002 Alpaca semen characteristics previous to a mating period. *Animal Reproduction Science* **72** 259-266

Garnica J, Achata R & Bravo PW (1993) Physical and biochemical characteristics of alpaca semen. *Animal Reproduction Science* **32** 85-90

Ghazi R 1981 Angioarchitectural studies of the utero-ovarian component in the camel (*Camelus dromedarius*). *Journal of Reproduction and Fertility* **61** 43-46.

Jones CJP, Abd-Elnaeim M, Bevilacqua E, Oliveira LV & Leiser R 2002 Comparison of uteroplacental glycosylation in the camel (*Camelus dromedarius*) and alpaca (*Lama pacos*). *Reproduction* **123** 115-126

Kafi M, Mesbah F, Nili H & Khalili A 2005 Chronological and ultrastructural changes in camel (*Camelus dromedarius*) oocytes during *in vitro* maturation. *Theriogenology* **63** 2458-2470

Khatir H, Anouassi A & Tibary A 2004 Production of dromedary (*Camelus dromedarius*) embryos by IVM and IVF and co-culture with oviductal or granulosa cells. *Theriogenology* **62** 1175-1185

Khatir H, Anouassi A & Tibary A 2005 In vitro and in vivo developmental competence of dromedary (*Camelus dromedarius*) embryos produced in vitro using two culture systems (mKSOMaa and oviductal cells). *Reproduction in Domestic Animals* **40** 245-249

Khatir H & Anouassi A 2006 The first dromedary (*Camelus dromedarius*) offspring obtained from in vitro matured, in vitro fertilized and in vitro cultured abattoir-derived oocytes. *Theriogenology* **65** 17127-1736

Khorchani T, Hammadi M & Moslani M 2005 Artificial nursing of camel calves: An effective technique for calves safeguard and improving herd productivity. In: Desertification combat and food safety; The added value of camel producers, Faye B and Esenov P (ed) NBato Science Series I, Vo 362, Pages 168-172

Lichtenwalner AB, Woods GL & Weber JA 1996 Seminal collection, seminal characteristics and pattern of ejaculation in llamas. *Theriogenology* **46** 293-305

Lichtenwalner AB, Woods GL & Weber JA 1997 Pattern of emission during copulation in male llamas. *Biology of Reproduction* **56** 298-298

Marie M & Anouassi A 1987 Induction of luteal activity and progesterone secretion in the nonpregnant one-humped camel (*Camelus dromedarius*). *Journal of Reproduction and Fertility* **80** 183-192

McKinnon AO, Tinson AH & Nation G 1994 Embryo transfer in dromedary camels. *Theriogenology* **41** 145-150

Miragaya MH, Aba MA, Capdevielle EF, Ferrer MS, Chaves MG, Rutter B & Aguero A 2004 Follicular activity and hormonal secretory profile in vicuna (*Vicugna vicugna*). *Theriogenology* **61** 663-671

Miragaya MH, Chaves MG & Agüero A 2006 Reproductive biotechnology in South American camelids. *Small Ruminant Research* **61** 299-310

Nagy P, Jutka J & Wernery U 2005 Incidence of spontaneous ovulation and development of the corpus luteum in non-mated dromedary camels (*Camelus dromedarius*). *Theriogenology* **64** 292-304

Nili H, Mesbah F, Kafi M & Esfahani MHN 2004 Light and transmission electron microscopy of immature camelus dromedarius oocyte. *Anatomia Histologia Embryologia* **33** 196-199.

Nowshari MA 2005 The effect of harvesting technique on efficiency of oocyte collection and different maturation media on the nuclear maturation of oocytes in camels (*Camelus dromedarius*). *Theriogenology* **63** 2471-2481

Nowshari MA, Ali SA & Saleem S 2005 Offspring resulting from transfer of cryopreserved embryos in camel (*Camelus dromedarius*). *Theriogenology* **63** 2513-2522.

Olivera L, Zago D, Leiser R, Jones C & Bevilacqua E 2003a Placentation in the alpaca *Lama pacos*. *Anatomy and Embryology* **207** 45-62

Olivera LVM, Zago DA, Jones CJP & Bevilacqua E 2003b Developmental changes at the materno-embryonic interface in early pregnancy of the alpaca, *Lamos pacos*. *Anatomy and Embryology* **207** 317-331.

Raggi LA, Ferrando G, Parraguez VH, MacNiven V & Urquieta B 1999 Plasma progesterone in alpaca (*Lama pacos*) during pregnancy, parturition and early postpartum. *Animal Reproduction Science* **54** 245-249

Ratto M, Berland M, Huanca W, Singh J & Adams GP 2005a In vitro and in vivo maturation of llama oocytes. *Theriogenology* **63** 2445-2457

Ratto MH, Huanca W, Singh J & Adams GP 2005b Local versus systemic effect of ovulation-inducing factor in the seminal plasma of alpacas. *Reproductive Biology and Endocrinology* **3** online paper 29.

Sansinena MJ, Taylor SA, Taylor PJ, Denniston RS & Godke R 2003 Production of nuclear transfer llama (*Lama glama*) embryos from in vitro matured llama oocytes. *Cloning and Stem Cells* **5** 191-198

Santiani A, Huanca W, Sapana R, Huanca T, Sepulveda N & Sanchez R 2005 Effects on the quality of frozen-thawed alpaca (*Lama pacos*) semen using two different cryoprotectants and extenders. *Asian Journal of Andrology* **7** 303-309

Sayed-Ahmed A, Rudas P & Bartha T 2005 Partial cloning and localisation of leptin and its receptor in the one-humped camel (*Camelus dromedarius*). *Veterinary Journal* **170** (2) 264-267

Sghiri A 1988 Evaluation des performances de reproduction d'un troupeau camelin (*Camelus dromedarius*) à Laayoune. Thèse de Doctorat Vétérinaire, Institut Agronomique et vétérinaire Hassan II, Rabat, Morocco.

Sghiri A & Driancourt MA 1999 Seasonal effects on fertility and ovarian follicular growth and maturation in camels (*Camelus dromedarius*). *Animal Reproduction Science* **55** 223-237

Skidmore JA, Allen WR & Heap RB 1994 Oestrogen synthesis by the peri-implantation conceptus of the one-humped camel (*Camelus dromedarius*). *Journal Reproduction and Fertility* **101** 363-367

Skidmore JA, Billah M & Allen WR 1996a The ovarian follicular wave pattern and induction of ovulation in the mated and non-mated one-humped camel (*Camelus dromedarius*). *Journal of Reproduction and Fertilility* **106** 185-192

Skidmore JA, Wooding FBP & Allen WR 1996b Implantation and early placentation in the one-humped camel (*Camelus dromedarius*). *Placenta* **17** 253-262

Skidmore JA, Starbuck GR, Lamming GE & Allen WR 1998 Control of luteolysis in the one-humped camel (*Camelus dromedarius*). *Journal of Reproduction and Fertility* **114** 201-209

Skidmore JA, Billah M, Allen WR & Short RV 2000 Modern reproductive methods to hybridize old and new world camelids: *Camelus dromedarius* x *Lama guanicoe*. *Reproduction in Domestic Animals* **Suppl. 6** 100-103

Skidmore JA, Billah M & Allen WR 2002 Investigation of factors affecting pregnancy rate after embryo transfer in the dromedary camel. *Reproduction, Fertility and Development* **14** 109-116

Skidmore JA, Billah M & Loskutoff NM 2005 Comparison of two different methods for the vitrification of hatched blastocysts from the dromedary camel (*Camelus dromedarius*). *Reproduction Fertility and Development* **17** 523-527

Tibary A & Anouassi A (1996) Ultrasonographic changes of the reproductive-tract in the female camel (*Camelus-Dromedarius*) during the follicular cycle and pregnancy. *Journal of Camel Practice and Research* **3** 71-90

Tibary A & Anouassi A 1997 Theriogenology in Camelidae: Anatomy, physiology, BSE, pathology and artificial breeding: Actes Editions, Institut Agronomique et Veterinaire Hassan II, 1997.

Tibary A 2001a Semen preservation in camelids. Proceedings of the Annual Conference of the Society for Theriogenology, September 12-15, 2001, Vancouver, BC, Canada, pp 369-378

Tibary A 2001b Fertilization, embryo and fetal development in camelids. Proceedings of the Annual Conference of the Society for Theriogenology, September 12-15, 2001, Vancouver, BC, Canada, pp 387-396

Tibary A 2001c Embryo transfer in camelids. Proceedings of the Annual Conference of the Society for Theriogenology, September 12-15, 2001, Vancouver, BC, Canada, pp 379-386

Tibary A, Anouassi A & Khatir H 2005a Update on reproductive biotechnologies in small ruminants and camelids. *Theriogenology* **64** 618-638

Tibary A, Anouassi A & Sghiri A 2005b Factors affecting reproductive performance of camels at the herd and individual level. In: Desertification combat and food safety; The added value of camel producers, Faye B and Esenov P (ed) NBato Science Series I, Vo 362, Pages 97-114.

Tibary A & Vaughan J 2006 Reproductive physiology and infertility in male South American camelids: A review and clinical observations. *Small Ruminant Research* **61** 283-298

Tingari MD & Moniem KA 1979 On the regional histology and histochemistry of the epididymis of the camel (*Camelus dromedarius*). *Journal of Reproduction and Fertility* **57** 11-20

Urquieta B, Cepeda R, Caceres JE, Raggi LA & Rojas JR 1994 Seasonal variation in some reproductive parameters of male vicuna in the High Andes in northern Chile. *Journal of Arid Environments* **26** 79-87

Urquieta B, Flores P, Munoz C, Bustos-Obregon E & Garcia-Huidobro J 2005 Alpaca semen characteristics under free and directed mounts during a mating period. *Animal Reproduction Science* **90** 329-339

Vaughan J & Tibary A 2006 Reproduction in female South American camelids: A review and clinical observations. *Small Ruminant Research* **61** 259-281

Vaughan JL, Macmillan KL & D'Occhio MJ 2004 Ovarian follicular wave characteristics in alpacas. *Animal Reproduction Science* **80** 353-361

Zhao XX, Huang YM & Chen BX 1994 Artificial insemination and pregnancy diagnosis in the Bactrian camel (*Camelus bactrianus*). *Journal of Arid Environments* **26** 61-65

Zhao XX, Li XL & Chen BX 2001 Isolation of ovulation-inducing factors in the seminal plasma of bactrian camel (*Camelus bactrianus*) by DEAE-cellulose chromatography. *Reproduction in Domestic Animals* **36** 177-181

State-of-the-art embryo technologies in cattle

P Lonergan

School of Agriculture, Food Science and Veterinary Medicine, College of Life Sciences, University College Dublin, Belfield, Dublin 4, Ireland

Over the past 30 years, basic and applied studies on classical and advanced embryo technologies have generated a vast literature on factors regulating oocyte and embryo development and quality. In addition, over this period, commercial bovine embryo transfer has become a large international business. It is well recognised that bovine embryos derived in vivo are of superior quality to those derived from in vitro maturation, fertilization and culture. Relatively little has changed in the techniques of producing embryos in vivo although there is increasing evidence of the importance of, for example, peripheral and follicular endocrine profiles for the subsequent developmental competence of the embryo. The in vitro production of ruminant embryos is a three-step process involving oocyte maturation, oocyte fertilization and in vitro culture. Only 30–40% of such oocytes reach the blastocyst stage, at which they can be transferred to a recipient or frozen for future use. We know now that the quality of the oocyte is crucial in determining the proportion of immature oocytes that form blastocysts while the post-fertilization culture environment has a major influence on the quality of the blastocyst. Use of sexed-sorted sperm in conjunction with in vitro embryo production is a potentially efficient means of obtaining offspring of the desired sex. Concerns regarding the use of sexed semen technology include the apparent lower fertility of sorted sperm, the lower survival of sorted sperm after cryopreservation and the reduced number of sperm that could be separated in a specified time period. Assessment of embryo quality is a challenge. Morphological assessment is at present the most popular method for embryo selection prior to transfer. Other non-invasive assessment methods include the timing of the first cleavage division which has been linked to developmental ability. Quantitative examination of gene expression is an additional valuable tool to assess the viability of cultured embryos. A substantial amount of evidence exists to demonstrate that the culture conditions to which the embryo is exposed, particularly in the post-fertilization period, can have perturbing effects on the pattern of gene expression in the embryo with potentially important long-term consequences. Collectively, in vivo and in vitro studies support the notion that the environment of the embryo is critical for its future. The identification and characterization of the short-term effects of in vitro culture

Corresponding author E-mail: pat.lonergan@ucd.ie

raises the question about long-term consequences and safety of assisted reproductive technologies. The impact of some of these technologies on animal production will be the subject of this review.

Introduction

Over the past 30 years, basic and applied studies on classical (superovulation, non-surgical recovery and transfer of cattle embryos) and advanced embryo technologies (in vitro embryo production, nuclear transfer, transgenesis) have generated a vast literature on factors regulating oocyte and embryo development and quality. In addition, over this period, commercial bovine embryo transfer has become a large international business; approximately 790,000 bovine embryos (550,000 in vivo-derived and 240,000 in vitro produced) were transferred in 2004, the latest year for which data are available (Thibier 2005).

Three different generations of embryo technology have been developed: (i) in vivo embryo production, (ii) in vitro embryo production and (iii) cloning by nuclear transfer and transgenics. The objectives of this review are to give the reader an overview of current technologies in a few selected areas of embryo technology and to discuss how such technologies may impact on animal production.

Embryo Technologies and Genetic Improvement

The rate of genetic improvement in most breeding programmes is controlled by four main factors: (i) the *selection intensity,* a measure of how choosy we are as breeders; the fewer animals we select based on their superior performance the faster the rate of genetic progress (ii) the *accuracy* with which we can predict the genetic merit of an individual animal; the greater the accuracy the greater the potential improvement (iii) the *genetic variation* in the particular trait in question; the greater the variation for a given trait the greater the scope we have as breeders to select animals which are far from the mean level of performance for the trait and (iv) the *generation interval,* a measure of how long it will take the selected animal(s) to contribute their superior genes to the next generation via their offspring. In domestic species this tends to be quite long (2 to 3 years). Coupled with these four parameters are the *selection differential* (the difference in performance in a particular trait between the selected animals and the overall group from which they are selected, clearly related to the selection intensity) and the *heritability* of the trait, the proportion of the superiority (i.e., the proportion of the selection differential) which, on average, is passed on to the offspring. Put another way, it is a measure of how much a trait is under genetic control or how similar offspring are to their parents in terms of performance in a particular trait.

Several reproductive biotechnologies can impact one or more of these parameters and in so doing improve the rate of genetic improvement (Fig. 1). These include more established technologies such as artificial insemination and multiple ovulation embryo transfer (MOET) as well as newer technologies of in vitro embryo production, cloning and transgenesis (Georges & Massey 1991; Lohuis 1995; Nicholas 1996; Cunningham 1999, Van Arendonk & Bijma 2003; Galli & Lazzari 2005). Advantages of MOET, for example, include higher female selection intensity and increased selection accuracy (more full and half-sib information). However, variability in individual animal response to superovulation and the low average number of transferable embryos recovered are still limiting factors. In vitro embryo production can potentially affect generation interval through its use on prepubertal donors; for example, modest success in the production of pregnancies from in vitro produced embryos derived from prepubertal calves has been reported (e.g., Taneja *et al.* 2000).

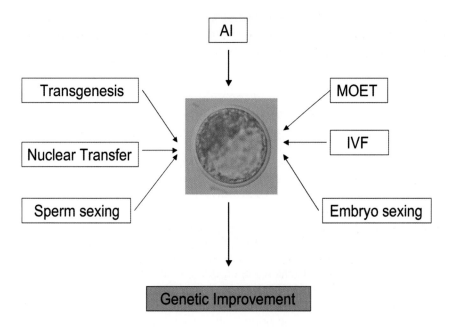

Fig. 1. The main reproductive technologies which can impact on the rate of genetic improvement.

MOET

Apart from artificial insemination, MOET is one of the longer established technologies impacting on animal breeding and reproduction. The techniques of MOET are well established and their use accounts for approximately 80% of cattle embryos transferred commercially (Thibier 2005). By increasing the reproductive rate of females MOET offers the opportunity to reduce the number of dams that need to be selected for the next generation while at the same time increasing the amount of information available on sibs for estimating breeding values. The biggest issue with respect to results is the huge variability in response of individual donors to superovulation. On average 4 to 6 transferable embryos are produced per treatment cycle, a yield which has not changed over the past 20 years (Hasler 2003); however, one-third of treated donors fail to respond, one-third yields 1 to 3 embryos and only the remaining one-third actually yields a significant number of embryos (Boland *et al.* 1991).

The maintenance of recipients is a major economic component of producing ET calves. If higher pregnancy rates are achieved by recipients, more efficient use of embryos would occur and financial savings could be realized through reductions in costs associated with transfer of embryos and maintenance of non-pregnant animals (Looney *et al.* 2006).

Assessing embryo quality

Embryo morphology assessment, with all its drawbacks, is at present the most widely used method for embryo selection prior to transfer for assisted reproduction in both domestic ruminants and humans. The high lipid content of ruminant embryos precludes the visualization of nuclear and nucleolar morphology in most cases, aspects which can easily be evaluated in the more transparent human or murine embryo. However, recent research focused on the correlation between bovine embryo morphology and ultrastructure, gene expression and cryotolerance, has provided evidence that much more can be deduced from morphological examination than previously thought (Van Soom *et al.* 2003). Morphological features such as colour of the blastomeres, the extent of compaction, kinetics of development and timing of blastocyst formation and expansion and diameter of the embryo at hatching can be linked with embryo quality. Other measures of embryo quality include non-invasive measures such as assessment of the timing of the first cleavage division after fertilization (Lonergan *et al.* 1999; Fair *et al.* 2004) and invasive measures such as examination of the pattern of transcript abundance in the blastocyst (Lonergan *et al.* 2003a,b; 2006).

In domestic species, there is a large body of evidence demonstrating that culture media can perturb gene expression in the developing embryo (see review by Wrenzycki *et al.* 2005). This is the case, not only when one compares in vitro and in vivo culture systems, but also when comparisons of different in vitro culture systems are made (see for example, Eckert & Niemann 1998; Natale *et al.* 2001; Rizos *et al.* 2002b; Rinaudo & Schultz 2004; Lonergan *et al.* 2003a;b; 2006). Apart from the culture medium used, the conditions of culture can also affect gene expression; bovine in vitro produced embryos respond to changes in oxygen concentrations by altering the expression of GLUT1 (Harvey *et al.* 2004) while in mice it has been reported that expression of GLUT1, GLUT3 and VEGF was significantly increased in embryos cultured under 2% vs 20% oxygen (Kind *et al.* 2004). The prolonged period of postfertilization culture of the embryo presents both challenges and opportunities for manipulation of gene expression and thereby protein expression and phenotype which should be exploited in order to take account of the changing needs of the developing embryo.

In vivo culture of embryos

Culture of bovine embryos in the oviduct of the ewe has been shown by several authors to be a suitable environment for the development of embryos from the zygote to blastocyst stage (Lonergan *et al.* 2003a;b) and even through the early stages of elongation (Rexroad & Powell 1999). Though not perfect, one advantage of this in vivo culture system is the ability to culture large numbers of embryos in a 'near in vivo' environment and in a cost effective manner. While the yield of blastocysts following such in vivo culture is not superior to that following culture in vitro, the quality of the blastocysts is significantly improved (Enright *et al.* 2000; Galli & Lazzari 1996; Lazzari *et al.* 2002; Rizos *et al.* 2002a;b; Lonergan *et al.* 2003a;b).

Until the recent establishment and efficient application of endoscopy as a means of performing oviduct transfer and collection of embryos in cattle (Wetscher *et al.* 2005), in vivo culture of in vitro matured or fertilized embryos in the homologous bovine oviduct was extremely difficult. Techniques developed by Besenfelder and colleagues have focused on accessing the bovine oviduct to perform comparative in vivo versus in vitro studies and finally tubal transfer and flushing were combined for in vivo culture of IVP-derived embryos (Besenfelder & Brem 1998; Besenfelder *et al.* 2001; Wetscher *et al.* 2005).

In vivo culture of zygotes derived by IVF could be one means of reducing the incidence of abnormally large offspring (Lazzari *et al.* 2002). Walmsley *et al.* (2004) examined the effect of transferring in vitro produced sheep embryos to recipient ewes on Day 2 versus Day 6 postfertilization on the incidence of oversized or abnormal lambs. Day 2 transfer of cleaved embryos did reduce but did not prevent the production of oversized or congenitally abnormal offspring that was likely attributable to the IVP system compared with transfer at Day 6.

Controlling gender

Determination of the sex of the embryo prior to transfer would offer a clear advantage to the end-user and thereby add value to the product. Up until the early-nineties much of the focus on controlling offspring gender in practice was based on establishing the sex of the embryo, usually by biopsy, prior to transfer. This has potentially deleterious implications for the likelihood of a pregnancy ensuing.

A number of recent reviews have addressed the use of sexed semen in cattle and sheep production (Seidel & Garner 2002; Seidel 2003a;b; Wheeler *et al.* 2006; Maxwell *et al.* 2004; Evans *et al.* 2004). While not an 'embryo technology' per se, the use of sex-sorted sperm in embryo production has massive potential for increasing efficiencies of animal production. Some of the concerns associated with the use of sexed semen include lower pregnancy rates, lower survival of sorted sperm after cryopreservation and number of sperm separated per unit time.

Numerous field trials have been conducted with single-ovulating heifers with fertility of low doses of sexed semen being about 60-90% of conventional doses of unsorted semen. The importance of heifer management to achieving comparable pregnancy rates has been highlighted (Seidel 2003b).

It is often the combination of two or more, rather than individual, technologies that make the biggest impact on practice. One appealing attribute of using sorted semen for in vitro embryo production is that much fewer sperm are need for IVF. Over the past decade or so sorted semen has been used for the production of embryos in vitro in both cattle (Lu *et al.* 1999) and sheep (Hollinshead *et al.* 2004; Morton *et al.* 2004).

For the dissemination of superior genes the use of sexed semen on superovulated donors of high genetic merit is desirable. In the study of Schenk *et al.* (2006), the number of embryos recovered from superovulated donors bred with sexed sperm was lower than that for donor females bred with non-sexed semen. However, the use of 20×10^6 sexed sperm per insemination dose resulted in similar numbers of transferable embryos of the desired sex compared to that for non-sexed sperm. Thus, use of sufficient numbers of sexed sperm in bovine superovulation and embryo transfer programmes can result in the production of more embryos of the desired sex with about half as many recipients, thus reducing production costs associated with embryos of undesirable sex.

In vitro embryo production

Techniques and procedures based on the use of in vitro fertilization technology offer significant benefits in relation to genetic improvement and reproductive management of domestic ruminants. Underlining the importance of this area and in recognition of some of the inherent problems facing users of this technology, a symposium was recently held in Orlando, Florida in conjunction with the 32nd Annual Conference of the International Embryo Transfer Society entitled 'Realizing the promise of IVF in cattle: Optimizing embryonic and fetal survival' at which many related topics were addressed including procedures for transferring a better embryo, errors in embryonic and fetal development, recipient management and cryopreservation (Hansen 2006).

In vitro, most studies terminate at the blastocyst or hatched blastocyst stage of development. This is a reflection of two things: (i) the blastocyst stage is, in cattle at least, the stage most routinely used for transfer or cryopreservation and (ii) in vitro systems do not support normal development of ruminant embryos through the post-hatching elongation stages. Artificial elongation has been induced in vitro using agarose gel tunnels (Vajta *et al.* 2000; Brandao *et al.* 2004). While such results are encouraging, the technology still requires considerable refinement; for example, the embryonic disk characterizing the pre-streak stage 1 is never established (Alexopoulos et al. 2005; Vejlsted et al. 2006)

For all IVF technologies, achieving economically viable success rates is still a problem. Less than 30% of cultured oocytes develop into transferable embryos. Wastage during pregnancy is also increased with pregnancy rates from IVF embryos typically 10% lower than those achieved with in vivo embryos. From a genetic improvement point-of-view, the use of ovaries collected from random females at local abattoirs offers little in the way of genetic superiority (Cunningham 1999).

The in vitro production of ruminant embryos is a three-step process involving in vitro oocyte maturation, in vitro fertilization and in vitro culture. In terms of efficiency, in cattle, approximately 90% of immature oocytes, generally recovered from follicles at unknown stages of the oestrous cycle, undergo nuclear maturation in vitro from prophase I to metaphase II (the stage at which they would be ovulated in vivo); about 80% undergo fertilization following insemination and cleave at least once, to the two-cell stage. However, only 30-40% of such oocytes reach the blastocyst stage, at which they can be transferred to a recipient or frozen for future use. Thus, the major fall-off in development occurs during the last part of the process (in vitro culture), between the 2-cell and blastocyst stages, suggesting that post-fertilization embryo culture is the most critical period of the process in terms of determining the blastocyst yield. There is considerable evidence demonstrating the quality of the oocyte is crucial in determining the proportion of immature oocytes that form blastocysts and that the post-fertilization culture environment, within certain limits, does not have a major influence on the capacity of the immature oocyte to form a blastocyst (see Rizos et al. 2002a). It would appear that once the oocyte has been removed from the follicle its ability to develop the blastocyst stage is more-or-less sealed and despite attempts at temporarily inhibiting resumption of meiosis to allow cytoplasmic maturation to proceed in vitro, thereby improving development (e.g. Lonergan et al. 2000; Sirard 2001) or modifications of maturation media, blastocyst yields in vitro using oocytes recovered from slaughtered animals rarely exceed 40% on a consistent basis. Therefore, attempts at increasing oocyte competence must be applied before removal from the follicle by manipulating follicle development (e.g., Blondin et al. 2002; Durocher et al. 2006)

Ovum Pick-Up (OPU)

OPU was first developed in cattle in the early nineties (Pieterse et al. 1991). Since then many studies have been carried out aimed at refining both technical (e.g., needle size, vacuum pressure) and biological (e.g., hormonal stimulation, frequency, operator skill) factors (reviewed by Merton et al. 2003; van Wagtendonk-de Leeuw 2006). The advent of OPU opened up significant potential for the application of IVF technology to animal breeding. Collecting the ovaries of slaughtered beef animals, while offering a fantastic source of raw material (oocytes and, after IVF, embryos) for basic research studies, had very little to offer in terms of genetic improvement; however, OPU allows the repeated access to the ovaries of high genetic merit cows and when coupled with semen from a high genetic merit sire, allows the feasibility of producing high quality embryos, genetically speaking, in large numbers. When the techniques of OPU and IVF are combined they are capable of producing 80-100 calves per donor cow per year albeit with large variation between donors (van Wagtendonk-de Leeuw 2006); this compares with typically one calf per cow per year with AI and approximately 20-25 calves per donor with MOET.

Cloning

Since the birth of Dolly, the first mammal to be cloned using the nucleus of a somatic cell from an adult animal, in 1997 (Wilmut *et al.* 1997) standard somatic cell nuclear transfer (SCNT) techniques have resulted in the birth of live offspring in 11 species (Vajta & Gjerris 2006). Although considerable variation exists the basic steps involved include (i) enucleation of an in vitro matured oocyte for use as recipient cytoplasm, (ii) insertion of donor cell/nucleus, and (iii) activation of the reconstructed embryo. A relatively new approach to nuclear transfer is the so-called handmade cloning technique which obviates the need for expensive micromanipulators. However, it does require more specialized culture conditions for reconstructed embryos as zona removal is an integral part of the process (Vajta *et al.* 2005).

The application of cloning techniques to animal breeding are numerous including the fact that many animals of the same genotype can theoretically be produced increasing the accuracy of evaluation (as each female can be evaluated on the average phenotypic performance of herself) and allowing the evaluation of post-slaughter traits in a cohort of clones, the exploitation of heterosis and the multiplication of transgenic animals (Cunningham 1999; Van Arendonk & Bijma 2003). Clearly, however, the testing of clones could be problematical, especially in the case of traits of low heritability, and the risks of a reduction in genetic variation need to be considered.

One of the biggest drawbacks of this technology is the inefficiency of the process. While blastocysts can be produced at rates comparable with those of routine IVF in cattle, a very limited percentage (0.5-5%) of the reconstructed embryos result in full term development. This is mainly due to a high frequency of postimplantation developmental arrest. Such losses are predominantly during the first trimester of pregnancy but can occur much later (Heyman *et al.* 2002). Heyman *et al.* (2002) monitored the evolution of pregnancy following the transfer of embryos derived from somatic cell cloning, embryonic cloning and IVP in order to detect the occurrence of late gestation losses and their frequency. On the basis of progesterone concentrations on Day 21, there were no significant differences in the percentages of initiated pregnancies between the groups (55.6-62.7%). Confirmed pregnancy rate by Day 35 using ultrasound scanning was significantly lower in the two somatic cloned groups (27.5-33.8%) compared with the embryonic clones (49.2%) and IVF embryos (52.9%). This pattern was maintained at Days 50, 70 and 90. The incidence of loss between Day 90 of gestation and calving was 43.7% for adult somatic clones and 33.3% for fetal somatic clones compared with 4.3% after embryonic cloning and 0% after IVP.

Long Term Consequences of Embryo Culture Conditions

Although normal calves can result from the transfer of embryos produced in vitro or by somatic cell nuclear transfer, these embryo technologies are more often associated with increased rates of abortion and abnormal development of the fetus and placenta (Bertolini & Anderson 2002; Farin *et al.* 2001; Kruip & den Daas 1997). Several studies have shown that embryos are sensitive to environmental conditions that can affect future growth and developmental potential, both pre- and postnatally (e.g., Young & Fairburn 2000; Fleming *et al.* 2004). It is well known that early ruminant embryos also exhibit sensitivity to their environment which may, after transfer, lead to the abnormal development, characterized by aberrant fetal and placental development, increased fetal myogenesis, dystocia, abnormal perinatal pulmonary activity and increased mortality in the early postnatal period (Walker *et al.* 1996; Sinclair *et al.* 2000). In addition, some of the effects, such as increased organ size, can persist into later life (McEvoy *et al.* 1998). These alterations in phenotype have been referred to as 'Large Calf Syndrome' or 'Large Offspring Syndrome' although a new term, 'abnormal offspring syndrome' has been proposed (Farin *et al.* 2006) to better represent the fact that abnormalities other than solely large size are observed.

It is worth bearing in mind that abnormalities in development arising from the environment to which the embryo is exposed are not limited to in vitro culture or manipulation. Maternal diet can impact preimplantation phenotype and long-term development. For example, high protein diets in sheep around conception have been associated with reduced viability and increased fetal and birth weights (McEvoy *et al.* 1997; 2001) as well as alterations in fetal development and physiology (Edwards & McMillen 2002; McMillen & Robinson 2005).

Collectively these in vivo and in vitro studies support the notion that the environment of the embryo is critical for its future. The identification and characterization of the short-term effects of in vitro culture raises the question about long-term consequences and safety of assisted reproductive technologies. For example, recent reports indicate that in vitro culture of mouse embryos can have irreversibly long-term consequences of postnatal development, growth, physiology, and behaviour in resulting offspring (Ecker *et al.* 2004; Fernandez-Gonzales *et al.* 2004)

Conclusions

As pointed out by Cunningham (1999), the baseline against which all new reproductive and genetic methods must be measured is natural service. The use of a stock bull by an individual farmer has a certain cost and convenience associated with it. For example, in Ireland AI usage in cattle has declined by approximately 20% over the past 5 years. The reasons for the decline are related to the labour availability required for AI and a perceived lack of benefit (Buckley *et al.* 2003). This is despite a recent objective analysis which demonstrated that daughters of relatively high Economic Breeding Index bulls leave on average 80 euro profit per lactation more than those of stock bulls (Berry *et al.* 2005). Therefore, new technologies, or refinement of existing ones, must take into account the demands of the end-user to ensure they are taken up in practice. As indicated by van Wagtendonk-de Leeuw (2006) not many breakthroughs are expected for OPU technology in the near future. Advances in IVP technology and our understanding of the factors controlling embryo development may allow the development of in vitro culture systems which can produce embryos of a quality similar to those produced in vivo which ultimately would lead to improved quality and welfare of subsequent offspring.

Acknowledgements

The author's research is funded in part by Science Foundation Ireland (Grant 02/IN1/B78). The opinions, findings and conclusions or recommendations expressed in this material are those of the author and do not necessarily reflect the views of the Science Foundation Ireland.

References

Alexopoulos NI, Vajta G, Maddox-Hyttel P, French AJ, & Trounson AO 2005 Stereomicroscopic and histological examination of bovine embryos following extended in vitro culture. *Reproduction Fertility and Development* **17** 799-808

Berry D, Cromie A, Coughlan S, Dillon P 2005 An objective comparison of artificial insemination and the stock bull in Irish dairy herds. Published online: *www.icbf.com.*

Bertolini M & Anderson GB 2002 The placenta as a contributor to production of large calves. *Theriogenology* **57** 181-187.

Besenfelder U & Brem G 1998 Tubal transfer of bovine embryos: a simple endoscopic method reducing long term exposure of in vitro produced embryos. *Theriogenology* **50** 739-745

Besenfelder U, Havlicek V, Mösslacher G & Brem G 2001 Collection of tubal stage bovine embryos by means of endoscopy. A technique report. *Theriogenology* **55** 837-845.

Boland MP, Goulding D & Roche JF 1991 Alternative gonadotrophins for superovulation in cattle. *Theriogenology* **35** 5-17.

Blondin P, Bousquet D, Twagiramungu H, Barnes F & Sirard MA 2002 Manipulation of follicular development to produce developmentally competent bovine oocytes. *Biology of Reproduction* **66** 38-43.

Brandao DO, Maddox-Hyttel P, Lovendahl P, Rumpf R, Stringfellow D, & Callesen H 2004 Post hatching development: a novel system for extended in vitro culture of bovine embryos. *Biology of Reproduction* **71** 2048-2055.

Buckley F, Mee J, Sullivan K, Evans R, Berry D, Dillon P 2003 Insemination factors affecting the conception rate in seasonal calving Holstein-Friesian cows. *Reproduction, Nutrition Development* **43** 543-555.

Cunningham EP 1999 The application of biotechnologies to enhance animal production in different farming systems. *Livestock Production Science* **58** 1-24.

Durocher J, Morin N & Blondin P 2006 Effect of hormonal stimulation on bovine follicular response and oocyte developmental competence in a commercial operation. *Theriogenology* **65** 102-115.

Ecker, DJ, Stein P, Xu Z, Williams CJ, Kopf GS, Biker WB, Abel T & Schultz RM 2004 Long-term effects of culture of preimplantation mouse embryos on behavior. *Proc Natl Acad Sci USA* **101** 1595-1600.

Eckert J & Niemann H 1998 mRNA expression of leukaemia inhibitory factor (Lif) and its receptor subunits glycoprotein 130 and Lif-receptor-beta in bovine embryos derived in vitro or in vivo. *Molecular Human Reproduction* **4** 957-965.

Edwards LJ & McMillen IC 2002 Impact of maternal undernutrition during the periconceptional period, fetal number and fetal sex on the development of the hypothalamo-pituitary adrenal axis in sheep during late gestation. *Biology of Reproduction* **66** 1562-1569.

Enright BP, Lonergan P, Dinnyes A, Fair T, Ward FA, Yang X Boland MP 2000 Culture of in vitro produced bovine zygotes in vitro vs in vivo: implications for early embryo development and quality. *Theriogenology* **54** 659-673.

Evans G, Hollinshead FK & Maxwell WMC 2004 Preservation and artificial insemination of sexed semen in sheep. *Reproduction Fertility and Development* **16** 455-464.

Fair T, Murphy M, Rizos D, Moss C, Martin F, Boland MP & Lonergan P 2004 Analysis of differential maternal mRNA expression in developmentally competent and incompetent bovine two-cell embryos *Molecular Reproduction and Development* **67** 136-144.

Farin PW, Crosier AE & Farin CE 2001 Influence of in vitro systems on embryo survival and fetal development in cattle. *Theriogenology* **55** 151-170.

Farin PW, Piedrahita & Farin CE 2006 Errors in development of fetuses and placentas from in vitro-produced bovine embryos. *Theriogenology* **65** 178-91.

Fernández-Gonzalez R, Moreira P, Bilbao A, Jiménez A, Pérez-Crespo M, Ramírez MA, Rodríguez De Fonseca F, Pintado B & Gutiérrez-Adán A 2004 Long-term effect of in vitro culture of mouse embryos with serum on mRNA expression of imprinting genes, development, and behavior. *Proceedings of the National Academy of Sciences of the United States of America* **101** 5880-5885.

Fleming TP, Kwong WY, Porter R, Ursell E, Fesenko I, Wilkins A, Miller DJ, Watkins AJ & Eckert JJ 2004 The embryo and its future. *Biology of Reproduction* **71** 1046-1054.

Galli C & Lazzari G 1996 Practical aspects of IVM/IVF in cattle. *Animal Reproduction Science* **42** 371-379.

Galli C & Lazzari G 2005 Embryo technologies in dairy cattle. 26th *European Holstein and Red Holstein Conference*, Prague.

Georges M & Massey JM 1991 Velogenetics, or the synergistic use of marker assisted selection and germline manipulation. *Theriogenology* **35** 151-159.

Hansen PJ 2006 Realizing the promise of IVF in cattle – an overview. *Theriogenology* **65** 119-125.

Harvey AJ, Kind KL, Pantaleon M, Armstrong DT & Thompson JG 2004 Oxygen-regulated gene expression in bovine blastocysts. *Biology of Reproduction* **71** 1108-1119.

Hasler JF 2003 The current status and future of commercial embryo transfer in cattle. *Animal Reproduction Science* **79** 245-264.

Heyman Y, Chavette-Palmer P, LeBourhis D, Camous S, Vignon X & Renard JP 2002 Frequency and occurrence of late-gestation losses from cattle cloned embryos. *Biology of Reproduction* **66** 6-13.

Hollinshead FK, Evans G, Evans KM, Catt SL, Maxwell WMC, O'Brien JK 2004 Birth of lambs of a predetermined sex after in vitro production of embryos using frozen-thawed sex-sorted and re-frozen-thawed ram spermatozoa. *Reproduction* **127** 557-568.

Kind KL, Collett RA, Harvey AJ & Thompson JG 2004 Oxygen-regulated expression of GLUT-1, GLUT-3 and VEGF in the mouse blastocyst. *Molecular Reproduction and Development* **70** 37-44.

Kruip TAM & den Daas JHG 1997 In vitro produced and cloned embryos: effects on pregnancy, parturition and offspring. *Theriogenology* **47** 43-52.

Lazzari G, Wrenzycki C, Herrmann D, Duchi R, Kruip T, Niemann H & Galli C 2002 Cellular and molecular deviations in bovine in vitro-produced embryos are related to the large offspring syndrome. *Biology of Reproduction* **67** 767-775.

Lohuis MM 1995 Potential benefits of bovine embryo manipulation technologies to genetic improvement programs *Theriogenology* **43** 51-60.

Lonergan P, Khatir H, Piumi F, Rieger D, Humblot P, Boland MP 1999 Effect of time interval from insemination to first cleavage on the developmental characteristics, sex ratio and pregnancy rate after transfer of bovine embryos. *Journal of Reproduction and Fertility* **117** 159-67.

Lonergan P, Dinnyes A, Fair T, Yang X & Boland MP 2000 Bovine oocyte and embryo development following meiotic inhibition with butyrolactone I. *Molecular Reproduction and Development* **57** 204-209.

Lonergan P, Rizos D, Gutierrez-Adan A, Fair T & Boland MP 2003a Oocyte and embryo quality: effect of origin, culture conditions and gene expression patterns. *Reproduction in Domestic Animals* **38** 259-267.

Lonergan P, Rizos D, Gutierrez-Adan A, Moreira P, Pintado B, de la Fuente J & Boland MP 2003b Temporal divergence in the pattern of messenger RNA expression in bovine embryos cultured from the zygote to blastocyst stage in vitro or in vivo. *Biology of Reproduction* **69** 1424-1431.

Lonergan P, Fair T, Corcoran D & Evans AC 2006 Effect of culture environment on gene expression and developmental characteristics in IVF-derived embryos. *Theriogenology* **65** 137-152.

Looney CR, Nelson JS, Schneider HJ & Forrest DW 2006 Improving fertility in beef cow recipients. *Theriogenology* **65** 201-209.

Lu KH, Cran DG & Seidel GE 1999 In vitro fertilization with flow-cytometrically sorted bovine sperm. *Theriogenology* **52** 1393-1405.

McEvoy TG, Robinson JJ, Aitken RP, Findlay PA, Robertson IS 1997 Dietary excesses of urea influence the viability and metabolism of preimplantation sheep embryos and may affect fetal growth among survivors. *Animal Reproduction Science* **47** 71-90.

McEvoy TG, Sinclair KD, Broadbent PJ, Goodhand KL & Robinson JJ 1998 Post-natal growth and development of Simmental calves derived from in vivo or in vitro embryos. *Reproduction Fertility and Development* **10** 459-464.

McEvoy TG, Robinson JJ, Ashworth CJ, Rooke JA & Sinclair KD 2001 Feed and forage toxicants affecting embryo survival and fetal development. *Theriogenology* **55** 113-129.

McMillen IC & Robinson JS 2005 Developmental origins of the metabolic syndrome: prediction, plasticity, and programming. *Physiological Reviews* **85** 571-633.

Maxwell WMC, Evans G, Hollinshead FK Bathgate R De Graff SP, Eriksson BM, Gillian L, Morton KM & O'Brien JK 2004 Integration of sperm sexing technology into the ART toolbox. *Animal Reproduction Science* **82-83** 79-95.

Merton JS, de Roos AP, Mullaart E, de Ruigh L, Kaal L, Vos PL, Dieleman SJ 2003 Factors affecting oocyte quantity in commercial application of embryo technologies in the cattle breeding industry. *Theriogenology* **59** 651-674.

Morton KM, Catt SL, Hollinshead FK, Maxwell WM & Evans G 2004 Production of lambs after the transfer of fresh and cryopreserved in vitro produced embryos from prepubertal lamb oocytes and unsorted and sex-sorted frozen-thawed spermatozoa. *Reproduction in Domestic Animals* **39** 454-61.

Natale DR, De Sousa PA, Westhusin ME & Watson AJ 2001 Sensitivity of bovine blastocyst gene expression patterns to culture environments assessed by differential display RT-PCR. *Reproduction* **122** 687-693.

Nicholas FW 1996 Genetic improvement through reproductive technology *Animal Reproductive Science* **42** 205-214.

Pieterse MC, Vos PLAM, Kruip TAM, Wurth YA, van Beneden ThH, Willemse AH Taverne MA 1991 Transvaginal ultrasound-guided follicle aspiration of bovine oocytes. *Theriogenology* **35** 19-24.

Rexroad CE & Powell AM 1999 The ovine uterus as a host for in vitro-produced bovine embryos. *Theriogenology* **52** 351-364.

Rinaudo P & Schultz RM 2004 Effects of embryo culture on global pattern of gene expression in preimplantation mouse embryos. *Reproduction* **128** 301-311.

Rizos D, Lonergan P, Ward F, Duffy P & Boland MP 2002a Consequences of bovine oocyte maturation, fertilization or early embryo development in vitro versus in vivo: implications for blastocyst yield and blastocyst quality. *Molecular Reproduction and Development* **61** 234-248.

Rizos D, Lonergan P, Boland MP, Arroyo-Garcia R, Pintado B, de la Fuente J, Gutierrez-Adan A 2002b Analysis of differential mRNA expression between bovine blastocysts produced in different culture systems: Implications for blastocyst quality. *Biology of Reproduction* **66** 589-595.

Schenk JL, Suh TK, Seidel Jr & GE 2006 Embryo production from superovulated cattle following insemination of sexed sperm *Theriogenology* **65** 299-307.

Seidel GE Jr & Garner DL 2002 Current status of sexing mammalian spermatozoa. *Reproduction* **124** 733-43.

Seidel GE Jr 2003a Sexing mammalian sperm—intertwining of commerce, technology, and biology. *Animal Reproduction Science* **79** 145-56.

Seidel GE Jr 2003b Economics of selecting for sex: the most important genetic trait. *Theriogenology* **59** 585-98.

Sinclair KD, Young LE, Wilmut I & McEvoy TG 2000 In-utero overgrowth in ruminants following embryo culture: lessons from mice and a warning to men. *Human Reproduction* **15**(Suppl. 5) 68-86.

Sirard MA 2001 Resumption of meiosis: mechanism involved in meiotic progression and its relation with developmental competence. *Theriogenology* **55** 1241-1254.

Taneja M, Bols PE, Van de Velde A, Ju JC, Schreiber D, Tripp MW, Levine H, Echelard Y, Riesen J, Yang X 2000 Developmental competence of juvenile calf oocytes in vitro and in vivo: influence of donor animal variation and repeated gonadotropin stimulation. *Biology of Reproduction* **62** 206213.

Thibier M 2005 International Embryo Transfer Society: Data Retrieval Committee Annual Report. *Embryo Transfer Newsletter* **22** 12-19.

Vajta G, Hyttel P & Trounson A 2000 Post-hatching development of in vitro produced bovine embryos on agar or collagen gels. *Animal Reproduction Science* **60-61** 208.

Vajta G & Gjerris M 2006 Science and technology of farm animal cloning: state of the art. *Animal Reproduction Science* **92** 211-230.

Vajta G, Kragh PM, Mtango NR & Callesen H 2005 Handmade cloning approach: potentials and limitations. *Reproduction, Fertility and Development* **17** 97-112.

Van Arendonk JAM & Bijma P 2003 Factors affecting commercial application of embryo technologies in dairy cattle in Europe – a modeling approach. *Theriogenology* **59** 635-649.

Van Soom A, Mateusen B, Leroy J & De Kruif A 2003 Assessment of mammalian embryo quality: what can we learn from embryo morphology? *Reproductive Biomedicine Online* **7** 664-70.

Van Wagtendonk-de Leeuw AM 2006 Ovum pick up and in vitro production in the bovine after use in several generations: A 2005 status. *Theriogenology* **65** 914-925.

Vejlsted M, Du Y, Vajta G & Maddox-Hyttel P 2006 Post-hatching development of the porcine and bovine embryo—defining criteria for expected development in vivo and in vitro. *Theriogenology* **65** 153-65.

Walker SK, Hartwich KM & Seamark RF 1996 The production of unusually large offspring following embryo manipulation: concepts and challenges. *Theriogenology* **45** 111-120.

Walmsley SC, Buckrell BC, Buschbeck C, Rumph N & Pollard JW 2004 Rate of abnormalities in lambs from in vitro produced embryos transferred on Day 2 compared with Day 6 postfertilization. *Theriogenology* **62** 195-206.

Wetscher F, Havlicek V, Huber T, Gilles M, Tesfaye D, Griese J, Wimmers K, Schellander K, Müller M, Brem G & Besenfelder U 2005 Intrafallopian transfer of gametes and early stage embryos for in vivo culture in cattle. *Theriogenology* **64** 30-40.

Wheeler MB, Rutledge JJ, Fischer-Brown A, VanEtten T, Malusky S, & Beebe DJ 2006 Application of sexed semen technology to in vitro embryo production in cattle. *Theriogenology* **65** 219-227.

Wilmut I, Schnieke AE, McWhir J, Kind AJ & Campbell KH 1997 Vital offspring derived from adult and adult mammalian cells. *Nature* **385** 810-813.

Wrenzycki C, Herrmann D, Lucas-Hahn A, Korsawe K, Lemme E & Niemann H 2005 Messenger RNA expression patterns in bovine embryos derived from in vitro procedures and their implications for development. *Reproduction, Fertility and Development* **17** 23-35.

Young LE & Fairburn HR 2000 Improving the safety of embryo technologies: possible role of genomic imprinting. *Theriogenology* **53** 627-648.

Nuclear reprogramming by somatic cell nuclear transfer – the cattle story

XC Tian[1], SL Smith[1], SQ Zhang[1,2], C Kubota[3], C Curchoe[1] F Xue[4], L Yang[1], F Du[1], L-Y Sung[1] and X Yang[1]

[1]Department of Animal Science/Center for Regenerative Biology, [4]Department of Molecular and Cellular Biology, University of Connecticut, Storrs, CT, USA; [2]College of Animal Science, South China Agricultural University, Guangzhou 510642, People's Republic of China; [3]Department of Veterinary Medicine, Kagoshima University, Kagoshima City, Kagoshima, Japan

Somatic cell nuclear transfer (cloning) returns a differentiated cell to a totipotent status; a process termed nuclear reprogramming. Nuclear transfer has potential applications in agriculture and biomedicine, but is limited by low efficiency. To understand the deficiencies of nuclear reprogramming, our research has focused on both candidate genes (imprinted and X-linked genes) and global gene expression patterns in cloned bovine embryos/offspring as compared to those generated by conventional reproduction. We found aberrant expression patterns of *H19* and *Igf2r* as well as X-linked genes in term cloned calves. The expression profiles of cloned blastocysts, however, closely resembled those of the naturally fertilized embryos but were considerably different from those of their nuclear donor cells. Our findings suggest that cloned embryos have undergone significant nuclear reprogramming by the blastocyst stage. However, it is possible that during re-differentiation in later development gene expression aberrancies occur. Additionally, small initial nuclear reprogramming errors may be manifested during subsequent development.

Introduction

A long-held dogma in developmental biology was that mammalian somatic cell differentiation was considered irreversible. Fig. 1 shows the landscape model of cell differentiation (Waddington 1940; Keeton & Gould 1984). It likens the process of mammalian cell differentiation as a ball rolling down a hill with many valleys. When the ball is on top of the hill, it can roll down through any valleys below; this represents the process of a totipotent cell that can differentiate into any tissue of the body. However, as the ball rolls passed an intersection, the available valleys for the ball to roll down become limited. When the ball reaches the bottom of the hill, it cannot move to another valley or back to the top of the hill. This model was used to illustrate a totipotent cell choosing among different developmental paths; when the cell's fate is partially determined its differentiation potential becomes limited. Once the cell is terminally differentiated, it can no longer trans-differentiate into another cell type or become totipotent again.

Corresponding author E-mail: Xiuchun.tian@uconn.edu

Fig. 1. The landscape model of mammalian cell differentiation (modified from Keeton & Gould 1984).

The success of cloning a whole animal using differentiated somatic cells, however, challenged this theory. During cloning, a differentiated somatic cell is injected into the oocyte's cytoplasm and a cloned embryo is created. The cloned embryo contains totipotent cells that can differentiate into any tissue type and result in a cloned animal (Fig. 2). The process of returning a differentiated somatic nucleus to a totipotent status is termed nuclear reprogramming. Currently, this process can only be accomplished by somatic cell nuclear transfer. During nuclear reprogramming, genes inactivated due to cell differentiation are subjected to re-activation, allowing the re-constructed cloned embryos to support development and generation of all tissue types in the cloned individual.

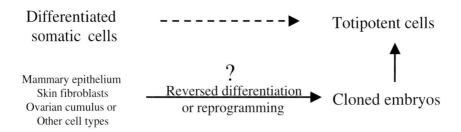

Fig. 2. Schematic drawing of the nuclear reprogramming process. Differentiated somatic cells used in nuclear transfer such as epithelial, fibroblast, cumulus or other cells can be made into cloned embryos. It has been shown that cloned embryos contain totipotent cells because embryonic stem cells can be readily generated from these embryos (Wang et al. *2005; Brambrink* et al. *2006; Wakayama* et al. *2006). Indirectly, the differentiated cells are reprogrammed to become totipotent cells by a yet unknown reversed differentiation process.*

Somatic cloning, however, is challenged with many problems, such as low efficiency, abnormally high rates of fetal death and abortions, premature birth is frequently observed in cloned pregnancies, neonatal death, and placental abnormalities to name a few (Hill *et al.* 1999; Kubota *et al.* 2000; Heyman *et al.* 2002; Xue *et al.* 2002). These observations suggest that nuclear transfer derived fetuses do not develop similarly to *in vivo* or *in vitro* fertilized embryos. These abnormalities are believed to result from incomplete re-activation of genes involved in embryonic development in the donor cells.

Epigenetics

The inactivation of genes during cell differentiation is believed to involve epigenetic modifications of chromatin. Epigenetics is defined as nuclear inheritance that is not based on differences in DNA sequence (Holliday 1987). It is best exemplified by different tissues comprised of cells expressing different proteins while having the same genetic makeup in an individual. Epigenetics is believed to involve differential DNA methylation, histone acetylation, chromatin configuration as well as other mechanisms. These epigenetic signals are stably transmitted during cell division but are reset in each generation in the gonads during fetal development (Goto & Monk 1998; Latham 1999). Therefore, epigenetic signals are not inheritable from one generation to the next but stably maintained within the generation.

To study gene re-activation and reprogramming by nuclear transfer, and to identify genes not expressed properly after cloning, we have employed two complementary approaches: 1) the candidate-gene approach to study individual gene expression; and 2) the gene-panning approach to study global gene expression.

Candidate gene approach

Two main epigenetic modifications of gene expression have been relatively well characterized: genetic imprinting and X chromosome inactivation. Both have been active areas of study in cloned animals because somatic cloning bypasses the natural process of parental specific erasure and re-establishment of epigenetic signals (occurs in the gonads). Cloning using somatic cells skips the gonads and epigenetic signal modifications. Genes that are subjected to epigenetic regulations are thus good candidates to study nuclear reprogramming. Additionally, clones of the same donor provide unique experimental materials in that they are genetically identical yet epigenetically different (Eggan *et al.* 2000; Xue *et al.* 2002). A thorough understanding of reprogramming of epigenetically regulated genes in cloned animals will also improve the young and promising technology by revealing the ideal conditions for complete reprogramming of the somatic nucleus.

Genetic imprinted genes

Genomic imprinting is an epigenetic phenomenon in which only one allele of a specific gene is expressed depending on its parental origin, mono-allelic expression (Latham 1999; Ferguson-Smith & Surani 2001). To date, more than 50 imprinted genes have been identified in the mouse and/or human (Dean *et al.* 2003) and many of them are involved in regulation of fetal growth. These genes are epigenetically modified in the gonads during natural reproduction. This is caused by differential "marks", in forms of differential methylation, established on the DNA of sperm and oocytes during gametogenesis. In nuclear transfer, however, both sets of chromosomes are derived from the same donor cell. It is therefore important to study how imprinted genes are regulated in cloned embryos/animals. Furthermore, many of the defects in large offspring syndrome (LOS) are similar to experimentally created imprinting disruptions (bi-allelic expression of imprinted genes) in mice and naturally occurring imprinting diseases in humans. Because most imprinted genes regulate fetal growth and many are essential for normal development, it is likely that some defects, especially LOS, and a portion of embryonic deaths, are caused by imprinting disruptions. We have chosen to study the *H19* and *IGF2R* genes in the cloned animals.

Imprinting status of the *H19* gene in clones

The *H19* gene encodes for an un-translated RNA molecule (Brannan *et al.* 1990) and is one of the best-studied imprinted genes in both the mouse and human. It is expressed from the maternal allele in both species with the paternal allele silent or nearly silent (Rachmilewitz *et al.* 1992; Bartolomei & Tilghman 1997). *H19* is expressed abundantly in the human placenta and in several embryonic tissues (Goshen *et al.* 1993). We identified a single nucleotide polymorphism (SNP) in exon 5 of bovine *H19*, and found that cattle produced by conventional breeding expressed the maternal allele of *H19* (Fig. 3a). In organs of three out of four deceased cloned calves, bi-allelic expression of *H19* was observed; supporting our hypothesis that imprinting disruption is present in cloned animals that suffered from developmental abnormalities at birth (Fig. 3b).

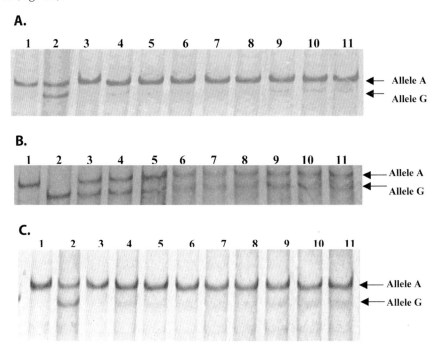

Fig. 3. SSCP images of the allele-specific expression pattern of the *H19* gene in cattle produced by natural reproduction and SCNT. a). Allelic expression of a beef calf: Lanes 1 and 2: Genotypes of a dam and her calf. The calf had two bands indicating the animal was heterozygous for the SNP while the dam only had one band (Allele A) indicating she was homozygous and the calf inherited the A allele from the maternal origin. Lanes 3-9: Expression pattern of *H19* in the calf's liver, kidney, heart, brain, lung, placenta, thymus, bladder, spleen. All organs were either predominantly or exclusively expressing the A allele, which was of maternal origin, indicating the *H19* is imprinted and maternally expressed. b). A representative SSCP image of the allele-specific expression pattern of the bovine *H19* gene in a deceased cloned calf showing bi-allelic expression of *H19*. Lanes 1 and 2: Genotypes of control animals homozygous for the *H19* SNP; Lane 3: genotype of the cloned animal, showing that she was heterozygous for the SNP; Lane 4: genotype of the donor cells; Lane 5: allelic expression of the donor cells, showing bi-allelic expression; Lanes 6-11: brain, heart, liver, lung, spleen and kidney of the cloned animal. c). An SSCP image of the allele-specific expression of *H19* in tissues of a cloned cow's offspring produced by artificial insemination. Lanes 1 and 2: genotypes of the clone's dam (the cloned cow) and her calf by natural reproduction; Lanes 3-11: allelic expression of the liver, kidney, heart, brain, lung, placenta, thymus, bladder, spleen of the clone's calf.

Interestingly, examination of the expression of *H19* in the offspring of a cloned animal produced by artificial insemination showed that the imprinting pattern in this animal was indistinguishable from those of control animals (Fig. 3c), suggesting that either imprinting disruptions in cloned animals are corrected through natural reproduction or that they are not present in healthy cloned animals capable of undergoing natural reproduction.

Insulin-like growth factor 2 receptor (*IGF2R*)

IGF2R, also called cation-independent mannose-6-phosphate receptor, was among the first imprinted genes discovered (Barlow *et al.* 1991). Species variations have been found for the imprinted status of this gene. It has been shown to be maternally expressed in the mouse (Willison 1991; Wutz & Barlow 1998) , sheep (Young *et al.* 2001; Young *et al.* 2003) , cattle (Killian *et al.* 2001b) and pig (Killian *et al.* 2001a) , but not in humans (Riesewijk *et al.* 1996; Wutz *et al.* 1998).

We analyzed allelic expression of *IGF2R* in placentas and organs of ten bovine clones derived from a 13-year-old cow. We found that the maternal *IGF2R* expression pattern of the donor cells was retained in the organs. In contrast, we found random preferential expression of either allele of *IGF2R* in the clones' placentas. Methylation analysis of the putative bovine imprinting control elements of *IGF2R* is underway. Our findings may indicate that independent epigenetic marks may exist for imprinting of *IGF2R*, and that nuclear reprogramming can erase those recognized by the placentas, but not by tissues from the epiblast.

Levels of expression of imprinted genes

By using real time reverse-transcription polymerase chain reaction (RT-PCR), we quantified the expression of the bovine *IGF2*, *IGF2R* and *H19* genes in eight major organs (brain, bladder, heart, kidney, liver, lung, spleen and thymus) of somatic cell cloned calves that died shortly after birth, in three tissues (skin, muscle and liver) of healthy clones that survived to adulthood, and in corresponding tissues of control animals from natural reproduction (Yang *et al.* 2005). We found that deceased bovine cloned calves exhibited abnormal expression of all three genes studied in various organs. Large variations in the expression levels of imprinted genes were also seen among these clones, which were produced from the same genetic donor. In surviving adult clones, however, the expression of these imprinted genes was largely normal, except for the expression of the *IGF2* gene in muscle, which was highly variable. Our data suggest that nuclear transfer can cause disruptions of expression of imprinted genes in bovine clones, possibly due to incomplete reprogramming of donor cell nuclei, and these abnormalities may contribute to the high neonatal mortality in cloned animals; clones that survived to adulthood, however, are not only physically healthy but also relatively normal at the molecular level (Yang *et al.* 2005).

X-linked genes

In mammals, males have one while females have two copies of the X chromosomes. This creates a situation in which there is unequal gene dosage between males and females. During evolution, this was solved by a process termed X chromosome inactivation (XCI) (Heard *et al.* 1997; Lyon 1999), the random transcriptional silencing of one of the two X chromosomes in somatic cells of females during early development. XCI occurs by the process of epigenetic modification, the inactivated X chromosome has hyper-methylated DNA and hypo-acetylated histones. Proper XCI is essential to embryonic development. Inactivation of both X chromosomes in mouse embryos leads to embryonic lethality, and having more than one active X chromosome is deleterious to extra-embryonic development and also causes early embryonic death in mice (Wang *et al.* 2001).

X chromosome inactivation occurs at the late blastocyst-stage in mice. As described in Fig. 4a, before fertilization, the egg carries an active X chromosome that is of maternal origin (Xam), while the sperm carries an inactive X of paternal origin (Xip). At the formation of the female zygote, both X chromosomes become active (XamXap). This state of activation persists through early blastocyst-stage. During late blastocyst-stage, the expression of the X-inactivation specific transcript (*Xist*) gene from one of the two X chromosomes, in a random fashion in the inner cell mass, leads to its inactivation (Xi). In extra-embryonic tissues, however, the paternal X chromosome is preferentially inactivated in the mouse, resulting in imprinted XCI. Once established, the inactive state of a particular X chromosome is epigenetically inherited throughout all subsequent cell divisions (Goto & Monk 1998).

A.

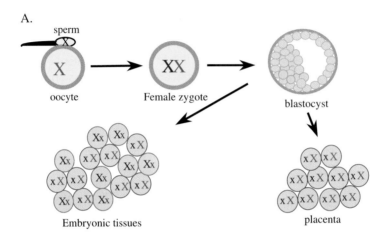

B.

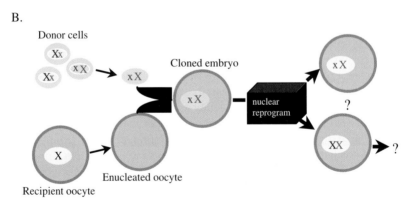

Fig. 4. XCI in early fertilized embryos (a). During natural fertilization, the sperm carries an inactive x (blue, lowercase) while the egg carries an active X (red, uppercase). Both X are active after formation of the female zygote (XX both uppercase). At the time of blastocyst formation, the inner cell mass randomly inactivates one X, either of paternal or maternal origin, resulting in random XCI. In the trophectoderm, which will become the placenta, the paternal X chromosome (blue) is preferentially inactivated, resulting imprinted XCI. b). Question of reprogramming of X chromosome inactivation in cloned embryos. During nuclear transfer, a somatic cell with a pre-existing active and inactive X chromosomes is transferred into an enucleated oocyte. After nuclear reprogramming, which is still largely a black box, it is unclear whether the inactivated X will become re-activated to result in two active X as in the naturally fertilized zygote, or the same pattern of XCI in the somatic donor cell will be maintained in the cloned animal.

In nuclear transfer, the cloned zygotes receive one active (Xa) and one inactive (Xi) X chromosome from the donor cell (Fig. 4b). This state of X inactivation is different from that in naturally fertilized female zygotes, in which both X are active. During nuclear reprogramming, it is unclear whether the inactivated X is re-activated or the pattern of XCI in the donor cell is maintained in the cloned animals. The first study of XCI reprogramming in cloned animals was conducted by Eggan *et al.* (2000) who reported that epigenetic marks on the somatic X chromosomes in mice were completely erased and then appropriately reestablished by the nuclear reprogramming process, leading to normal random XCI in the cloned embryos. The question remains whether or not this is universal across all species. We conducted an extensive study on XCI in cattle and established patterns of XCI in female cattle from natural reproduction (Xue *et al.* 2002). By following the allele specific expression of the X-linked house-keeping gene monoamine oxidase type A (*MAOA*), we found that, as in the mouse, XCI in cattle somatic cells is also random and is paternally imprinted in the placenta (Fig. 5). In cloned calves that died shortly after birth, however, expression of both alleles of the *MAOA* gene was observed, suggesting aberrant XCI. We also conducted a series of studies to examine the normalcy of XCI reprogramming in 9 full-term calves cloned from different cell types using 10 X-linked genes sampled from various available organs. We examined allele specific expression of *MAOA*, and the expression of 9 additional X-linked genes, in major organs or in skin and blood of these 9 full-term cloned XX calves. Surprisingly, we found aberrant expression patterns in 9 out of 10 X-linked genes in all deceased clones. Inactivation of both alleles of several X-linked genes was observed in organs of all 5 deceased clones. Interestingly, the transducin (beta) like 1 (*TBL1*) a gene known to escape XCI in humans and mice (Bassi *et al.* 1999; Carrel *et al.* 1999) was expressed in all organs of these clones.

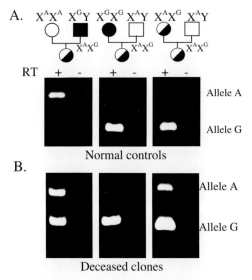

Fig. 5. Allele-specific expression of *MAOA* and expression of *XIST* in bovine placenta. a) three pedigrees showing the inheritance of *MAOA* gene during natural reproduction. circle = female, square = male, X^A = X chromosome carrying the A allele (clear), X^G = X chromosome carrying the G allele (filled), half-filled circle = heterozygous female. The three informative daughters are heterozygous for the *MAOA* gene, and the banding patterns of *MAOA* in their placentas demonstrate mono-allelic expression of the maternal allele of this gene. RT was added (+) or omitted (-) as a control for DNA contamination in RT-PCR. b) RTPCR-RFLP of *MAOA* showing expression of both alleles A and G in the placentas of deceased clones CA, E2 and I.

Consistent with observations of aberrant XCI in internal organs, we also found random XCI in the placenta of all deceased clones examined (Xue *et al.* 2002). Placental abnormalities have been reported in both live and deceased cloned calves (Hill *et al.* 1999; Hill *et al.* 2000).

The aberrant XCI in bovine clones may have resulted from incomplete erasure of the epigenetic marks on the X chromosomes of the somatic donor cells during nuclear reprogramming, which in turn may lead to only partial reactivation of the Xi or silencing of both X chromosomes prior to XCI. Upon differentiation, those epigenetic marks already present on the X chromosomes of the cloned embryo may interfere with the ones further imposed during XCI in clone development, ultimately leading to the observed aberrant expression patterns of X-linked genes.

The interesting finding that *TBL1*, a gene that escapes XCI, was properly expressed in all clones may indicate that regions of the X chromosomes not subjected to XCI, thus not epigenetically modified, are less affected by events involved in nuclear reprogramming. These data are consistent with the abnormal levels of DNA methylation found in cloned embryos and fetuses discussed earlier.

The global gene expression approach

Candidate gene expression studies can only analyze a handful of genes at a time; therefore, the study of global gene expression at early embryonic stages is a powerful approach to study nuclear reprogramming. We used a 7,872 cattle cDNA microarray to compare gene expression profiles between cloned and control blastocyst-stage embryos (Smith *et al.* 2005). This microarray was primarily derived from the bovine placental and spleen cDNA libraries, and was able to detect the expression of approximately 3,500 genes in the early embryos. In conjunction with linear amplification, individual NT embryos were compared to 1) their donor fibroblast cells; and 2) to control embryos created *in vivo* by artificial insemination (AI) and *in vitro* fertilization (IVF). Genes that differed by $\geq$ two fold (ANOVA $P <$ 0.05 with False Discovery Rate (FDR) correction) in relative expression levels were considered differentially expressed.

Surprisingly, the NT embryos' gene expression profiles were drastically different from those of the donor cells (Fig. 6a). A total of 1,546 genes were differentially expressed, representing 29% of the total genes analyzed (n = 5,356). Among these, 751 were up-regulated and 795 were down-regulated in the donor cells versus the NT embryos. Gene Ontology (GO) analysis revealed that over-represented categories among genes up-regulated in the NT embryos were: "carrier activity," "mitochondrial inner membrane," "primary active transporter activity," "RNA splicing" and "ion transporter activity."

Because early embryos such as those generated from NT are expected to contain totipotent cells, we analyzed the expression of 94 genes on the microarray known to be highly enriched in human and mouse ES cells (Ivanova *et al.* 2002; Ramalho-Santos *et al.* 2002; Sato *et al.* 2003; Abeyta *et al.* 2004). Genes previously characterized as ES cell-specific: *ODC1*, *PECAM1* and *CCNE1* (Kelly *et al.* 2000) and an additional 20 genes had significantly higher expression in the NT embryos compared to the differentiated donor cells, suggesting that the enucleated bovine oocytes reprogrammed the differentiated fibroblast nuclei to totipotency. Additionally, the dissimilarity of gene expression profiles between the donor cells and NT embryos indicates that significant nuclear reprogramming of the donor cell nuclei is evident at the blastocyst stage after cloning.

Another surprising observation was that the gene expression profiles of the NT and AI embryos were more similar than those of the IVF and AI embryos (Fig. 6b). The correlation coefficient between the NT embryos and AI embryos was 0.808, but was only 0.714 between the IVF and AI embryos. One of the most interesting findings was that substantially less variation was found among individual NT embryos, with the correlation coefficient being 0.838, as opposed to 0.733 observed among IVF embryos and 0.812 among the AI embryos. This exhibited lower variability among the NT embryos indicates that by the blastocyst stage NT embryos of the same donor animal behaved similarly at the molecular level, while AI and IVF embryos are more variable due to different genetic backgrounds.

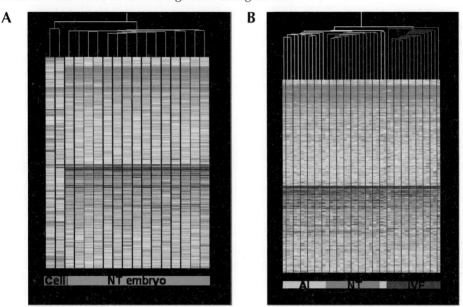

Fig. 6. Hierarchical cluster of global gene expression profiles. a) Cluster of donor cells (pink) and NT embryos (blue) reveals a difference of 84.2% in gene expression over 5,356 genes. b) Cluster of AI (yellow), NT (blue) and IVF (red) embryos illustrates more similarity between AI and NT embryos, with IVF embryos clustering as the outlier (n = 5174 genes). Color indicates the normalized expression values (sample: standard reference). Red equals high expression in the sample compared to the standard reference; yellow, equal expression; and green, low expression.

Even though the expression profiles between the NT embryos and the *in vivo* produced AI embryos displayed the most similarity, 50 out of 5,174 analyzed genes were identified as differentially expressed. Among these differentially expressed genes, eight were differentially expressed in the NT embryos when compared to both the AI and IVF embryos; this could be the result of a specific effect of nuclear reprogramming or it is possible that these genes were expressed from regions of the donor DNA still yet incompletely reprogrammed. Similarly, 17 genes were differentially expressed in the AI embryos versus both NT and IVF embryos; these genes may be important for the high developmental potential of AI embryos. Lastly, 25 genes were differentially expressed only between the AI and NT embryos. Some of these uniquely expressed genes (*COL4A1, DUSP6, FOLR1, MEIS2, MITF* and *TFAP2A*) are involved in development and down-regulated in the NT embryos. Thus, these genes could be potential candidates for perpetuating the abnormal development and mortality observed in NT fetuses.

Although many of abnormalities in cloned animals suggest imprinting disruptions (Mann *et al.* 2003; Ogawa *et al.* 2003), the expression of the 20 out of the 21 imprinted genes on the microarray were similarly expressed in the AI, IVF and NT embryos. Only *CD81*, a gene imprinted in the mouse placenta (Lewis *et al.* 2004), was differentially expressed between the NT and AI embryos. The observation indicates that the other 20 imprinted genes examined were either properly reprogrammed in the NT embryos or that imprinting has not been established in bovine embryos at this stage. The former possibility is more likely because eleven of the imprinted genes studied—CD81, COPG2, DCN, GNAS, GRB10, IGF2R, MEST, PEG3, PLAGL1, SDHD and SGCE—were significantly differentially expressed between the donor cells and the NT embryos. Interestingly, the differentially expressed imprinted genes were not located on the same bovine chromosomes, suggesting that there was no clustering effect of imprinting reprogramming.

Aberrant expression of X-linked genes has been previously reported in bovine NT embryos and the tissues of deceased clones (Wrenzycki et al. 2002; Xue et al. 2002). Interestingly, no X-linked genes (n = 123 analyzed genes) were identified as differentially expressed between the NT and AI embryos. Previously in the bovine, it has been determined that by day 14-15, XCI is completely established in vivo (De La Fuente et al. 1999). Therefore, it is possible that XCI was not complete in the blastocyst-stage embryos examined here.

During normal bovine preimplantation embryonic development, dramatic methylation reprogramming takes place. These precise events could be difficult to recapitulate after nuclear transfer and indeed bovine NT embryos have been shown to be abnormally hypermethylated (Bourc'his et al. 2001; Kang et al. 2001; Dean et al. 2003) . We therefore sought to study genes that are involved in methylation regulation: ATF7IP, DMAP1, DNMT2, DNMT3A, DNMT3B, FOS, MBD4, MIZF and p66alpha. These genes were not differentially expressed among the three embryo types. This was consistent with our observation of similar expression of imprinted and X-linked genes between NT and normal AI embryos, and further indicates that the methylation regulation involved in nuclear reprogramming is not deficient in the NT embryos. Additional support for this observation came from the finding that both the de novo methyltransferases, DNMT3A and DNMT3B, were very highly and consistently expressed in the AI, NT and IVF embryos. This level of expression was not seen in the donor cells, signifying that these de novo methyltransferases were properly reprogrammed in the NT embryos at the blastocyst stage.

Chromatin remodeling and modification is crucial for mammalian development and efficient nuclear reprogramming (Li 2002). Twenty-six genes associated with chromatin modification and epigenetic regulation were examined: ARID1A, ASF1A, BAT8, BAZ1B, CHD4, CHRAC1, CPA4, CTCF, CUGBP1, HDAC1, 2, 3, 7A, L3MBTL, MLL3, MSL3L1, MYST1, 4, RBM14, RPS6KA5, SET07, SIRT5, SMARCA5, SMARCC1, SMARCD3 and TRIM28. No differential expression was seen among the AI, IVF and NT embryos. However in the NT embryos and donor cell comparison, ASF1A, BAZ1B, HDAC1, MLL3, RPS6KA5 and TRIM28 were up-regulated in the NT embryos and HDAC7A and SMARCD3 were up-regulated in the donor cells. This indicates that proper reprogramming of genes important for chromatin modification took place.

Conclusions

In summary, our data on global gene expression documented that the NT embryos' gene expression profiles were vastly different from those of their donor cells and look a great deal like those of AI embryos, suggesting that reprogramming of the differentiated somatic cell by the oocyte cytoplasm is effective and relatively complete by the blastocyst stage. This conclusion is supported by three lines of evidence: 1) blastocyst development rates of NT embryos are similar to those of IVF embryos, suggesting NT embryos are as competent as embryos fertilized by natural gametes in regards of preimplantation development; 2) embryonic stem cell (ESC) lines can be derived from NT embryos (ntESC) with high efficiency, suggesting NT embryos are reprogrammed and contain totipotent cells; 3) ntESC have similar global gene expression patterns as ESC's derived from fertilized embryos (Brambrink et al. 2006; Wakayama et al.

2006). Combined with our results obtained by the candidate gene approach in cloned fetuses and neonates, we hypothesize that the commonly observed low developmental efficiency of NT embryos is potentially caused by abnormal gene reprogramming during post-implantation fetal/placental development (i.e., gene re-differentiation). This hypothesis is supported by at least 2 lines of evidence: 1) the majority of the failure in NT embryo development occurs after implantation and/or around placentation (Heyman *et al.* 2002); 2) the small number of differentially expressed genes found in our microarray study were mainly involved in tissue differentiation/development, but not in pre-implantation development. Further research is required to determine if the aberrant expression of these genes at the blastocyst stage are magnified downstream in development.

Acknowledgments

This work is supported by funds from NIH and USDA to XCT and XY. The microarray study was conducted in collaboration with Drs. Harris A. Lewin and Robin E. Everts of the University of Illinois, Urbana-Champaign.

References

Abeyta MJ, Clark AT, Rodriguez RT, Bodnar MS, Pera RA & Firpo MT 2004 Unique gene expression signatures of independently-derived human embryonic stem cell lines. *Human Molecular Genetics* **13** 601-608.

Barlow DP, Stoger R, Herrmann BG, Saito K & Schweifer N 1991 The mouse insulin-like growth factor type-2 receptor is imprinted and closely linked to the Tme locus. *Nature* **349** 84-7.

Bartolomei MS & Tilghman SM. 1997. Genomic imprinting in mammals. *Annual Review of Genetics* **31** 493-525.

Bassi MT, Ramesar RS, Caciotti B, Winship IM, De Grandi A, Riboni M, Townes PL, Beighton P, Ballabio A & Borsani G 1999 X-linked late-onset sensorineural deafness caused by a deletion involving OA1 and a novel gene containing WD-40 repeats. *American Journal of Human Genetics* **64** 1604-1616.

Bourc'his D, Le Bourhis D, Patin D, Niveleau A, Comizzoli P, Renard JP & Viegas-Pequignot E 2001 Delayed and incomplete reprogramming of chromosome methylation patterns in bovine cloned embryos. *Current Biology* **11** 1542-1546.

Brambrink T, Hochedlinger K, Bell G & Jaenisch R 2006 ES cells derived from cloned and fertilized blastocysts are transcriptionally and functionally indistinguishable. *Proceedings of the National Academy of Sciences of the United States of America* **103** 933-938.

Brannan CI, Dees EC, Ingram RS & Tilghman SM 1990 The product of the H19 gene may function as an RNA. *Molecular and Cellular Biology* **10** 28-36.

Carrel L, Cottle AA, Goglin KC & Willard HF 1999 A first-generation X-inactivation profile of the human X chromosome. *Proceedings of the National Academy of Sciences of the United States of America* **96** 14440-14444.

De La Fuente R, Hahnel A, Basrur PK & King WA 1999 X inactive-specific transcript (Xist) expression and X chromosome inactivation in the preattachment bovine embryo. *Biology of Reproduction* **60** 769-775.

Dean W, Santos F & Reik W 2003 Epigenetic reprogramming in early mammalian development and following somatic nuclear transfer. *Seminars in Cell & Developmental Biology* **14** 93-100.

Eggan K, Akutsu H, Hochedlinger K, Rideout W, 3rd, Yanagimachi R & Jaenisch R 2000 X-Chromosome inactivation in cloned mouse embryos. *Science* **290** 1578-1581.

Ferguson-Smith AC & Surani MA 2001 Imprinting and the epigenetic asymmetry between parental genomes. *Science* **293** 1086-1089.

Goshen R, Rachmilewitz J, Schneider T, de-Groot N, Ariel I, Palti Z & Hochberg AA 1993 The expression of the H-19 and IGF-2 genes during human embryogenesis and placental development. *Molecular Reproduction and Development* **34** 374-379.

Goto T & Monk M 1998 Regulation of X-chromosome inactivation in development in mice and humans. *Microbiology and Molecular Biology Reviews* **62** 362-378.

Heard E, Clerc P & Avner P 1997 X-chromosome inactivation in mammals. *Annual Review of Genetics* **31** 571-610.

Heyman Y, Chavatte-Palmer P, LeBourhis D, Camous S, Vignon X & Renard JP 2002 Frequency and occurrence of late-gestation losses from cattle cloned embryos. *Biology of Reproduction* **66** 6-13.

Hill JR, Burghardt RC, Jones K, Long CR, Looney CR, Shin T, Spencer TE, Thompson JA, Winger QA & Westhusin ME 2000 Evidence for placental abnormality as the major cause of mortality in first-trimester somatic cell cloned bovine fetuses. *Biology of Reproduction* **63** 1787-1794.

Hill JR, Roussel AJ, Cibelli JB, Edwards JF, Hooper NL, Miller MW, Thompson JA, Looney CR, Westhusin ME, Robl JM & Stice SL 1999 Clinical and pathologic features of cloned transgenic calves and fetuses (13 case studies). *Theriogenology* **51** 1451-1465.

Holliday R 1987 The inheritance of epigenetic defects. *Science* **238** 163-170.

Ivanova NB, Dimos JT, Schaniel C, Hackney JA, Moore KA & Lemischka IR 2002 A stem cell molecular signature. *Science* **298** 601-604.

Kang YK, Koo DB, Park JS, Choi YH, Chung AS, Lee KK & Han YM 2001 Aberrant methylation of donor genome in cloned bovine embryos. *Nature Genetics* **28** 173-177.

Keeton W & Gould J *1984 Biological Science. New York, NY: WW Norton and Company, Inc.*

Kelly DL & Rizzino A 2000 DNA microarray analyses of genes regulated during the differentiation of embryonic stem cells. *Molecular Reproduction and Development* **56** 113-123.

Killian JK, Buckley TR, Stewart N, Munday BL & Jirtle RL 2001a Marsupials and Eutherians reunited: genetic evidence for the Theria hypothesis of mammalian evolution. *Mammalian Genome* **12** 513-517.

Killian JK, Nolan CM, Wylie AA, Li T, Vu TH, Hoffman AR & Jirtle RL 2001b Divergent evolution in M6P/IGF2R imprinting from the Jurassic to the Quaternary. *Human Molecular Genetics* **10** 1721-1728.

Kubota C, Yamakuchi H, Todoroki J, Mizoshita K, Tabara N, Barber M & Yang X 2000 Six cloned calves produced from adult fibroblast cells after long-term culture. *Proceedings of the National Academy of Sciences of the United States of America* **97** 990-995.

Latham KE 1999 Epigenetic modification and imprinting of the mammalian genome during development. *Current Topics in Developmental Biology* **43** 1-49.

Lewis A, Mitsuya K, Umlauf D, Smith P, Dean W, Walter J, Higgins M, Feil R & Reik W 2004 Imprinting on distal chromosome 7 in the placenta involves repressive histone methylation independent of DNA methylation. *Nature Genetics* **36** 1291-1295.

Li E 2002 Chromatin modification and epigenetic reprogramming in mammalian development. *Nature Reviews. Genetics* **3** 662-673.

Lyon MF 1999 X-chromosome inactivation. *Current Biology* **9** R235-237.

Mann MR, Chung YG, Nolen LD, Verona RI, Latham KE & Bartolomei MS 2003 Disruption of imprinted gene methylation and expression in cloned preimplantation stage mouse embryos. *Biology of Reproduction* **69** 902-914.

Ogawa H, Ono Y, Shimozawa N, Sotomaru Y, Katsuzawa Y, Hiura H, Ito M & Kono T 2003 Disruption of imprinting in cloned mouse fetuses from embryonic stem cells. *Reproduction* **126** 549-557.

Rachmilewitz J, Gileadi O, Eldar-Geva T, Schneider

T, de-Groot N & Hochberg A 1992 Transcription of the H19 gene in differentiating cytotrophoblasts from human placenta. *Molecular Reproduction and Development* **32** 196-202.

Ramalho-Santos M, Yoon S, Matsuzaki Y, Mulligan RC & Melton DA 2002 "Stemness": transcriptional profiling of embryonic and adult stem cells. *Science* **298** 597-600.

Riesewijk AM, Schepens MT, Welch TR, van den Berg-Loonen EM, Mariman EM, Ropers HH & Kalscheuer VM. 1996. Maternal-specific methylation of the human IGF2R gene is not accompanied by allele-specific transcription. *Genomics* **31** 158-166.

Sato N, Sanjuan IM, Heke M, Uchida M, Naef F & Brivanlou AH 2003 Molecular signature of human embryonic stem cells and its comparison with the mouse. *Developmental Biology* **260** 404-413.

Smith SL, Everts RE, Tian XC, Du F, Sung LY, Rodriguez-Zas SL, Jeong BS, Renard JP, Lewin HA & Yang X 2005 Global gene expression profiles reveal significant nuclear reprogramming by the blastocyst stage after cloning. *Proceedings of the National Academy of Sciences of the United States of America* **102** 17582-17587.

Waddington C *1940 The temporal course of gene reactions, organisers and genes. Cambridge, UK: Cambridge University Press.*

Wakayama S, Jakt ML, Suzuki M, Araki R, Hikichi T, Kishigami S, Ohta H, Van Thuan N, Mizutani E, Sakaide Y, Senda S, Tanaka S, Okada M, Miyake M, Abe M, Nishikawa SI, Shiota K & Wakayama T 2006 Equivalency of Nuclear Transfer-Derived Embryonic Stem Cells to those Derived from Fertilized Mouse Blastocyst. *Stem Cells* **24** 2023-2033.

Wang J, Mager J, Chen Y, Schneider E, Cross JC, Nagy A & Magnuson T 2001 Imprinted X inactivation maintained by a mouse Polycomb group gene. *Nature Genetics* **28** 371-375.

Willison K 1991 Opposite imprinting of the mouse Igf2 and Igf2r genes. *Trends in Genetics* **7** 107-109.

Wrenzycki C, Lucas-Hahn A, Herrmann D, Lemme E, Korsawe K & Niemann H 2002 In vitro production and nuclear transfer affect dosage compensation of the X-linked gene transcripts G6PD, PGK, and Xist in preimplantation bovine embryos. *Biology of Reproduction* **66** 127-134.

Wutz A & Barlow DP 1998 Imprinting of the mouse Igf2r gene depends on an intronic CpG island. *Molecular and Cellular Endocrinology* **140** 9-14.

Wutz A, Smrzka OW & Barlow DP 1998 Making sense of imprinting the mouse and human IGF2R loci. *Novartis Foundation Symposium* **214** 251-259; discussion 260-263.

Xue F, Tian XC, Du F, Kubota C, Taneja M, Dinnyes A, Dai Y, Levine H, Pereira LV & Yang X 2002 Aberrant patterns of X chromosome inactivation in bovine clones. *Nature Genetics* **31** 216-220.

Yang L, Chavatte-Palmer P, Kubota C, O'Neill M, Hoagland T, Renard JP, Taneja M, Yang X & Tian XC 2005 Expression of imprinted genes is aberrant in deceased newborn cloned calves and relatively normal in surviving adult clones. *Molecular Repro-duction and Development* **71** 431-438.

Young LE, Fernandes K, McEvoy TG, Butterwith SC, Gutierrez CG, Carolan C, Broadbent PJ, Robinson JJ, Wilmut I & Sinclair KD 2001 Epigenetic change in IGF2R is associated with fetal overgrowth after sheep embryo culture. *Nature Genetics* **27** 153-154.

Young LE, Schnieke AE, McCreath KJ, Wieckowski S, Konfortova G, Fernandes K, Ptak G, Kind AJ, Wilmut I, Loi P & Feil R 2003 Conservation of IGF2-H19 and IGF2R imprinting in sheep: effects of so-matic cell nuclear transfer. *Mechanisms of Development* **120** 1433-1442.

Gene expression analysis of single preimplantation bovine embryos and the consequence for developmental potential

NT Ruddock-D'Cruz[1], VJ Hall[2], RT Tecirlioglu[3] and AJ French[4]

[1]Centre for Reproduction and Development, Monash University, Victoria, Australia, [2]Neuronal Survival Unit, Wallenberg Neuroscience Centre, Lund University, Sweden, [3]Monash Immunology and Stem Cell Laboratories, Monash University, Victoria, Australia, [4]Stemagen Co., La Jolla, California, USA.

Preimplantation embryo development typically involves sequential morphological events connecting embryonic cleavage, morula compaction and blastocyst formation, and occurs in parallel with transcriptional regulation, specifically, the maternal to embryonic transition. The underlying homeostatic and metabolic mechanisms governing embryo development are influenced by both genetic and epigenetic factors that respond to environmental stimuli and may impact development during later gestational and fetal growth. There is a renewed interest in the identification and characterization of developmentally important genes during embryonic and fetal development. Perturbations in gene expression, resulting from environmental conditions, can have serious consequences on further embryonic development, homeostasis and disease pathogenesis. The bovine embryo is, however, capable of tolerating and adapting to a wide range of conditions, although little is known of the molecular fingerprint required for oocyte maturation, fertilization and development to term. The genomic revolution united with promising new technologies offer greater opportunity to elucidate the mechanisms behind this well-orchestrated biological process. This paper reviews the current literature on gene expression in the bovine embryo with reference to environmental interference and the development of new technologies to observe this biological process. Defining the difference in molecular signalling between in vivo and in vitro systems will undoubtedly improve the safety and efficiency of assisted reproductive technologies. The future challenge is to devise culture conditions that mimic the changing environment required by developing embryos to allow the correct temporal and spatial expression of a cohort of developmental genes in a manner similar to that seen in vivo.

---

[1]Corresponding author E-mail: afrench@stemagen.com

Introduction

The initiation of mammalian embryogenesis is regulated by a complex network involving the oocyte genome, transcriptome and proteome. In the absence of new transcription (Davidson 1986), completion of the first meiotic and mitotic cell cycles, gamete reprogramming (oocyte and sperm) and the maternal to embryonic transition rely on transcripts and proteins made during oocyte growth, as well as signal transduction events associated with maturation, ovulation and fertilization (Knowles *et al.* 2003). During preimplantation development in the mouse, an estimated 15,700 genes are expressed (Stanton *et al.* 2003), and it is likely that a similar number will be expressed in other mammalian species. This preimplantation period culminates in a synchronous and intricate discourse between the embryo and the receptive uterus, resulting in implantation and the pathway to further embryonic development (Wang & Dey 2006).

In vitro production (IVP) technologies provide an alternative source of oocytes and embryos for both research and routine embryo transfer. Historically, IVP success has been primarily gauged on the morphological assessment of the preimplantation embryo. However, while the true developmental competence of any given embryo is a continuum that proceeds throughout its lifecycle, the attainment of full term development is a critical first milestone. This process particularly in a uniparous animal, such as the bovine, is restricted by the long gestational interval and the reliance on intensive recipient management programs, which requires significant capital investment. The relevance is no more apparent then when IVP embryos upon transfer to recipients show similar implantation rates when compared to *in vivo* embryos but then undergo significant embryonic and fetal losses (Reichenbach *et al.* 1992). A proportion (up to 30%) of the surviving animals also show increased birth weight and other anatomical abnormalities that have been described as the large offspring syndrome (Holm *et al.* 1996; Young *et al.* 1998; Renard *et al.* 1999; Niemann & Wrenzycki 2000; Sinclair *et al.* 2000). While bovine preimplantation embryos appear capable of tolerating and adapting to wide ranging environment stimuli, a relative short exposure to sub-optimal *in vitro* conditions can initiate a range of downstream consequences (Wrenzycki *et al.* 2004).

The development of expression analysis techniques to examine the cohort of essential and developmentally important genes during early mammalian development provides a useful method to assess the normality of embryo development and to allow *in vitro* culture and assisted reproductive technologies to be examined and improved without the requirement for extensive *in vivo* testing. The interplay between the effect of microenvironment modifications and epigenetic alternations at early stages of development suggest that limited *in vivo* testing will still be required and not suppressed from future animal studies.

The consequence of temporal or spatial and qualitative or quantitative shifts in gene expression patterns can influence the well-orchestrated events controlling resumption of meiosis, initiation of embryo cleavage, maternal to embryonic transition, and cellular differentiation during and well beyond blastocyst formation.

What is becoming increasingly apparent is that the intrinsic quality of the oocyte plays a key factor in the overall success of these events (Lonergan *et al.* 2003a). However, the molecular fingerprint of an oocyte that is capable of undergoing maturation, fertilization and supporting development to term is virtually unknown.

Defining this profile will likely require comparative gene expression studies involving single embryo analyses, to account for the variability between embryos, with groups of embryos in the same environment where the outcome is to increase the mean behaviour of the group. It would appear however that many of the genes analysed (see Table 1) show consistent expression patterns as a consequence of environmental conditions in both the single and pooled embryo studies. Modifications to culture conditions would necessitate a balance to the envi-

Table 1. Gene expression in bovine oocytes and preimplantation embryos

Gene	Expression Profile (Graph Relative Transcript Abundance) in bovine oocytes and embryos									Observations (Single (S) or Pooled (P) oocytes or embryos for analyses)	Reference
	Follicle	Oocyte	Zygote	2-cell	4-cell	8-cell	16-cell	Morula	Blast.		
Acrogranin										Aberrant expression in SCNT. Bl Only (S)	(Hall et al. 2005b)
Bax	n.a	+	+	+	+	n.a.	n.a.	n.a.	+	High in poor quality oocyte and embryos (Ref 2, (P)) and (Ref 1, (S))	(Yang & Rajamahendran 2002; Li et al. 2006)
Bcl2	n.a.	+	+	+	+	n.a.	n.a.	n.a.	+	High in good quality oocyte and embryos (Ref 2, (P)) and (Ref 1, (S))	(Yang & Rajamahendran 2002; Li et al. 2006)
bFGF									+	Up Regulated in IVP Embryo. Bl Only (S)	(Lazzari et al. 2002)
BMP15	+	+	+	+	+	+	-	-	-	(P)	(Pennetier et al, 2004)
Cdx2									+	Aberrant expression in SCNT. Bl Only (S)	(Hall et al. 2005b)
Chop-10									+	(S)	(Li et al. 2006)
Cu/Zn-SOD									+	Up Regulated in IVP blastocysts. Bl Only (S)	(Lazzari et al. 2002)
Cx31									+	Bl Only (P)	(Rizos et al. 2004)
Cx43				_(bar graph)_	_(bar graph)_	_(bar graph)_	_(bar graph)_		+	(Ref 1, (P)) and (Ref 2, (S))	(Gutierrez-Adan et al. 2001; Li et al. 2006)
Dc II								+	+	Variable expression in IVP morula and blastocyst (Ref 2, (P) and Ref 1, (S))	(Wrenzycki et al. 2001a ; Knijn et al. 2005)
Dc III								+	+	Up regulation in IVP morula (P)	(Wrenzycki et al. 2001a)
Dlk1	n.a.	-	-	-	-	-	-	-	+	(S)	(Ruddock et al. 2004)
Dnmt1									+	Up regulation in IVP blastocyst. Bl Only (S)	(Li et al. 2006)
Dnmt3a									+	Gene expression variation according to IVP system. Up regulation in SCNT Blastocysts. Bl Only (P)	(Wrenzycki & Niemann 2003)

Table 1. Contd.

Gene	Follicle	Oocyte	Zygote	2-cell	4-cell	8-cell	16-cell	Morula	Blast.	Observations (Single (S) or Pooled (P) oocytes or embryos for analyses)	Reference
E-cad								+	+	Decrease in IVP morula (Ref4 (P), Ref 3 (S), Ref 1 (P) and Ref 2 (S))	(Wrenzycki et al. 2001a;b; Rizos et al. 2004; Li et al. 2006)
Eomes									+	BI Only (S)	(Hall et al. 2005b)
ErbB3									+	BI Only (S)	(Hall et al. 2005b)
ERR2									+	Aberrant expression in SCNT. BI Only (S)	(Hall et al. 2005b)
FGF2	n.a.	+	n.a.	+	+	+	n.a.	+	+	(S)	(Daniels et al. 2000; 2001)
FGF4	n.a.	−	n.a.	−	−	−	n.a.	+	+	Decreased expression resulting from SCNT procedure (S)	(Daniels et al. 2000; 2001)
FGFr2	n.a.	+	n.a.	+	+	−	n.a.	−	+	(S)	(Daniels et al. 2000; 2001)
G6PD								+	+	Skewed sex ratio in IVP. Higher expression in Female IVP embryos (S)	(Wrenzycki et al. 2002)
GDF-9	+	+	+	+	+	+	+	−	−	(Ref 1 (P) and Ref 2 (S))	(Pennetier et al. 2004 ; Ruddock et al. 2004)
Glut-1				+	+	+	+	+	+	Down regulated in IVP morula and blastocyst (Ref 2 (P), Ref 1, 3-4 (S))	(Wrenzycki et al. 2001a; Bertolini et al. 2002; Li et al. 2006; Oliveira et al. 2006)
Glut-3									+	Up regulated expression in IVP Blastocysts. BI Only (S)	(Lazzari et al. 2002)
Glut-4									+	Up regulated expression in IVP Blastocysts. BI Only (S)	(Lazzari et al. 2002)
Glut-5					(bar graph)	(bar graph)	(bar graph)			(P)	(Gutierrez-Adan et al. 2001)
Glut-8									+	Up regulated expression in IVP Blastocysts. BI Only (S)	(Knijn et al. 2005)

Table 1. Contd.

Gene	Expression Profile (Graph Relative Transcript Abundance) in bovine oocytes and embryos									Observations (Single (S) or Pooled (P) oocytes or embryos for analyses)	Reference
	Follicle	Oocyte	Zygote	2-cell	4-cell	8-cell	16-cell	Morula	Blast.		
Gnas	n.a.	–	–	–	–	+	+	+	+	Detected following MET(S)	(Ruddock et al. 2004)
Gp130	n.a.	+	n.a.	+	+	–	n.a.	+	+	Ref 2-3 (S) and Ref 1 (S and P)	(Eckert & Niemann 1998; Daniels et al. 2000; 2001)
Grb10	n.a.	+	+	+	+	+	+	+	+	(S)	(Ruddock et al, 2004)
Hand1									+	Bl Only (S)	(Hall et al. 2005b)
Hsp70.1	n.a.	–	–	–	–	–	–	+	+	Up regulation in IVP and SCNT embryos (S)	(Lazzari et al. 2002; Li et al. 2006)
IFN-ζ	n.a.	–	–	–	–	–	–	+	+	(Ref 2 (S) and Ref 1 (P))	(Rizos et al. 2004; Li et al. 2006)
IGF-1R				*(bar graph)*	*(bar graph)*	*(bar graph)*	*(bar graph)*			(Ref 1 (P) and Ref 2 (S))	(Gutierrez-Adan et al. 2001; Li et al. 2006)
IGF-IIR									+	Down regulated in female IVP blastocysts Bl Only (P)	(Bertolini et al. 2002)
IGF-I									+	Bl Only(P)	(Bertolini et al. 2002)
IGF-II				*(bar graph)*	*(bar graph)*	*(bar graph)*	*(bar graph)*			Down regulated in IVP blastocysts (Ref 1-2 (P) and Ref 3 (S))	(Gutierrez-Adan et al. 2001; Bertolini et al. 2002; Li et al. 2006)
Il-6	n.a.	–	n.a.	–	–	–	+	+	+	Decreased expression resulting from donor cell and SCNT procedure (S)	(Daniels et al. 2000; 2001)

Table 1. Contd.

Gene	Expression Profile (Graph Relative Transcript Abundance) in bovine oocytes and embryos									Observations (Single (S) or Pooled (P) oocytes or embryos for analyses)	Reference
	Follicle	Oocyte	Zygote	2-cell	4-cell	8-cell	16-cell	Morula	Blast.		
Jam		▮	▮	▮	▮	▮	▮	▮	▮	Intracellular tight junction implicated in blastocyst formation (S)	(Miller et al. 2003)
Lamin B	–	+	+	+	+	+	+	+	+	GV Only, Not detected in MII (S)	(Hall et al. 2005a)
Lamin A/C	+	+	+	+	+	–	–	+	+	GV Only, Not detected in MII	(Hall et al. 2005a)
LIF	+	+	+	+	n.a.	+	+	+	+	(S) and (P)	(Eckert & Niemann 1998)
LR-ß	+	+	+	+	+	+	+	+	+	(S) and (P)	(Eckert & Niemann 1998)
Mash-2									+	Down regulated in IVP blastocysts. Variable expression in blastocysts from different SCNT methods. Bl Only (S)	(Wrenzycki et al. 2001b)
Mater	+	+	+	+	+	–	–	–	–	(P)	(Pennetier et al. 2004)
Mest (iso 1)	n.a.	–	–	–	–	–	–	–	–	(S)	(Ruddock et al. 2004)
Mest (iso 2)	n.a.	+	+	+	+	+	+	+	+	(S)	(Ruddock et al. 2004)
MnSOD				▮	▮	▮	▮			(Ref 1 (P) and Ref 2 (S))	(Gutierrez-Adan et al. 2001; Li et al. 2006)
MRJ	n.a.	–							+	Bl Only (S)	(Hall et al. 2005b)
Ndn	n.a.	+	–	–	–	+	+	+	+	Detected following MET (S)	(Ruddock et al. 2004)
Nnat			+	+	–	–	–	–	+	Allelic expression differences (S)	(Ruddock et al. 2004)

Table 1. Contd.

Gene	Expression Profile (Graph Relative Transcript Abundance) in bovine oocytes and embryos	Observations (Single (S) or Pooled (P) oocytes or embryos for analyses)	Reference
Occludin	Follicle, Oocyte, 8-cell, 16-cell, Morula, Blast.	Intracellular tight junction implicated in blastocyst formation (S)	(Miller *et al.* 2003)
Oct4	Follicle +, Oocyte +, Zygote +, 2-cell +, 4-cell +, 8-cell +, 16-cell +, Morula +, Blast. +	Expression increase after MET (Ref 1-3 (S))	(Daniels *et al.* 2000; 2001; Kurosaka *et al.* 2004)
OOP1	Follicle, Oocyte, 2-cell, 4-cell	Oocyte-specific marker with two splice variants- specifically expressed in bovine (P)	(Tremblay *et al.* 2006)
Pan ZO-1	Follicle, Oocyte, 8-cell, 16-cell, Morula, Blast.	Intracellular tight junction implicated in blastocyst formation (S)	(Miller *et al.* 2003)
Pan ZO-2	Follicle, Oocyte, 8-cell, 16-cell, Morula, Blast.	Intracellular tight junction implicated in blastocyst formation (S)	(Miller *et al.* 2003)

Table 1. Contd.

Gene	Follicle	Oocyte	Zygote	2-cell	4-cell	8-cell	16-cell	Morula	Blast.	Observations (Single (S) or Pooled (P) oocytes or embryos for analyses)	Reference
PGK								+	+	IVP morula and female embryos transcribe more. IVP blastocysts have higher expression (S)	(Wrenzycki et al. 2002)
Plako								+	+	Down regulated in IVP morula and blastocysts (S)	(Wrenzycki et al. 2001a)
Rex1									+	Bl Only (S)	(Hall et al. 2005b)
Sgce	n.a.	+	+	+	+	+	+	−	−	Potential role in oocyte fertilization and/or cleavage (S)	(Ruddock et al. 2004)
SOX				▥	▥	▥	▥			Higher expression in IVP Blastocysts (P)	(Gutierrez-Adan et al. 2001; Rizos et al. 2002a)
XIAP						+	+	+	+	Decreased expression in IVP blastocyts (P)	(Knijn et al. 2005)
Xist	n.a.	−	−	−	−	+	+	+	+	Detected following MZT. Increased expression in SCNT embryos (S)	(Wrenzycki et al. 2002; Ruddock et al. 2004)
ZAR1	▥	▥	▥	▥	▥	▥	▥	▥		Changes in expression around MET (P)	(Brevini et al. 2004)
ZO-1α+	▥	▥				▥	▥	▥	▥	Intracellular tight junction implicated in blastocyst formation (P)	(Brevini et al. 2004; Pennetier et al. 2004)

Table 1. Contd.

Gene	Expression Profile (Graph Relative Transcript Abundance) in bovine oocytes and embryos									Observations (Single (S) or Pooled (P) oocytes or embryos for analyses)	Reference
	Follicle	Oocyte	Zygote	2-cell	4-cell	8-cell	16-cell	Morula	Blast.		
ZO-2ß						▓	▓	▓	▓	Intracellular tight junction implicated in blastocyst formation (S)	(Miller *et al.* 2003)

+ expressed, - not expressed, n.a - not analysed, Bl Only - blastocyst stage examined

ronmental response in the individual embryo against those required for a group of embryos, that's even if correlations between morphological data and gene expression profiling are to be established. Another way to discern this variability would be to analyse gene expression of blastomeres following embryo biopsy. These analyses could be used to modify culture conditions to the single embryo according to how it responds to the environment, although the variability of gene expression between individual (or few) blastomeres is not known.

The continual development of highly sensitive techniques will enable the quantitative profiling of transcriptomes and proteomes, which are necessary to regulate and coordinate events from folliculogenesis to early embryo development and are vital to improving the efficiency of *in vitro* production systems (IVP) and assisted reproductive technologies. To date, research has focussed on embryonic cleavage rates (Lonergan *et al.* 2000; Ward *et al.* 2001; Comizzoli *et al.* 2003; Holm *et al.* 2003), developmental arrest (Yang & Rajamahendran 2002), developmental competence of *in vitro* produced (IVP) embryos (Thompson 1997; Holm & Callesen 1998; Enright *et al.* 2000), embryo manipulation (Wrenzycki *et al.* 2001b), culture (Wrenzycki *et al.* 2001a) and metabolism (Khurana & Niemann 2000; Thompson 2000), embryo genomic activation (De Sousa *et al.* 1998a; Memili & First 2000), embryo sex (Avery *et al.* 1992; Xu *et al.* 1992; Gutierrez-Adan *et al.* 2001), oocyte quality (Lonergan *et al.* 2003a), methylation status (Bourc'his *et al.* 2001), protein synthesis (De Sousa *et al.* 1998b), species differences (Wrenzycki *et al.* 2002), transcript abundance (Watson *et al.* 2000) and the functional organization of the nucleus and nucleolus (remodelling and reprogramming) (Hall *et al.* 2005a ; Corcoran *et al.* 2006).

The degree to which embryo culture or manipulation influences gene expression and the downstream consequences are beginning to be revealed. This review will briefly examine gene expression studies in the bovine preimplantation embryo and the development of new methodologies to elucidate optimal conditions for improving the developmental competence of embryos generated from a variety of IVP systems.

In-Vitro production in the bovine

Despite ongoing improvements, the full potential of the IVP production system remains hampered by the overall quality of the embryo when compared to those derived *in vivo*. This is graphically demonstrated in Fig. 1 where embryo viability until weaning from *in vivo* and *in vitro* production systems are shown from this group over a 5 year period (2000-2005).

Differences between *in vivo* and *in vitro* derived embryos have been reviewed extensively (Thompson 1997; Enright *et al.* 2000; Niemann & Wrenzycki 2000; Lonergan *et al.* 2003a) and the disparity between *in vitro* and *in vivo* embryos can be categorized to either the proportion that reach the blastocyst stage of development or survival following embryo transfer. Around 35% of *in vitro* matured and fertilized (IVM/IVF) oocytes reach the blastocyst stage and around 40% survive to term following transfer when contrasted with over 70% and 60% for the *in vivo* counterparts, respectively (Thompson 1997; Holm & Callesen 1998). Manipulation of IVP embryos for the purpose of Somatic Cell Nuclear Transfer (SCNT) reduces the embryo viability even further (Wrenzycki *et al.* 2004).

The source of variation has been attributed to a variety of causes including the processes of *in vitro* maturation, site of and effect (quality) of sperm at fertilization, chromosome imbalance, polyspermy, variation in pronuclei formation, incorrect morula compaction/ blastocyst formation and sensitivity to cooling and freezing due to elevated lipid content (Thompson 1997; Holm & Callesen 1998; Rizos *et al.* 2002b).

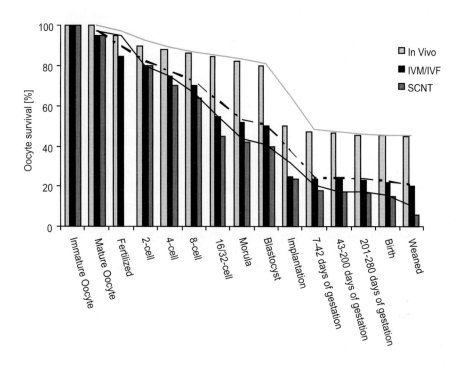

Fig. 1: The developmental competence of bovine embryos derived from *In Vivo* and *In Vitro* production (IVM/IVF and SCNT) systems.

Gene expression analysis in bovine oocytes and embryos

A variety of methods are available for investigating gene expression (see below) and their sensitivity can detect qualitative and quantitative changes to transcriptional regulation during oocyte growth, maturation and embryonic development. Data from these analyses has established molecular expression profiles that may be a more accurate indicator of developmental competence when compared to morphological and blastocyst development observations (Leibfried & First 1979; de Loos *et al.* 1992). The morphological classification and grading of bovine embryos (Lindner & Wright 1983) with minor modifications (Hasler 2001), and subsequently adopted by the International Embryo Transfer (www.iets.org), is the standard to which all embryos are currently described. This practice is used extensively in both research and commercial activities (Thompson 1997; Holm & Callesen 1998; Hyttel 2001). However it has also become increasingly apparent, particularly with the development of SCNT, that morphological data does not correlate solely with developmental competence (see Fig. 1). To battle this discrepancy, much research has recently focused on gene expression profiling in individual oocytes and embryos.

The literature precludes that over 100 genes to date have been associated with developmentally important processes in bovine preimplantation embryos from *in vivo, in vitro* produced, such as genes involved in compaction/cavitation, metabolism, transcription/ translation, DNA methylation and histone modification, oxidative stress, response to or production of growth factors, cytokine signalling, cell cycle regulation and apoptosis. The development of SCNT technology adds a further level of complexity to gene expression studies where the interrelationship between incomplete somatic cell reprogramming and environmental constraints imposed by the in vitro production system need to be distinguished.

These categories of genes have been evaluated in bovine preimplantation embryos from *in vivo, in vitro* produced and SCNT embryos, for reviews see (Corcoran *et al.* 2005; Niemann & Wrenzycki 2000; Lonergan *et al.* 2001; 2006; Niemann *et al.* 2002; Wrenzycki *et al.* 2004; Wrenzycki *et al.* 2005a;b). Many genes in IVP (and SCNT) embryos have displayed aberrant expression patterns compared with their *in vivo* counterparts. Both genetic and epigenetic mechanisms (methylation and histone modifications) are thought to be involved in the differences in gene expression, regardless of the fact that the developing embryo appears capable of enduring substantial dysregulation of both imprinted and non-imprinted genes (Humpherys *et al.* 2001; Reik *et al.* 2001). There are many steps associated with the IVP system including maturation, fertilization, and culture and in addition to this, various manipulations are undertaken for the production of SCNT embryos, all of which have the potential to further alter gene expression patterns in the developing embryo. These noted alterations in gene expression have been associated with the type of medium (maturation, fertilization and culture) and various additive components (growth factors, serum, etc). Interestingly, epigenetic modifications seem more prevalent in imprinted genes (Blondin *et al.* 2000; Ruddock *et al.* 2004). Many of these differences are clearly established after 1 day of culture, underling the sequential interactions between the environment and gene expression (Rizos *et al.* 2002a;b; 2003; Lonergan *et al.* 2003b). Embryo manipulations involved in the production of SCNT embryos have also resulted in varied gene expression patterns and may be associated with oxidative stress, impaired trophoblastic function, DNA methylation and X chromosome inactivation (Wrenzycki *et al.* 2004). The establishment of diagnostic techniques using these predictive values of aberrantly expressed genes as markers for embryo quality and viability is a critical step towards improving the efficiency of each system. One of the first applications of a limited microarray in the bovine showed 18 genes from a subset of intermediate-filament protein, metabolic, lysosomal-related, stress related and major histocompatibility complex class I were differentially expressed between IVF and SCNT blastocysts. (Gutierrez-Adan *et al.* 2001).

A summary of genes detected in single and pooled bovine oocyte /embryo samples is provided in Table 1. However while a considerable body of work has been amassed on the expression of specific genes and the role of genetic and epigenetic reprogramming, there is a need for greater understanding of the relationships between altered phenotypes, changes in both genetic code and epigenetic patterns within the genome, and alterations in mRNA or protein expression profiles and the downstream consequence for embryos generated from a variety of IVP manipulations. Direct interventions are now required to modify specific epigenetic characteristics to correlate with the biological processes associated with developmental competence, including the correct mRNA and protein expression profiles (Lazzari *et al.* 2002; Fernandez-Gonzalez *et al.* 2004; Farin *et al.* 2006).

Gene expression as an indicator of developmental competence

A number of studies have shown similar rates of ATP production, glucose metabolism, pyruvate uptake and utilization between *in vivo* and *in vitro* produced embryos (Thompson 1997; 2000). However, lower total cell counts in the blastocyst and skewed ratios between inner cell mass and trophectoderm cells in IVP embryos highlight that significant developmental differences exist, which may be reflected at the molecular and cellular level. Gene expression comparisons between *in vivo, in vitro* and SCNT embryos have identified a number of genes associated with developmental competence. These include the relative abundance of transcripts encoding the α1 subunit of Na1/K1-ATPase (De Sousa *et al.* 1998c), impaired cavitation and blastocyst formation induced by a decrease in expression of connexin (*Cx43*) in the IVP blastocyst (Rizos *et al.* 2002a); absence of FGF4 expression in SCNT blastocysts (Daniels *et al.* 2000; 2001), altered expression of lamin A/C and B following fusion and early cleavage that was restored by morula stage in SCNT embryos when compared to IVF (Hall *et al.* 2005a). The developmental competence of oocytes and embryos is influenced by the high expression of Bcl-2 gene (anti-apoptotic) and low expression of the Bax gene (pro-apoptotic), with the ratio of Bcl-2 to Bax expression considered a marker of embryo survival (Yang & Rajamahendran 2002). A recent study examined 16 candidate genes associated with rapid cleavage in bovine two-cell embryos, three genes (YEAF, IDH, H2A) were differentially expressed in the early cleaving embryo and have the potential to be markers of developmental competence (Dode *et al.* 2006).

The majority of these studies report only steady state mRNA levels in gene expression pathways; however, RNA biogenesis is a central multi-step process that must balance message fidelity against steady-state levels of the mature RNA. An equally important area in the regulation of gene expression in mammals is the role of transcription turnover (Milligan *et al.* 2002). However, relatively little is known of the consequence of environmental conditions to affect transcription turnover and mRNA half-life leading to altered gene expression pathways in the embryo. Few studies, in this under investigated area, show altered gene expression levels in the embryo are due to changes in transcriptional activity.

Gene expression analysis strategies and DNA amplification

A variety of methods which are suitable for detecting gene transcription in preimplantation embryos are provided in Table 2. Conventionally, methods for detecting gene transcripts have included Northern Blotting, In Situ hybridization and RNAse Protection Assay. Limitations include low overall sensitivity and the need for a high complementary DNA (cDNA) copy number for successful analysis. Given the relative scarcity of mammalian preimplantation embryos and the requirement to pool (100) embryos for sufficient RNA, that is not applicable for analyses of rare constructs or for quantitative measurement in the individual embryo (Lechniak 2002).

Polymerase chain reaction

In contrast, the development of RT–PCR has enabled the detection of mRNAs from low yields of RNA obtained from single embryos (Bustin *et al.* 2005). However, the requirement to reverse transcribe RNA into cDNA is influenced by a number of variables and even a small shift in amplification efficiency may lead to exponential differences in the final PCR product (Gilliland *et al.* 1990; Nicoletti & Sassy-Prigent 1996; Lechniak 2002).

The development of competitive (Auboeuf & Vidal 1997) and semi-quantitative, non-competitive RT-PCR (Saric & Sydney 1997) strategies have addressed variability in amplification efficiency. The first involves simultaneous co-amplification of target RNA with added exogenous synthetic RNA that competes with available reagents and primers. The constant ratio between the two types of RNA allows the initial concentration (quantitation) of target RNA to be determined. PCR products are distinguished by small deletions, modifications to restriction sites or the addition of unrelated motifs to the synthetic RNA (Nicoletti & Sassy-Prigent 1996). The second involves co-amplification of the target RNA and unaffected endogenous (GAPDH, ß-actin, rRNA) or exogenous rabbit (ß-globulin) RNA using separate primers. Amplification in

Table 2. Applicable RNA quantitation strategies for the bovine preimplantation embryo

Method	Source	Benefits/ Limitations	Reference
Northern	mRNA	**Ben.:** mRNA size, integrity, alternative splicing **Limit.:** Low sensitivity/ relatively high complementary DNA (cDNA) copy number necessary for successful analysis.	(Bilodeau-Goeseels & Schultz 1997)
RNAse protection assay	mRNA	**Ben.:** Screening of similar size mRNA **Limit.:** Low sensitivity/ relatively high complementary DNA (cDNA) copy number necessary for successful analysis.	(Prediger 2001)
In Situ Hybridisation	mRNA	**Ben.:** Localization of transcript within analysed tissue **Limit. :** Low sensitivity/ relatively high complementary DNA (cDNA) copy number necessary for successful analysis.	(Jin & Lloyd 1997)
RT-PCR	mRNA/cDNA	**Ben.:** Qualitative amplification from a single cell. Semi – quantitative techniques available using competitive and non-competitive RNA. **Limit.:** Variables (primers, polymerase secondary RNA structure) influence efficiency of RT and PCR amplification that do not permit quantitation.	(Bustin et al. 2005)
qRT-PCR	mRNA/cDNA	**Ben.:** multiplex amplification with a single reaction. Quantitation of initial amounts of target sequence. No gel electrophoresis required. **Limit.:** Careful design or primers for amplification process and greater attention to optimization of PCR conditions.	(Gibson et al. 1996) SYBR Green (www.strategene.com) Molecular Beacons (www.molecular-beacons.org) TaqMan® (www.invitrogen.com)
Microarrays	cDNA arrays	**Ben.:** No sequence analyses required for the array, Product of interest identified by sequence analysis after array. **Limit.:** Requires plasmids from cDNA banks, Difficult maintaining PCR products. Alternative splice variants not distinguished.	(Duggan et al. 1999; Winzeler et al. 1999; Lou et al. 2001)
	Short Oligonucleotides: (25-base) in situ synthesis	**Ben.:** Commercially available. Widely supported. Up to 20,000 genes. **Limit.:** Difficult to incorporate new sequence data	Affymetrix (www.affymetrix.com) Febit (www.febit.de)
	Long (50-80 base) Oligonucleotides	**Ben.:** Design arrays in silico no cDNA banks required. More specificity in hybridization, more constant hybridization. Distinguish between alternative spliced mRNA conditions. Produce arrays with > 25,000 elements. Low and high density arrays **Limit.:** Requires completion of genome sequence. Design of oligonucleotides is of up most importance.	Combimatrix (www.combimatrix.com) (Blanchard & Hood 1996; Hughes & Shoemaker 2001) Clonetech (http://www.clonetech.com Compugen Http://www.cgen.com MWG Biotech Http://www.mwg-iotech.com Operon http://www.operon.com
	Genome-Wide SNP:	**Ben.:** 48,000 bovine SNP markers on single array. Near Commercial reality **Limit.:** Reliant on bovine genome sequence. Higher economic cost.	Illumina http://www.illumina.com

the linear range for the two primers allows the target RNA PCR products to be normalized against control products (Gutierrez-Adan *et al.* 2000).

A further modification is Real-time PCR which detects mRNA during the exponential phase of amplification and avoids the necessity for subsequent gel analysis (Bustin *et al.* 2005). The assay uses unique florescent reporters (Molecular beacons (Marras *et al.* 2006), SYBR Green (Morrison *et al.* 1998) and TaqMan® probes (Heid *et al.* 1996)), which emit signals in direct proportion to the PCR product during each round of amplification. While care must be taken to optimize both the RNA isolation and RT–PCR conditions, it is possible to detect as few as 400 copies of mRNA (Gibson *et al.* 1996). Internal, exogenous RNA (rabbit globin) and endogenous RNA standards (GAPDH, ß-actin and rRNA) are required during the amplification process to normalize the RNA levels and to justify the observed variations. The application, including use of various primers, detection limitations and benefits of each of the different systems have been described in detail (Lechniak 2002; Wrenzycki *et al.* 2004).

Microarray for analysis of gene expression

The combination and convergence of a variety of technologies led to the development of microarrays in the 1990s. This technology is being continually refined and its widespread adoption has indicated that DNA microarrays are to become the main technological workhorse for gene expression studies (Barrett & Kawasaki 2003; Kunz *et al.* 2004; Taylor *et al.* 2004; Ginsberg 2005). The sensitivity of the assay is high, with reports suggesting that this methodological approach is sensitive enough to detect the presence of one mRNA copy per cell (Barrett & Kawasaki 2003).

The sequencing of the >3 billion nucleotides in the human genome, and its extensive characterization suggest it is comprised of approximately 30,000-40,000 genes. This wealth of genomic information permits researchers to study thousands of genes simultaneously in the cell or tissue type of choice. Subsequently, the (micro) genomic revolution of sequencing is being extended to many different species, including the bovine (see Bovine genome sequencing project paragraph thereafter). Several microarray platforms (see Table 2) are highly suited to the study of gene expression in the bovine preimplantation embryo. Of note are the cDNA arrays (robotic printing), Short Oligonucleotides (25 bases, *in situ* synthesis) and Long Oligonucleotide arrays (40-80 base pairs, *in situ* synthesis and robotic printing) (Barrett & Kawasaki 2003).

The release of the bovine draft sequence and genome assembly has enabled the recent development of high-quality bovine SNP marker arrays to map quantitative trait loci (QTLs) which will result in a tool of significant value to cattle breeding. DNA differences detected through genotyping will expand gene discovery for better meat and milk quality and production, disease susceptibility and the discovery of elements responsible for phenotypic variation, growth and development. With further development it is likely to be applicable to elucidating the relationships between different populations of embryos.

While the application of microarray technologies continues to increase exponentially, it is perhaps cautionary to note that few attempts have been made to replicate and/or compare mRNA data across different platforms. In fact, some comparative cross-platform (and inter-platform) analyses show considerable variations between analyses of the same tissue (Kothapalli *et al.* 2002; Kuo *et al.* 2002). In addition, other methods (RT-PCR etc) have not been used to confirm results. In an effort to address the standardization of both experiments and controls, guidelines have been published to reduce confusion surrounding interpretation and replication of microarray data (Brazma *et al.* 2001; Ball *et al.* 2002; Stoeckert *et al.* 2002).

Bovine genome sequencing project

The bovine sequencing project is an international effort to sequence the genome of the cow (*Bos taurus*). The bovine genome is similar in size to the genomes of humans and other mammals, containing approximately 3 billion DNA base pairs (Gibbs *et al.* 2004).

The collaboration aims to produce a 15-fold coverage of the bovine genome and allow detailed tracking of the DNA differences between a number of cattle breeds to assist discovery of traits for better meat and milk production and to model human disease. The first draft of the bovine genome sequence has been deposited into free public databases for use by biomedical and agricultural researchers around the globe (GenBank (www.ncbi.nih.gov/Genbank) at NIH's National Center for Biotechnology Information (NCBI), EMBL Bank (www.ebi.ac.uk/embl/index.html) at the European Molecular Biology Laboratory's Nucleotide Sequence Database and the DNA Data Bank of Japan (www.ddbj.nig.ac.jp).

Global Gene Expression using cDNA microarray technology

The wealth of sequence data will allow researchers to capitalize on gene expression studies through the flexibility in choosing appropriate and specific arrays (Kurimoto *et al.* 2006; Vige *et al.* 2006). The utility of the microarray platform in preimplantation embryonic development has recently been demonstrated in both the mouse and bovine.

In the mouse, global changes in gene expression during folliculogenesis (primordial to large antral follicles) were examined for pathways that accompany the acquisition of meiotic and developmental competence (Hosack *et al.* 2003). The highest degree of up-regulation and down-regulation of gene expression (one third of transcripts exhibited a two fold change in relative abundance) was observed between the primordial to primary follicle transition (Pan *et al.* 2005). Subsequent transitions were about 10-fold less. Of particular interest in the primordial to primary follicle transition, was the increased or decreased transcriptional activity of specific regions (predominantly) on selected chromosomes. A phenomenon not observed at later stages of oocyte or embryo development.

The changes in global patterns of gene expression during *in vitro* maturation of an oocyte from either a primordial or secondary follicle stage displayed only a 4% and 2% difference when compared to the *in vivo* counterparts, respectively. Additional findings revealed an over-representation of genes involved in transcription of the commonly mis-expressed genes (1%) (Pan *et al.* 2005).

Global patterns of gene expression that surround the development of preimplantation mouse embryos have been examined with cDNA microarrays (Hamatani *et al.* 2004; Rinaudo & Schultz 2004; Zeng *et al.* 2004; Wang S. *et al.* 2005; Zeng & Schultz 2005). Results have confirmed previous analyses showing similarities between oocytes and 1-cell embryos, most likely due to the inheritance of mRNA from the oocyte, with major reprogramming of gene expression associated with maternal to embryonic transition and a period of mid-preimplantation gene activation, which precedes the dynamic morphological and functional changes that occur between the morula and blastocyst stage of development (Hamatani *et al.* 2004). Further analyses during this maternal to embryonic transition revealed a network of genes associated with *Myc* and its role in genome activation and reprogramming of gene expression, and *Hdac1* role in the repression of gene expression through chromatin-mediated changes (Zeng *et al.* 2004; Zeng & Schultz 2005). Global gene expression patterns in preimplantation embryos have been altered by the type of culture medium. Microarray analysis has shown that genes involved in protein synthesis, cell proliferation and transporter function are down-regulated as a consequence of exposure to *in vitro* culture medium (Rinaudo & Schultz 2004; Wang *et al.* 2005).

Microarray analyses have been applied to bovine embryo development (Sirard *et al.* 2005) and recently a 7872 bovine cDNA microarray was use to compare the global gene expression profiles of individual bovine SCNT blastocyst with their donor (somatic) cell and embryos derived from IVF and artificial insemination. The gene expression profile of an individual SCNT blastocyst indicated that it had undergone significant nuclear reprogramming from its original somatic profile (donor cell) so that it unexpectedly resembled a naturally fertilized AI embryo, more so than IVF embryo. Further analysis is required as to whether the consequence of early-stage reprogramming errors (1% of genes examined) have significant downstream effects to embryonic and fetal development (Smith *et al.* 2005). Another more recent study comparing IVF and SCNT blastocysts across 5000 cDNAs could only detect a difference in expression of *KRT18*, and *SLC16A1*. Further examination of transcript levels could not distinguish an NT from an IVF embryo. The unpredictability of gene expression on a global background of multiple gene expression changes argues for a predominantly stochastic nature of reprogramming errors (Somers *et al.* 2006). The dynamic changes during global activation of the embryonic genome have been recently examined using the Affymetrix bovine-specific DNA microarray. In the MII oocytes, genes controlling DNA methylation and metabolism were up-regulated. While during embryonic genomic activation (8-cell), those genes essential to the regulation of transcriptoin, chromatin-structure adhesion and signal transduction were up-regulated. Changes in gene expression during these critical development time points is expected to provide unique chromatin structures that maintain totipotency during embryogenesis and permit lineage-specific differentiation during post-implantation development. The consequence of this dynamics has many implications for a number of assisted reproductive technologies (Misirlioglu *et al.* 2006).

Future developments

While techniques that detect aberrant gene expression at the level of RNA in the single embryo continue to improve, little is known of the consequence at the level of the proteome. There is a need to develop diagnostic non-destructive solutions that permit the selection of viable embryos following *in vitro* culture and assisted reproductive techniques (ie Somatic Cell Nuclear Transfer). Several promising strategies are emerging and their application in cell biology is eagerly awaited. Of particular note is the development of fluorescent techniques that may enable researchers to determine the functions of proteins in individual cells.

Traditionally, florescent organic dyes conjugated with antibodies to the protein of interest have allowed imaging of biological events in individual fixed and permeabilized cells. However, new classes of florescent probes have been developed from synergistic developments in molecular biology, organic chemistry and material science that permit multiple functional analyses (gene expression, protein trafficking, biochemical signals) in single cells using non destructive imaging without significantly perturbing endogenous protein function (Giepmans *et al.* 2006). Two techniques amongst others could be compatible with protein expression studies in the oocyte and developing embryo.

Quantum dots (QD) are inorganic nanocrystals with intense and sustained fluorescence at specific wavelengths. Specialized coatings have made QD water soluble (to prevent quenching) and allow conjugation of protein targeting molecules (antibodies). The final size (QD + biomolecule) prevents easy transport across intact cellular membranes, limiting application to permeabilized cells or extracellular or endocytosed proteins. Further refinements addressing the size limitation using single-molecule optoelectronics have also now been developed (Lee *et al.* 2005).

Genetic Tagging or the construction of genetically encoded fluorophores covalently fused to specific cytoplasmic or surface proteins by spontaneous attachment or enzyme ligation also show potential for biological application. The most developed is the tetracysteine-biarsenical system which modifies a target protein to identify it from the many other proteins inside live cells. This specific protein is then fluorescently stained by membrane permeable non-fluorescent dye molecules that attach with picomolar affinity (Griffin *et al.* 1998). Small dithiol antidotes are added simultaneously to reduce binding, toxicity and perturbations to endogenous proteins and their function (Martin *et al.* 2005).

Various modifications to these fluorescent approaches have allowed the study of single proteins or complex endogenous pathways of proteins in a single cell (Sachs *et al.* 2005), protein localization, diffusion and trafficking (Lidke *et al.* 2004), dynamic and conformational changes in spatiotemporal resolution (Wallrabe & Periasamy 2005), protein-protein interactions (Gaietta *et al.* 2002), and by using Chromophore-activated light inactivation (CALI) to inactivate a protein with greater spatiotemporal control than possible with genetic KO or RNA interference approaches (Jay & Sakurai 1999).

Conclusions

In the bovine, as in the mouse and other species, the developmental competence of the preimplantation embryo can be influenced by the environment to which it is exposed. While it is capable of tolerating various sub-optimal conditions, the timing and severity of aberrant gene expression can initiate a range of downstream consequences for fetal and postnatal growth and development. Two critical conditions affect the severity of gene expression. They are the inherent quality of the oocyte which affects its ability to fertilise and form a blastocyst and *in vitro* culture effects which may impair the quality of the blastocyst. The integration of new and emerging genomic technologies allows researchers to profile the transcription patterns of thousands of genes in single oocytes and embryos following *in vitro* maturation, fertilization, manipulation and culture. Microarray analyses, when properly standardized, will provide unprecedented levels of information at the molecular level and offer greater opportunity to elucidate the requirements necessary for developmentally competent oocytes and/or embryos. In addition, the availability of new non-invasive florescent technologies that permit the examination of proteins endogenous pathway is eagerly awaited. The potential benefits arising from this knowledge include the ability to select viable embryos using non-invasive, quantitative assays of markers for developmental competence and ultimately the manufacture of *in vitro* culture conditions that properly mimic the correct temporal and spatial expression of a cohort of developmental genes in a manner similar to that seen *in vivo*.

References

Auboeuf D & Vidal H 1997 The use of the reverse transcription-competitive polymerase chain reaction to investigate the in vivo regulation of gene expression in small tissue samples. *Analytical Biochemistry* **245** 141-148.

Avery B, Jorgensen CB, Madison V & Greve T 1992 Morphological development and sex of bovine in vitro-fertilized embryos. *Molecular Reproduction and Development* **32** 265-270.

Ball CA, Sherlock G, Parkinson H, Rocca-Sera P, Brooksbank C, Causton HC, Cavalieri D,

Gaasterland T, Hingamp P, Holstege F et al. 2002 Standards for microarray data. *Science* **298** 539.

Barrett JC & Kawasaki ES 2003 Microarrays: the use of oligonucleotides and cDNA for the analysis of gene expression. *Drug Discovery Today* **8** 134-141.

Bertolini M, Beam SW, Shim H, Bertolini LR, Moyer AL, Famula TR & Anderson GB 2002 Growth, development, and gene expression by in vivo- and in vitro-produced day 7 and 16 bovine embryos. *Molecular Reproduction and Development* **63** 318-328.

Bilodeau-Goeseels S & Schultz GA 1997 Changes in

the relative abundance of various housekeeping gene transcripts in in vitro-produced early bovine embryos. *Molecular Reproduction and Development* **47** 413-420.

Blanchard AP & Hood L 1996 Sequence to array: probing the genome's secrets. *Nature Biotechnology* **14** 1649.

Blondin P, Farin PW, Crosier AE, Alexander JE & Farin CE 2000 In vitro production of embryos alters levels of insulin-like growth factor-II messenger ribonucleic acid in bovine fetuses 63 days after transfer. *Biology of Reproduction* **62** 384-389.

Bourc'his D, Le Bourhis D, Patin D, Niveleau A, Comizzoli P, Renard JP & Viegas-Pequignot E 2001 Delayed and incomplete reprogramming of chromosome methylation patterns in bovine cloned embryos. *Current Biology* **11** 1542-1546.

Brazma A, Hingamp P, Quackenbush J, Sherlock G, Spellman P, Stoeckert C, Aach J, Ansorge W, Ball CA, Causton HC et al. 2001 Minimum information about a microarray experiment (MIAME)-toward standards for microarray data. *Nature Genetics* **29** 365-371.

Brevini TA, Cillo F, Colleoni S, Lazzari G, Galli C & Gandolfi F 2004 Expression pattern of the maternal factor zygote arrest 1 (Zar1) in bovine tissues, oocytes, and embryos. *Molecular Reproduction and Development* **69** 375-380.

Bustin SA, Benes V, Nolan T & Pfaffl MW 2005 Quantitative real-time RT-PCR—a perspective. *Journal of Molecular Endocrinology* **34** 597-601.

Comizzoli P, Urner F, Sakkas D & Renard JP 2003 Upregulation of glucose metabolism during male pronucleus formation determines the early onset of the s phase in bovine zygotes. *Biology of Reproduction* **68** 1934-1940.

Corcoran D, Fair T & Lonergan P 2005 Predicting embryo quality: mRNA expression and the preimplantation embryo. *Reproductive Biomedicine Online* **11** 340-348.

Corcoran D, Fair T, Park S, Rizos D, Patel OV, Smith GW, Coussens PM, Ireland JJ, Boland MP, Evans AC & Lonergan P 2006 Suppressed expression of genes involved in transcription and translation in in vitro compared with in vivo cultured bovine embryos. *Reproduction* **131** 651-660.

Daniels R, Hall V & Trounson AO 2000 Analysis of gene transcription in bovine nuclear transfer embryos reconstructed with granulosa cell nuclei. *Biology of Reproduction* **63** 1034-1040.

Daniels R, Hall VJ, French AJ, Korfiatis NA & Trounson AO 2001 Comparison of gene transcription in cloned bovine embryos produced by different nuclear transfer techniques. *Molecular Reproduction and Development* **60** 281-288.

Davidson E 1986 *Gene activity in early development*, New York: Academic Press.

de Loos F, van Maurik P, van Beneden T & Kruip TA 1992 Structural aspects of bovine oocyte maturation in vitro. *Molecular Reproduction and Development* **31** 208-214.

De Sousa PA, Caveney A, Westhusin ME & Watson AJ 1998a Temporal patterns of embryonic gene expression and their dependence on oogenetic factors. *Theriogenology* **49** 115-128.

De Sousa PA, Watson AJ, Schultz GA & Bilodeau-Goeseels S 1998b Oogenetic and zygotic gene expression directing early bovine embryogenesis: a review. *Molecular Reproduction and Development* **51** 112-121.

De Sousa PA, Westhusin ME & Watson AJ 1998c Analysis of variation in relative mRNA abundance for specific gene transcripts in single bovine oocytes and early embryos. *Molecular Reproduction and Development* **49** 119-130.

Dode MA, Dufort I, Massicotte L & Sirard MA 2006 Quantitative expression of candidate genes for developmental competence in bovine two-cell embryos. *Molecular Reproduction and Development* **73** 288-297.

Duggan DJ, Bittner M, Chen Y, Meltzer P & Trent JM 1999 Expression profiling using cDNA microarrays. *Nature Genetics* **21** 10-14.

Eckert J & Niemann H 1998 mRNA expression of leukaemia inhibitory factor (LIF) and its receptor subunits glycoprotein 130 and LIF-receptor-beta in bovine embryos derived in vitro or in vivo. *Molecular Human Reproduction* **4** 957-965.

Enright BP, Lonergan P, Dinnyes A, Fair T, Ward FA, Yang X & Boland MP 2000 Culture of in vitro produced bovine zygotes in vitro vs in vivo: implications for early embryo development and quality. *Theriogenology* **54** 659-673.

Farin PW, Piedrahita JA & Farin CE 2006 Errors in development of fetuses and placentas from in vitro-produced bovine embryos. *Theriogenology* **65** 178-191.

Fernandez-Gonzalez R, Moreira P, Bilbao A, Jimenez A, Perez-Crespo M, Ramirez MA, Rodriguez De Fonseca F, Pintado B & Gutierrez-Adan A 2004 Long-term effect of in vitro culture of mouse embryos with serum on mRNA expression of imprinting genes, development, and behavior. *Proceedings of the National Academy of Sciences of the United States of America* **101** 5880-5885.

Gaietta G, Deerinck TJ, Adams SR, Bouwer J, Tour O, Laird DW, Sosinsky GE, Tsien RY & Ellisman MH 2002 Multicolor and electron microscopic imaging of connexin trafficking. *Science* **296** 503-507.

Gibbs R, Weinstock G, Kappes S, Schook L, Skow L & Womack J 2004 Bovine Genomic Sequencing Initiative - Cattle-izing the Human Genome.

Gibson UE, Heid CA & Williams PM 1996 A novel method for real time quantitative RT-PCR. *Genome Research* **6** 995-1001.

Giepmans BN, Adams SR, Ellisman MH & Tsien RY 2006 The fluorescent toolbox for assessing protein location and function. *Science* **312** 217-224.

Gilliland G, Perrin S, Blanchard K & Bunn HF 1990 Analysis of cytokine mRNA and DNA: detection and quantitation by competitive polymerase chain reaction. *Proceedings of the National Academy of Sci-*

ences of the United States of America **87** 2725-2729.

Ginsberg SD 2005 RNA amplification strategies for small sample populations. *Methods* **37** 229-237.

Griffin BA, Adams SR & Tsien RY 1998 Specific covalent labeling of recombinant protein molecules inside live cells. *Science* **281** 269-272.

Gutierrez-Adan A, Oter M, Martinez-Madrid B, Pintado B & De La Fuente J 2000 Differential expression of two genes located on the X chromosome between male and female in vitro-produced bovine embryos at the blastocyst stage. *Molecular Reproduction and Development* **55** 146-151.

Gutierrez-Adan A, Lonergan P, Rizos D, Ward FA, Boland MP, Pintado B & de la Fuente J 2001 Effect of the in vitro culture system on the kinetics of blastocyst development and sex ratio of bovine embryos. *Theriogenology* **55** 1117-1126.

Hall VJ, Cooney MA, Shanahan P, Tecirlioglu RT, Ruddock NT & French AJ 2005a Nuclear lamin antigen and messenger RNA expression in bovine in vitro produced and nuclear transfer embryos. *Molecular Reproduction and Development* **72** 471-482.

Hall VJ, Ruddock NT & French AJ 2005b Expression profiling of genes crucial for placental and preimplantation development in bovine in vivo, in vitro, and nuclear transfer blastocysts. *Molecular Reproduction and Development* **72** 16-24.

Hamatani T, Carter MG, Sharov AA & Ko MS 2004 Dynamics of global gene expression changes during mouse preimplantation development. *Developmental Cell* **6** 117-131.

Hasler JF 2001 Factors affecting frozen and fresh embryo transfer pregnancy rates in cattle. *Theriogenology* **56** 1401-1415.

Heid CA, Stevens J, Livak KJ & Williams PM 1996 Real time quantitative PCR. *Genome Research* **6** 986-994.

Holm P, Walker SK & Seamark RF 1996 Embryo viability, duration of gestation and birth weight in sheep after transfer of in vitro matured and in vitro fertilized zygotes cultured in vitro or in vivo. *Journal of Reproduction and Fertility* **107** 175-181.

Holm P & Callesen H 1998 In vivo versus in vitro produced bovine ova: similarities and differences relevant for practical application. *Reproduction, Nutrition, Development* **38** 579-594.

Holm P, Booth PJ & Callesen H 2003 Developmental kinetics of bovine nuclear transfer and parthenogenetic embryos. *Cloning and Stem Cells* **5** 133-142.

Hosack DA, Dennis G, Jr., Sherman BT, Lane HC & Lempicki RA 2003 Identifying biological themes within lists of genes with EASE. *Genome Biology* **4** R70.

Hughes TR & Shoemaker DD 2001 DNA microarrays for expression profiling. *Current Opinion in Chemical Biology* **5** 21-25.

Humpherys D, Eggan K, Akutsu H, Hochedlinger K, Rideout WM, 3rd, Biniszkiewicz D, Yanagimachi R & Jaenisch R 2001 Epigenetic instability in ES cells and cloned mice. *Science* **293** 95-97.

Hyttel P 2001 Nucleolus formation in pre-implantation cattle and swine embryos. *Italian Journal of Anatomy and Embryology* **106** 109-117.

Jay DG & Sakurai T 1999 Chromophore-assisted laser inactivation (CALI) to elucidate cellular mechanisms of cancer. *Biochimica et Biophysica Acta* **1424** M39-48.

Jin L & Lloyd RV 1997 In situ hybridization: methods and applications. *Journal of Clinical Laboratory Analysis* **11** 2-9.

Khurana NK & Niemann H 2000 Effects of oocyte quality, oxygen tension, embryo density, cumulus cells and energy substrates on cleavage and morula/blastocyst formation of bovine embryos. *Theriogenology* **54** 741-756.

Knijn HM, Wrenzycki C, Hendriksen PJ, Vos PL, Zeinstra EC, van der Weijden GC, Niemann H & Dieleman SJ 2005 In vitro and in vivo culture effects on mRNA expression of genes involved in metabolism and apoptosis in bovine embryos. *Reproduction, Fertility, and Development* **17** 775-784.

Knowles BB, Evsikov AV, de Vries WN, Peaston AE & Solter D 2003 Molecular control of the oocyte to embryo transition. *Philosophical Transactions of the Royal Society of London. Series B, Biological Sciences* **358** 1381-1387.

Kothapalli R, Yoder SJ, Mane S & Loughran TP, Jr 2002 Microarray results: how accurate are they? *BMC Bioinformatics* **3** 22.

Kunz M, Ibrahim SM, Koczan D, Scheid S, Thiesen HJ & Gross G 2004 DNA microarray technology and its applications in dermatology. *Experimental Dermatology* **13** 593-606.

Kuo WP, Jenssen TK, Butte AJ, Ohno-Machado L & Kohane IS 2002 Analysis of matched mRNA measurements from two different microarray technologies. *Bioinformatics* **18** 405-412.

Kurimoto K, Yabuta Y, Ohinata Y, Ono Y, Uno KD, Yamada RG, Ueda HR & Saitou M 2006 An improved single-cell cDNA amplification method for efficient high-density oligonucleotide microarray analysis. *Nucleic Acids Research* **34** e42.

Kurosaka S, Eckardt S & McLaughlin KJ 2004 Pluripotent lineage definition in bovine embryos by Oct4 transcript localization. *Biology of Reproduction* **71** 1578-1582.

Lazzari G, Wrenzycki C, Herrmann D, Duchi R, Kruip T, Niemann H & Galli C 2002 Cellular and molecular deviations in bovine in vitro-produced embryos are related to the large offspring syndrome. *Biology of Reproduction* **67** 767-775.

Lechniak D 2002 Quantitative aspect of gene expression analysis in mammalian oocytes and embryos. *Reproductive Biology* **2** 229-241.

Lee TH, Gonzalez JI, Zheng J & Dickson RM 2005 Single-molecule optoelectronics. *Accounts of Chemical Research* **38** 534-541.

Leibfried L & First NL 1979 Characterization of bovine follicular oocytes and their ability to mature in vitro. *Journal of Animal Science* **48** 76-86.

Li X, Amarnath D, Kato Y & Tsunoda Y 2006 Analysis

of development-related gene expression in cloned bovine blastocysts with different developmental potential. *Cloning and Stem Cells* **8** 41-50.

Lidke DS, Nagy P, Heintzmann R, Arndt-Jovin DJ, Post JN, Grecco HE, Jares-Erijman EA & Jovin TM 2004 Quantum dot ligands provide new insights into erbB/HER receptor-mediated signal transduction. *Nature Biotechnology* **22** 198-203.

Lindner GM & Wright RW, Jr 1983 Bovine embryo morphology and evaluation. *Theriogenology* **20** 407-416.

Lonergan P, Gutierrez-Adan A, Pintado B, Fair T, Ward F, Fuente JD & Boland M 2000 Relationship between time of first cleavage and the expression of IGF-I growth factor, its receptor, and two housekeeping genes in bovine two-cell embryos and blastocysts produced in vitro. *Molecular Reproduction and Development* **57** 146-152.

Lonergan P, Rizos D, Ward F & Boland MP 2001 Factors influencing oocyte and embryo quality in cattle. *Reproduction, Nutrition, Development* **41** 427-437.

Lonergan P, Rizos D, Gutierrez-Adan A, Fair T & Boland MP 2003a Oocyte and embryo quality: effect of origin, culture conditions and gene expression patterns. *Reproduction in Domestic Animals* **38** 259-267.

Lonergan P, Rizos D, Kanka J, Nemcova L, Mbaye AM, Kingston M, Wade M, Duffy P & Boland MP 2003b Temporal sensitivity of bovine embryos to culture environment after fertilization and the implications for blastocyst quality. *Reproduction* **126** 337-346.

Lonergan P, Fair T, Corcoran D & Evans AC 2006 Effect of culture environment on gene expression and developmental characteristics in IVF-derived embryos. *Theriogenology* **65** 137-152.

Lou XJ, Schena M, Horrigan FT, Lawn RM & Davis RW 2001 Expression monitoring using cDNA microarrays. A general protocol. *Methods in Molecular Biology* **175** 323-340.

Marras SA, Tyagi S & Kramer FR 2006 Real-time assays with molecular beacons and other fluorescent nucleic acid hybridization probes. *Clinica Chimica Acta* **363** 48-60.

Martin BR, Giepmans BN, Adams SR & Tsien RY 2005 Mammalian cell-based optimization of the biarsenical-binding tetracysteine motif for improved fluorescence and affinity. *Nature Biotechnology* **23** 1308-1314.

Memili E & First NL 2000 Zygotic and embryonic gene expression in cow: a review of timing and mechanisms of early gene expression as compared with other species. *Zygote* **8** 87-96.

Miller DJ, Eckert JJ, Lazzari G, Duranthon-Richoux V, Sreenan J, Morris D, Galli C, Renard JP & Fleming TP 2003 Tight junction messenger RNA expression levels in bovine embryos are dependent upon the ability to compact and in vitro culture methods. *Biology of Reproduction* **68** 1394-1402.

Milligan L, Forne T, Antoine E, Weber M, Hemonnot B, Dandolo L, Brunel C & Cathala G 2002 Turn-

over of primary transcripts is a major step in the regulation of mouse H19 gene expression. *EMBO Reports* **3** 774-779.

Misirlioglu M, Page GP, Sagirkaya H, Kaya A, Parrish JJ, First NL & Memili E 2006 Dynamics of global transcriptome in bovine mature oocytes and preimplantation embryos. *Proceedings of the National Academy of Sciences of the United States of America* **103** 18905-18910.

Morrison TB, Weis JJ & Wittwer CT 1998 Quantification of low-copy transcripts by continuous SYBR Green I monitoring during amplification. *BioTechniques* **24** 954-958, 960, 962.

Nicoletti A & Sassy-Prigent C 1996 An alternative quantitative polymerase chain reaction method. *Analytical Biochemistry* **236** 229-241.

Niemann H & Wrenzycki C 2000 Alterations of expression of developmentally important genes in preimplantation bovine embryos by in vitro culture conditions: implications for subsequent development. *Theriogenology* **53** 21-34.

Niemann H, Wrenzycki C, Lucas-Hahn A, Brambrink T, Kues WA & Carnwath JW 2002 Gene expression patterns in bovine in vitro-produced and nuclear transfer-derived embryos and their implications for early development. *Cloning and Stem Cells* **4** 29-38.

Oliveira AT, Lopes RF & Rodrigues JL 2006 Gene expression and developmental competence of bovine embryos produced in vitro with different serum concentrations. *Reproduction in Domestic Animals* **41** 129-136.

Pan H, O'Brien M J, Wigglesworth K, Eppig JJ & Schultz RM 2005 Transcript profiling during mouse oocyte development and the effect of gonadotropin priming and development in vitro. *Developmental Biology* **286** 493-506.

Pennetier S, Uzbekova S, Perreau C, Papillier P, Mermillod P & Dalbies-Tran R 2004 Spatio-temporal expression of the germ cell marker genes MATER, ZAR1, GDF9, BMP15,andVASA in adult bovine tissues, oocytes, and preimplantation embryos. *Biology of Reproduction* **71** 1359-1366.

Prediger EA 2001 Detection and quantitation of mRNAs using ribonuclease protection assays. *Methods in Molecular Biology* **160** 495-505.

Reichenbach HD, Liebrich J, Berg U & Brem G 1992 Pregnancy rates and births after unilateral or bilateral transfer of bovine embryos produced in vitro. *Journal of Reproduction and Fertility* **95** 363-370.

Reik W, Dean W & Walter J 2001 Epigenetic reprogramming in mammalian development. *Science* **293** 1089-1093.

Renard JP, Chastant S, Chesne P, Richard C, Marchal J, Cordonnier N, Chavatte P & Vignon X 1999 Lymphoid hypoplasia and somatic cloning. *Lancet* **353** 1489-1491.

Rinaudo P & Schultz RM 2004 Effects of embryo culture on global pattern of gene expression in preimplantation mouse embryos. *Reproduction* **128** 301-311.

Rizos D, Lonergan P, Boland MP, Arroyo-Garcia R, Pintado B, de la Fuente J & Gutierrez-Adan A 2002a Analysis of differential messenger RNA expression between bovine blastocysts produced in different culture systems: implications for blastocyst quality. *Biology of Reproduction* **66** 589-595.

Rizos D, Ward F, Duffy P, Boland MP & Lonergan P 2002b Consequences of bovine oocyte maturation, fertilization or early embryo development in vitro versus in vivo: implications for blastocyst yield and blastocyst quality. *Molecular Reproduction and Development* **61** 234-248.

Rizos D, Gutierrez-Adan A, Perez-Garnelo S, De La Fuente J, Boland MP & Lonergan P 2003 Bovine embryo culture in the presence or absence of serum: implications for blastocyst development, cryotolerance, and messenger RNA expression. *Biology of Reproduction* **68** 236-243.

Rizos D, Gutierrez-Adan A, Moreira P, O'Meara C, Fair T, Evans AC, Boland MP & Lonergan P 2004 Species-related differences in blastocyst quality are associated with differences in relative mRNA transcription. *Molecular Reproduction and Development* **69** 381-386.

Ruddock NT, Wilson KJ, Cooney MA, Korfiatis NA, Tecirlioglu RT & French AJ 2004 Analysis of imprinted messenger RNA expression during bovine preimplantation development. *Biology of Reproduction* **70** 1131-1135.

Sachs K, Perez O, Pe'er D, Lauffenburger DA & Nolan GP 2005 Causal protein-signaling networks derived from multiparameter single-cell data. *Science* **308** 523-529.

Saric T & Sydney AS 1997 Semiquantitive RT-PCR: enhancement of assay accuracy and reproducibility. *BioTechniques* **22** 630-636.

Sinclair KD, Young LE, Wilmut I & McEvoy TG 2000 In-utero overgrowth in ruminants following embryo culture: lessons from mice and a warning to men. *Human Reproduction* **15 Supplement 5** 68-86.

Sirard MA, Dufort I, Vallee M, Massicotte L, Gravel C, Reghenas H, Watson AJ, King WA & Robert C 2005 Potential and limitations of bovine-specific arrays for the analysis of mRNA levels in early development: preliminary analysis using a bovine embryonic array. *Reproduction, Fertility, and Development* **17** 47-57.

Smith SL, Everts RE, Tian XC, Du F, Sung LY, Rodriguez-Zas SL, Jeong BS, Renard JP, Lewin HA & Yang X 2005 Global gene expression profiles reveal significant nuclear reprogramming by the blastocyst stage after cloning. *Proceedings of the National Academy of Sciences of the United States of America* **102** 17582-17587.

Somers J, Smith C, Donnison M, Wells DN, Henderson H, McLeay L & Pfeffer PL 2006 Gene expression profiling of individual bovine nuclear transfer blastocysts. *Reproduction* **131** 1073-1084.

Stanton JA, Macgregor AB & Green DP 2003 Gene expression in the mouse preimplantation embryo. *Reproduction* **125** 457-468.

Stoeckert CJ, Jr., Causton HC & Ball CA 2002 Microarray databases: standards and ontologies. *Nature Genetics* **32 Supplement** 469-473.

Taylor TB, Nambiar PR, Raja R, Cheung E, Rosenberg DW & Anderegg B 2004 Microgenomics: Identification of new expression profiles via small and single-cell sample analyses. *Cytometry A* **59** 254-261.

Thompson JG 1997 Comparison between in vivo-derived and in vitro-produced pre-elongation embryos from domestic ruminants. *Reproduction, Fertility, and Development* **9** 341-354.

Thompson JG 2000 In vitro culture and embryo metabolism of cattle and sheep embryos - a decade of achievement. *Animal Reproduction Science* **60-61** 263-275.

Tremblay K, Vigneault C, McGraw S, Morin G & Sirard MA 2006 Identification and characterization of a novel bovine oocyte-specific secreted protein gene. *Gene* **29** 29.

Vige A, Gallou-Kabani C, Gross MS, Fabre A, Junien C & Jais JP 2006 An oligonucleotide microarray for mouse imprinted genes profiling. *Cytogenetic and Genome Research* **113** 253-261.

Wallrabe H & Periasamy A 2005 Imaging protein molecules using FRET and FLIM microscopy. *Current Opinion in Biotechnology* **16** 19-27.

Wang H & Dey SK 2006 Roadmap to embryo implantation: clues from mouse models. *Nature Reviews. Genetics* **7** 185-199.

Wrenzycki C & Niemann H 2003 Epigenetic reprogramming in early embryonic development: effects of in-vitro production and somatic nuclear transfer. *Reproductive Biomedicine Online* **7** 649-656.

Wang S, Cowan CA, Chipperfield H & Powers RD 2005 Gene expression in the preimplantation embryo: in-vitro developmental changes. *Reproductive Biomedicine Online* **10** 607-616.

Ward F, Rizos D, Corridan D, Quinn K, Boland M & Lonergan P 2001 Paternal influence on the time of first embryonic cleavage post insemination and the implications for subsequent bovine embryo development in vitro and fertility in vivo. *Molecular Reproduction and Development* **60** 47-55.

Watson AJ, De Sousa P, Caveney A, Barcroft LC, Natale D, Urquhart J & Westhusin ME 2000 Impact of bovine oocyte maturation media on oocyte transcript levels, blastocyst development, cell number, and apoptosis. *Biology of Reproduction* **62** 355-364.

Winzeler EA, Schena M & Davis RW 1999 Fluorescence-based expression monitoring using microarrays. *Methods in Enzymology* **306** 3-18.

Wrenzycki C, Herrmann D, Keskintepe L, Martins A, Jr., Sirisathien S, Brackett B & Niemann H 2001a Effects of culture system and protein supplementation on mRNA expression in pre-implantation bovine embryos. *Human Reproduction* **16** 893-901.

Wrenzycki C, Wells D, Herrmann D, Miller A, Oliver J, Tervit R & Niemann H 2001b Nuclear transfer protocol affects messenger RNA expression patterns in cloned bovine blastocysts. *Biology of Reproduction* **65** 309-317.

Wrenzycki C, Lucas-Hahn A, Herrmann D, Lemme E, Korsawe K & Niemann H 2002 In vitro production and nuclear transfer affect dosage compensation of the X-linked gene transcripts G6PD, PGK, and Xist in preimplantation bovine embryos. *Biology of Reproduction* **66** 127-134.

Wrenzycki C, Herrmann D, Lucas-Hahn A, Lemme E, Korsawe K & Niemann H 2004 Gene expression patterns in in vitro-produced and somatic nuclear transfer-derived preimplantation bovine embryos: relationship to the large offspring syndrome? *Animal Reproduction Science* **82-83** 593-603.

Wrenzycki C, Herrmann D, Lucas-Hahn A, Gebert C, Korsawe K, Lemme E, Carnwath JW & Niemann H 2005a Epigenetic reprogramming throughout preimplantation development and consequences for assisted reproductive technologies. *Birth Defects Research. Part C, Embryo Today* **75** 1-9.

Wrenzycki C, Herrmann D, Lucas-Hahn A, Korsawe K, Lemme E & Niemann H 2005b Messenger RNA expression patterns in bovine embryos derived from

in vitro procedures and their implications for development. *Reproduction, Fertility, and Development* **17** 23-35.

Xu KP, Yadav BR, King WA & Betteridge KJ 1992 Sex-related differences in developmental rates of bovine embryos produced and cultured in vitro. *Molecular Reproduction and Development* **31** 249-252.

Yang MY & Rajamahendran R 2002 Expression of Bcl-2 and Bax proteins in relation to quality of bovine oocytes and embryos produced in vitro. *Animal Reproduction Science* **70** 159-169.

Young LE, Sinclair KD & Wilmut I 1998 Large offspring syndrome in cattle and sheep. *Reviews of Reproduction* **3** 155-163.

Zeng F & Schultz RM 2005 RNA transcript profiling during zygotic gene activation in the preimplantation mouse embryo. *Developmental Biology* **283** 40-57.

Zeng F, Baldwin DA & Schultz RM 2004 Transcript profiling during preimplantation mouse development. *Developmental Biology* **272** 483-496.

Gene expression in elongating and gastrulating embryos from ruminants

I Hue, SA Degrelle, E Campion and J-P Renard

INRA, UMR 1198; ENVA; CNRS, FRE 2857, Biologie du Développement et Reproduction, Jouy en Josas, F-78350, France

In ruminants, more than 30% of the embryonic losses observed after artificial insemination (AI) have an early origin, coincident with a marked elongation of the trophoblast which occurs before implantation. Several observations provide clear evidence that early elongation of the conceptus relies on cell multiplication, cell growth and cell shape remodeling. Recent results indicating an intense multiplication of a non-fully differentiated trophoblast, which still expresses some epiblast genes, has to be considered at the onset of elongation. It has also been shown in the last two years that general metabolism and protein trafficking are characteristic of the onset of elongation whereas cellular interactions, cell to cell signaling and cell adhesion become predominant at the end of elongation. Accordingly, expression of most of the single genes identified so far increases during elongation and is related to the establishment of embryo-maternal exchanges before implantation. However, not much is known of what controls the induction of the elongation process or the coordinated development of the embryonic and extra-embryonic tissues. This review highlights new information on this developmental phase and summarizes the views on the complex cross-talk among molecules which might govern conceptus development and lead to successful implantation.

Introduction

After fertilisation and cleavage, the mammalian embryo reaches the blastocyst stage. This stage occurs after *in vivo* or *in vitro* development and is characterised by the morphological distinction between two cell types: the inner cell mass (or ICM) and the trophectoderm, even though molecular distinctions between these lineages appear earlier (Rossant 2004). In ruminants, the blastocyst is a sphere of 150-200 μm in diameter which contains approximately 250-300 cells (8 days after mating or insemination). A few days later, the sphere has enlarged, the number of cells is 10 times higher and the blastocyst has evolved to an ovoid shape (Day 11 in sheep, Day 12 in cows). According to Chang (1952) the ICM evolves into a germinal disc and the endoderm cells underlying the disc migrate under the trophectoderm to form an extra-embryonic endoderm layer. On Day 12 in sheep and Day 14 in cows, the blastocyst has a tubular shape, and the germinal disc is a true embryonic disc that is flat and open to the uterine environment since the trophoblast layer (called polar trophoblast or Rauber's layer) which covers the ICM

Corresponding author E-mail: isabelle.hue@jouy.inra.fr

until this stage has disappeared. Subsequently, the mural trophoblast lineage further elongates and gives rise to a filamentous-shaped conceptus (Day 14 in sheep, Day 16 to 18 in cows) composed of differentiated embryonic and extra-embryonic tissues. The extra-embryonic tissues include the chorion, formed by fusion of trophectoderm with migrating extra-embryonic mesoderm, and the growing yolk sac composed of extra-embryonic endoderm and mesoderm. Growth and differentiation processes of embryonic and extra-embryonic lineages are referred to as gastrulation and elongation, respectively. Finally, the trophoblast forms loose cellular contacts with the endometrial layer of the uterus, which marks the onset of implantation (Day 15 in sheep, Day 19 in cows; Guillomot 1995).

This review focuses on the process of conceptus elongation without comments on trophoblast lineage differentiation or implantation which were reviewed recently (Roberts *et al.* 2004; Spencer & Bazer 2004). We will concentrate on recent molecular data and comment on elongation models to gain insights into this complex process, mostly studied as an essential step in establishing functional embryo-maternal communications (Wolf *et al.* 2003; Imakawa *et al.* 2004; Klein *et al.* 2006). Maternal recognition of pregnancy is indeed essential for the uterus to switch from a cyclic to a pregnant state as underlined by the high rate of embryonic loss before implantation (Anderson 1978; Dunne *et al.* 2000; Goff 2002). This review will mainly describe gene expression patterns in ruminants, but occasionally comments on results obtained from studies of porcine conceptuses which also elongate and gastrulate before implantation.

Elongation and extra-embryonic development

Elongation defines the exponential growth of bovine, ovine and porcine blastocyst which occurs before implantation, with rapid evolution in their shapes from spherical to ovoid, tubular and finally filamentous forms (see Fig. 1). Though simple in its description, elongation is a complex process which does not have a clear starting point. Depending on species, elongation starts when conceptuses are at a spherical to slightly ovoid stage (pig) or an ovoid to slightly tubular stage (sheep, cow), on Day 11 (pig, sheep) or Day 12 (cow) post-insemination and at diverging blastocyst sizes: 1 to 2 mm (sheep, cow) or 9 to 10 mm (pig). It seems that these parameters (Wintenberger-Torres & Flechon 1974; Geisert *et al.* 1982; Betteridge & Flechon 1988), though extremely useful when collecting the embryos, are not adequate to establish a clear starting point to understand it at the molecular level. Therefore, we define the 150-200 μm freshly hatched blastocyst as the developmental stage for beginning elongation, the spherical-ovoid transition as the induction phase of elongation, the tubular-filamentous transition as a step of elongation maintenance and implanting stages as a phase of elongation arrest.

When starting elongation at the early blastocyst stage, cell proliferation likely constitutes the initiating event of the process. The clear increase in cell numbers initially observed between spherical and ovoid stages (Chang 1952; Wintenberger-Torres & Flechon 1974) has been confirmed using several molecular approaches. Analysing the transition in development of bovine blastocysts between Days 7 and 14, Ushizawa *et al.* (2004) identified 680 genes for which expression increased and 26 genes that were down-regulated including those encoding for oncogenes, tumor suppression, cell cycle control and apoptosis (see Fig. 1). Similarly, cell cycle and cell proliferation related genes represented 29% of the annotated repertoire we established at the ovoid stage (Degrelle *et al.* 2005) while a small set of genes preferentially expressed at the ovoid stage included a marker of cell proliferation in human cancers (Nap1L1). However, since none of these transcripts has been localised *in situ* at spherical or ovoid stages, the molecular and cellular basis of this proliferation process remains poorly characterised.

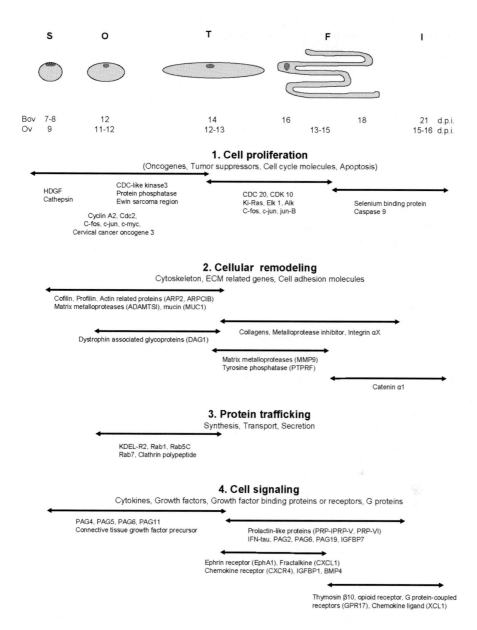

Fig. 1. Gene expression across elongation in ruminants. Some of the genes known to be expressed across elongation are named here together with the functions they relate to. Differential expression patterns are symbolised by horizontal arrows encompassing the appropriate stages: spherical (S), ovoid (O), tubular (T), filamentous (F) or implanted (I). These sets of genes originate from the following reports: Ushizawa *et al.* 2004; Cammas *et al.* 2005; Degrelle *et al.* 2005; Xavier *et al.* 1997; ovoid cDNA library (I Hue & SA Degrelle, unpublished observations). The conceptuses are schematically drawn according to their shape: light grey represents extra-embryonic tissues, dark grey: embryonic tissues. The developmental timing is scaled in days post insemination (or d.p.i) where day 0 marks the onset of oestrus.

In addition to cell proliferation, the onset of elongation involves a cellular reorganisation based upon changes in cell morphology, density or migration. At the spherical stage, the trophecto-derm is a layer of flattened cells and the endoderm cells which migrate out from the inner cell mass, start underlining the trophectoderm layer to give rise to the extra-embryonic endoderm. At the ovoid stage, these endoderm cells are flattened too, except at the level of the nucleus, whereas the trophectoderm cells are cuboidal and show numerous signs of phagocytosis (Betteridge & Flechon 1988). In the pig, changes in the cellular organisation of both trophecto-derm and endoderm layers have been put forward to explain the quick transitions observed in the size and shape of elongating conceptuses. Indeed, within a thin band of cells called "elon-gation zone", trophectoderm cells exhibited alterations of microfilaments and junctional com-plexes whereas endoderm cells formed numerous filopodia (Geisert et al. 1982). No similar observations of an "elongation zone" (surrounding the ICM and extending from the disc to distal ends of the conceptus) have been reported for ruminants but genes related to cell migra-tion and cell remodelling have been identified in bovine and ovine conceptuses (Fig. 1). Similarly, genes involved in protein trafficking (which includes synthesis, secretion and trans-port) were highly expressed in sheep as previously revealed on transmission electron micro-graphs by the presence of rough endoplasmic reticulum or endocytic vesicles (Betteridge & Flechon 1988). Cell differentiation and remodelling have also been explored at "late" stages of elongation and associated with implantation (Guillomot 1995; Spencer et al. 2004; Ushizawa et al. 2004; Cammas et al. 2005).

Last but not least, the elongation process has also been defined as an essential step in the establishment of pregnancy through an active cross-talk between the conceptus and the uterus well ahead of implantation. Cell signalling molecules have thus been identified accordingly (Fig. 1) and reported as increasingly expressed throughout the whole process or preferentially expressed at "late" elongated stages.

Despite the obvious roles of cell proliferation, cellular remodelling and cell signalling in the elongation process, most of the genes related to those functions were reported as "expressed in bovine or ovine elongating conceptuses" with no formal proof of a direct role in the induction, maintenance or arrest of the elongation process. In the absence of direct arguments through in vitro or in vivo knock-down approaches, it thus seems difficult to define the genes which really play the key roles in this complex process.

Recent molecular data based on the analysis of the transcriptome provided also genes or Expressed Sequence Tags (EST), the function of which is not elucidated so far. At first glance, this only lengthened the list of such genes we already had. For example, genes encoding Retinol Binding Proteins (RBP: Trout et al. 1991; Dore et al. 1994; Yelich et al. 1997a), Tropho-blast Kunitz Domain Proteins (TKDP: MacLean et al. 2003; MacLean et al. 2004; Chakrabarty et al. 2006) or Pregnancy Associated Glycoproteins (PAG: Green et al. 2000; Hughes et al. 2003; Klisch et al. 2005) were isolated during elongation and further characterised as interest-ing gene families (TKDP, PAG) or important molecules for the elongation of porcine concep-tuses (RBP). However their precise roles remain incompletely understood. Conversely, some of the genes identified through the use of transcriptomic approaches (arrays or SAGE: Serial Analysis of Gene Expression) brought novel perspectives on extra-embryonic development due to the functions they have outside of the elongation process. Among them: (i) an unusual chemokine (CX3CL1; sheep) which could mediate inflammatory reaction or cell adhesion; (ii) a Wnt inhibitor (Dickkopf1; cow) known to be expressed in embryonic tissues during gastrula-tion or in uterine tissues prior to implantation (Idkowiak et al. 2004; Bauersachs et al. 2006); and (iii) two developmental factors potentially regulated by estrogens: Stratifin and Midkine (pig; Blomberg et al. 2006).

Elongation and embryonic development

Coincident with elongation, growth and differentiation of extra-embryonic tissues, embryonic development proceeds quickly from an inner cell mass, to a germinal disc and ultimately to an embryonic disc which undergoes progressive growth and differentiation (see Fig. 2). These developmental steps have been described extensively over the past decades (Chang 1952; Greenstein *et al.* 1958; Betteridge & Flechon 1988; Vejlsted *et al.* 2006a), but there is little information on the molecular basis of gastrulation in ruminants or ungulates. Molecular markers have been isolated to characterise embryonic poles (anterior versus posterior), embryonic layers (ectoderm, endoderm or mesoderm) and specific structures appearing during gastrulation (e.g., primitive streak), but these markers have mostly been used to identify the stage of development of the embryonic disc. Expression patterns for goosecoid and brachyury were thus reported for ungulates using whole-mount *in situ* hybridisation and used as early markers for the primitive endoderm or the mesoderm, respectively in order to stage the embryos (Meijer *et al.* 2000; van de Pavert *et al.* 2001; Hue *et al.* 2001). Recent studies using either RNA or antibodies as probes revealed additionally interesting specificities in the expression patterns of key gastrulation genes in ungulates as compared to their patterns in the mouse (Guillomot *et al.* 2004; Flechon *et al.* 2004; Degrelle *et al.* 2005; Vejlsted *et al.* 2005; Vejlsted *et al.* 2006b; Blomberg le *et al.* 2006). To our knowledge however, no study has been conducted in ruminants to analyse the interactions between embryonic and extra-embryonic tissues as recently documented in the mouse where reciprocal inductive interactions are involved in (i) defining the antero-posterior pattern of the embryo (Beck *et al.* 2002; Richardson *et al.* 2006) and (ii) preventing precocious differentiation of the extra-embryonic ectoderm (Guzman-Ayala *et al.* 2004).

Therefore, we decided to study these interactions by taking advantage of the ease of non-surgical recovery of bovine conceptuses prior to implantation. We thus selected bovine conceptuses according to their size (ovoid: 1-20 mm, tubular: 50-60 mm, early filamentous: 100-150 mm) and compared the gene expression profiles of each elongation stage (ovoid, tubular and filamentous) while screening a bovine array of about 10 000 cDNA (10K array). A hierarchical clustering based on a set of differentially expressed genes between ovoid, tubular and early filamentous conceptuses revealed two main gene clusters, as well as two main developmental groups (Fig. 3). A major difference appeared between the ovoid versus the tubular and filamentous stages which confirmed on a large repertoire (10K array) that gene expression profiles between tubular and early filamentous conceptuses were not statistically different, as initially observed (1K array; Degrelle *et al.* 2005). In addition to the differences in the size of the conceptus or the extra-embryonic layers formed at those 3 elongation stages, another important feature distinguished them. At both tubular and filamentous stages, the embryonic disc is flat with a clear antero-posterior axis, whereas at the ovoid stage the embryonic disc is a non polarised germinal disc. When related to these morphological observations, the two gene clusters suggest that the transcriptional pattern which distinguishes an ovoid stage from tubular and filamentous stages is, in fact, a pre-gastrulating versus a gastrulating signature. The identity of the genes within these 2 clusters needs now to be explored and understood in the light of embryonic-extra-embryonic interactions. However, the existence of these clusters already extends previous observations where some of the key genes for the embryonic-extra-embryonic interactions in the mouse have been identified in ruminant extra-embryonic tissues (furin or PACE1: Degrelle *et al.* 2005; BMP4 for bone morphogenic protein-4: Cammas *et al.* 2005). Nevertheless, this does not mean necessarily that similar gene networks are involved in establishing early embryonic polarities in ruminant and in rodent extra-embryonic tissues. This topic awaits further studies in ruminants (I Hue, unpublished observations), but reassures us in proposing that the ovoid stage is a critical step in the transition between a spherical and a tubular shape, and also between an ICM and an epiblast developmental stage in bovine conceptuses.

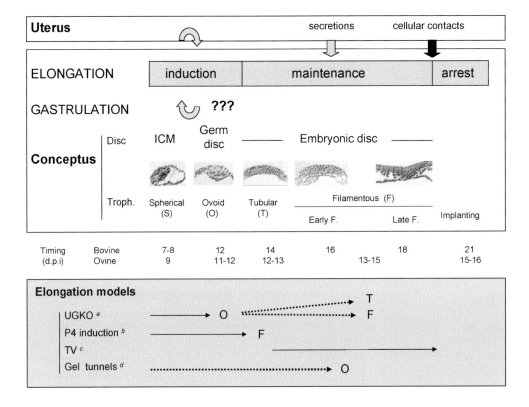

Fig. 2. From a puzzle to an integrated model? In ruminants, the elongation of the conceptus defines the exponential growth of the blastocyst as well as the evolution in its size and shape from a spherical (S) to an ovoid (O), a tubular (T) and finally a filamentous (F) form. These elongating stages also correspond to different stages of embryonic development from an inner cell mass (or ICM) to a germinal disc and finally an embryonic disc. Four main models have been established so far to study what governs the induction and maintenance of the elongation process. Among them: the UGKO model in sheep and the P4 induction model (in sheep and cow) helped analysing the induction phase ( ). The excision of the disc from tubular conceptuses to get trophoblastic vesicles (or T.V., in sheep and cow) demonstrated that uterine secretions maintain the elongation process once properly induced ( ). The culture of spherical blastocysts in gel tunnels confirmed that physical constraints do not fully mimic elongation (in cows). In each model, the elongation stage reached by the blastocyst is given by the appropriate letter: O, T or F. When most of the blastocysts reached the same stage a plain arrow is used but when they reached a stage or another, a dotted arrow is used. However, the events leading to the coordination of elongation and gastrulation have not been cleared so far in ruminants ( ). On the other hand, the loose contacts formed between the conceptus and the uterus clearly mark ( ) the arrest of elongation and the onset of implantation (I). The developmental timing is scaled in days post insemination (or d.p.i) where day 0 marks the onset of oestrus. Italics refer to original reports: (a) Gray *et al.* 2001; 2002, (b) Satterfield *et al.* 2006, (c) Heyman *et al.* 1984; Flechon *et al.* 1986, (d) Vajta *et al.* 2004; Vejlsted *et al.* 2006a.

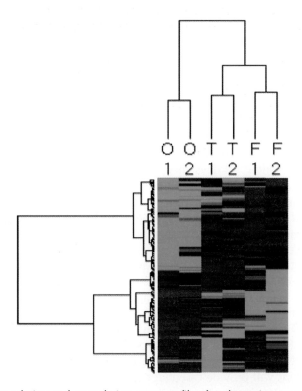

Fig. 3. Pre-gastrulating and gastrulating gene profiles for elongating extra-embryonic tissues. Hierarchical clustering of 239 genes which were differentially expressed between extra-embryonic tissues from ovoid, tubular and early filamentous conceptuses (statistically significant with AnovArray; Hennequet-Antier *et al.* 2005). The elongating stages were defined according to the size of the conceptuses (ovoid (O): 1-20 mm, tubular (T): 50-60 mm, filamentous (F): 100-150 mm) and each probe (O, T, F) corresponded to a pool of four conceptuses. Two independent probes per elongation stage (named: 1, 2) were hybridised to the 10K array. This array will be described in a forthcoming paper (SA Degrelle, unpublished observations). Within the cluster, highly expressed genes are in red and low abundance genes in green. The sizes of the branches within the tree indicate distances between gene sets or elongation stages.

Elongation and uterine environment

Conceptuses of sheep and cattle produce proteins which exert an antiluteolytic effect by inhibiting uterine pulsatile release of PGF2α. As a consequence, continued secretion of progesterone by the corpora lutea stimulates and maintains endometrial functions necessary for growth and elongation of pre-implanting conceptuses (Thatcher *et al.* 1984; Spencer & Bazer 2004). Since the first gene identified in the trophoblast encoded IFN-tau (Farin *et al.* 1990), endometrial genes regulated by IFN-tau have been identified in sheep and cattle together with genes induced by other secretory products of the conceptus in pregnant ruminants (Song *et al.* 2005; Gray *et al.* 2006; Klein *et al.* 2006). Despite the pivotal role of IFN-tau in the embryo-maternal dialog (Thatcher *et al.* 1989; Bazer & Spencer 2006), other cytokines, hormones, growth factors and growth factor receptors or prostaglandins and prostaglandin receptors (Kliem *et al.* 1998; Burghardt *et al.* 2002; Wolf *et al.* 2003; Ashworth *et al.* 2006; Cammas *et al.* 2006), exert important functions. However, to go one step further in the understanding of the complex signalling pathways between the conceptus and the uterus, one now needs to analyse for temporal and spatial (cell specific) gene expression profiles both before and after elongation of the conceptus.

Reciprocally, blastocyst growth and elongation depend on uterine secretions as evidenced by increased embryonic losses in documented cases of conceptus-uterus asynchrony (Pope 1988; Barnes 2000). This has been confirmed and extended using two experimental models (Fig. 2); one that precludes uterine gland secretions, i.e, the Uterine Gland Knock Out (UGKO) ewe model (Gray *et al.* 2001; Gray *et al.* 2002) and the other that advances uterine secretory activity, i.e, the early progesterone-induced ewe (Satterfield *et al.* 2006) and cow (Garrett et al. 1988) models. Interestingly, 14 days after mating, UGKO ewes presented no conceptus (4 of 8 ewes), growth-retarded conceptuses (3 of 8 ewes) or filamentous conceptuses (1 of 8 ewes), versus 4/4 ewes with filamentous conceptuses in untreated control ewes (Fig. 2). These results demonstrated a direct consequence of lack of uterine glands on conceptus development, that is, no uterine glands and no conceptuses, moderate gland density and retarded to normal conceptuses. On the contrary, exogenous progesterone administration 1.5 days after mating led to increased blastocyst diameters on day 9 and advanced the timing of elongation (filamentous stages on Day 12 already). Unfortunately, not much is known about conceptus development in the ewe model, but secretion of IFN-tau was advanced in cows. There remain many questions, e.g., how does a growth-retarded sheep conceptus (developed in UGKO ewes) appear morphologically when no INF-tau is present in the corresponding uterine flushes? Conversely, in the P4-induced model is the advanced elongation process perfectly similar to that for normal conceptuses? Undoubtedly, gene expression profiling studies on embryonic and extra-embryonic tissues in such situations will provide essential data on the genes or developmental cascades involved in successful uterus/conceptus cross-talk.

In vitro elongation

Mimicking the elongation process *in vitro* has long been a challenge, but has provided many interesting features and questions (Fig. 2). The first experimental approach was to collect *in vivo* elongated conceptuses at the tubular stage (sheep: Day 12; cow: Day 14) cut them into pieces on both sides of the embryonic disc, cultivate them *in vitro* or transfer them back *in utero*. In both species, it appeared that the resulting "trophoblastic vesicles" survived, but did not elongate further *in vitro* whereas they could elongate and produce IFN-tau during 5 day period following transfer back into a synchronous uterine environment (Heyman *et al.* 1984; Flechon *et al.* 1986). As such, it seems that the elongation process and IFN-tau secretion require the uterine environment, but not the embryonic disc (Fig. 2). However, in the absence of any additional results, one does not know whether such trophoblast elongation resembles a normal transition of tubular to filamentous conceptuses.

The second approach to initiating the elongation process *in vitro* was to incubate spherical blastocysts in gel tunnels containing glucose-rich medium (Vajta *et al.* 2004). Compared to cultures in Petri dishes where blastocysts only formed spheres of increasing diameters, cultures in gel tunnels induced a constrained elongation in most blastocysts although very few survived until Day 16 (2/67). The size of the longest blastocyst obtained was about 12 mm which is the size of an ovoid blastocyst developed *in vivo* (1-20 mm; Day 12). Nevertheless, in the few healthy-looking blastocyts (2/67), the inner cell mass had developed, a second cell layer had formed and completed development (Brandao *et al.* 2004), but there was no regression of the Rauber's layer and no formation of a true embryonic disc (Vejlsted *et al.* 2006a). Moreover, many blastocysts degenerated during culture, even those which grew in Petri dishes, suggesting that the medium used may not have met the metabolic needs of these blastocysts. In light of recent results concerning differential and constitutive gene expression during elongation of ruminant conceptuses, optimising a medium that can (i) sustain protein, nucleic acid, carbohydrate and lipid metabolism of the embryos and (ii) accommodate associated catabolism, seems challenging. Even more challenging are efforts *in vitro* to mimic the uterine environment which is just being decrypted through proteomic studies (Berendt *et al.* 2005).

Elongation: hypotheses and open questions

Based on the models and results presented in this review, three basic features can describe what we know about elongation: (i) it depends on uterine secretions; (ii) it is tightly coordinated with growth and differentiation of the embryonic disc; and (iii) it is not likely to be solely dependent on a physical constraint such as the shape of the uterus. This knowledge is, however, difficult to integrate in a simple model of elongation induction, maintenance and arrest. This is partly due, in our opinion, to: (i) confusion around the start point of elongation and lack of understanding of the mechanisms which induce it; (ii) absence of biological parameters to define the phases of induction or maintenance of conceptus development; and (iii) partial datasets available *in vivo, in vitro* and in experimental models since the uterus or the conceptus, but not both, were usually studied. To gain insights into this simple view of induction, maintenance and arrest of elongation, we will summarize what is known about each phase of the conceptus development.

Induction: After mating, uteri of UGKO ewes maintained conceptuses until Day 9 (spherical stage, 900μm), the conceptuses could not elongate in UGKO ewes, but could in normal ewes. Also, conceptuses from normal ewes did not elongate in uteri of UGKO ewes. Therefore, the mechanism for induction of conceptus elongation is not inherent within the conceptus. However, between the onset of elongation and the parameters that define the ovoid stage: (i) a shape is defined by conceptus length that is about 1.3 times its width (Grealy *et al.* 1996); (ii) a specific programming of the ruminant trophoblast results in its expression of epiblast genes (Degrelle *et al.* 2005); (iii) an intense proliferation of the mural lineage occurs; and (iv) a gene expression profile for extra-embryonic tissues which define pre-gastrulating features. The question is what do the uterine secretions induce? Do they interfere with or influence embryonic-extra-embryonic interactions in any way?

Maintenance: Similar to the situation of normal sheep blastocysts in UGKO ewes, trophoblastic vesicles derived at tubular stages do not elongate *in vitro,* but do so *in vivo,* indicating that induction of conceptus elongation depends on uterine signalling. However, we do not know which components of uterine secretions induce the elongation process or which components of these secretions maintain the conceptus. Do they maintain (i) an elongated shape which does not result solely from physical constraints imposed by the uterine wall since equine conceptuses (though confined as well in long uterine horn) remain spherical from Day 6 to Day 22 (reviewed in Allen 2001) (ii) an ongoing proliferation process (iii) ongoing IFN-tau secretion and/or (iv) a gastrulating signature in tubular and filamentous extra-embryonic tissues?

Arrest: We have known for decades that elongation stops at apposition, the onset of the implanting process (Guillomot 1995; Spencer *et al.* 2004). However, embryonic development does not stop concomitantly, since neural folds and somites appear at that time as if this phase of elongation arrest initiates another stage to achieve synchrony between pre-placental and fetal tissues. These events, however, are outside the focus of this review.

Defining similarities and differences between elongating processes in conceptuses from sheep, cow and pig could help us understand the basic core mechanisms responsible for elongation of conceptuses. Unfortunately, this has not been studied so far although a number of reports on molecular data are available for pigs (Yelich *et al.* 1997b; Wilson *et al.* 2000; Ross *et al.* 2003; Blomberg *et al.* 2005; Lee *et al.* 2005). At present, we have some pieces of the elongation puzzle, but must continue searching for the missing cues until an integrated model can be developed and validated.

Conclusions

Although known for decades, mechanisms responsible for elongation of conceptuses are far from being understood. This is likely due to: i) the numerous interactions with the uterus and the embryonic disc so that only well synchronised processes give rise to a successful pregnancy; ii) the complexity of the process: early processes and late processes which might be distinct but interdependent on each other; iii) cross-talk between cell layers which compose the extra-embryonic tissues; and iv) the inability, to date, to mimic or dissect this complex process *in vitro*.

Acknowledgments

The authors gratefully acknowledge A. Hernandez[a], R. Everts[a], H. Lewin[a], C. Tian[b] and J. Yang[b] for their contribution to the construction of the bovine10K array ([a]University of Urbana, Illinois, USA; [b]University of Storrs, Connecticut, USA), the UCEA from Bressonvilliers for the access to *in vivo* developed bovine conceptuses and M. Guillomot (INRA, VMR1198) for helpful discussions and critical reading of the manuscript.

References

Allen WR 2001 Fetomaternal interactions and influences during equine pregnancy. *Reproduction* **121** 513-527.

Anderson LL 1978 Growth, protein content and distribution of early pig embryos. *The Anatomical Record* **190** 143-153.

Ashworth MD, Ross JW, Hu J, White FJ, Stein DR, Desilva U, Johnson GA, Spencer TE & Geisert RD 2006 Expression of porcine endometrial prostaglandin synthase during the estrous cycle and early pregnancy, and following endocrine disruption of pregnancy. *Biology of Reproduction* **74** 1007-1015.

Barnes FL 2000 The effects of the early uterine environment on the subsequent development of embryo and fetus. *Theriogenology* **53** 649-658.

Bauersachs S, Ulbrich SE, Gross K, Schmidt SEM, Meyer HHD, Wenigerkind H, Vermehren M, Sinowatz F, Blum H & Wolf E 2006 Embryo-induced transcriptome changes in bovine endometrium reveal species-specific and common molecular markers of uterine receptivity. *Reproduction* **132** 319-331.

Bazer FW & Spencer TE 2006 Methods for studying interferon tau stimulated genes. *Methods in Molecu-*

lar Medicine **122** 367-380.

Beck S, Le Good JA, Guzman M, Ben Haim N, Roy K, Beermann F & Constam DB 2002 Extraembryonic proteases regulate Nodal signalling during gastrulation. *Nature Cell Biology* **4** 981-985.

Berendt FJ, Frohlich T, Schmidt SE, Reichenbach HD, Wolf E & Arnold GJ 2005 Holistic differential analysis of embryo-induced alterations in the proteome of bovine endometrium in the preattachment period. *Proteomics* **5** 2551-2560.

Betteridge KJ & Flechon JE 1988 The anatomy and physiology of pre-attachment bovine embryos. *Theriogenology* **29** 155-187.

Blomberg le A, Long EL, Sonstegard TS, Van Tassell CP, Dobrinsky JR & Zuelke KA 2005 Serial analysis of gene expression during elongation of the peri-implantation porcine trophectoderm (conceptus). *Physiological Genomics* **20** 188-194.

Blomberg le A, Garrett WM, Guillomot M, Miles JR, Sonstegard TS, Van Tassell CP & Zuelke KA 2006 Transcriptome profiling of the tubular porcine conceptus identifies the differential regulation of growth and developmentally associated genes. *Molecular Reproduction and Development* **73** 1491-1502.

Brandao DO, Maddox-Hyttel P, Lovendahl P, Rumpf R, Stringfellow D & Callesen H 2004 Post hatching development: a novel system for extended in vitro culture of bovine embryos. *Biology of Reproduction* **71** 2048-2055.

Burghardt RC, Johnson GA, Jaeger LA, Ka H, Garlow JE, Spencer TE & Bazer FW 2002 Integrins and extracellular matrix proteins at the maternal-fetal interface in domestic animals. *Cells, Tissues, Organs* **172** 202-217.

Cammas L, Reinaud P, Dubois O, Bordas N, Germain G & Charpigny G 2005 Identification of differentially regulated genes during elongation and early implantation in the ovine trophoblast using complementary DNA array screening. *Biology of Reproduction* **72** 960-967.

Cammas L, Reinaud P, Bordas N, Dubois O, Germain G & Charpigny G 2006 Developmental regulation of prostacyclin synthase and prostacyclin receptors in the ovine uterus and conceptus during the periimplantation period. *Reproduction* **131** 917-927.

Chakrabarty A, Green JA & Roberts RM 2006 Origin and evolution of the TKDP gene family. *Gene* **373** 35-43.

Chang MC 1952 Development of bovine blastocyst with a note on implantation. *The Anatomical Record* **113** 143-161.

Degrelle SA, Campion E, Cabau C, Piumi F, Reinaud P, Richard C, Renard JP & Hue I 2005 Molecular evidence for a critical period in mural trophoblast development in bovine blastocysts. *Developmental Biology* **288** 448-460.

Dore JJ, Roberts MP & Godkin JD 1994 Early gestational expression of retinol-binding protein mRNA by the ovine conceptus and endometrium. *Molecular Reproduction and Development* **38** 24-29.

Dunne LD, Diskin MG & Sreenan JM 2000 Embryo and foetal loss in beef heifers between day 14 of gestation and full term. *Animal Reproduction Science* **58** 39-44.

Farin CE, Imakawa K, Hansen TR, McDonnell JJ, Murphy CN, Farin PW & Roberts RM 1990 Expression of trophoblastic interferon genes in sheep and cattle. *Biology of Reproduction* **43** 210-218.

Flechon JE, Guillomot M, Charlier M, Flechon B & Martal J 1986 Experimental studies on the elongation of the ewe blastocyst. *Reproduction, Nutrition, Development* **26** 1017-1024.

Flechon JE, Degrouard J & Flechon B 2004 Gastrulation events in the prestreak pig embryo: ultrastructure and cell markers. *Genesis* **38** 13-25.

Garrett JE, Geisert RD, Zavy MT, Morgan GL 1988 Evidence for maternal regulation of early conceptus growth and development in beef cattle. *Journal of Reproduction and Fertility* **84** 437-446.

Geisert RD, Brookbank JW, Roberts RM & Bazer FW 1982 Establishment of pregnancy in the pig: II. Cellular remodeling of the porcine blastocyst during elongation on day 12 of pregnancy. *Biology of Reproduction* **27** 941-955.

Goff AK 2002 Embryonic signals and survival. *Reproduction in Domestic Animals* **37** 133-139.

Gray CA, Taylor KM, Ramsey WS, Hill JR, Bazer FW, Bartol FF & Spencer TE 2001 Endometrial glands are required for preimplantation conceptus elongation and survival. *Biology of Reproduction* **64** 1608-1613.

Gray CA, Burghardt RC, Johnson GA, Bazer FW & Spencer TE 2002 Evidence that absence of endometrial gland secretions in uterine gland knockout ewes compromises conceptus survival and elongation. *Reproduction* **124** 289-300.

Gray CA, Abbey CA, Beremand PD, Choi Y, Farmer JL, Adelson DL, Thomas TL, Bazer FW & Spencer TE 2006 Identification of endometrial genes regulated by early pregnancy, progesterone, and interferon tau in the ovine uterus. *Biology of Reproduction* **74** 383-394.

Grealy M, Diskin MG & Sreenan JM 1996 Protein content of cattle oocytes and embryos from the two-cell to the elongated blastocyst stage at day 16. *Journal of Reproduction and Fertility* **107** 229-233.

Green JA, Xie S, Quan X, Bao B, Gan X, Mathialagan N, Beckers JF & Roberts RM 2000 Pregnancy-associated bovine and ovine glycoproteins exhibit spatially and temporally distinct expression patterns during pregnancy. *Biology of Reproduction* **62** 1624-1631.

Greenstein JS, Murray RW & Foley RC 1958 Observations on the morphogenesis and histochemistry of the bovine preattachment placenta between 16 and 33 days of gestation. *The Anatomical Record* **132** 321-341.

Guillomot M 1995 Cellular interactions during implantation in domestic ruminants. *Journal of Reproduction and Fertility, Supplement* **49** 39-51.

Guillomot M, Turbe A, Hue I & Renard JP 2004 Staging of ovine embryos and expression of the T-box genes Brachyury and Eomesodermin around gastrulation. *Reproduction* **127** 491-501.

Guzman-Ayala M, Ben-Haim N, Beck S & Constam DB 2004 Nodal protein processing and fibroblast growth factor 4 synergize to maintain a trophoblast stem cell microenvironment. *Proceedings of the National Academy of Sciences of the United States of America* **101** 15656-15660.

Hennequet-Antier C, Chiapello H, Piot K, Degrelle S, Hue I, Renard JP, Rodolphe F & Robin S 2005 AnovArray: a set of SAS macros for the analysis of variance of gene expression data. *BMC Bioinformatics* **6** 150.

Heyman Y, Camous S, Fevre J, Meziou W & Martal J 1984 Maintenance of the corpus luteum after uterine transfer of trophoblastic vesicles to cyclic cows and ewes. *Journal of Reproduction and Fertility* **70** 533-540.

Hue I, Renard JP & Viebahn C 2001 Brachyury is expressed in gastrulating bovine embryos well ahead of implantation. *Development Genes and Evolution* **211** 157-159.

Hughes AL, Green JA, Piontkivska H & Roberts RM 2003 Aspartic proteinase phylogeny and the origin of pregnancy-associated glycoproteins. *Molecular Biology and Evolution* **20** 1940-1945.

Idkowiak J, Weisheit G, Plitzner J & Viebahn C 2004 Hypoblast controls mesoderm generation and axial patterning in the gastrulating rabbit embryo. *Development Genes and Evolution* **214** 591-605.

Imakawa K, Chang KT & Christenson RK 2004 Pre-implantation conceptus and maternal uterine communications: molecular events leading to successful implantation. The *Journal of Reproduction and Development* **50** 155-169.

Klein C, Bauersachs S, Ulbrich SE, Einspanier R, Meyer HH, Schmidt SE, Reichenbach HD, Vermehren M, Sinowatz F, Blum H, Wolf E 2006 Monozygotic twin model reveals novel embryo-induced transcriptome changes of bovine endometrium in the preattachment period. *Biology of Reproduction* **74** 253-264.

Kliem A, Tetens F, Klonisch T, Grealy M & Fischer B 1998 Epidermal growth factor receptor and ligands in elongating bovine blastocysts. *Molecular Reproduction and Development* **51** 402-412.

Klisch K, De Sousa NM, Beckers JF, Leiser R & Pich A 2005 Pregnancy associated glycoprotein-1, -6, -7, and -17 are major products of bovine binucleate trophoblast giant cells at midpregnancy. *Molecular Reproduction and Development* **71** 453-460.

Lee SH, Zhao SH, Recknor JC, Nettleton D, Orley S, Kang SK, Lee BC, Hwang WS & Tuggle CK 2005 Transcriptional profiling using a novel cDNA array identifies differential gene expression during porcine embryo elongation. *Molecular Reproduction and Development* **71** 129-139.

MacLean JA, 2nd, Chakrabarty A, Xie S, Bixby JA, Roberts RM & Green JA 2003 Family of Kunitz proteins from trophoblast: expression of the trophoblast Kunitz domain proteins (TKDP) in cattle and sheep. *Molecular Reproduction and Development* **65** 30-40.

MacLean JA, 2nd, Roberts RM & Green JA 2004 Atypical Kunitz-type serine proteinase inhibitors produced by the ruminant placenta. *Biology of Reproduction* **71** 455-463.

Meijer HA, Van De Pavert SA, Stroband HW & Boerjan ML 2000 Expression of the organizer specific homeobox gene goosecoid (gsc) in porcine embryos. *Molecular Reproduction and Development* **55** 1-7.

Pope WF 1988 Uterine asynchrony: a cause of embryonic loss. *Biology of Reproduction* **39** 999-1003.

Richardson L, Torres-Padilla ME & Zernicka-Goetz M 2006 Regionalised signalling within the extraembryonic ectoderm regulates anterior visceral endoderm positioning in the mouse embryo. *Mechanisms of Development* **123** 288-296.

Roberts RM, Ezashi T & Das P 2004 Trophoblast gene expression: transcription factors in the specification of early trophoblast. *Reproductive Biology and Endocrinology* **2** 47.

Ross JW, Ashworth MD, Hurst AG, Malayer JR & Geisert RD 2003 Analysis and characterization of differential gene expression during rapid tropho-blastic elongation in the pig using suppression subtractive hybridization. *Reproductive Biology and Endocrinology* **1** 23.

Rossant J 2004 Lineage development and polar asymmetries in the peri-implantation mouse blastocyst. *Seminars in Cell & Developmental Biology* **15** 573-581.

Satterfield MC, Bazer FW & Spencer T 2006 Progesterone regulation of preimplantation conceptus growth and galectin 15 (LGALS15) in the ovine uterus. *Biology of Reproduction* **75** 289-296.

Song G, Spencer TE & Bazer FW 2005 Cathepsins in the ovine uterus: regulation by pregnancy, progesterone, and interferon tau. *Endocrinology* **146** 4825-4833.

Spencer TE & Bazer FW 2004 Uterine and placental factors regulating conceptus growth in domestic animals. *Journal of Animal Science* **82 E-Suppl** E4-13.

Spencer TE, Johnson GA, Bazer FW & Burghardt RC 2004 Implantation mechanisms: insights from the sheep. *Reproduction* **128** 657-668.

Thatcher WW, Bartol FF, Knickerbocker JJ, Curl JS, Wolfenson D, Bazer FW & Roberts RM 1984 Maternal recognition of pregnancy in cattle. *Journal of Dairy Science* **67** 2797-2811.

Thatcher WW, Hansen PJ, Gross TS, Helmer SD, Plante C & Bazer FW 1989 Antiluteolytic effects of bovine trophoblast protein-1. *Journal of Reproduction and Fertility, Supplement* **37** 91-99.

Trout WE, McDonnell JJ, Kramer KK, Baumbach GA & Roberts RM 1991 The retinol-binding protein of the expanding pig blastocyst: molecular cloning and expression in trophectoderm and embryonic disc. *Molecular Endocrinology* **5** 1533-1540.

Ushizawa K, Herath CB, Kaneyama K, Shiojima S, Hirasawa A, Takahashi T, Imai K, Ochiai K, Tokunaga T, Tsunoda Y, Tsujimoto G, Hasizume K 2004 cDNA microarray analysis of bovine embryo gene expression profiles during the pre-implantation period. *Reproductive Biology and Endocrinology* **2** 77.

Vajta G, Alexopoulos NI & Callesen H 2004 Rapid growth and elongation of bovine blastocysts in vitro in a three-dimensional gel system. *Theriogenology* **62** 1253-1263.

van de Pavert SA, Schipper H, de Wit AA, Soede NM, van den Hurk R, Taverne MA, Boerjan ML & Stroband HW 2001 Comparison of anterior-posterior development in the porcine versus chicken embryo, using goosecoid expression as a marker. *Reproduction, Fertility and Development* **13** 177-185.

Vejlsted M, Avery B, Schmidt M, Greve T, Alexopoulos N & Maddox-Hyttel P 2005 Ultrastructural and immunohistochemical characterization of the bovine epiblast. *Biology of Reproduction* **72** 678-686.

Vejlsted M, Du Y, Vajta G & Maddox-Hyttel P 2006a Post-hatching development of the porcine and bovine embryo—defining criteria for expected development in vivo and in vitro. *Theriogenology* **65** 153-165.

Vejlsted M, Offenberg H, Thorup F & Maddox-Hyttel P 2006b Confinement and clearance of OCT4 in the porcine embryo at stereomicroscopically defined stages around gastrulation. *Molecular Reproduction and Development* **73** 709-718.

Wilson ME, Sonstegard TS, Smith TP, Fahrenkrug SC & Ford SP 2000 Differential gene expression during elongation in the preimplantation pig embryo. *Genesis* **26** 9-14.

Wintenberger-Torres S & Flechon JE 1974 Ultrastructural evolution of the trophoblast cells of the pre-implantation sheep blastocyst from day 8 to day 18. *Journal of Anatomy* **118** 143-153.

Wolf E, Arnold GJ, Bauersachs S, Beier HM, Blum H, Einspanier R, Frohlich T, Herrler A, Hiendleder S, Kolle S, Prelle K, Reichenbach HD, Stojkovic M, Wenigerkind H, Sinowatz F 2003 Embryo-maternal communication in bovine - strategies for deciphering a complex cross-talk. *Reproduction in Domestic Animals* **38** 276-289.

Xavier F, Lagarrigue S, Guillomot M & Gaillard-Sanchez I 1997 Expression of c-fos and jun protooncogenes in ovine trophoblasts in relation to interferon-tau expression and early implantation process. *Molecular Reproduction and Development* **46** 127-137.

Yelich JV, Pomp D & Geisert RD 1997a Detection of transcripts for retinoic acid receptors, retinol-binding protein, and transforming growth factors during rapid trophoblastic elongation in the porcine conceptus. *Biology of Reproduction* **57** 286-294.

Yelich JV, Pomp D & Geisert RD 1997b Ontogeny of elongation and gene expression in the early developing porcine conceptus. *Biology of Reproduction* **57** 1256-1265.

Fetal-maternal interactions during the establishment of pregnancy in ruminants

TE Spencer[1,2], GA Johnson[2], FW Bazer[1,2] and
RC Burghardt[2]

[1]Center for Animal Biotechnology and Genomics, Departments of Animal Science and [2]Veterinary
Integrative Biosciences, Texas A&M University, College Station, Texas, 77843 USA

This review integrates established information with new insights into molecular and physiological mechanisms responsible for events leading to pregnancy recognition, endometrial receptivity, and implantation with emphasis on sheep. After formation of the corpus luteum, progesterone acts on the endometrium and stimulates blastocyst growth and elongation to form a filamentous conceptus (embryo/fetus and associated extraembryonic membranes). Recurrent early pregnancy loss in the uterine gland knockout ewe model indicates that endometrial epithelial secretions are essential for peri-implantation blastocyst survival and growth. The elongating sheep conceptus secretes interferon tau (IFNT) that acts on the endometrium to inhibit development of the luteolytic mechanism by inhibiting transcription of the estrogen receptor alpha (ESR1) gene in the luminal (LE) and superficial ductal glandular (sGE) epithelia, which prevents estrogen-induction of oxytocin receptors (OXTR) and production of luteolytic prostaglandin F2-alpha pulses. Progesterone downregulates its receptors (PGR) in LE and then GE, correlating with a reduction of anti-adhesive MUC1 (mucin glycoprotein one) and induction of secreted LGALS15 (galectin 15) and SPP1 (secreted phosphoprotein one), that are proposed to regulate trophectoderm growth and adhesion. IFNT acts on the LE to induce WNT7A (wingless-type MMTV integration site family member 7A) and to stimulate LGALS15, CTSL (cathepsin L), and CST3 (cystatin C), which may regulate conceptus development and implantation. During the peri-implantation period, trophoblast giant binucleate cells (BNC) begin to differentiate from mononuclear trophectoderm cells, migrate and then fuse with the uterine LE as well as each other to form multinucleated syncytial plaques. Trophoblast giant BNC secrete chorionic somatomammotropin (CSH1 or placental lactogen) that acts on the endometrial glands to stimulate their morphogenesis and differentiated function. The interactive, coordinated and stage-specific effects of ovarian and placental hormones regulate endometrial events necessary for fetal-maternal interactions and successful establishment of pregnancy.

Corresponding author E-mail: tspencer@tamu.edu

Introduction

Establishment of pregnancy includes pregnancy recognition signaling for maintenance of a functional corpus luteum, and endometrial differentiation and function for uterine receptivity to implantation of the conceptus (embryo/fetus and associated extraembryonic membranes). The coordinated actions of progesterone and placental hormones regulate fetal-maternal inter-actions required for establishment and maintenance of pregnancy. This review integrates established information with new insights into molecular and physiological mechanisms re-sponsible for events leading to pregnancy recognition, endometrial receptivity, and implanta-tion with emphasis on sheep.

Overview of pregnancy establishment and conceptus implantation

Establishment of pregnancy in domestic ruminants (sheep, cattle, goats) begins at the blasto-cyst stage and involves coordinate pregnancy recognition signaling and conceptus implanta-tion (Fig. 1). According to Guillomot (1995), the phases of implantation are designated as: (1) shedding of the zona pellucida; (2) pre-contact and blastocyst orientiation; (3) apposition; (4) adhesion; and (5) endometrial invasion. All of these phases occur in domestic ruminants, but endometrial invasion is very limited (see Figs. 1 and 2).

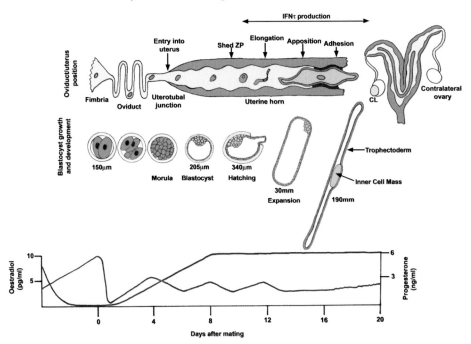

Fig. 1. Early pregnancy events in sheep. This schematic illustrates changes in embryo/blastocyst/conceptus development after fertilization in relation to position in the female reproductive tract and circulating levels of ovarian steroid hormones. Fertilization occurs in the oviduct and morula stage embryos enter into the uterus on Day 4-5. Blastocysts form by Day 6, hatch from the zona pellucida on Day 8, transition from spherical to tubular forms by Day 11 and then elongate to filamentous conceptuses between Days 12 and 16. Elongation of conceptuses marks the beginning of implantation which involves apposition and transient attachment (Days 12 to 15) and firm adhesion by Day 16. By Day 17, the filamentous conceptus occupies the entire ipsilateral horn and has elongated through the uterine body into the contralateral horn.

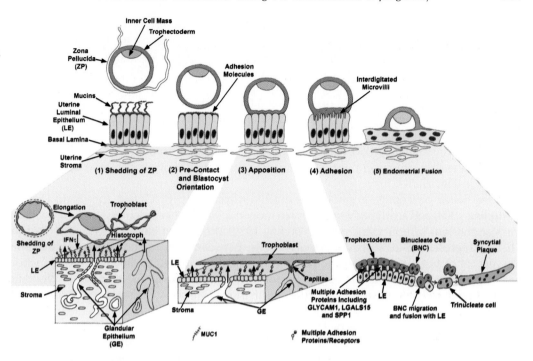

Fig. 2. Phases of blastocyst implantation in sheep. Pre-attachment blastocysts undergo shedding of the zona pellucida (Phase 1) and pre-contact blastocyst orientation (Phase 2). The first phase of implantation is shedding of the zona pellucida on Day 8 that exposes the trophectoderm to begin the second pre-implantation phase wherein blastocyst orientation occurs on Days 9 to 11 as it transitions from a spherical to a tubular conceptus and migrates to the middle region of the uterine horn ipsilateral to the corpus luteum. The anti-adhesive MUC1 present on endometrial LE prevents contact of trophectoderm and adhesive receptors such as integrins. Histotroph, secreted from endometrial LE and GE under the influence of progesterone, nourishes the developing blastocysts. Phase 3 includes apposition and transient attachment. After Day 11, the tubular blastocyst elongates to form a filamentous conceptus. During this period, expression of MUC1 declines on LE to expose constitutively expressed integrins on LE and trophectoderm. Apposition between trophectoderm and endometrial LE occurs followed by formation of trophoblast papillae that extend into superficial ducts of uterine glands. Elongation of conceptuses likely requires apposition and transient attachment of trophectoderm to endometrial LE. Firm adhesion of trophectoderm to endometrial LE occurs during Phase 4 between Days 15 and 16. Available evidence indicates that several molecules (GLYCAM1, LGALS15, SPP1) secreted by the endometrial epithelia interact with receptors (integrins and glycoconjugates) on the apical surface of trophectoderm and LE to facilitate adhesion. Invasion of the endometrium during Phase 5 involves formation of binucleate cells (BNC) that begin differentiate from mononuclear trophoblast cells between Days 14 and 16 and then migrate to and fuse with LE to form multinucleated syncytial plaques. Although the BNC are inherently invasive, they do not cross the basal lamina between the LE and stratum compactum stroma.

In sheep (Fig. 1), morula-stage embryos enter the uterus on Day 4-5 and the blastocyst, formed by Day 6, contains an inner cell mass and a blastocoele or central cavity surrounded by a monolayer of trophectoderm (Guillomot 1995; Spencer *et al.* 2004a). After hatching from the zona pellucida on Day 8, blastocysts develop into a tubular form by Day 11, and then elongate on Day 12 to 10 cm or more in length by Day 14 and achieve a length of 25 cm or more by Day 17. The elongated or filamentous conceptus is composed mainly of extraembryonic trophecto-derm. Hatched blastocysts and trophoblastic vesicles do not elongate *in vitro,* but do so when transferred into the uterus of either sheep or cows (Heyman *et al.* 1984; Flechon *et al.* 1986). Elongation of the blastocyst is critical for developmentally regulated production of interferon tau (IFNT), the pregnancy recognition signal, and for implantation (Farin *et al.* 1989; Guillomot *et al.* 1990; Gray *et al.* 2002). Between Days 9 and 14, close association between conceptus trophectoderm and endometrial luminal epithelium (LE) is achieved and followed first by adhe-sion and then interdigitation of cytoplasmic projections of trophectoderm cells and microvilli of the LE assures firm attachment in both the caruncular and intercaruncular areas by Day 16 of pregnancy (Fig. 2).

In the sheep conceptus, trophoblast giant binucleate cells (BNC) first appear on Day 14 (Wooding 1984) and are thought to arise from mononuclear trophectoderm cells by consecutive nuclear divisions without cytokinesis (Wooding 1992) also termed mitotic polyploidy. By Day 16, BNC represent 15-20% of trophectoderm cells. Between Days 16 and 24, the LE transforms to syncytial plaques which appear to result from migration of BNC to the microvillar junction and then fusion with individual LE cells to produce trinucleate fetomaternal hybrid cells (Wooding 1984). Continued BNC migration and fusion with trinucleate cells, together with displacement and/or death of the remaining uterine LE, apparently produces the multinucleated syncytial plaques that cover the caruncles by Day 24. The syncytial plaques appear to be linked by tight junctions and limited in size to 20-25 nuclei. This caruncular syncytium, in which no nuclear division has been reported, expands in area during formation and maintenance of cotyledons, presumably deriving nuclei from continued BNC migration and fusion. The BNC have at least two main functions: (1) formation of a hybrid fetomaternal syncytium for successful implanta-tion and subsequent cotyledonary growth in the placentome; and (2) synthesis and secretion of protein and steroid hormones such as CSH1 (chorionic somatomammotropin hormone 1; alias placental lactogen)), PAGs (pregnancy associated glycoproteins), and progesterone (Wooding 1992). As will be discussed later, placental CSH1 binds to prolactin receptors (PRLR) unique to uterine GE and stimulates their growth and differentiated functions during pregnancy (Spencer *et al.* 2004c).

Pregnancy recognition signaling by IFNT

Maternal recognition of pregnancy is the physiological process whereby the conceptus signals its presence to the maternal system and prolongs the lifespan of the corpus luteum (CL) (Bazer *et al.* 1991). In ruminants, IFNT is the pregnancy recognition signal secreted by the elongating conceptus (Roberts *et al.* 1999). In addition to its antiluteolytic actions, IFNT acts on the endometrium to induce or enhance expression of a number of genes (IFNT-stimulated genes or ISGs) that are hypothesized to regulate uterine receptivity and conceptus development during implantation (Hansen *et al.* 1999; Spencer *et al.* 2004a).

Inhibition of the endometrial luteolytic mechanism

Sheep experience uterine-dependent oestrous cycles until establishment of pregnancy (Spen-cer & Bazer 2004). The oestrous cycle is dependent on the uterus, because it releases prostag-landin F2 alpha (PGF) in a pulsatile manner to induce luteolysis during late diestrus. As illus-trated in Fig. 3, the luteolytic pulses of PGF are produced by the endometrial LE and superficial ductal glandular epithelium (sGE) in response to oxytocin binding to oxytocin receptors (OXTR)

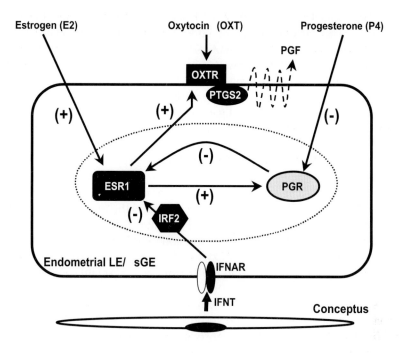

Fig. 3. Schematic illustrating hormonal regulation of the endometrial luteolytic mechanism and antiluteolytic effects of the conceptus on the endometrium in the ovine uterus. During estrus and metestrus, OXTR are present on LE/sGE because circulating estrogens increase expression of ESR1 and OXTR. The PGR are present, but circulating levels of progesterone are inadequate to activate PGR and suppress expression of ESR1 and OXTR. During early diestrus, endometrial ESR1 and estrogen are low and circulating progesterone increases with maturation of the CL so that to activate, via PGR, suppression of expression of ESR1 and OXTR for 8 to 10 days. Continuous exposure of the endometrium to progesterone eventually results in down-regulation of *PGR* in endometrial LE/sGE by Days 11 to 12 of the estrous cycle to end the progesterone block to *ESR1* and *OXTR* expression. Thus, *ESR1* expression increases on Days 11 and 12 post-estrus followed by estrogen induction of OXTR on LE/sGE on Days 13 and 14. The increase in OXTR is facilitated by increasing secretion of estrogen by ovarian follicles. In both cyclic and pregnant sheep, OXT is released from the posterior pituitary and CL beginning on Day 9. In cyclic ewes, OXT binds to OXTR on LE/sGE to induce release of luteolytic pulses of PGF that leads to CL regression through a PTGS-dependent pathway. In pregnant sheep, IFNT is synthesized and secreted by the elongating conceptus between Days 11 to 12 and 21 to 25 of pregnancy. IFNT binds to Type I IFN receptors (IFNAR) on endometrial LE/sGE to inhibit transcription of the *ESR1* gene through a signaling pathway involving IRF2. These antiluteolytic actions of IFNT on the *ESR1* gene prevent ESR1 expression and, therefore, the ability of estrogen to induce expression of OXTR required for pulsatile release of luteolytic PGF. E2, Estrogen; ESR1, estrogen receptor; IFNAR, Type I IFN receptor; IFNT, interferon tau; IRF2, interferon regulatory factor two; OXT, oxytocin; OXTR, oxytocin receptor; P4, progesterone; PGF, prostaglandin F2α; PGR, progesterone receptor; PTGS2, prostaglandin-endoperoxide synthase 2 (prostaglandin G/H synthase and cyclooxygenase)

on those epithelia (McCracken *et al.* 1999). Expression of the *OXTR* is regulated by progesterone and estrogen. Progesterone acts via progesterone receptors (PGR) to "block" expression of oestrogen receptor alpha (ESR1) and OXTR in LE/sGE between Days 5 and 11 after which time progesterone downregulates PGR in LE/sGE after Day 11 and GE after Day 13 in cyclic ewes which allows rapid increases in expression of *ESR1* on Days 12 to 13 and then *OXTR* on Day 14 (see (Spencer & Bazer 2004)). ESR1, presumably activated by oestrogen from ovarian follicles or possibly growth factors from the stroma, stimulates transcription of the *OXTR* gene. Two GC-rich SP1 binding sites, located within 140 bp of the translational start site, regulate basal activity of the ovine *OXTR* promoter as well as liganded ESR1 induction of promoter activity (Fleming et al. 2006). Oxytocin, secreted from as early as Day 9 from posterior pituitary and/or the CL, binds to OXTR to induce pulsatile release of PGF between Days 14 and 16, which elicits structural and functional regression of the CL.

IFNT is synthesized and secreted by conceptus trophectoderm between Days 10 and 20 or so of gestation in sheep (Roberts et al. 1999), acts on endometrial LE/sGE to inhibit transcription of the *ESR1* gene and, therefore, *OXTR* gene, which abrogates the luteolytic mechanism required for uterine production of luteolytic pulses of PGF (Wathes & Lamming 1995). IFNT does not inhibit either basal production of PGF or expression of *PTGS2* (prostaglandin-endoperoxide synthase 2; alias COX-2), a rate-limiting enzyme in prostaglandin production, in endometrial LE/sGE of pregnant ewes (Charpigny et al. 1997; Kim et al. 2003b). Rather, IFNT indirectly inhibits *OXTR* gene transcription by silencing *ESR1* gene transcription in the endometrial LE/sGE. The promoter/enhancer region of the ovine *ESR1* gene contains four IFN regulatory factor elements (IRFEs) and one IFN-stimulated response element (ISRE) that bind IFN regulatory factor two (IRF2) (Fleming et al. 2001), a transcriptional repressor expressed specifically in LE/sGE of the ovine uterus that increases in abundance from Day 10 to Day 16 in pregnant, but not cyclic ewes (Choi et al. 2001). IFNT and IRF2 inhibit transcriptional activity of the ovine *ESR1* promoter *in vitro*, but the precise cellular and molecular mechanism remains to be determined (Fleming et al. 2001). However, IFNT does not affect activity of the ovine *OXTR* gene promoter nor inhibit liganded ESR1 induction of *OXTR* promoter activity (Fleming et al. 2006). Thus, available results from sheep support the hypothesis that the antiluteolytic effects of IFNT are to silence *ESR1* gene transcription in LE/sGE subsequent to PGR down-regulation, which precludes induction of *OXTR* gene expression and uterine release of luteolytic pulses of PGF. In cows, the conceptus, presumably via IFNT, inhibits expression of endometrial OXTR before suppressing ESR1, suggesting mechanistic differences between cattle and sheep in terms of the endometrial luteolytic mechanism and pregnancy recognition signaling (Robinson et al. 2001).

IFNT-stimulated genes (ISGs) in the endometrium

IFNT induces or increases expression of several ISGs in the endometrium that are hypothesized to be important for conceptus implantation (Hansen et al. 1999; Spencer et al. 2004a). Since expression of ISGs increases in a stage-specific manner within the endometria of diverse species, including domestic animals, laboratory rodents, primates and humans during early pregnancy, they may be universally important in establishment of uterine receptivity to conceptus implantation.

Bovine endometrial, ovine endometrial and human 2fTGH fibroblast cells have been used to determine that IFNT activates the classical JAK-STAT-IRF (janus kinase-signal transducer and activator of transcription-interferon regulatory factor) signaling pathway used by other Type I IFNs (see (Stark et al. 1998) and Fig. 4). Many classical ISGs, such as *ISG15* (ISG15 ubiquitin-like modifier; alias IFI15 or UCRP), are expressed in LE/sGE of the ovine uterus on Days 10 or 11 of the cycle and pregnancy, but are undetectable in LE/sGE by Days 12 to 13 (Johnson et al.

1999c) (see Table 1 and Fig. 4). In response to IFNT from elongating conceptuses, *ISG15* is induced in the stratum compactum stroma and GE by Days 13 to 14, and expression extends to the stratum spongiosum stroma, deep glands, and myometrium of the ovine uterus by Days 15 to 16 of pregnancy (Johnson *et al.* 1999c; Johnson *et al.* 2000a). As IFNT production by the conceptus declines, expression of ISGs also declines, but some remain abundant in endometrial stroma and GE on Days 18 to 20 of pregnancy. Similar temporal and spatial alterations in *ISG15* expression occurs in the bovine uterus during early pregnancy (Johnson *et al.* 1999a). Indeed, numerous ISGs are induced or stimulated in the endometrium during conceptus elongation in both cattle and sheep (Klein *et al.* 2005; Gray *et al.* 2006).

Curiously, *in vivo* studies revealed that many classical ISGs (*STAT1, STAT2, ISGF3G (IFN-stimulated gene factor 3 gamma), B2M (beta-2-microglobulin), GBP2 (guanylate binding protein 2), IFI27 (interferon, alpha-inducible protein 27), IFIT1 (interferon-induced protein with tetratricopeptide repeats 1), ISG15, MIC (MHC class I polypeptide),* and *OAS (2',5'-oligoadenylate synthetases 1, 2 and 3)* are not induced or increased by IFNT in LE/sGE of the sheep uterus (Johnson *et al.* 1999c; Choi *et al.* 2001; Johnson *et al.* 2001b; Choi *et al.* 2003; Kim *et al.* 2003a). This finding was initially surprising, because all endometrial cell types express IFNAR1 (interferon (alpha, beta and omega) receptor 1) and IFNAR2 subunits of the common Type I IFN receptor (Rosenfeld *et al.* 2002). Available results indicate that IRF2, a potent transcriptional repressor of ISGs, is expressed specifically in LE/sGE and represses transcriptional activity of ISRE- and IRFE-containing promoters (Choi *et al.* 2001). Thus, IRF2 in LE/sGE is proposed to restrict IFNT induction of many ISGs to stroma and GE of the ovine uterus (see Fig. 4 and Table 1). In fact, all components of ISGF3 (*STAT1, STAT2, ISGF3G*) and the other studied ISGs (*B2M, GBP2, IFI27, IFIT1, ISG15, MIC, OAS*) contain ISREs in their promoters. The silencing of *MIC* and *B2M* genes in endometrial LE/sGE during pregnancy recognition and establishment may be a critical mechanism preventing immune rejection of the conceptus allograft (Choi *et al.* 2003). Given that the critical signaling components of the JAK-STAT signaling system (STAT1, STAT2, ISGF3G) are not expressed in the endometrial LE/sGE, IFNT must utilize a non-classical STAT1-independent cell signaling pathway to regulate expression of genes in endometrial LE (Fig. 4). Transcriptional profiling of human U3A (STAT1 null) cells and ovine endometrium treated with IFNT were used to discover novel ISGs in the endometrial LE/sGE during pregnancy establishment including *WNT7A* (wingless-type MMTV integration site family, member 7A), *LGALS15* (lectin, galactoside-binding, soluble, 15), *CTSL* (cathepsin L) and *CST3* (cystatin C) (Kim *et al.* 2003a; Song *et al.* 2005; Gray *et al.* 2006; Song *et al.* 2006b). The expression patterns of classical and novel ISGs are summarized in Table 1, and proposed biological roles of novel epithelial ISGs in conceptus implantation will now be discussed.

WNT7A. During early pregnancy, *WNT7A* mRNA is detected on Day 10, undetectable on Day 12, and is the only gene found to be induced by IFNT between Days 12 and 14 of pregnancy specifically in LE/sGE (Kim *et al.* 2003a). The WNT family (19 genes in human) includes many highly conserved and secreted glycoproteins that regulate cell and tissue growth and differentiation during embryonic development and play a central role in coordinating uterine-conceptus interactions required for implantation in mice and perhaps humans (Mohamed *et al.* 2005). WNT7A from the endometrial LE may activate canonical signaling pathways in the trophectoderm to stimulate proliferation and differentiation. Further, WNT7A may have autocrine actions on LE/sGE to regulate expression of target genes important for uterine receptivity and conceptus implantation, such as *LGALS15, CST3* and *CTSL*.

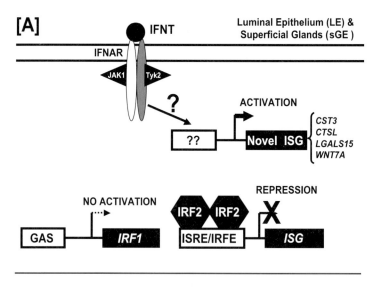

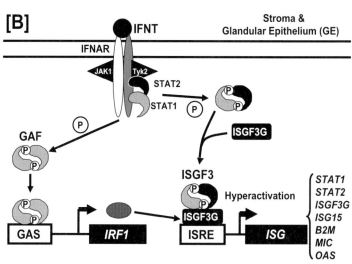

Fig. 4. Schematic illustrating current working hypothesis on IFNT signaling in the endometrium of the ovine uterus. IFNT, produced by developing conceptuses of ruminants, binds to IFNAR present on cells of the ovine endometrium. In LE/sGE, IFNT is prevented from activating ISGs by IRF2 (see upper panel [A]). IRF2, a potent and stable transcriptional repressor present in the nucleus, increases during early pregnancy in LE/sGE. The continual presence of IRF2 inhibits ISRE- and IRFE-containing target genes (*STAT1*, *STAT2*, *ISGF3G*, *BMG*, *ISG15*, *MHC*, *OAS*) through direct ISRE and IRFE binding and coactivator repulsion. Furthermore, critical factors in the JAK-STAT-IRF pathway (STAT1, STAT2, and ISGF3G) are not present, resulting in the absence of ISGF3 or IRF1 transcription factors necessary to stimulate ISGs. However, IFNT does activate an unknown cell signaling pathway that results in induction of *WNT7A* and stimulation of *CST3*, *CTSL* and *LGALS15* specifically in LE/sGE. In cells of the stroma and middle to deep GE (see lower panel [B]), IFNT-mediated association of IFNAR subunits facilitates cross-phosphorylation and activation of JAK, which in turn phosphorylates the receptor and creates a docking site for STAT2. STAT2 is then phosphorylated, thus creating a docking site for STAT1 which is

then phosphorylated. STAT1 and STAT2 are then released from the receptor and can form two transcription factor complexes. ISGF3 is formed by association of a STAT1-2 heterodimer and ISGF3G in the cytoplasm, translocates to the nucleus, and transactivates genes containing an ISRE(s), such as *STAT1, STAT2, ISGF3G, BMG, ISG15, MHC* and *OAS.* GAF is formed by STAT1 homodimers, which translocates to the nucleus and transactivates genes containing a GAS element(s), such as *IRF1.* IRF1 can also bind and transactivate ISRE-containing genes as well an IRFE-containing genes. The simultaneous induction of *STAT2* and *ISGF3G* by IFNT appears to shift transcription factor formation from GAF towards predominantly ISGF3. Therefore, IFNT activation of the JAK-STAT-IRF signal transduction pathway allows for constant formation of ISGF3 and GAF transcription factor complexes and hyperactivation of ISG expression. B2M, beta-2-microglobulin; CST3, cystatin C; CTSL, cathepsin L; GAF, gamma activated factor; GAS, gamma activation sequence; IFNAR, Type I IFN receptor; IFNT, interferon tau; IRF1, interferon regulatory factor one; IRF2, interferon regulatory factor two; IRFE, IRF-response element; ISGF3G, interferon-stimulated transcription factor 3, gamma 48kDa; ISG15, (ISG15 ubiquitin-like modifier; alias IFI15 or UCRP); ISRE, IFN-stimulated response element; JAK, janus kinase; LGALS15, galectin-15; MIC, MHC class I polypeptide-related sequence; OAS, 2',5'-oligoadenylate synthetases; STAT1, signal transducer and activator of transcription 1, 91kDa; STAT2, signal transducer and activator of transcription 2, 113kDa; WNT7A, wing-less-type MMTV integration site family, member 7A.

CTSL and CST3. *CTSL* and *CST3* are induced by progesterone in LE/sGE between Days 10 and 12 and are further increased by IFNT (Song *et al.* 2005; 2006b). Cathepsins are peptidases that can degrade extracellular matrix, catabolize intracellular proteins, process prohormones, and regulate uterine receptivity for implantation and trophoblast invasion in several mammals (Salamonsen 1999). CST3 is an inhibitor of CTSL. A balance of proteases and their inhibitors is likely required to modify the glycocalyx on endometrial LE and trophoblast during apposition and adhesion phases of implantation (Carson *et al.* 2000).

LGALS15. Similar to *CTSL* and *CST3*, *LGALS15* is induced by progesterone in LE/sGE between Days 10 and 14 and is further increased by IFNT (Gray *et al.* 2004). Galectins are proteins with a conserved carbohydrate recognition domain that bind beta-galactosides, thereby cross-linking glycoproteins as well as glycolipid receptors on the surface of cells, such as integrins, and initiating biological responses (Yang & Liu 2003). LGALS15, originally termed ovgal11, was originally identified in ovine intestinal epithelium as being induced in response to infection by the nematode parasite *Haemonchus contortus* (Dunphy *et al.* 2000). Interestingly, LGALS15 is the 14K protein from sheep endometrium initially characterized as a progesterone-modulated protein associated with crystalline inclusion bodies in uterine epithelia and conceptus trophectoderm (Kazemi *et al.* 1990). LGALS15 is implicated in conceptus implantation (Spencer *et al.* 2004a), because functional studies of other galectins have implicated these proteins in cell growth, differentiation and apoptosis as well as in cell adhesion, chemoattraction and migration (Yang & Liu 2003). Indeed, some galectin family members are involved in both innate and adaptive immune responses and participate in the activation or differentiation of immune cells.

CXCL10. The only ISG with a reported biological effect on trophectoderm growth and adhesion is chemokine (C-X-C motif) ligand 10 (CXCL10; alias IP-10) (Nagaoka *et al.* 2003a; Nagaoka *et al.* 2003b). CXCL10 is a member of the C-X-C chemokine family that regulates multiple aspects of inflammatory and immune responses primarily through chemotactic activity toward subsets of leukocytes. *CXCL10* mRNA was localized to monocytes in the subepithelial stroma of pregnant, but not cyclic uteri of sheep. Whether IFNT directly regulates *CXCL10* in the monocytes or simply attract the monocytes to the endometrium remains to be determined. In the ovine uterus, CXCL10 appeared on Day 17 in the uterine lumen, and the CXCR3 recep-

Table 1. Temporal roadmap of progesterone and IFNT regulated genes during establishment of pregnancy in sheep[1]

| | Day of pregnancy | | | | |
	10/11	12/13	14/15	16/17	18/20
Conceptus	IFNT	IFNT	**IFNT**	**IFNT**	IFNT
				CSH1	CSH1
Uterine	GLYCAM1	GLYCAM1	**GLYCAM1**	**GLYCAM1**	
Lumen	LGALS15	LGALS15	**LGALS15**	**LGALS15**	
Protein			SPP1	SPP1	n.d.[2]
		CTSL	**CTSL**	**CTSL**	
		CST3	**CST3**	**CST3**	
			CXCL10	CXCL10	
LE	PGR				
	MUC1	**MUC1**	MUC1	MUC1	
		GLYCAM1	**GLYCAM1**	**GLYCAM1**	**GLYCAM1**
		LGALS15	**LGALS15**	**LGALS15**	**LGALS15**
		CTSL	**CTSL**	**CTSL**	CTSL
		CST3	**CST3**	CST3	CST3
	WNT7A		WNT7A	WNT7A	WNT7A
		PTGS2	PTGS2	PTGS2	PTGS2
	IRF2	IRF2	**IRF2**	IRF2	IRF2
				G1P3	G1P3
	ISG15	ISG15			
	MIC	MIC			
	B2M	B2M			
GE	PGR	PGR			
	PRLR	PRLR	PRLR	PRLR	PRLR
		SPP1	SPP1	**SPP1**	**SPP1**
			CTSL	**CTSL**	CTSL
			CST3	**CST3**	CST3
					SERPIN
					STC1
				GRP	GRP
			STAT1	**STAT1**	STAT1
			STAT2	**STAT2**	STAT2
			ISGF3G	**ISGF3G**	ISGF3G
			IRF1	**IRF1**	IRF1
		GBP2	GBP2	**GBP2**	GBP2
		IFIT1	IFIT1	**IFIT1**	IFIT1
			IFI27	**IFI27**	IFI27
			ISG15	**ISG15**	ISG15
			MIC	**MIC**	MIC
			B2M	**B2M**	**B2M**
ST	PGR	PGR	PGR	**PGR**	**PGR**
			STAT1	**STAT1**	**STAT1**
			STAT2	**STAT2**	**STAT2**
			ISGF3G	**ISGF3G**	**ISGF3G**
			IRF1	IRF1	**IRF1**
		GBP2	**GBP2**	**GBP2**	**GBP2**
		IFIT1	IFIT1	**IFIT1**	**IFIT1**
			ISG15	**ISG15**	**ISG15**
			MIC	**MIC**	**MIC**
			B2M	**B2M**	**B2M**

[1]Relative abundance of mRNAs across days of pregnancy are indicated: gray = low; normal = moderate; and **bold** = high).

[2]Not determined (n.d.) due to inability to flush intact conceptuses from the uterine lumen on Days 18-20 of pregnancy.

tor was localized to trophoblast cells. Subsequently, recombinant CXCL10 was shown to stimulate migration of trophoblast cells and promote their adhesion to fibronectin, as well as increase expression of integrins α5, αV, and ß3 subunit mRNAs in trophoblast cells (Nagaoka *et al.* 2003b; Imakawa *et al.* 2006). Integrins are essential for conceptus implantation (Burghardt *et al.* 2002) and will be discussed later.

Functional role of the endometrial epithelia in blastocyst growth and elongation

All mammalian uteri contain endometrial epithelia that synthesize and secrete or transport a complex mixture of amino acids, ions, glucose, enzymes, growth factors, hormones, transport proteins and other substances termed histotroph (Bazer 1975). During the pre-attachment period, nutrition of the conceptus depends on uterine secretions. The epithelial cells of the uterine lumen are highly secretory during implantation, and the trophectoderm exhibits intense pinocytotic activity which increases as the conceptus develops (Guillomot 1995). Therefore, factors supporting growth of pre- and peri-implantation blastocysts and elongating conceptuses are thought to be obtained primarily from uterine histotroph. This hypothesis is supported by results from studies of asynchronous uterine transfer of embryos and trophoblast vesicles (Lawson *et al.* 1983; Flechon et al. 1986) and from studies of uterine gland knockout (UGKO) ewes (Gray *et al.* 2001b; Gray *et al.* 2002).

The UGKO ewe model is produced by continuous administration of a synthetic, non-metabolizable progestin to neonatal ewes from birth to postnatal day 56 (Gray *et al.* 2000a). This inappropriate exposure to a progestin permanently ablates differentiation and development of the endometrial glands from LE and produces an UGKO phenotype without apparent alterations in development of myometrium, or other Müllerian duct-derived female reproductive tract structures, or function of the hypothalamic-pituitary-ovarian axis (Gray *et al.* 2000b; Gray *et al.* 2001a). The UGKO endometrium is devoid of glands with markedly reduced LE surface area. UGKO ewes exhibit recurrent early pregnancy loss as blastocyst fails to elongate, and transfer of blastocysts from uteri of control ewes into uteri of timed recipient UGKO ewes does not ameliorate this defect (Gray *et al.* 2001b). Morphologically normal blastocysts are present in uterine flushes of bred UGKO ewes on Days 6 and 9 post-mating, but not on Day 14 (Gray *et al.* 2001b; Gray *et al.* 2002) when uteri contain either no conceptus or a severely growth-retarded tubular conceptus. These results demonstrate that histotroph from endometrial epithelia are required for peri-implantation blastocyst survival and elongation in sheep.

Defects in blastocyst survival and elongation in UGKO ewes are not due to alterations in expression of steroid receptors, anti-adhesive mucin glycoprotein one (MUC1), adhesive integrins on the endometrial LE, or responsiveness of the endometrium to IFNT (Gray *et al.* 2000a; Gray *et al.* 2002). However, uterine flushes from Day 14 bred UGKO ewes contain either very low amounts or undetectable amounts of secreted phosphoprotein one (SPP1 or osteopontin) and glycosylated cell adhesion molecule one (GLYCAM1) proteins which are adhesion proteins secreted by the LE and GE that are relatively abundant in uterine histotroph (Gray *et al.* 2002). Therefore, the reduction or absence of adhesion proteins from endometrial GE is a likely cause of recurrent pregnancy loss in UGKO ewes. Other essential, but as yet undefined, components of histotroph are undoubtedly absent or reduced in the uteri of infertile UGKO ewes.

Progesterone, adhesion molecules and implantation

The cellular and molecular mechanism(s) regulating trophectoderm outgrowth during blastocyst elongation, although not well understood, are hypothesized to require progesterone-dependent uterine secretions as well as apposition and transient attachment of the trophoblast to the LE that is mediated by factors of endometrial origin (see Fig. 2).

Progesterone regulation of pre-implantation blastocyst growth and elongation

As the hormone of pregnancy, progesterone stimulates and maintains endometrial functions necessary for conceptus growth, implantation, placentation and development to term (Bazer *et al.* 1979; Geisert *et al.* 1992; Spencer & Bazer 2002; Spencer *et al.* 2004b). Circulating concentrations of progesterone in early pregnancy affect blastocyst survival and growth during early pregnancy (Mann & Lamming 1999). In dairy cattle, successful establishment of pregnancy and rapid blastocyst development occurs in cows with an early rise in progesterone after ovulation and mating (Mann & Lamming 2001). Exogenous progesterone from Days 2 to 5 and Days 5 to 9 enhanced conceptus development and size in heifers on Day 14 and cows on Day 16, respectively (Garrett *et al.* 1988; Mann *et al.* 2006); however, exogenous progesterone from Days 12 to 16 did not affect conceptus development in cows on Day 16 (Mann *et al.* 2006). Heifers and ewes with lower concentrations of progesterone in the early luteal phase had retarded conceptuses that secreted less IFNT (Nephew *et al.* 1991; Mann & Lamming 2001). Indeed, advancement of conceptus development by administration of progesterone during metestrus and early diestrus has been described also in sheep (Kleemann *et al.* 1994).

The mechanisms whereby progesterone stimulates blastocyst survival and growth are not known, but presumed to be mediated by histotroph (Geisert *et al.* 1992). Uterine-derived growth factors, including CSF2 (colony stimulating factor 2 (granulocyte-macrophage)), FGF2 (fibroblast growth factor 2 (basic)), IGF1 (insulin-like growth factor one), and IGF2, stimulate IFNT production by cultured conceptuses and isolated mononuclear trophectoderm cells (Ko *et al.* 1991; Imakawa *et al.* 1993; Michael *et al.* 2006); however, these effects could be either direct or indirect due to effects on cell proliferation. As noted previously, progesterone acts on the endometrium to induce a number of epithelial genes (*CST3, CTSL, GLYCAM1, LGALS15, SPP1*) in a stage-specific manner that are hypothesized to regulate conceptus development during the peri-implantation period of pregnancy. Paradoxically, progesterone induction of those genes appears to be via the loss of PGR in the endometrial epithelia as discussed next.

PGR regulation and endometrial gene expression

In most mammalian uteri, PGR are expressed in endometrial epithelia and stroma during the early to mid-luteal phase, allowing direct regulation (induction or repression) of genes by progesterone. However, continuous exposure of the endometrium to progesterone negatively regulates PGR expression in LE and GE (see Table 1). In the ovine uterus, PGR protein is not detectable in LE and GE after Days 11 and 13 of pregnancy, respectively, but can be detected in the uterine stroma and myometrium throughout gestation (Spencer *et al.* 2004). The paradigm of loss of PGR in uterine epithelia immediately prior to implantation is common in domestic ruminants and across mammals (Carson *et al.* 2000), strongly suggesting that the loss of PGR alters the program of gene expression in the endometrial LE and then GE (see Table 1). PGR loss in those epithelia is determined by timing of the post-ovulatory rise in progesterone and requires continuous exposure to progesterone, which in sheep is at least 8 days. Thus, an earlier increase in circulating progesterone advances the timing of PGR loss from uterine epithelia. Indeed, PGR loss in the endometrial LE is strongly associated with a reduction in expression of the anti-adhesive MUC1 and induction of *LGALS15*, GLYCAM1, *CST3* and *CTSL*.

In GE, PGR loss is associated with induction of *SPP1*, *STC1* (stanniocalcin) and SERPIN (ovine uterine serpin peptidase inhibitor; alias uterine milk proteins). Indeed, inhibition of progesterone action by an anti-progestin prevents PGR down-regulation and, in turn, progesterone induction of GLYCAM1, *LGALS15*, *CTSL*, *CST3*, *SPP1*, *STC1* and *SERPINs*. Several candidate adhesion factors that mediate blastocyst implantation under the influence of progesterone and, for some IFNT, will now be discussed briefly except for LGALS15 that was discussed previously (see (Burghardt *et al.* 2002; Johnson *et al.* 2003; Spencer *et al.* 2004a)).

Mucin glycoprotein one (MUC1)

As the blastocyst approaches the endometrial LE, it encounters the glycocalyx which includes MUC1, a large transmembrane mucin glycoprotein expressed at the apical surface of epithelia in reproductive tracts of several species (Brayman *et al.* 2004). Expression of the glycoproteins MUC1 and MUC4 on uterine LE may block accessibility of trophoblast integrin receptors to their ligands for cell-cell and cell-extracellular matrix (ECM) adhesion necessary for initial stages of implantation (Carson *et al.* 2000; Burghardt *et al.* 2002). Removal of MUC1 from LE after Day 15 of pregnancy in sheep, which correlates to loss of PGR, may be necessary to expose other glycoproteins involved in adhesion between trophoblast and LE (Table 1). Given that the mucins contain large amounts of glycans that may be potentially recognized by blastocysts or secreted animal lectins such as LGALS15, they may be involved in the apposition phase of implantation.

Integrins

Integrins are a family of heterodimeric intrinsic transmembrane glycoprotein receptors that mediate cellular differentiation, motility and adhesion (Giancotti & Ruoslahti 1999). They play a dominant role in interactions with ECM to transduce cellular signals in uterine epithelial cells and conceptus trophoblast (Burghardt *et al.* 2002; Johnson *et al.* 2003). The central role of integrins in the adhesion cascade leading to implantation is binding ECM ligand(s) to induce cytoskeletal reorganization, stabilize adhesion, and mediate cell migration, proliferation and differentiation through numerous cell signaling pathways (Aplin 1997; Burghardt *et al.* 2002). During the peri-implantation period of pregnancy in ewes, integrin subunits α (v, 4 and 5) and ß (1, 3 and 5) are constitutively expressed on the apical surfaces of both conceptus trophectoderm and endometrial LE (Johnson *et al.* 2001a). This suggests that receptivity to implantation does not involve changes in either temporal or spatial patterns of integrin expression, but may require expression of other glycoproteins and ECM proteins such as LGALS15, SPP1 and oncofetal fibronectin which are ligands for heterodimers of these integrins (Johnson *et al.* 2003). More detail on integrins and implantation can be found elsewhere (Aplin 1997; Burghardt *et al.* 2002; Johnson *et al.* 2003; Spencer *et al.* 2004a).

Glycosylated cell adhesion molecule one (GLYCAM1)

GLYCAM1, a sulfated glycoprotein secreted by the endothelium, mediates leukocyte-endothelial cell adhesion (Lasky *et al.* 1992) by functioning as a carbohydrate ligand for the lectin domain of leukocyte cell surface selectin (SELL or L-selectin) in the lymphoid system (Rosen 1993). Ligation of SELL by GLYCAM1 activates ß1 and ß2 integrins and promotes firm adhesion to ECM components, such as fibronectin (Hwang *et al.* 1996). In humans, trophoblast SELL appears to be responsible for interactions with the uterine epithelium that are considered critical for implantation and establishing pregnancy (Genbacev *et al.* 2003). The temporal and spatial patterns of expression of GLYCAM1 in cyclic and pregnant ovine uteri implicates it as a potential regulator of implantation (Spencer *et al.* 1999a) (see Table 1). In pregnant ewes, the relative amount of immunoreactive GLYCAM1 in uterine flushings is low on Days 11 and 13, but abundant on Days 15 and 17. Thus, a GLYCAM1-like protein may be a secretory product of the endometrial epithelium and/or conceptus trophoblast. Patterns of distribution observed for immunoreactive GLYCAM1 in the endometrial epithelium, combined with proposed functions for lymphoid GLYCAM1, suggest that this mucin glycoprotein may be involved in conceptus-maternal interactions during the peri-implantation period of pregnancy in sheep (Spencer *et al.* 2004a).

Secreted phosphoprotein one (SPP1)

SPP1 is a member of the Small Integrin-Binding Ligand, N-Linked Glycoprotein (SIBLING) family of related ECM proteins recognized as key players in a number of diverse processes such as bone mineralization, cancer metastasis, cell-mediated immune responses, inflammation, angiogenesis, and cell survival (Sodek *et al.* 2000; Johnson *et al.* 2003). During the peri-implantation period of pregnancy in sheep, *SPP1* mRNA is first detected in the endometrial glands of some ewes by Day 13 and is present in all glands by Day 19 (Johnson *et al.* 1999b). In the uterine lumen, SPP1 protein appears on Day 15 and is found at the trophectoderm-LE interface throughout gestation, suggesting that it plays a key role in adhesion of the trophectoderm to LE via integrin receptors (Johnson *et al.* 2001a). Ovine trophectoderm and LE cells show evidence of integrin receptor activation and cytoskeletal reorganization in response to SPP1 binding *in vitro* (Johnson *et al.* 2001a). Progesterone induces expression of *SPP1* in endometrial glands, and this requires loss of PGR (Spencer *et al.* 1999b; Johnson *et al.* 2000b). SPP1 is hypothesized to serve as a bifunctional bridging ligand that mediates adhesion between LE and trophectoderm essential for implantation and placentation in sheep (Johnson *et al.* 2003).

Progesterone and placental hormone (IFNT, CSH1) regulation of uterine gland morphogenesis and secretory function

During early pregnancy, the ovine uterus is exposed sequentially to oestrogen, progesterone, IFNT, and CSH1 which is proposed to initiate and maintain endometrial gland morphogenesis and differentiated secretory functions (see (Spencer *et al.* 1999b; Spencer & Bazer 2002; Spencer *et al.* 2004b) for review). The placentae of a number of species, including rodents, humans, nonhuman primates and sheep, secrete hormones structurally related to pituitary PRL (prolactin) and GH (growth hormone) that are termed CSH1 (alias placental lactogen) (Soares 2004). Ovine CSH1 is produced by trophoblast giant BNC from Days 15 to 16 of pregnancy which is coordinate with onset of expression of *SERPIN*, *SPP1*, *GRP* (gastrin-releasing peptide), and *STC1* (Ing *et al.* 1989; Whitley *et al.* 1998; Stewart *et al.* 2000; Song *et al.* 2006a), which are excellent markers for GE differentiation and secretory function during pregnancy in sheep. A homodimer of the PRLR, as well as a heterodimer of PRLR and GHR (growth hormone receptor), transduce signals by ovine CSH1 (Gertler & Djiane 2002). In the ovine uterus, *PRLR* gene expression is unique to GE (Cassy *et al.* 1999; Stewart *et al.* 2000). Temporal changes in circulating levels of CSH1 are correlated with endometrial gland hyperplasia and hypertrophy and increased production of SERPIN and SPP1 during pregnancy. The sequential exposure of the pregnant ovine endometrium to estrogen, progesterone, IFNT, and CSH1 appears to be required to activate and maintain endometrial remodeling, secretory function of GE and uterine growth during gestation. Chronic treatment of ovariectomized ewes with progesterone induces expression of *SPP1*, *SERPIN* and *STC1* by GE (Moffatt *et al.* 1987; Spencer *et al.* 1999b; Johnson *et al.* 2000b; Song *et al.* 2006a). However, intrauterine infusions of CSH1 further increases endometrial *SPP1*, *SERPIN* and *STC1* mRNA, but only when ewes receive progesterone and intrauterine infusions of IFNT between Days 11 and 20 (Spencer *et al.* 1999b). The effects of IFNT may be attributed, in part, to stimulation of PRLR in GE (Martin *et al.* 2004). These results indicate that placental hormones play key roles in stimulating endometrial gland morphogenesis and differentiated functions during pregnancy that are required for conceptus growth and development.

Conclusions

During the past decade, knowledge of mechanisms and factors regulating fetal-maternal interactions during establishment of pregnancy has increased in sheep and cattle. Transcriptional profiling studies are now accelerating the pace of discovery; however, our knowledge of cellular and molecular mechanisms governing fetal-maternal interactions and, in particular, progesterone actions and trophectoderm growth and differentiation remain very limited. Results from studies of rodents strongly suggests that implantation involves a multiplicity of receptor-ligand interactions that are organized into a combinatorial cascade . Therefore, individual and integrative roles of adhesion factors must be mechanistically determined using *in vivo, ex vivo* and *in vitro* experimental models. Pregnancy loss in ruminants is greatest during the period of pregnancy recognition and establishment prior to placentation (Mann & Lamming 2001). Therefore, a more complete understanding of key molecules and signal transduction pathways that regulate fetal-maternal interactions during establishment and maintenance of pregnancy can be used to diagnose and identify the cause(s) of recurrent pregnancy loss and improve pregnancy rates and reproductive efficiency in domestic animals and humans.

Acknowledgments

We thank our colleagues and the present and past members of our laboratories who contributed to the research presented in this review. Due to space limitations, many primary references could not be included in the manuscript. This work was supported by National Research Initiative Competitive Grants 2001-02259, 2005-35203-16252 from the USDA Cooperative State Research, Education and Extension Service, and NIH Grants HD32534 and P30 ES09106.

References

Aplin JD 1997 Adhesion molecules in implantation. *Reviews of Reproduction* **2** 84-93.

Bazer FW 1975 Uterine protein secretions: Relationship to development of the conceptus. *Journal of Animal Science* **41** 1376-1382.

Bazer FW, Roberts RM & Thatcher WW 1979 Actions of hormones on the uterus and effect on conceptus development. *Journal of Animal Science* **49** 35-45.

Bazer FW, Thatcher WW, Hansen PJ, Mirando MA, Ott TL & Plante C 1991 Physiological mechanisms of pregnancy recognition in ruminants. *Journal of Reproduction and Fertility* **43** 39-47.

Brayman M, Thathiah A & Carson DD 2004 MUC1: A multifunctional cell surface component of reproductive tissue epithelia. *Reproductive Biology and Endocrinology* **2** 4.

Burghardt RC, Johnson GA, Jaeger LA, Ka H, Garlow JE, Spencer TE & Bazer FW 2002 Integrins and extracellular matrix proteins at the maternal-fetal interface in domestic animals. *Cells Tissues Organs* **171** 202-217.

Carson DD, Bagchi I, Dey SK, Enders AC, Fazleabas AT, Lessey BA & Yoshinaga K 2000 Embryo implantation. *Developmental Biology* **223** 217-237.

Cassy S, Charlier M, Guillomot M, Pessemesse L & Djiane J 1999 Cellular localization and evolution of prolactin receptor mRNA in ovine endometrium during pregnancy. *FEBS Letters* **445** 207-211.

Charpigny G, Reinaud P, Tamby JP, Creminon C, Martal J, Maclouf J & Guillomot M 1997 Expression of cyclooxygenase-1 and -2 in ovine endometrium during the estrous cycle and early pregnancy. *Endocrinology* **138** 2163-2171.

Choi Y, Johnson GA, Burghardt RC, Berghman LR, Joyce MM, Taylor KM, Stewart MD, Bazer FW & Spencer TE 2001 Interferon regulatory factor-two restricts expression of interferon- stimulated genes to the endometrial stroma and glandular epithelium of the ovine uterus. *Biology of Reproduction* **65** 1038-1049.

Choi Y, Johnson GA, Spencer TE & Bazer FW 2003 Pregnancy and interferon tau regulate MHC class I and beta-2-microglobulin expression in the ovine uterus. *Biology of Reproduction* **68** 1703-1710.

Dunphy JL, Balic A, Barcham GJ, Horvath AJ, Nash AD & Meeusen EN 2000 Isolation and characterization of a novel inducible mammalian galectin. *Journal of Biological Chemistry* **275** 32106-32113.

Farin CE, Imakawa K & Roberts RM 1989 In situ localization of mRNA for the interferon, ovine trophoblast protein-1, during early embryonic development of the sheep. *Molecular Endocrinology* **3** 1099-1107.

Flechon JE, Guillomot M, Charlier M, Flechon B & Martal J 1986 Experimental studies on the elongation of the ewe blastocyst. *Reproduction, Nutrition, Development* **26** 1017-1024.

Fleming JA, Choi Y, Johnson GA, Spencer TE & Bazer FW 2001 Cloning of the ovine estrogen receptor-alpha promoter and functional regulation by ovine interferon-tau. *Endocrinology* **142** 2879-2887.

Fleming JG, Spencer TE, Safe SH & Bazer FW 2006 Estrogen regulates transcription of the ovine oxytocin receptor gene through GC-rich SP1 promoter elements. *Endocrinology* **147** 899-911.

Garrett JE, Geisert RD, Zavy MT, Gries LK, Wettemann RP & Buchanan DS 1988 Effect of exogenous progesterone on prostaglandin F2 alpha release and the interestrous interval in the bovine. *Prostaglandins* **36** 85-96.

Geisert RD, Morgan GL, Short EC, Jr. & Zavy MT 1992 Endocrine events associated with endometrial function and conceptus development in cattle. *Reproduction, Fertility, and Development* **4** 301-305.

Genbacev OD, Prakobphol A, Foulk RA, Krtolica AR, Ilic D, Singer MS, Yang ZQ, Kiessling LL, Rosen SD & Fisher SJ 2003 Trophoblast L-selectin-mediated adhesion at the maternal-fetal interface. *Science* **299** 405-408.

Gertler A & Djiane J 2002 Mechanism of ruminant placental lactogen action: molecular and in vivo studies. *Molecular Genetics and Metabolism* **75** 189-201.

Giancotti FG & Ruoslahti E 1999 Integrin signaling. *Science* **285** 1028-1032.

Gray C, Bartol FF, Taylor KM, Wiley AA, Ramsey WS, Ott TL, Bazer FW & Spencer TE 2000a Ovine uterine gland knock-out model: effects of gland ablation on the estrous cycle. *Biology of Reproduction* **62** 448-456.

Gray CA, Taylor KM, Bazer FW & Spencer TE 2000b Mechanisms regulating norgestomet inhibition of endometrial gland morphogenesis in the neonatal ovine uterus. *Molecular Reproduction and Development* **57** 67-78.

Gray CA, Bazer FW & Spencer TE 2001a Effects of neonatal progestin exposure on female reproductive tract structure and function in the adult ewe. *Biology of Reproduction* **64** 797-804.

Gray CA, Taylor KM, Ramsey WS, Hill JR, Bazer FW, Bartol FF & Spencer TE 2001b Endometrial glands are required for preimplantation conceptus elongation and survival. *Biology of Reproduction* **64** 1608-1613.

Gray CA, Burghardt RC, Johnson GA, Bazer FW & Spencer TE 2002 Evidence that absence of endometrial gland secretions in uterine gland knockout ewes compromises conceptus survival and elongation. *Reproduction* **124** 289-300.

Gray CA, Adelson DL, Bazer FW, Burghardt RC, Meeusen EN & Spencer TE 2004 Discovery and characterization of an epithelial-specific galectin in the endometrium that forms crystals in the trophectoderm. *Proceedings of the National Academy of Sciences of the USA* **101** 7982-7987.

Gray CA, Abbey CA, Beremand PD, Choi Y, Farmer JL, Adelson DL, Thomas TL, Bazer FW & Spencer TE 2006 Identification of endometrial genes regulated by early pregnancy, progesterone, and interferon tau in the ovine uterus. *Biology of Reproduction* **74** 383-394.

Guillomot M, Michel C, Gaye P, Charlier N, Trojan J & Martal J 1990 Cellular localization of an embryonic interferon, ovine trophoblastin and its mRNA in sheep embryos during early pregnancy. *Biology of the Cell* **68** 205-211.

Guillomot M 1995 Cellular interactions during implantation in domestic ruminants. *Journal of Reproduction and Fertility* **49** 39-51.

Hansen TR, Austin KJ, Perry DJ, Pru JK, Teixeira MG & Johnson GA 1999 Mechanism of action of interferon-tau in the uterus during early pregnancy. *Journal of Reproduction and Fertility* **54** 329-339.

Heyman Y, Camous S, Fevre J, Meziou W & Martal J 1984 Maintenance of the corpus luteum after uterine transfer of trophoblastic vesicles to cyclic cows and ewes. *Journal of Reproduction and Fertility* **70** 533-540.

Hwang ST, Singer MS, Giblin PA, Yednock TA, Bacon KB, Simon SI & Rosen SD 1996 GlyCAM-1, a physiologic ligand for L-selectin, activates beta 2 integrins on naive peripheral lymphocytes. *Journal of Experimental Medicine* **184** 1343-1348.

Imakawa K, Helmer SD, Nephew KP, Meka CS & Christenson RK 1993 A novel role for GM-CSF: enhancement of pregnancy specific interferon production, ovine trophoblast protein-1. *Endocrinology* **132** 1869-1871.

Imakawa K, Imai M, Sakai A, Suzuki M, Nagaoka K, Sakai S, Lee SR, Chang KT, Echternkamp SE & Christenson RK 2006 Regulation of conceptus adhesion by endometrial CXC chemokines during the implantation period in sheep. *Molecular Reproduction and Development* **73** 850-858.

Ing NH, Francis H, McDonnell JJ, Amann JF & Roberts RM 1989 Progesterone induction of the uterine milk proteins: major secretory proteins of sheep endometrium. *Biology of Reproduction* **41** 643-654.

Johnson GA, Austin KJ, Collins AM, Murdoch WJ & Hansen TR 1999a Endometrial ISG17 mRNA and a related mRNA are induced by interferon-tau and localized to glandular epithelial and stromal cells from pregnant cows. *Endocrine* **10** 243-252.

Johnson GA, Spencer TE, Burghardt RC & Bazer FW 1999b Ovine osteopontin: I. Cloning and expression of messenger ribonucleic acid in the uterus during the periimplantation period. *Biology of Reproduction* **61** 884-891.

Johnson GA, Spencer TE, Hansen TR, Austin KJ, Burghardt RC & Bazer FW 1999c Expression of the interferon tau inducible ubiquitin cross-reactive protein in the ovine uterus. *Biology of Reproduction* **61** 312-318.

Johnson GA, Spencer TE, Burghardt RC, Joyce MM & Bazer FW 2000a Interferon-tau and progesterone regulate ubiquitin cross-reactive protein expression in the ovine uterus. *Biology of Reproduction* **62** 622-627.

Johnson GA, Spencer TE, Burghardt RC, Taylor KM, Gray CA & Bazer FW 2000b Progesterone modulation of osteopontin gene expression in the ovine uterus. *Biology of Reproduction* **62** 1315-1321.

Johnson GA, Bazer FW, Jaeger LA, Ka H, Garlow JE, Pfarrer C, Spencer TE & Burghardt RC 2001a Muc-1, integrin, and osteopontin expression during the implantation cascade in sheep. *Biology of Reproduction* **65** 820-828.

Johnson GA, Stewart MD, Gray CA, Choi Y, Burghardt RC, Yu-Lee LY, Bazer FW & Spencer TE 2001b Effects of the estrous cycle, pregnancy, and interferon tau on 2',5'- oligoadenylate synthetase expression in the ovine uterus. *Biology of Reproduction* **64** 1392-1399.

Johnson GA, Burghardt RC, Bazer FW & Spencer TE 2003 Osteopontin: roles in implantation and placentation. *Biology of Reproduction* **69** 1458-1471.

Kazemi M, Amann JF, Keisler DH, Ing NH, Roberts RM, Morgan G & Wooding FB 1990 A progesterone-modulated, low-molecular-weight protein from the uterus of the sheep is associated with crystalline inclusion bodies in uterine epithelium and embryonic trophectoderm. *Biology of Reproduction* **43** 80-96.

Kim S, Choi Y, Bazer FW & Spencer TE 2003a Identifica-tion of genes in the ovine endometrium regulated by interferon tau independent of signal transducer and activator of transcription 1. *Endocrinology* **144** 5203-5214.

Kim S, Choi Y, Spencer TE & Bazer FW 2003b Effects of the estrous cycle, pregnancy and interferon tau on expression of cyclooxygenase two (COX-2) in ovine endometrium. *Reproductive Biology and Endocrinology* **1** 58.

Kleemann DO, Walker SK & Seamark RF 1994 Enhanced fetal growth in sheep administered progesterone during the first three days of pregnancy. *Journal of Reproduction and Fertility* **102** 411-417.

Klein C, Bauersachs S, Ulbrich SE, Einspanier R, Meyer HHD, Schmidt SEM, Reichenbach H-D, Vermehren M, Sinowatz F, Blum H et al. 2006 Monozygotic twin model reveals novel embryo-induced transcriptome changes of bovine endometrium in the pre-attachment period. *Biology of Reproduction* **74** 253-264.

Ko Y, Lee CY, Ott TL, Davis MA, Simmen RC, Bazer FW & Simmen FA 1991 Insulin-like growth factors in sheep uterine fluids: concentrations and relationship to ovine trophoblast protein-1 production during early pregnancy. *Biology of Reproduction* **45** 135-142.

Lasky LA, Singer MS, Dowbenko D, Imai Y, Henzel WJ, Grimley C, Fennie C, Gillett N, Watson SR & Rosen SD 1992 An endothelial ligand for L-selectin is a novel mucin-like molecule. *Cell* **69** 927-938.

Lawson RA, Parr RA & Cahill LP 1983 Evidence for maternal control of blastocyst growth after asynchronous transfer of embryos to the uterus of the ewe. *Journal of Reproduction and Fertility* **67** 477-483.

Mann GE & Lamming GE 1999 The influence of progesterone during early pregnancy in cattle. *Reproduction in Domestic Animals* **34** 269-274.

Mann GE & Lamming GE 2001 Relationship between maternal endocrine environment, early embryo development and inhibition of the luteolytic mechanism in cows. *Reproduction* **121** 175-180.

Mann GE, Fray MD & Lamming GE 2006 Effects of time of progesterone supplementation on embryo development and interferon-tau production in the cow. *Veterinary Journal* **171** 500-503.

Martin C, Pessemesse L, de la Llosa-Hermier MP, Martal J, Djiane J & Charlier M 2004 Interferon-{tau} upregulates prolactin receptor mRNA in the ovine endometrium during the peri-implantation period. *Reproduction* **128** 99-105.

McCracken JA, Custer EE & Lamsa JC 1999 Luteolysis: a neuroendocrine-mediated event. *Physiological Reviews* **79** 263-323.

Michael DD, Alvarez IM, Ocon OM, Powell AM, Talbot NC, Johnson SE & Ealy AD 2006 Fibroblast Growth Factor-2 Is Expressed by the Bovine Uterus and Stimulates Interferon-Tau Production in Bovine Trophectoderm. *Endocrinology* **147** 3571-3579.

Moffatt J, Bazer FW, Hansen PJ, Chun PW & Roberts RM 1987 Purification, secretion and immunocytochemical localization of the uterine milk proteins, major progesterone-induced proteins in uterine secretions of the sheep. *Biology of Reproduction* **36** 419-430.

Mohamed OA, Jonnaert M, Labelle-Dumais C, Kuroda K, Clarke HJ & Dufort D 2005 Uterine Wnt/beta-catenin signaling is required for implantation. *Proceedings of the National Academy of Sciences of the USA* **102** 8579-8584

Nagaoka K, Nojima H, Watanabe F, Chang KT, Christenson RK, Sakai S & Imakawa K 2003a Regulation of blastocyst migration, apposition, and initial adhesion by a chemokine, interferon gamma-inducible protein 10 kDa (IP-10), during early gestation. *Journal of Biological Chemistry* **278** 29048-29056.

Nagaoka K, Sakai A, Nojima H, Suda Y, Yokomizo Y, Imakawa K, Sakai S & Christenson RK 2003c A chemokine, interferon (IFN)-gamma-inducible protein 10 kDa, is stimulated by IFN-tau and recruits immune cells in the ovine endometrium. *Biology of Reproduction* **68** 1413-1421.

Nephew KP, McClure KE, Ott TL, Dubois DH, Bazer FW & Pope WF 1991 Relationship between variation in conceptus development and differences in estrous cycle duration in ewes. *Biology of Reproduction* **44** 536-539.

Roberts RM, Ealy AD, Alexenko AP, Han CS & Ezashi T 1999 Trophoblast interferons. *Placenta* **20** 259-264.

Robinson RS, Mann GE, Lamming GE & Wathes DC 2001 Expression of oxytocin, oestrogen and progesterone receptors in uterine biopsy samples throughout the oestrous cycle and early pregnancy in cows. *Reproduction* **122** 965-979.

Rosen SD 1993 Ligands for L-selectin: where and how many? *Research in Immunology* **144** 699-703; discussion 754-662.

Rosenfeld CS, Han CS, Alexenko AP, Spencer TE & Roberts RM 2002 Expression of interferon receptor subunits, IFNAR1 and IFNAR2, in the ovine uterus. *Biology of Reproduction* **67** 847-853.

Salamonsen L 1999 Role of proteases in implantation. *Reviews of Reproduction* **4** 11-22.

Soares MJ 2004 The prolactin and growth hormone families: pregnancy-specific hormones/cytokines at the maternal-fetal interface. *Reproductive Biology and Endocrinology* **2** 51.

Sodek J, Ganss B & McKee MD 2000 Osteopontin. *Critical Reviews in Oral Biology and Medicine* **11** 279-303.

Song G, Spencer TE & Bazer FW 2005 Cathepsins in the ovine uterus: regulation by pregnancy, progesterone, and interferon tau. *Endocrinology* **146** 4825-4833.

Song G, Bazer FW, Wagner GF & Spencer TE 2006a Stanniocalcin (STC) in the endometrial glands of the ovine uterus: Regulation by progesterone and placental hormones. *Biology of Reproduction* **74** 913-922.

Song G, Spencer TE & Bazer FW 2006b Progesterone and interferon tau regulate cystatin C (CST3) in the endometrium. *Endocrinology* **147** 3478-3483.

Spencer TE, Bartol FF, Bazer FW, Johnson GA & Joyce MM 1999a Identification and characterization of glycosylation-dependent cell adhesion molecule 1-like protein expression in the ovine uterus. *Biology of Reproduction* **60** 241-250.

Spencer TE, Gray A, Johnson GA, Taylor KM, Gertler A, Gootwine E, Ott TL & Bazer FW 1999b Effects of recombinant ovine interferon tau, placental lactogen, and growth hormone on the ovine uterus. *Biology of Reproduction* **61** 1409-1418.

Spencer TE & Bazer FW 2002 Biology of progesterone action during pregnancy recognition and maintenance of pregnancy. *Frontiers in Bioscience* **7** d1879-1898.

Spencer TE & Bazer FW 2004 Conceptus signals for establishment and maintenance of pregnancy. *Reproductive Biology and Endocrinology* **2** 49.

Spencer TE, Johnson GA, Bazer FW & Burghardt RC 2004a Implantation mechanisms: insights from the sheep. *Reproduction* **128** 657-668.

Spencer TE, Johnson GA, Burghardt RC & Bazer FW 2004b Progesterone and placental hormone actions on the uterus: insights from domestic animals. *Biology of Reproduction* **71** 2-10.

Stark GR, Kerr IM, Williams BR, Silverman RH & Schreiber RD 1998 How cells respond to interferons. *Annual Review of Biochemistry* **67** 227-264.

Stewart MD, Johnson GA, Gray CA, Burghardt RC, Schuler LA, Joyce MM, Bazer FW & Spencer TE 2000 Prolactin receptor and uterine milk protein expression in the ovine endometrium during the estrous cycle and pregnancy. *Biology of Reproduction* **62** 1779-1789.

Wathes DC & Lamming GE 1995 The oxytocin receptor, luteolysis and the maintenance of pregnancy. *Journal of Reproduction and Fertility* **49** 53-67.

Whitley JC, Shulkes A, Salamonsen LA, Vogiagis D, Familari M & Giraud AS 1998 Temporal expression and cellular localization of a gastrin-releasing peptide-related gene in ovine uterus during the oestrous cycle and pregnancy. *Journal of Endocrinology* **157** 139-148.

Wooding FB 1984 Role of binucleate cells in fetomaternal cell fusion at implantation in the sheep. *American Journal of Anatomy* **170** 233-250.

Wooding FB 1992 Current topic: the synepitheliochorial placenta of ruminants: binucleate cell fusions and hormone production. *Placenta* **13** 101-113.

Yang RY & Liu FT 2003 Galectins in cell growth and apoptosis. *Cellular and Molecular Life Sciences* **60** 267-276.

The effects of maternal nutrition around the time of conception on the health of the offspring

MH Oliver, AL Jaquiery, FH Bloomfield and JE Harding

Liggins Institute, University of Auckland, Auckland, New Zealand

The incidence of prematurity, diabetes and cardiovascular disease have been increasing in both the developed and developing world. Increasing numbers of human studies suggest that these serious health outcomes may have developmental origins originating from nutritional deficits in the periconceptional period, with maternal nutrition around the time of conception now shown to have important effects on the length of gestation, trajectory of fetal growth and on postnatal growth and health. Biomedical research using the pregnant sheep has been widely employed to gain a deeper understanding of the underlying mechanisms involved. There is growing awareness that this field of research has major implications for the livestock production industry. From our own studies on sheep we have evidence that maternal undernutrition during the periconceptional period results in altered fetal hypothalamic-pituitary-adrenal axis (HPAA) development, an increased rate of premature birth, altered fetal pancreatic function, insulin signalling and amino acid metabolism, and also alterations in maternal adaptation to pregnancy. We are currently studying the postnatal consequences of these changes. Other research groups have shown that restricted nutrition of sheep in the early part of pregnancy alters postnatal muscle development, fat deposition, cardiovascular regulation and HPAA function. One aim of this review is to illustrate how biomedical research using animals such as the sheep has been used to gain a better understanding of the consequences of reduced maternal nutrition during the periconceptional period. We suggest that there are equally important consequences of this research for the livestock production industries.

Introduction

Maternal nutrition, and particularly undernutrition, has been recognised as an important influence on the growth and metabolism of the offspring. A large amount of research has focused on the effects of maternal undernutrition in late pregnancy. However the effects of undernutrition around the time of conception are now a major focus of both biomedical and agricultural research. There is increasing epidemiological evidence from human populations to suggest that poor maternal nutrition around the time of conception or in early pregnancy may lead to a higher incidence of premature birth (Rayco-Solon et al. 2005) and an increased risk of developing hypertension, heart disease and obesity later in adult life (Roseboom et al. 2001a; Gluckman

Corresponding author E-mail: m.oliver@auckland.ac.nz

et al. 2005). In the sheep, maternal nutrition around the time of conception has an important influence on the development of the conceptus in terms of its growth, physiology (Oliver et al. 2005) and the length of gestation (Bloomfield et al. 2003). The implications for livestock production parameters such as fat and muscle development are an important area of current research.

The focus of this review is to discuss the importance of maternal nutrition at the time of conception on fetal development, offspring survival, and for postnatal growth and development. Other factors, such as the hormonal environment at the time of conception are also recognised to have profound effects on the development of the offspring (Kleemann et al. 2001). Similarly, experiments using in vitro culture of embryos have indicated powerful effects of the local nutrient environment, leading to alterations in cell number and allocation to different embryonic cell lines and hence altered fetal and placental growth (Barnes 2000). It is beyond the scope of this review to discuss these other periconceptional factors except where they shed light on possible mechanistic aspects of nutritional effects. Rather, we will concentrate on the effects of maternal undernutrition around the time of conception on fetal growth, fetal hypothalamic-pituitary adrenal axis (HPAA) development, insulin secretion and signalling, regulatory physiology, maternal adaptation to pregnancy and the possible postnatal consequences of all of these factors. We have recently demonstrated that late gestation twin fetal sheep of ewes that were well nourished throughout pregnancy demonstrate a similar physiological phenotype to singleton fetuses of ewes that were undernourished around the time of conception. As twinning is an important aspect of commercial sheep breeding, we will also overview the consequences of twinning for many of these outcomes.

Fetal growth

Early studies of the relationship between fetal growth and later disease risk came from cohort studies of British populations born in the 1920s (Hales et al. 1991). Careful follow-up revealed that men and women of lower birth weight had higher blood pressure, increased risk of coronary heart disease (Fall et al. 1995) and impaired glucose tolerance as adults (Phipps et al. 1993; Law 1996). These initial findings by Barker and colleagues were soon replicated in many other populations, establishing a strong and consistent link between size at birth and the incidence of diabetes (Fall et al. 1998; Ravelli et al. 1998) and cardiovascular disorders in later life (Barker, 2000). However it was less clear whether it was small birth size itself, or the cause of that small size, that was most important in determining outcome.

One cause of small size at birth in many animal species is maternal nutritional intake inadequate for fetal demands. In human pregnancy, maternal undernutrition, if present, is often chronic and confounded by other factors such as poverty and infection, both before and after birth. The effect of different timings of severe undernutrition in a previously well nourished human population have been described in a cohort subjected to the Dutch Hunger winter. Towards the end of world war II the occupying German forces imposed embargos of food supplies in response to continued Dutch resistance efforts, and severe famine resulted for 5 months before the lifting of the blockade by the Allied forces. Very good data were collected on the timing and nature of the food rationing, and also most importantly on the babies born during the period. Those babies have now been followed into adult life, providing a unique insight into the effect of maternal undernutrition at different periods of pregnancy on the size and health of the offspring. Babies of women exposed to famine conditions only in late pregnancy had reduced size at birth, and impaired glucose tolerance as adults (Ravelli et al. 1998). However babies of women who conceived during the famine, that often extended well into the first trimester,were of normal size at birth, but as adults were at increased risk of coronary heart disease (Roseboom *et al.* 2001b), hyperlipidaemia (Roseboom *et al.* 2000) and obesity (Ravelli *et al.* 1999). These and many other studies strongly suggest that events around the time of conception can have life long effects on the offspring, without necessarily affecting size at birth. Furthermore, altered size at birth does not appear to lie on the causal pathway between reduced nutrition in early pregnancy and its consequences before or after birth.

We have examined the effects of maternal undernutrition during different periods of pregnancy in sheep, where fetal growth trajectory can be monitored from day to day using surgically implanted catheters. In late gestation, rapidly growing fetuses slowed their growth promptly in response to maternal undernutrition designed to reduce maternal blood glucose concentrations by 30-40% (Harding 1997). However, fetuses whose mothers had also been moderately undernourished around the time of conception (60d before to 30d after mating to produce a 10-15% reduction in maternal weight) grew more slowly in late gestation, and were able to continue this slow growth trajectory in the face of further severe maternal undernutrition in late gestation. These data suggest that both trajectory of fetal growth, and also the feto-placental capacity to adapt to a late gestation nutritional insult, are determined by events around the time of conception. Intriguingly, although trajectory of fetal growth was affected by periconceptional undernutrition, gross fetal size (weight and length) were not different at the end of the experiment, consistent with the human data from the Dutch Hunger Winter. However, the fetuses of periconceptionally undernourished ewes had altered body composition, with proportionately larger hearts, livers and kidneys (Harding 1997). Others have recently demonstrated that when maternal intake is reduced by 50% from day 30 to 70 of pregnancy there is in an increased ratio of oxidative to glycolytic muscle fibre type in 2 week old lambs (Fahey *et al.* 2005). Similar nutritional restriction in early pregnancy has also been demonstrated to increase fat mass in postnatal sheep (Symonds *et al.* 2005). Thus nutrition in early pregnancy may also determine important aspects of both prenatal growth and body composition that may not necessarily be reflected in altered size at birth.

Growth rate of twins in late gestation is less than that of singletons and is often discordant within the twin pair. If fetal growth has important associations with fetal and postnatal development, studying twins may provide insights into these associations. We recently presented data that demonstrate complex interactions in twin sheep between periconceptional nutrition, twinning and fetal growth responses to maternal undernutrition in late gestation (Bloomfield *et al.* 2005). In response to a 3 day maternal fast in late gestation, the heavier sheep fetus of a twin pair whose mother was well nourished around conception does not reduce its growth rate, thereby behaving in a similar manner to a singleton fetus whose mother was undernourished during the periconceptional period. The lighter twins of both the periconceptionally well nourished and undernourished groups do slow their growth dramatically in response to maternal fasting in late gestation. We speculate that the placenta serving the lighter twin may have less functional capacity to maintain fetal growth. In contrast, the heavier twin of the periconceptionally undernourished group initially slows its growth in response to maternal fasting in late gestation in a similar manner to the lighter, undernourished twin, but then returns to its original growth rate. The placenta serving the heavier twin presumably has better placental functional capacity to adapt to maternal restriction and is therefore able to allow the fetus to maintain its original growth rate. These conclusions are supported by changes in the circulating concentrations of glucose, lactate, urea and amino acids in the fetus during the course of nutritional manipulations (Bloomfield *et al.* 2005). We suggest that these data support two important concepts: firstly, that aspects of twin growth and metabolism are also determined around the time of conception and, secondly, that there may be important interactions between periconceptional events and subsequent placental functional capacity. This may be causal in determining the relative size of twins within a pair, but may also determine the ability of the placenta to supply the fetus with nutrients in the face of reduced maternal nutrient supply. The concept of placental capacity being determined in early pregnancy is supported by elegant data from Kwong et al. who demonstrated that the proportion of inner cell mass which is allocated to the trophoblast and thus the placenta can be affected by the prevailing nutrient environment in the periconceptional period (Kwong et al. 2000), thereby emphasizing the critical nature of this period of pregnancy.

The consequences of periconceptional undernutrition for postnatal growth and development of singleton and twin sheep are now under study. Preliminary data from our laboratory show that in the period from birth to 3 months, weight gain was greater in singleton lambs from ewes undernourished in the periconceptional period (Fig. 1) despite the fact that milk intake was not increased. Insulin like growth factor 1 (IGF-1), an important marker of growth in young animals, was also higher in the same lambs throughout the same period. However, weight at 4 months of age was similar in both groups, suggesting an altered pattern of early postnatal growth after periconceptional undernutrition. Twinning obscured any effect of periconceptional undernutrition on weight gain and IGF-1 levels. The long term effects of periconceptional undernutrition on important production parameters such as the development of muscle and fat requires further investigation.

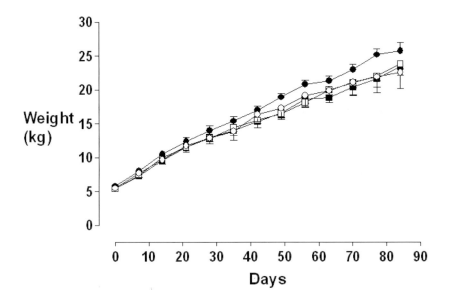

Figure 1. Weights of lambs from birth until weaning (means ± SEM). Singleton lambs born to ewes undernourished from 60 d before, to 30 d after mating (closed circles, n = 10) increased in weight faster than singleton lambs born to ewes well fed during the same period (open circles, n = 6, $p < 0.05$, multiple linear regression). There was no difference between twin lambs born to ewes undernourished from 60 d before, to 30 d after mating (closed squares, n = 13) and those born to ewes well fed during the same period (open squares, n = 21).

Timing of birth and HPAA development

We have also reported that moderate maternal undernutrition in the periconceptional period (individual feed intake adjusted to reduce maternal weight by 10-15% from 60 d before until

30 d after mating) reduces the mean gestation length of singleton-bearing ewes by 8 d (Bloomfield *et al.* 2003). None of these preterm lambs survived, despite being appropriately sized for gestational age and being born indoors. More detailed physiological tests and molecular studies in a parallel cohort killed in late gestation revealed that periconceptional undernutrition had led to precocious activation of the fetal HPAA, resulting in premature elevation of fetal plasma ACTH, cortisol (Bloomfield *et al.* 2004a) and prostaglandins PGE2 and PGFM (Kumarasamy *et al.* 2005), thus resulting in earlier onset of parturition (Fig 2.). Subsequent studies in human populations have also suggested that nutrition before and in early pregnancy affects the length of gestation, with maternal dieting, eating disorders (Cnattingius *et al.* 1998; Sollid *et al.* 2004) and early pregnancy vomiting being associated with an increased risk of preterm birth (Dodds *et al.* 2006).

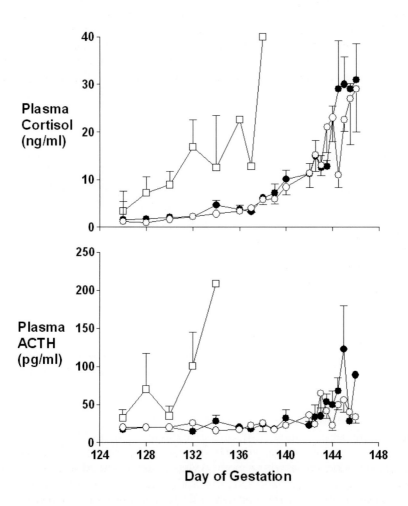

Figure 2. Fetal plasma cortisol and ACTH concentrations from 125 d until delivery. Lambs born prematurely to ewes undernourished from 60 d before, to 30 d after mating (open squares, n = 5) had an early rise in plasma cortisol and ACTH concentrations. Term lambs born to ewes underfed from 60 d before, to 30 d after mating (closed circles, n = 5) or ewes well fed during the same period (open circles, n = 8) had later increases in plasma cortisol and ACTH concentrations (both $p < 0.01$, ANOVA).

These findings have important implications for the livestock industry. Early birth of an appropriately sized lamb may not be recognised within the paradigm of normal farm-based breeding practices, and subsequent death may be ascribed to causes in the immediate environment at birth rather than nutrition of the mother five or more months earlier. Such effects may perhaps be reflected in the high lamb mortality rates that are a regular feature of the pastoral sheep farming industry in New Zealand. It is also widely recognised that multiple lambs are born smaller and earlier making them more vulnerable.

Apart from an increased risk of prematurity, precocious activation of the HPAA has wider implications for the postnatal health, growth and performance of the production animal. Cardiovascular and metabolic homeostasis are profoundly influenced by central HPPA activity and also by expression of glucocorticoid receptors (GR) and 11ß -hydroxysteroid dehydrogenase (11ß-HSD) isoforms in the peripheral tissues. The action of 11ß-HSD-1 is to activate cortisol from cortisone while 11ß-HSD-2 works in the other direction. Others have shown that early gestation undernutrition results in increased fat deposition in postnatal lambs (Symonds *et al.* 2005) and it is likely that this could be mediated by changes in the expression and interaction of leptin, GR and 11ß-HSD isoforms in adipose tissues (Budge *et al.* 2005). Increased adipose expression of GR and 11ß-HSD-1, but decreased expression of 11ß-HSD-2, is observed in obese humans (Budge *et al.* 2005). Leptin levels in the circulation are strongly correlated with fat mass and chronic administration of the hormone in rats results in increased adipose expression of GR and 11ß-HSD-1, and decreased expression of 11ß-HSD-2 (Gnanalingham *et al.* 2005).

We have recently shown that ACTH and cortisol responses of 10 month old female sheep to a combined CRH and AVP challenge are inversely related to birth weight in both singletons and twins, but the birth size effect was greater in twins (Bloomfield *et al.* 2004b). These findings raise the possibility that HPAA function in relation to size at birth may have important implications for both animal production and carcass composition of twins.

Insulin secretion and signalling

The development of impaired glucose tolerance, and ultimately frank diabetes, is thought to arise from a combination of impaired insulin secretion and insulin resistance, although the relative contributions of these two phenomena are not always clear. The human epidemiological studies have been primarily focused on insulin resistance (Phillips *et al.* 1994; Osmond & Barker 2000), although there is also clear evidence of pancreatic beta cell dysfunction (Hales *et al.* 1991; Fall *et al.* 1998). Our own sheep studies demonstrate that both insulin secretion (Oliver *et al.* 2001) and insulin signalling (Buckley *et al.* 2005) may be influenced by periconceptional undernutrition, and that the early changes are evident before birth.

In the fetus, as in the adult, pancreatic beta cells secrete insulin in response to a variety of stimuli including glucose and amino acids such as arginine (Fowden 1980). Pancreatic maturation in the late gestation fetus is characterised by increasing responsiveness to glucose (Aldoretta *et al.* 1998). Late gestation fetuses whose mothers were undernourished in the periconceptional period appear to have advanced pancreatic maturation, as evidenced by exaggerated insulin response to glucose but not arginine challenge (Oliver *et al.* 2001). This apparent accelerated maturation is consistent with the precocious development of the HPAA, described above, suggesting that nutritional status around the time of conception sets the trajectory of growth and maturation of a number of organ systems. It would be surprising, therefore, if these changes did not lead to altered postnatal physiology, but the results of such studies are not yet available.

One possible mechanism underlying the effects of undernutrition involves altered supply of specific nutrients critical to developing organ systems. Taurine may be one such nutrient; a semi essential amino acid that plays a key role in pancreatic cell development. Rats fed a low protein diet (9% vs 18% in controls) during pregnancy have low circulating taurine levels and give birth to offspring with impaired pancreatic beta cell function (Cherif *et al.* 1998). Insulin secretion can be normalised in the offspring by supplementing the maternal low protein diet with taurine during pregnancy. In sheep, the initial depression in maternal plasma concentrations of taurine during periconceptional undernutrition is followed by an elevation in maternal and fetal levels in late gestation, long after the mother has been returned to a normal diet. In the same fetuses pancreatic maturation is advanced, suggesting a possible link between fetal taurine levels and pancreatic development (Oliver *et al.* 2001).

The other component of impaired glucose tolerance is insulin resistance. In the rat, maternal protein restriction (Snoeck *et al.* 1990) or prenatal exposure to glucocorticoids of maternal or exogenous origin (Nyirenda *et al.* 2001) result in insulin resistance in the adult offspring. Paradoxically there is often an initial increase in insulin sensitivity in the offspring of exposed mothers in early life followed by a progressive deterioration as they age (Petry *et al.* 2000). The initial increase in insulin sensitivity is associated with elevations in the message and protein expression of key insulin signalling proteins. Consistent with this, we have evidence that levels of protein kinase-ζ (PK-ζ), a key insulin signalling protein, are elevated in the muscle of late gestation sheep fetuses whose mothers were undernourished around the time of conception (Buckley *et al.* 2005). The postnatal consequences of advanced pancreatic maturation and altered insulin sensitivity during fetal life are currently under investigation.

Physiological and metabolic regulation in the fetal/placental unit

One mechanism by which periconceptional undernutrition may result in altered fetal growth and maturation may involve altered development of the placenta and hence feto-placental metabolic interactions. In rats, a maternal low protein diet in the 4.5 d period before implantation results in altered distribution of cells between the inner cell mass and trophectoderm of the blastocyst, and hence to the future embryo and placenta respectively (Kwong *et al.* 2000). Such embryos also show subtle alterations in substrate metabolism with altered glucose consumption and lactate production. Altered substrate metabolism is also evident in late gestation in fetal sheep after periconceptional undernutrition. Plasma lactate concentrations are higher while blood oxygen content and pH are lower than in fetuses from well nourished ewes, and the feto-placental unit appears to convert a greater proportion of available glucose into lactate (Oliver *et al.* 2005). This may represent a means of conserving carbohydrate in the fetal compartment.

It is tempting to speculate that altered balance of the development of oxidative and glycolytic pathways may be reflected in the finding that 2 week old lambs whose mothers were fed 50% of requirement from day 30 to 70 have reduced fast twitch, or glycolytic muscle fibres and more slow twitch, or oxidative fibres (Fahey *et al.* 2005).

Another example of altered feto-placental metabolism after periconceptional undernutrition involves serine and glycine metabolism. Glycine is an important substrate for the growing fetus, being essential for the synthesis of such critical end products as DNA, RNA, heme, bile salts, collagen and glutathione (Jackson 1991). In postnatal life there is considerable capacity for glycine synthesis, but during pregnancy demand may exceed maternal synthetic capacity, especially when nutrition is poor (Jackson *et al.* 1997). In both human and sheep pregnancy, glycine and serine do not cross the placenta in appreciable amounts, and so the feto-placental unit must synthesise most of its requirements by complex interconversion pathways between serine and glycine in the placenta and the fetal liver (Cetin *et al.* 1991). We have found that these pathways are perturbed in late gestation fetuses of ewes who had been undernourished in the periconceptional period. The ratio of serine to glycine in the fetal circulation was increased, in association with decreased activity in the placenta of one of the enzymes involved in serine and glycine interconversion, serine hydroxymethyltransferase (SHMT) and increased SHMT activity in the fetal liver (Fig 3.) (Thorstensen et al. 2005). The long term postnatal consequences of this imbalance remain to be determined.

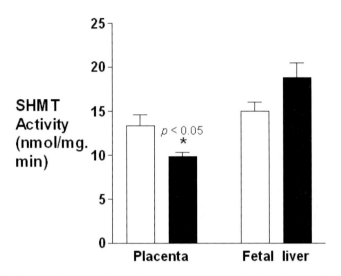

Figure 3. Placental and fetal hepatic serine hydroxymethyltransferase (SHMT) activity in 131 d sheep fetuses. Placentas from ewes underfed from 60 d before, to 30 d after mating (dark bars, n = 10) had lower SHMT activities than those from ewes well fed during the same period (open bars, n=9, $p < 0.05$, unpaired T-test). In contrast, SHMT activity tended to be higher in livers of fetuses whose mothers were underfed from 60 d before, to 30 d after mating ($p = 0.07$).

Altered maternal adaptation to pregnancy

Much of the research concerning the relationship between size at birth and disease risk in adulthood has focused on the growth, physiology and endocrinology of the developing conceptus. There has been much less attention paid to possible alterations in maternal physiology and endocrinology that may give rise to such effects. In the case of periconceptional undernutrition, we have evidence from our sheep studies that several maternal hormonal and metabolic regulatory systems are altered and remain affected long after correction of the original nutritional deficit.

Excess exposure to maternal glucocorticoid has long been suggested to be the mediating factor in nutritionally-induced adaptive changes in fetal and postnatal development. This could happen in several ways. Firstly, undernutrition may be a stressor resulting in an elevation in maternal glucocorticoids, which then cross the placenta and lead to fetal exposure to excess glucocorticoids (Benediktsson et al. 1993; Cleasby et al. 2003). Secondly, activity of 11-ßHSD 2, the enzyme that normally inactivates glucocorticoids, could be reduced in the placenta as a result of maternal undernutrition (Lesage et al. 2001) so that more maternal glucocorticoid reaches the fetus. Thirdly, 11-ßHSD-1 the enzyme that activates cortisone to cortisol (or corticosterone in rodents), could be increased in the placenta or fetus in response to maternal undernutrition (Seckl *et al.* 1995), again resulting in increased fetal exposure to active gluco-corticoid. Rats fed a low protein diet (9% vs 18% in controls) during pregnancy exhibit in-creased corticosterone levels and give birth to pups of reduced birth weight, reduced glucose tolerance and elevated blood pressure as adults (Langley-Evans 1997a; Seckl & Meaney 2004). These same findings can be reproduced by administration of synthetic glucocorticoids or by direct inhibition of 11ß-HSD-2, and prevented by maternal adrenalectomy to prevent the rise in maternal glucocorticoids in response to nutritional stress (Langley-Evans 1997b). Similarly, synthetic glucocorticoid administration on day 28 of pregnancy in sheep results in hypertension of the offspring (Dodic *et al.* 1998), and reduced birth size in human pregnancy is associated with decreased placental activity of 11ß-HSD-2 (Hofmann *et al.* 2001).

We hypothesised that the effects of periconceptional undernutrition in sheep may likewise be mediated by the elevation of maternal glucocorticoid levels in response to the "stress" of undernutrition. However contrary to our hypothesis, we found that there was a marked de-crease in maternal plasma cortisol and ACTH concentrations during prolonged but moderate undernutrition (Bloomfield *et al.* 2004a). Similar findings have been reported elsewhere in slightly different undernutrition experiments (Edwards & McMillen 2001; Bispham *et al.* 2003). We went on to show that undernutrition was associated with an increased ACTH response to AVP+CRH challenge but a decreased cortisol response, suggesting that the suppression of the maternal HPAA by periconceptional undernutrition may result in adrenal ACTH resistance (Jaquiery *et al.* 2006). Interestingly we found that even after 20 d of refeeding (day 50 of pregnancy), that maternal adrenal mRNA levels of the ACTH receptor, steroidogenic acute regulatory protein (a key regulator of cholesterol transport in steroid synthesis) and P450c17, the rate limiting enzyme in cortisol production, were all still reduced and circulating maternal cortisol levels were also still low. Thus it is possible that the accelerated maturation of fetal physiology and endocrinology that we observed may result from development in a low, rather than high, maternal glucocorticoid environment during the first half of pregnancy. However placental 11-ßHSD 2 activity was also decreased in these animals, raising the possibility that the fetus could be exposed to excess maternal glucocorticoids later in pregnancy if maternal adrenal function recovered while placental enzyme activity remained low. Further studies in late gestation are currently under way to address this question.

Other aspects of maternal endocrinology were also markedly perturbed after periconceptional undernutrition. Progesterone levels were reduced from 10 d after mating until 80-90 d of ges-tation (Bloomfield *et al.* 2004a). This may reflect impaired function of the corpus luteum, and potentially delayed transfer of progesterone production to the placenta in mid pregnancy. Oth-ers have shown that elevation of maternal progesterone levels for the first 3 d of pregnancy have a profound impact on local hormone production and embryonic, fetal and placental growth (Kleemann *et al.* 2001). Plasma placental lactogen (PL) concentrations were reduced in periconceptionally undernourished ewes for the last half of pregnancy (Bloomfield *et al.* 2004a). It is suggested that PL in the maternal circulation promotes lipolysis (Thordarson *et al.* 1987) and insulin resistance (Ryan & Enns 1988), thereby favouring transfer of nutrients to the pla-centa and fetus. Thus impaired production of PL may be another mechanism whereby periconceptional undernutrition may perturb fetal growth and metabolism in late gestation. Our finding that the feto-placental unit in these animals may convert a greater fraction of glucose to lactate as a means of carbohydrate conservation (Oliver *et al.* 2005) may also be consistent with this.

The development of moderate insulin resistance is a normal physiological adaptation to pregnancy, thought to contribute to sparing of maternal glucose supply to meet the high feto-placental demand. In human pregnancy, impaired development of insulin resistance is associ-ated with impaired fetal growth (Catalano *et al.* 1995; Bernstein *et al.* 1997; Caruso *et al.*

1998). In sheep, we have found that maternal undernutrition from 60 d before to 30 d after conception impaired development of insulin resistance of glucose and fat metabolism in the mother at 65 d gestation, and that the effects on fat metabolism were still present at 120d gestation (Jaquiery *et al.* 2005). Preliminary data suggest that impaired development of pregnancy associated insulin resistance does not occur if the ewe is well fed until mating and only subjected to undernutrition for the first 30 d of pregnancy. The mechanisms underlying the development of insulin resistance in pregnancy, and thus their impairment with periconceptional undernutrition, are not yet clear. However the changes in progesterone and placental lactogen described above may both be contributory (Ryan & Enns 1988), and other mechanisms remain to be explored.

Consequences of periconceptional undernutrition for the postnatal animal

Human epidemiological studies suggest that cardiovascular disease, hyperlipidaemia and obesity are all increased in adults whose mothers were exposed to undernutrition in early pregnancy (Roseboom *et al.* 2001a). In a normal pastoral farm setting, premature birth of domestic livestock due to periconceptional undernutrition would result in neonatal death, precluding a postnatal outcome. There are a limited number of studies that have investigated postnatal outcomes in sheep after maternal undernutrition in early to mid pregnancy. Hansen and colleagues demonstrated that maternal undernutrition from mating until day 70 of pregnancy resulted in lambs with moderate hypertension (Hawkins *et al.* 2000) and an enhanced cortisol response to ACTH both in late gestation (Hawkins *et al.* 2001) and in early postnatal life. Similar regimes of maternal undernutrition result in postnatal lambs with increased fat mass, leptin mRNA expression (Symonds *et al.* 2005) and an increase in the ratio of oxidative/slow twitch to glycolytic fast twitch muscle fibres in two week old lambs (Fahey *et al.* 2005). In the rat maternal low protein diet in the pre-implantation period resulted in reduced birth weight and postnatal hypertension (Kwong *et al.* 2000).

There are also some preliminary data emerging suggesting altered postnatal physiology in twins compared to singletons. Juvenile twin lambs, up to 6 months of age, appear to be more insulin sensitive than singletons and this is related to being the lighter of the twin pairs (Clarke *et al.* 2000). We did not find any difference in post pubertal glucose tolerance in post-pubertal twin lambs, and this may reflect the switch from increased sensitivity to progressive resistance referred to earlier. However, post pubertal female twins do have increased HPAA activity compared to singletons, and once again the lighter of a twin pair had the higher activity (Bloomfield *et al.* 2004b). Data are also emerging from human studies suggesting differences between twins and singletons. For example, insulin sensitivity is reduced in 7 year old twins compared to singletons (Jefferies *et al.* 2004) and glucose tolerance in adult Danish twins is reduced in the lighter of the twin pairs (Poulsen & Vaag 2006). Unfortunately most studies in twins suffer from the lack of an appropriate control group and from subject selection; more well designed studies in both humans and animals are urgently needed to confirm or refute these preliminary observations.

The above findings may have important implications for the livestock industry. Premature birth and reduced neonatal survival are very costly problems for the industry. Alterations in muscle differentiation, insulin secretion and signalling, and HPAA regulation at the hormonal and tissue level are also likely to have important consequences for health, growth, composition and productivity of farm animals. Many studies have manipulated maternal nutrition during pregnancy, beginning well after conception and often also well after implantation. Our studies suggest that nutrition before and immediately after mating may have important consequences for the growth and development of the offspring, its survival at birth, and its subsequent health and productivity.

Conclusions

Maternal undernutrition during the periconceptional period has important effects on the growth and development of the offspring *in utero* and there are an increasing number of epidemiological and experimental studies that suggest there are profound effects on the health of those offspring after birth. The length of pregnancy can be perturbed by poor maternal nutrition during the periconceptional period and there are also increased risks of obesity, diabetes and heart disease in postnatal life. The challenge in medical science is to develop better understanding of optimum nutrition in the periconceptional period and to develop methods for detecting and treating perturbed development to reduce adverse consequences. In the livestock production industry, a better understanding of the impact of maternal nutrition during the periconceptional period on the growth, health and production characteristics of the offspring will allow an opportunity to intervene in a way that is unparalleled in human medicine.

Acknowledgements

We would like to acknowledge the valuable input of our colleagues Bernhard Breier, Alexandra Buckley, John Challis, Nina DeBoo, Peter Gluckman, Paul Hawkins and Murray Mitchell. Our studies would not have been possible without the excellent technical assistance of Christine Keven, Toni Mitchell, Samantha Rossenrode, Eric Thorstensen and Pierre Van Zjil. Our research is supported by the Health Research Council of New Zealand, National Research Centre for Growth and Development, New Zealand Lottery Board and The Lion Foundation.

References

Aldoretta PW, Carver TD & Hay WW, Jr 1998 Maturation of glucose-stimulated insulin secretion in fetal sheep. *Biology of the Neonate* 73 375-386.

Barker DJ 2000 In utero programming of cardiovascular disease. *Theriogenology* 53 555-574.

Barnes FL 2000 The effects of the early uterine environment on the subsequent development of embryo and fetus. *Theriogenology* 53 649-658.

Benediktsson R, Lindsay RS, Noble J, Seckl JR & Edwards CR 1993 Glucocorticoid exposure in utero: new model for adult hypertension. *Lancet* 341 339-341.

Bernstein IM, Goran MI & Copeland KC 1997 Maternal insulin sensitivity and cord blood peptides: relationships to neonatal size at birth. *Obstetrics and Gynecology* 90 780-783.

Bispham J, Gopalakrishnan GS, Dandrea J, Wilson V, Budge H, Keisler DH, Broughton Pipkin F, Stephenson T & Symonds ME 2003 Maternal endocrine adaptation throughout pregnancy to nutritional manipulation: consequences for maternal plasma leptin and cortisol and the programming of fetal adipose tissue development. *Endocrinology* 144 3575-3585.

Bloomfield FH, Oliver MH, Hawkins P, Campbell M, Phillips DJ, Gluckman PD, Challis JR & Harding JE 2003 A periconceptional nutritional origin for non-infectious preterm birth. *Science* 300 606.

Bloomfield FH, Oliver MH, Hawkins P, Holloway AC, Campbell M, Gluckman PD, Harding JE & Challis JR 2004a Periconceptional undernutrition in sheep accelerates maturation of the fetal hypothalamic-pituitary-adrenal axis in late gestation. *Endocrinology* 145 4278-4285.

Bloomfield FH, Oliver, MH, Harding, JE 2004b The importance of being a twin: within twin pair analysis of glucose tolerance and hypothalamic-pituitary-adrenal axis responsiveness. In *Medical Sciences Congress of New Zealand* pp. C8. Queenstown.

Bloomfield FH, Rumball C, Oliver MH, Jaquiery AL, Harding JE 2005 Periconceptional undernutrition and twin size affect both growth and metabolic responses of twin sheep to an acute maternal fast in late gestation. In *International Congress of Developmental Origins of Health and Disease. Pediatric Research* 58(5) P3-032, Toronto.

Buckley AJ, Oliver MH, Bloomfield FH, Harding JE 2005 Upregulated expression of PKCz in fetal skeletal muscle of periconceptionally undernourished ewes. In *3rd International Congress of Developmental Origins of Health and Disease. Pediatric Research* **58(5)** P1-038, Toronto.

Budge H, Gnanalingham MG, Gardner DS, Mostyn A, Stephenson T & Symonds ME 2005 Maternal nutritional programming of fetal adipose tissue development: long-term consequences for later obesity. *Birth Defects Research Part C Embryo Today* **75** 193-199.

Caruso A, Paradisi G, Ferrazzani S, Lucchese A, Moretti S & Fulghesu AM 1998 Effect of maternal carbohydrate metabolism on fetal growth. *Obstetrics and Gynecology* **92** 8-12.

Catalano PM, Drago NM & Amini SB 1995 Maternal carbohydrate metabolism and its relationship to fetal growth and body composition. *American Journal of Obstetrics and Gynecology* **172** 1464-1470.

Cetin I, Fennessey PV, Quick AN, Jr., Marconi AM, Meschia G, Battaglia FC & Sparks JW 1991 Glycine turnover and oxidation and hepatic serine synthesis from glycine in fetal lambs. *American Journal of Physiology* **260** E371-378.

Cherif H, Reusens B, Ahn MT, Hoet JJ & Remacle C 1998 Effects of taurine on the insulin secretion of rat fetal islets from dams fed a low-protein diet. *Journal of Endocrinology* **159** 341-348.

Clarke L, Firth K, Heasman L, Juniper DT, Budge H, Stephenson T & Symonds ME 2000 Influence of relative size at birth on growth and glucose homeostasis in twin lambs during juvenile life. *Reproduction Fertility and Development* **12** 69-73.

Cleasby ME, Kelly PA, Walker BR & Seckl JR 2003 Programming of rat muscle and fat metabolism by in utero overexposure to glucocorticoids. *Endocrinology* **144** 999-1007.

Cnattingius S, Bergstrom R, Lipworth L & Kramer MS 1998 Prepregnancy weight and the risk of adverse pregnancy outcomes. *New England Journal of Medicine* **338** 147-152.

Dodds L, Fell DB, Joseph KS, Allen VM & Butler B 2006 Outcomes of pregnancies complicated by hyperemesis gravidarum. *Obstetrics and Gynecology* **107** 285-292.

Dodic M, May CN, Wintour EM & Coghlan JP 1998 An early prenatal exposure to excess glucocorticoid leads to hypertensive offspring in sheep. *Clinical Science (Lond)* **94** 149-155.

Edwards LJ & McMillen IC 2001 Maternal undernutrition increases arterial blood pressure in the sheep fetus during late gestation. *Journal of Physiology* **533** 561-570.

Fahey AJ, Brameld JM, Parr T & Buttery PJ 2005 The effect of maternal undernutrition before muscle differentiation on the muscle fiber development of the newborn lamb. *Journal of Animal Science* **83** 2564-2571.

Fall CH, Osmond C, Barker DJ, Clark PM, Hales CN, Stirling Y & Meade TW 1995 Fetal and infant growth and cardiovascular risk factors in women. *British Medical Journal* **310** 428-432.

Fall CH, Stein CE, Kumaran K, Cox V, Osmond C, Barker DJ & Hales CN 1998 Size at birth, maternal weight, and type 2 diabetes in South India. *Diabetic Medicine* **15** 220-227.

Fowden AL 1980 Effects of arginine and glucose on the release of insulin in the sheep fetus. *Journal of Endocrinology* **85** 121-129.

Gluckman PD, Hanson MA, Spencer HG & Bateson P 2005 Environmental influences during development and their later consequences for health and disease: implications for the interpretation of empirical studies. *Proceedings: Biological Sciences* **272,** 671-677.

Gnanalingham MG, Mostyn A, Webb R, Keisler DH, Raver N, Alves-Guerra MC, Pecqueur C, Miroux B, Symonds ME & Stephenson T 2005 Differential effects of leptin administration on the abundance of UCP2 and glucocorticoid action during neonatal development. *American Journal of Physioliology, Endocrinology and Metabolism* **289** E1093-1100.

Hales CN, Barker DJ, Clark PM, Cox LJ, Fall C, Osmond C & Winter PD 1991 Fetal and infant growth and impaired glucose tolerance at age 64. *British Medical Journal* **303** 1019-1022.

Harding JE 1997 Periconceptual nutrition determines the fetal growth response to acute maternal undernutrition in fetal sheep of late gestation. *Prenatal and Neonatal Medicine* **2** 310-319.

Hawkins P, Hanson MA & Matthews SG 2001 Maternal undernutrition in early gestation alters molecular regulation of the hypothalamic-pituitary-adrenal axis in the ovine fetus. *Journal of Neuroendocrinology* **13** 855-861.

Hawkins P, Steyn C, Ozaki T, Saito T, Noakes DE & Hanson MA 2000 Effect of maternal undernutrition in early gestation on ovine fetal blood pressure and cardiovascular reflexes. *American Journal of Physiology* **279** R340-348.

Hofmann M, Pollow K, Bahlmann F, Casper F, Steiner E & Brockerhoff P 2001 11 beta-hydroxysteroid dehydrogenase (11 beta-HSD-II) activity in human placenta: its relationship to placental weight and birth weight and its possible role in hypertension. *Journal of Perinatal Medicine* **29** 23-30.

Jackson AA 1991 The glycine story. *European Journal of Clinical Nutrition* **45** 59-65.

Jackson AA, Persaud C, Werkmeister G, McClelland IS, Badaloo A & Forrester T 1997 Comparison of urinary 5-L-oxoproline (L-pyroglutamate) during normal pregnancy in women in England and Jamaica. *British Journal of Nutrition* **77** 183-196.

Jaquiery AL, Oliver MH, Bloomfield FH, Connor KL, Challis JR & Harding JE 2006 Fetal exposure to excess glucocorticoid is unlikely to explain the effects of periconceptional undernutrition in sheep. *Journal of Physioliology* **572** 109-118.

Jaquiery AL, Oliver MH, Buckley A, Harding JE 2005 Effect of periconceptional undernutrition on insulin sensitivity at 65 days gestation in singleton bearing pregnant ewes. In *3rd International Congress of Developmental Origins of Health and Disease. Pediatric Research* **58(5)** P1-053, Toronto.

Jefferies CA, Hofman PL, Knoblauch H, Luft FC, Robinson EM & Cutfield WS 2004 Insulin resistance in healthy prepubertal twins. *Journal of Pediatrics* **144** 608-613.

Kleemann DO, Walker SK, Hartwich KM, Fong L, Seamark RF, Robinson JS & Owens JA 2001 Fetoplacental growth in sheep administered progesterone during the first three days of pregnancy. *Placenta* **22** 14-23.

Kumarasamy V, Mitchell MD, Bloomfield FH, Oliver MH, Campbell ME, Challis JR & Harding JE 2005 Effects of periconceptional undernutrition on the initiation of parturition in sheep. *American Journal of Physiology* **288** R67-72.

Kwong WY, Wild AE, Roberts P, Willis AC & Fleming TP 2000 Maternal undernutrition during the preimplantation period of rat development causes blastocyst abnormalities and programming of postnatal hypertension. *Development* **127** 4195-4202.

Langley-Evans SC 1997a Hypertension induced by foetal exposure to a maternal low-protein diet, in the rat, is prevented by pharmacological blockade of maternal glucocorticoid synthesis. *Journal of Hypertension* **15** 537-544.

Langley-Evans SC 1997b Intrauterine programming of hypertension by glucocorticoids. *Life Sciences* **60** 1213-1221.

Law CM 1996 Fetal and infant influences on non-insulin-dependent diabetes mellitus (NIDDM). *Diabetic Medicine* **13** S49-52.

Lesage J, Blondeau B, Grino M, Breant B & Dupouy JP 2001 Maternal undernutrition during late gestation induces fetal overexposure to glucocorticoids and intrauterine growth retardation, and disturbs the hypothalamo-pituitary adrenal axis in the newborn rat. *Endocrinology* **142** 1692-1702.

Nyirenda MJ, Welberg LA & Seckl JR 2001 Programming hyperglycaemia in the rat through prenatal exposure to glucocorticoids-fetal effect or maternal influence? *Journal of Endocrinology* **170** 653-660.

Oliver MH, Hawkins P, Breier BH, Van Zijl PL, Sargison SA & Harding JE 2001 Maternal undernutrition during the periconceptual period increases plasma taurine levels and insulin response to glucose but not arginine in the late gestational fetal sheep. *Endocrinology* **142** 4576-4579.

Oliver MH, Hawkins P & Harding JE 2005 Periconceptional undernutrition alters growth trajectory and metabolic and endocrine responses to fasting in late-gestation fetal sheep. *Pediatric Research* **57** 591-598.

Osmond C & Barker DJ 2000 Fetal, infant, and childhood growth are predictors of coronary heart disease, diabetes, and hypertension in adult men and women. *Environmental Health Perspectives* **108** Suppl 3 545-553.

Petry CJ, Ozanne SE, Wang CL & Hales CN 2000 Effects of early protein restriction and adult obesity on rat pancreatic hormone content and glucose tolerance. *Hormone and Metabolic Research* **32** 233-239.

Phillips DI, Barker DJ, Hales CN, Hirst S & Osmond C 1994 Thinness at birth and insulin resistance in adult life. *Diabetologia* **37** 150-154.

Phipps K, Barker DJ, Hales CN, Fall CH, Osmond C & Clark PM 1993 Fetal growth and impaired glucose tolerance in men and women. *Diabetologia* **36** 225-228.

Poulsen P & Vaag A 2006 The intrauterine environment as reflected by birth size and twin and zygosity status influences insulin action and intracellular glucose metabolism in an age- or time-dependent manner. *Diabetes* **55** 1819-1825.

Ravelli AC, van der Meulen JH, Michels RP, Osmond C, Barker DJ, Hales CN & Bleker OP 1998 Glucose tolerance in adults after prenatal exposure to famine. *Lancet* **351** 173-177.

Ravelli AC, van Der Meulen JH, Osmond C, Barker DJ & Bleker OP 1999 Obesity at the age of 50 y in men and women exposed to famine prenatally. *American Journal of Clinical Nutrition* **70** 811-816.

Rayco-Solon P, Fulford AJ & Prentice AM 2005 Differential effects of seasonality on preterm birth and intrauterine growth restriction in rural Africans. *American Journal of Clinical Nutrition* **81** 134-139.

Roseboom TJ, van der Meulen JH, Osmond C, Barker DJ, Ravelli AC & Bleker OP 2000 Plasma lipid profiles in adults after prenatal exposure to the Dutch famine. *American Journal of Clinical Nutrition* **72** 1101-1106.

Roseboom TJ, van der Meulen JH, Ravelli AC, Osmond C, Barker DJ & Bleker OP 2001a Effects of prenatal exposure to the Dutch famine on adult disease in later life: an overview. *Molecular and Cellular Endocrinology* **185** 93-98.

Roseboom TJ, van der Meulen JH, van Montfrans GA, Ravelli AC, Osmond C, Barker DJ & Bleker OP 2001b Maternal nutrition during gestation and blood pressure in later life. *Journal of Hypertension* **19** 29-34.

Ryan EA & Enns L 1988 Role of gestational hormones in the induction of insulin resistance. *Journal of Clinical Endocrinology and Metabolism* **67** 341-347.

Seckl JR, Benediktsson R, Lindsay RS & Brown RW 1995 Placental 11 beta-hydroxysteroid dehydrogenase and the programming of hypertension. *Journal of Steroid Biochemistry and Molecular Bioliogy* **55** 447-455.

Seckl JR & Meaney MJ 2004 Glucocorticoid programming. *Annals of the New York Academy of Sciences* **1032** 63-84.

Snoeck A, Remacle C, Reusens B & Hoet JJ 1990 Effect of a low protein diet during pregnancy on the fetal rat endocrine pancreas. *Biology of the Neonate* **57** 107-118.

Sollid CP, Wisborg K, Hjort J & Secher NJ 2004 Eating disorder that was diagnosed before pregnancy and pregnancy outcome. *American Journal of Obstetrics and Gynecology* **190** 206-210.

Symonds ME, Budge H, Stephenson T & Gardner DS 2005 Experimental evidence for long-term programming effects of early diet. *Advances in Exerimental*

Medicine and Biology **569** 24-32.

Thordarson G, McDowell GH, Smith SV, Iley S & Forsyth IA 1987 Effects of continuous intravenous infusion of an ovine placental extract enriched in placental lactogen on plasma hormones, metabolites and metabolite biokinetics in non-pregnant sheep. *Journal of Endocrinology* **113** 277-283.

Thorstensen EB, Van Zijl P, Oliver MH, Harding JE 2005 Periconceptional undernutrition perturbs serine and glycine metabolism in the pregnant sheep. In *The 9th annual congress of the Perinatal Society of Australia and New Zealand.*

Flock differences in the impact of maternal dietary restriction on offspring growth and glucose tolerance in female offspring

BE Burt[1,2], BW Hess[1,2], PW Nathanielsz[1,3] and SP Ford[1,2]

[1]The Center for the Study of Fetal Programming, Laramie, WY 82071; [2]Department of Animal Science, University of Wyoming, Laramie, WY 82071; [3]Department of Obstetrics and Gynecology, University of Texas Health Sciences Center, San Antonio, TX 78229

Variable impacts of in-utero programming stimuli on postnatal offspring development suggest that genotype may play a role in this response. In this study, ewes from two flocks of similar breeding but adapted for 6-8 generations to one of two markedly different production environments were utilized (Baggs ewes - nomadic lifestyle and limited nutrition; UW ewes - sedentary lifestyle and adequate nutrition). Ewes from each flock were fed 50% (nutrient restricted) or 100% (control) National Research Council (NRC) requirements between day 28 and 78 of gestation; some ewes in each dietary group were then necropsied. Remaining ewes were fed 100% NRC requirements from day 79 to term. Weights of singleton female fetuses were reduced ($P < 0.05$) in nutrient restricted UW ewes compared to control UW ewes on day 78. Two month old ewe lambs from nutrient restricted UW ewes had greater ($P < 0.05$) baseline glucose concentrations, and exhibited greater ($P < 0.05$) glucose and insulin concentrations to an intravenous glucose bolus than lambs from control UW ewes. From 4 to 12 months of age, ewe lambs from nutrient restricted UW ewes were heavier ($P < 0.05$) than lambs from control UW ewes. In contrast, no differences in fetal weight, baseline glucose, glucose and insulin concentration to an intravenous glucose bolus, or body weight were observed for nutrient restricted and control Baggs ewes. These data suggest that a multigenerational adaptation of ewes to different production systems impacts their ability to protect their fetus against a bout of early to mid-gestational nutrient restriction.

Introduction

Significant data has accumulated in recent years establishing an association between suboptimal intrauterine and post-natal environments and the development of adult diseases including coronary heart disease, hypertension, stroke, type II diabetes and dyslipidemia and neurologic disorders (Barker and Osmond 1988; Barker et al. 1989; Barker 1998; Gluckman and Hanson 2004). These observations have been confirmed in numerous animal studies across multiple species (Moss et al. 2001; Bertram and Hanson 2001; Mathews et al. 2002; Gluckman and Hanson 2004).

Corresponding author E-mail: spford@uwyo.edu

Recently, Vonnahme et al. (2006) reported that Rambouillet/Columbia cross ewes from the same two flocks as utilized in this study and subjected to the same nutrient restriction, exhibited flock specific differences in placentomal growth, development, and efficiency in response to maternal nutrient restriction from early to mid-gestation. Specifically, Vonnahme et al. (2006) reported that in ewes adapted to harsh range conditions and limited nutrient availability (Baggs ewes), placentomes advanced from type A to more efficient types B, C, and D by day 78 of gestation. Further, this placentomal conversion by Baggs ewes in the face of a bout of undernutrition was shown to support normal fetal growth and maintain fetal blood glucose and amino acid concentrations at normal levels (Wu et al. 2005). Conversely, ewes not accustomed to a harsh environment or to periods of nutrient restriction (UW ewes) failed to convert type A placentomes to types B, C, and D by day 78 and exhibited decreased fetal blood glucose and intrauterine growth restriction (IUGR; Vonnahme et al. 2006). In addition, Kwon et al. (2004) reported that essential amino acids were markedly reduced in the blood of these nutrient restricted UW ewes when compared to UW ewes fed to requirement. Classification of placentomal type is based on the scheme of Vatnick et al. (1991) and depends on placentome appearance as follows: 1) caruncular tissue completely surrounding the cotyledonary tissue (type A), 2) cotyledonary tissue beginning to grow over the surrounding caruncular tissue (type B), 3) flat placentomes with caruncular tissue on one surface and cotyledonary tissue on the other (type C), and 4) everted placentomes resembling bovine placentomes (type D). As placentomes progress from type A through type D, they increase in size, and exhibit a greater capillary volume and arteriolar density (Ford et al. 2004), resulting in increased placentomal blood flow (Ford et al. 2006). This adaptive mechanism among different but genetically similar, populations could provide valuable insight as to the etiology of the metabolic syndrome X. This syndrome is linked to IUGR and is marked by glucose tolerance, insulin resistence, obesity and hypertension, predisposing an individual to cardiovascular disease in later life (Latini et al. 2004; Reaven 2005).

Maternal undernutrition has been shown to induce insulin resistance (a precursor to type II diabetes mellitus) in a variety of species including the rat, sheep, and human (Fowden and Hill 2001; and Schwitzgebel 2001; Simmons et al. 2001). The goal of this project was to compare growth patterns and insulin and glucose responses of singleton female lambs born to nutrient restricted and control Baggs and UW ewes to a glucose challenge.

Methods

Animals

All animal procedures were approved by the University of Wyoming Animal Care and Use Committee. Ewes of similar breeding (Rambouillet/Columbia cross), age (4-5 years old) and parity (2-3 lamb crops) were obtained from two different flocks for use in this study. The first flock located near Baggs, Wyoming was adapted over 6-8 generations (~ 30 years) to a nomadic existence, grazing a land mass of ~ 250 miles/year which ranged from desert terrain to high mountain pastures with very limited nutritional supplementation (Baggs ewes). The second flock was also maintained for 6-8 generations (~ 30 years) by the University of Wyoming, and in contrast to the Baggs ewes, had a relatively sedentary lifestyle and consumed a diet from birth that always met or exceeded their dietary requirements (UW ewes). Over the course of two years, 60 Baggs ewes and 60 UW ewes were utilized for this study.

Ewes were checked for estrus twice daily and bred to an intact ram of the same breeding as the ewes at first exhibition of estrus and 12 h later (first day of mating = day 0). Animals were fed alfalfa hay at a rate of ~ 2% of their estimated body weight from 60 days before mating to day 20 post-mating. On day 20 post-mating, ewes were weighed so that individual diets could be provided on a metabolic body weight basis (body weight $^{0.75}$). The control ration consisted of a pelleted beet pulp (93.5% dry matter [DM], 79% total digestible nutrients [TDN], and 10.0% crude protein). Daily rations were delivered on a DM basis to meet the TDN required for maintenance of a pregnant ewe (NRC, 1985). A mineral-vitamin mixture (51.43% sodium triphosphate, 47.62% potassium chloride, 0.39% zinc oxide, 0.06% cobalt acetate, and 0.50% ADE vitamin premix [8,000,000 IU vitamin A, 800,000 IU vitamin D3, and 400,000 IU vitamin E per pound; amount of vitamin premix was formulated to meet the vitamin A requirements]) was included with the beet pulp pellets to meet requirements. On day 21 post-mating, all UW and Baggs ewes were placed in individual pens and fed the control diet satisfying 100% of NRC requirements for the early gestational ewe (NRC 1985). On day 28 post-mating, ewes within each flock were randomly assigned in equal numbers to either remain on the control diet at 100% NRC requirements (control) or were fed the control diet at 50% NRC requirements (nutrient restricted). On day 45 post-mating, pregnancy was confirmed by ultrasonography (Ausonics Microimager 1000 sector scanning instrument; Ausonics Pty Ltd, Sydney, Australia). At weekly intervals from day 28 of gestation until parturition, all ewes were weighed and daily rations adjusted for body weight gain for control or loss for nutrient restricted. On day 79 of gestation, 10 ewes in each of the four treatment groups carrying singleton pregnancies were necropsied and fetal weight and crown rump lengths were determined for female fetuses. All remaining nutrient restricted UW and Baggs ewes were re-alimented to the control diet on day 79, and thereafter all remaining ewes from both flocks were fed to meet NRC requirements for the early gestational ewe (NRC 1985) until day 104 of gestation. From day 105 of gestation until the day of parturition, the rations fed to all pregnant ewes were increased according to NRC guidelines for late gestational ewes (NRC 1985).

Parturition and postnatal procedures

Lambing was allowed to proceed naturally in all ewes, which were given free choice access to good quality alfalfa hay thereafter. Immediately after birth, lambs were towel dried and morphometric data (birth weight, crown-rump length, abdominal and thoracic circumferences, biparietal diameter, and right and left humerus lengths) were collected and recorded for all singleton ewe lambs. Singleton ewe lambs born to each of 8 control UW ewes and 7 control Baggs ewes, as well as 7 nutrient restricted UW and 8 nutrient restricted Baggs ewes were selected for inclusion in this study to avoid potential differences due to pregnancy status (singleton and twin). Prior to 14 days of age, ewe lambs were tail-docked as per Federation of Animal Science Societies recommendations (FASS 1999). All lambs were given free access to a standard commercially available creep feed (Lamb Creep B30 w/Bovatec; Ranch-Way Feeds, Ft. Collins, CO) from birth to weaning. At 120 ± 2 days of age, ewe lambs were weaned and placed in an outdoor housing facility with shelter and ad libitum water. Lambs were fed a diet formulated to maintain a daily gain of 0.23 kg according to NRC requirements for replacement ewes (NRC 1985). The diet consisted of a 65% alfalfa hay and 35% corn mixture. In order to satisfy NRC phorphorus requirements, the corn mixture was fortified with sodium triphosphate mineral mix such that 99.25% of the mixture was corn and 0.75% was sodium triphosphate. Lambs were fed this diet until 8 months of age at which time they were switched to a diet consisting of ad libitum access to an alfalfa hay diet.

Glucose Tolerance Test (GTT)

All selected singleton ewe lambs were removed from their mothers at 63 ± 1 days of lactation, for a 12 h period, then weighed and jugular veins were catheterized without anesthesia (Terumo® Surflash 18 gauge x 2.5 inches long, Ann Arbor, MI), and using aseptic procedures. Lambs were allowed approximately 60 min to recover before initiating a glucose tolerance test. At -15 and -5 min before administration of a bolus injection of glucose (0.25 g/kg of body weight in 20s, 50% dextrose solution; Vedco; St. Joseph, MO), 5 mL blood samples were drawn from the venous catheter and 2mL placed into tubes containing heparin and sodium fluoride (2.5 mg/mL ; Sigma) and the remaining 3 mL placed into a vacutainer tube with no anticoagulant (5mL; Sigma, St. Louis, MO) to establish baseline values of glucose and insulin, respectively. Additional blood samples were collected at 2, 5, 10, 15, 30, 60, and 120 min after glucose injection. Catheters were flushed with heparinized saline following glucose infusion and after each blood sampling. Heparinized blood samples were stored on ice until centrifuged at 3000 x g for 10 min, and plasma stored at -80°C until analysis; while the nonheparinized samples were allowed to clot for 24 h at 4°C before centrifugation at 3000 x g for 10 min, and serum stored at -80°C until subsequent analysis.

Hormone assays

Glucose was analyzed using the Infinity™ (ThermoTrace Ltd, Cat. # TR15498; Melbourne, Australia) colorimetric assay modified in the following manner; plasma was diluted 1:5 in dH_2O, and $10\mu L$ of diluted plasma was added to $300\mu L$ reagent mix. All samples were run in triplicate, and sample analysis was completed using multiple assays. The intra-assay and inter-assay CVs were 5% and 7%, respectively. Insulin was measured by RIA in accordance with manufacturer recommendations (Coat-A-Count®, Diagnostic Products Corporation, Los Angeles, CA) and completed using two assays. The intra-assay CV was $< 5 \%$, while the inter-assay CV was $< 3\%$. Fasted glucose and insulin baseline values were ascertained from the -15 and -5 min samples.

Calculations and statistics

All data are presented as means $\pm$ SEM and significance was accepted when $P < 0.05$, with $P < 0.10$ considered a trend. Area under the curve (AUC) was determined for insulin and glucose using the trapezoidal rule with GraphPad Prism software (Version 3, GraphPad). Weight gain was calculated as the change in body weight between two consecutive measurements. Statistical comparisons between groups (nutrient restricted vs. control) were completed using independent t-test, split-plot ANOVA where appropriate (SAS V8.2, SAS Inst. Inc., Cary, NC). Post hoc analysis was performed as indicated with a LSD test.

Results

Maternal and fetal data

On day 28 of gestation, body weights of Baggs and UW ewes were similar averaging 75.23 ± 3.05 kg. At the end of the treatment period on day 78 of gestation, control Baggs and UW ewes had increased $7.21 \pm 0.70\%$ in body weight, while nutrient restricted Baggs and UW ewes had lost $8.32 \pm 0.51\%$ of their day 28 weight. Weights and crown rump lengths of day 78 singleton fetuses from nutrient restricted UW ewes were reduced (P < 0.05) when compared to those of control UW ewes and nutrient restricted and control Baggs ewes which were similar (Table 1). At term, control and nutrient restricted-realimented Baggs and UW ewes reached similar weights averaging 94.04 ± 4.20 kg.

Table 1. Body weight (BWT) and crown rump length (CRL) of singleton female fetuses on day 78 of gestation in nutrient restricted and control UW and Baggs ewes.

Measurements	UW control lambs (n = 5)	UW nutrient restricted lambs (n = 4)	Baggs control lambs (n = 4)	Baggs nutrient restricted lambs (n = 5)
BWT (g)	301 ± 11[a]	225 ± 7[b]	294 ± 9[a]	289 ± 12[a]
CRL (cm)	22.8 ± 0.4[a]	20.1 ± 0.4[b]	22.4 ± 0.5[a]	22.2 ± 0.6[a]

Values are expressed as the mean ± SEM. Values with different superscripts differ (P < 0.05).

Neonatal morphometric data and postnatal growth

No differences were observed at birth in the lambs from nutrient restricted and control Baggs ewes for any of the morphometric measurements taken (Table 2). In contrast, while the lambs from nutrient restricted and control UW ewes showed no differences in birth weight (5.2 ± 0.3 vs. 5.4 ± 0.2 kg ; P > 0.05), abdominal girth was reduced (P < 0.05) and thoracic girth tended to be reduced (P < 0.10) in offspring from nutrient restricted UW ewes when compared to offspring from control UW ewes.

Table 2. Neonatal morphometric data taken immediately after birth and prior to the first nursing of singleton female lambs.

	UW Control	UW NR	Baggs Control	Baggs NR
CRL (cm)	55.5 ± 0.9[a]	57.3 ± 0.9[a]	56.1 ± 0.8[a]	57.7 ± 0.8[a]
BPD (cm)	17.1 ± 0.5[a]	16.7 ± 0.5[a]	18.8 ± 0.8[a]	19.5 ± 0.5[a]
RHL (cm)	15.5 ± 0.3[a]	15.5 ± 0.2[a]	16.4 ± 0.9[a]	16.7 ± 0.3[a]
LHL (cm)	15.8 ± 0.2[a]	15.5 ± 0.2[a]	16.5 ± 0.9[a]	16.9 ± 0.3[a]
TG (cm)	41.2 ± 0.5[a]	39.9 ± 0.6[a,*]	42.2 ± 1.3[a]	43.0 ± 0.5[a]
AG (cm)	39.4 ± 0.6[a]	37.0 ± 0.7[b]	40.0 ± 0.8[a]	42.2 ± 1.0[a]
BWT (kg)	5.2 ± 0.2[a]	5.4 ± 0.2[a]	5.9 ± 0.6[a]	5.8 ± 0.3[a]

Mean ± SEM with different superscripts differ (P < 0.05). * Mean ± SEM differ (P < 0.10). Morphometric measurements: crown rump length (CRL); bi-parietal diameter (BPD); right humerus length (RHL); left humerus length (LHL); thoracic girth (TG); abdominal girth (AG); birth weight (BWT).

There was no difference in the weights of ewe lambs from nutrient restricted and control Baggs ewes from birth to one year of age (Fig. 1). In contrast, by four months of age UW lambs from nutrient restricted UW ewes were heavier (P < 0.05; Fig. 2) than lambs from control UW ewes, and these lambs remained heavier (P < 0.05) than lambs from control UW ewes until weighing ceased at one year of age.

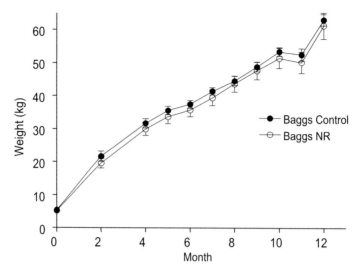

Fig. 1. Body weight response in female lambs born to Baggs ewes fed to satisfy requirements throughout gestation (●, n = 7) and to ewes fed 50% of requirements between day
28 through day 78 of gestation then fed to satisfy requirements thereafter (O, n = 8).
Values are expressed as mean ± SEM.

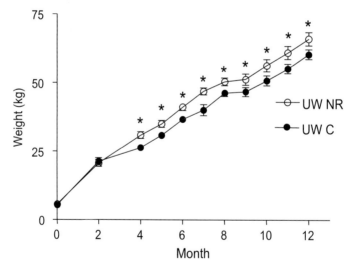

Fig. 2. Body weight response in female lambs born to UW ewes fed to satisfy requirements
throughout gestation (●, n = 8) and to ewes fed 50% of requirements between day 28
through day 78 of gestation then fed to satisfy requirements thereafter (O, n = 7). Astericks
indicate significant differences (P < 0.05). Values are expressed as mean ± SEM.

Two month glucose tolerance test

Baseline glucose levels in systemic blood averaged 82.4 ± 8.1 mg/dL for lambs of nutrient
restricted Baggs ewes and 76.8 ± 7.2 mg/dL for lambs from control Baggs ewes, which were
similar. In contrast, lambs from nutrient restricted UW ewes had higher (P < 0.05) average

baseline glucose concentrations than lambs from control UW ewes which averaged 90.7 ± 4.8 mg/dL and 71.3 ± 4.0 mg/dL, respectively. Peak glucose concentrations were attained at two min following the administration of the glucose bolus in lambs from all Baggs and UW ewes regardless of dietary group (Figs. 3 and 4). Peak concentrations of glucose in the offspring of control and nutrient restricted Baggs ewes averaged 226.64 ± 15.86 mg/dL and 243.4241 ± 15.25858 mg/dL respectively, which did not differ (Fig. 3). Similarly, no treatment differences in glucose concentration were observed throughout the remainder of the glucose response curve in Baggs lambs. In contrast, peak glucose concentrations of lambs from nutrient restricted UW offspring reached higher concentrations (P < 0.05) than those of lambs from control UW ewes which averaged 251.98 ± 22.52 mg/dL and 205.64 ± 16.15 mg/dL, respectively (Fig. 4). Additionally, glucose concentrations remained higher (P < 0.05) in the lambs from nutrient restricted UW ewes than lambs from control UW ewes for 10 min after bolus administration. Further, the total area under the curve (AUC) for glucose tended (P < 0.09) to be higher in the lambs from nutrient restricted UW ewes than in the lambs from control UW ewes, while no differences were observed between the two Baggs dietary groups. Glucose concentration of lambs from both control and nutrient restricted Baggs and UW ewes returned to baseline concentrations within the two hour testing period.

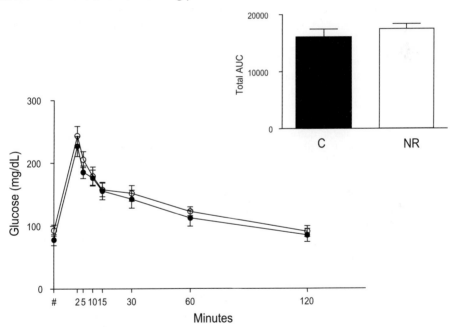

Fig. 3. Glucose concentrations before and after a glucose challenge (0.25 g/kg of 50% dextrose solution intravenously) at 63 ± 1 day of post- natal age in female lambs born to Baggs ewes fed to satisfy requirements throughout gestation (●, n = 7) and to Baggs ewes fed 50% of requirements between d 28 through 78 of gestation then day fed to satisfy requirements thereafter (O, n = 8). The (#) symbol on the X axis denotes pre-injection concentrations (average of -15 and -5 time points). Area under the curve (AUC; μIU min^{-1}) depicted in the inset. Values are expressed as mean ± SEM.

Baseline insulin values averaged 1.29 ± 0.23 pg/mL for the lambs from Baggs control ewes and 1.27 ± 0.31 pg/mL for the lambs from nutrient restricted Baggs ewes, which did not differ (Fig. 5). Similarly, no significant differences were found in average baseline insulin concentrations

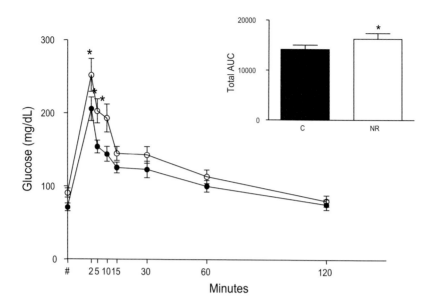

Fig. 4. Glucose concentrations before and after a glucose challenge (0.25 g/kg of 50% dextrose solution intravenously) at 63 ± 1 day of post-natal age in female lambs born to UW ewes fed to satisfy requirements throughout gestation (●, n = 8) and to UW ewes fed 50% of requirements between d 28 through day 78 of gestation then fed to satisfy requirements thereafter (O, n = 8). The (#) symbol on the X axis denotes pre-injection concentrations (average of -15 and -5 time points). Area under the curve (AUC; μIU min^{-1}) depicted in the inset was greater in nutrient restricted than control fed offspring. The asterisk indicates significant difference (P < 0.01). Values are expressed as mean ± SEM.

between lambs from control and nutrient restricted UW ewes, which averaged 1.26 ± 0.35 pg/ mL and 1.80 ± 0.35 pg/mL, respectively (Fig.e 6). Peak insulin levels were reached at five to ten minutes after glucose administration in all groups (Figs. 5 and 6). In the Baggs control lambs, peak insulin values were 17.12 ± 4.53 pg/mL, and in the nutrient restricted Baggs lambs they averaged 13.24 ± 2.68 pg/mL, but this difference was not significant (P = 0.43). Insulin concentrations in the blood of lambs from both control and nutrient restricted Baggs ewes had returned to baseline by 120 min. (Fig. 5). While peak insulin values were reached at the same time interval after glucose administration in the lambs from UW ewes (Fig. 6), significant treatment differences were observed. The offspring from control UW ewes reached peak insulin values of 10.80 ± 2.05 pg/ml, while the nutrient restricted animals obtained average peak values of 18.65 ± 4.21 pg/mL, nearly double their control cohorts (Fig. 6; P < 0.05). Further, insulin levels remained higher (P < 0.05) for 60 min after the glucose bolus in the lambs from nutrient restricted UW ewes when compared to lambs from control UW ewes. Again, insulin concentration of lambs from control and nutrient restricted UW ewes returned to baseline concentrations within the two hour testing period.

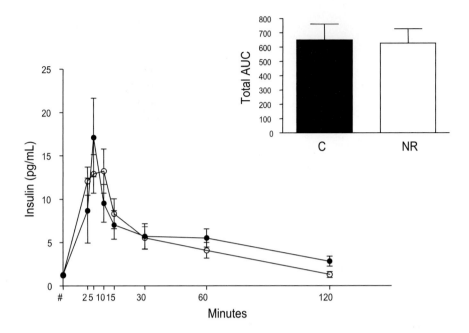

Fig. 5. Insulin concentrations before and after a glucose challenge (0.25 g/kg of 50% dextrose solution intravenously) at 63 ± 1 day of post-natal age in female lambs born to Baggs ewes fed to satisfy requirements throughout gestation (●, n = 7) and to Baggs ewes fed 50% of requirements between d 28 through day 78 of gestation then fed to satisfy requirements thereafter (O, n = 8). The (#) symbol on the X axis denotes pre-injection concentrations (average of -15 and -5 time points). Area under the curve (AUC; μIU min⁻¹) depicted in the inset. Values are expressed as mean ± SEM.

Discussion

To our knowledge, this is the first study in a livestock species demonstrating that ewes of similar breeding adapted over several generations to markedly different management systems alters the impact of maternal nutrient restriction on fetal growth as well as postnatal growth and endocrine status of their offspring. Specifically, female fetuses gestated by nutrient restricted UW ewes exhibited markedly reduced weights on day 78 of gestation when compared to female fetuses gestated by control UW ewes, while no reduction in fetal weight was observed in nutrient restricted Baggs ewes. Consistent with these findings, we have previously reported that this model of early to mid-gestational nutrient restriction results in decreased concentrations of glucose, essential amino acids and polyamines in the blood of fetuses gestated by nutrient restricted UW ewes, but not in the fetuses of nutrient restricted Baggs ewes on day 78 (Vonnahme et al. 2003; Kwon 2004; Wu et al. 2005). Much data has accrued that links IUGR to the metabolic syndrome X (Barker and Clark 1997; Latini et al. 2004; Stocker et al. 2005). The syndrome symptoms include impaired glucose and lipid metabolism, hypertension, and insulin resistance (Reaven 2005). Further, maternal undernutrition from early to mid-pregnancy in ewes has been shown to affect the hypothalamic-pituitary-adrenal axis and cardiovascular function of their lambs in the absence of altered birth weight (Hawkins et al. 2000; Gilbert et al. 2005).

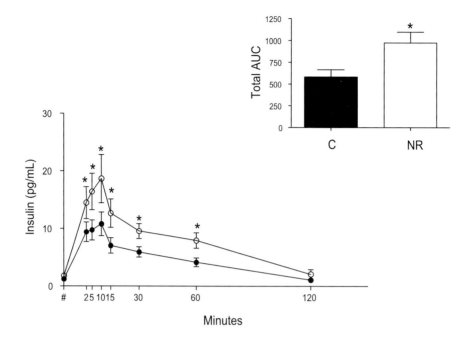

Fig. 6. Insulin concentrations before and after a glucose challenge (0.25 g/kg of 50% dextrose solution intravenously) at 63 ± 1 day of post-natal age in female lambs born to UW ewes fed to satisfy requirements throughout gestation (●, n = 8) and to ewes fed 50% of requirements between d 28 through day 78 of gestation then fed to satisfy requirements thereafter (O, n = 7). The (#) symbol on the X axis denotes pre-injection concentrations (average of -15 and -5 time points). Area under the curve (AUC; μIU min^{-1}) depicted in the inset was greater in nutrient restricted than control fed offspring. The asterisk indicates significant difference (P < 0.05). Values are expressed as mean ± SEM.

Because the weights of fetuses of nutrient restricted UW ewes were reduced compared to fetuses gestated by control UW ewes on day 78, and the weights of lambs from nutrient restricted and control UW ewes were similar at birth, an increased fetal growth rate must have occurred after realimentation in nutrient restricted UW ewes (see Fig. 7). Several studies have proposed a strong connection between IUGR, postnatal "catch-up growth" and the incidence of obesity and type 2 diabetes mellitus (Desai and Hales 1997; Bertin et al. 1999; Ericksson et al. 1999; Simmons et al. 2001; Hales and Ozanne 2003). The additional consequences, if any, of the observed pre-natal acceleration of fetal growth rate in the offspring of nutrient restricted-realimented UW ewes observed in this study is at present unknown.

The current observations show that a significant divergence in body weight occurred between the lambs of control and nutrient restricted UW ewes but not Baggs ewes between 2 and 4 months of age. This period includes the time at which the animals were suckling and allowed free access to creep feed. While we have no measurements of actual feed consumption during this interval, it is possible that consumption of milk and or consumption of creep feed may have differed between the lambs from nutrient restricted and control UW ewes. During this period, the lambs from nutrient restricted UW ewes also demonstrated an increased insulin response measured as area under the curve to a glucose tolerance test. In response to this initial glucose tolerance test at 63 days of age (pre-weaning), the lambs from nutrient restricted UW ewes presented a reduced capacity to clear glucose from the bloodstream as well as hyper-secretion of insulin in

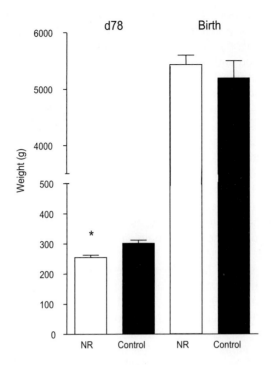

Fig. 7. Day 78 fetal weights and birth weights of lambs from nutrient restricted (NR) and control UW ewes. The asterisk indicates significant difference (P < 0.05). Values are expressed as mean ±SEM.

comparison to the control UW lambs. These data suggest that maternal undernutrition during early to mid-gestation triggers a hyper-responsive pancreatic insulin secretion during the pre-weaning period in response to a glucose rise. The augmented insulin secretion that we observed during this time period may underlie the divergence in growth rate observed, as it has been demonstrated previously that insulin is a potent anabolic hormone in the growing lamb. (Wolffe *et al.* 1989).

As previously stated, our group has recently reported that differences in placental efficiency may mediate the fetal growth rate differences seen in nutrient restricted UW ewes versus nutrient restricted Baggs ewes (Vonnahme *et al.* 2006). Results from several laboratories have demonstrated an increased conversion of type A placentomes to types B, C, and D placentomes in undernourished ewes when compared to their well-fed counterparts, however, this conversion has been thought to occur only in late gestation in association with the exponentially growing fetus (Crowe *et al.* 1996; Hoet and Hanson 1999; Osgerby *et al.* 2002; Osgerby *et al.* 2004). The ability of nutrient restricted Baggs ewes, but not UW ewes, to increase placentomal conversion and thus the size and vascularity of placentomes in response to early to mid-gestational undernutrition, would be expected to result in increased oxygen and nutrient availability to the fetus, thus preventing IUGR and associated health problems of offspring in later life. This conclusion is consistent with data presented in this study which demonstrated that nutrient restricted UW ewes experienced IUGR, as well as increased growth rates, hyperglycemia, and insulin resistance in their female offspring. In contrast, no differences were observed in fetal growth of nutrient restricted versus control Baggs ewes, or in the growth rates, blood glucose or insulin concentrations in their female offspring.

Simmons et al. (2001) used uterine artery ligation to induce IUGR in late gestation rats. They showed that restricted juvenile offspring developed accelerated growth rates, obesity, insulin resistance, and hyperglycemia. Further, as these restricted animals aged, insulin elevations progressively declined and animals actually experienced an inability to handle glucose. This may have been a consequence of an approximate 60% decrease in ß-cell mass and subsequent ß-cell failure (Simmons et al. 2001). Heywood et al. (2004) also found that maternal protein restriction in rats decreases the glucokinase activity of pancreatic beta cells of the offspring, causing decreased glucose tolerance and possible trend towards diabetes. Using a similar model, Lopes da Costa et al. (2004) demonstrated increased insulin secretion and Glut-2 transporters in pancreatic islets in young male neonatal rats. Gardner et al. (2005) reported that undernutrition (50% of requirements) of pregnant ewes during the last 37 days of gestation, but not during the first 30 days, resulted in glucose intolerance in the resulting offspring at 12 months of age. The impacts of periconceptional undernutrition on ovine fetal pancreatic function have also been demonstrated previously, as Oliver et al. (2001) reported that feeding 60% of requirements from 60 days before to 30 days after conception magnified fetal insulin release to a glucose challenge in late gestation.

In the past few years, several laboratories have evaluated the impact of maternal nutrient restriction from early to mid gestation on fetal and placental development in the sheep (Clarke et al. 1998; Heasman et al. 1999; Osgerby et al. 1999 and 2002; Vonnahme et al. 2003) and have reported markedly different results. The reasons for these differences may relate to differences in ewe breed, age, or parity, but may also relate to differences in nutritional and/or environmental exposures either before or after birth or even to alterations in the epigenetic state of the fetal genome. An understanding of how the two flocks of ewes utilized in this study evolved in a relatively short time frame to be either "sensitive to feed restriction" or "resistent to feed restriction" could be critically relevant to our understanding of factors dictating nutrient requirements during gestation in livestock species. Buhimschi et al (2001) reported that the levels of hypertension, intrauterine growth restriction, and fetal death to nitric oxide inhibition during pregnancy were significantly different in the same strain of rats purchased from different commercial suppliers. These researchers concluded that these differences may represent subtle genetic or environmental differences between the two outbred colonies resulting in dramatic variations in response to nitric oxide inhibition. Further, Knight et al. (2005) reported markedly different impacts of a 30% maternal nutrient restriction from day 6.5 to 18.5 of gestation in two strains of mice on fetal and placental size and offspring growth These studies suggest that within or between breed (or strain) differences exist in the susceptability of the conceptus to a variety of maternal stressors.

Conclusion

These data suggest that multigenerational adaptation of ewes of similar breeding to differing production systems impacts their ability to protect their fetus against a bout of early to mid-gestational nutrient restriction. More specifically, ewes adapted to harsh range conditions and limited nutrient availability maintained fetal growth in the face of an early gestational nutrient restriction, thereby preventing IUGR and thus the associated health problems of their offspring in adult life. In contrast, ewes adapted to less harsh conditions, and supplied with a diet which always met or exceeded their dietary requirements, failed to maintain fetal growth in the face of a similar bout of nutrient restriction resulting in IUGR, as well as hyperglycemia, and insulin resistance in their offspring, conditions which are linked to health problems in later life. Additional research is urgently needed to gain an understanding of the physiologic basis of these adaptation-induced differences in susceptibility to gestational nutrient restriction in ruminant species.

Acknowledgements

The authors wish to acknowledge support from University of Wyoming BRIN P20 RR16474, and INBRE P20 RR016474-04, and NIH HD 21350, and to thank Carole Hertz and Myrna Miller for laboratory and technical assistance and Reed Davidson and Mark Grant for animal care and management.

References

Barker DJ 1998 In utero programming of chronic disease. *Clinical Science* (London) **95** 115-128.

Barker DJ & Clark PM 1997 Fetal undernutrition and disease in later life. *Reviews of Reproduction* **2** 105-112.

Barker DJ & Osmond C 1988 Low birth weight and hypertension. *British Medical Journal* **297** 134-135.

Barker DJ, Winter PD, Osmond C, Margetts B & Simmonds SJ 1989 Weight in infancy and death from ischaemic heart disease. *Lancet* **2** 577-580.

Buhimschi IA, Shi SQ, Saade GR & Garfield RE 2001 Marked variation in response to long-term nitric oxide inhibition during pregnancy in outbred rats from two different colonies. *American Journal of Obstetrics & Gynecology* **184** 686-693.

Bertin E, Gangnerau MN, Bailbe D & Portha B 1999 Glucose metabolism and beta-cell mass in adult offspring of rats protein and/or energy restricted during the last week of pregnancy. *American Journal of Physiology* **277** E11-E17.

Bertram CE & Hanson MA 2001 Animal models and programming of the metabolic syndrome. *British Medical Bulletin.* **60** 103-121.

Clarke L, Heasman L, Juniper DT & Symonds ME 1998 Maternal nutrition in early-mid gestation and placental size in sheep. *British Journal of Nutrition* **79** 359-364.

Crowe C, Hawkins P, Saito T, Stratford LL, Noakes DA & Hanson MA 1996 Nutritional plane in early pregnancy: effect on fetal and placental size in late gestation. *Proceedings Fetal and Neonatal Physiology Society* pp 11.

Desai M & Hales CN 1997 Role of fetal and infant growth in programming metabolism in later life. *Biology Review of Cambridge Philosophy Society* **72** 329-348.

Eriksson JG, Forsen T, Tuomilehto J, Winter PD, Osmond C & Barker DJP 1999 Catch-up growth in childhood and death from coronary heart disease: longitudinal study. *Britsih Medical Journal* **318** 427-431.

FASS 1999 Sheep and goat husbandry. In *Guide for the Care and Use of Agricultural Animals in Agricultural Research and Teaching*, First Revised Edition, pp 67-71. Savoy, Illinois: Federation of Animal Science Societies.

Ford SP, Nijland MJ, Miller MM, Hess BW & Nathanielsz PW 2006 Maternal undernutrition advanced placentomal type, in association with increased placentomal size, and cotyledonary (COT) blood flow. *Journal of the Society for Gynecologic Investigation.*

Supplement **13** 212A.

Ford SP, Vonnahme KA, Drumhiller MC, Reynolds LP, Nijland MJ & Nathanielsz PW 2004 Arteriolar density and capillary volume increase as placentomes advance from type A through type D developmental stages. *Proceedings Society for the Study of Reproduction* pp 509.

Fowden AL & Hill DJ 2001 Intra-uterine programming of the endocrine pancreas. *British Medical Bulletin* **60** 123-142.

Gardner DS, Tingey K, Van Bon BWM, Ozanne SE, Wilson V, Dandrea J, Keisler DH, Stephenson T and Symonds ME 2005 Programming of glucose-insulin metabolism in adult sheep after maternal undernutrition. *American Journal of Physiology; Regulatory Integrated Comparative Physiology* **289** R947-R954.

Gilbert JS, Lang AL, Grant AR & Nijland MJ 2005 Maternal nutrient restriction in sheep: hypertension and decreased nephron number in offspring at 9 months of age. *Journal of Physiology* **565.1** 137-147.

Gluckman PD & Hanson MA 2004 Developmental origins of disease paradigm: a mechanistic and evolutionary perspective. *Pediatric Research* **56** 311-317.

Hales CN & Ozanne SE 2003 The dangerous road of catch-up growth. *Journal of Physiology* **547** 5-10.

Hawkins P, Steyn C, McGarrigle HH, Calder NA, Saito T, Strarford LL, Noakes DE & Hansona MA 2000 Cardiovascular and hypothalamic-pituitary-adrenal axis development in late gestation fetal sheep and young lambs following modest maternal nutrient restriction in early gestation. *Reproduction, Fertility & Development* **12** 443-456.

Heasman L, Clarke L, Stephenson TJ & Symonds ME 1999 The influence of maternal nutrient restriction in early to min-pregnancy on placental and fetal development in sheep. *Proceedings of the Nutrition Society* **58** 283-288.

Heywood WE, Mian N, Milla PJ & Lindley KJ 2004 Programming of defective rat pancreatic beta-cell function in offspring from mothers fed a low-protein diet during gestation and the suckling periods. *Clinical Science (London)* **107** 37-45.

Hoet JJ & Hanson MA 1999 Intrauterine nutrition: its importance during critical periods for cardiovascular and endocrine development. *Journal of Physiology* **514** 617-627.

Knight B, Pennell C, Lye S & Lunenfeld S 2005 Strain differences in the impact of maternal dietary restriction on fetal growth, pregnancy and postnatal development in mice. *Proceedings DOHaD Meeting* pp1031.

Kwon H, Ford SP, Bazer FW, Spencer TE, Nathanielsz PW, Nijland MJ, Hess BW & Wu G 2004 Maternal nutrient restriction reduces concentrations of amino acids and polyamines in ovine maternal and fetal plasma and fetal fluids. *Biology of Reproduction* **71** 901-908.

Latini G, De Mitri B, Del Vecchio A, Chitano G, De Felice C & Zetterstrom R 2004 Foetal growth of kidneys, liver and spleen in intrauterine growth restriction: "programming" causing "metabolic syndrome" in adult age. *ACTA Paediatrica* **93** 1559-1560.

Lopes Da Costa C, Sampaio De Freitas M & Sanchez Moura A 2004 Insulin secretion and GLUT-2 expression in undernourished neonate rats. *Journal of Nutrition & Biochemistry* **15** 236-241.

Matthews SG, Owen D, Banjanin S & Andrews MH 2002 Glucocorticoids, hypothalamo-pituitary-adrenal (HPA) development, and life after birth. *Endocrine Research* **28** 709-718.

Moss TJ, Sloboda DM, Gurrin LC, Harding R, Challis JR & Newnham JP 2001 Programming effects in sheep of prenatal growth restriction and glucocorticoid exposure. *American Journal of Physiology; Regulatory Integrated Comparative Physiology* **281** R960-R970.

NRC 1985 Nutrient requirements and signs of deficiency. In *Nutrient Requirements of Sheep*, edn 6, pp 2-25. The National Academies Press, Washington, DC.

Oliver MH, Hawkins P, Breier BH, Van Zijl PL, Sargison SA & Harding JE 2001 Maternal undernutrition during the periconceptual period increases plasma taurine levels and insulin response to glucose but not arginine in the late gestational fetal sheep. *Endocrinology* **142** 4576-4579.

Osgerby JC, Gadd TS & Wathes DC 1999 Expression of insulin-like growth factor binding protein-1 (IGFBP-1) mRNA in the ovine uterus throughout the estrous cycle and early pregnancy. *Journal of Endocrinology* **162** 279-287.

Osgerby JC, Wathes DC, Howard D & Gadd TS 2002 The effect of maternal undernutrition on ovine fetal growth. *Journal of Endocrinology* **173** 131-141.

Osgerby JC, Wathes DC, Howard D, & Gadd TS 2004 The effect of maternal undernutrition on the placental growth trajectory and the uterine insulin-like growth factor axis in the pregnant ewe. *Journal of Endocrinology* **182** 89-103.

Reaven GM 2005 Why syndrome X? From Harold Himsworth to the Insulin Resistence Syndrome. *Cell Metabolism* **1** 9-14.

Schwitzgebel VM 2001 Programming of the pancreas. *Molecular & Cellular Endocrinology* **185** 99-108.

Simmons RA, Templeton LJ & Gertz SJ 2001 Intrauterine growth retardation leads to the development of type 2 diabetes in the rat. *Diabetes* **50** 2279-2286.

Stocker CJ, Arch JRS & Cawthorne MA 2005 Fetal origins of insulin resistence and obesity. *Proceedings of the Nutrition Society* **64** 143-151.

Vatnick, I, Schoknecht, PA, Darrigrand, R & Bell, AW 1991. Growth and metabolism of the placenta after unilateral fetectomy in twin pregnant ewes. *Journal of Developmental Physiology* **15** 351-356.

Vonnahme KA, Hess BW, Hansen TR, McCormick RJ, Rule DC, Moss GE, Murdoch WJ, Nijland MJ, Skinner DC, Nathanielsz PW & Ford SP 2003 Maternal undernutrition from early- to mid-gestation leads to growth retardation, cardiac ventricular hypertrophy, and increased liver weight in the fetal sheep. *Biology of Reproduction* **69** 133-140.

Vonnahme KA, Hess BW, Nijland MJ, Nathanielsz PW & Ford SP 2006 Placentomal differentiation may compensate for nutrient restriction in ewes adapted to harsh range conditions. *Journal of Animal Science* **84** 3451-3459.

Wolffe JE, Dobbie PM & Petric DR 1989 Anabolic effects of insulin in growing lambs. *Quarterly Journal of Experimental Physiology* **74** 451-463.

Wu G, Shi W, Spencer TE, Hess BW, Nathanielsz PW & Ford SP 2005 Production system under which ewes are selected alters nutrient availability to the fetus in response to early pregnancy undernutrition. *Proceedings American Society of Animal Science*, Cincinnati, OH, p.297.

The developmental origins of health and disease: current theories and epigenetic mechanisms

KD Sinclair[1], RG Lea[2], WD Rees[3] and LE Young[4]

[1]School of Biosciences, University of Nottingham, Sutton Bonington, LE12 5RD, UK; [2]School of Veterinary Medicine, University of Nottingham, Sutton Bonington, LE12 5RD, UK; [3]Rowett Research Institute, Bucksburn, Aberdeen, AB21 9SB, UK; [4]School of Human Development, University of Nottingham, Nottingham, NG7 2UH, UK

The retrospective cohort studies of David Barker and colleagues during the late 1980s established the principle that the incidence of certain adult diseases such as stroke, type 2 diabetes and dyslipidaemia may be linked to *in utero* development. Later termed the "Developmental Origins of Health and Disease (DOHaD)" hypothesis, there have been several more recent attempts to explain this phenomenon. Although a general conceptual framework has been established to explain how mechanisms may have evolved to facilitate rapid adaptations to changing ecological conditions, it doesn't identify the actual mechanisms responsible for such effects. Extensive covalent modifications to DNA and related proteins occur from the earliest stages of mammalian development. These determine lineage-specific patterns of gene expression and so represent the most plausible mechanisms by which environmental factors can influence development during the life course. In providing a contemporary overview of chromatin modifications during early mammalian development, this review highlights both the complexity and our current lack of understanding of how epigenetic alterations may contribute to *in utero* programming. It concludes by providing some thoughts to future research endeavours where the emphasis should be on bettering our understanding of epigenesis and devising more thoughtful experimental approaches that focus on specific environmental factors in appropriate animal and cellular models.

Introduction

Clinical interest in the concept that late onset diseases can originate from events occurring *in utero* arose from the initial retrospective cohort studies of Barker and Osmond (1988) and Barker *et al.* (1989), who assessed relationships between size at birth, hypertension and ischaemic heart disease in adult humans. Further studies established inverse relationships between birth weight and the incidence of stroke, type 2 diabetes and dyslipidaemia, and these phenomena, often collectively referred to as Syndrome X or metabolic syndrome, led to the "fetal origins (later to be called developmental origins) of health and disease" or "DOHaD" hypothesis (Barker 1995). However, the central tenet of this hypothesis, that low birth weight is associ-

Corresponding author E-mail: Kevin.Sinclair@nottingham.ac.uk

ated with elevated blood pressure or hypertension in adults, has since been challenged (e.g. Huxley *et al.* 2002). It transpires that reduced size at birth, which is only a proxy for sub-optimal *in utero* conditions, is not necessarily associated with increased disease risk in adults. Nevertheless, the general consensus based on an overwhelming body of epidemiological evidence among different human populations and from direct-interventionalist studies with animals supports the general concept that sub-optimal *in utero* conditions can predispose to late-onset disease in adults (Langley-Evans 2006).

In the decade or so that followed the initial observations of Barker and colleagues a number of hypotheses were proposed to explain this phenomenon. Many attempted to support this hypothesis by setting it in a broader 'evolutionary' context. Numerous animal-based studies have also been conducted to determine the effects of exposure to specific environmental factors, at different stages of gestation, on indices of long-term health. The purpose of the present article is not to provide an exhaustive overview of this topic, which has been provided elsewhere (McMillen & Robinson 2005), but rather to develop and critically assess some of the emerging concepts and related theories, whilst defining the role of heritable epigenetic modifications to DNA and associated proteins. Although studies in ruminants feature, the emphasis is on investigations which have provided unique conceptual insights into early mammalian development and some of the regulatory mechanisms where, in keeping with the original ideas of David Barker and colleagues, the bias is towards the effects of physiologically relevant alterations to maternal nutrition.

DOHaD: Current controversies and related concepts

Experimentally induced exposure to a number of environmental factors during *in utero* development is known to modify post-natal physiology. Although studies have been conducted in the mouse, guinea pig, and domesticated species such as the pig and horse, the favoured animal models to date have been the rat and sheep. Studies with these species have revealed that maternal insults during pregnancy can alter subsequent cardiovascular and metabolic function in offspring in the absence of effects on birth weight (Table 1). Indeed, these observations have contributed to the current debate as to whether or not birth weight *per se* has any causal effect on subsequent physiological function in offspring (Lucas *et al.* 1999). An emerging concept from this debate is that a period of accelerated or 'compensatory' growth, that frequently accompanies episodes of growth restriction, may underlie many of the adverse effects on cardiovascular function and metabolism observed during adulthood (Singhal & Lucas 2004). Importantly, although initially proposed to defend the observation that catch-up growth during childhood in low-birth weight infants leads to cardiovascular disease (CVD) in adults, this hypothesis can be applied to any specific period during pre- and post-natal development.

Thrifty phenotype and Predictive Adaptive Responses

Much of the current thinking in the field (including that above) stems from the 'thrifty phenotype' hypothesis, originally proposed by Hales & Barker (1992), which attempts to explain the link between fetal growth and the metabolic syndrome. This hypothesis proposes that poor nutrition in early life (particularly during fetal development) leads to permanent changes in glucose-insulin metabolism. Thus offspring malnourished during pregnancy or infancy are ill-equipped to cope with high calorie diets in later life. This concept has been expanded by some to consider such adaptations as a means by which phenotypic modifications can be induced within a single generation in order to best accommodate prevailing or anticipated

Table 1. Examples of experimentally induced perturbations during pregnancy leading to altered physiology.

Intervention	Species	Stage of gestation	Physiological outcomes Pre-natal	Post-natal	Reference
A. Non-nutritional interventions					
(i) With intra-uterine growth restriction					
Assisted reproduction (e.g. IVF)	Mice	< 20 %	↓ fetal:placental weight ratio	↑ growth, ↑ adiposity	Sjöblom *et al.* (2005)
Carunclectomy	Sheep	14 - 20 %	Hypoxia, hypoglycaemia	N/D	Owens *et al.* (1989)
Maternal hyperthermia	Sheep	25 - 95 %	Hypoxia, hypoglycaemia, ↓ ponderal index	↑ neonatal mortality	Anthony *et al.* (2003)
Utero-placental embolization	Sheep	80 - 100 %	↑ mean arterial bp	↓ mean arterial bp	Louey *et al.* (2000)
Glucocortiocoid administration	Rat/Sheep	67 - 100 %	↑ ventricular wall thickness, - heart rate and bp,	↑ bp, hypertension, hyperglycaemia, hyperinsulinaemia	Seckle (2001)
Temporarily induced anaemia	Sheep	80 - 95 %	↓ haematocrit, ↓ blood O2	↓ adult weight, ↑ coronary conductance	Davis *et al.* (2002)
Between breed embryo transfer	Horse	0 - 100 %	↑↓ in placental and birth weights	Cross dependent ↑ ↓ arterial bp and baroreflex sensitivity	Giussani *et al.* (2003)
B. Nutritional interventions					
(i) With intra-uterine growth restriction					
Low protein diet	Rat	0 - 20 %	↓ blastomere number	↑ post-weaning growth, systolic bp (males)	Kwong *et al.* (2000)
(ii) Without intra-uterine growth restriction					
Low protein diet (9% casein)	Rat	0 - 33 %		↑ weaning weight, systolic bp (males)	Langley-Evans *et al.* (1996)
		34 - 67 %	↑ birth weight	↑ systolic bp	
		68 -100 %		↑ weaning weight (males), ↑ systolic bp	
		0 - 100 %		↑ weaning weight (males), ↑ systolic bp	
Lard enriched diet	Rats	0 - 100 %	N/D	↑ systolic, diastolic bp (females)	Khan *et al.* (2003)
Global nutrient restriction	Sheep	-40 - 5 %	↑ arterial bp and rate pressure product in twins	N/D	Edwards & McMIlien (2002)
	Sheep	0 - 20 %	N/D	↑ pulse pressure, ↓ pulse pressure product, blunted baroreflex sensitivity to Ang II	Gardner *et al.* (2004)
	Sheep	20 - 55 %	N/D	↓ nephron number, ↑ mean arterial bp	Gilbert *et al.* (2005)
	Sheep	70 - 80 %	N/D	Altered hypothalamic-piuitary-adrenal axis	Bloomfield *et al.* (2003)

↑, ↓ = increase/decrease relative to Controls; N/D = not determined; bp = blood pressure; Ang II = angiotensin II

environmental circumstances (Bateson 2001; Gluckman & Hanson 2004). Citing numerous examples within the animal kingdom, these authors distinguish such responses (termed Predictive Adaptive Responses [PARs]) from those that confer an immediate advantage or arise from disrupted or teratogenic development as a consequence of severe environmental challenges either during pregnancy or infancy. This hypothesis draws some support from long term follow-up studies of human populations subjected to famine during the second world war, for example, in The Netherlands during the winter of 1944-45 (Ravelli *et al.* 1998) and during the siege of Leningrad from 1941 to 1944 (Stanner *et al.* 1997); and from animal studies, including some of those listed in Table 1.

Intergenerational transmission, Lamarckian inheritance and epigenetics

An apparent limitation of the hypothesis proposed by Gluckman & Hanson (2004) is that it is based on the premise that the *in utero* environment can provide the necessary cues for the fetus to predict future (i.e. post-natal) environmental conditions. Whilst it is understandable that appropriate *in utero* responses to circadian and seasonal cues may confer immediate post-natal advantages to the neonate, particularly for short-lived and altricial mammalian species, the longer term advantages, particularly for larger animals, are less apparent. Indeed, for long-lived species, such as the human and ruminant, such responses would seem improbable, given that they have to transcend seasonal fluctuations in day length, climate and nutrient provision. Nevertheless, one population based study in humans conducted with subjects in both the northern and southern hemispheres demonstrated a clear effect of season of birth on longevity (autumn born babies live longer), which the authors (Doblhammer & Vaupel 2001) attributed to *in utero* nutrition.

An attempt to explain the evolutionary significance of PARs was made by Kuzawa (2005), who considered the phenomenon in the context of life history theory. This is a branch of evolutionary biology which postulates that many physiological traits are linked to key maturational and reproductive characteristics that define the life course. Compelling evidence was presented that nutrient provision during pregnancy (particularly to the female fetus) conveys information reflecting the nutritional environments experienced by matrilineal ancestors. Termed 'Inter-generational Phenotypic Inertia' this hypothesis predicts that traits such as birth weight would have an inter-generational component; a prediction supported by data from the Dutch famine of 1944-1945 (Lumey 1992). However, such a mechanism is also expected to minimise the influence of immediate or short-term fluctuations in nutrient provision, typically associated with seasonal or other stochastic factors, in favour of longer-term strategic goals reflecting changing ecological conditions. That is, according to this model, the effects of PARs within a single generation would be minimal, but would accumulate across generations. Furthermore, the hypothesis predicts (and some evidence is presented) that the 'buffering effect' afforded by this mechanism primarily benefits the female fetus, so that the male fetus may be more susceptible to factors experienced by the mother during pregnancy (Kuzawa 2005). Indeed, in a nutritional re-alignment study with rats which had been offered a low protein diet for 12 generations the following observations were made during nutritional realignment: (i) there was a graded response in terms of increased birth weight and cognitive abilities which depended on the timing of protein re-alignment (*in utero* > from birth > from 4 weeks), and (ii) further incremental improvements in residual physical and behavioural outcomes were observed over the following two generations (Stewart *et al.* 1980). Importantly, the effects of both the initial 12 generations of protein restriction and of dietary re-alignment were greater in male offspring. The full significance of this gender-effect specifically in the context of DOHaD, however, remains to be confirmed. The available evidence is inconclusive as unfortunately few experiments have considered long-term intergenerational effects, and experiments have often been conducted with insufficient power to determine gender effects within a single generation.

Nevertheless, such phenomena evoke the ideas of the French naturalist Jean-Baptiste Lamark (1744-1829), who proposed that the inheritance of acquired characteristics is the hereditary mechanism by which changes in physiology acquired during the lifetime of an organism are transmitted to its offspring. Previously refuted in favour of Darwin's theories on natural selection, the emerging field of epigenetics, whilst in no way contradicting the laws of natural selection, has provided a potential mechanism for Larmark's ideas on the inheritance of acquired traits. The environmental lability of acquired epigenetic modifications, particularly during the earliest stages of development (discussed later), provides a mechanistic basis for predictive adaptive responses and transgenerational inheritance.

Effects on fertility

Investigations into the long-term consequences of the war-time famine in The Netherlands revealed that undernutrition during gestation can lead to a number of disease states in later life, the nature and severity of which are dependent on the stage of gestation at the time of insult (Painter *et al.* 2005). A striking observation from studies with this cohort, however, was that the subsequent fertility of offspring undernourished during pregnancy was unaffected (Lumey 1998). Although there are reported effects on ovulation rate among non-obese adolescent girls born small for gestational age (effects associated with central adiposity and dyslipidaemia (Ibáñez *et al.* 2002), it remains to be seen if these effects persist to normal reproductive age. Furthermore, whilst it is clear that ovulation rate and/or litter size in adult ewes can be reduced by undernutrition during fetal development (Rae *et al.* 2002) or during the pre-pubertal period (Rhind *et al.* 1998), there seems to be little effect on male fertility and no effect on the incidence of barreness among ewes; that is, it seems that fertility is not affected.

These observations are consistent with current theories on PARs and the transgenerational effects discussed earlier. Teleologically, it makes sense that a polyovular species, faced with a long-term decline in nutrient provision, would attempt to reduce ovulation rate. Failure to ovulate, which impedes the initiation of pregnancy, could only be an emergency measure of short-term significance, unsustainable in the longer term. It would also seem probable for long-lived species faced with several generations of nutritional impoverishment that, in addition to a reduction in mature size (Kuzawa 2005), puberty would be delayed and reproductive senescence advanced. Although there is a considerable body of evidence to support a delay in the onset of puberty under such circumstances in humans (Frisch 1994), at present there is limited evidence to support the notion that growth restriction during late gestation or during early childhood advances the onset of menopause (Cresswell *et al.* 1997; Hardy & Kuh 2002). Regrettably, this is an area where there appears to be a dearth of suitable animal models (Roof *et al.* 2005; Danilovich & Ram Sairam 2006). Given the evolutionary theories concerning ageing discussed next it would seem that rodents are not best suited for the study of reproductive senescence in humans.

Evolutionary theories of ageing

Many of the existing evolutionary theories on ageing intricately link fecundity to mature size and lifespan, and so go some way to underpin many of the current hypptheses on the origins of adult health and disease (Holliday 2005). The 'accumulated mutations' theory first proposed by PB Medawar in 1952, for example, proposes that deleterious genetic mutations that are manifest at a young age would be selected against whereas those displayed in late (post-reproductive) life would not, and so would accumulate across generations. A related theory, the 'antagonistic pleiotropy' theory (proposed GC Williams in 1957), suggests that genes exist which confer an advantage early in life (e.g. increased fecundity or cancer resistance) but which later advance the onset of age-related senescence. To date, however, this has been mainly supported from studies with the fruit fly and nematode with only limited evidence from the mouse (Leroi et al. 2005). More recently a third theory has attracted greater interest. The 'disposable soma' theory links the energy costs of maintenance to extrinsic mortality (e.g. predation), fecundity and longevity (Kirkwood 2005). The theory considers trade-offs between maintenance and reproduction, so that small mammals that develop and reproduce early, and have larger litters, have shorter lifespans. In contrast, in the absence of factors contributing significantly to extrinsic mortality, natural selection favours species that are better adapted to their environment, so that mature size increases, reproductive rates are reduced, reproduction occurs later and longevity is increased. This theory explains the rather paradoxical scenario where caloric restriction throughout life enhances longevity; a feat achieved through reductions in fecundity. The effect in rodents (a species utlised extensively in investigations into the DOHaD hypothesis) has been estimated to be 10 times greater than that in humans (Phelan & Rose 2005); demonstrating the extent to which these small mammals invest in reproduction (or annual fecundity rate), also calculated to be 10 times greater in rodents than in humans.

In pursuit of immortality: the germline and the soma

Primordial germ cells (PGCs) in mammals are derived from the inner-cell mass and are segregated from the somatic lineage during the early stages of gastrulation (Matsui & Okamura 2005). Destined to differentiate into gametes, these specialised cells are unique in that they give rise to the lineage responsible for the transfer of the genome from one generation to another. Relieved from the responsibility of transgenerational inheritance at an early stage in metazoan evolution, the surrounding soma was able to specialise into the many other cell types found in multicellular organisms but, in so doing, these cells incurred a cost; that is they became expendable. In his 'deprivation syndrome paradigm', Heininger (2002) explained that during early metazoan evolution a common strategy in response to environmental stress was to redirect resources from the soma towards the germline (e.g. the dying fruiting body and germ-like spores of Dictyostelium). Ageing, it was argued, evolved as a counter measure on the part of the soma to postpone its inevitable demise.

It follows that the distinction between the germ line and the surrounding soma in higher organisms, and the conflict that exists between them, is central to the current thinking on the evolution of ageing and age-related disease, under-pinning the ideas presented earlier about the trade-off between somatic maintenance and reproductive effort in a world of finite resources (Kirkwood 2005). This evolutionary conflict led Heininger (2002) to state that "death is no altruistic suicide but is imposed on the soma by the "enemy within", the germ cells".

The case for DOHaD

Some of the hypotheses described in this section predict that, at least in viviparous species, environmental influences from the earliest stages of development can alter key physiological processes during the life course. Some of these modifications may be adaptive responses, some may not, but all could contribute either directly or indirectly to DOHaD. Set in the context of life history theory, the hypothesis of 'trans-generational phenotypic inertia' further predicts that these effects accumulate gradually over successive generations. The 'disposable soma' theory links reproductive effort to extrinsic mortality and resistance to 'environmental stress', and forms the basis of the concept that ageing, and age-related diseases, are an inevitable consequence of the efforts of the soma to evade death; which in an evolutionary context is seen as a 'germ-cell triggered' event.

The importance of understanding these latter hypotheses are two-fold. Firstly, it may yet transpire that germ cells are uniquely resistant to in utero 'programming' effects, which have so clearly been demonstrated in somatic lineages; so that traits such as fertility may be unaffected by factors such as maternal nutrition during pregnancy. In this regard, the onset of reproductive senescence will be an important future area of study. Secondly, as many of the post-natal health-related outcomes programmed *in utero* are displayed during 'post-reproductive' life, so they may be intricately linked to the mechanisms that extend life; that is, these mechanisms may also be 'programmed' *in utero*.

Although compelling, evidence supporting many of these hypotheses is largely correlative and the mechanistic basis for the observed effects remains to firmly established. The period of *in utero* development, however, would appear to be particularly sensitive to alterations in the maternal environment. This is perhaps not surprising given that, of the estimated 43 cell divisions that occur during the course of human development, 37 have been predicted to take place *in utero* (Freitas 1999), with 8 occurring during the pre-implantation period. It follows that as each round of cell division is accompanied by a round of DNA replication, so the mechanistic basis for *in utero* programming of adult health and disease may reside in subtle alterations to ongoing chromatin or epigenetic modifications taking place during the normal processes of cellular differentiation and lineage commitment.

Epigenetics: the chemical basis for change

The modern use of the term epigenetics is to describe heritable changes in gene function that occur without an alteration in DNA sequence. These changes arise principally as a consequence of specific covalent modifications to DNA and associated histone proteins which act in concert with chromatin structure to define the transcriptome associated with a specific cell lineage. Epigenetic inheritance is therefore defined as the transmission of such modifications from a cell (or multicellular organism) to its descendants without any alteration in nucleotide sequence.

Histone modifications

The nucleosome is the functional unit of chromatin and consists of an octomeric complex of core histone proteins (H2A, H2B, H3 and H4), around which is wrapped 146 bp of DNA. A single histone H1 polypeptide interacts with an additional 20 bp of DNA as it enters and leaves the nucleosomal core. The negatively charged DNA is usually very tightly associated with the positively charged N-terminal histone tails which protrude from the core. Specific amino acid residues on these tails are targets for a number of enzyme-catalysed post-translation modifications which affect their charge and function (Table 2). This, in turn, can affect the configuration of the chromatin resulting in either a tightly bound heterochromatic configuration or a more open euchromatic configuration. These modifications can also serve to direct the binding of transcription factors and DNA damage repair proteins.

Table 2. Examples of post-translational histone modifications in mammalian cells.

Histone modification	Histone and residue	Modifying enzyme	Demodifying enzyme	Function	Species	Reference
Acetylation	H3K14	GCN5/PCAF	HDAC1	Transcriptional activation	Hn	Yamagoe *et al.* (2003)
	H4K?	HAT1/GCN5	HDAC 1, 2, 3, 7	Activation/ Repression	Bov	McGraw *et al.* (2003)
Phosphorylation	H3S10	Aurora-B-Kinase	?	Transcriptional activation	Ms	Fischle *et al.* (2005)
	H3T11	Dlk/ZIP Kinase	?	Mitosis	Hn/rat	Preuss *et al.* (2003)
	H2AX	PI3 Kinase	PP2A	DNA damage repair	Hn	Chowdhury *et al.* (2005)
Methylation	H3K4	Set9	LSD1	Transcriptional activation	Ms/Hn	Francis *et al.* (2005) Shi *et al.* (2004)
	H3K9	SUV39HI	?	DNA methylation/ heterochromatin	Hn	Snowden *et al.* (2002)
	H3R17	CARM1	PAD4	Transcriptional activation	Hn	Wang *et al.* (2004)
	H4R3	PRMT1	PAD4	Transcriptional activation	Hn	Wang *et al.* (2004)
Ubiquitylation	H2AK199	Ring1b	?	X-inactivation in trophoblast	Ms	Fang *et al.* (2004)
Sumoylation	H4	UBC9	ULP proteases?	Repression via HDAC and HP1	Hn	Shiio & Eisenman (2003)

Histone (H); lysine (K); arginine (R); serine (S); threonine (T); human (Hn); mouse (Ms); bovine (Bov)

The post-translational modifications that occur to these histone proteins have been best studied in yeast (*S. cerevisiae* and *S. pombe*), the fruit fly (*D. melanogaster*) and in *Arabidopsis* (Bannister & Kouzarides 2005). Of these modifications histone acetylation is perhaps the best understood. This modification involves acetyl substitution of the ε-amino group of lysine (K) (e.g. H3K9 or H4K12) which results in a more acidic (less positive) state. These reactions are catalysed by members of one of five families of histone acetyltransferases (HATs) (Kouzarides 1999); the reverse reactions being catalysed by histone deacetylase (HDAC) of which there are are 17 known mammalian members belonging to three classes.

Histone methylation occurs on non-acetylated lysine and arginine residues, and the methylation of H3 and H4 has attracted most attention where the outcomes can either be stimulatory or inhibitory depending on the particular lysine and arginine residue modified. Methylation of lysine and arginine residues is directed by histone methyltransferase (HMT) and protein arginine methyltransferase (PRMT) family members respectively; the former can direct S-adenosylmethionine (SAM) mediated methylation of lysine residues (Lee *et al.* 2005). In contrast to histone methyltransferases, the identification of histone demethylases is comparatively recent, and so our understanding of how these enzymes operate is incomplete (Bannister & Kouzarides 2005). A key feature of these post-translational modifications, however, is that they seldom act in unison but there exists considerable 'cross-talk' between these processes (Fischle *et al.* 2003). What's more, these histone modifications are intricately linked to covalent DNA modifications to collectively define the transcriptome. Details of such interactions are only now emerging but they highlight a complex bi-directional interplay, for example, between histone acetylation, methylation and DNA methylation (Fuks 2005).

DNA methylation

DNA methylation involves the addition of a methyl group to the number 5 carbon atom of the cytosine pyrimidine ring. This covalent modification is targeted to CpG dinucleotides which are recognised by DNA methyltransferase enzymes (DNMTs) (Goll & Bestor 2005). It has been estimated that around 75% of CpG dinucleotides in the human genome are methylated and are to be found in non-translated and heterochromatic regions; the remaining 25% unmethylated sites being located in specific 1 to 2 kb regions known as CpG islands (Caiafa & Zampieri 2005). These CpG islands, in turn, are frequently found in the promoter regions of genes. Among its numerous roles, DNA methylation is thought to serve in tissue-specific gene expression, X-chromosome inactivation, genomic imprinting and the silencing of transposable elements which constitute around 45% of the human genome (Lander *et al.* 2001). A key feature of DNA methylation, currently thought to distinguish it from many other chromatin modifications, is that it is normally faithfully restored during each cell cycle so that methylation patterns established during the earliest stages of development can persist in adult somatic cells. At present, the temporal means by which DNA methylation is targeted to specific regions of the genome is poorly understood, although it is thought to involve protein-protein interactions between the DNMTs and transcriptional repressor proteins, chromatin insulators and RNA-mediated interference (Klenova *et al.* 2002; Klose & Bird 2006). Methylated DNA in turn attracts methyl-CpG-binding proteins and these interact with other repressor proteins, including HDACs, to form transcriptional repressor complexes.

Although DNA methylation is thought to be an irreversible reaction, catalysed exclusively by members of the DNMT family of methyltransferases, it has long been recognised that active-demethylation of paternally-derived single copy genes takes place in the mammalian zygote (Oswald *et al.* 2000); and there is emerging, albeit controversial, evidence that DNA methylation in other cell types may also be fashioned by demethylase activity (Santos *et al.* 2002; Vairapandi 2004). The universal methyl donor SAM has been shown to inhibit demethylase activity (Detich *et al.* 2003), and so intracellular depletion of SAM may lead to a reduction in DNA methylation by reducing the provision of labile methyl groups and by increasing DNA demethylase activity.

Methyl cycle metabolism

The metabolic cycles which supply methyl groups for DNA and histone methylation involve a complex series of interactions between a number of different intracellular metabolites (Stipanuk 2004) (Fig. 1). The two principal methyl donors in animal metabolism are betaine (trimethyl glycine), a metabolite of choline, and SAM, a metabolite of methionine. SAM is a universal donor of methyl groups providing one-carbon moieties for phospholipid biosynthesis, creatine production and protein synthesis in addition to DNA methylation. In donating its methyl group to an acceptor, SAM is converted to S-adenosyl homocysteine (SAH). The subsequent loss of the adenosine residue produces the non-protein amino acid homocysteine (hcy), a central intermediate in the metabolism of sulphur in all animals. Homocysteine in turn can either be catabolised to cystathionine in the trans-sulphuration pathway or remethylated to form methionine. In mammals, two separate enzymes catalyse the remethylation step to complete a cycle which enables other molecules, notably serine and glycine, to act as methyl donors. One enzyme (*methionine synthase*, MTR) utilises a derivative of folic acid, N5-methyltetrahydrofolate as the methyl donor. The reaction is mediated by the coenzyme, methylcobalamin, derived from dietary vitamin B12 (cyanocobalamin). The activity of this reaction is therefore determined by the availability of two B vitamins, namely folic acid and cyanocobalamin. An alternative re-methylation reaction using betaine as the cofactor is also available to convert hcy to methionine. As betaine is derived from the breakdown of choline, the activity of this enzyme (*betaine-hyc methyl transferase*; BHMT) is partly determined by the availability of choline in the diet.

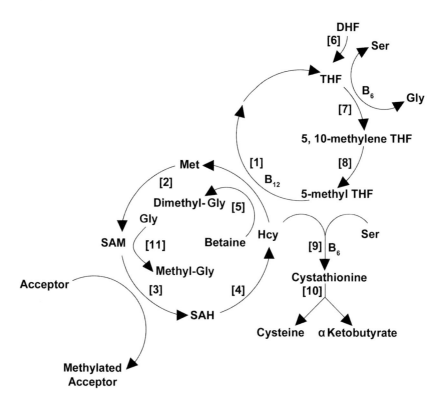

Fig. 1. Combined methionine and folate cycles. Metabolites: Methionine (Met), S-adenosyl methionine (SAM), S-adenosyl homocysteine (SAH), homocysteine (Hcy), dihydrofolate (DHF), tetrahydrofolate (THF), serine (Ser), glycine (Gly), methylcobalamin (B12), pyridoxal phosphate (B6). Enzymes: *methionine synthase* [1], *methionine adenosyl trasnferase* [2], SAM-dependent methyltransferases [3], *SAH-hydrolase* [4], *betaine-homocysteine methyltransferase* [5], *dihydrofolate reductase* [6], *serine hydroxymethyl transferase* [7], *5,10-methylene THF reductase* [8], *cystathionine b-synthase* [9], *cystathionine g-lyase* [10], *glycine methyltransferase* [11].

In contrast to rats, the total capacity of *choline oxidase* and BHMT is much lower than MTR in the adult sheep (Table 3). This reflects differences in the metabolism of the two species. Adult ruminants derive only a limited quantity of the precursors of transmethylation from the diet as those that are present are degraded by rumen microbial action. The transition from pre- (neonatal) to post-ruminant states in ruminants results in a decline in choline and creatine availability leading to a modification in single carbon metabolism. Nevertheless, both human subjects and sheep exhibit strikingly similar rates of methionine recycling, despite differences in dietary methyl group provision (Lobley *et al.* 1996).

Table 3. Methyl cycle enzyme activity (adapted from Snoswell & Xue, 1987)

Enzyme	Specific activity (pmol/min/mg protein)	
	Sheep	Rats
Choline oxidase	294	7980
Betaine Hcy methyl transferase	130	725
Methionine synthase	203	55
Glycine methyltransferase	25	7098

Metabolic responses to methyl nutrients are also known to differ between individuals as a result of polymorphic variation in at least four enzymes involved in these cycles. For example, the 677C to T polymorphism in *methylene tetrahydrofolate reductase (MTHFR)* reduces enzyme activity and is highly prevalent among Caucasians (5-15% for the TT and 40-50% for the CT genotypes respectively). Such individuals require high intakes of folic acid to reduce plasma hcy concentrations (de Bree *et al.* 2003). Given the central role of MTR in ruminant single-carbon metabolism, it is interesting to note that the A to G polymorphism at position 2756 of this gene is common in human subjects (~ 40% and 5% for AG and GG genotypes respectively; Sharp & Little 2004).

Epigenetic programming during development

Given that the genetic code does not vary between cell types, the transition to multicellularity and the division between the germline and the soma necessitated the evolution of epigenetic mechanisms (such as the methylation of DNA) that would determine lineage-specific gene expression (Jablonka 1994). Recent studies using restriction landmark genome-scanning (RLGS) have identified numerous tissue-specific differentially methylated regions (DMRs) within CpG islands in both germ and somatic cells of differing lineages in mice (Shiota *et al.* 2002; Kremenskoy *et al.* 2003). The temporal complexity of the epigenetic mechanisms that direct somatic cell lineage specification was further illustrated by Rupp *et al.* (2002) who considered the interactions that exist between myogenic regulatory proteins and chromatin modifying enzymes during the differentiation of skeletal muscle. Both MyoD and Myf5 are expressed during somitic myogenesis under the influence of Sonic hedgehog and Wnt-1, and serve to induce mesodermal precursor cells to differentiate into myoblasts, as well as activating down-stream target genes such as myogenin (Maltin *et al.* 2001). MyoD expression is associated with a loss of methylation at a conserved distal enhancer element known to mediate its activation. The DNA binding activity of MyoD is regulated by both HATs and HDAC and, together with Myf5, MyoD is believed to initiate target gene transcription by recruiting mammalian SW1/SNF complexes (members of the trithorax group of proteins associated with euchromatin). It is interesting, therefore, that temporal shifts in Myf5 protein expression in proliferating ovine fetal muscle cells should be induced by temporarily exposing Day 3 pre-implantation embryos to an advanced uterine environment (Maxfield *et al.* 1998); suggesting that events occurring during the earliest stages of post-fertilisation development can alter down-stream gene expression and cellular proliferation (discussed later).

Epigenetic mechanisms are also known to participate in the specification and formation of primordial germ cells (Allegrucci *et al.* 2005a). Crucially, these cells undergo additional dramatic epigenetic modifications during gametogenesis that serve to erase epimutations, re-establish totipotency and establish sex specific imprints. Yet further important modifications occur following fertilisation in the period leading up to zygotic gene activation; modifications which may herald the initiation of the genetic conflict that is thought to exist between the sexes (Moore & Reik 1996) and/or that are associated with zygotic activation and preparation for implantation.

Epigenetic modifications following fertilisation

Epigenetic programming during pre-implantation development has been extensively studied in the mouse, where the paternally derived genome is actively (in the absence of DNA replication) demethylated during the first cell cycle, consistent with early male pronuclear and zygotic gene transcription in this species (Schultz, 2002). Recent studies using immunoflourescent staining targeted to heterochromatic regions in one-cell mouse embryos have revealed that, as the sperm nucleus decondenses, it acquires H3-K9 acetylated and mono-methylated histones as the DNA becomes progressively demethylated (Santos *et al.* 2005). In contrast, the maternal pronucleus retains high levels of H3 di- and tri-methylated K9 and K27, which these authors suggest may serve to protect DNA methylation in the maternal genome at this time.

Whilst similar patterns of global DNA demethylation of the paternal genome have been observed in the rat, cow and human zygote, there are significant differences in the rabbit and sheep zygote (Beaujean *et al.* 2004a). Through the use of interspecies intracytoplasmic sperm injection, these authors were able to demonstrate that the more extreme loss of paternal demethylation in the mouse compared to the sheep male pronucleus arises as a consequence of (1) the enhanced resistance of sheep sperm to demethylation and (2) the greater demethylating capacity of the mouse ooplasm (Beaujean *et al.* 2004b). Subsequent cell divisions during the early pre-implantation period are associated with further losses of DNA methylation which again are more extreme in the mouse than either the cow or sheep genomes (Beaujean *et al.* 2004a). Although the functional significance of these species-related differences in DNA methylation are unclear, they may partly be due to pre-existing levels of methylation in the male pronucleus at the time of syngamy, or to differences in the timing of embryonic gene activation (discussed by Young & Beaujean, 2004). There are also species differences in the expression patterns of *DNMTs* in oocytes and pre-implantation embryos (Golding & Westhusin 2003; Vassena *et al.* 2005) which could partially account for these effects.

Epigenetic modifications in germ cells

Primordial germ cells in mammals are believed to be specified from the epiblast during the early stages of grastrulation. Most extensively studied in the mouse, their specification has been found to involve complex interactions between maternally inherited and zygotically induced molecules in a multi-step process (Matsui & Okamura 2005). In addition to epigenetic modifications associated with germline specification during this period, PGCs undergo further unique epigenetic alterations believed to coincide with their migration across the genital ridge (reviewed by Allegrucci *et al.* 2005a). From around 10.5 days *post coitum* in the mouse, single copy imprinted and non-imprinted genes are actively demethylated, thereby ensuring that epigenetic modifications that accumulate in somatic cells with age are erased through the germline. In contrast, the demethylation of repetitive elements is more protracted, variable and generally less complete; a process hypothesised to be a protective mechanism designed to minimise transposition and aberrant expression of adjacent genes (Hajkova *et al.* 2002).

Sex-specific remethylation of PGCs then takes place in the gonad. However, the timing of *de novo* methylation differs between the male and female germlines, at least in the mouse. Remethylation in the male, and the re-establishment of gametic imprints, commences in mitotically dividing (pre-meiotic) spermatogonial stem cells (Rousseaux *et al.* 2005). In contrast, the timing of *de novo* methylation in the female germ cell is less certain, but methylation acquisition at a number of imprinted loci is known to generally commence later, during the growth phase of oocytes arrested at the diplotene stage of the first meiotic division (Lucifero *et al.* 2002). Importantly, these events are known to be immediately preceded by an increase in transcript expression for key DNMTs (La Salle *et al.* 2004).

Epigenetic basis of DOHaD

Existing theories provide a conceptual framework to explain how, in viviparous species, mechanisms have evolved during the course of evolution to facilitate rapid adaptations to changing ecological conditions; and that these may form the basis for the developmental origins of adult health and disease. From the preceding discussion it is apparent that extensive covalent modifications to DNA and related proteins occur from the earliest stages of mammalian development to determine lineage-specific patterns of gene expression; and so these represent the most plausible mechanisms by which environmental factors such as diet can influence development during the life course. Indeed, both the nature and extent of the modifications which occur normally during gametogenesis and early embryogenesis suggest that these may be particularly vulnerable periods during which epimutations can be induced. However, it is also apparent that our understanding of how these epigenetic processes may underpin the environmentally-induced phenotypic modifications discussed earlier is far from complete. The following discussion highlights some of the major findings of recent years with some thoughts on future directions.

Inducing epigenetic modifications

In addition to the epigenetic modifications that occur as a part of normal development, changes can also be induced by environmental factors although often the mechanisms by which these effects arise are unclear. For example, short term exposure of mammalian zygotes to non-physiological in vitro culture environments can lead to epigenetic alterations in global DNA methylation in mice (Shi & Haaf 2002) and genomic imprinting in sheep (Young *et al.* 2001). Recently, transient exposure of gestating female rats to specific endocrine disrupting compounds during the period of fetal gonadal sex determination was shown to decrease spermatogenic capacity and increase male infertility in offspring; effects associated with altered DNA methylation at several loci (Anway *et al.* 2005). Also, in altricial mammals, infant licking and grooming by the mother during the first week of life can reversibly modify DNA methylation and histone acetylation at the glucocorticoid receptor promoter, thereby altering glucocorticoid receptor (GR) expression (Meaney & Szyf 2005). Dietary induced epigenetic effects have also been reported in mice following weaning. For example, synthetic diets and diets deficient in folic acid, vitamin B12, methionine and choline offered to weanlings each caused a persistent loss of imprinting of *Igf2* in the kidney relative to controls (Waterland *et al.* 2006).

Of particular relevance to the DOHaD hypothesis are the observations that alterations to maternal diet during pregnancy and/or lactation can also lead to epigenetic modifications in DNA methylation associated with varied phenotypic outcomes. For example, viable yellow agouti (Avy) mice harbour a transposable element (responsible for ectopic *agouti* transcription) at the agouti locus, the epigenetic status of which is metastable. Recently, Waterland & Jirtle (2003) demonstrated that dietary supplementation with the methyl donors folate, vitamin B12, choline and betaine throughout gestation and lactation can increase CpG methylation at this site, modifying the expression of the agouti gene which lies adjacent to the transposable element. Significantly, these authors observed that the A^{vy} methylation status of tissues derived from the three germ layers of the early embryo were affected indicating that the effects are likely to have been induced during early embryo development. Subsequent studies with genistein (the major isoflavone in soy) offered to gestating female mice at physiological levels also found alterations in coat colour of heterozygous viable agouti offspring toward the mottled agouti phenotype (Dolinoy *et al.* 2006). Moreover, these effects, which were associated with a similar increase in methylation at the agouti locus, persisted into adulthood and reduced the incidence of obesity, a phenotype commonly associated with ectopic *agouti* expression. Significantly, the increase in DNA methylation at this locus occurred in the absence of any effect on either SAM or SAH (Fig. 1), indicating a dietary-induced epigenetic mechanism independent of single carbon metabolism.

Recent studies in the rat, using protein restricted diets similar to those employed in previous studies with this species (Table 1), assessed the methylation status and expression of the GR and peroxisomal proliferator-activated receptor (PPAR) genes in weanlings (Lillycrop *et al.* 2005). Methylation-sensitive PCR revealed a significant reduction in CpG methylation in the promoter regions of both PPARa and GR which, in each case, was associated with a significant increase in gene expression. Importantly, the inclusion of 1mg/kg folic acid in the protein restricted diet reversed the effects of protein restriction on DNA methylation and the expression of both genes. Although epigenetic mechanisms were not determined, Jackson *et al.* (2002) was able to demonstrate that the non-essential amino acid glycine (which plays a key role in methionine and SAM metabolism [Fig. 1] *via* the actions of *glycine methyltransferase* [Table 3]) was able to reverse the effects of protein restriction during pregnancy on offspring blood pressure in Wistar rats. These studies serve to further demonstrate that epigenetic effects may be involved in changes of phenotype due to the macronutrient composition of the maternal diet.

Transgenerational epigenetic inheritance

The most significant observation from the study of Anway *et al.* (2005), cited earlier, was that both epigenetic and phenotypic effects observed in the F1 generation persisted in all subsequent generations studied (i.e. F1 through to F4) indicating that, in the absence of genetic mutations, phenotypic effects can be transmitted to successive generations *via* epigenetic inheritance. Clues that such inheritance patterns may exist in mammals were provided earlier when the altered phenotypes and patterns of gene expression observed in mouse nucleocytoplasmic hybrids were found to be present in their offspring (Roemer *et al.* 1997). There is also evidence of epigenetic transgenerational inheritance at the viable yellow agouti (Avy) locus where, in the absence of alterations to the nutritional environment, the phenotype of the offspring has been found to resemble that of the dam; indicating incomplete erasure of epigenetic modifications through the female germline (Morgan *et al.* 1999). Evidence of paternal as well as maternal transmission of epigenetic modifications in mice was later provided for the *axin-fused* (*Axin*[Fu]) allele (involved in determining embryonic axis formation in vertebrates) which is also associated with a retrotransposon; the methylation state of this allele in mature sperm closely resembled that of somatic cells of the animal suggesting this also was not completely erased through the germline (Rakyan *et al.* 2003).

The significance of these observations is that they establish the principle of transgenerational epigenetic inheritance in mammals, previously only observed in plants and fission yeast (Chong & Whitelaw 2004). None of these examples, however, could be considered to meet the criteria of PARs described by Gluckman & Hanson (2004) or the inter-generational effects predicted by the phenotypic inertia model of Kuzawa (2005). Nevertheless, they demonstrate that environmentally induced epigenetic modifications, which may lead to adult disease, can be transmitted to future generations through the germ line.

Future directions

The foregoing discussion highlights both the complexity of epigenetic mechanisms and our current lack of understanding of how they may contribute to *in utero* programming. It does, however, provide some guidance for future research endeavours. The evidence contained within this article indicates that both nutritional and non-nutritional factors contribute to the DOHaD. To date, the former has relied largely on two experimental paradigms that involve either calorie or protein restriction. In contrast, ongoing collaborative investigations between the authors' laboratories are assessing the effects of deficiencies in specific micronutrients (i.e. methionine, choline, folate, vitamin B12 and B6) that are known to directly influence single carbon metabolism. The clinical significance of these nutrients is that their consumption is acknowledged to be highly variable in most Western societies and also in developing countries (e.g. Stabler & Allen 2004). The focus in future, therefore, should be directed towards understanding the effects of specific dietary nutrients and environmental factors. Further detailed studies should recognise (i) that the epigenetic regulatory mechanisms involved in environmentally induced gene disregulation encompass complex interactions between non-transcribed RNAs and covalently altered proteins as well as DNA, and (ii) that the susceptibility of individuals to environmentally induced epimutations may have a genetic base. The significance of this latter point is that the metabolic responses which can epigenetically modify chromatin vary between individuals as a consequence of polymorphic variation in relevant enzymes. Although it will be necessary to establish such relationships in model animal species, direct studies in human cells will be of greatest value. To this end we have begun to identify key polymorphisms in enzymes directing the combined methionine/folate cycles within human embryonic stem cells, which serve as a model of the human embryo (Allegrucci *et al.* 2005b). Complementary studies in appropriate animal models will allow such investigations to take place at later stages of gestation and confirm their significance with respect to the onset of adult disease. They would also permit the investigation of transgenerational inheritance.

Acknowledgements

Original research is supported by a cooperative agreement from the National Institutes of Health, NICHD (U01-HD044638) awarded to Sinclair, Lea, Rees and Young, and the Biotechnology and Biological Sciences Research Council (BBS/B/06164) awarded to Young and Sinclair.

References

Allegrucci C, Thurston A, Lucas E & Young L 2005a Epigenetics and the germline. *Reproduction* **129** 137-149.

Allegrucci C, Denning CN, Burridge P, Steele W,

Sinclair KD & Young LE 2005b Human embryonic stem cells as a model for nutritional programming: an evaluation. *Reproductive Toxicology* **20** 353-367.

Anthony RV, Scheaffer AN, Wright CD & Regnault TR 2003 Ruminant models of prenatal growth restriction. *Reproduction Supplement* **61** 183-194.

Anway MD, Cupp AS, Uzumcu M & Skinner MK 2005 Epigenetic transgenerational actions of endocrine disruptors and male fertility. *Science* **308** 1466-1469.

Bannister AJ & Kouzarides T 2005 Reversing histone methylation. *Nature* **436** 1103-1106.

Barker DJ & Osmond C 1988 Low birth weight and hypertension. *British Medical Journal* **297** 134-135.

Barker DJ, Winter PD, Osmond C, Margetts B & Simmonds SJ 1989 Weight in infancy and death from ischaemic heart disease. The *Lancet* **2** 577-580.

Barker DJ 1995 The Wellcome Foundation Lecture, 1994. The fetal origins of adult disease. *Proceedings of the Biological Sciences* **262** 37-43.

Bateson P 2001 Fetal experience and good adult design. *International Journal of Epidemiology* **30** 928-934.

Beaujean N, Hartshorne G, Cavilla J, Taylor J, Gardner J, Wilmut I, Meehan R & Young L 2004a Non-conservation of mammalian preimplantation methylation dynamics. *Current Biology* **14** R266-R267.

Beaujean N, Taylor JE, McGarry M, Gardner JO, Wilmut I, Loi P, Ptak G, Galli C, Lazzari G, Bird A, Young LE & Meehan RR 2004b The effect of interspecific oocytes on demethylation of sperm DNA. *Proceedings of the National Academy of Sciences U.S.A* **101** 7636-7640.

Bloomfield FH, Oliver MH, Giannoulias CD, Gluckman PD, Harding JE & Challis JR 2003 Brief undernutrition in late-gestation sheep programs the hypothalamic-pituitary-adrenal axis in adult offspring. *Endocrinology* **144** 2933-2940.

Caiafa P & Zampiera M 2005 DNA methylation and chromatin structure: the puzzling CpG islands. *Journal of Cellular Biochemistry* **94** 257-265.

Chong S & Whitelaw E 2004 Epigenetic germline inheritance. *Current Opinions in Genetics and Development* **14** 692-696.

Chowdhury D, Keogh MC, Ishii H, Peterson CL, Buratowski S & Lieberman J 2005 gamma-H2AX dephosphorylation by protein phosphatase 2A facilitates DNA double-strand break repair. *Molecular Cell* **20** 801-809.

Cresswell JL, Egger P, Fall CHD, Osmond C, Fraser RB & Barker DJP 1997 Is the age of menopause determined in-utero? *Early Human Development* **49** 143-148

Danilovich N & Ram Sairam M 2006 Recent female mouse models displaying advanced reproductive aging. *Experimental Gerontology* **41** 117-122.

Davis L, Roullet JB, Thornburg KL, Shokry M, Hohimer AR & Giraud GD 2002 *Journal of Physiology* **547** 53-9.

de Bree A, Verschuren WM, Bjorke-Monsen AL, van der Put NM, Heil SG, Trijbels FJ & Blom HJ 2003 Effect of the methylenetetrahydrofolate reductase 677C—>T mutation on the relations among folate intake and plasma folate and homocysteine concentrations in a general population sample. *American Journal Clinical Nutrition* **77** 687-693.

Detich N, Hamm S, Just G, Knox JD & Szyf M 2003 The methyl donor S-Adenosylmethionine inhibits active demethylation of DNA: a candidate novel mechanism for the pharmacological effects of S-Adenosylmethionine. *Journal Biological Chemistry* **278** 20812-20820.

Doblhammer G & Vaupel JW 2001 Lifespan depends on month of birth. *Proceedings National Academy Sciences U.S.A* **98** 2934-2939.

Dolinoy DC, Weidman JR, Waterland RA & Jirtle RL 2006 Maternal genistein alters coat color and protects Avy mouse offspring from obesity by modifying the fetal epigenome. *Environmental Health Perspectiv* **114** 567-572.

Edwards LJ & McMillen IC 2002 Periconceptional nutrition programs development of the cardiovascular system in the fetal sheep. *American Journal Physiology, Regulation, Integration and Comparative Physiology* **283** R669-R679.

Fang J, Chen T, Chadwick B, Li E & Zhang Y 2004 Ring1b-mediated H2A ubiquitination associates with inactive X chromosomes and is involved in initiation of X inactivation. *Journal Biological Chemistry* **279** 52812-52815.

Fischle W, Wang Y & Allis CD 2003 Histone and chromatin cross-talk. *Current Opinion Cell Biology* **15** 172-183.

Fischle W, Tseng BS, Dormann HL, Ueberheide BM, Garcia BA, Shabanowitz J, Hunt DF, Funabiki H & Allis CD 2005 Regulation of HP1-chromatin binding by histone H3 methylation and phosphorylation. *Nature* **438** 1116-1122.

Francis J, Chakrabarti SK, Garmey JC & Mirmira RG 2005 Pdx-1 links histone H3-Lys-4 methylation to RNA polymerase II elongation during activation of insulin transcription. *Journal Biological Chemistry* **280** 36244-36253.

Freitas RA Jr. 1999. Cytometrics In: Nanomedicine, Vol. I: Basic Capabilities, Landes Bioscience, Georgetown, TX, USA.

Frisch RE 1994 The right weight: body fat, menarche and fertility. *Proceedings of the Nutrition Society* **53** 113-129.

Fuks F 2005 DNA methylation and histone modifications: teaming up to silence genes. *Current Opinion Genetics andDevelopment* **15** 490-495.

Gardner DS, Pearce S, Dandrea J, Walker R, Ramsay MM, Stephenson T & Symonds ME 2004 Peri-implantation undernutrition programs blunted angiotensin II evoked baroreflex responses in young adult sheep. *Hypertension* **43** 1290-1296.

Gilbert JS, Lang AL, Grant AR & Nijland MJ 2005 Maternal nutrient restriction in sheep: hypertension and decreased nephron number in offspring at 9 months of age. *Journal of Physiology* **565** 137-147.

Giussani DA, Forhead AJ, Gardner DS, Fletcher AJ, Allen WR & Fowden AL 2003 Postnatal cardiovascular function after manipulation of fetal growth by embryo transfer in the horse. *Journal of Physiology* **547** 67-76.

Gluckman PD & Hanson MA 2004 The developmental origins of the metabolic syndrome. *Trends in Endocrinology and Metabolism* **15** 183-187.

Golding MC & Westhusin ME 2003 Analysis of DNA (cytosine 5) methyltransferase mRNA sequence and expression in bovine preimplantation embryos, fetal and adult tissues. *Gene Expression Patterns* **3** 551-558.

Goll MG & Bestor TH 2005 Eukaryotic cytosine methyltransferases. *Annual Reviews in Biochemistry* **74** 481-514.

Hajkova P, Erhardt S, Lane N, Haaf T, El-Maarri O, Reik W, Walter J & Surani MA 2002 Epigenetic reprogramming in mouse primordial germ cells. *Mechanisms of Development* **117** 15-23.

Hales CN & Barker DJ 1992 Type 2 (non-insulin-dependent) diabetes mellitus: the thrifty phenotype hypothesis. *Diabetologia* **35** 595-601.

Hardy R & Kuh D 2002 Does early growth influence timing of the menopause? Evidence from a British birth cohort. *Human Reproduction* **17** 2474-2479.

Heininger K 2002 Aging is a deprivation syndrome driven by a germ-soma conflict. *Ageing Research Reviews* **1** 481-536.

Holliday R 2005 Ageing and the extinction of large animals. *Biogerontology* **6** 151-156.

Huxley R, Neil A & Collins R 2002 Unravelling the fetal origins hypothesis: is there really an inverse association between birthweight and subsequent blood pressure? *Lancet* **360** 659-665.

Ibanez L, Potau N, Ferrer A, Rodriguez-Hierro F, Marcos MV & De ZF 2002 Anovulation in eumenorrheic, nonobese adolescent girls born small for gestational age: insulin sensitization induces ovulation, increases lean body mass, and reduces abdominal fat excess, dyslipidemia, and subclinical hyperandrogenism. *Journal of Clinical Endocrinology and Metabolism* **87** 5702-5705.

Jablonka E 1994 Inheritance systems and the evolution of new levels of individuality. *Journal of Theological Biology* **170** 301-309.

Jackson AA, Dunn RL, Marchand MC & Langley-Evans SC 2002 Increased systolic blood pressure in rats induced by a maternal low-protein diet is reversed by dietary supplementation with glycine. *Clinical Science (London)* **103** 633-639.

Khan IY, Taylor PD, Dekou V, Seed PT, Lakasing L, Graham D, Dominiczak AF, Hanson MA & Poston L 2003 Gender-linked hypertension in offspring of lard-fed pregnant rats. *Hypertension* **41** 168-75.

Kirkwood TB 2005 Understanding the odd science of aging. *Cell* **120** 437-447.

Klenova EM, Morse HC, III, Ohlsson R & Lobanenkov VV 2002 The novel BORIS + CTCF gene family is uniquely involved in the epigenetics of normal biology and cancer. *Seminars in Cancer Biology* **12** 399-414.

Klose RJ & Bird AP 2006 Genomic DNA methylation: the mark and its mediators. *Trends in Biochemical Science* **31** 89-97.

Kouzarides T 1999 Histone acetylases and deacetylases in cell proliferation. *Current Opinion in Genetics and Development* **9** 40-48.

Kremenskoy M, Kremenska Y, Ohgane J, Hattori N, Tanaka S, Hashizume K & Shiota K 2003 *Biochemisty and Biophysics Research Communications* **311** 884-890.

Kuzawa CW 2005 Fetal origins of developmental plasticity: Are fetal cues reliable predictors of future nutritional environments? *American Journal of Human Biology* **17** 5-21.

Kwong WY, Wild AE, Roberts P, Willis AC & Fleming TP 2000 Maternal undernutrition during the preimplantation period of rat development causes blastocyst abnormalities and programming of postnatal hypertension. *Development* **127** 4195-4202.

La Salle S, Mertineit C, Taketo T, Moens PB, Bestor TH & Trasler JM 2004 Windows for sex-specific methylation marked by DNA methyltransferase expression profiles in mouse germ cells. *Developmental Biology* **268** 403-415.

Lander ES, Linton LM, Birren B, et al. 2001. Initial sequencing and analysis of the human genome. *Nature* **409** 860-921.

Langley-Evans SC, Welham SJ, Sherman RC & Jackson AA 1996 Weanling rats exposed to maternal low-protein diets during discrete periods of gestation exhibit differing severity of hypertension. *Clinical Science (London)* **91** 607-615.

Langley-Evans SC 2006 Developmental programming of health and disease. *Proceedings of the Nutrition Society* **65** 97-105.

Lee DY, Teyssier C, Strahl BD & Stallcup MR 2005 Role of protein methylation in regulation of transcription. *Endocrine Reviews* **26** 147-70.

Leroi AM, Bartke A, De BG, Franceschi C, Gartner A, Gonos ES, Fedei ME, Kivisild T, Lee S, Kartaf-Ozer N, Schumacher M, Sikora E, Slagboom E, Tatar M, Yashin AI, Vijg J & Zwaan B 2005 What evidence is there for the existence of individual genes with antagonistic pleiotropic effects? *Mechanisms of Ageing and Development* **126** 421-429.

Lillycrop KA, Phillips ES, Jackson AA, Hanson MA & Burdge GC 2005 Dietary protein restriction of pregnant rats induces and folic acid supplementation prevents epigenetic modification of hepatic gene expression in the offspring. *Journal of Nutrition* **135** 1382-1386.

Lobley GE, Connell A & Revell D 1996 The importance of transmethylation reactions to methionine metabolism in sheep: effects of supplementation with creatine and choline. *British Journal of Nutrition* **75** 47-56.

Louey S, Cock ML, Stevenson KM & Harding R 2000 Placental insufficiency and fetal growth restriction lead to postnatal hypotension and altered postnatal growth in sheep. *Pediatric Research* **48** 808-814.

Lucas A, Fewtrell MS & Cole TJ 1999 Fetal origins of

adult disease-the hypothesis revisited. *British Medical Journal* **319** 245-249.

Lucifero D, Mertineit C, Clarke HJ, Bestor TH & Trasler JM 2002 Methylation dynamics of imprinted genes in mouse germ cells. *Genomics* **79** 530-538.

Lumey LH 1992 Decreased birthweights in infants after maternal in utero exposure to the Dutch famine of 1944-1945. *Paediatrics Perinatology and Epidemiology* **6** 240-253.

Lumey LH 1998 Reproductive outcomes in women prenatally exposed to undernutrition: a review of findings from the Dutch famine birth cohort. *Proceedings of the Nutrition Society* **57** 129-135.

Maltin CA, Delday MI, Sinclair KD, Steven J & Sneddon AA 2001 Impact of manipulations of myogenesis in utero on the performance of adult skeletal muscle. *Reproduction* **122** 359-374.

Matsui Y & Okamura D 2005 Mechanisms of germ-cell specification in mouse embryos. *BioEssays* **27** 136-143.

Maxfield EK, Sinclair KD, Dunne LD, Broadbent PJ, Robinson JJ, Stewart E, Kyle DG & Maltin CA 1998 Temporary exposure of ovine embryos to an advanced uterine environment does not affect fetal weight but alters fetal muscle development. *Biology of Reproduction* **59** 321-325.

McGraw S, Robert C, Massicotte L & Sirard MA 2003 Quantification of histone acetyltransferase and histone deacetylase transcripts during early bovine embryo development. *Biology of Reproduction* **68** 383-389.

McMillen IC & Robinson JS 2005 Developmental origins of the metabolic syndrome: prediction, plasticity, and programming. *Physiological Reviews* **85** 571-633.

Meaney MJ & Szyf M 2005 Environmental programming of stress responses through DNA methylation: life at the interface between a dynamic environment and a fixed genome. *Dialogues in Clinical Neuroscience* **7** 103-123.

Moore T & Reik W 1996 Genetic conflict in early development: parental imprinting in normal and abnormal growth. *Reviews of Reproduction* **1** 73-77.

Morgan HD, Sutherland HG, Martin DI & Whitelaw E 1999 Epigenetic inheritance at the agouti locus in the mouse. *Nature Genetics* **23** 314-318.

Oswald J, Engemann S, Lane N, Mayer W, Olek A, Fundele R, Dean W, Reik W & Walter J 2000 Active demethylation of the paternal genome in the mouse zygote. *Current Biology* **10** 475-478.

Owens JA, Falconer J & Robinson JS 1989 Glucose metabolism in pregnant sheep when placental growth is restricted. *American Journal of Physiology* **257** R350-R357.

Painter RC, Roseboom TJ & Bleker OP 2005 Prenatal exposure to the Dutch famine and disease in later life: an overview. *Reproductive Toxicology* **20** 345-352.

Phelan JP & Rose MR 2005 Why dietary restriction substantially increases longevity in animal models but won't in humans. *Ageing Research Reviews* **4**

339-350.

Preuss U, Landsberg G & Scheidtmann KH 2003 Novel mitosis-specific phosphorylation of histone H3 at Thr11 mediated by Dlk/ZIP kinase. *Nucleic Acids Research* **31** 878-885.

Rae MT, Kyle CE, Miller DW, Hammond AJ, Brooks AN & Rhind SM 2002 The effects of undernutrition, in utero, on reproductive function in adult male and female sheep. *Animal Reproduction Science* **72** 63-71.

Rakyan VK, Chong S, Champ ME, Cuthbert PC, Morgan HD, Luu KV & Whitelaw E 2003 Transgenerational inheritance of epigenetic states at the murine Axin(Fu) allele occurs after maternal and paternal transmission. *Proceedings National Academy of Science U.S.A* **100** 2538-2543.

Ravelli AC, van der Meulen JH, Michels RP, Osmond C, Barker DJ, Hales CN & Bleker OP 1998 Glucose tolerance in adults after prenatal exposure to famine. The *Lancet* **351** 173-177.

Rhind SM, Elston DA, Jones JR, Rees ME, McMillen SR & Gunn RG 1998 Effects of restriction of growth and development of Brecon Cheviot ewe lambs on subsequent lifetime reproductive performance *Small Ruminant Research* **30** 121-126

Roemer I, Reik W, Dean W & Klose J 1997 Epigenetic inheritance in the mouse. *Current Biology* **7** 277-280.

Roof KA, Hopkins WD, Izard MK, Hook M & Schapiro SJ 2005 Maternal age, parity, and reproductive outcome in captive chimpanzees (Pan troglodytes). *American Journal of Primatology* **67** 199-207.

Rousseaux S, Caron C, Govin J, Lestrat C, Faure AK & Khochbin S 2005 Establishment of male-specific epigenetic information. *Gene* **345** 139-153.

Rupp RA, Singhal N & Veenstra GJ *2002 When the embryonic genome flexes its muscles.* European Journal of Biochemistry **269** 2294-2299.

Santos F, Hendrich B, Reik W & Dean W 2002 Dynamic reprogramming of DNA methylation in the early mouse embryo. *Developmental Biology* **241** 172-182.

Santos F, Peters AH, Otte AP, Reik W & Dean W 2005 Dynamic chromatin modifications characterise the first cell cycle in mouse embryos. *Developmental Biology* **280** 225-236.

Schultz RM 2002 The molecular foundations of the maternal to zygotic transition in the preimplantation embryo. *Human Reproduction Update* **8** 323-331.

Seckl JR 2001 Glucocorticoid programming of the fetus; adult phenotypes and molecular mechanisms. *Molecular and Cellular Endocrinology* **185** 61-71.

Sharp L & Little J 2004 Polymorphisms in genes involved in folate metabolism and colorectal neoplasia: a HuGE review. *American Journal of Epidemiology* **159** 423-443.

Shi W & Haaf T 2002. Aberrant methylation patterns at the two-cell stage as an indicator of early developmental failure. *Molecular Reprodion and Development* **63** 329-334.

Shi Y, Lan F, Matson C, Mulligan P, Whetstine JR,

Cole PA, Casero RA & Shi Y 2004 Histone demethylation mediated by the nuclear amine oxidase homolog LSD1. *Cell* **119** 941-953.

Shiota K, Kogo Y, Ohgane J, Imamura T, Urano A, Nishino K, Tanaka S & Hattori N 2002 Epigenetic marks by DNA methylation specific to stem, germ and somatic cells in mice. *Genes and Cells* **7** 961-969.

Shiio Y & Eisenman RN 2003 Histone sumoylation is associated with transcriptional repression *Proceedings of the National Academy of Sciences U S A* **100** 13225-13330.

Singhal A & Lucas A 2004 Early origins of cardiovascular disease: is there a unifying hypothesis? The *Lancet* **363** 1642-1645.

Snoswell AM & Xue GP 1987 Methyl group metabolism in sheep. *Comparative Biochemistry and Physiology B* **88** 383-394.

Snowden AW, Gregory PD, Case CC & Pabo CO 2002 Gene-specific targeting of H3K9 methylation is sufficient for initiating repression *in vivo*. *Current Biology* **12** 2159-2166.

Sjöblom C, Roberts CT, Wikland M & Robertson SA 2005 Granulocyte-macrophage colony-stimulating factor alleviates adverse consequences of embryo culture on fetal growth trajectory and placental morphogenesis. *Endocrinology* **146** 2142-2153.

Stabler SP & Allen RH 2004 Vitamin B12 deficiency as a worldwide problem. *Annual Reviews of Nutrition* **24** 299-326.

Stanner SA, Bulmer K, Andres C, Lantseva OE, Borodina V, Poteen VV & Yudkin JS 1997 Does malnutrition in utero determine diabetes and coronary heart disease in adulthood? Results from the Leningrad siege study, a cross sectional study. *British Medical Journal* **315** 1342-1348.

Stewart RJC, Sheppard H, Preece R & Waterlow JC 1980 The effect of rehabilitation at different stages of development of rats marginally malnourished for ten to twelve generations. *British Journal Nutrition* **43** 403-412

Stipanuk MH 2004 Sulfur amino acid metabolism: Pathways for production and removal of homocysteine and cysteine. *Annual Reviews of Nutrition* **24** 539-577

Vairapandi M 2004 Characterization of DNA demethylation in normal and cancerous cell lines and the regulatory role of cell cycle proteins in human DNA demethylase activity. *Journal Cellular Biochemistry* **91** 572-583.

Vassena R, Dee SR & Latham KE 2005 Species-dependent expression patterns of DNA methyltransferase genes in mammalian oocytes and preimplantation embryos. *Molecular Reproduction and Development* **72** 430-436.

Wang Y, Wysocka J, Sayegh J, Lee YH, Perlin JR, Leonelli L, Sonbuchner LS, McDonald CH, Cook RG, Dou Y, Roeder RG, Clarke S, Stallcup MR, Allis CD & Coonrod SA 2004 Human PAD4 regulates histone arginine methylation levels via demethylimination. *Science* **306** 279-283.

Waterland RA & Jirtle RL 2003 Transposable elements: targets for early nutritional effects on epigenetic gene regulation. *Molecular and Cellular Biology* **23** 5293-5300.

Waterland RA, Lin J-L, Smith CA & Jirtle RL 2006 Post-weaning diet affects genomic imprinting at the insulin-like growth factor 2 (*Igf2*) locus. *Human Molecular Genetics* **15** 705-716.

Yamagoe S, Kanno T, Kanno Y, Sasaki S, Siegel RM, Lenardo MJ, Humphrey G, Wang Y, Nakatani Y, Howard BH & Ozato K 2003 Interaction of histone acetylases and deacetylases in vivo. *Molecular and Cellular Biology* **23** 1025-1033.

Young LE, Fernandes K, McEvoy TG, Butterwith SC, Gutierrez CG, Carolan C, Broadbent PJ, Robinson JJ, Wilmut I & Sinclair KD 2001 Epigenetic change in IGF2R is associated with fetal overgrowth after sheep embryo culture. *Nature Genetics* **27** 153-154.

Young LE & Beaujean N 2004 DNA methylation in the preimplantation embryo: the differing stories of the mouse and sheep. *Animal Reproduction Science* **82-83** 61-78.

Poster Abstracts

Poster author index

1. Muscle fibroblasts, myogenic precursors and differentiated myotubes result in similar cattle cloning efficiency

AL Green, DN Wells & B Oback

Reproductive Technologies, AgResearch Ruakura, Hamilton, New Zealand

It has been postulated that mammalian nuclear transfer (NT) cloning efficiency is inversely correlated with the donor cell differentiation status[1]. However, previous comparative data sets addressing this issue were confounded by the use of donor cells from different genetic backgrounds, sex, cell cycle stages or cloning methodologies applied. In order to conclusively correlate cloning efficiency with differentiation, we have directly compared epigenetically distinct but genetically identical donors within the myogenic lineage. Donor cells were isolated from cryopreserved rump muscle of a male Day (D) 150 bovine fetus. Cells were cultured on matrigel-coated plates in DMEM/F12 containing either 1% or 10% FCS for 24 to 144 hours. Every 24 hours, cells were fixed and the proportion of the total number of Hoechst 33342-positive cells expressing the following antigens was quantified by immunofluorescence microscopy: PAX7, MYOD, MYF5, MYOGENIN, DESMIN and 5-Bromo-2-deoxyuridine (BrdU) after a 30 min and 24 h labelling period. For each antigen, a minimum of 150 cells were counted per replicate (n = 4-11). Based on their antigen and proliferation profile in vitro, donor cells serum-starved for 24, 96 and 120 hours were classified as myogenic precursors (MPC), myotubes (MT) and muscle-derived fibroblasts (MF), respectively. Individual mononucleated cells were selected for zona-free NT as described in cattle[2]. There was a clear effect of differentiation status on development to Grade 1-2 compacted morulae/blastocysts (M/B$^{1-2}$) on Day 7 (118/520 = 23% vs. 93/873 = 11% vs. 66/174 = 38% for MPC vs. MT vs. MF, respectively, P< 0.001, n = 5-11). For each donor cell type, the expression of the following transcripts was measured by RT-PCR in i) donor cells selected for NT, ii) recipient oocytes, iii) NT couplets, iv) NT reconstructs, v) NT 2-cell embryos, vi) NT blastocysts and vii) control in-vitro fertilized (IVF) blastocysts: 18S rRNA and ß-actin (housekeeping), Pax7, MyoD, Myf5, Myogenin and MRF4 (muscle-specific). While 18S and ß-actin continued to be expressed in all samples analysed, the muscle-specific transcription factor mRNAs were all silenced immediately before or after NT and remained undetectable in single NT embryos and IVF controls. After embryo transfer of single B$^{1-2}$, we found no significant difference in pregnancy rate at D35 and development to weaning between the cell types (14/22 = 64% vs. 8/23 = 35% vs. 10/22 = 46% and 3/22 = 14% vs. 3/23 = 9% vs. 4/22 = 18%, for MPC vs. MT vs. MF, respectively). In conclusion, two donor cell subpopulations of divergent differentiation status within the same somatic lineage showed similar transcriptional reprogramming up to the blastocyst stage and no difference in cloning efficiency.

We would like to thank our cloning team and farm staff for excellent technical assistance. This work was funded by the New Zealand Foundation for Research, Science and Technology and AgResearch.

[1]Oback & Wells, 2002. Cloning Stem Cells 4:147-168.
[2]Oback & Wells, 2003. Cloning Stem Cells 5:243-256.

2. Serum progesterone and estrone concentrations in cattle pregnant with AI-, IVP- or nuclear transfer-generated embryos

CJ Morrow, MC Berg, RM McDonald, DN Wells, AJ Peterson & RSF Lee

AgResearch Ltd, Ruakura, Hamilton, New Zealand

Abnormal placentation, pregnancy failure and hydroallantois are associated with somatic cell nuclear transfer (SCNT) in cattle. Identification of diagnostic markers for abnormal placentation in early gestation would permit therapeutic intervention.

We report on maternal serum progesterone and estrone concentrations from individual cows between Days 35–130 of gestation in AI (n = 11), IVP (n = 20) and SCNT (n = 25) generated pregnancies. Resulting fetuses were either clones (SCNT) or offspring (AI/IVP) of a donor Holstein bull from which the donor cells (AESF-1) were derived.

Serum samples were collected on Days 35, 50, 70, 100 and 130 of gestation. Placental and fetal morphological data were collected following slaughter between Days 135-163 of gestation. After slaughter, pregnancy outcomes for the SCNT group were retrospectively classified as: failed (failing spontaneously between Days 50-135; n = 11); hydrops (greater than 10L combined amniotic and allantoic fluid at post mortem, n = 8) and viable (pregnant and apparently normal at time of recovery between Days 135-163, n = 6). Serum concentrations were analysed by Anova on log transformed data.

The SCNT viable group had serum progesterone concentrations significantly higher than the IVP and AI group on Day 35 (14.7 ± 1.3, 10.4 ± 1.4 and 8.4 ± 0.5 ng/ml for SCNT, IVP and AI respectively; $P < 0.05$). Serum progesterone in the SCNT viable group continued to be significantly higher than the IVP group on Day 50 and the AI group on Day 70. However, serum progesterone concentrations were not significantly different among the groups of cows carrying SCNT embryos that failed, developed hydrops or appeared viable at post-mortem.

Serum estrone concentrations were lower in cows that subsequently aborted a SCNT fetus (SCNT failed) than in the AI and IVP groups on Day 50 ($P < 0.05$) and the AI group on Day 100 (23.8 ± 5.0 v. 41.0 ± 5.2 pg/ml; $P < 0.05$). In all other groups, serum estrone concentrations were on average 2.4-fold higher (range 1.3-3.7) on Day 130 than on Day 100, reflecting gain of steroidogenic function. There were no significant differences in serum estrone concentrations among the remaining SCNT, IVP and AI groups until Day 130. On Day 130 the SCNT viable cows had significantly lower serum estrone concentrations (67.7 ± 18.5 pg/ml) than AI (105.9 ± 12.1 pg/ml; $P < 0.05$) but not IVP (88.7 ± 7.8 pg/ml) or SCNT hydrops (88.8 ± 17.8 pg/ml) cows.

Maternal serum progesterone concentrations in SCNT-carrying cows were elevated before Day 100 and serum estrone concentrations tended to be lower in SCNT pregnancies that spontaneously aborted. Further investigation of serum steroid concentrations and the potential of serum estrone as a diagnostic marker to predict abnormal placentation and/or pregnancy failure is warranted.

3. Fertility of donor ewes following surgical collection of embryos

CAG Diaz & E Emsen

Ataturk University, Department of Animal Science, Erzurum, Turkey

Embryo collection in the small ruminant species is a surgical procedure although non-surgical techniques have been the subject of considerable research. The cervix of sheep and goats is very difficult to penetrate, especially 6-7 days after heat, and commercial success in non-surgical embryo collection is limited to goats. Surgical procedures allow for the recovery of a high percentage of embryos. However, because of the surgical trauma and resulting adhesions, surgical recovery can be repeated only a few times. The interval between recovery and subsequent pregnancy may vary in different breeds, recovery procedures, superovulation regimes, season etc. In certain species (e.g. hamsters) it is known that low pregnancy rates after recovery are mainly due to trauma to the endometrium [1].

The fertility rate of fat-tailed Awassi ewes used previously as donors in an embryo flushing program was evaluated by artificial insemination in the late breeding season (January). Estrus was synchronized by means of fluorogestone acetate sponges and ewes were superovulated with porcine FSH (six decreasing doses of 4, 4, 3, 3, 2, 2 mg i.m. at 12h intervals, starting 24h before sponge withdrawal). Donors were inseminated via laparoscopy with frozen-thawed semen (50 million live spermatozoa post-thawing) and subjected to surgical embryo recovery. Six (n = 17) or 9 weeks (n = 16) following the surgical embryo recovery, donor ewes were artificially inseminated and pregnancy rate determined. Donor ewes were not super-ovulated in the subsequent insemination but synchronized with vaginal sponge (30mg FGA) for 12 days and 600 I.U. eCG were injected i.m. at sponge withdrawal.

While pregnancy rate was low (27%) in donors inseminated at 6 weeks following the surgical embryo recovery, more donors ($P < 0.05$) established a successful pregnancy when inseminated 9 weeks after embryo recovery (70%). The superovulation response (at least 3 Corpora Lutea) had no significant effect ($P > 0.05$) on subsequent fertility rate. One principal reason for pregnancy failure 6 weeks after surgical embryo recovery could be due to the uterus failing to provide adequate support for optimal conceptus development. Rebreeding donors 9 weeks after surgical embryo recovery provided a practical means of increasing the fertility of the donor ewes. These preliminary results show that the duration between embryo recovery and subsequent insemination played an important role in the fertility of donor ewes. When scheduling the rebreeding of donor ewes with a short breeding season, such as Awassi (104 days), a longer duration (3 weeks more than usual) between recovery and insemination should be taken into account in multiple ovulation and embryo transfer programs.

---

[1] Jarosz & Dukelow, 1990. Zoological Sci 7:85-91.

4. Lambing rates in multiparous ewes and ewe lambs after frozen-thawed laparoscopic artificial insemination

CAG Diaz

Ataturk University, Department of Animal Science, Erzurum, Turkey

Successful development, breeding, and lambing of ewe lambs is one of most important tasks of the shepherd. Ewe lambs can be mated successfully without detrimental affects on subsequent reproductive performance providing they achieve a threshold body weight within the breeding season[1]. However, ewe lambs that are to be artificially inseminated with frozen-thawed semen do not always have acceptable pregnancy rates compared to mature ewes. Many management factors affect successful pregnancy of ewe lambs. Because numerous factors influence conception rates among ewe lambs, it is possible for some sheep producers to get as many as 95 to 100 percent of their ewe lambs pregnant, while others only get 10 to 40 percent pregnant.

The objective of this study was to determine lambing rates of adult ewes and ewe lambs inseminated with frozen-thawed semen. A total of 26 cyclic Awassi ewes, involving multiparous ewes (n = 13) and ewe lambs (n = 13) were used in this study. The ewe lambs were about 17 months old at the beginning of the present study. They were of the generally recommended breeding size (65% of expected adult body weight). The experiment was conducted during autumn. All females were treated with 30 mg FGA in impregnated polyurethane sponges. Pessaries were inserted deep into the vagina and left in place for 14 days. At sponge withdrawal, both groups of females received an intramuscular injection of 600 IU PMSG. PGF2 alpha was injected i.m. 24h prior to PMSG injection. Animals were inseminated via laparoscopy using frozen-thawed semen (100 million motile spermatozoa after thawing) at 50-55h after PMSG injection.

There was a significant difference ($P < 0.05$) in lambing rates between multiparous ewes (43%) and ewe lambs (21%). Prolificacy was similar for both multiparous ewes (1.3) and ewe lambs (1.5). Adult ewes and ewe lambs produced lambs of similar birth weights (3.5kg and 3.6kg, respectively). However, all of the lambs born were successfully raised by the multiparous ewes, whereas only 57% of lambs born from the ewe lambs reached weaning age (75 days of age; $P < 0.05$).

Ewe lambs begin cycling somewhat later than mature ewes of the same breed. Therefore, the opportunity to get ewe lambs inseminated is lower than with mature ewes. In conclusion, this study showed that ewe lambs have lower pregnancy rates, and lower survivability in their offspring, following laparoscopic artificial insemination performed early in the breeding season. Pregnancy rates in ewe lambs could be improved by using fresh diluted semen instead of frozen-thawed semen and with insemination delayed to the mid-breeding season.

[1]Gaskins *et al.* 2005. J Anim Sci 83:1680-1689.

5. A preliminary evaluation on the efficiency of supplementing a GnRH-agonist in the superovulatory response of Boer Goat does

JPC Greyling[a], KC Lehloenya[a], S Grobler[b] & LMJ Schwalbach[a]

[a]Department of Animal, Wildlife & Grassland Sciences, University of the Free State, P.O. Box 339, Bloemfontein 9300, South Africa and [b]Department of Obstetrics and Gynaecology, University of the Free State, P.O. Box 339, Bloemfontein 9300, South Africa

The presence of a dominant follicle at the onset of a superovulation has been reported to decrease the ovarian response in small ruminants. One way of improving superovulation yields in small ruminants might therefore be by the induction of a low LH concentration, as dominance is established in the presence of high blood LH concentrations. The pattern of LH secretion has been modified with the aid of exogenous gonadotrophin-releasing hormone (GnRH) antagonists or agonists in small ruminants MOET protocols. When a GnRH antagonist is used as a pre-treatment during the progestagen treatment phase of a MOET programme, a decline in LH concentration and high ovulation rate could be recorded. The aim of this preliminary study was to evaluate preliminary the efficiency of supplementing a GnRH-agonist (GnRHa) to the FSH treatment in the superovulatory protocol for Boer goat does during the natural breeding season.

Oestrus was synchronised in 21 Boer goat does with CIDRs (Phamacia & Upjohn, Auckland, New Zealand) placed for 17 days. Starting on Day 7 of CIDR insertion, does were superovulated with FSH alone (control group, n = 10:CG) or in combination with a GnRHa (Leuprolide, Lucrin, NL, Ch: treatment group, n = 11: TG), for 7 days. A total of 200 mg FSH-p (Folltropin, Vetrepharm)/doe was administered i.m. in 7 doses at 12h intervals, starting 48 h prior to CIDR removal (the first dose being 50 mg and all others 30 mg). Laparoscopic inseminations with fresh diluted Boer goat semen were performed 36h and 48h following CIDR removal. Embryos were flushed 6 days following the second AI.

The oestrus response (80 vs 100%), onset of oestrus (25.5 ± 7.4 vs 24.0 ± 7.4h) and duration of the induced oestrous period (26.9 ± 4.0 vs 19.6 ± 5.5h) for the control and GnRHa treatment groups, respectively, did not differ significantly. There was no significant difference in the mean total ova (unfertilised ova and embryos) collected, embryos, and unfertilised ova per donor between treatments. The mean degenerated embryos /donor tended to be higher in the GnRHa treated does (6.6 ± 4.2), compared to the control group (1.7 ± 1.5). The mean transferable embryos per donor tended to be lower in GnRHa treated goats (3.4 ± 2.7) compared, to the control does (9.3 ± 6.1).

From these results it can be concluded that the oestrous response, onset of oestrus and to a lesser extend the duration of the induced oestrus does not seem to be affected by GnRH agonist treatment. The GnRH agonist also had a detrimental effect on the total number of ova collected (fertilised and non-fertilised) and reduced the number of transferable embryos by increasing the total number of degenerated embryos.

6. Ovarian follicular response to repeat stimulation with FSH and oocyte recovery and quality in Brahman heifers treated with a GnRH agonist

A Kenny[a], N Phillips[b] & MJ D'Occhio[a]

[a]School of Animal Studies and [b]School of Veterinary Science, The University of Queensland, Gatton Campus, Queensland, Australia 4343

The aim in the present study was to ascertain whether removal of ovarian follicular dominance by downregulation of endogenous gonadotrophin secretion with a GnRH agonist would increase follicular responses to weekly cycles of FSH stimulation and ovum pick-up (OPU) in heifers. Peripubertal Brahman (*Bos indicus*) crossbred heifers were randomly assigned on live weight (LW) to 2 groups: Control (n = 8, 295 ± 5 kg LW), no treatment; GnRH agonist (n = 8, 296 ± 4 kg LW), received deslorelin in a controlled release implant.[1] During wk 7 after implanting, follicles ≥ 3 mm were ablated from both ovaries in all heifers by ultrasound-guided follicular aspiration (5 MHz endovaginal sector transducer equipped with 18 G aspiration needle). Five days later (Week 1, Day 0), follicular status was recorded (Aloka SSD-500, 7.5 MHz transducer) and heifers were placed on FSH (Folltropin-VÒ, Bioniche A/Asia) treatment as follows: Day 0, 20 mg A.M. and 10 mg P.M.; Day 1, 10 mg A.M. Oocytes were aspirated from ≥ 3 mm follicles by standard OPU on Day 2, A.M. The cycle of FSH stimulation and OPU, starting 5 days after the previous OPU, was repeated for Wk 2, 3 and 4. The size of the largest follicle at the start of FSH treatment was smaller (P < 0.01) in heifers treated with deslorelin compared with controls from Wk 1 (7.2 ± 0.8 and 11.0 ± 0.8 mm) to Wk 4 (6.1 ± 1.0 and 12.0 ± 1.5 mm) and the number of follicles ≥ 4 mm was less in deslorelin treated heifers from Wk 1 (3.3 ± 1.0 and 6.6 ± 0.4, P < 0.05) to Wk 4 (1.8 ± 0.9 and 7.3 ± 1.0, P < 0.01). Heifers treated with deslorelin appeared to have a greater number of 2-3 mm follicles at the start of FSH but this was difficult to accurately quantify. The number of follicles (≥ 3 mm) aspirated for each heifer did not differ (P > 0.05) between deslorelin and control heifers from Wk 1 (18.0 ± 4.2 and 12.3 ± 3.8, respectively) to Wk 8 (14.7 ± 2.2 and 18.7 ± 6.0, respectively). The average number of oocytes collected each wk and total number for Wk 1-4 were similar for deslorelin treated (65 and 262) and control (64 and 257) heifers. There were no apparent differences in oocyte quality between deslorelin and control heifers from Wk 1-4. For the 2 groups combined, oocyte quality was distributed as Grade A, 6%; Grade B, 17%; Grade C, 52%; with the remaining 25% denuded, expanded or atretic. The rationale for downregulating of gonadotrophin secretion in the present study was to restrict follicle growth and remove inter-follicular dominance, thus potentially enhancing follicular and oocyte responses to FSH stimulation before OPU in heifers. Treatment with deslorelin restricted follicle growth but this did not increase the number of follicles available for aspiration after stimulation with FSH, the number of oocytes recovered, or the quality of oocytes. This finding was interpreted to suggest that differences between follicles in the capacity to respond to FSH are already established at the early antral stage. The influence of inter-follicular dominance on the capacity of individual follicles to respond to FSH could therefore be regarded as relatively minor compared with the potential acquired during early follicular development. Strategies to enhance the follicular response to exogenous FSH may need to consider the period of development before the early antral stage.

We gratefully acknowledge Bioniche Animal Health A/Asia Pty for generous supply of Folltropin V® and Mr Tom Copely, "Salty", Crows Nest, Queensland, for the provision of Brahman crossbred heifers.

[1]D'Occhio *et al.* Anim Reprod Sci 60-61:433-442.

456

7. A culture procedure for assessing the viability of thawed embryos prior to transfer

DA Contreras[a], JG Avila[b], MA Aspron[c], NA Moreno[d] & CS Galina[a]

[a]Departamento de Reproducción, [b]Departamento de Rumiantes, Facultad de Medicina Veterinaria y Zootecnia, Universidad Nacional Autónoma de México, [c]Escuela de Medicina Veterinaria, Facultad de Ciencias Naturales, Universidad Autónoma de Querétaro and [d]Departamento de Biología Celular y Fisiología, Instituto de Investigaciones Biomédicas, UNAM, México

Frozen embryos have been stored in developing countries for many years with little information as to whether these embryos are viable upon thawing. The aim of this study was to evaluate a culture system as a non-invasive means of assessing the viability of recently thawed embryos prior to transfer.

Forty five "Holstein x Brahman" embryos were collected 7 days after insemination from 20 synchronized and superovulated cows. Embryos were classified as good n = 15 (compact blastomeres, amber color, no damage in the zona pellucida), fair n = 15 (extruded blastomeres, cellular debris in the perivitelline space) and poor n = 15 (marked degeneration, small cell mass, underdevelopment, abundant cellular debris and dark color) and frozen in 10% of glycerol. After five months of storage, embryos were thawed and incubated with controlled humidity and 5% CO_2 at 37° C in McCoy© medium.

Morphology and stage of development were monitored for a period of 24 hours to register the development every 30 minutes the first 2 h, thereafter every hour. During the culture period, four embryos of each classification were separated at 4 h, another four at 12 h and the remaining seven at 24 hours and the degree of apoptosis was determined using the TUNEL technique.

Descriptive statistics were used to assess the data, and analysis of variance to compare the number of apoptotic cells. Embryos of good and fair quality did not suffer major changes in their development even after 7 h of incubation, whereas poor quality embryos experienced changes as early as 2 h after incubation. Good quality embryos invariably showed fewer apoptotic cells than those of fair and poor quality ($P < 0.05$) suggesting embryo culture to be useful in assessing viability and confirming the quality of thawed embryos previously stored in liquid nitrogen, prior to transfer.

Acknowledgements: Financial support was provided by UNAM-PAPIIT IN201903.

8. Effect of using different concentrations of permeating cryoprotectants and rehydration methods on the cryopreservation of dromedary camel embryos (*Camelus dromedarius*)

JA Skidmore[a], M Billah[a] & NM Loskutoff[b]

[a]*Camel Reproduction Centre, Dubai, UAE and* [b]*Center for Conservation & Research, Henry Doorly Zoo, Omaha, Nebraska 68107 USA*

Embryos (n = 60) were recovered from the uteri of donor camels by non-surgical uterine lavage on day 6 (n = 20) or 7 (n = 40) after ovulation. They were examined morphologically and those of Grades 1 or 2 were exposed to either 5% (n = 40) or 10% (n = 20) ethandiol (EG; v:v) in holding media (HM) for 10 min before being loaded individually into 0.25 ml straws and placed directly into the Embryo Freezing machine at −7°C. The straws were seeded after 1 min and held for a further 10 min to equilibrate before cooling at -0.5°C/min to -30°C, at which time they were plunged into liquid nitrogen.

The embryos were subsequently thawed by holding in air for 8 secs, followed by swirling in a 32°C water bath for 2 min before all embryos were rehydrated in one of two ways. They were either expelled initially into 1) HM containing 0.2M sucrose for 5 min. (day 7, 10% n = 10; day 7, 5% n = 10; day 6, 5% n = 8) or 2) HM containing 0.5M sucrose for 2.5 min, followed by HM with 0.25M sucrose for 2.5 min (day 7, 10% n = 10; day 7, 5% n = 10; day 6, 5% n = 12) before transferring them into HM without sucrose prior to transfer. All embryos were transferred into recipient camels on day 6 after ovulation and pregnancy was diagnosed by ultrasonography of the uterus between days 18 and 20 of gestation.

Of the embryos transferred to recipient camels after exposure to 10% EG and rehydrated in HM + 0.2M sucrose (n = 10), 4 resulted in viable fetuses and those rehydrated using 0.5M/0.25M sucrose (n = 10) resulted in 3 pregnancies, but these 3 had resorbed by day 23. However, none of the day 7 embryos exposed to 5% EG resulted in a pregnancy regardless of the method of rehydration, whereas 2/8 (25%) day 6 embryos rehydrated in HM + 0.2M sucrose and 3/12 (25%) day 6 embryos rehydrated in HM + 0.5M/0.25M sucrose did result in viable fetuses.

These results show that it is possible to cryopreserve camel embryos using slow-cooling methods and 10% ethandiol as cryoprotectant although further work is required to improve pregnancy rates. However, it would appear that 5% ethandiol only provides enough cryoprotection for the smaller day 6 embryos.

The project was kindly sponsored by His Highness Shaikh Mohammed Bin Rashid Al Maktoum, Vice President and Prime Minister of the UAE and Ruler of Dubai.

458

9. A simple technique for ruminant cervical dilatation dilation using topical application of cloprostenol

MC Berg & RSF Lee

AgResearch, Ruakura Research Centre, Hamilton, New Zealand

A simple, inexpensive technique, using a commonly used animal drug, has been developed to facilitate the recovery of bovine fetuses prior to fetal death, the termination of hydroallantoic pregnancies and the use of large-bore catheters to non-surgically flush and recover elongation stage embryos. Previously, a combination of prostaglandins (cloprostenol or $PGF_{2\alpha}$ and PGE_2) and oxytocin has been used to allow first trimester fetal recovery within a few hours after fetal death.[1] Fetal tissues, recovered prior to fetal death, enable anatomical and histological examinations and recovery of intact DNA for molecular analysis. Viable fetal tissues are also valuable for the isolation of primary cell lines which can be used as donor cells for somatic cell nuclear transfer (SCNT) or as starting material for genetic modification for nuclear transfer transgenesis.[2]

A conventional Cassou ET pistolette and sheath, with a syringe fitted to a luer hub replacing the plunger, is used to place twice daily 1 ml doses of cloprostenol (250 μg/ml) into the cervical canal. These non-luteolytic doses act locally to soften the cervix and copious mucus production is evident after the first application. Typically, a modified catheter (7 mm I.D. endotracheal tube) can be passed 12 hours after the second dose through the cervix into the body of the uterus to allow full flushing to recover elongation stage embryos. Additional doses are needed to allow the cervix to soften further for fetal recoveries. A foley catheter with a fully inflated cuff placed in the body of the uterus is useful to prevent a very young fetus from slipping out prematurely if the cervix softens faster than expected. Because the pistolette can easily traverse the cervix, care must be taken not to perforate the allantoic membrane if an intact conceptus is to be recovered. After four doses, most cervices will be dilated up to 1-2 fingers width in diameter and younger than Day 70 fetuses may be passed with intact membranes into the vagina. Two or more additional doses may be needed for larger fetuses to pass through the cervix. When the cervix is dilated to four fingers width, it is usually soft enough to manually dilate further, reach inside the uterus to rupture membranes and extract the fetus.

This technique has been used successfully to deliver and recover fetuses from hydroallantoic pregnancies (up to 20% of SCNT pregnancies) where the fetuses may be overgrown with grossly distended abdomens. A dose of long-acting dexamethasone (5 mg/ml) administered 5 days prior to this treatment has enabled the extraction of >Day 150 hydroallantoic SCNT fetuses prior to fetal death. These older fetuses are then euthanized with a small dose of sodium pentobarbital. Atonal uteri and retained membranes necessitate careful treatment following fetal extraction from hydroallantoic pregnancies. These recipient dams usually make complete recoveries and have successful subsequent pregnancies. This technique has also been successful in recovering Day 120 cervine mummified fetuses resulting from SCNT pregnancies.

[1]Lavoir & Betteridge, 1996. J Reprod and Fertil 106:95-100.
[2]Wells *et al.* 2003. Theriogenology 59:45-49.

10. The effect of external genital stimulation on uterine contractions and conception rate in cattle

<u>M Ono</u>[a], M Matsui[a], S Koseki[b], S Nakata[c], T Osawa[d], & Y-I Miyake[a]

[a]Dept of Clinical Vet. Sci, Obihiro University of Agric. and Vet. Med., Obihiro, [b]Tokachi District Agricultural Mutual Aid Association, Obihiro, [c]Nemuro District Agricultural Mutual Aid Association, Bekkai and [d]Dept of Clinical Vet. Sci, Iwate University, Morioka, Japan

An understanding of the physiological effects in cow management at artificial insemination (AI) can improve reproductive efficiency. Clitoral stimulation at AI has been reported to increase conception rate in cows[1,2]. Previous report showed that various manual stimuli could increase uterine contractility[3]. However, its mechanism has not been fully understood. The first experiment was conducted to assess the effect of stimulation of the external genitals on the uterine contractions in the cow. Experiment 2 was conducted to assess the effect of labial stimulation after AI on conception rate in the cows and heifers.

In experiment 1, uterine contractions were measured in 14 Holstein-Friesian cows by using a tubal insufflation device. The methods of external genital stimulation were clitoral stimulation (3sec) and labial stimulation (15sec). The manual stimulation was applied in the estrous phase and luteal phase. After clitoral stimulation (3sec), uterine contractions were observed in 3 out of 22 cows (13.6%) at estrous phase and 1 out of 17 cows (5.9%) at luteal phase. Labial stimulation (15sec) induced uterine contractions in 8 out of 9 cows (88.9%) at estrous phase and 1 out of 9 cows (11.1%) at luteal phase. At estrous phase, labial stimulation induced uterine contractions in significantly higher ($p < 0.01$) percentage of cows than clitoral stimulation. Uterine contractions by labial stimulation were observed in significantly higher ($p < 0.01$) percentage of cows at estrous phase compared with cows at luteal phase.

In experiment 2, labia of 297 cows and 156 heifers were stimulated for 15 seconds after AI, and 240 cows and 139 heifers were assigned to the control group. The conception rate in cows receiving labial stimulation (35.0%) was similar to that in cows not receiving stimulation (32.1%). In heifers, the conception rates were not significantly different between the animals with or without stimulation; 68.6% or 64.7%, respectively. When we analyzed the effect of labial stimulation on the conception rate of semen from two different bulls that were used for inseminating > 10 heifers per group, one bull showed a significant difference ($p < 0.05$) between stimulated group (15/18, 83.3%) and control group (5/13, 38.5%), and another one showed no difference between stimulated group (60/85, 70.6%) and control group (58/84, 69.0%). In heifers not receiving labial stimulation, the conception rate tend to be lower ($p < 0.10$) in August (15/28, 53.5%) than in June (29/38, 76.3%), but this tendency was not observed in heifers receiving labial stimulation during August (20/31, 79.5%) and June (35/44, 64.5%).

The present study clearly demonstrated that labial stimulation at estrus induced uterine contractions. Our results suggested that labial stimulation may prevent reduction in fertility during summer months. We propose that external genital stimulation may be applied to improve fertility after AI under the heat stress condition.

[1]Randel et al. 1975. J Anim Sci 40:1119-1123.
[2]Short et al. 1979. J Anim Sci 49:647-650.
[3]Hays et al. 1953. Am J Physiol 172:553-556

11. The use of a modified laparoscopic technique for recovery of oocytes in young goats

MF Cordeiro, PA Di Filippo, RGS Dória, DPM Dias, CAG Beretta, GD Pavão, G Ribeiro, ALL Duarte & WRR Vicente

Departamento de Medicina Veterinária Preventiva e Reprodução Animal, Universidade Estadual Paulista, São Paulo, Brasil

Amongst the surgical techniques, laparoscopy has gained prominence for being less invasive and faster, therefore reducing the stress suffered by the animal[1]. Techniques for the attainment of oocytes *in vivo* for use in basic research and the production of embryos *in vitro* in goats have not developed with the same speed as others biotechnologies[2]. The objective of this study was to evaluate a modified technique for follicular aspiration by laparoscopy in young goats.

Eight female goats with ages varying from 4 to 6 months of age were each submitted for six sessions to harvest oocytes, totaling 48 laparoscopies. Trocars were introduced in the abdomen (cranial region to udder) for the insertion of the endoscope and non-traumatic forceps. The abdomen was inflated with CO_2, using a total pressure of 5mmHg. For follicular aspiration a 16G peripheral venous access catheter (BD AngiocathÒ) was attached to a vacuum system. Firstly, one ovary was grasped in the non-traumatic forceps and positioned next to the lateral abdominal wall. Then the catheter was inserted into the cavity, next to the ovary to be aspirated. At end of the session, the ovaries were bathed with PBS plus heparin in order to prevent future adhesions. The aspirated follicular fluid was examined with a stereomicroscope and the recovered oocytes classified according to quality.

The total number of oocytes recovered from each doe averaged 23. The average number of oocytes recovered per doe per session was considered low (3.88). This may, have been partly attributable to the number of punctured follicles (458) being less than the number of observed follicles (615), as well as to the bore of the needle (16G) that may have limited the aspiration of small follicles (around 5mm in diameter). Although the pressure used in the vacuum still caused denudation of some oocytes, it did not overly compromise total quality (12.37% grade I; 36.56% grade II; 24.19% grade III and 26.88% grade IV). Formation of adhesions between the structures of the genital system and other organs was not observed in the great majority of the females. The use of the catheter allowed access to the ovaries without presenting the risk of injury of adjacent organs given that the procedure required a little time to complete, and was carried out away from other organs. Moreover, the plastic case protected the tip of the needle while the ovary was manipulated, being displayed only at the moment follicular aspiration. The punctures in the abdominal cavity healed within three days. On average, the laparoscopic procedure required only 35 minutes per doe.

This laparoscopic technique for follicular aspiration is simple and safe, and can be used a number of times in the same female. However, adjustments in the technique must be made to further improve the proportion and the quality of the recovered oocytes.

Acknowledgements: To the Fundação de Amparo à Pesquisa do Estado de São Paulo - FAPESP for the financial support; Paulo Sérgio da Silva and Edilson Gabriel da Silva Júnior for assistance with the handling, the feeding and the maintenance of the animals.

[1]Kühholzer *et al.* 1997. Theriogenology 48:545-550.
[2]Baldassarre *et al.* 2002. Theriogenology 57:275-284.

12. Effects of equine Chorionic Gonadotrophin (eCG) on ovarian dynamics in Nelore cows

MACM Bergamaschi[a], WRR Vicente[a], RT Barbosa[b], R Machado[b], PS Baruselli[c], MM Alencar[b] & M Binelli[c]

[a]Departamento de Reprodução Animal, Universidade Estadual Paulista (UNESP), Jaboticabal, Brazil, [b]Embrapa Pecuária Sudeste, São Carlos and [c]Departamento de Reprodução Animal, Universidade São Paulo (USP), Brazil

All protocols of timed artificial insemination involve some sort of hormone treatment to synchronize ovulation. It has been demonstrated that progestagen methods give better results[1]. In addition, the administration of equine chorionic gonadotrophin (eCG) at progestagen withdrawal improves conception rates in beef cattle[2]. This study aimed to evaluate the effects of eCG on follicular dynamics, as well as corpus luteum (CL) development and function after an ovulation synchronization protocol.

A group of 16 mature, synchronized (CRESTAR®), lactating Nelore (*Bos taurus indicus*) cows were randomly allotted to receive either 400 IU of eCG at implant withdrawal (G_{eCG}; n = 8) or remain as controls (G_C; n = 8). Ultrasound *per rectum* evaluation of ovaries was conducted daily, from implant removal up to the following ovulation (a complete estrous cycle). Simultaneously, blood samples were taken to determine the plasma concentrations of progesterone ([P4]). Data were analyzed by GLM of the SAS program.

The length of the 2nd wave of follicular growth (FG) was shorter (P < .05) in G_{eCG}, even though traits assessed of the 1st and 3rd waves were not affected (P > .05) by eCG. As a result, the interval between the synchronized and the subsequent natural ovulation was similar (P > .05) between groups. Preovulatory follicle diameter of the synchronized cycle was not different (P > .05) between G_C (13.38 ± .39 mm) and G_{eCG} (12.72 ± .39 mm). The highest [P4] occurred on the 11th day of the estrous cycle for both groups. However, [P4] of G_{eCG} (8.15 ± .64 ng/mL) was higher (P < .05) than G_C (6.37 ± .64 ng/mL). G_C presented three 2-wave (37.5%) and five 3-wave cows (62.5%) in comparison with one 2-wave (12.5%) and seven 3-wave (87.5%) cows in the G_{eCG}. However, this difference was not significant (P > .05).

eCG has FSH-like activity and promotes follicular growth and maturation, ovulation of the dominant follicle and further development of the subsequent CL. In addition, eCG binds to follicular and luteal gonadotrophin receptors and causes an increase in the number of large luteal cells, which are responsible for some 80% of the progesterone synthesized by CL[3]. Indeed, G_{eCG} cows showed increased [P4] from the mid luteal phase (D11) up to the following natural ovulation. The luteotrophic effects of eCG reduced the length of the 2nd wave of follicular growth perhaps by the high P4 concentrations lowering gonadotrohin secretion. A lack of gonadotrophin stimulation would likely retard follicular growth and induce precocious atresia. We conclude that eCG given at progestagen removal affected follicular dynamics of the subsequent estrous cycle and promoted increases in the [P4] from the mid luteal phase.

[1]Odde, 1990. Journal of Animal Science 60-61:713-723.
[2]Baruselli *et al.* 2004. Animal Reproduction Science 82-83:479-86
[3]Niswender *et al.* 1985. Recent Progress in Hormone Research 41:101.

13. A comparison of staining media for sex-sorting cooled dairy bull spermatozoa

L Underwood[a], R Bathgate[a], L Gillan[b] WMC Maxwell[a] & G Evans[a]

[a]ReproGen, Faculty of Veterinary Science, The University of Sydney, NSW 2006, Australia and
[b]Sydney IVF, Sydney, NSW 2000, Australia

There is increasing interest among scientists and dairy producers in sex-sorting of cooled, transported (or frozen-thawed) bull spermatozoa. This would allow the sexing of spermatozoa from bulls located at a distance from the sorting facility. The aim of this study was to determine the best medium for staining cooled dairy bull spermatozoa for optimum resolution during sex-sorting.

Semen (n = 3 bulls) was collected, diluted 1:1 (semen: diluent) with a 20% egg yolk tris-citrate-fructose medium, cooled to 5°C and transported from Victoria to Sydney over an 18 hour period in a temperature-controlled (5°C) foam box (Minitube Australia; Smythes Creek, Australia). Semen was centrifuged through PureSperm gradients (NidaCon, Sweden; 600g; 20min). The lower gradient was resuspended in modified Androhep (AH; Minitube Australia, Smythes Creek, Australia) and centrifuged (300g; 10min). The sperm pellet was diluted to 200 x 10^6 sp/ml with modified Androhep (AH), Tris-citrate-fructose (TCF), Sperm TALP (ST) or XY Staining TALP (XY), each containing 90μM Hoechst 33342 (Molecular Probes[TM], Invitrogen[TM]; Eugene, Oregon, USA; 1 hour, 34°C). Samples were diluted 1:1 (v: v) with their respective medium + 4% egg yolk and assessed by SX MoFlo® (DakoCytomation, CO, USA) for percent non-viable, percent live sperm correctly orientated, maximum sorting speed (spermatozoa/second), and subjective visual resolution of the populations of X- and Y-chromosome-bearing spermatozoa. Samples were also assessed for acrosome damage using FITC-PNA stain immediately prior to sorting. Additional cooled, transported semen (n = 3 bulls) was diluted in each medium to 40 x 10^6 sperm/ml and incubated at 34°C for 2 hours. Objective motility was assessed by CASA (HTM-IVOS, Hamilton-Thorne, USA) at 0, 1 and 2 hours. The experiment was replicated three times and analysed by ANOVA.

Percentage of non-viable spermatozoa in samples stained in AH, TCF and XY were significantly lower than those stained in ST (8.1% ± 0.38, 8.1% ± 0.38, 9.0% ± 0.56 vs. 19.2% ± 1.48, respectively; P< 0.05). Proportion of live spermatozoa correctly orientated, maximum sorting speed, and proportion of spermatozoa with intact acrosomes was significantly higher in samples stained in AH, TCF and XY than those stained in ST (49.8% ± 1.77, 48.5% ± 2.17, 47.3% ± 1.74 vs. 21.7% ± 2.8; 3167 sp/s ± 141, 3076 sp/s ± 241, 2978 sp/s ± 215 vs. 840 sp/s ± 161.4; 97.1% ± 0.57, 97% ± 0.42, 97.4% ± 0.60 vs. 95.2% ± 0.85, respectively; P< 0.05). Visual resolution between the sperm populations was superior in samples stained in AH and XY compared to TCF and ST. Objective motility at 0, 1 and 2 hours was significantly higher in samples diluted in AH compared to TCF, ST and XY (0 hours: 91.2% ± 2.15 vs. 84.3% ± 2.14, 86.6% ± 2.4, 83.8% ± 2.46; 1 hour: 90.8% ± 1.39 vs. 81.4% ± 2.88, 82.3% ± 2.03, 77.8% ± 2.98; 2 hours: 88.9% ± 1.77 vs. 80.2% ± 2.88, 75.3% ± 3.62, 68.8% ± 5.13, respectively; P< 0.05).

These results demonstrate that staining in AH and XY resulted in better sorting parameters and visual resolution than the other diluents examined for staining cooled, transported bull sperm. Incubating semen in AH resulted in higher motility than the other diluents examined. Further experiments will be conducted to determine if there is a post-thaw difference between staining cooled, transported bull sperm in AH and XY media after sex-sorting and freezing.

Research supported by The Geoffrey Gardiner Foundation, XY Inc., and Total Livestock Genetics.

14. Low-density egg yolk lipoprotein and its effect on fertility in the bovine

G McMillan, C Pitt, P Shannon & ZZ Xu

Livestock Improvement Corporation, Hamilton, New Zealand

It is well known that egg yolk is an effective protecting agent for bovine spermatozoa during *in vitro* storage or cryopreservation and that the low-density lipoprotein component of egg yolk confers the majority of this protection. Previously, we showed that semen processed in diluent containing the water-soluble fraction of egg yolk had both increased *in vitro* survival time and non-return rate (NRR) after artificial insemination (AI) compared with semen processed in diluent containing whole egg yolk (WEY)[1]. However, the complexity of the purification procedures prevented its application in commercial semen production. The objective of the present study was to evaluate the performance of a water-soluble lipoprotein fraction (SPL) that had been extracted from egg yolk using a simplified procedure.

Egg yolk was mixed with water at a ratio of 1:4. After stirring for 2 h at 4 °C, the mixture was centrifuged for 1 h at 18,000 g to remove the water-insoluble fraction. The supernatant was harvested and dialysed against 50 volumes of water for 6 days at 4°C in dialysis sacs with a molecular weight cut off of 12,000 Da. Water was changed every 48 h. Following dialysis, the retentate was harvested, re-suspended in equal volume of a citrate buffer, and the protein content determined. The SPL fraction was used for diluent preparation at a level that achieved similar lipoprotein concentration in the final diluent to that achieved by 5% whole egg yolk.

In *in vitro* studies, semen from the same ejaculates was processed to a final concentration of 10×10^6 sperm per mL in caprogen diluents prepared using either WEY or SPL and the total survival hours during incubation at 37°C was compared. Sperm processed in SPL diluent survived 14 h longer (P < 0.01) than sperm in WEY diluent. Using survival hours during incubation as a criterion, we showed that SPL protected sperm against cold shock with equal or better potency compared with WEY. When antibiotics were added to the re-suspended medium, the SPL fraction maintained its beneficial effects over WEY through 60 days of storage at 4°C.

To study if the observed benefits of SPL over WEY during in vitro incubation was reflected in improvement in NRR after AI, 16 ejaculates from 11 bulls of known fertility were split and processed in diluents containing either 5% WEY or equivalent amount of SPL to a final concentration of 4 or 8×10^6 sperm per mL, which translated into sperm dose rates of 1 and 2×10^6 per insemination. The semen straws were randomly allocated to herds throughout New Zealand and inseminated on Day 3 after processing. No significant difference in NRR between the WEY and the SPL treatments at either dose rate was detected. The NRR was 66.7% (n = 10,463) for WEY and 65.4% (n = 9,640) for SPL at the 1×10^6 sperm dose rate. Corresponding values were 67.1% (n = 7,950) for WEY and 67.0% (n = 8,021) for SPL at the 2×10^6 sperm dose rate.

In summary, despite the *in vitro* results showing that the water-soluble fraction of egg yolk was superior to whole egg yolk in supporting sperm survival during incubation and protecting sperm against cold shock, fertility of semen after insemination was not improved. The reasons for the lack of effect of SPL over WEY on fertility and for disagreement with results from our previous study[1] are not known. There appeared to be bull and breed differences in their responses to the treatments.

[1]Vishwanath *et al.* 1992. Animal Reproduction Science 29:185-194.

15. Kinematics of sex-sorted ram spermatozoa in response to seminal plasma

SP de Graaf[a], T Leahy[a], L Gillan[b], G Evans[a], WMC Maxwell[a] & JK O'Brien[a]

[a]Faculty of Veterinary Science, The University of Sydney, NSW 2006, Australia and [b]Sydney IVF, Sydney, NSW, 2000, Australia

The influence of seminal plasma on ram spermatozoa is well documented proving beneficial in restoring certain aspects of function following cryopreservation[1]. Sex-sorted spermatozoa are known to have altered kinematics and membrane status, which may affect their fertility[2]. This has stimulated the investigation of additives, such as seminal plasma in the present study, which may stabilise or reverse the functional changes observed in sex-sorted sperm.

Semen was collected from three Merino rams and split into two parts for sex-sorting and freezing (SF) or, in the case of the control, standard freezing (Control) as previously described[2]. SF and Control samples were thawed (37°C, 2 min) and diluted in Androhep (Minitube Australia, Smythes Creek, Australia) containing either 0, 5, 10 or 20% ram seminal plasma (frozen-thawed) before incubation at 37°C for 6 h. Sperm kinematics (HTM-IVOS; Hamilton-Thorne, USA) were assessed at 0, 3 and 6 h post-thaw.

The total motility ($\pm$ SEM) of Control sperm (averages pooled over 6 h) was significantly higher ($P < 0.05$) in the presence of 10% (52.6 $\pm$ 4.07%) and 20% (55.0 $\pm$ 3.99%) seminal plasma than in its absence (43.1 $\pm$ 4.03%), but did not differ significantly from Control sperm diluted with 5% seminal plasma (46.6 $\pm$ 4.2%). Conversely, the motility of SF samples incubated without seminal plasma (64.7 $\pm$ 4.82%) was superior ($P < 0.05$) to those supplemented with 10% (57.8 $\pm$ 5.84%) and 20% (30.5 $\pm$ 6.96%) seminal plasma, but not 5% (63.3 $\pm$ 4.17%) seminal plasma. The average path velocity (VAP) of Control samples did not differ ($P > 0.05$) with the supplementation of 0% (94.9 $\pm$ 5.34 μs^{-1}), 5% (92.3 $\pm$ 5.15 μs^{-1}), 10% (92.8 $\pm$ 3.74 μs^{-1}) or 20% (90.7 $\pm$ 4.19 μs^{-1}) seminal plasma. However, the VAP of SF sperm significantly decreased ($P < 0.05$) in the presence of both 10% (68.7 $\pm$ 6.55 μs^{-1}) and 20% (38.0 $\pm$ 8.23 μs^{-1}) seminal plasma when compared to 0% (85.8 $\pm$ 5.33 μs^{-1}) or 5% (79.5 $\pm$ 4.40 μs^{-1}) seminal plasma.

While these results lend further weight to the suggestion that seminal plasma is able to improve the function of frozen-thawed non-sorted ram sperm, the reverse appears to be true for sex-sorted sperm. The reason for this deleterious effect is unclear, but may be due to alteration in sperm surface proteins from the mechanical and dilution stress of sex-sorting, increasing the sensitivity of sperm to the effect of seminal plasma. Further research is required to determine the precise nature of this functional change and the component(s) of seminal plasma responsible for these positive and/or negative effects.

Research supported by XY, Inc.

[1]Maxwell & Johnson, 1999. Theriogenology 52: 1353-1362.
[2]Hollinshead et al. 2003. Reproduction, Fertility and Development 15: 351-359.

16. The effect of Trolox addition to egg yolk-Tris extender on the motility and membrane integrity of frozen-thawed ram spermatozoa

CC Sicherle[a], MS Maia[b], SD Bicudo[a], RE Green[a], DB Sousa[a] & HC Azevedo[c]

[a]Department of Animal Reproduction and Veterinary Radiology, UNESP, Botucatu, São Paulo, Brazil and [b]Embrapa Semi Árido, Petrolina, Pernambuco, Brazil, [c]Embrapa Tabuleiros Costeiros, Sergipe, Alagoas, Brazil

The aim of this study was to evaluate the effect of a water-soluble antioxidant, Trolox (6-hydroxy-2,5,7,8-tetramethylchroman-2-carboxilic acid) added to an extender on the post thaw motility and membrane integrity (MI) of ram spermatozoa. The semen was collected from six rams (two ejaculates per ram) and it was diluted to a final concentration of $400x\ 10^6$ sperm/mL in egg yolk-Tris extender[1] containing Trolox (0, 50, 100, 150 or $200\mu M/10^8$ spermatozoa). The sperm total motility (TM) was evaluated using a computer-assisted sperm analysis (CASA*) and the MI was verified using the fluorescent stains propidium iodide and carboxyfluorescein diacetate[2]. The addition of Trolox at either 50 or $100\mu M$ did not improve the TM significantly when compared to the control samples (58.64, 51.45 vs 54.45%). Nevertheless when the antioxidant was added at 150 and at $200\mu M$, a significant decrease ($P< 0.05$) on TM was observed if compared to the $50\mu M$ concentration (38.73 and 36.81 vs 58.64%). The addition of Trolox at 150 or $200\mu M$ to the extender reduced ($P<0.05$) the progressive motility (PM) when compared to the control samples (18.18, 17.73 vs 34.45%). The percentage of rapid spermatozoa (RAP) was lower when the antioxidant was used in the concentration of $200\mu M$ (28.81%) when compared to the control samples (50.36%). The other spermatic kinetic parameters as average path velocity, curvilinear velocity and straight-line velocity (VAP, VCL and VSL) did not differ significantly among the groups. The addiction of Trolox did not present any significant effect ($P>0.05$) on MI among the samples from all groups. The percentages of MI spermatozoa were: 38.18, 39.27, 37.91, 37.36 and 31.1% for the control samples, and for the Trolox samples at $50\mu M$, $100\mu M$, $150\mu M$ and $200\mu M$, respectively. In conclusion, the addition of Trolox in the concentrations utilized did not improve any of the parameters studied: total motility, membrane integrity, VAP, VCL and VSL when compared to the control samples. However, when it was added at the concentrations of 150 and 200mM, a deleterious effect on progressive motility parameters was noted. It can be suggested that the addition of Trolox at high concentrations is toxic to ram spermatozoa.

[1] Maia et al. 2005. Acta Scientiae Veterinariae 33 (Supl 1): s311.
[2] Harrison & Vickers, 1990. J Reprod Fert 88: 343-352.
* UltiMate, Sperm Analyzer – Hamilton Thorne Biosciences- IMV

17. Effect of Trolox addition on the motility and membrane integrity of ram spermatozoa with high and low freezability

MS Maia[a], CC Sicherle[b], SD Bicudo[b], DB Sousa[b] & HC Azevedo[c]

[a]Embrapa Semi Árido, Petrolina, Pernambuco, Brazil, [b]Department of Animal Reproduction and Veterinary Radiology, UNESP, Botucatu, São Paulo, Brazil and [c]Embrapa Tabuleiros Costeiros, Sergipe, Alagoas, Brazil

The aim of this study was to evaluate the effect of a water-soluble antioxidant, Trolox (6-hydroxy-2,5,7,8-tetramethylchroman-2-carboxilic acid) added to an extender on the post thaw motility and membrane integrity of ram spermatozoa with high and low freezability. Semen was collected from six rams (two ejaculates per ram) and diluted to a final concentration of 400×10^6 sperm/mL in egg yolk-Tris extender[1] containing Trolox (50, 100, 150 or $200\mu M/10^8$ spermatozoa) or control (0). The semen samples with low freezability (n = 6) had average post thaw sperm motility of ≤35%. The samples with high freezability (n = 6) were those that had an average sperm motility > 35%. Sperm motility was evaluated using a computer-assisted sperm analysis (CASA) and membrane integrity was verified using the fluorescent stains propidium iodide and carboxyfluorescein diacetate. The total motility (TM), progressive motility (PM), average path velocity (VAP), straight-line velocity (VSL), curvilinear velocity (VCL), rapid movement (RAP), lateral displacement of the head (ALH) and membrane integrity (MI) were significantly higher (P< 0.05) in high freezability semen compared to the low freezability semen, TM: 56.77 vs 29.84%; MP: 35.6 vs 10.96%; VAP: 123.52 vs 99.29μm/s; VSL: 107.62 vs 86.28μm/s; VCL: 189.84 vs 155.63μm/s; RAP: 50.65 vs 25.04%; ALH: 6.88 vs 6.06μm; MI: 40.66 vs 29.12% of intact cells, respectively. The addition of Trolox at 0, 50, 100, 150 and 200μM had a significant (P<0.05) effect on TM, PM and RAP in the high freezability semen samples. The evaluated parameters from the above cited samples with respective concentrations were; TM: 66.57, 71.14, 62.71, 43.57 and 39.85%, PM: 43.2, 48.7, 40.4, 23.71 and 21.0%; RAP: 61.5, 65.7, 56.4, 38.8 and 30.7%. Trolox concentrations above 100μM/10^8 spermatozoa induced a decrease in sperm motility parameters. A toxic effect for bovine embryos was observed when a high concentration of Trolox was used (above 400μM)[2] and the high Trolox concentrations (150 and 200μM) used in the present study were possibly toxic for ram spermatozoa. Nevertheless, low freezability spermatozoa seem to be less vulnerable to Trolox. The low freezability semen showed no significant effect (P>0.05) of the Trolox addition on the sperm motility parameters. In this group, the TM were: 29.4, 33.6, 30.4, 28.8 and 27.0%; MP: 15.6, 12.2, 10.0, 8.4 and 8.6% for 0, 50, 100, 150 and 200μM of Trolox, respectively. Lipid peroxidation in the ram spermatozoa was negatively correlated with sperm motility.[3] The spermatozoa with low freezability in the present study were possibly undergoing high oxidative stress due to the excessive generation of reactive oxygen species, which are known to originate from dead and defective spermatozoa. In this case, the addition of an antioxidant may have been insufficient to overcome the effect of the reactive oxygen species.

[1]Maia et al. 2005. Acta Scientiae Veterinariae 33 (Supl 1):s311.
[2]Feugang et al. 2004. Theriogenology 61:71-90.
[3]Kasimanickam et al. 2005. Theriogenology, in press.

18. Simultaneous evaluation of plasma membrane integrity, acrosome reaction and mithocondrial membrane potential in ram spermatozoa by a combination of fluorescent probes

HC Azevedo[a], SD Bicudo[b], DB Sousa[b], MS Maia[b] & CC Sicherle[b]

[a]Embrapa Tabuleiros Costeiros, Aracaju, Brazil and [b]Faculdade de Medicina Veterinária e Zootecnia da Universidade Estadual Paulista, Brazil

The purpose of this study was to assess sperm quality by the combined use of fluorescent probes which can determine simultaneously the plasma membrane integrity, acrosome reaction and mitochondrial membrane potential in ram spermatozoa. One ejaculate was collected from each of 25 Santa Inês rams. The semen samples were diluted at 32°C with a Tris extender[1] to obtain a final concentration of 400 x 10^6 spermatozoa/mL. The diluted semen was packaged in 0.25mL straws, cooled in an automatic freezer (TetakonÒ-TK 3000) programmed to refrigerate the semen at 0.25°C/minute down to 5°C and to maintain this temperature for 120 minutes[2]. Samples of fresh and diluted/cooled semen were submitted to an evaluation of sperm membranes by simultaneous assessment of plasma membrane integrity, acrosome reaction and mitochondrial membrane potential according to the technique described by Celeghini (2005)[3] with some modifications. A sample of fresh or diluted/cooled semen was diluted in 500μL of X-Cell® previously heated to 37°C, keeping the sperm concentration at 4 x 10^6 spermatozoa/mL. To the sample the following were added: 1.5 μL of propidium iodide (PI – 0.5 mg/mL in PBS), 1.5 μL of 5,5',6,6'tetrachloro-1,1',3,3'- tetraethilbenzimidazolyl carbocyanine iodide (JC-1 - 153 M in DMSO) and 25 μL of *Pisum sativum* agglutinin conjugated with fluorescein isothiocyanate (FITC-PSA - 100 g/mL in 1 % sodium azide in PBS). The mixture was homogenized whilst protected from the light and after 8 minutes of 37°C incubation, 10μL of this were removed, placed on a previously heated slide (37°C) and covered with a coverslip. A microscope with epifluorescence illumination was used to read at 1000x magnification under oil. A total of 100 cells were counted for each slide and distributed in eight categories. The averages of spermatozoa with intact plasma membranes measured by PI in fresh and diluted/cooled semen were 85.8 and 59.4% respectively. The averages of cells with intact acrosomes measured by FITC-PSA in fresh and diluted/cooled semen were 92.4 and 76.0% respectively. The proportions of cells with high and low mitochondrial membrane potential determined by JC-1 were 79.6 and 20.4% in fresh semen and 19.8 and 80.2% in diluted/cooled semen respectively. The procedure to assess simultaneously the plasma membrane integrity, acrosome reaction and mitochondrial membrane potential combining fluorescent probes such as IP, JC-1 and FITC-PSA is suitable for application to and interpretation of fresh and diluted/cooled ram semen.

Acknowledgements: FAPESP for the project financial support; EMBRAPA and UNESP for staff support.

[1]Maia *et al.* 2005. Acta Scientiae Veterinariae 33, Suppl. 1 :311.
[2]Azevedo *et al.* 2005. Abstract 129-177, XVI Brazilian Congress of Animal Reproduction.
[3]Celeguini, ECC, 2005. PhD. Dissertation, University of São Paulo, Brazil, 186p.

19. Determination of capacitation in ram spermatozoa using a modified chlortetracycline fluorescence assay

HC Azevedo[a], SD Bicudo[b], MS Maia[b], DB Sousa[b] & CC Sicherle[b]

[a]Embrapa Tabuleiros Costeiros, Aracaju, Brazil and [b]Faculdade de Medicina Veterinária e Zootecnia da Universidade Estadual Paulista, Brazil

The chlortetracycline test (CTC) has been applied to ram spermatozoa for classifying the cellular population into three categories: uncapacitated with intact acrosome (F), capacitated with intact acrosome (B) and acrosomal reaction sperm (AR)[1]. The purpose of this study was to test some modifications of CTC technique for a better evaluation of both the capacitation and acrosomal reaction in recently ejaculated and diluted/cooled ram spermatozoa. Ejaculates were collected from 25 Santa Inês rams and processed according to Azevedo et al. (2005)[2]. The semen samples were diluted at 32°C with a Tris extender to obtain a final concentration of 400 x 10^6 spermatozoa/mL, packaged in 0.25mL straws and cooled in an automatic freezer (Tetakon®-TK 3000) programmed to refrigerate the semen at 0.25°C/minute to 5°C and to maintain this temperature for 120 minutes. Semen samples with 24 x 10^6 spermatozoa were diluted in 1000 μL (q.s.p.) of PBS previously heated at 37°C and submitted to centrifugation at 900 g for 4 minutes to remove the seminal plasma or extender. The sperm pellet was then resuspended in 150 μL of PBS and an aliquot of 10 μL was mixed with 10 μL of 1mM CTC prepared daily in a stock solution containing 20mM of Tris, 130mM of NaCl and 5mM of L-Cysteine (pH 7.8). The mixture was slowly homogenized for 20 seconds and then it was added to 10 μL of 1% of glutaraldehyde solution prepared in 2M of Tris and corrected with HCl (pH 7.8). A sample of 10 μL of this suspension was placed on a previously heated slide (37°C) and mixed with 10iL of 0.22 M 1,4-diazabicyclo[2.2.2]octane (DABCO) solution prepared in a medium composed of PBS and glycerol in a 1:9 proportion. The mixture was covered with coverslips, compressed firmly with absorbent paper to remove the fluid excess, sealed with colorless enamel and stored at 4°C in the dark. The samples were evaluated within 1 hour using a microscope with epifluorescence illumination at 1000x magnification under oil. A total of 100 cells were counted for each slide and distributed into 3 categories: F, B and AR. The centrifugation allowed a clear visualization principally of diluted/cooled semen and all CTC patterns were found in all samples. The averages of spermatozoa classified as F, B and AR were respectively 78.3, 12.2 and 9.5% in fresh semen and 58.9, 27.9 and 13.2% in diluted/cooled semen. A typical distribution of the CTC pattern was verified in fresh semen samples with more than 75% of uncapacitated acrosome intact cells[3]. The frequency of the spermatozoa with F pattern was higher than observed by Gillan et al. (1997)[1]. The data obtained in this work agreed with the ones found in the literature and validate the modifications made to the CTC technique for its use in fresh and diluted/cooled ram semen.

Acknowledgements: FAPESP for the project financial support; EMBRAPA and UNESP for staff support.

[1]Gillan et al. 1997. Reproduction Fertility and Development 9:481-487.
[2]Azevedo et al. 2005. Abstract of XVI Brazilian Congress of Animal Reproduction 129-177.
[3]Pérez et al. 1996. Theriogenology 46:131-140.

20. Heterospermic mixture of three Hereford bulls' semen does not improve fertility compared to homospermic semen

H Voges[a], E Newey[a], J Melis[a] & R Vishwanath[b]

[a]Livestock Improvement Corporation, Private Bag 3016, Hamilton, New Zealand and [b]AgResearch Ruakura, Private Bag 3123, Hamilton, New Zealand

The insemination of a female with admixtures of semen from more than one male is called heterospermic insemination. Heterospermic insemination has been reported to increase overall fertility in some species. While most studies in cattle have been equivocal[1], a number of trials have demonstrated improved conception of up to 11%. The aim of this experiment was to conduct a large-scale trial and determine unequivocally whether heterospermic semen conferred any fertility advantage over homospermic semen.

Semen was collected from three Hereford bulls (H1, H2 & H4) over several months. After pooling ¼ of the ejaculates of each of the three bulls on collection day, the heterospermic and remaining matching homospermic semen batches were processed and frozen in 0.25ml French ministraws at 60×10^6 sperm/ml. The semen was despatched to dairy herds across New Zealand over three consecutive mating seasons (2001-2003) as part of Livestock Improvement's Beef Pak artificial insemination (AI) product. Cow calving and culling information was collected in the subsequent seasons. Herds with incomplete calving records were excluded from the data, as were heifer matings and AI outside the principal spring mating season (October-December). Insemination and calving data for matching batches from all 3 bulls and the corresponding heterospermic batch was available for 15 collection dates. Finally, all cows that calved between 268 and 292 days after AI with trial semen were considered pregnant (**P**) to that insemination. Cows that calved > 292 days post-AI and cows with an "empty" ("MT") cull code were considered not pregnant (**NP**). All remaining culled cows and cows calving < 262 days post-AI were also excluded from the final dataset.

Over 3500 inseminations from each of the control bulls and the heterospermic semen batches qualified as either P or NP. The final dataset thus included 14,532 inseminations carried out in 1205 herds, making this the largest trial to assess the overall fertility effect of heterospermic insemination in cattle. In addition to bull/heterospermic treatment, collection date and AI season, cow location, breed and age were recorded. The calving to mating interval (prior to trial insemination) was also calculated. The data was analysed using a generalized linear models procedure with calving rate (CR) as outcome variable.

While all variables had highly significant univariate effects on CR, even distribution across herds throughout NZ ensured that the bull/heterospermic treatment was generally unaffected by the other measurements. The GLM analysis including bull/heterospermic treatment, cow breed and treatment X breed interaction was highly significant for all factors ($p < 0.001$). Bulls H1 and H2 had significantly better CR than H4 (57.9% & 55.9% vs 50.3% respectively; $p < 0.001$). The heterospermic CR (54.6%) did not differ significantly from H1 and H2, but was significantly higher than bull H4 ($p < 0.001$). Jersey cows (> 75% Jersey) had lower a CR than both Holstein-Friesian (> 75% H-F) and mixed breed cows at 51.1% vs 55.6% and 55.3% respectively ($p < 0.01$). The results of this large-scale study clearly demonstrate that mixing semen from several bulls did not improve fertility over homospermic semen batches.

[1]Revell, 1993. The Veterinary Record 133 (1):20.

21. Cryopreservation of epididymal alpaca sperm in pellets and straws frozen on dry ice or in liquid nitrogen vapour

KM Morton, R Bathgate, G Evans & WMC Maxwell

Centre for Advanced Technologies in Animal Genetics and Reproduction (ReproGen), Faculty of Veterinary Science, The University of Sydney, Australia 2006

Alpaca ejaculates contain a highly viscous seminal plasma which traps sperm and makes handling, dilution and cryopreservation difficult. As a consequence, the protocol for the cryopreservation of alpaca sperm is not well established. The use of epididymal sperm avoids the viscous seminal plasma and provides a model for establishing protocols for the cryopreservation of ejaculated semen. Epididymal sperm can also be used to establish gene banks for endangered camelids and deceased males of high genetic value. Despite this, there are limited reports on the frozen storage of epididymal alpaca sperm[1,2] and the development of freeze and thawing protocols are still in their early stages.

Epididymides and testes were obtained from males after castration and transported to the laboratory (approx. 6 h). Sperm were recovered by mincing the epididymides with a scalpel and allowed to swim out for 30 mins into Androhep (AH; Minitube, Germany). Sperm were frozen using a modified Westendorf method[3] with a cooling extender (11 % lactose supplemented with 20 % egg yolk) and a freezing extender consisting of cooling extender supplemented with 9 % glycerol and 1.5 % sodium dodecyl sulfate. Briefly, sperm suspensions were centrifuged (300 g; 10 mins), the pellet resuspended to a final volume of 2 mL with cooling extender and cooled to 4°C over 2 hours (-0.14°C/min). Cooled sperm were further diluted with 1.0 mL freezing extender and frozen as pellets (250 μL) on dry ice, or loaded into 0.25 mL or 0.5 mL straws and frozen on dry ice or in liquid nitrogen vapour (10 cm above liquid nitrogen for 10 mins then at 2 cm for 5 mins). Motility and acrosome integrity (FITC-PNA) were recorded after sperm harvest, cooling and at 0 and 3 h post-thaw (35°C). Statistically significant differences were determined by ANOVA and means compared using LSD.

Sperm motility after harvest was 37.5 $\pm$ 11 % and was similar after cooling to 4°C. Motility immediately (0 h) post-thaw was higher for sperm frozen as pellets (27.5 $\pm$ 8.5 %), than in 0.25 mL (6.3 $\pm$ 2.4 %) or 0.5 mL (6.3 $\pm$ 4.7 %) straws (on dry ice), or 0.25 mL (10.0$\pm$3.5%) and 0.5 mL (11.7 $\pm$ 4.4 %) straws (in liquid nitrogen vapour). Motility at 3 h post-thaw was similar between groups (< 5 %). Acrosome integrity was similar immediately after harvest and cooling to 4°C (89.3 $\pm$ 3.8 %) and did not differ between groups at 0h (range: 81.0 - 91.0 %) and 3 h (range: 78.3 - 83.0 %) post-thaw.

Pellet freezing resulted in superior motility compared with 0.25 and 0.5 mL straws frozen either on dry ice or in liquid nitrogen vapour. Immediately post-thaw pellet frozen sperm retained most of their original motility, but motility declined by 3 hours post-thaw. While these results are encouraging, further research aimed at improving the post-thaw motility and longevity of epididymal alpaca sperm is required.

Supported by Rural Industries Research and Development Corporation (RIRDC) and Australian Alpaca Association (AAA).

[1]Bravo *et al.* 2000. Proc Int Congress Anim Reprod 2:92.
[2]Morton *et al.* 2006. Proc European Soc Dom Anim Reprod (submitted).
[3]Bathgate *et al.* 2006. Reprod Domest Anim 41:68-73.

22. Effect of alpaca age on testes growth and development, sperm production and post-thaw sperm survival: preliminary results

KJ Bailey, KM Morton, R Bathgate, G Evans & WMC Maxwell

Centre for Advanced Technologies in Animal Genetics and Reproduction (ReproGen), Faculty of Veterinary Science, The University of Sydney, NSW, 2006, Australia

Alpacas have a naturally low reproductive rate and long generation interval. Male alpacas reach sexual maturity between the ages of 2 and 3 years[1], and are selected for breeding at 3 years with testes length of ≥ 3 cm[1]. Selecting males for breeding based on testis size at 14 months of age may result in the use of males with superior sperm production[1] and increased pregnancy rates after natural mating. The current project aims to examine the growth and development of alpaca testes, and to identify possible relationships between age and testis size (length and width), testis development, sperm concentration and post-harvest and post-thaw motility.

Testes and epididymides were collected at castration from male alpacas (n = 18) with an average age of 19.4 ± 0.9 months (range 10 – 25 months). Testes were measured (length and width) and sperm extracted from the epididymides and pellet-frozen[2]. Sperm concentration and motility were assessed post-harvest (pre-freeze) and post-thaw. Testicular development was assessed histologically. Briefly, whole testes were fixed in 10 % buffered formalin and processed through ethanol and xylol before embedding in paraffin wax. Tissue was sectioned at 5 μm, mounted and stained with Whitlock's Haematoxylin and alcohol eosin. Correlation and regression analyses were calculated using Genstat.

Average testis length and width was 3.5 ± 0.2 cm and 2.4 ± 0.1 cm, respectively. Sperm concentration was 55.1 ± 18.1 x10^6 per mL but this varied considerably between males (range: 0 – 265 x 10^6 per mL). Post-harvest motility was 23.7 ± 5.8 % and post-thaw motility was 12.1 ± 4.2 %. Alpaca age was highly related to both testes length (P = 0.0039), and testes width (P = 0.0094). Testis length and width were highly related (P = 0.0021). There was no relationship between age and sperm concentration, pre-freeze or post-thaw motility. Testis length was not related to sperm concentration (P > 0.05).

These results demonstrate the relationship between alpaca age and testis growth (length and width). Interestingly, there was no relationship between age and testis development or sperm concentration. Motile sperm were harvested from a 15 month male, while three males over 20 months of age produced no sperm suggesting individual variation in testicular development and timing of sperm production. Selecting males with testes length ≥ 3 cm does not provide a reliable indicator of sperm production. Further research is required to elucidate the patterns of testicular growth and development in male alpacas, and to determine the optimal breeding age, and testis size of male alpacas.

Supported by RIRDC and the Australian Alpaca Association.

[1]Galloway, 2000. Proc Aust Alpaca Ass 1:21-23.
[2]Morton et al. 2006. Proc European Soc Dom Anim Reprod (submitted).

23. Birth of offspring of pre-determined sex after artificial insemination of frozen-thawed, sex-sorted and re-frozen-thawed ram spermatozoa

SP de Graaf[a], G Evans[a], WMC Maxwell[a], DG Cran[b] & JK O'Brien[a]

[a]Faculty of Veterinary Science, The University of Sydney, NSW 2006, Australia and [b]XY (UK) Ltd.
Scottish Agricultural College, Aberdeen, Scotland

Accurate predetermination of sex in mammals is made possible by the use of sexed sperm, sorted in a modified flow cytometer into those bearing an X or a Y chromosome[1]. However, the application of this technology to the breeding of livestock and wildlife is limited when the sorter is located a long distance from the male(s) and/or female(s). In an attempt to overcome this problem, it was postulated that sperm which had been previously frozen could be thawed, sex-sorted and re-frozen to be later used for IVF and embryo transfer or artificial insemination. This method has been successfully used for *in vitro* sheep embryo production[2] but has yet to be tested in artificial insemination (AI) programs. The following experiment was conducted to determine the fertility of frozen-thawed, sex-sorted, and re-frozen-thawed ram sperm after AI.

Semen was collected from three Merino rams and split into two parts for sex-sorting and freezing (SF) or, in the case of the control, standard freezing (Control). Portions of frozen semen from the control group were then thawed, processed for sex-sorting and re-frozen (FSF). Intrauterine inseminations were conducted using commercial laparoscopic techniques in 292 mature Merino ewes in synchronised oestrus. Ewes allocated semen from each treatment group received 15×10^6 motile sperm ($Control_{15}$, SF_{15} & FSF_{15}), with the exception of a second control group which received 50×10^6 motile sperm/ewe ($Control_{50}$). Pregnancy was diagnosed at Day 54 by real-time cutaneous ultrasound, and the number of lambs born per ewe was determined at lambing.

The percentage of ewes lambing after insemination was similar for $Control_{15}$ (36/74; 48.6%), SF_{15} (35/76; 46.1%) and FSF_{15} (26/72; 36.1%) groups (P > 0.05). A higher percentage of ewes produced lambs in the $Control_{50}$ (38/70; 54.3%) than the FSF_{15} group (P < 0.05). There was no fetal loss between Day 54 of pregnancy and lambing. The number of lambs born/ewe ($\pm$ SEM) did not differ (P > 0.05) between $Control_{50}$ (1.5 $\pm$ 0.08), $Control_{15}$ (1.6 $\pm$ 0.09), SF_{15} (1.6 $\pm$ 0.08) and FSF_{15} (1.7 $\pm$ 0.10) groups. Fifty-one of the 55 (92.7%) lambs derived from fresh, sex-sorted frozen-thawed spermatozoa were of the predicted sex. Similarly, 41/43 (95.3%) lambs derived from frozen-thawed, sex-sorted, re-frozen-thawed spermatozoa were of the correct sex.

The lower pregnancy and lambing rate of ewes inseminated with FSF_{15} sperm, as compared to $Control_{50}$ sperm, may be due to the advanced maturation status of FSF sperm and a consequent reduced lifespan inside the female reproductive tract.

These are the first recorded births of live pre-sexed offspring from any species inseminated with frozen-thawed, sex-sorted and re-frozen-thawed sperm, and demonstrate the potential of this technique for application in livestock or wildlife species where animals remain distant from a sex-sorting facility.

Research supported by XY, Inc.

[1]Johnson & Pinkel, 1986. Cytometry. 7:268-273.
[2]Hollinshead *et al.* 2004. Reproduction 127:557-568.

24. Biospeckle technique evaluation of frozen bovine semen using two different laser wavelengths

PHA Carvalho[a], JB Barreto Filho[a], RODS Rossi[a], GF Rabelo[b] & RA Braga Jr[b]

[a]Veterinary Department, Animal Reproduction Laboratory and [b]Optics and Laser Laboratory, Federal University of Lavras, Lavras, Minas Gerais, Brazil

Artificial insemination centers can greatly benefit from laser light and biospeckle evaluation techniques when it comes to eliminating light microscopy shortcomings. The incidence of different wavelengths on biologically active materials such as semen generates successive interference images known as biospeckle. Processing of these images generates an index called inertia moment (IM) which quantifies the activity or viability of the illuminated material.[1]

This study was aimed at testing the sensibility of two different sources of laser light (red laser of Helium-Neonium with 632 nm and 10 mW and green laser of Argonium with 543 nm and 10 mW) for evaluation of frozen bovine semen by the biospeckle technique and comparing the IM generated for each lit sample with the velocity and motility parameters evaluated by light microscopy.

Seventy-six straws with 0.5 ml of bovine semen were thawed and evaluated microscopically as to velocity and motility. Immediately, each sample was lit by either kind of laser light (Helium –Neonium , n = 56; Argonium n = 20). Biospeckle data were captured by a "charge coupled detector" camera and transformed into matrixes (spatial temporal speckle – STS and occurrence matrix – OM) by a specific software. A material presenting a high biologic activity generates an OM in which the pixels representing the kinetics scatter along the diagonal axis of the matrix. The IM was calculated by a function able to quantify the scattering of the pixels of the OM. Statistic data were analysed by Spearman correlation considering microscopically evaluated velocity and individual motility as compared to biospeckle generated IM.

Illumination performed with red and green laser light resulted in an IM ranging from 78 to 311 and 60 to 111, respectively for samples with low and high kinetic standard. A positive correlation ($p < 0.0001$) was found between IM and velocity ($r = 0.59$; $r = 0.80$) and IM and motility ($r = 0.67$; $r = 0.76$) for the samples lit by red and green laser light, respectively.

Results suggest that laser light can be an useful tool for the objective evaluation of semen quality, and that Argonium green laser is more sensitive and consistent in detecting the biologic activity and viability of frozen bovine semen.

[1]Arizaga, 1999. Optics & Laser Technology 4-5:1-7.

25. Easy-*in vitro* fertilization of bovine oocytes

R Urrego[a-b], A Tarazona[b], M Olivera[b] & O Camargo[a-b]

[b]Grupo de Fisiología y Biotecnología de la Reproducción Animal, Facultad de Ciencias Agrarias, Universidad de Antioquia and [a]Grupo de Biotecnología Animal, Universidad Nacional de Colombia - Sede Medellín, Colombia

This study was conducted to evaluate an *in vitro* fertilization (IVF) procedure that avoids the washing-selection-centrifugation of sperm before the encounter with oocytes. A special dish was devised as follows: in the middle of each of the 4-wells of a conventional plastic dish (Nunc, Roskilde) another little ($\approx$50 µl) glass dish was placed (0.3 mm-tall x 0.8 mm-diameter), in such a way that when the larger plate is filled with 700 µl of IVF medium (Fert–TALP), the inner well is also filled up by overflowing.

Ten *in vitro* matured oocytes (10) are placed into the inner well, and 30 µl of thawed semen are placed on the bottom of the outer well. After one hour the medium from the outer well is removed, and the well is incubated for 16 hours. Embryos were cultured in a CR1aa media for 7 days. As a control, a conventional IVF, using the classical Percoll selection was carried out.

The rates of cleavage (70.8 vs. 73.1) and the proportion of oocytes reaching morula and blastocyst stages (18.8 vs. 20.3) by the "easy-IVF" were similar to the results obtained by the classical method ($P < 0.05$). These results indicate that the two methods yield similar results but the "easy-IVF" definitely saves time, human effort, reagents, and reduces the manipulation of the sperm with potential benefits towards improving in vitro reproduction procedures.

26. Effect of centrifugation on plasma membrane and DNA of spermatozoa of bovine

R Urrego[a-b], J Rios[b], M Olivera[a] & O Camargo[a-b]

[a]Grupo de Fisiología y Biotecnología de la Reproducción Animal, Universidad de Antioquia and
[b]Grupo de Biotecnología Animal, Universidad Nacional de Colombia -Sede Medellín, Colombia

Most of protocols of *in vitro* bovine embryo production involve at lest one washing-centrifugation step to clear and select motile sperm cells of thawed semen, both by swim-up and by Percoll. It is accepted that this procedures leads to reactive oxygen species overproduction affecting the performance of sperm cells and eventually embryo development.

The aim of this study was to evaluate the effects of different times of centrifugation on the integrity of the plasma membrane and DNA of the bovine sperm cells, by means of hypoosmotic test (HOST) and assay comet, respectively. The sperm cells were centrifuged to 700 g for 10, 30 and 45 min. Sperm that was not centrifuged was used as negative control while sperm cells exposed to hydrogen peroxide (H_2O_2) to a final concentration of 200 mM, was used as a positive one.

The results of this study indicate that while the integrity of the plasma membrane was not affected by the centrifugation, DNA were affected in an independent manner of centrifugation times ($p < 0.05$). Analysis of variance and Student's t-test were used for statistical analysis of the variables and sample means studied, three replicates were carried out and each treatment was represented in each replicate. These results indicate that centrifugation does not cause injurious effects on plasma membrane of bovine sperm cells but it certainly tends to cause harm to sperm DNA as assessed by fragmentation with undetermined effects on performance of embryo development. Additionally, some degree of DNA fragmentation in sperm cells not centrifuged (negative control) suggested that DNA of bovine sperm cells could be affected by other factors, probably the freezing procedure.

27. Acrosome reaction in cryopreserved bovine semen

AM Tarazona[a], ZT Ruiz[a], Sebastián Isaza[b] & M Olivera[a]

[a]Reproducción Fisiología y Biotecnología ,Facultad de Ciencias Agrarias Universidad de Antioquia and [b]Reprovet Colombia

Examining acrosome reaction (AR) is very useful in bovine reproductive programs. AR complements common semen parameters indicative of fertility in frozen/thawed semen.

We evaluated different protocols to induce the AR *in vitro*. Semen samples from a bull with excellent fertility tested *in vivo* and *in vitro* was used as a positive control. Each sample was divided in aliquots: one to measure spontaneous AR and the other to measure it after induction with four different protocols (2mg/ml heparin, 10% v/v Bovine Calf Serum (BCS), 10μM Calcium ionophore (CI), and BCS plus heparin). The basic medium was Tyrodes, Albumin, Lactate, Pyruvate (TALP) in all groups. The negative control was TALP medium alone. Samples were incubated for 3 hours at 39°C in an atmosphere of 5% CO_2. Three evaluations using *Pisum sativum*-FITC were performed per treatment (at 1, 2 and 3 hours), and three response patterns were differentiated: complete acrosome reaction (CAR), partial acrosome reaction (PAR) and intact acrosome (IA).

All treatments were capable of inducing acrosome reaction but the time course varied: CI induced AR in 1 hour, BCS plus heparin required 2 hours and BCS and heparin induce AR only after 3 hours. This set of protocols was used to probe a bull with normal spermatozoal parameters (800-1200 million/ml concentration, >80% motility and >80% normal morphology) but with a history of low pregnancy rates (<10%). When the AR was induced the results were: CAR 12%; PAR 23%, IA 66%.

We believe then that there is a possible relationship between AR failure and low fertility of frozen semen and conclude that testing for AR is an excellent tool to check bulls with normal sperm parameters but low fertility.

28. Effect of supplementation with bypass fat on maternal ability in beef cows

J Angulo[b], L Mahecha[a] , M Cerón[a], C Giraldo[b] & M Olivera[b]

[a]Grica and [b]Fisiología y Biotecnología de la Reproducción, Facultad de Ciencias Agrarias Universidad de Antioquia, Colombia

We measured the effect of supplementing by-pass fatty acids during the suckling period (7 months with the permanent presence of the calf), in 30 Angus X Zebu (AxZ) and 18 Zebu (Z) cows. Weight and body condition of the mothers and weight of the calves as well as reproductive parameters such as uterine involution, presence of the first dominant postpartum follicle and pregnancy rate at first detected estrus were recorded. Supplementation was stopped at weaning, and during the following 75 days heat detection and pregnancy diagnoses were performed. The cows and the calves grazed on brachiaria with green dry matter (DGM) varying from 1,587 Kg DGM/ha during the rainy period (August-November) to 175 Kg DGM/ha in the transition and the dry period (December-April). The data corresponding to each breed group were analyzed independently using a mixed model with repeated measures within individuals. Supplemented Z cows (n = 12) as well as their controls (n = 6) lost weight throughout the experimental period but the supplemented Z cows maintained a better body condition up to the fourth month (p < 0.05).

Supplemented AXZ cows (n = 20) gained weight during the first three months and kept a better body condition (p < 0.01) than their controls. Thereafter, supplemented AXZ cows lost weight and body condition (p < 0.05) but at a reduced rate compared to their controls. No liveweight differences were observed among the calves. There was no observed differences in uterine involution for any groups.   However, the first dominant follicle was detected sooner in supplemented cows (23 d postpartum) then in control cows (28 d postpartum). The proportion of supplemented cows detected in estrus while suckling calves was low (7 of 32), indicating that the by-pass fatty acid supplementation did not overcome the suckling effect.   Of the treated cows that showed estrus and were bred, 5/6 became pregnant. The majority of supplemented cows (17/26) were observed in estrus during the 75 day-period after weaning, but interestingly only 4/17 became pregnant. Thus, feeding by-pass fatty acids during calf-suckling in primiparous cows improves body condition, but did not overcome the low level of reproductive performance in this type of animal under extensive range management conditions.

29. Feeding heifers with non-protein nitrogen can affect oocyte competence to reach the hatched blastocyst stage after in vitro fertilization

FA Ferreira[a], RG Gomez[b], D Carlino[b], YF Watanabe[b] & PHM Rodrigues[c]

[a]Department of Animal Reproduction, College of Veterinary Medicine, Universidade de São Paulo (FMVZ-USP), São Paulo, SP, Brazil, [b]Grupo Vitrogen®, Cravinhos, SP, Brazil and [c]Department of Animal Nutrition and Production, FMVZ-USP, Pirassununga, SP, Brazil

The hypothesis that feeding cows with high levels of crude protein, associated with high levels of plasma urea nitrogen (PUN), may exert deleterious effects on their fertility has been studied by many researchers (reviewed by Butler, 1998)[1]. In this way, supplementing cows with urea, a nitrogenous compound, can reduce their reproductive performance. Our objective was to investigate the effect of feeding heifers with an excessive nitrogen content diet on the competence of their oocytes to develop during in vitro embryo production (IVEP). Forty bovine crossbred heifers were used twice (n=80) during an eight week trial, designed in randomized blocks, in two treatments: with urea (U) or without urea (NU). Every other week, two groups of five animals each were allocated to the treatments and the diets (corn silage + concentrate mixture, 1.4 kg of dry matter and 0.88 kg of dry matter/animal/day, respectively) were offered to the animals, once a day, during six days. The two concentrate mixtures had the same composition (finely ground corn, soybean meal, salt and minerals) except for urea, present only in the U group's diet (66 g of urea/kg DM of concentrate mixture). On the last day of each experimental period, the heifers were submitted to transvaginal ovum pick up (TOPU) and had their blood sampled twice (at fasting and 3 hours post-feeding) for later determination of PUN. Only viable oocytes (grade I- homogeneous ooplasm and compact cumulus cells with many layers; grade II- fewer cumulus cells than grade I; grade III- heterogeneous ooplasm with few cumulus cells) were used for IVEP, which was performed according to standard procedures, considering day 0 the fertilization day. ANOVA was used to analyse embryo data and repeated measures ANOVA to analyse PUN data. Despite a significant PUN rise in the U group (mean ± SEM: 22.12 mg/dl ± 0.86 vs. 31.31 mg/dl ± 1.13, for NU vs. U respectively, $P<0.0001$; there was no interaction between treatments and different times of blood sampling, $P=0.1355$), no treatment effects were observed, either on cleavage ratio assessed at day 3 (mean ± SEM: 0.57 ± 0.03 vs. 0.59 ± 0.03, for NU vs. U respectively, $P=0.7312$) or in blastocyst ratios, assessed at day 6 (0.26 ± 0.03 vs. 0.22 ± 0.037, for NU vs. U respectively, $P=0.3374$), at day 7 (0.33 ± 0.03 vs. 0.34 ±0.03, for NU vs. U, respectively, $P=0.8443$) or at day 9 (0.34 ± 0.03 vs. 0.37 ± 0.04, for NU vs. U respectively, $P=0.5914$). However, there was a significant treatment effect on the hatched blastocyst (day 11)/total blastocyst (day 9) ratio (mean ± SEM: 0.83 ± 0.052 vs. 0.64 ± 0.069 for NU vs. U, respectively, $P=0.0407$). This might mean a possible influence of high PUN on oocyte gene expression during follicle development affecting late embryonic development. In conclusion, the effect of feeding urea on reproduction in heifers seems to reduce the embryo's capacity to hatch from the zona pellucida and would be expected to impede further development in vivo.

Acknowledgements to FAPESP and CAPES for grants, Grupo Vitrogen®, DVM. Márcio L. Ferraz, Dr. Flávio V. Meirelles, Dr. Paula M. Meyer, Dr. João A. Negrão and Laboratory Technicians Ari de Castro, Gilson de Godoy, Sandra Oliveira and Simi Aflalo.

[1]Butler, 1998. J Dairy Sci 81:2533-2539.

30. Quantity and quality of oocytes recovered by ovum pick up from heifers fed with urea are not affected by high levels of plasma urea nitrogen

FA Ferreira[a], RG Gomez[b], D Carlino[b], YF Watanabe[b] & PHM Rodrigues[c]

[a]Department of Animal Reproduction, College of Veterinary Medicine, Universidade de São Paulo (FMVZ-USP), São Paulo, SP, Brazil, [b]Grupo Vitrogen®, Cravinhos, SP, Brazil and [c]Department of Animal Nutrition and Production, FMVZ-USP, Pirassununga, SP, Brazil

Feeding cows with urea as an additional and cheap source of non-protein nitrogen may damage their reproductive performance, since it leads to a raise in plasma urea nitrogen (PUN). Rhoads *et al.* (2006)[1] suggested that a potential harmful effect of PUN could affect either oocyte or early embryo development. This study aimed to verify if feeding heifers with urea would influence the total quantity and quality of oocytes recovered by transvaginal ovum pick up (TOPU). Forty bovine crossbred heifers were used twice (n = 80) during an eight week trial, designed in randomized blocks, in two treatments: with urea (U) or without urea (NU). Each experimental period lasted six days, in which two groups of five animals each were allocated to the treatments. Diets, composed of corn silage (1.4 kg of dry matter/animal/day) and concentrate mixture (0.88 kg of dry matter/animal/day), were offered daily to the animals during the whole experimental period. The two concentrate mixtures had the same composition (finely ground corn, soybean meal, salt and minerals) except for urea, present only in the U groups' diet (66 g of urea/kg of dry matter of concentrate mixture). On the last day of each experimental period, heifers were submitted to TOPU and had their blood sampled twice (at fasting and 3 hours post-feeding) for later determination of PUN. All recovered oocytes were counted and classified as viable (grade I- homogeneous ooplasm and compact cumulus cells with many layers; grade II- fewer cumulus cells than grade I; grade III- heterogeneous ooplasm with few cumulus cells) and unviable (degenerated, atretic and expanded). ANOVA was used to analyse oocyte data and repeated measures ANOVA to analyse PUN data. A significant treatment effect regarding PUN data was observed (mean $\pm$ SEM: 22.12 mg/dl $\pm$ 0.86 vs. 31.31 mg/dl $\pm$ 1.13, for NU vs. U, respectively, $P < 0.0001$; there was no interaction between treatments and different times of blood sampling, $P = 0.1355$). However, no treatment effects were observed in oocyte data regarding the total number recovered (mean $\pm$ SEM: 9.15 $\pm$ 0.82 vs. 8.82 $\pm$ 0.95, for NU vs. U, respectively, $P = 0.5027$), number of viable oocytes (mean $\pm$ SEM: 7.03 $\pm$ 0.82 vs. 6.31 $\pm$ 0.83, for NU vs. U, respectively, $P = 0.3552$) and viable/total ratio (mean $\pm$ SEM: 0.74 $\pm$ 0.02 vs. 0.70 $\pm$ 0.02, for NU vs. U, respectively, $P = 0.3180$) or the number of oocytes classified in different quality grades (mean $\pm$ SEM: GI- 0.73 $\pm$ 0.28 vs. 0.64 $\pm$ 0.24, GII- 0.98 $\pm$ 0.24 vs. 0.56 $\pm$ 0.15, GIII- 5.33 $\pm$ 0.51 vs. 5.10 $\pm$ 0.62, Degenerated- 1.2 $\pm$ 0.18 vs. 1.49 $\pm$ 0.26, Atretic- 0.76 $\pm$ 0.13 vs. 0.79 $\pm$ 0.14, Expanded- 0.15 $\pm$ 0.07 vs. 0.23 $\pm$ 0.07, for NU vs. U, respectively, $P > 0.05$). Therefore, any deleterious effect of PUN on cattle reproduction was not evident in terms of oocyte growth, at least not regarding the quantity and quality of oocytes recovered by TOPU.

Acknowledgements to FAPESP and CAPES for grants, Grupo Vitrogen®, DVM. Márcio L. Ferraz, Dr. Flávio V. Meirelles, Dr. Paula M. Meyer, Dr. João A. Negrão and Laboratory Technicians Ari de Castro, Gilson de Godoy, Sandra Oliveira and Simi Aflalo.

[1]Rhoads *et al.* 2006. Anim Reprod Sci 91 (1-2):1-10.

31. Nutritional supplementation improves the reproductive response of female goats to the male effect in extensive conditions

MA De Santiago-Miramontes[a], FG Véliz[a], R Rivas-Muñoz[a], JD Hernández-Bustamante[a], M Muñoz-Gutiérrez[c], RJ Scaramuzzi[b], & JA Delgadillo[a]

[a]Centro de Investigación en Reproducción Caprina, Departamento de Ciencias Médico Veterinarias, Universidad Autónoma Agraria Antonio Narro, Torreón, Coahuila, Mexico, [b]Department of Veterinary Basic Sciences, Royal Veterinary College, London, UK and [c]Departamento de Biología de la Reproducción, Universidad Autónoma Metropolitana Iztapalapa, 09340 Mexico City, Mexico

We investigated if short-term nutritional supplementation improved sexual activity of anoestrous female goats subjected to the male effect in extensive grazing conditions in sub-tropical Mexico (26° N). Two groups of 25 female goats were used with two sexually active bucks per group. One group received no supplementation (NS), while the other group was supplemented (S) for seven days before the introduction of the males with 290 g of rolled corn, 140 g of soya bean and 950 g of alfalfa hay per animal per day. Oestrus was recorded daily and ovulation rate was determined by ultrasonography 5 and 18 days after introduction of the bucks.

Over the first 5 days following introduction of the bucks, more females displayed oestrus in the S (23/25) than in the NS group (15/25; $P < 0.01$) and the proportion of females ovulating increased (22/25 vs. 16/25, S and NS respectively; $P < 0.05$). Ovulation rate of does ovulating over the first 5 days was higher ($P < 0.05$) in S (1.6 ± 0.2) compared to NS (1.0 ± 0.2) group. However, between days 6 and 15, the supplementation did not ($P > 0.05$) affect the proportion of females in oestrus (S: 23/25; NS: 21/25) or the number of females ovulating (S: 21/25; NS: 20/25) or their ovulation rate (S: 1.44 ± 0.1; NS: 1.36 ± 0.1 ovulations).

We conclude that seven days of nutritional supplementation increased the proportion of does displaying oestrus and stimulated ovulation rate of female goats exposed to the male effect. The nutritional effect was immediate and affected the first ovulation in response to the male effect but the stimulation was transient and these improvements were not observed at the second ovulation 6 to 15 days after the introduction of the bucks.

32. Nutrition improves the mating potential of Merino rams

C Viñoles[a], J Olivera[b], J Gil[c], S Fierro[d], I De Barbieri[a] & F Montossi[a]

[a]National Research Institute for Agriculture, km 386, Ruta 5, Tacuarembó, Uruguay, [b]Faculty of Veterinary Medicine, Paysandú, Uruguay, [c]DILAVE, Paysandú, Uruguay and [d]Undergraduate student at the Faculty of Veterinary Medicine

Feeding rams with lupins increases testicular size and sperm production[1]. Although the scrotal circumference and the serving capacity determine the mating potential of a ram[2], it is not clear if nutrition can improve the mating potential. The aim of this study was to evaluate the effect of different sources of energy and protein on the reproductive potential of young Merino rams.

In March 2005, thirty-two 17-month old fine Merino rams were divided in 4 groups ($n = 8$) homogeneous in body condition, live weight, scrotal circumference and serving capacity. The experimental design was a factorial with two factors: type of pasture (native (NP) or improved (IP) pasture) and supplement (with (+S) or without (-S) supplement). Rams grazed continuously during 14 weeks in a NP (30 ha, stocking rate adjusted to 4 animals/ha) and on an IP (*Lotus corniculatus* cv. INIA Draco; 4 ha, stocking rate: 8 animal/ha). The supplement (0.75% of the live weight) was a mixture of sorghum (70%) and soybean meal (30%). Live weight and scrotal circumference were evaluated every 2 weeks, and the serving capacity every 4 weeks using the Laborde test[3]. All animals were castrated on week 14 and the weight of testicles and epididymis registered. Variables were analysed using analysis of variance in Genstat and the mixed procedure in SAS. Values were considered significant if $P < 0.05$.

Live weight and scrotal circumference were affected by pasture and supplement over time ($P < 0.001$). From week 8, rams grazing on IP-S (62 ± 1.0 kg), IP+S (63 ± 1.0 kg) and NP+S (62 ± 1.0 kg) were heavier than those grazing on NP-S (59 ± 1.0 kg). From week 10, the scrotal circumference was larger in IP-S (30 ± 0.4 cm), IP+S (30 ± 0.4 cm) and NP+S (31 ± 0.4 cm) than in NP-S (29 ± 0.4 cm). The serving capacity (4.3 ± 0.4 and 3.1 ± 0.4 services/40 minutes), thus mating potential (76.6 ± 6.3 and 57 ± 6.3 ewes/ram) decreased from March to June ($P < 0.01$). The supplement maintained the mating potential elevated (+S: 74.3 ± 5.4 and -S: 61.7 ± 6.3 ewes/ram; $P = 0.052$). The weight of testicles and epididymis was affected by pasture and supplement ($P < 0.05$), while the interaction between the two factors affected testicular weight ($P < 0.05$). On week 14, rams grazing on NP-S had lighter testicles (209 ± 17 g) and epididymis (44 ± 2 g) compared to rams grazing on IP-S (284 ± 17 g and 53 ± 2 g), IP+S (291 ± 17 g and 58 ± 2 g) and NP+S (283 ± 17 g and 56 ± 2 g, for testicles and epididymis respectively; $P < 0.05$).

A high plane of nutrition improves mating potential probably due to the actions of insulin and IGF-I on the responsiveness of testicular tissue to gonadotrophins, since the effect of nutrition on LH and FSH concentrations is acute[1].

The author's whish to thank the staff of INIA's Research Station and the students in thesis from the Faculty of Veterinary Medicine for they're invaluable help.

[1]Blache *et al.* 2002. Reproduction 59: 219-233.
[2]Blockey & Wilkins, 1984. In: Reproduction in sheep: 53-58.
[3]Ibarra *et al.* 2000. Small Rumin Res 37: 165-169.

33. Are high energy diets safe for breeding rams?

LMJ Schwalbach, N Bester, HJ van der Merwe, KC Lehloenya & JPC Greyling

Department of Animal, Wildlife and Grassland Sciences, University of The Free State, P.O. Box 339, Bloemfontein 9300, South Africa

In South Africa it is common to feed high-energy diets to breeding rams for various reasons. Rams are often fed high-energy diets during intensive growth performance test trials and in preparation for shows and sales - as fat rams are usually sold at higher prices than lean rams of similar quality at auctions. However, very little is known about the effect of such diets on the semen characteristics of rams. Anedoctal evidence suggests that the feeding of high-energy diets and the over-conditioning of rams reduces semen quantity, quality and freezing ability.

A study to evaluate the effects of high dietary energy levels on the semen quantity, quality and freezibility of ram semen was conducted during the natural breeding season. Twenty-four 11 to 12 month old Dorper rams were divided into two similar groups of 12 rams each and fed at two energy levels: Low Energy (LE; 6.52 MJ ME/kg DM) and High Energy (HE; 9.39 MJ ME/kg DM) for 127 days. The body weight of the rams was monitored weekly and at the end of the trial period, semen was collected from each ram (with the aid of an artificial vagina) and cryopreserved using a one-step dilution (1 + 4) technique using Salomon's medium containing 5% glycerol. The fresh and frozen (post-thawed) semen samples were evaluated using standard laboratory techniques for quality (overall and progressive sperm motility, percentage live and normal sperm cells) and the results (for both fresh and post-thawed semen), compared statistically between the two treatment groups, using ANOVA procedures of SAS.

The rams in the HE group grew significantly ($P < 0.05$) faster (ADG 229 vs 112 g/d), were heavier at the end of the trial (BW 71 vs 56.5 kg) and recorded a larger mean scrotal circumference (35.5 vs 29.5 cm). Ejaculate volume (1.2 vs 0.8 ml), overall motility (77 vs 84%), sperm concentration (1.83 vs 1.85 x10^9 sperm/ml) and perentage live and normal cells (64 and 91% vs 70 and 96%, respectively) were not statistically different between the two treatment groups. Similarly no significant differences were recorded between the two groups in the frozen semen samples. Overall motility (57 vs 60%), and percentage live sperm (50 vs 60%) of the frozen semen (post thawing) were lower in the HE than in the LE group respectively, although these differences were also not statistically significant.

It could be concluded that conditioning yearling Dorper rams for a period of 4 months with a diet containing up to 9.39 MJ ME/kg DM does not seem to have any detrimental effect on their semen quantity, quality and freezing ability. However, further research on the effect of HE diets (at even higher concentrations and for longer periods of time) on the quantity, quality and feezing ability is warranted to evaluate the use of such diets to condition rams intended for breeding or for semen collection (and freezing) for artificial insemination.

34. Changes in plasma ghrelin in response to a glucose challenge are associated with changes in plasma fatty acid and insulin, but not glucose

JR Roche[a], AJ Sheahan[a], L Chagas[a], DP Berry[b] & RC Boston[c]

[a]Dexcel, Hamilton, New Zealand, [b]Teagasc Moorepark, Fermoy, Co. Cork, Ireland and [c]Dept of Clinical Studies, University of Pennsylvania, USA

Dry matter intake is probably the first limiting factor for both production and reproduction in domesticated ruminants. Plasma concentrations of the orexigenic hormone ghrelin increase during fasting and fall following feeding, but the mechanisms of control are largely unknown. Schaller et al.[1] reported that postprandial changes in ghrelin were not a result of either changes in glucose or insulin. We hypothesised that the effect is related to the insulin-mediated decline in plasma fatty acids following a meal, rather than associated changes in either glucose or insulin.

Ten grazing Holstein-Friesian dairy cows of two divergent genetic strains and offered either 0, 3 or 6 kg DM of high starch concentrates (~ 70% starch) were subjected to an intravenous glucose challenge (300 mg D-glucose/kg body weight) at on average day 21 of lactation. Prior to infusion and at regular intervals following infusion plasma glucose, insulin, non-esterified fatty acids (NEFA) and ghrelin concentrations were monitored. All metabolites and hormones were positively skewed and optimal transformations were determined using Box-Cox methodology. In an effort to isolate associations amongst these metabolic patterns generalised linear models were used and the significance of the relationship determined using the F-test. All metabolites and hormones were treated as continuous.

Glucose levels rose from a mean ($\pm$ standard deviation) of 3.3 ($\pm$0.41) to 14.6 ($\pm$1.23) mmol/L within two minutes of infusion, declining to baseline concentrations within 90 minutes. Plasma insulin concentrations were quadratically associated with blood glucose ($r^2 = 0.88$; $P < 0.001$), peaking at 27.2 ($\pm$7.54) microU/L 10 ($\pm$4.1) minutes after infusion before declining less rapidly to baseline levels (2.1 $\pm$0.52 microU/L) after 90 minutes. Glucose's insulin-mediated suppression of lipolysis reduced plasma NEFA concentrations until 29 ($\pm$14.8) minutes post infusion, after which NEFA concentrations rose rapidly. A quadratic regression on insulin explained 36% of the variation in NEFA.

Plasma ghrelin concentrations declined post-infusion before reaching a nadir at 22 ($\pm$9.7) minutes post-infusion. An inverse quadratic relationship was evident between plasma ghrelin and glucose ($r^2 = 0.40$; $P < 0.001$) and insulin ($r^2 = 0.34$; $P < 0.001$). In comparison, a positive quadratic association existed with plasma NEFA ($r^2 = 0.37$; $P < 0.001$), with plasma ghrelin rising with increasing NEFA at a proportionally greater rate at increasing concentrations. Following the inclusion of the effect of strain and feeding treatment as well as the significant interaction between feeding treatment and NEFA, the r-square increased to 0.58.

A multivariate analysis of ghrelin revealed that following adjustment for strain and feed system, a quadratic regression on NEFA explained a greater proportion of the remaining variation in plasma ghrelin than insulin, while blood glucose concentration was not significant in the model. Although NEFA, glucose and insulin are interrelated, the multivariate analysis suggests that NEFA plays a more important role in the regulation of ghrelin production than either glucose or insulin.

[1]Schaller et al. 2003. Diabetes 52:16-20.

35. Effect of varied prepartum diets on postpartum plasma IGF-1 and interval to first ovulation in Holstein-Friesian dairy cows

TE Moyes[a], CR Stockdale[b] & KL Macmillan[a]

[a]Department of Veterinary Science, The University of Melbourne, Werribee, Victoria, Australia and
[b]Primary Industries Research Victoria, Kyabram, Victoria, Australia

Plasma concentrations of IGF-1 can be decreased by restricting feed intake or increased by offering high energy diets to lactating dairy cows[1,2]. Given the positive association between IGF-1 and fertility[2], the aim of the experiment was to investigate the extent to which nutritional manipulation of IGF-1 during the prepartum period was associated with variation in postpartum concentrations of IGF-1 and the interval to the first ovulation (IFO). Seventy-two cows were each fed one of 3 diets during the last 3 to 4 weeks before expected calving date in August or September of 2002. The 3 diets were: i) a standard total mixed ration (TMR) (81.1 MJME/cow/day); ii) the TMR + 4 kg of grain concentrate (94.5 MJME/cow/day; high energy); and iii) the TMR + 3.5 kg of soybean meal (115.9 MJME/cow/day; high protein). After calving, cows were offered a common diet of 35 to 40 kg DM pasture/cow/day with 6kgDM grain concentrates and monitored for at least 10 weeks into lactation. Plasma IGF-1 concentrations were measured in each cow on a weekly basis throughout the experiment. The IFO was estimated using dates of observed oestrus as well as plasma progesterone profiles. Concentrations of plasma IGF-1 were high at the commencement of the prepartum treatment period (~ 160ng/mL) before decreasing rapidly from 2 weeks before calving through the first week postpartum (~ 65ng/mL), and then increasing gradually for the remainder of the experimental period (~ 100ng/mL by Week 7 of lactation). Manipulation of diet in the prepartum period altered plasma concentrations of IGF-1 during that time with maximum weekly average concentrations of 167.0 ± 12.9, 173.2 ± 12.7 and 204.7 ± 11.7 ng/ml for Diets 1 to 3 respectively. Diet 3 also delayed the prepartum decline in IGF-1. These differences in concentrations of IGF-1 in the prepartum period did not carry over into early-lactation as postpartum plasma IGF-1 concentrations were unaffected by prepartum diet. Neither did the prepartum diets affect the average IFO (Diet 1: 45.0 ± 4.8 days; Diet 2: 50.2 ± 4.6 days; Diet 3: 44.7 ± 4.6 days, p=0.636). There was a negative relationship between plasma IGF-1 concentrations measured at the start of the AI program and the IFO for individual cows (r = -0.31, p = 0.013). Even though manipulating the diet in the prepartum period was sufficient to alter plasma concentrations of IGF-1, there was no carry-over affect on either the postpartum concentrations of IGF-1 or the IFO. This suggests that energy and protein intakes during the experimental period of this study were not factors that affected the duration of anoestrus. The fact that differences in postpartum concentrations of IGF-1 among individual cows were significantly related to IFO even though they were not related to prepartum diet suggests that the relationship may be genetic and could explain the limited effectiveness of some high energy prepartum diets on subsequent reproductive performance.

[1]Spicer, et al. 1990. J Dairy Sci 73:929-937.
[2]Lucy, et al. 1992. Anim Production 54:323-331.

36. Metabolic responses to a glucose challenge in grazing primiparous dairy cows

LM Chagas[a], PJS Gore[a], D Blache[b], RC Boston[c] and GA Verkerk[a]

[a]Dexcel Limited, Private Bag 3221, Hamilton, New Zealand, [b]The University of Western Australia, Nedlands, Australia and [c]Dept. of Clinical Studies, University of Pennsylvania, USA.

The postpartum period of dairy cows is characterized by metabolic adaptations of the cows to the energy demands of lactation. This involves the mobilization of adipose tissue which results in the release of NEFA into the blood stream[1]. Lactating cows are thought to exhibit insulin resistance by decreasing the sensitivity of adipose and muscle tissues to insulin[2]. Cows calving with good body condition mobilize more fat tissue and have greater NEFA levels than cows calving with low body condition[3]. We hypothesized that heifers with unrestricted access to pasture in early lactation would have increased fat mobilization and NEFA that is associated with increased insulin resistance after calving.

Twenty-four primiparous Holstein-Friesian cows (2 yr old) were managed to calve with an average body condition score (BCS) of 5.25 (from 4.5 to 6.0) on a 1 to 10 scale. After calving half of the group was offered unrestricted pasture (fully fed, FF) and the other half received a restricted pasture allowance (RES). The restriction was similar to those to be found on normal New Zealand commercial dairy farms. Allocation to experimental treatments was random and balanced for body weight and genetic merit for milk production. Two weeks after calving, animals received an intravenous infusion of glucose (300 mg D-glucose/kg LW) and blood samples were collected at 0, 2, 3, 4, 5, 6, 7, 8, 9, 10, 12, 14, 16, 18, 20, 22, 24, 26, 28, 30, 32, 36, 40, 45, 50, 60, 65, 70, 75, 80, 90, 120, 150, 180, 210 and 240 minutes relative to the time of the infusion for measurement of glucose, insulin, GH and NEFA.

There were no differences in the glucose response pattern by treatment ($P > 0.05$). There was an increase in insulin in both groups but the FF group had a lower secretion of insulin than the RES group ($P < 0.05$). The FF group responded with a higher secretion of GH than the RES group ($P < 0.05$). The FF also has an earlier nadir (lipolysis resumed sooner) than the RES group. This recovery of NEFA level to the pre challenge baseline was faster in the FF than the RES group. These results suggest that the RES have a greater degree of insulin resistance and this is associated with the differences in GH and NEFA between the groups. The role of insulin resistance associated with diet in postpartum in dairy cows needs to be further investigated in relationship to the GH and metabolism.

This research was funded by the Foundation for Science, Research and Technology, New Zealand.

[1]Emery et al. 1964. J Dairy Sci 47:1074-1079.

[2]Cronjé, 2000. Ruminant Physiology: Digestion, metabolism, growth and reproduction, C. P.B., Editor. CAB International. p. 409-422.

[3]Chagas et al. 2006. J Dairy Sci 89:1981-1989.

37. The effect of mono propylene glycol treatment on the expression of GHR1A, GHRtotal and IGF-I mRNA in the liver of postpartum dairy cows

LM Chagas[a], G Graham[b], PJS Gore[a], D Blache[b] & G Verkerk[a]

[a]Dexcel Limited, Private Bag 3221, Hamilton, New Zealand and [b]The University of Western Australia, Nedlands, Australia

A major source of infertility in New Zealand dairy herds is prolonged postpartum anoestrous intervals (PPAI). Current technologies to resolve declining reproductive performance are hormonal interventions to induce cyclicity and calving. Our previous study demonstrated that supplementing Friesian heifers in low body condition at calving with monopropylene glycol (MPG) after calving substantially reduced PPAI[1]. However, the drenching regimen used was not cost effective. By identifying the mechanism by which MPG reduces PPAI, strategies for using 'targeted nutrition' may be developed to overcome anoestrum. This study examined the effects of MPG on metabolically important hepatic genes pre- and postpartum.

Holstein-Friesian heifers (n = 49) that conceived to a synchronised first insemination were used. They were managed to achieve a body condition score of 5.25 (from 4.5 to 6.0) at calving. After calving they were split into three groups: unrestricted access of pasture (fully fed, FF), restricted pasture plus 250 ml of MPG twice a day (RES +), and restricted (RES). The restriction was at a similar level to commercial dairy farms. On weeks -1, 1, and 4, a liver sample was collected from a subset of animals (n = 6 per group) to determine the mRNA expression of growth hormone receptor 1A (GHR1A), total growth hormone receptor (GHRtotal) and insulin-like growth factor-I (IGF-I). RNA was extracted from 100mg of liver using trizol (Invitrogen) according to the manufacturer's directions. Total RNA (2 ug) was reverse transcribed using the Superscript III first strand cDNA synthesis kit (Invitrogen) according to the manufacturer's directions. Real-time PCR was performed on a Corbett Rotorgene 3000. Dual labelled fluorescent probes and standards were synthesised for cattle GHR1A, GHRtotal and IGF-I (Proligo). Results were analysed using two-way ANOVA.

GHR1A mRNA expression was lower at week 4 in all groups ($P < 0.05$). The expression of GHR1A was lower ($P < 0.05$) in the RES than the FF group at week 4 but not in weeks −1 and 1. There was no interaction between time and treatment (week -1 FF 42.50 $\pm$ 4.69 pg/ml, RES 35.05 $\pm$ 7.81 pg/ml, RES+ 31.47 $\pm$ 4.28 pg/ml; week 1 FF 54.54 $\pm$ 10.27 pg/ml, RES 53.82 $\pm$ 13.44 pg/ml, RES+ 36.50 $\pm$ 7.58 pg/ml; week 4 FF 26.34 $\pm$ 5.18, RES 12.07 $\pm$ 3.07 pg/ml, RES+ 19.91 $\pm$ 5.92 pg/ml). GHRtotal mRNA expression was lower postpartum ($P < 0.05$) with no effect of treatment or interaction between time and treatment. IGF-I mRNA expression was lower postpartum ($P < 0.05$) with no effect of treatment or interaction between time and treatment (week -1 FF 6.25 $\pm$ 1.20 pg/ml, RES 6.84 $\pm$ 1.26 pg/ml, RES+ 7.89 $\pm$ 3.34 pg/ml; week 1 FF 3.16 $\pm$ 0.81 pg/ml, RES 1.82 $\pm$ 0.57 pg/ml, RES+ 2.76 $\pm$ 0.91 pg/ml; week 4 FF 4.57 $\pm$ 0.80, RES 3.05 $\pm$ 1.03 pg/ml, RES+ 2.58 $\pm$ 0.49 pg/ml). In conclusion, this study suggested that the reduction of PPAI by MPG is not mediated by changes in liver sensitivity to growth factors or production of IGF-1. However, MPG might have a long-term effects on the expression of GHR1A, GHRtotal and IGF-I which will be elucidated by investigating changes in expression at later time-points postpartum.

This research was funded by the Foundation for Science, Research and Technology, New Zealand.

[1]Chagas *et al.* 2003. ANZ Dairy Vets Conference, 215-220.

38. Impaired fertility associated with negative energy balance in cows is accompanied by changes in hepatic gene expression of the insulin-like growth factor (IGF) family

MA Fenwick[a], DG Morris[b], D Kenny[c], J. Murphy[d], R Fitzpatrick[b] & DC Wathes[a]

[a]Reproduction, Genes and Development Group, Royal Veterinary College, London, UK, [b]Teagasc Research Centre, Athenry, Co. Galway, Ireland, [c]Faculty of Agriculture, University College Dublin, Ireland and [d]Teagasc Moorepark, Co. Cork, Ireland

The growth hormone-IGF axis functions to maintain nutritional homeostasis. In high yielding dairy cows, modulation of GH-IGF signaling in the liver during lactation leads to excessive mobilization of tissue reserves and precipitates a state of negative energy balance (NEB). The reduced systemic IGF-I and release of metabolic byproducts during NEB correlates with impaired fertility in these cows. This study determined how NEB alters the endocrine response of hepatic IGF secretion by examining candidate and global gene expression in the liver during the early *post partum* period.

Multiparous Holstein-Friesian cows (n = 24) were blocked 2 weeks prior to calving by parity, body condition score and previous lactation yield. Cows were randomly allocated to 2 treatments designed to produce either mild (MNEB) or severe negative energy balance (SNEB). MNEB cows were fed *ad lib* grass silage supplemented with 8 kg day[-1] concentrate and milked once daily. SNEB cows were restricted to 25 kg day[-1] silage with 4 kg day[-1] concentrate and milked thrice daily. Six cows showing extremes in EB from each group (calculated using the French NE system) were culled in week 2 of lactation when liver and blood samples were taken for analysis. Hepatic RNA expression was examined with a 23K bovine oligonucleotide microarray (Affymetrix) and real time RT-PCR was used to quantitate differences in hepatic gene expression for all members of the IGF family and relevant steroid receptors. Mean energy balance on the day of slaughter for MNEB and SNEB groups was -1.8 ± 0.96 UFL/day and -6.6 ± 1.01 UFL/day respectively ($P < 0.05$). Metabolic differences between the two groups were confirmed by a comparative increase in non-esterified fatty acids (NEFA) and ß-hydroxybutyrates (BHB) and reduced glucose in SNEB cows (all $P < 0.01$). For total IGF-I expression, plasma protein and hepatic mRNA was reduced in SNEB cows compared with MNEB cows ($P < 0.01$). Data from both levels of expression correlated highly across all animals ($r = 0.905$; $P < 0.01$) confirming PCR assay validity, and the liver as a major endocrine source of IGF. Normalised microarray data revealed a total of 113 genes that were differentially expressed by at least 2-fold in SNEB compared with MNEB livers (LPE test; $P < 0.05$). Of these genes, 24% were related to metabolism, including members of the IGF family, and a large proportion (9%) could be attributed to immune response and stress. By real time RT-PCR, SNEB cows exhibited reduced hepatic synthesis of IGF-1R, IGF-2R, IGF binding proteins (IGFBPs) -3,-5,-6, IGFBP acid labile subunit, and growth hormone receptor 1A variant ($P < 0.02$ respectively), while IGFBP-2 was elevated ($P < 0.01$). Liver expression of IGF-II, IGFBP-1, -4, and receptors for insulin (A/B), oestrogen (α/ß) and glucocorticoid-α were unaffected by EB. Taken together these results demonstrate that SNEB affects the synthesis and secretion of IGF-I, and the synthesis of key IGFBPs known to modulate the bioavailability of circulating IGF-I. A more complete transcriptome analysis indicates that other physiological processes likely contribute to a loss of reproductive efficiency in high yielding dairy cows.

This work was funded by the Wellcome Trust, UK. No. 072315

39. Postpartum anoestrous intervals in relation to somatotropic hormones and metabolic profiles in mature dairy cows managed on pasture with or without grain supplement during the transition period

<u>CR Burke</u>, JR Roche, PW Aspin & JM Lee

Dexcel Limited, Private Bag 3221, Hamilton, New Zealand

Extended postpartum anoestrous intervals (PPAI) present a substantial challenge to obtaining optimal reproductive performance in seasonal calving dairy herds. Age, breed and nutrition are well known factors that influence PPAI. The manner in which nutrition influences PPAI is complex, but is at least partially mediated via signalling pathways involving the somatotropic axis (see [1]Robinson *et al.*, 2006). Key hormones of this axis (GH, IGF and insulin) are responsive to body energy reserves, dietary energy intake and diet composition.

In the current study, a 2 x 2 factorial arrangement examined PPAI in cows with pre- and post-calving diets that were iso-energetic, but varying in the ratio of structural to non-structural carbohydrates. At 36 ± 8.7 d prepartum, cows were assigned iso-energetic diets (114 MJ ME/cow/d) of either pasture/pasture silage (prePast; n = 34) or similar supplemented with 3 kg DM/cow/d barley-maize concentrate (preGrain; n = 34). After calving, cows within each prepartum diet were assigned iso-energetic diets (179 MJ ME/cow/d) of either pasture/pasture silage (postPast; n = 34) or similar supplemented with 5 kg DM/cow/d concentrate (postGrain; n = 34). Blood samples were collected weekly from 3 weeks before calving (d. 0) until resumption of oestrous cycles postpartum. Additional samples were collected at calving and then daily for 4 days. Progesterone concentrations were determined in milk samples collected twice weekly to determine PPAI (first sustained elevation > 3 ng/ml).

Type of diet prepartum did not affect PPAI (P = 0.7), but postGrain cows had reduced (P < 0.05) PPAI (28.4 ± 1.8 d) compared with postPast cows (36.4 ± 2.5 d). There was no interaction between pre- and postpartum diets (P = 0.7). Diet did not affect body condition score, but degree of negative energy balance may have been greater in postGrain cows, as suggested by greater plasma NEFA concentrations on d. 21 and 35 postcalving. As compared with postPast, cows fed grain after calving had consistently increased (P < 0.01) concentrations of IGF-1 from d. 2 to 35 (9.4 vs. 10.2; sed; 0.7), while levels of GH and insulin were not consistently different (P > 0.1). Univariate regression analyses of data pooled within treatments found PPAI to be influenced (P < 0.05) by body condition score at any timepoint, NEFA on d. 0, 14 and 28, insulin and glucose on d. 0, IGF-1 on d. 0 to 3 and urea on d. 3; but not by GH or beta-hydroxy butyrate. However, none of these factors by themselves could account for more than 10% of the variability in PPAI ($r^2 < 0.2$).

The 8-day shortening of PPAI in postpartum grain-supplemented cows was not associated with improved body condition, and occurred in spite of an apparently greater negative energy balance in early lactation. However, postpartum grain supplementation consistently increased plasma IGF-1, and IGF-1 was a consistent predictor of PPAI within treatments. Several other predictors of PPAI were identified with the significance of these mostly limited to the immediate transition period. These findings suggest that fortification of a pasture-based diet with non-structual carbohydrates provided a nutritional cue that advanced the recoupling of the somatotropic axis after calving, an event known to beneficially influence subsequent reproductive events.

[1]Robinson *et al.* 2006. Anim Feed Sci Technol 126:259-276.

40. The effect of n-3 polyunsaturated fatty acids (PUFAs) on luteal steroidogenesis in the ewe

EC Chin[a], J Nadaffy[a], JS Brickell[a], ZC Cheng[a], LM Hodges[a], EL Sheldrick[b], APF Flint[b], DC Wathes[a] & DRE Abayasekara[a]

[a]Reproduction, Genes and Development Group, Royal Veterinary College, London UK and [b]Animal Physiology Group, University of Nottingham, Nottingham, UK

Increased dietary intake of polyunsaturated fatty acids (PUFAs) has been shown to modulate many aspects of female reproduction including ovulation, ovarian cyclicity and gestational length[1,2]. It has been suggested that these changes are principally mediated through alterations in the pattern of prostaglandin synthesis. Prostaglandins of the 2 –series derived from the n-6 PUFA, arachidonic acid, exert diametrically opposing effects on luteal steroidogenesis - prostaglandin E_2 (PGE_2) stimulates whilst prostaglandin $F_{2\alpha}$ ($PGF_{2\alpha}$) inhibits luteal progesterone (P_4) output. Moreover, increased intake of either n-6 or n-3 PUFAs in the diet leads to profound changes in the circulating levels of P_4 in dairy cows[3]. The mechanisms underlying these alterations in steroid secretion remain unelucidated. Therefore, in this study we have investigated the hypothesis that changing the composition of polyunsaturated fatty acids in the sheep diet affects luteal steroid synthesis by altering the expression of enzymes involved in steroid and prostaglandin synthesis. Two groups of Welsh Mountain ewes (n = 8 per group) were individually fed either a control diet or a diet high in α-linolenic acid (LNA; 18:3 n-3) for 4 weeks. Blood samples were collected daily throughout the oestrous cycle (day 0 to 17) and day I to 4/5 of the subsequent oestrous cycle for measurement of progesterone by specific radioimmunoassay (RIA). Additional sampling was carried out every 2h from days 13 to 16 for measurement of prostaglandin 13,14-dihydro-15 keto prostaglandin $F_{2\alpha}$ (PGFM) by RIA. All ewes were slaughtered on day 4/5 of the subsequent cycle when ovarian tissue was collected. Ovarian tissue was used for the analysis of proteins (western blotting) involved in the regulation of steroidogenesis i.e. cytochrome P450 dependent cholesterol side chain cleavage (P450cscc), 3 beta hydroxysteroid dehydrogenase/isomerase (3ß-HSD) and steroidogenic acute regulatory protein (StAR). The high n-3 diet affected plasma P_4 with levels being significantly increased on day 15 compared to control and significantly decreased compared to control on day 4/5 of the subsequent cycle. The overall mean concentration as well as the number pulses of plasma PGFM secreted on day 14 was decreased in n-3 PUFA fed ewes. StAR protein expression was significantly inhibited in n-3 PUFA-fed ewes ($P < 0.05$). Expression of P450cscc and 3ßHSD were unaffected by the diet. We conclude that the high n-3 PUFA diet delayed luteolysis though inhibiting $PGF_{2\alpha}$ synthesis. Observations (data not shown) that n-3 PUFAs inhibited the expression of cyclooxygenase-2 (COX-2) in bovine endometrial cell cultures supports this particular notion. Whilst this is the most likely explanation, we cannot exclude the possibility that PUFAs influenced steroidodogenesis by modifying the expression of StAR directly. Support for this alternative explanation is provided by our findings in the ovine adrenal cortex, where n-3 PUFAs in vivo and in vitro modified StAR expression and cortisol secretion without apparently affecting COX-2 expression or prostaglandin secretion.

This work was supported by The Wellcome Trust and BBSRC.

[1]Abayasekara & Wathes, 1999. Prostaglandins Leukotrienes and Essential Fatty Acids 61:275-287.
[2]Robinson et al. 2002. Reproduction 124:119-131.
[3]Elmes et al. 2005. Journal of Physiology 562:583-592.

41. Uterine and hepatic mRNA expression for insulin and IGF-I signaling intermediaries, end-products, and repressors in lactating Holstein cows

JP Meyer, ML Rhoads & MC Lucy

University of Missouri-Columbia, Missouri, USA

Negative energy balance occurs when energy demands exceed energy intake. During negative energy balance body energy stores (i.e., triglycerides within adipocytes) are used to fulfill systemic energy needs. Elevated plasma fatty acids inhibit glucose uptake into tissues.[1] This response created by negative energy balance causes systemic insulin resistance.[2] Intense selection for milk production has lowered dairy fertility. This decrease in fertility may be associated with insulin resistance and decreased glucose uptake within reproductive tissues.[2]

The objective was to measure uterine and liver mRNA for insulin and IGF-I signaling intermediaries, end-products, and repressors. Uterine and liver biopsies (n=8) were collected from 4 Holstein cows on day 60 after calving. Total cellular RNA was isolated and reverse transcribed into cDNA. The insulin signaling mRNA amount was measured by using quantitative reverse transcriptase polymerase chain reaction (RTPCR) with Syber Green detection. The PCR reactions were performed and fluorescence was quantified with the ABI PRISM 7500 Sequence Detector (Applied Biosystems, Foster City, CA). Analyses of amplification plots were performed with the Sequence Detection Software (Applied Biosystems). Bovine primers were developed to measure mRNA for acetyl-CoA carboxylase (Acc), Akt1, Akt2, Bclx, phosphatase 14 (Dusp14), eukaryotic translation initiation factor 2 binding protein 1 (Eif2bp1), Elk1, ERas, Ets1, Ets2, glucose-6-phosphatase (G6pc), Grb2, hexokinase 2 (Hk2), IGF-I, IGF binding protein 5 (Igf1bp5), IGF-I receptor (Igf1r), insulin receptor (Insr), insulin receptor substrate 1 (Irs1), Irs2, Mapk1, Mapk3, phosphoenolpyruvate carboxykinase 2 (Pck2), 3-phosphoinositide dependent protein kinase-1 (Pdpk1), PI3-kinase (p85-alpha subunit) (Pik3r1), PPAR gamma, coactivator 1 alpha (Ppargc1a), Pten, protein tyrosine phosphatase 1b (ptp1b), Raf-1, ubiquitin (Ubq), uncoupling protein 2 (Ucp2), Ucp3, and Vegf. Each gene product was PCR amplified and sequenced to verify its identity. All 32 mRNA were detected in liver but only 24 out of 32 were detected in uterus (Acc, Bclx, ERas, Ets2, G6pc, Insr, Pdk1, and Ucp3 targets were not detected in uterus). For detected mRNA, the effects of tissue and cow were significant (P< 0.001) for mRNA amount [expressed as the average PCR cycle at which the amplification reached threshold (C_T)]. Across all expressed genes, the average C_T was 6.3 cycles less for liver (25.8 ± 0.2) compared with uterus (32.1 ± 0.2). These results suggest a 30-fold increase in expression of pathway genes in liver compared with uterus. Genes expressed at a similar level for liver and uterus were Irs2, Pten, and Elk1. Differences in gene expression between individual cows were less than the differences between tissues. Cow differences were nonetheless highly significant (P< 0.001) with a C_T difference of 3 between highest and lowest expressing cows (approximately 8-fold difference in gene expression).

These data demonstrate that insulin and IGF signaling pathway components differ for liver and uterus. Pathway genes are coordinately expressed (i.e., there appears to be global control of gene expression). Individual cows differ for mRNA expression across numerous pathway genes and those differences may encode differential insulin sensitivity in individual cows.

[1]Roith & Zick, 2001. Diabetes Care 24:588-597.
[2]Lucy, 2004. Proc NZ Soc Animal Pract 64:19-23.

42. Regulation of mRNA expression for the bovine ovarian IGF system in *post partum* negative energy balance

S Llewellyn[a], R Fitzpatrick[b], DA Kenny[c] RJ Scaramuzzi[a] & DC Wathes[a]

[a]*Department of Veterinary Basic Sciences, Royal Veterinary College, Potters Bar, Herts, UK,* [b]*Teagasc Research Centre, Athenry, Co. Galway, Ireland and* [c]*Faculty of Agriculture, University College Dublin, Ireland*

Post partum negative energy balance (NEB) in dairy cattle is associated with a delayed return to ovarian cyclicity. This effect may be, at least in part, mediated by alterations to members of the systemic and ovarian insulin-like growth factor (IGF) family. This study tested the hypothesis that NEB during the early *post partum* period in dairy cows alters mRNA expression for members of the follicular IGF system.

Twelve multiparous cows were blocked two weeks prior to expected calving according to parity, body condition score, and previous lactation yield, and randomly allocated to 2 treatments designed to produce mild (mNEB) or severe negative energy balance (sNEB). Mild NEB cows were fed *ad lib* grass silage with 8 kg day[-1] of a 21% crude protein dairy concentrate and milked once daily; sNEB were fed 25 kg day[-1] silage with 4 kg day[-1] concentrate and milked thrice daily. Calculations of energy balance were based on the French NE system[1]. Cows were slaughtered on day 6 or 7 of the first follicular wave after calving and the ovary opposite to that containing the dominant follicle was collected and frozen at $-$80°C. Expression of follicular mRNA to IGF-II, type 1 IGF receptor (IGF-1R), and IGF binding proteins (IGFBP)-1 to $-$6 mRNA was determined by *in situ* hybridisation and quantified by measuring optical density (OD) units as described previously[2]. Blood samples were analysed for ß-hydroxybutyrate (BHB), non-esterified fatty acids (NEFA), glucose and IGF-I.

Circulating concentrations of glucose and IGF-I were higher ($P < 0.05$) in the mNEB cows while BHB and NEFA were higher ($P < 0.05$) in the sNEB cows. Follicular IGF-II mRNA was localised to theca cells where expression decreased with atresia and increasing follicle size (healthy follicles only). Expression of IGF-IR mRNA was detected in granulosa cells and was unaffected by treatment, follicle size or health. Severe NEB decreased granulosa IGFBP-2 mRNA compared with mNEB (ODs: sNEB 0.11 ± 0.008, mNEB 0.16 ± 0.008; [$P < 0.001$]). Thecal IGFBP-4 mRNA and granulosa IGFBP-5 mRNA expression was higher in very small (VS, 1-2.5 mm) than small (S, 2.5-5 mm) follicles for mNEB (IGFBP-4 ODs: VS 0.22 ± 0.017, S 0.15 ± 0.017; [$P = 0.01$]; IGFBP-5 ODs: VS 0.85 ± 0.163, S 0.31 ± 0.225; [$P = 0.1$]) but not sNEB (IGFBP-4 ODs: VS 0.19 ± 0.012, S 0.20 ± 0.011; IGFBP-5 ODs: VS 0.35 ± 0.167, S 0.36 ± 0.175) cows. Concentrations of follicular IGFBP-5 mRNA in VS atretic follicles were positively correlated with plasma glucose concentrations ($r = 0.673$, $P < 0.05$) and negatively correlated with plasma BHB and NEFA concentrations (BHB: $r = -0.763$, $P < 0.05$; NEFA: $r = -0.664$, $P = 0.051$). IGF-IR mRNA was negatively correlated with serum amyloid-alpha ($r = -0.738$, $P < 0.01$) and IGFBP-2 mRNA in S atretic follicles was positively correlated with plasma urea ($r = 0.744$, $P < 0.05$). Thecal IGFBP-6 mRNA concentrations were unaffected by treatment, but changed in relation to follicle size. In conclusion, *post partum* negative energy balance alters ovarian gene expression for the IGF system, which is associated, in part, with changes in plasma metabolites. This may influence both ovarian activity and fertility.

[1]Jarrige, 1989. Eds J. Agabriel, P. Champciaux and C. Espinasse. CNERTA, Dijon, France.
[2]Perks *et al.* 1999. Journal of Reproduction and Fertility 116:157-165.

43. Influence of pre-weaning diet and sire on growth rates of replacement dairy heifers

JS Brickell[a], N Bourne[a], ECL Bleach[b] & DC Wathes[a]

[a]Reproduction, Genes and Development Group, Royal Veterinary College, London, UK and [b]Writtle College, Chelmsford, UK

A short herd lifespan is a significant economic loss to the dairy industry, with infertility a major cause. We have shown that the pre-pubertal IGF-I concentrations are related to growth rate and can predict IGF-I around first calving, when low levels are associated with a delayed resumption of ovarian cyclicity[1]. This study investigated the effects of sire and pre-weaning diet in terms of growth rates and IGF-I concentrations leading up to the first service period at 15 months. A group of consecutively born Holstein type heifers ($n = 75$) with 5 sires ($n = 8$ to 23 offspring each) were recruited from one farm (CEDAR, Reading), subdivided into 3 equal groups according to birth weight and dam parity and fed different pre-weaning diets until 6 weeks of age; warm ad-libitum milk (W, mean daily intake 14.7 litres), cold ad-libitum milk (C, mean daily intake 11.1 litres), and restricted milk (R, 2 litres twice daily). The subsequent post weaning management was the same across groups. Size parameters (body weight (BW), height at withers (HT), crown-rump length (CRL), and heart girth), and plasma IGF-I were measured at 1 month, pre-pubertally (6 months), and pre-service (14 months). Data were analysed by 2 way ANOVA and Pearson correlations.

Across the groups no calves died during the first 6 months, but 3 died between 6-14 months. All size parameters at 1 month were significantly affected by both diet (W > C > R $P < 0.05$-0.001) and sire ($P < 0.05$-0.001). By 6 months there was still significant variation in size parameters between the groups (W > C > R for BW, HT, Girth; $P < 0.001$), but no sire effect. Conversely at 14 months diet was no longer significant but there was a significant sire effect on CRL ($P < 0.01$). Some growth rate parameters during the first 6 months were significantly affected by both diet (R > W > C Girth; $P < 0.01$) and sire (HT, Girth; $P < 0.05$). There was also a significant diet effect on the girth increase during the first 14 months (R > W > C $P = 0.015$), and the BW increase between 6-14 months (R > W > C $P = 0.06$). This is indicative of higher post weaning compensatory growth in the calves that were initially restricted.

The IGF-I concentration at 1 month also showed significant variation with diet (W > C > R $P < 0.001$), however IGF-I concentrations at 6 and 14 months did not vary between the groups. There was no sire effect on IGF-I concentration at any of the time points. IGF-I concentration at 6 months was significantly correlated with growth rate during the first 6 months (BW, Girth; $P < 0.05$), and growth rate between 6-14 months (BW, CRL, Girth; $P < 0.05$).

A total of 72 heifers reached the first service period; 5 of these did not conceive and 1 aborted (2 per milk group). Age at conception was not significantly affected by diet or sire, however there were trends ($P = 0.1$) for both diet (W < R < C, range 1.7 – 2.3 serves) and sire (range 1.3 – 2.4 serves) influences on the number of services required for conception at 15 months, with a significant diet by sire interaction ($P = 0.015$).

In conclusion, we have shown that the pre-weaning diet can influence the IGF-I concentration and growth rate of heifers during the rearing period. The prevalent sire effects on size at 1 month were lost by 6 months of age with the exception of the girth increase over this period. Sire did, however, appear to influence heifer fertility. Further studies will determine if growth rate and IGF-I concentration can predict future fertility or yield.

Funded by DEFRA, MDC and Volac International.

[1]Taylor et al. (2004) Journal of Endocrinology 180: 63-75.

44. A short grazing period on *Lotus uliginosus* cv. Maku can increase ovulation rate in Corriedale ewes

GE Banchero & G Quintans

National Institute of Agricultural Research (INIA), Ruta 50, Km 12. 70000, Uruguay

Short periods of supplementation with lupins, which are high in rumen degradable protein and energy, increased ovulation rate (OR) in ewes. This response was attained with only 4 days of supplementation[1]. Smith[2] studied the nutrients responsible for this increase in OR and reported that ewes consuming similar levels of energy presented higher OR as protein increased. Although there are no lupins in Uruguay, it is possible that high protein from improved pastures may have the same effect. Banchero *et al.*[3] examined the response in OR of sheep grazing Lotus Maku (LM, PC= 15 to 24%). Preliminary results indicated that there was no effect of type of pasture on OR. However, the ewes that grazed LM for 13 to 14 days tended (P=0.08) to have more double-ovulations than ewes grazing native pastures (42% *vs.* 25%). The present experiment tested the hypothesis that ewes grazing LM would have a higher OR than ewes grazing native pasture and this effect would be enhanced when the ewes were supplemented with an energy rich maize grain diet. Oestrous cycles in a mob of 265 mature Corriedale ewes (45.6 ± 0.63 kg liveweight (LW) and 2.1 ± 0.04 units of body condition) were synchronised (Day 0 = synchronised ovulation) and ewes were randomly assigned to one of four treatments. Ewes from treatment 1 were allowed to graze a native pasture (CP = 7.8%) which was offered at 12 kg dry matter (DM)/100 kg LW from Day 2 to Day 20. Ewes of treatment 2 were offered the same pasture plus a supplement of 500 g of whole maize (CP = 6.9%) per animal per day from Day 4 to 13. Ewes from treatment 3 were allowed to graze Lotus Maku (CP= 20%) which was offered at 12 kg dry matter (DM)/100 kg LW from Day 2 to day 20 and then drafted to a native pasture. Ewes of treatment 4 were offered the same pasture plus a supplement of 500 g of whole maize per animal per day from Day 4 to 13. The second ovulation was detected using 10 fertile rams which were painted to detect days of service. Seven days after (Day 25), the numbers of corpora lutea were determined by laparoscopy to calculate the ovulation rate of all ewes. Ewes grazing LM had higher OR than ewes grazing native pasture (1.35 *vs.* 1.21, P< 0.05) and the supplement did not improve the OR but improved the pregnancy rate (90% and 82% for supplemented and unsupplemented ewes, P=0.10). Hence, a short period of feeding with LM has the potential to increase OR and this would have a tremendous benefit under extensive grazing condition.

[1]Stewart & Oldham, 1986. Proc An Prod Austr 16:367.
[2]Smith, 1985. Genetics of Reproduction in Sheep. pp 349.
[3]Banchero *et al.* 2003. 12th World Corriedale Congress Uruguay, pp 111.

45. Lamb vigour of Merino ewes in high and low body condition with or without a lupin supplement during the last two weeks of pregnancy

G Banchero & G Quintans

National Institute of Agricultural Research (INIA), Ruta 50 km 12, 70000, Uruguay

Nutrition of the ewe during pregnancy affects the behaviour of the lamb during the neonatal period. Undernutrition of the dam may lead to birth of lambs with low birth weight and poor vigour[1]. On the other hand, large and heavy lambs that can cause prolonged or difficult birth are also predisposed to poor vigour after birth. The aim of this experiment was to study the vigour of lambs born to ewe in high or low body condition that have been supplemented or not with lupins (*Lupinus angustifolius*) during the last 15 days of pregnancy. The experiment was a 2 x 2 factorial experiment with body condition (high or low) and supplementation (+ or -) as factors. Sixty two 3-year old Merino ewes carrying single lambs were used. Ewe body condition was manipulated from day 80 of pregnancy when ewes were randomly assigned to two condition scores groups (high and low) and fed with a different amount of the same feed to achieve high and low body condition scores (more than 2.5 vs 1.5; 5-point scale) by day 135 of pregnancy. Fourteen days before lambing, half of each body condition group was supplemented with 1 kg/head/day of lupin grain until parturition. Observations on lamb behaviour were then recorded for each lamb at 10-second intervals during the first hour of life. Lambs were tagged and weighted immediately after behavioural observations finished. High body condition ewes achieved the target of ≥ 2.5 and were 0.8 condition score higher (P < 0.05) than low condition score ewes (2.5 $\pm$ 0.07 vs 1.7 $\pm$ 0.07 CS) and 6 kg heavier (P < 0.05) at day 135 of pregnancy when the supplementation started (64.3 $\pm$ 1.2 vs 58.4 $\pm$ 1.3kg). Body condition and supplementation of the dam had no effect on the time the lamb attempted to stand, the time successfully stood or the total time standing during the first hour of life. However, lambs belonging to ewes in good body condition at day 135 of pregnancy attempted to suck earlier than lambs belonging to ewes in low body condition (22.3 $\pm$ 2.24 vs 29.9 $\pm$ 2.71 min; P < 0.05). They also were able to suck more quickly after birth (at 29.9 $\pm$ 2.54 vs 39.5 $\pm$ 3.08 min; P < 0.05) and for longer (5.8 $\pm$ 0.80 vs 3.5 $\pm$ 0.78 min; P < 0.05) than the lambs belonging to mothers in poor body condition. Supplementation did not affect the time the lambs attempted to suck, time from birth to successfully sucking and time sucking. The body condition or supplementation of the ewe did not affect the birth weight of the lambs or their growth up to 20 days of age. Lambs mothered by ewes in high body condition attempted to suck and successfully sucked earlier and spent 64% more time sucking during their first hour of life than those lambs mothered by ewes in low body condition. Generally, larger lambs are more vigorous than smaller lambs at least when dystocia is not a problem but, in this case, the difference in sucking activity in lambs cannot be attributed to a higher birth weight since lamb birth weights did not differ between ewes in high and low body condition (4.9 $\pm$ 0.12 vs 4.8 $\pm$ 0.11kg). Nonetheless, it may be partially attributed to the different quantity of food the ewes received during gestation when feeding to achieve high and low body condition. This difference in nutrients would not have increased the lambs' birth weight because when the treatments were imposed, the placenta was probably completely developed but could have affected the nutrients, especially glucose, that the lamb received during the last weeks of gestation. On the other hand, 2 weeks of supplementation prior to lambing seems not to be enough to increase the lamb activity.

[1]Alexander *et al.* 1959. Australian Veterinary Journal 35:433-441.

46. Effects of feeding palm oil fatty acids on blood metabolites in early lactating cows

A Heravi Moussavi[a], H Grailli[b] & M Danesh Mesgaran[a]

[a]Excellence Center for Animal Science, Ferdowsi University of Mashhad, Mashhad 91775-1163, Iran
and [b]Rajaii Agriculture College, Neishabour, Iran

During early lactation, dry matter intake lags behind the nutrient requirements for milk production and, consequently, lipids, proteins, and minerals are mobilized from body stores to support milk production. For this reason, the addition of fat sources to the diet, such as Ca salts of fatty acids, may be useful to overcome limitations in energy supplies. The use of Ca salts of fatty acids in the diets of dairy cows in early lactation to increase energy intake is well documented[1] but the effects of palm oil fatty acids on plasma metabolites have not been studied in early lactation cows. Because the effects of dietary fat are influenced by diet and stage of lactation[2], this study was conducted to characterize further the effects of dietary fatty supplementation on plasma concentrations of glucose, urea nitrogen (PUN), cholesterol and triglycerides in early lactating cows.

From d 5 to 70 postpartum (PP), sixteen cows were fed isonitrogenous diets containing 0 (n = 8) or 2% (n = 8) palm oil fatty acids (POFA; Bergafat; Berg + Schmidt GmbH, Hamburg, Germany). No differences were detected at calving in the body weight or body condition score of cows randomly assigned to each diet. Blood samples were collected weekly from d 1 to 70 PP via venipuncture of coccygeal vessels prior to the morning feeding and plasma was separated using centrifugation. Plasma metabolites were analyzed by enzymatic colorimetric assays using procedures modified from available kits. The data were analyzed using the MIXED procedure of SAS (2001) for a completely randomized design with repeated measures. The model contained the effects of treatment, week of treatment, cow within treatment and the interaction of treatment by week. The overall effect of treatment was tested using cow within treatment as the error term. Least squares means are reported throughout, and significance was declared at $P < 0.05$.

Plasma glucose concentrations were similar among diets (70.32 and 72.56 $\pm$ 1.65 mg/dl, respectively for control and POFA groups). The effect of time (week of treatment) was significant ($P < 0.05$) and plasma glucose increased over the time. Diet had no effect on PUN (14.04 and 13.57 $\pm$ 0.3 mg/dl, control and POFA group respectively), neither did PUN change over time. Plasma cholesterol concentrations were higher ($P < 0.01$) in cows fed POFA (167.68 and 206.11 $\pm$ 8.4 mg/dl, for control and POFA groups respectively). The effect of time was significant ($P < 0.001$) with plasma cholesterol increasing throughout the study. Diet had a significant impact on plasma triglycerides concentrations ($P < 0.05$) with the POFA group having decreased concentrations when compared to the control group (70.84 and 63.78 $\pm$ 1.9 mg/dl, for control and POFA groups respectively). Plasma triglycerides concentrations did not change significantly over time.

Results from this experiment demonstrate the increase in plasma cholesterol and decrease in plasma triglycerides in early lactation cows when feeding palm oil fatty acids.

The authors would like to thank the Excellence Center for Animal Science of Ferdowsi University of Mashhad for supporting this study.

[1]Chouinard et al. 1998. J Dairy Sci 81:471-481.
[2]Khorasani & Kennelly, 1998. J Dairy Sci 81:2459-2468

47. In sheep, a maternal diet enriched in absorbable polyunsaturated fatty acids (PUFA) skews the sex of conceptuses towards males

MP Green[a,b], LD Spate[a], TE Parks[a], K Kimura[a,c], CN Murphy[a], JE Williams[a], MS Kerley[a], JA Green[a], DH Keisler[a] & RM Roberts[a,d]

[a]Department of Animal Sciences, [d]Christopher S. Bond Life Sciences Center, University of Missouri, Columbia MO 65211, USA and [b]The Liggins Institute, University of Auckland, Auckland, New Zealand and [c]National Institute of Livestock and Grassland Science, Reproductive Physiology Lab, Tochigi 329-2793, Japan

There are numerous reports of sex ratio variation in mammals in relation to factors such as food availability, competition for resources and environmental cues, although such data remain controversial. In deer, does with better body condition scores (BCS), generally produce more sons than daughters, an outcome consistent with the sex allocation theory of Trivers & Willard[1]. There have been few studies in livestock, however, that have examined whether nutrition of the mother can affect sex ratio under controlled experimental conditions despite the fact that an ability to influence the ratio of sons to daughters born could have potential financial benefits in some agriculture systems. Anecdotally, prior to the present study we had noted an unexpectedly high number of male lambs born to ewes in the University of Missouri flock over two successive years. The hypothesis underpinning the present experiment was that the energy content of the diet fed to ewes around the time of conception could influence the sex ratio of offspring born. Non-parous Romanov crossbred ewes (n = 44), a breed with high fecundity, were divided into two weight-matched groups and assigned either a carbohydrate-rich (C) or a diet approximately five-fold richer in absorbable PUFA (F) from four weeks prior to breeding until d13 post-estrus. The diets formulated were isonitrogenous and contained a similar mass of metabolizable energy. Estrous cycles were synchronized and ewes bred to rams not fed either of the two test diets. On d13 post-estrus, conceptuses were flushed from the uteri, their development and dimensions recorded, and placed individually into culture. After 4 and 24h, the interferon-tau (IFNT) content of each medium was assayed and the sex of the conceptuses determined by PCR. After surgery, all ewes were fed the C diet for six weeks. In phase 2 of the experiment, all ewes were fed the F diet to negate any possibility of parity order influencing sex ratio. An identical schedule to phase 1 was followed. Weekly blood samples were collected to determine the plasma concentration of glucose, insulin, IGF-I, leptin and non-esterified fatty acids (NEFA). Plasma progesterone was measured daily for two weeks after breeding. In total, 129 conceptuses were recovered. Maternal diet did not affect conceptus size but conceptus sex differed ($P < 0.001$) from the expected 1:1 ratio in ewes fed the F diet, as 69% of the conceptuses were males. The C diet did not cause a significant shift in sex ratio. Diet did not affect plasma concentrations of insulin, IGF-I, glucose or leptin either overall or in early pregnancy. NEFA concentrations were, however, significantly ($P < 0.05$) higher in ewes on the F diet. Progesterone concentrations rose more slowly and were overall lower in ewes on the F than the C diet. Diet had no affect on IFNT production, but female conceptuses, and large conceptuses produced more ($P < 0.05$) IFNT than male or small conceptuses respectively. A greater maternal BCS also promoted increased IFNT production ($P < 0.05$). These results suggest that increasing the amount of PUFA fed to ewes in the period prior to and around conception can lead to an increase in male conceptuses and alters the concentration of at least one key hormone, namely progesterone.

Supported by USDA/CSREES/NRI Grant 2001-35203-10693.

[1]Trivers & Willard, 1973. Science 179:90-92.

48. The effect of short-term lupin grain supplements on plasma glucose and insulin and follicular fluid glucose in cyclic ewes

A Somchit[a,c], BK Campbell[d], M Khalid[b], NR Kendall[d] & RJ Scaramuzzi[a]

Depts [a]Veterinary Basic Science & [b]Veterinary Clinical Sciences, Royal Veterinary College, London, UK, [c]Dept Anim Husb, Chulalongkorn University, Bangkok, Thailand and [d]Div of Obstetrics & Gynaecology, University of Nottingham, UK

The interaction between nutrition and reproduction has long been known to have important implications for reproduction in farm animals[1]. Lupin grain supplementation for 4-6 days in the late luteal phase can stimulate folliculogenesis and so increase ovulation rate at the next ovulation. Dietary supplementation with lupin grain is associated with effects on the biokinetics of glucose and the blood concentrations of several metabolic hormones. Insulin has general effects on cellular function including enhanced glucose uptake and increased cellular proliferation and insulin and glucose together influence ovarian function directly[2]. The aim of this study was to determine the effect of lupin grain supplementation during the late luteal phase of the oestrous cycle on plasma and follicular fluid concentrations of glucose and plasma concentrations of insulin.

Sixteen, cyclic Welsh Mountain ewes were used. The ewes were randomly divided by weight into 2 groups: control (n = 8) and lupin-fed (n = 8). Ewes in the lupin-fed group were fed lupin grain (500g/ewe/day) for 5 days during late luteal phase. All ewes had free access to hay and water. Daily blood samples were taken from all ewes before and twice daily during treatment, plasma was harvested and stored at -20°C. At the end of the experiment, animals were humanely killed, their ovaries immediately collected and all ovarian follicles > 1.0 mm in diameter were dissected and classified as small (> 1.0-3.5 mm) and large (> 3.5 mm). Follicular fluid was collected and stored at -20°C. Plasma and follicular fluid samples were assayed for glucose (hexokinase assay) and plasma samples were assayed for insulin (radioimmunoassay).

Average weight did not differ between groups during the experiment. The numbers of large and small follicles, although higher in the lupin-fed group, this increase was not statistically significant. The plasma insulin concentrations from the start of lupin feeding until slaughter were higher in the lupin-fed group (P = 0.007). Before the start of lupin feeding the mean ($\pm$ SEM) plasma glucose concentrations in the lupin and control groups were 50.2 $\pm$ 2.00 and 51.2 $\pm$ 1.70 mg/dl. After the start of lupin feeding, the concentration of plasma glucose increased in the lupin-fed group (P = 0.012). The mean follicular fluid concentration of glucose in follicles < 3.5 mm in diameter in lupin-fed group (161 $\pm$ 44.7 mg/dl) was higher (P = 0.01) than in the control group (36.7 $\pm$ 5.75 mg/dl). The mean follicular fluid concentrations of glucose in follicles $\geq$ 3.5 mm in lupin-fed and control groups were 67.2 $\pm$ 4.92 and 90.9 $\pm$ 10.8 mg/dl respectively (P = 0.052).

In conclusion, short-term supplementation with lupin grain can increase blood insulin and glucose concentrations and follicular fluid glucose concentrations in cyclic ewes. This direct metabolic effect in the follicle is associated with stimulated follicular growth.

A Somchit is supported by a scholarship from the Royal Thai Government.

[1]Scaramuzzi et al. 2006. Nutrition Reproduction and Development (In Press).
[2]Downing et al. 1999. Journal of Endocrinology 163:531-541.

49. Early embryonic development in Holstein cows fed diets enriched in saturated or unsaturated fatty acids

G Thangavelu[a], MG Colazo[b], DJ Ambrose[a,b], M Oba[a], EK Okine[a] & MK Dyck[a]

[a]University of Alberta and [b]Alberta Agriculture Food and Rural Development, Edmonton, AB, Canada

The present objective was to determine embryonic development in Holstein cows fed diets enriched in saturated or unsaturated fatty acids. We hypothesized that a diet enriched in α-linolenic acid would enhance embryonic development over diets enriched in linoleic acid or saturated fatty acids. Cycling, lactating, Holstein cows (n = 24) were assigned randomly but equally to 1 of 3 diets containing hydrogenated tallow (TAL, high in palmitic acids), whole flax seed (FLX, high in α-linolenic acid), or sunflower seed (SUN, high in linoleic acid). Rations were formulated to provide 750 g supplemental fat/cow/d in all dietary groups. Diets were isonitrogenous and estimated energy intake was similar across diets. Ultrasound-guided follicular ablation (FA) was performed 5 d after ovulation in all cows. Two d after FA (i.e., expected day of follicular wave emergence), FSH treatment (300 mg total, in 8 divided doses, over 4 d; Folltropin-V, Bioniche Animal Health) was initiated and 25 mg dinoprost (Lutalyse, Pfizer Animal Health) was administered along with the 7th and 8th FSH treatments. One d after the last FSH treatment, 25 mg of porcine LH (Lutropin-V, Bioniche Animal Health) was given and cows were artificially inseminated (AI) 12 and 24 h later, with frozen-thawed semen of one bull. Embryos were collected nonsurgically 7 d after AI, evaluated (MZ7.5, Leica Microsystems) and categorized into unfertilized, degenerate, morula, blastocyst, or expanded blastocyst. Numbers of follicles ≥ 9 mm at pLH treatment, and that of CL at flushing, were determined by transrectal ultrasonography. Sixty-one transferable embryos were stained as previously described[1], mounted in glycerol and visualized under a 2-photon microscope system (Zeiss NLO510). Cell counts were obtained using the spot-counting function of Imaris software (Bitplane AG, Zurich, Switzerland). Data were analyzed by GLM procedures of SAS. Recovery rates (number of ova & embryos/number of CL) for cows fed TAL, FLX and SUN were 43, 40, and 54%, respectively (P > 0.05). Numbers of follicles, CL, total ova & embryos, or transferable embryos did not differ among diets (P > 0.05; overall means ± SEM, 18.5 ± 1.9, 11.4 ± 1.3, 5.3 ± 0.9, and 3.4 ± 0.7, respectively). Fertilization rate was numerically higher (P = 0.17) in cows fed TAL (93.3%), than in those fed FLX (84.4%) or SUN (77.5%). Cell number was affected (P < 0.001) by diet and stage of embryos, and their interactions (P = 0.05). Embryos collected from cows fed TAL (77.1 ± 3.9) had lower cell numbers (P < 0.003) than those from cows fed FLX (93.4 ± 3.3) or SUN (97.2 ± 3.5). Cell numbers of morulae (68.8 ± 3.2) did not differ (P > 0.10) among treatment groups. However, blastocysts of cows fed TAL had fewer cells (P < 0.05; 77.5 ± 5.3) than those of cows fed SUN (93.7 ± 5.0) but not different (P = 0.16) from those of cows fed FLX (88.6 ± 5.7). Differences were most evident in the expanded blastocyst stage; embryos of cows on FLX and SUN diets had higher (P < 0.02) cell numbers than those of cows on TAL (115.4 ± 5.7, 132.3 ± 7.6, and 89.3 ± 8.7 cells, respectively). Cell numbers of expanded blastocysts tended to differ (P = 0.08) between FLX and SUN cows. Our hypothesis was not directly supported, but findings indicate that embryo development is enhanced in Holstein cows fed unsaturated fatty acids compared to those fed saturated fatty acids.

Research supported by Alberta Agricultural Funding Consortium, Alberta Milk, Pfizer Animal Health, and Vetoquinol Canada Inc.

[1]Thouas et al. 2001. Reprod Biomed Online 3:25-29.

50. Conception rates and pregnancy losses in dairy cows fed a diet supplemented with rolled flaxseed

DJ Ambrose[a], CT Estill[b], MG Colazo[a], JP Kastelic[c] & R Corbett[a]

[a]Alberta Agriculture Food & Rural Development, Edmonton, AB, Canada, [b]Oregon State University, Corvallis, OR, USA and [c]Agriculture & Agri-Food Canada, Lethbridge, AB, Canada

In previous studies with limited numbers of cattle, supplementing the diet with flaxseed (a rich source of alpha linolenic acid) improved fertility in dairy cows[1,2]. The objective of the present study was to determine the repeatability of these findings in a large (1286 cows) commercial herd of Holstein cows. In the early postpartum period, cows were randomly assigned to receive a total mixed ration supplemented with (n = 156, FLX) or without (n = 147, CON) rolled flaxseed. Diets were isonitrogenous, isocaloric, and contained the same percent fat on a dry matter basis. Diets began on average 32.0 ± 15.6 d (mean $\pm$ SD) after calving and cows were timed-inseminated (TAI) at least 28 d after initiation of diets. Ovarian status was presynchronized with 25 mg of dinoprost (Lutalyse, Pfizer Animal Health) given twice, 14 d apart. Starting 14 d after the 2nd dinoprost treatment, cows received 100 μg of GnRH (Cystorelin, Merial) on D −10 and −1, 25 mg of dinoprost on D −3 (all treatments given IM), and were TAI on D 0, by 1 of 2 technicians. In a subset of cows (n = 226), plasma progesterone concentrations were determined on blood samples collected on D -10 and -3 (to determine cylicity status), and on D -3, 0 and 7 (to determine synchronization response). Diets continued for 31 d after TAI when pregnancy diagnosis was performed by transrectal ultrasonography, with re-examination (transrectal palpation) of pregnant cows on D 94 (to detect pregnancy losses from D 31 to 94). Data were analyzed by CATMOD procedures of SAS. The initial model included all potential variables; thereafter, any variable with a P value > 0.15 was excluded and the data were reanalyzed. When main effects were significant, their interactions were tested. The proportion of cycling cows was 81.6%. Conception rates at D 31 were numerically lower in cows fed FLX compared to those fed CON (28.2 vs 42.9 %, $P = 0.16$; overall, 35.3%, 107/303). Cows that responded to the synchronization protocol had higher ($P < 0.001$) conception rates (51.1 %, 47/92) than those that did not respond (25.4%, 34/134); when the latter were removed from the analysis, conception rates were lower ($P = 0.04$) in FLX than in CON cows (35.3%, 30/85 vs 51.2%, 43/84). Conception rates at D 31 were affected by AI technician ($P < 0.02$; 24.4% [22/90] vs 40.5% [85/210] for Technicians 1 and 2, respectively). Diet did not affect ($P = 0.35$) conception rates at D 94 (25.6%, 40/156 and 36.7%, 54/147 for FLX and CON diets, respectively). However, cows that responded to the synchronization protocol had higher ($P < 0.001$) conception rate at D 94 (43.5%, 40/92) than those that did not respond (22.4%, 30/134). Overall, 12.2% of the pregnancies were lost between D 31 and 94. Diets did not ($P = 0.42$) influence pregnancy loss, with 9 out of 63 pregnancies lost in CON, and 4 out of 44 lost in FLX. In conclusion, although previous studies[1,2] reported higher conception rates in cows fed flaxseed, diets did not significantly affect conception rates in the present study when all cows were considered. Furthermore, when cows not responding to the synchronization treatment were excluded, conception rates were significantly lower in cows fed FLX relative to those fed a CON diet.

Research supported by Agriculture Funding Consortium, Alberta Milk, Merial, and Pfizer Animal Health.

[1]Petit *et al.* 2001. Can J Anim Sci 81:263.
[2]Ambrose *et al.* 2003. J Dairy Sci 86, Suppl 1:187.

51. Whole sunflower seed supplementation after parturition hastens the reinitiation of ovarian activity in primiparous but not in multiparous grazing Holstein dairy cows

AF Mendoza[a], D Crespi[a], AF La Manna[a], MA Crowe[b] & D Cavestany[a]

[a]Nat Agriculture Research Institute (INIA), Colonia, Uruguay and [b]School of Agriculture, Food Science & Vet Medicine, University College Dublin, Ireland

Polyunsaturated fatty acids fed to dairy cows can modulate different reproductive processes, like follicular growth, due to changes in the dynamics of hormones like IGF-I or precursors for steroidogenesis like cholesterol[1], although information in dairy cows in grazing conditions is scarce. We designed an experiment to evaluate the effect of the addition of whole sunflower seeds to the diets of grazing Holstein dairy cows (n = 48) during the first 2 months postpartum (PP) on the fate of the first follicular wave (FFW) and the interval from parturition to first ovulation. Treatments were: 0 (C), 0.7 (L), and 1.4 (H) kg/cow/day of whole sunflower seed, and were asigned to 24 primiparous and 24 multiparous Holstein cows. The diet was complemented with grass-legume pastures, whole-plant wheat silage, and concentrates, and was designed to be isoenergetic and isonitrogenous (6.7 MJ NEL/kg DM, 18.5 % CP). Ovaries were examined by ultrasonography three times a week from day 8 PP until ovulation or regression of the dominant follicle of the FFW, and thereafter once a week until ovulation and formation of a corpus luteum. The diameter of the dominant follicle (DDF) during the FFW was recorded and the concentration of plasma IGF-I and cholesterol at that moment were measured. Number of cows ovulating during the FFW (OV1), interval from parturition to first ovulation (DOV), DDF, IGF-I and cholesterol, were determined by PROC GLM and PROC FREQ (SAS), with treatment, parity and its interaction as the main effects. Pearson correlation coefficients were calculated to establish relationships between reproductive variables, IGF-I, and cholesterol. Milk production and body condition score (scale of 1 to 5) were recorded weekly.

A significant interaction was detected between OV1 and DOV ($P < 0.05$). While in treatments L and H, 7/8 and 6/8 primiparous cows ovulated the first postpartum follicular wave, only 1/8 did it in treatment C ($P < 0.05$), whereas no differences were detected in multiparous cows ($P > 0.05$). DOV was 31.3, 16.4 and 19.8 days for the primiparous cows of treatments C, L, and H, respectively ($P < 0.05$), and 22.8, 24.2 and 26.0 for multiparous cows of treatments C, L, and H, ($P > 0.1$). The DDF in the FFW was not affected by the treatments ($P > 0.1$). Plasma IGF-I concentration was not affected by fat supplementation or parity ($P > 0.1$), and was not related to the onset of luteal activity or the characteristics of the FFW ($P > 0.1$). Plasma cholesterol was not affected by treatments or parity ($P > 0.1$), and was not related to any reproductive variable ($P > 0.1$). Fat corrected milk (4 %) production and body condition score (scale 1 to 5) during the experimental period was 23.1, 24.3 and 24.4 kg and 2.36, 2.33 and 2.37, for treatments C, L and, H, respectively, and were not affected by sunflower seeds supplementation ($P > 0.1$).

Whole sunflower seeds supplementation hastened the resumption of ovarian activity after parturition in primiparous dairy cows, which is the category with a longer anoestrous period in grazing conditions[2], but not in multiparous cows. These results could not be explained by alterations of plasma IGF-I or cholesterol concentrations, so other mechanisms that link fat supplementation with a faster resumption of PP ovarian activity must be evaluated in the future.

[1]Staples et al. 1998. J Dairy Sci 81:856-871.
[2] Meikle et al. 2004. Reproduction 127:727-737.

52. Comparison between the addition of a progesterone releasing device for 7 or 9 days in a modified Ovsynch/TAI protocol

D Cavestany[a], A Bentancur[b] & G Grasso[b]

[a]National Agricultural Research Institute (INIA), Colonia, & Department of Reproduction, Veterinary College, Uruguay and [b]DVM. Private practitioners

The development of an estrus synchronization protocol that allowed insemination at a fixed time (TAI), avoiding the need of estrus detection[1] was an important achievement in reproductive management programs. However, lack of synchrony of ovulation in heifers, and thus lower pregnancy rates has been a major constraint for the application of this protocol in this category[2]. In order to overcome this we tried, in 85 Holstein heifers (25.5 ± 0.5 months old and 406 ± 7 Kg body weight), a modified Ovsynch/TAI protocol with the addition of a progesterone releasing vaginal device (Terapress®) for 7 days between the first GnRH and the $PGF_{2\alpha}$ treatment ($n = 43$), or for 9 days between the first and the second GnRH treatments ($n = 42$). Treatment protocol was: Day 0: GnRH (0.250 mg of Gonadorelin, Fertagyl, Intervet, Montevideo, Uruguay); Day 7: $PGF_{2\alpha}$ (0.150 ml of d-cloprostenol, Enzaprost, Biogénesis, Montevideo, Uruguay); Day 9: GnRH as in Day 0; Day 10: (16 hours after the GnRH) TAI. An intravaginal progesterone releasing device with 1 g of natural progesterone (Terapress®, Biogénesis, Montevideo, Uruguay) was inserted at Day 0, and removed at Day 7 at the same time of the $PGF_{2\alpha}$ treatment (Ovsynch + P4) or at Day 9, at the same time as the GnRH treatment (Ovsynch9 + P4). Heat detection was done twice a day from Day 0 to Day 9, and heifers in estrus were inseminated 12 hours later. Pregnancy diagnosis was done by ultrasonography at day 33 after service. Data was analyzed by logistic regression (SAS).

The addition of the progesterone device suppressed the ovulations until removal, but in group Ovsynch + P4 47% of the heifers were detected in estrus between Day 7 and Day 9 and were inseminated. In group Ovsynch9 + P4, as the progesterone source was removed at the time of the second GnRH treatment, no heats were observed and all heifers were TAI. Pregnancy rate after TAI for the Ovsynch + P4 group was 47.4% and for the Ovsynch9 + P4 38.1% ($P > 0.1$). Within the Ovsynch + P4 group, conception rate for heifers inseminated after a detected heat (before the second GnRH treatment) was 52.6%, not statistically different to the 47.4% pregnancy rate of the heifers bred at a timed AI ($P > 0.1$).

Conception rate of heifers inseminated after a detected heat was lower than the $> 60\%$ expected, in comparison to related reports[2], so in comparison, the overall pregnancy rate of 41.2% in the TAI heifers was considered acceptable in this situation. It must be noted that, if estrous detection had not been performed and all heifers were bred at the appointed time, pregnancy rate in this protocol, and under these field conditions, would have been much lower. Notwithstanding the overall herd fertility in this trial, the maintenance of the progesterone source until the day after the TAI, suppressed premature heats and ovulation without altering the fertility, and makes this variation appealing.

[1]Pursley et al. 1995. Theriogenology 44:915-923.
[2]Ambrose et al. 2005. Theriogenology 64:1457-1474.

53. Cytological changes of cervical discharge during ovulation synchronization programs using an intra-vaginal progesterone-releasing device

MR Ahmadi[a], S Nazifi[a], J Sajedianfard[b], G Moattari[c]

Departments of [a]Clinical Studies, [b]Physiology and [c]Graduate School of Veterinary Medicine, Shiraz University, Shiraz, Iran

Synchronization of estrus in cattle can facilitate the use of artificial insemination by reducing the time and cost spent on estrous detection compared to cattle entering estrus spontaneously. Therefore, providing a tool for determination of estrus is beneficial. The objective of this study was to determine cytological changes in cervical mucus during the induction of estrus with controlled intra-vaginal drug release (CIDR) devices in dairy cows. Sixty post-partum healthy Holstein Frisian dairy cows were selected from a commercial dairy farm around Shiraz, in southern Iran. All cows used were 80 ± 10 days post-partum with no clinical signs of endometritis evident during routine examination performed to monitor their fertility. Records of estrous behavior, palpation of growing corpora lutea and response to the induction of estrus were used to determine the date of estrus. Cows were synchronized using one of five protocols, namely, Ovsynch (n = 10; 15 μg GnRH 10 days before AI, PGF2α 3 days before AI, 15 μg GnRH 16 hours before AI), CIDR + OV (n = 18; controlled intra-vaginal drug release device + Ovsynch: as the Ovsynch protocol with the addition of a CIDR place vaginally at the time of the first GnRH injection and removed at the time of PGF2a administration), S-CIDR + OV (n = 9; skeleton of CIDR + Ovsynch protocol- as the CIDR + OV protocol except the CIDR contained no progesterone), CIDR + E2 (n = 16; CIDR placed vaginally and 4 mg estradiol benzoate injected 10 days prior to AI, CIDR removed and PGF2α administered 3 days before AI, 1 mg estradiol benzoate administered 24 hours before AI) and S-CIDR + E2 (n = 7, as for CIDR + E2 except the CIDR contained no progesterone).

Blood samples were collected from the coccygeal vein on days -10, -3, 0 and 12 relative to AI from all groups. Cervical mucus discharges were collected on same days and analyzed cytologically. There were significant differences ($p < 0.05$) between percentages of neutrophil densities in cervical mucosa at different days in group S-CIDR + E2. This variation was not observed in other groups. There were significant positive correlations between progesterone and estradiol concentrations and neutrophil and epithelial cell percentages in S-CIDR + OV and S-CIDR + E2 groups. However, the correlation was not significant in other groups. There were significant differences ($P < 0.05$) between neutrophil proportion in cervical mucus smears of the five synchronization methods on day zero or day of insemination. There were no differences among groups in neutrophil percentages 10 or 3 days before AI or 12 days after AI. However, on the day of AI, neutrophil percentages were significantly higher ($P < 0.05$) in the S-CIDR + OV and S-CIDR + E2 groups (28.89 ± 43.36 and 39.00 ± 48.71, respectively) compared to the Ovsynch group (0.6 ± 1.58). Values observed in the CIDR + OV and CIDR + E2 were intermediate at 5.61 ± 13.73 and 14.56 ± 28.66, respectively. Although we could not show the effect of hormonal changes on the variation of cell population in each group clearly, comparing neutrophil proportion on day zero among the synchronization methods used supported the effect of hormonal changes and mechanical effects of CIDR on cellular population of cervical mucus. Additional studies are required to further determine the hormonal and mechanical effects of CIDR devices on the uterus.

54. Validation of a synchronization protocol for beef cows in Brazil

RL Valarelli[a] & JLM Vasconcelos[b]

aPfizer Animal Health, São Paulo, SP, Brazil and bFMVZ- UNESP, Botucatu, SP, Brazil

In 2005 a protocol was developed for synchronization of beef cows[1]. That protocol promoted a 50.6% conception rate after timed artificial insemination (TAI) in Nelore cows and consisted of the insertion of an intravaginal P4 device (CIDR 1.9g P4, Pfizer, Brazil) plus an injection of 2 mg of estradiol benzoate (Estrogin, Farmavet, Brazil) on day 0. On day 7, 12.5 mg of dinoprost (Lutalyse, Pfizer, Brazil) was administered. Forty eight hours later (day 9) CIDR was removed, 0.5 mg of estradiol cipionate (ECP, Pfizer, Brazil) was injected and the calves were removed until finishing TAI, that was performed 48 hours after CIDR removal.

This study aimed to validate the protocol described above. Data of 7527 TAI on beef cows during 2005/2006 breeding season, after synchronization with this protocol was analyzed by logistic regression.

Conception rate at TAI was 48.9% (3681/7527) and was affected ($P < 0.001$) by category, breed and body score condition (BSC). The conception rate differed between dry cows, primiparous and multiparous [39.4% (137/348), 46.3% (937/2022) and 50.5% (2607/5157)]. Crossbred cows showed higher conception rate at TAI when compared to *Bos taurus* and Nelore cows [62.3% (964/1547), 53% (150/283) and 45.1% (2567/5697)]. Cows with lower BSC showed lower conception rate at TAI [BSC 2.5: 41.9% (687/1638); BSC 2.75: 50.3% (831/1653); BSC 3.0: 49.4% (1006/2035); BSC 3.25: 49,2% (386/785); BSC 3.5: 54.5% (771/1416)]. The variable number of times that CIDR was used before (0 vs 1 vs 2) did not affected conception rate at TAI.

These results validated the protocol for the synchronization of beef cows and factors like category, breed and BSC must be considered before implementing a TAI program in beef cows.

[1]Meneghetti *et al.* 2005. A Hora Veterinária 147:25-27.

55. The effects of ovarian function on estrus synchronization with PGF in dairy cows

A Waldmann[a,b], J Kurykin[a], Ü Jaakma[a], T Kaart[a], M Aidnik[a], M Jalakas[a], L Majas[a] & P Padrik[a]

[a]Institute of Veterinary Medicine and Animal Sciences, Estonian University of Life Sciences, Tartu, Estonia and [b]Institute of General and Molecular Pathology, University of Tartu, Estonia

This investigation was undertaken to test the hypothesis that ovarian function prior to, or following synchronization treatments, may account for poor estrous synchrony and low pregnancy rates following timed AI (TAI) in dairy cows. The study was conducted on an 1800 cow commercial dairy herd with average annual milk production of 8400 kg per cow. A total of 108 cows were treated twice at a 14 day interval with 25 mg im of PGF (Dinolytic, Pharmacia & Upjohn Co.) to synchronize estrus beginning 70 ± 1.4 (mean $\pm$ SE) days after calving. Single inseminations were performed 80 to 82 h after the second PGF treatment, without regard to signs of behavioural estrus. Milk samples were taken every 2 days starting at the time of the first PGF treatment until day 48 after TAI for measurement of P4 concentrations. At the time of the second PGF treatment and TAI the ovaries of cows were visualized with a real-time B-mode diagnostic ultrasound scanner (Honda, HS-120, Japan), and size of follicles and CL were recorded. Test day milk yield and milk composition data were obtained from the Estonian Agricultural Registers and Information Centre (Tartu, Estonia). For statistical comparisons cows were grouped according to ovarian response to PGF treatment. The following groups were formed: normal (expected) response to PGF treatment, failure of luteolysis, delayed ovulation after luteolysis, and anestrus. Statistical comparisons in milk P4 data, size of ovarian structures, and lactation performance data (milk yield, milk composition, fat to protein ratio at milk recordings) between the groups were conducted using GLM Procedure of SAS (SAS Institute Inc., Cary, NC, USA). Milk P4 profiles revealed that anestrus, failure of luteolysis following treatment with PGF, and delayed ovulation following luteolysis were the main reasons for low pregnancy rate with TAI. Twenty eight percent of the treated cows could not conceive during the 7-day interval after the second injection of PGF because they did not ovulate during that time. Pregnancy rates 24 and 47 days after TAI in cows that responded as expected to the synchronization treatment were 62 % and 54 %, respectively. Pregnancy was precluded in nonresponsive cows. The largest follicle at the time of TAI in cows experiencing late embryonic mortality was smaller (P = 0.02) than in cows that successfully maintained pregnancy suggesting that follicular size in cows undergoing TAI after synchronization by PGF might affect late embryonic loss. Response to PGF was associated with variation in milk yield, milk composition, and concentration of milk P4. Anestrous cows had a higher percentage of milk fat (P < 0.05) and higher fat to protein ratio (P < 0.01), and cows that did not undergo luteolysis had higher milk yield (P < 0.05) and lower percentage of milk protein (P < 0.05) than cows that responded to PGF treatment. Cows that did not undergo luteolysis and cows that did not ovulate following luteolysis had lower milk P4 during the luteal phase preceding the second PGF injection (P < 0.01 and P < 0.05, respectively). Results suggest that a primary reason for low pregnancy rates in dairy cows after administration of PGF and TAI is discrepancies in ovarian function at the time of, or following treatment. Further studies are needed to clarify whether P4 and milk test day data (milk yield, milk fat, milk protein, fat to protein ratio in milk) can be used to assign cows to estrous synchronization protocols according to postpartum ovarian function. Financial support of the work was provided by the Estonian Science Foundation (Grants 4822, 6065), the Estonian Ministry of Agriculture, and TARPM 0421.

56. Development and evaluation of a steroid-based method for synchronization of ovulation and timed AI in repeat-breeder lactating dairy cows in Thailand

S Sangsritavong[a] & MC Wiltbank[b]

[a]Biotec Central Research Unit Thailand Science Park, Prathumthani, Thailand and [b]Department of Dairy Science University of Wisconsin-Madison, WI 53706 USA

Our understanding of the bovine estrous cycle and the physiology of follicular wave dynamics have expanded greatly in recent years, primarily with the advent of ultrasonography to monitor the dynamics of ovarian structures. This knowledge has accelerated the development of timed AI protocols such as Ovsynch that allow AI without the need for detection of estrus. Due to high prices for GnRH in many parts of the world and the desire to develop more efficient protocols for timed AI, many recent studies have utilized other hormonal treatments, particularly steroid hormones in designing protocols for synchronizing ovulation. This abstract describes the follicular wave dynamics (Experiment 1) and the conception rates (Experiment 2) during a steroid-based synchronization protocol developed for repeat-breeding dairy cattle in Thailand. In Experiment 1 an intravaginal progesterone-releasing device (CIDR) was inserted into the vagina of 24 lactating dairy cows at random stages of the estrous cycle. At the same time, cows were treated with progesterone (50 mg) and estradiol-17ß (2 mg) by i.m. injection (Day 0 of protocol). On Day 6 cows were treated with 25 mg $PGF_{2\alpha}$ and the CIDR was removed on Day 7. One day later (Day 8) cows were treated with estradiol benzoate (1 mg). Cows had follicular sizes determined by ultrasonography on a daily basis from Day 0 until Day 8 and every 6 h after the estradiol benzoate injection (Day 8) until ovulation. Days to new follicular wave emergence, ovulation times, and sizes of preovulatory follicles were recorded. Emergence of a new follicular wave occurred at 3.67 +/-1.09 days after estradiol-17ß + progesterone treatment (Day 0). A total of 79.2% (19/24) of the cows ovulated to the estradiol benzoate treatment with an ovulatory follicle size of 13.3 +/-1.6 mm. Ovulation time was 77.5 +/-6.8 h. after CIDR removal. In Experiment 2, a total of 1390 repeat-breeding dairy cows (open cows with $\geq$3 previous inseminations) had ovulation synchronized by the protocol outlined in Experiment 1 and received timed AI at 60 hr after CIDR removal. Conception rates during each month of the year were recorded and compared to normal cows bred to a natural estrus (breeding by AM/PM rule) in the same herds (n = 184 farms). Overall conception rates for the two groups was greater (P = 0.03) for the timed AI group (32.0%; 455/1390) than for cows bred to a natural estrus (28.2%; 653/2315). During the summer months there was a greater (P = 0.003) conception rate in cows bred to the synchronized ovulation and timed AI (29.6%; 104/351) than to the natural estrus (19.5 %; 152/779). Surprisingly, there was also an overall greater (P = 0.08) percentage of female calves born to the cows bred to the timed AI program (58.2%; 225/387) than cows bred to a natural estrus (48.7%; 279/572). Thus, this synchronization protocol appears to produce a good synchronization of follicular waves and sufficient synchronization of time of ovulation to produce conception rates in repeat-breeder dairy cows that are similar or sometimes better than conception rates in normal cows bred to a natural estrus. The intriguing observation of improved conception rates during summer heat stress by using a steroid-based synchronization and timed AI protocol needs to be confirmed with future randomized trials using this protocol. In addition, the dramatic increase in percentage of cows with female offspring after this timed AI protocol is an intriguing but mechanistically-unexplained observation that is difficult to ignore.

57. Detection of ovulation failure permits early resynchronization and increase number of pregnant cows in 28 days

JLM Vasconcelos[a] & PHM Garcia[b]

[a]FMVZ-UNESP, Botucatu, SP, Brazil and [b]Fazenda São João, Inhauma, MG, Brazil

Ovulation rate after detected estrus was reported to be 84.8% in lactating Holstein cows[1]. The aim of this study was to evaluate if early detection of ovulation failure (by presence of CL between days 7 to 13 after AI) with early resynchronization increases pregnancy rates within 28 days in lactating Holstein cows.

The trial was conducted at a dairy farm located in Inhauma, MG, Brazil in 2005. Lactating Holstein cows (n = 3876) producing 26.7 ± 8.3 kg of milk/d with 213.7 ± 130.6 DIM were bred by AI (12 h after heat detection) or by Timed-AI (TAI). The TAI protocol consisted of an intravaginal P4 device (CIDR 1.9mg Pfizer Animal Health, Brazil), plus an injection of GnRH (1mL Fertagyl Intervet) on the first day. After 7 days, the CIDR was removed and an injection of PGF2α (5mL Lutalyse Pfizer Animal Health, Brazil) was administered, followed by an injection of Estradiol Cypionate (0.5mL ECP Pfizer Animal Health, Brazil) 24 h later. Cows were inseminated 48 h after ECP. Cows were divided in two groups: those in Group 1 (G1; n = 2989) were not evaluated for the presence of CL and those in Group 2 (G2; n = 887) were evaluated for the presence of CL by ultrasound (US). In G2, 597 cows were on day 7 after timed AI and 290 cows were between days 7 and 13 after AI. The cows in G2 detected without CL were resynchronized with the protocol above. Pregnancy diagnosis was performed by US between 28 to 34 days after AI. Cows that showed heat between AI and pregnancy diagnosis were inseminated again. The data were analyzed by a chi-square test.

Early evaluation of presence of CL by US did not affect conception rates: 23% (687/2989) in G1 and 23.8% (211/887) in G2. In the cows evaluated for the presence of CL, the incidence of CL absence in those inseminated after heat detection was 24.5% (71/290) and in those bred by TAI it was 24.1% (144/597). Cows with a CL had a conception rate of 28.3% (62/219) after AI and 32.9% (149/453) after TAI. Cows without a CL at evaluation were resynchronized and the conception rate was 25.6% (55/215). Heat detection in the period of 28 days and conception rate were, respectively, 49.9% (1148/2302) and 23.6% (271/1148) for G1 and 70.1 (323/461) and 20.7% (67/323) for G2. Heat detection was higher (P < 0.01) in G2, probably due to the cows without a CL being resynchronized. The percentage of pregnant cows in 28 days was higher (P < 0.01) in G2: 37.5% (211 + 55 + 67 / 887) than in G1: 32.2% (687 + 271 / 2989).

These data show that ovulation failure could be a factor that negatively affects conception rates in lactating Holstein cows. Early diagnosis of open cows (absence of CL by US) did not affect conception and, associated with early resynchronization (EARLY-RESYNCH), increases pregnancy rate.

---

[1]Vasconcelos et al. 2006. Theriogenology 65:192-200.

58. Fourteen days of temporal calf removal reduced postpartum anoestrous without compromising maternal bond or calves' normal growth

G Quintans, D Negrín & C. Jimenez de Aréchaga

National Institute of Agricultural Research (INIA), Ruta 8 km 281, 33000, Uruguay

The detrimental effect of suckling on the duration of post-partum anoestrous (PPA) has been extensively reported. The maternal-offspring bond is generally considered to be the major cause of delayed ovulation in beef cows. Under extensive conditions and in primiparous cows in moderate to low body condition, premature weaning was effective shortening the PPA while the use of nose plates was unable to maintain normal oestrous cycles after induced ovulation[1]. In an exploratory study, calves were completely removed from their mothers for 10 days and the maternal bond was not broken[2]. The aim of the present study was to determine the influence of long-term calf separation (14 days) on reproductive performance and on calf growth in primiparous cows under range conditions. This experiment involved 60 primiparous Angus x Hereford cows with a post-calving live weight of (mean $\pm$ S.E.M.) 351 $\pm$ 4.76 kg and body-condition score (BCS) of 4.3 $\pm$ 0.06 units (1 = thin, 8 = fat). The cows remained with their calves until 70.7 $\pm$ 1.3 days postpartum (designated Day 0), when they were assigned to one of three treatments: i) calves remained with and had free access to their dams (C; n = 20), ii) calves fitted with nose plates for 14 days while remaining with their mothers (NP, n = 20), iii) calves were completely removed from their dams for 14 days and then returned to their dams (CR, n = 20). During the period of separation (CR) calves were offered alfalfa hay, water on an *ad libitum* basis and a total of 9.5 kg of a high protein (16% crude protein) supplement, offered in an increased sequential way. The mating period lasted 75 days commencing 15 days before the onset of treatments (Day -15). Cows were naturally mated using two bulls of proven fertility. Cow live weight and BCS were recorded at calving and at monthly intervals. Calves were weighed at calving, at the start of the experiment and at biweekly intervals. Calves were completely weaned at 6 months of age. Two pregnancy diagnosis by ultrasonography were done to report early (first 45 days of mating period) and total pregnancy rate. Detailed observations for signs of oestrous behaviour were conducted twice a day (6 am and 6 pm) for an hour, during the mating period. Two cows from the C group, one from the NP and three from the CR group had ovulated before treatments began and were subsequently eliminated from data analysis. General Linear Model procedure and Fisher´s exact test were used. The interval (days) from the onset of treatments (Day 0) to the first oestrous was shorter ($P < 0.05$) in CR (6 ± 4.4) and NP (10 ± 4.1) than in C cows (38 ± 4.3). The early pregnancy rate was higher ($P < 0.05$) in CR cows (94%) than in C cows (55%) but not different in NP (74%) and the final pregnancy rate was also higher ($P < 0.05$) in CR cows (94%) than in C cows (61%) but not different in NP (79%). The mean daily live weight of calves (kg/a/d) was lower ($P < 0.05$) in NP (0.412 ± 0.02) and in CR groups (0.398 ± 0.02) than in C calves (0.579 ± 0.02) and the weaning weight was higher ($P < 0.05$) in C calves than in NP and CR calves (158 ± 3.7, 140 ± 3.3 and 143 ± 3.4 kg respectively). Fourteen days of calf separation did not break the maternal bond, calves had a lower but acceptable weaning weight and cows presented higher pregnancy rate than control cows.

[1]Quintans & Vázquez, 2002. Proc. 6th Internat. Ruminants Reprod Symp (A65).
[2]Quintans *et al.* 2003. Proc .IX World Conference Anim Production pp 219.

59. Comparison of the reproductive performance of tropically adapted heifers treated with three different protocols for synchronization of estrus

NJ Phillips[a], C Hockey[a], D Boothby[b], ST Norman[a] & MR McGowan[a]

[a]School of Veterinary Science, University of Queensland, Brisbane, Australia and [b]Beef Breeding Services, Queensland Department of Primary Industries and Fisheries, Brisbane, Australia

Previous studies[1] conducted in northern Australia have shown that the pregnancy rates following fixed-timed insemination (FTAI) are significantly higher in tropically adapted heifers treated with progestogen/progesterone protocols to synchronise estrus compared to those treated with a double prostaglandin $F_{2\alpha}$ ($PGF_{2\alpha}$) protocol. The objective of this study was to compare the pregnancy rates of heifers treated with progesterone and double $PGF_{2\alpha}$ protocols but inseminated following observed signs of estrus.

225 Santa Gertrudis heifers weighing an average of 411kg, aged 21mths, BCS 3.5, and with palpably normal reproductive tracts were stratified according to weight and then assigned to one of three treatment groups (n = 72-77 heifers each). Group 1 heifers received two intramuscular injections of $PGF_{2\alpha}$ (500µg/ml cloprostenol, Estroplan, Parnell Labs, Australia) 10 days apart. Group 2 heifers received 1mg oestradiol benzoate (OB; 1mg/ml, Ciderol, Genetics Australia, Australia) and an intravaginal progesterone implant (CIDR, 1.9g, Pfizer); at day-8 the CIDR was removed and each heifer received 250µg/ml $PGF_{2\alpha}$. Group 3 heifers received 1mg OB and a CIDR; the CIDR was removed at day-14 and 500µg/ml $PGF_{2\alpha}$ was administered to each heifer at day-24. Heifers were observed for signs of oestrus twice daily (am/pm) for at least one hour during the period 36 to 80 hours after the final treatment. Heifers seen standing to be mounted and those whose mount detector (Estrus Alert™) coating was >75% erased were recorded as being in estrus. These heifers were AI'd within 12 hours of the first in-oestrus observation; those not on heat by 80 hours were FTAI'd. All heifers were pregnancy tested by rectal palpation 9 weeks later. The significance of differences in submission and pregnancy rates between treatment groups were determined using chi-square.

There was no significant difference in the overall pregnancy rates (includes FTAI'd heifers) between groups 1 (41%), 2 (43%) and 3 (38%). Submission rates (proportion of heifers seen on heat and AI'd during the observation period) for groups 1, 2 and 3 were 63% (48/76), 85% (61/72) and 78% (60/77), respectively, with significant difference identified between groups 1 and 2 (p-value 0.003) and groups 1 and 3 (p-value 0.045). The pregnancy rates for heifers AI'd to observed oestrus in Groups 1, 2 and 3 were 60, 51 and 42%, respectively, with the difference between groups 1 and 3 approaching significance ($P < 0.06$). The insemination protocol used in this study (oestrous detection and insemination for 2 days followed by FTAI of remaining heifers) resulted in an overall pregnancy rate for the double $PGF_{2\alpha}$ group similar to that achieved by those treated with a standard CIDR protocol (Group 2). The marked difference in pregnancy rate for heifers AI'd at observed oestrus following treatment with $PGF_{2\alpha}$ (Group 1 v's Group 3) cannot be readily explained, and will be investigated in future trials.

[1]Kerr et al. 1991. Therio 36:129-141.

60. Changes in vaginal electrical resistance during different stages of the reproductive cycle in tropically adapted beef heifers

CD Hockey[a], ST Norman[a], JM Morton[a], D Boothby[b], NJ Phillips[a] & MR McGowan[a]

[a]School of Veterinary Science, University of Queensland, Brisbane, Australia and [b]Beef Breeding Services, Queensland Department of Primary Industries and Fisheries, Brisbane, Australia

An important limitation to conducting large scale AI programmes in beef cattle is the difficulty of efficiently and accurately identifying oestrous animals. Studies have shown that the vaginal electrical resistance (VER) of cattle during oestrus is significantly lower than in diestrus[1]. However, the majority of research has been conducted in dairy cattle and little VER work has been done on tropically adapted beef cattle. The purpose of this trial was to determine the difference in VER of tropically adapted beef heifers during different stages of the reproductive cycle and to evaluate whether measurements of VER using a commercially available probe (Ovatec probe®, Probe Diagnostics, Inc.) could be used to improve heat detection in a large scale AI programme.

225 Santa Gertrudis heifers of similar age and weight (average 411kg, 21 mths and BCS 3.5 on a scale of 5) were randomly divided into three groups of similar sample size (n = 72-77) and subjected to three oestrus synchronisation protocols. Group 1 were injected twice intramuscularly (IM) with 500µg of PGF2α (Estroplan®, cloprostenol 250µg/ml), on days 14 and 24 of the experiment; Group 2. were given an injection of 1mg of oestradiol benzoate (ODB; Ciderol® 1mg/ml) IM and had a CIDR® (1.9g progesterone) inserted on day 16, on day 24 the CIDR was removed and the heifers received an injection of 250µg of PGF2α IM; Group 3 were given an injection with 1mg of ODB IM and had a CIDR inserted on day 0, on day 14 the CIDR was removed, and on day 24 they were given an injection of 500µg of PGF2α. All heifers had Estrus Alert™ patches placed on them on day 24 and were observed for signs of oestrus on days 26 and 27. Animals were considered to be in oestrus if they were observed standing whilst being mounted at least once during an observation period of one hour, twice daily am/pm, or had > 75% of their Estrus Alert™ rubbed off within the previous observation period. Oestrous heifers were inseminated within 12 hours of first observation. Immediately prior to AI a VER measurement was taken according to the Ovatec probe® manufacturer's instructions. Group 1 had VER values recorded on day 24 at the time of their second PGF2α injection. At 3 and 9 weeks following insemination all groups had VER measurements taken. At 9 wks following insemination the heifers were rectally palpated for pregnancy.

Heifers presumed in oestrus (n = 167, 75%) showed a VER at time of insemination of 78.66 ± 10.18 and this value was not different (p < 0.05) to the VER at time of insemination for only the heifers that became pregnant (VER = 78.97 ± 9.58, n = 91). VER of heifers presumed in dioestrus (Group 1 at time of last PGF2α treatment, VER = 114 ± 20.63, n = 76) was significantly different to those presumed in oestrus (p < 0.01). Using these values a ROC curve was plotted for diagnosis of dioestrus with an area under curve = 0.9702. Hence measurements of VER using the Ovatec probe® are highly discriminatory and may be useful to determine oestrus in tropically adapted beef heifers. Values of VER for pregnant heifers at 3 and 9 wks post AI (VER = 120.47 ± 23.21 and 123.99 ± 28.11 respectively) were not significantly different to values for presumed dioestrus (p < 0.05). Thus a single VER measurement is not a good indicator of pregnancy.

[1]Leidl & Stolla, 1976. Theriogenology 6:237-249.

61. Estrous detection in Holstein heifers using a radio-telemetric pedometer under different rearing conditions

M Sakaguchi[a], R Fujiki[b], K Yabuuchi[b], Y Takahashi[b] & M Aoki[a]

[a]Department of Animal and Grassland Sciences, National Agricultural Research Center for Hokkaido Region and [b]Graduate School of Veterinary Medicine, Hokkaido University, Sapporo, Japan

Rapid progress in genetics and management in the dairy industry has resulted in increased milk production per cow, and metabolic demands for more milk negatively impact reproductive function of postpartum cows. It was demonstrated that cows having the longest intervals from calving to first ovulation produced the most milk and also had prolonged intervals to first estrous activity[1]. However, the delayed first ovulation did not always prolonged the days open, and first inseminations at first estrus had the possibility to produce greater pregnancy rates than those at subsequent periods of estrus[1,2]. Thus, the exact detection of estrus is important to improve reproductive performance of high yielding dairy cows. Therefore, a preliminary evaluation was made of the efficiency and accuracy of a new radiotelemetric pedometer system for estrous detection using Holstein heifers reared in pasture, an open paddock, or a tie-stall barn.

Data were collected from total 45 estrous cycles of 15 heifers (11-15 months old) fitted with a pedometer (Gyuho, COMTEC, Japan) on both the neck and hind leg. An additional pedometer was attached to the front leg of each heifer reared in a tie-stall barn. The number of steps per hour counted by each pedometer was automatically transferred to a computer. The latest 24h-total value (L-value) at each 1h-interval time point was compared with the mean value of past days (P-mean) as a reference period (RP, 3-7 days). When the ratio of L-value to P-mean exceeded a threshold level, estrus was deemed to have occurred and the detection date and time was recorded. After finishing each trial, the activity data were downloaded and retrospectively reanalyzed to determine true positive (TP) and false positive alerts at different criteria (threshold levels and RP). Efficiency and accuracy were evaluated by the ratio of TP to total visually observed estrus and by the ratio of TP to total pedometer alerts, respectively.

The number of steps counted by hind leg-pedometer was better correlated with that counted by the visual observation as compared to by neck pedometer under the pasture and paddock rearing conditions. In the pasture condition, the neck pedometer detected 80-100 % of 10 estrous activities at 1.2-1.3 fold of threshold with 3-7 days of RP but the accuracy was only 20-32%. The hind leg pedometer detected estrus of pastured heifers with 80-100 % efficiency and 89-100% accuracy at 1.4-1.7 fold threshold and 5-7 days RP. For 12 estrus activities of heifers under the paddock condition, efficiency and accuracy of detection were 92 % and 61-65% (1.3 fold threshold, 6-7 days RP) by the neck pedometer and 92% and 92-100% (1.6-1.7 fold, 4-7 days) by the hind leg pedometers, respectively. The heifers housed in a tie-stall barn showed 23 estrus activity events and the neck pedometer detected 61-92% of them with 33-44% accuracy (1.2-1.3 fold, 3-7 days). The hind and front leg pedometers detected 70-91% of estrus events with 60-88% accuracy and 74-96% with 50-83% at 1.3-1.4 fold threshold and 3-7 days RP, respectively. These results indicate that the neck pedometer is relatively useful for estrous detection of heifers reared in a paddock and that the leg pedometer can detect estrous activities efficiently and accurately in a variety of rearing conditions. The most suitable threshold setting for estrous detection of this system depends on the bodily location of the pedometer and rearing conditions.

[1]Sakaguchi et al. 2004. J Dairy Sci 87:2114-2121.
[2]Sakaguchi et al. 2006. Vet Rec (in press).

62. A novel approach to induce the absence of a large follicle at the start of FSH treatment improving superovulatory response in goats: Day 0 Protocol

A Menchaca[a], M Vilariño[a] & E Rubianes[b]

[a]Laboratorio de Fisiología de la Reproducción, Facultad Veterinaria, UdelaR and [b]Dpto de Producción Animal y Pasturas, Facultad Agronomía, UdelaR, Montevideo, Uruguay

Higher superovulatory response is obtained in goats when gonadotropin treatment begins in absence of a dominant follicle[1]. A "Day 0 Protocol" was designed to begin the FSH administration soon after ovulation (Day 0), at the moment of emergence of wave 1 in the absence of a dominant follicle[2]. The present work compares Day 0 Protocol to a Traditional Protocol in multiparous Alpine goats during the breeding season (May-July, Uruguay, 34° SL). In the Day 0 Protocol (n = 20), the first follicular wave was synchronized by insertion of a CIDR-G (0.3 g progesterone, InterAg, New Zealand) before FSH administration as described previously[3]. The FSH treatment started 84 hours after CIDR withdrawal. Traditional treatment (n = 18) began with the insertion of a CIDR-G for 11 days, with FSH and one dose of a PGF2α analogue being given on day 9. In both groups, superovulatory treatment consisted of six decreasing doses of FSH (12 h apart, 8.8 mg NIADDK oFSH-17, Ovagen, ICPbio, Auckland, NZ). With Day 0 Protocol, two half doses of PGF2α analogue (each 80 μg) were given together with the last two FSH doses. One dose of GnRH analogue (8.0 μg, buserelin acetate, Receptal, Hoechst, Argentina) was given 24 hours after the first half dose of PGF2α or 24 hours after CIDR withdrawal for Day 0 Protocol and Traditional treatment, respectively. Timed artificial insemination using frozen semen was performed 15 and 24 hours later. Follicular characteristics at the onset of FSH treatment were determined by transrectal ultrasonography (7.5 MHz, Aloka 500, Japan). The number of corpora lutea (CLs) was determined by laparoscopy at time of embryo collection, seven days after insemination. Data were compared using χ^2 or ANOVA after $\log + 1$ transformation.

Premature regression of CLs occurred in 9/18 goats with Traditional treatment against only 3/20 with Day 0 Protocol (P < 0.05). At the onset of FSH treatment, the diameter of the largest follicle was smaller in animals receiving the Day 0 Protocol (4.3 ± 0.3 mm) than using Traditional treatment (6.6 ± 0.4 mm; P < 0.001), and the number of small follicles (< 4.0 mm) was higher for Day 0 Protocol (11.8 ± 1.1 vs 7.2 ± 0.7; respectively; P < 0.001). The number of CLs were higher for Day 0 Protocol than Traditional treatment (9.6 ± 0.6 vs 6.3 ± 0.8, respectively; P < 0.01). The number of transferable embryos showed a tendency to be higher (4.9 ± 0.7 vs 2.6 ± 0.5; P < 0.1) and the transferable/total embryos rate was higher (84/114: 74% vs 23/43: 54%; P < 0.05) for Day 0 Protocol compared to Traditional treatment, respectively.

These results confirm the effectiveness of the Day 0 Protocol which begins a superovulatory treatment in the absence of a large follicle, and suggest this method may provide a better embryo yield per goat compared to the Traditional method.

[1]Menchaca, Pinczak & Rubianes, 2002. Theriogenolgy 58:1713-1721.
[2]Rubianes & Menchaca, 2003. Anim Reprod Sci 78:271-287.
[3]Menchaca & Rubianes, 2004. Reprod Fertil Develop 16:403-414.

63. Endocrine response and time of ovulation induced after a Short Progesterone Treatment associated with eCG or estradiol benzoate in goats

A Menchaca[a] & E Rubianes[b]

[a]Laboratorio de Fisiología de la Reproducción, Facultad Veterinaria, UdelaR and [b]Dpto de Producción Animal y Pasturas, Facultad Agronomía, UdelaR, Montevideo, Uruguay

Short Progesterone Treatment associated with eCG is an effective method to obtain a high pregnancy rate after a single Timed Artificial Insemination in goats[1]. However, the endocrine response to Short Progesterone Treatment has not yet been described. In addition, the repeated use of eCG to induce ovulation could promote inadequate response associated to a high immunogenic effect. Estradiol benzoate (EB) has been used in cattle to induce ovulation, but it has not been evaluated in goats as another option to avoid the use of eCG. In the present study, we used Short Progesterone Treatment while comparing efficacy of eCG or EB to synchronize ovulation, measuring progesterone, estradiol 17ß and LH serum concentrations. Twenty three multiparous Alpine dairy goats were used during the breeding season (May-June, 34° South Latitude, Uruguay). Short Progesterone Treatment consisted of 5 days intravaginal treatment with a CIDR-G insert (0.3 g progesterone, InterAg, New Zealand) together with one dose of a PGF2α analogue (160 µg delprostenate, Universal Lab, Montevideo, Uruguay) at CIDR insertion. Thereafter, either one dose of 250 IU eCG (Novormon, Syntex, Buenos Aires, Argentina) was administered at CIDR withdrawal (eCG group; n = 7); or an im dose of 200 µg of EB was given 24 hours after CIDR withdrawal (EB group; n = 8); or the goats remained as a control group receiving neither eCG nor EB (n = 8). Blood samples were collected during 120 hours after CIDR withdrawal for determination of serum hormones by RIA. Progesterone and estradiol-17ß concentrations were measured at 12 hour intervals and LH levels at 4 hour intervals. Timing of ovulation was determined by transrectal ovarian ultrasonography using a 7.5 MHz probe (Aloka 500, Tokio, Japan). Data were compared by ANOVA.

In 21/23 goats progesterone concentrations declined significantly from CIDR withdrawal (1.8 $\pm$ 1.8 ng/ml) until 12 hours later (0.2 $\pm$ 0.1 ng/ml; P < 0.01). The maximum concentrations of estradiol-17ß were higher in the EB group (76.9 $\pm$ 24.6 pmol/l) than in the control group (41.8 $\pm$ 9.0 pmol/l; P < 0.01), and were intermediate in the eCG group (70.3 $\pm$ 32.5 pmol/l). The LH peak occurred earlier in the eCG and EB group (38.4 $\pm$ 2.0 hours and 41.0 $\pm$ 4.1 hours, respectively) than in the control group (46.3 $\pm$ 5.1 hours; P < 0.05). Ovulation was induced in 5/7 eCG goats and 8/8 EB goats and occurred earlier (58.8 $\pm$ 2.7 hours and 63.0 $\pm$ 5.6 hours, respectively) than in 7/8 responding control goats (70.3 $\pm$ 8.3 hours; P < 0.05).

In conclusion, Short Progesterone Treatment induced adequate levels of progesterone, allowing administration of estradiol-17ß or LH to synchronize ovulation in dairy goats. Estradiol benzoate given 24 hours after CIDR withdrawal synchronized LH peak and ovulation in 100% of treated goats. These results demonstrate that EB is an efficacious alternative to eCG for inducing ovulation and allowing Timed Artificial Insemination in goats.

[1]Rubianes & Menchaca, 2003. Anim Reprod Sci 78:271-287.

64. GnRH and eCG associated with a progesterone treatment increased pregnancy rate after FTAI in prepubertal heifers

A Menchaca[a,b], M Vilariño[a] & D Ibarra[c]

[a]Departamento de Fisiología, Facultad de Veterinaria, UdelaR, [b]Syntex SA and [c]Departamento de Reproducción, Facultad de Veterinaria, UdelaR, Montevideo, Uruguay

Fixed-Time Artificial Insemination (FTAI) after a progesterone-oestradiol benzoate treatment has been used to achieve an acceptable pregnancy rate (i.e. higher than 50%) in postpartum anoestrous cows[1]. However, in anoestrous heifers (i.e. prepubertal heifers), this same protocol does not result in an acceptable pregnancy rate. The objective of this study was to evaluate the inclusion of eCG and/or GnRH for improving pregnancy rates in prepubertal heifers submitted to FTAI following a progesterone-OB protocol.

A factorial 2x2 design was applied using 486 Aberdeen Angus and Hereford heifers 14 to 16 months old. Heifers had an individual body weight ≥240 kg, and a body condition score (Mean ± SD) of 5.1 ± 0.3 (1 to 8 scale). Prepubertal status was defined by the absence of oestrous behaviour during the previous 20 days before the experiment began and was confirmed by transrectal ovarian ultrasonography (Aloka 500, Tokio, Japan) at the start of the treatment. All of the heifers received 2 mg of OB (Syntex, Buenos Aires, Argentina) at the same time as insertion of an intravaginal bovine device (DIB, 1 g of progesterone, Syntex, Buenos Aires, Argentina). At DIB withdrawal, 7 days later, heifers either received or did not receive an injection of 300 IU eCG (Novormon, Syntex, Buenos Aires, Argentina). Heifers within each of these two groups then received either 1 mg of OB 24 hours after DIB withdrawal (OB group, n = 127; OB + eCG group, n = 118), or 50 µg of GnRH analogue (lecirelina acetate, Gonasyn, Syntex, Buenos Aires, Argentina) 48 hours after DIB withdrawal (GnRH group, n = 118; GnRH + eCG group, n = 123). Fixed-time AI was performed 52 to 56 hours after DIB withdrawal and pregnancy rate was determined by ultrasonography 45 days later. Data were analysed using logistic regression.

Pregnancy rate was improved with the use of GnRH (118/241, 49.0%) vs OB (98/245, 40.0%; $P < 0.05$) and also with the use of eCG (120/241, 49.8%) vs no eCG (97/241, 40.3%; $P < 0.05$). The combination of GnRH + eCG resulted in a pregnancy rate of 55.3% (68/123), which was higher ($P < 0.05$) than the OB group (36.2%, 46/127; $P < 0.05$). Pregnancy rates were intermediate for the GnRH and OB + eCG groups (42.4%, 50/118 and 44.1%, 52/118; respectively).

In the present study, GnRH and eCG administered after a progesterone treatment resulted in a higher pregnancy rate than the OB treatment. The implication is that an acceptable (> 50 %) pregnancy rate with FTAI in prepubertal beef heifers is achievable using 7 days of progesterone treatment with 2 mg of OB at DIB insertion, 300 IU of eCG at DIB withdrawal and GnRH 48h after DIB withdrawal.

[1]Bó et al. 2003. Anim Reprod Sci 78:307-326.

65. Day 0 Protocol for superovulation in sheep: ovulatory response and embryo yield

A Menchaca[a], M Vilariño[a], S Kmaid[b], A Pinczak[b] & JM Saldaña[b]

[a]Laboratorio de Fisiología de la Reproducción, Facultad de Veterinaria, UdelaR and [b]Private Assesor, Montevideo, Uruguay

A new protocol to initiate the FSH treatment in absence of a large follicle has been developed in goats[1]. In sheep, most of the studies agree that the physiological moment that ensures the absence of a large follicle is found immediately after ovulation (Day 0), when the large follicle of the previous cycle has just ovulated[2]. The Day 0 Protocol begins the FSH treatment soon after ovulation thereby stimulating the first follicular wave. In the present study we determined the ovulation rate and embryo yield obtained with Day 0 Protocol compared to traditional superovulatory treatment on sheep. A total of 44 multiparous Merino ewes were treated during the breeding season (October, 40° NL, China). For the Day 0 Protocol (n = 22), the FSH treatment was started after a short interval of progesterone treatment[2] to synchronize ovulation using a CIDR-G (0.3 g progesterone, InterAg, Hamilton, New Zealand) for 6 days. An im dose of prostaglandin F2α analogue (PGF2α, 160 μg delprostenate, Glandinex, Universal Lab, Montevideo, Uruguay) and 340 IU of eCG (Pregnecol, Bioniche, Ontario, Canada) were administered at the time of CIDR insertion and withdrawal, respectively. An analogue of GnRH (8.0 μg, buserelin acetate, Receptal, Hoechst, Buenos Aires, Argentina) was given 36 hours after CIDR withdrawal. The first FSH dose in Day 0 Protocol was administered 84 hours after CIDR withdrawal. Traditional superovulatory treatment (n = 22) consisted of 14 days of progesterone exposure using two CIDR for 7 days each, beginning the FSH treatment two days before progesterone treatment finished. Both groups received the same superovulatory treatment using 8 decreasing FSH doses (12 h apart, 240 mg NIH-FSH-P1, Folltropin, Bioniche, Ontario, Canada). In Day 0 Protocol two half doses of a PGF2α analogue were given with the last two FSH doses. In Traditional treatment, PGF2α was administered with the first FSH dose. One dose of GnRH analogue was injected 24 hours after the first half dose of PGF2α or 24 hours after CIDR withdrawal for Day 0 Protocol and Traditional treatment, respectively. Laparoscopic artificial insemination was performed 15 hours after GnRH dose. Ovulation rate was checked and embryos were collected six days later. Data were compared using ANOVA after $\log + 1$ transformation.

The number of CL per treated donors was higher in Day 0 Protocol than in Traditional treatment (13.5 ± 1.4 vs 10.1 ± 1.1, respectively; $P < 0.05$). Total collected structures (10.3 ± 1.2 vs 8.0 ± 0.9) and transferable embryos (7.9 ± 1.4 vs 5.9 ± 1.1) showed a tendency to be higher using Day 0 Protocol than Traditional treatment ($P < 0.1$).

These results show a higher ovulation rate using Day 0 Protocol on sheep, as has been demonstrated in goats[2]. This novel treatment produced two more embryos per donor; however, other studies should be performed to confirm this tendency in the embryo yield.

[1]Rubianes & Menchaca, 2003. Anim Reprod Sci 78:271-287.
[2]Menchaca & Rubianes, 2004. Reprod Fertil Dev 16:403-413.

66. Effects of suckling inhibition with nose plates associated to a timed insemination protocol in primiparous beef cows

T de Castro[a], A Menchaca[bc], R Bonino[c] & A Peñagaricano[c]

Departments of [a]Animal Reproduction and [b]Physiology, Veterinary Faculty, Montevideo, and [c]Syntex SA, Uruguay

The use of Fixed-Time Artificial Insemination (FTAI) is increasing in bovine breeding programs raised under rangeland conditions[1]. Suckling inhibition (i.e. using nose plates on calves) is a low cost tool that is being implemented by the farmers to improve pregnancy rates in postpartum beef cows. To study the effects of suckling inhibition associated with a progesterone-estradiol benzoate (EB) based treatment for FTAI, a trial was conducted at San Sebastian farm, Florida Dept., Uruguay (34° SL) using 193 primiparous beef cows (121 Hereford, 62 Hereford x Aberdeen Angus and 11 Hereford x Limousin). Animals were kept together grazing on rangeland conditions. Before starting treatments, ovarian status was evaluated by transrectal ultrasonography using a 7.5 MHz probe (Aloka 500, Japan). Cows were in a mean ($\pm$SEM) BCS of 4.5 ± 0.04 (1-8 scale) and ultrasonography showed that 9/193 (4.7%) cows had a CL, 125/193 (64.8%) had follicles $\geq$8mm (no CL) and 59/193 (30.6%) had follicles <8mm (no CL). Animals were divided in two homogeneous groups according to breed, BCS, ovarian status and days postpartum. All the cows received on Day 0 (60–90 d postpartum) a DIB (bovine intravaginal device, 1 g progesterone, Syntex, Argentina) and 2 mg of EB (Syntex, Argentina) were administered. On Day 8 DIBs were removed and cows received 150 µg of D(+) cloprostenol (Ciclase, Syntex, Argentina) and 400 IU of eCG (Novormon, Syntex, Argentina). On Day 9, 1 mg of EB was administered. All cows were fixed-time inseminated at 52-56 hours after device removal with frozen thawed evaluated semen from the same bull. In one group of animals (NP, n = 95) nose plates were placed on calves on Day 0 in order to prevent suckling and left for 10 d. Nose plates were removed at the moment of FTAI. In the other group (S, n = 98), dams were permanently suckled by their calves. Pregnancy was diagnosed by transrectal ultrasonography 32 d after FTAI. Results were analyzed using Chi-square test.

Conception rates were similar in both groups of animals, 58/95 (61.1%) and 54/98 (55.1%) for NP and S respectively (P = 0.4). Results show that although a large proportion of the cows were not cycling (postpartum anestrous), progesterone-EB based treatment yielded an overall conception rate of 58% (112/193). Although no statistically significant differences were found between treatment groups, further trials with larger number of animals are needed to confirm the usefulness of application of nose plates associated with FTAI protocols.

[1]Bó et al. 2003. Anim Reprod Sci 78:307-326.

67. GnRH-induced gonadotropin surge and ovulation in heifers: Effect of day of cycle

JA Atkins, DC Busch, JF Bader, DH Keisler & MF Smith

Division of Animal Sciences, University of Missouri, Columbia, MO, USA

Increasing the proportion of heifers inseminated artificially is dependent upon the development of an effective fixed-time artificial insemination (TAI) protocol that results in acceptable pregnancy rates. A TAI protocol that has been used in beef heifers and cows includes an injection of GnRH on day -9 (GnRH1), prostaglandin $F_{2\alpha}$ (PGF) on day -2, and an injection of GnRH on day 0 (GnRH2; day of insemination). Pregnancy rates following the preceding protocol have consistently been lower in heifers compared to cows in part due to a reduced proportion of heifers ovulating following GnRH1 (day -9). Our laboratory previously reported that only 42% of pubertal heifers ovulated following GnRH1, which may be due to absence of a dominant follicle, presence of a dominant follicle that is unable to ovulate (atretic), and(or) inadequate gonadotropin secretion in response to GnRH1.

To examine the effect of day of the cycle on GnRH1-induced LH release and ovulation, 43 pubertal heifers were assigned by age and weight to one of five treatment groups based on day of the cycle when GnRH1 was administered (day 2, 5, 10, 15, or 18; day 0 = estrus) or to a control group (PGF on day 7 and GnRH two days later; day 0 = estrus). The preceding days of the estrous cycle were chosen to represent different physiological stages at which a dominant follicle may or may not be present while the control group was designed to ensure the presence of a dominant follicle and measure the maximum amount of LH released in response to GnRH1. Day of the cycle affected the proportion of heifers ovulating after GnRH1 (0/7[a], 8/8[b], 0/6[a] 5/8[c], and 5/8[c] of the day 2, 5, 10, 15, and 18 heifers respectively; [abc]P < .05) and GnRH2 (3/7[a], 8/8[b], 5/6[ab], 1/8[c], and 2/8[c] of the day 2, 5, 10, 15, and 18 heifers, respectively; [abc]P < .05). Each control heifer ovulated (6/6) after GnRH1. Excluding the control group, total percentage ovulating to GnRH1 and GnRH2 was 49% and 52%, respectively. Peak LH concentrations occurred 90 minutes after GnRH1 treatment and were highest in the control heifers followed by heifers in the day 18, 5, 15, 10, and 2 groups (113[a], 84[b], 60[b], 36[c], 24[c], and 10[d] ng/mL, respectively; [abcd]P < .05). Total LH released (arbitrary units) during the four hours following GnRH1 was highest in the control heifers (13,642[a]) followed by heifers in the day 18 (9,054[b]), 5 (5,774[bc]), 15 (4,672[c]), 10 (2,548[c]), and 2 (915[d]; respectively; [abcd]P < .05) groups. Heifers that ovulated in response to GnRH1 had more total LH released (P < .03) than heifers that did not ovulate; however this was confounded by day of the cycle. Total LH secreted following GnRH1 treatment was not different in heifers that did or did not ovulate in the day 15 and 18 heifers. Serum concentrations of estradiol followed a similar pattern as the total LH released and were highest in the control heifers followed by heifers in the day 18, 5, 15, 10, and 2 groups (9.4[a], 4.3[b], 3.7[b], 1.5[c], 1.4[c], 0.9[c] pg/mL, respectively; [abc]P < .05).

In summary, day of the cycle affected the proportion of heifers ovulating and total LH released in response to GnRH1. It is unclear whether the low ovulatory response of heifers to GnRH1 is due to an inadequate amount of LH released and(or) the stage of the follicular wave at GnRH1 injection due to confounding effects of day of cycle and response to GnRH1.

68. Effect of progesterone content in a vaginal insert on pregnancy rates in *Bos indicus* cross-bred beef heifers inseminated at a fixed time

D Pincinato[a], L Cutaia[abc], LC Peres[a] & GA Bo[ab]

[a]*Instituto de Reproduccion Animal Cordoba (IRAC), [b]Universidad Catolica de Cordoba and [c]Syntex S.A. J.L. de Cabrera 106, X5000GVD, Cordoba, Argentina*

We have previously shown that a vaginal progesterone-releasing device impregnated with 0.5 or 1.0 g of progesterone (P4) resulted in comparable pregnancy rates to fixed-time artificial insemination (FTAI) protocols in dairy cows and heifers and dry beef cows[1]. An experiment was designed to valuate pregnancy rates in cross-bred beef heifers treated with new or once-used P4-releasing devices impregnated with 0.5 or 1 g of P4 and FTAI. Cross-bred *Bos indicus* heifers (n = 239), 20 to 26 months of age and BCS between 2.5 and 3.5 (1 to 5 scale) were randomly allocated in 1 of 4 treatment groups in a 2 x 2 factorial design. On Day 0, all heifers received 2 mg of estradiol benzoate (EB, Syntex, Argentina) i.m. and were divided to receive a new or a once-used (i.e. used previously for 8 days) P4 device impregnated with 0.5 g of P4 (DIB 0.5; Syntex) or a P4 device impregnated with 1 g of P4 (DIB, Syntex). On Day 8, DIB devices were removed and all heifers received 150 µg D (+) cloprostenol (Ciclase, Syntex) i.m. On Day 9, all heifers received 1 mg EB and were FTAI 52 to 56 h after DIB removal. Ultrasonography was performed on Day 0 to determine the presence of a CL (all heifers had a CL) and 30 d after FTAI to determine pregnancy rates. Data were analyzed by logistic regression. There were no significant differences (P > 0.6) in pregnancy rates in heifers treated with new (26/60, 43.3%) or once-used DIB devices (27/59, 45.0%) impregnated with 1 g of P4, or those treated with new DIB devices impregnated with 0.5 g of P4. However, pregnancy rates were significantly lower (P < 0.003) in heifers treated with once-used DIB devices impregnated with 0.5 g of P4 (12/60, 20%). It was concluded that treatment of cross-bred zebu heifers with P4 devices containing 1 or 0.5 g of P4 results in comparable pregnancy rates when new devices are used. However, pregnancy rates are significantly affected with DIB devices impregnated with 0.5 g of P4 that were previously used.

[1]Cutaia *et al.* 2006. . Reprod Fertil Dev 18:114-115.

69. Effect of time of prostaglandin F2α administration on pregnancy rates in *Bos indicus* cross-bred beef heifers treated with progesterone vaginal devices and inseminated at a fixed time

L Cutaia[abc], LC Peres[a], D Pincinato[a] & GA Bó[ab]

[a]*Instituto de Reproducción Animal Córdoba (IRAC),* [b]*Universidad Católica de Córdoba and* [c]*Syntex S.A. J.L. de Cabrera 106, X5000GVD, Cordoba, Argentina*

We have previously shown that a vaginal progesterone (P4) releasing device impregnated with 0.5 or 1.0 g of P4 resulted in comparable pregnancy rates to fixed-time artificial insemination (FTAI) protocols in beef and dairy cows and dairy heifers[1]. Furthermore, induction of early luteolysis, by giving prostaglandin F2α (PGF2α) at the time of insertion of a P4-releasing device, hastened follicular wave emergence and increased the size of the ovulatory follicle in cycling beef cows[2]. An experiment was designed to test the hypothesis that lowering circulating levels of P4 during follicle wave emergence and growth would increase pregnancy rates to FTAI in *Bos indicus* cross-bred beef heifers treated with P4-releasing devices. The effects of a half dose of PGF2α at the time of insertion and removal of P4-vaginal devices impregnated with 0.5 or 1 g of P4 was compared to that of a full dose of PGF2α at device removal. Crossbred *Bos indicus* heifers (n = 239), 20 to 26 months of age and BCS between 2.5 and 3.5 (1 to 5 scale) were randomly allocated in 1 of 4 treatment groups in a 2 x 2 factorial design. On Day 0, all heifers received 2 mg of estradiol benzoate (EB, Syntex, Argentina) i.m. and were divided to receive either a P4-device impregnated with 0.5 g of P4 (DIB 0.5; Syntex) or a P4-device impregnated with 1 g of P4 (DIB, Syntex). Heifers in each group were subdivided to receive 75 µg of D(+) cloprostenol (Ciclase, Syntex) i.m. at the time of DIB insertion (Day 0) and again at DIB removal (Day 8) or a single treatment of 150 µg D(+) cloprostenol on Day 8. All heifers received 1 mg EB on Day 9 and were FTAI 52 to 56 h after DIB removal. Ultrasonography was performed on Day 0 to determine the presence of a CL and 30 d after FTAI to determine pregnancy rates. Data were analyzed by logistic regression. There were no significant differences ($P > 0.6$) in pregnancy rates in heifers treated with DIB devices impregnated with 1 g of P4 (134/239, 56.1%) and those treated with DIB devices impregnated with 0.5 g of P4 (133/243, 54.7%). Pregnancy rates were significantly higher ($P < 0.05$) in heifers treated with PGF2α at the time of insertion of a DIB device impregnated with 1 g of P4 (74/118, 62.7%) than in those treated with PGF2α at the time of DIB removal (60/121, 49.6%). However, the time of PGF2α treatment did not affect pregnancy rates among heifers treated with DIB impregnated with 0.5 g of P4 (70/124, 56.5% vs 63/119, 52.6% for heifers treated with PGF2α on Days 0 and 8 and those treated with PGF2α on Day 8, respectively). Finally, the presence or absence of a CL significantly affected ($P < 0.01$) pregnancy rates (241/407, 59.2% and 23/63, 36.5% for heifers with or without a CL on Day 0, respectively). It was concluded that early luteolysis increased pregnancy rates in cross-bred *Bos indicus* heifers treated with P4-devices containing 1 g of P4, but not in those treated with P4 devices containing 0.5 g of P4.

[1]Cutaia *et al.* 2006. Repro Fertil Dev 18:114-115.
[2]Bo *et al.* 2004. ICAR, Abstracts 1:110.

70. Starting a GnRH-PGF$_2$α based estrus synchronization protocol by day 21 postpartum can induce normal ovarian activity within a 60-d voluntary waiting period in dairy cows

C Amaya-Montoya[a], M Matsui[a], C Kawashima[b], G Matsuda[a], M Ono[a], B Devkota[a], K Kida[c], A Miyamoto[b] & Y-I Miyake[a]

[a] Department of Clinical Vet. Sci., [b]Graduate School of Anim. and Food Hygiene and [c]Field Center of Anim. Sci. and Agric., Obihiro University of Agric. and Vet. Med., Obihiro, Hokkaido, Japan

In dairy cows, resumption of ovarian activity within 3 wk post partum (pp) significantly favors early conception by increasing the number of cycles before first insemination[1,2]. Moreover, anestrous dairy cows express estrus, ovulation and reduction of subsequent abnormal cycles following exposure to progesterone (P$_4$) within the first month pp[3]. To evaluate the efficiency of an estrous synchronization protocol during pp, the ovarian response and the subsequent cyclicity were examined.

Lactating Holstein cows (n = 21) received GnRH, followed by PGF$_{2\alpha}$ 7-d later. Cows were treated with GnRH on d 21 (G21, n = 8) or around d 37 (G37, n = 13) pp. Untreated cows served as control (C, n = 40). Blood samples were collected throughout the pp and within the protocol, and assayed for P$_4$. Ovarian morphology within the protocol was monitored daily by transrectal ultra-sonography. Endocrine and morphology evaluations at GnRH treatment confirmed cows as having (ACL, n = 12) or not having (ICL, n = 9) active corpora lutea (CL). All cows ovulated in response to GnRH. At the time of PGF$_{2\alpha}$ treatment, ICL cows had larger induced CL than ACL cows ($p < 0.01$), but similar P$_4$ levels. ICL cows had larger synchronized ovulatory follicles than ACL cows ($p < 0.05$). Ovulation rate after PGF$_{2\alpha}$ was similar for G21 and G37 (85.7% and 76.9%), but greater in ACL than in ICL cows (91.7% vs. 62.5%).

To investigate the effect of early ovulation pp, the ovarian cyclicity after GnRH- PGF$_{2\alpha}$ treatment was evaluated. Commencement of luteal activity within 3 wk pp (CLA) was confirmed by plasma P$_4$ levels. In C group, the cows without CLA (n = 27) showed delayed occurrence of typical cyclicity than those with CLA (n = 13) (55.8 d vs. 41.3 d; $p < 0.01$). Cows without CLA treated with GnRH-PGF$_{2\alpha}$ at d 37 pp (n = 7) also showed delayed occurrence of ovarian cyclicity (55.5 d). However, treatment with GnRH-PGF$_{2\alpha}$ at d 21 pp accelerated the start of cyclicity (35.3 d) in the cows without CLA (n = 3). In the cows with CLA, GnRH-PGF$_{2\alpha}$ treatment at d 21 pp (n = 3) had no effect on the occurrence of ovarian cyclicity when compared to C cows with CLA (36.6 d vs. 41.3 d, respectively). Regardless of CLA, mean cycle length after GnRH-PGF$_{2\alpha}$ protocol in G21 (21.5 d) and G37 (22.2 d) were comparable to the one between the 2nd and 3rd ovulation in C cows with and without CLA (25.1 d and 21.7 d, respectively). Abnormal cycles were reduced in G21 and G37 cows when compared to those of C cows (33.3%, 25% and 50% respectively).

In conclusion, starting a GnRH-PGF$_{2\alpha}$ based protocol in dairy cows by 21 d pp can synchronize estrus and ovulation, and complete the resumption of normal cyclicity within a 60-d voluntary waiting period despite a first ovulation within 3 wk pp. These findings suggest that the plasma P$_4$ rise following an induced ovulation within 3wk pp may enhance the resumption of normal ovarian cyclicity.

[1]Darwash et al. 1997. Anim Sci 65:9-16.
[2]Senatore et al. 1996. Anim Sci 62:17-23.
[3]Roche et al. 1992. Anim Reprod Sci 28:371-378.

71. Bovine viral diarrhea virus (BVDV) in bovine fetuses collected from slaughterhouses in Presidente Prudente region, Sao Paulo State, Brazil

CN Camargo[a], NTC Galleti[a], E Stefano[a], LH Okuda[a], LC Vianna[c], C Del Fava[a], EM Pituco[a] & DU Mehnert[b]

[a]Instituto Biológico. Av.Cons.Rodrigues Alves 1252,CEP 04014-002,São Paulo/SP/Brazil, [b] Departamento de Microbiologia - Instituto de Ciências Biomédicas – USP and [c] Universidade do Oeste Paulista - UNOESTE

BVDV is one of the main viral agents causing embryonic and fetal death in bovines. The consequences of infection may be abortion, congenital defects, growth impairment, and the birth of persistently infected (PI) animals. Due to the unknown epidemiological situation of the infection in the country, the present trial aimed to evaluate the occurrence of BVDV in fetuses of zebu heifers. Fetuses were collected at slaughterhouses serving 26 cities in the western region of Sao Paulo state, Brazil, between May and December, 2003. Herds analyzed were mostly growth and finishing ones, and there was an intense movement of animals among farms. A total of 274 fetuses in different stages of development were necropsied. Organs (brain, liver, spleen, kidney, lungs and placenta) and fetal serum were collected, as well as serum of the mothers. The presence of antibodies was analyzed by virus neutralization (VN) in the serum of the mothers, whereas in fetal serum, ELISA (Idexx Laboratories™) was used due to toxicity of the sample. Fetal serum samples were also analyzed for the presence of the virus by an ELISA kit for BVDV antigen (Idexx Laboratories™). Viral RNA extracted from the brain of fetuses was reverse transcribed and subjected to nested multiplex PCR using primers of the genomic region of protein NS5B, which amplifies 360 pb for BVDV 1 and 604 pb for BVDV 2. Results of VN demonstrated a frequency equal to 40.51 % (111/274) in reactive mothers. Total frequency of fetal infection was 6.2 % (17/274), derived from the sum of the subtotals of positive samples obtained by the three techniques (antibody by indirect ELISA: 2.18 % (6/274), ELISA antigen: 2.18 % (6/274) and PCR: 1.82 % (5/274)). Therefore, it is recommended that both serological and molecular tests are used in order to increase the sensitivity in BVDV diagnosis because they complement each other. All positive fetuses were born from seroreactive heifers. It should be emphasized that the age of the mothers is a risk factor for primary infection with BVDV as the heifers are more susceptible to the disease after the disappearance of colostral antibodies (six months of age) until two years of age. Nested multiplex enabled the characterization of four samples of BVDV serotype 1 (1.46%) and one of BVDV 2 (0.36%) from the brain samples, emphasizing the importance of the production of vaccines containing both serotypes. The frequency of maternal antibodies was similar to that of the national mean. The present study enables the conclusion that the virus is disseminated in the region studied and there is a great risk of BVDV infection in the herds due to the intense movement of animals. Therefore, studies that evaluate the economical impact produced by BVDV should be undertaken in order to determine control measures in the herds of this region.

72. Parturition duration and birthing difficulty in twin and triplet lambs

JM Everett-Hincks, KG Dodds, KJ Knowler & JI Kerslake

AgResearch Invermay, Puddle Alley, Private Bag 50034, Mosgiel, New Zealand

The primary objective of the study was to investigate the relationship between observations of parturition behaviour, localised subcutaneous oedema (as determined by post mortem findings) and lamb survival. A flock of mixed age Coopworth ewes were monitored from mating through to lamb weaning at AgResearch's Woodlands farm in 2004 to investigate birthing difficulties in twin and triplet lambs. Parturition behaviour was recorded for 131 ewes and their 303 lambs (twins n = 180; triplets n = 123). Parturition duration was recorded as a lamb trait and adjusted for the random maternal effect (litter effect) using the MIXED procedure in SAS. Lamb parturition times were log transformed and were reported as back transformed values and SEMs were back transformed to multiplicative values.

Twin lambs and first born triplet lambs that did not survive to three weeks of age took at least twice as long to be born than lambs that survived (parturition duration was recorded from the time the feet were observed for the first born lamb and for lambs that died versus lambs that survived to three weeks of age: first born twin lambs 27 x/÷ 1.6 minutes vs 10 x/÷ 1.2 minutes, P < 0.05; first born triplets 32 x/÷ 1.6 minutes vs 7 x/÷ 1.3 minutes, P < 0.10 and second born twins 107 x/÷ 1.8 vs 36 x/÷ 1.2, P < 0.05). Parturition durations were greater than three times longer for all lambs diagnosed with moderate to severe localised subcutaneous oedema (n = 17) compared to those that did not have localised oedema (n = 9) at post mortem (50 x/÷ 1.7 minutes vs 14 x/÷ 1.5 minutes, P < 0.05).

This study showed that birthing difficulties determined by prolonged parturition were associated with increased lamb losses and the incidence of localised subcutaneous oedema for dead lambs. Further research is required with a larger dataset to verify the findings and to better understand the factors contributing to prolonged parturition for larger litters. It is necessary to examine if the mechanisms which cause prolonged parturition, birthing difficulties and localised subcutaneous oedema, operate differently over a range of physiological and management conditions for different ewe ages and whether there are genetic influences. With this information appropriate management techniques can be identified and appropriate animal breeding programmes formulated to reduce birthing difficulties (dystocia) and lamb mortality rates.

73. Effect of vitamin E /selenium supplementation prior to mating on ewe reproduction

E Emsen, M Yaprak, CA Gimenez Diaz

Ataturk University, Department of Animal Science, Erzurum, Turkey

It is widely documented that vitamins and minerals play an important role in the growth of animals and their physiological functions, as well as their reproductive performance. Segerson and Ganapathy[1] recorded significantly more fertilized ova, significantly more uterine contractions and significantly more ewes with large numbers of sperm attached to the zona pellucida due to improved sperm mobility when ewes (marginal blood selenium status 0.05 mg/ml) were given injections of 10 mg Se and 136 IU vitamin E before mating. The objectives of this study were to evaluate the effects of vitamin E plus Se administration on fertility, litter size, birth weight of lambs and survivability of lambs at weaning.

Sixty Awassi and 40 Redkaraman ewes were equally divided into two groups (Vit E plus Se and Control) to evaluate the effect of supplemental vitamin E (300 IU) plus selenium (2mg as sodium selenite), given daily for 14 days prior to mating, on reproductive performance. Vitamin E + Se were administrated by adding into the drinking water of animals in treatment groups. There were no differences between treatment and control groups in any reproductive trait excepting lamb survival. The number of lambs averaged 1.17 and 1.14 in treatment and control groups respectively. At weaning, survival of lambs was increased ($P < 0.05$) by vitamin E + Se treatment, treated ewes weaned approximately 10% more lambs per ewe mated.

Estrus and fertility results obtained in this study were lower than those of obtained by Gabryszuk and Klewiec[2] who injected Polish Merino ewes with 5 ml sodium selenite plus 250 mg vitamin E four weeks before mating. The reason for the lower fertility compared to Polish Merino ewes might be explained by the administration method where group supplementation was utilized so that amount of vitamin E + Se consumption per ewe was difficult to determine, and also breed differences. Similar fertility and prolificacy values between control and treatment groups suggest that the vitamin E and Se status of ewes in control groups was adequate for reproductive performance or that overall reproduction is not effected by vitamin E and Se supplementation prior to mating. Likewise, these data also indicate that a high concentration (about 10 times NRC required concentration) of vitamin E and Se did not adversely affect reproductive performance.

Data obtained in this study indicate that vitamin E and Se did not significantly increase reproduction in Awassi and Redkaraman ewes. The remarkable result of this research was the increase in survivability of lambs born from supplemented ewes.

[1]Segerson & Ganapathy, 1981. J Anim Sci 51:386-394.
[2]Gabryszuk & Klewiec, 2002. Small Ruminant Research 43:127-132.

74. Hormonal control of fat-tailed sheep reproduction

E Emsen, M Kutluca & CAG Diaz

Ataturk University, Department of Animal Science, Erzurum, Turkey

Estrus synchronization (ES) in goats and sheep is achieved by control of the luteal phase of the estrous cycle, either by providing exogenous progesterone or by inducing premature luteolysis. The traditional product of choice for ES in goats and sheep is the intravaginal sponge impregnated with progestagen (e.g., flurogestone acetate or methyl acetoxyprogesterone) for 9 to 19 d followed by PMSG injected 48 to 0 h from sponge removal[1].

The aim of this study was to determine the effects of exogenous hormone treatments on fertility of Awassi, Red Karaman and Tuj ewes, and survivability and performance of their lambs. In this research, 59 Awassi, 42 Red Karaman and 44 Tuj ewes and their lambs were used. Each breed of animals was randomly allocated into two groups; one being the experimental group and the other the control group. Estrus was synchronized by means of fluorogestone acetate (30mg FGA) intravaginal sponges and eCG (600 I.U.) was injected i.m. at the time of sponge withdrawal to the experimental group on October 15th (breeding season in the region).

The interval from sponge removal to the onset of estrus was significantly different ($P < 0.01$) between treatment groups. Lambing rate of Awassi, Red Karaman and Tuj ewes treated with FGA + PMSG and control were 85.7% and 77.4%; 91.3% and 84.2%; 66.7% and 95.2% respectively. Prolificacy significantly increased following treatment with eCG from 104 to 163% in the Awassi, from 119 to 138% in the Red Karaman and from 110% to 143% in the Tuj ewes. Progestagen sponge and eCG treatment resulted in an extra 7.4 (Awassi), 2.6 (Red Karaman) and 4.2 (Tuj) kg meat being produced per ewe, compared to the control groups. Birth and weaning weights of the lambs from three breeds of fat-tailed ewes were found to be different ($P < 0.01$) for the treatment (3.2kg) and control (3.8kg) groups. There was no significant differences in lamb survival rates at weaning (75 days of age) in the three breeds (100% versus 83% and 94% versus 89% and 89% versus 86% in the Awassi, Red Karaman and Tuj control and treated ewes, respectively).

The response of Awassi ewes to 600 IU eCG was generally higher than those for the Red Karaman and Tuj ewes. This is most likely due to the fact that the Awassi is a more prolific breed than the other two fat-tailed breeds. On the other hand, Awassi lambs were recorded with higher mortality rates compared to Tuj and Red Karaman lambs. The main reason for this could be that the Red Karaman and Tuj sheep is a local breed of the region and is better adapted to unfavorable environmental factors detrimentally effect survivability rate. It is concluded that intravaginal sponges (FGA) can be used for estrus synchronization and mating in Awassi, Red Karaman and Tuj fat-tailed ewes during the breeding season and that treatment with eCG following progestagen treatment has the potential of increasing the twinning rate. This can provide a more uniform lamb crop for sale to the consumer and extra meat produced per ewe.

[1]Wildeus, 1999. Current concepts in synchronization of estrous: sheep and goat. Available: http://www.asas.org/JAS/symposia/proceedings/0016.pdf

75. Interactions between temperament and reproductive performance

D Blache[a], K.Hart[ab], A Chadwick[a], F Sebe[c], T St Jorre de St Jorre[a], P Poindron[c], R Nowak[c] & D Fergusson[d]

[a]School of Animal Biology, The University of Western Australia,, WA, Australia, [b]Department of Agriculture Western Australia, [c]UMR 6175 INRA-CNRS-Université de Tours-Haras Nationaux, 37380 Nouzilly, France and [d]CSIRO Livestock Industries, FD McMaster Lab, NSW, Australia

Temperament can be defined as the "fearfulness" and the reactivity of an animal in response to human contact and/or novel or threatening environments[1]. Temperament is moderately heritable ($h^2 \sim 0.26$ to 0.40) in different breeds of sheep. Using a selection index based on two behavioural tests, one measuring the reactivity to human presence and the other measuring the reactivity to isolation, Merino sheep have been selected for "calmness" and "nervousness" over 15 years at the University of Western Australia. The ewes from the selection lines have been naturally joined to a single sire (1:30 ratio, no synchronisation, no flushing) under various climatic and nutritional conditions at the Allandale experimental farm (WA). Data obtained over the last few years show that the selection for calmness has a positive/beneficial impact at different stages of the reproductive cycle.

Inexperienced calm ewes are more proceptive and receptive to rams than ewes selected for nervousness[2]. The selection does not seem to have an impact on the male as neither their sexual behaviour nor the quality of the sperm produced was affected. The number of twins detected by scanning at around 50 days of gestation is double in the calm than in the nervous strain. The effect of temperament on ovulation rate or on embryo survival is not known yet. Behavioural and physiological differences between the two lines lead to a higher lamb mortality rate in the nervous line than in the calm line. In fact, the calm lambs are less agitated and less vocal in response to isolation during the first days of life. Maternal behaviour and rearing ability are better in the calm ewes[1]. The lower reactivity of calm mothers and their lambs to stressors might decrease the chance of separation during the very critical post-natal period. The production of both colostrum and milk over the post-partum period is similar between the two lines. However, the calm lambs seem to be better fed because the colostrum from calm ewes is less viscous than that from nervous ewes, making it easier to ingest, and the milk from calm mothers contains more protein[3]. This better nutritional resource is translated into a higher bodyweight at 18 months of age in both males and females of the calm line while their birth weight does not differ from nervous lambs. The impact of these differences on body weight at the age of puberty has not been investigated yet.

Selection for calmness has the potential to increase the reproductive efficiency of a flock because the calm ewes and lambs are less sensitive to disturbance. In addition, the responses in colostrum and milk quality to selection for calmness might affect basic metabolic systems that seem to lead to a better management of the energy. This theory needs further testing.

This work was partly supported by Meat and Livestock Australia.

[1]Murphy, 1999. PhD Thesis, The University of Western Australia.
[2]Gelez et al. 2003. Appl Anim Behav Sci 84:81-87.
[3]Sart et al. 2004. Proc Aust Soc Anim Prod 25:307.

76. The effects of size and age at first calving on subsequent fertility in dairy cows

N Bourne[a], A Swali[a], AK Jones[b], S Potterton[b] & D.C. Wathes[a]

[a]Reproduction, Genes and Development Group, Royal Veterinary College, London, UK and
[b]Centre for Dairy Research, Reading, UK

The decline in dairy cow fertility over the past 30 years has major economic, welfare, genetic and environmental consequences. A significant number of potential replacement heifers never reach their first lactation due to peri-natal mortality, death as young stock or failure to conceive and about one third of heifers born complete only a single lactation. For those cows entering the milking herd, infertility is the biggest cause of culling with a current rate in the UK of approximately 18%.

This study investigated the effects of growth rates and age at first calving on subsequent fertility. Holstein-Friesian heifer calves were recruited at birth on a single farm (n = 134), and monitored until the end of their second lactation. Size parameters were measured at birth, 3, 6, 9 and 15 months (the start of the service period for animals to calve at 2 years old); and weight and body condition score were measured before and after each calving. Fertility parameters were recorded at all stages (nulliparous (NP), primiparous (PP2) and multiparous (MP)). Milk progesterone profiles were used to estimate the interval to commencement of luteal activity (CLA), length of first luteal phase and to identify cows with normal or abnormal profiles. Of the 134 calves recruited, 12 (9%) died or were culled before calving. A further 10 animals did not conceive in the first service period and therefore calved for the first time as three year olds (PP3). Of these, 6 were culled in early lactation thus only 4 were subsequently reserved. Data were analysed to compare: (a) fertility at each stage, (b) fertility in cows calving for the first time as 2 and 3 year olds and (c) fertility in relation to size.

The number of services per conception were: NP 1.6 ± 0.09, n = 112; PP2 2.4 ± 0.18, n = 78 and PP3 2.3 ± 0.75, n = 4. The calving to conception intervals were: PP2 144 ± 9.8, and PP3 238 ± 99.3 days (P=0.113). MP fertility data are not yet complete. Following their first calving, cows had significantly shorter days to CLA (PP2 19.5 ± 1.13^a, MP 26.5 ± 1.60^b, PP3 23.2 ± 13.67; a<b P<0.001) and the length of the first luteal phase was significantly longer (PP2 19.5 ± 0.66^b, MP 15.8 ± 1.33^a, PP3 35.8 ± 13.32^c days, a<b<c P<0.001) than following the second calving. These parameters were not correlated in individuals (P>0.1). A greater proportion of PP2 cows had normal progesterone profiles compared with MP (66/104, 63% versus 26/84, 31%, P<0.001), although the first progesterone profile was not predictive of the second (P=0.3). All four PP3 cows had abnormal progesterone profiles (P<0.02). Furthermore, only 2/10 PP3 cows conceived again in their first lactation (20%), compared with 106/112 (95%) of the PP2 cows. With regards size parameters, the animals which failed to conceive at 15 months were lighter at 9 months of age than those which did conceive (259 $\pm$ 2.4 n = 112 versus 233 $\pm$ 8.1 kg n=10, P=0.011). PP3 cows were much heavier and had a higher BCS immediately pre calving than PP2 cows (weight PP3 778 $\pm$ 30.5 kg, PP2 619 $\pm$ 5.2 kg, P=0.003; BCS PP3 3.9 $\pm$ 0.15, PP2 2.2 $\pm$ 0.06, P<0.001).

In conclusion, poor juvenile growth rate depresses fertility of NP heifers. If such animals are kept over to calve as 3 year-olds their outcome is very poor. Fertility generally declined with age, possibly in relation to increasing milk yields and more disease prevalence.

Funded by DEFRA and Milk Development Council.

77. Effects of transportation stress, handling stress and flunixin meglumine approximately 13 d after AI on pregnancy establishment

TW Geary[a], MD MacNeil[a] & RP Ansotegui[b]

[a]USDA Agriculture Research Service, Miles City, Montana, USA and [b]Department of Animal and Range Science, Montana State University, Bozeman, USA

Embryonic mortality in cattle represents a significant loss to the beef and dairy industries. Transportation of cattle at critical times after AI has been reported to decrease pregnancy rates, presumably through increased embryonic loss. Our objective was to determine effects of transportation or handling stress with or without a single injection of the prostaglandin inhibitor Flunixin Meglumine (FM; 1.1 mg/kg BW, i.m.) ~ 13 d after AI on pregnancy establishment. Four experiments were conducted using estrus-synchronized heifers (n = 1,725) and cows (n = 1,278). Technicians and AI sires were equally represented across treatments within locations and experiments. Bulls were introduced following treatment. Pregnancy was diagnosed 28 to 50 d after AI using ultrasonography. In Exp 1, suckled beef cows (n = 224) within two locations and beef heifers (n = 259) at one location were divided within location to receive 4 h transportation stress (TS), TS + FM (TSFM), control (CON) or CON + FM (CONFM) treatment. The CONFM treatment was omitted at one cow location due to insufficient numbers (n = 97). Analyses of data for cows and heifers in Exp 1 that received TS, TSFM, and CON treatments revealed a tendency for TSFM females (74%) to have higher (P < 0.07) AI pregnancy rates than TS (64%) or TS + CON females combined (66%). Analyses of data from the two locations at which cows and heifers received TS, TSFM, CON, or CONFM treatments as a 2x2 factorial revealed higher (P = 0.05) AI pregnancy rates among FM treated cows and heifers (71%) compared to non-FM females (61%). The AI pregnancy rates of TS (59%), CON (64%), TSFM (70%), and CONFM (73%) were not different (P > 0.10). Serum collected from females at 0, 2 and 4 h after treatment onset revealed decreased prostaglandin F metabolite for FM compared to non-FM females and increased cortisol for TS compared to non-TS females at the 2 and 4 h blood samplings. In Exp 2, beef heifers (n = 1,219) were divided within five locations to receive FM or no further treatment (Control). At insemination, heifers were divided into two similar pastures or pens and ~ 13 d later, one group of heifers within each location was worked through an animal handling facility to administer FM treatment. There was no location by treatment interaction (P > 0.10) on AI pregnancy rates, so data were pooled. Pregnancy rates to AI were reduced (P < 0.025) among heifers receiving FM (66%) compared to control heifers (72%). In Exp 3, suckled beef cows (n = 719) were assigned within two locations to receive FM or no further treatment (Control) ~ 13 d after AI. At insemination, Control and FM cows were divided into separate pastures and only FM cows were handled after AI. There was no location by treatment interaction (P > 0.10) so data were pooled. Pregnancy rates to AI did not differ (P > 0.10) between FM (57%) and Control cows (59%). In Exp 4, beef heifers (n = 247) and suckled beef cows (n = 335) from one location were assigned at AI to receive FM or Control treatment ~ 13 d later. In Exp. 4, all cows and heifers were handled through a working facility but only half of each age group received FM treatment. Pregnancy rates to AI between FM (45%) and Control cows (42%) or FM (56%) and control (55%) heifers were not different (P > 0.10). We conclude that transportation and handling stress ~ 13 d after AI decreases pregnancy establishment and that this decrease may be related to the degree of stress perceived. Administration of FM (1.1 mg/kg BW) ~ 13 d after AI minimized the loss, but did not improve pregnancy establishment in beef cows and heifers.

78. Should indigenous South African goat breeds be selected for increased litter size? Some practical implications

KC Lehloenya, LMJ Schwalbach, JPC Greyling

University of the Free State, P.O. Box 339, Bloemfontein 9300, South Africa

Indigenous goat breeds in South Africa, namely the Boer and the unimproved indigenous feral goat play a major role in both commercial and subsistence production systems. These breeds are locally adapted and therefore often more productive than exotic breeds under local harsh environmental farming conditions. For these reasons, indigenous goat breeds need to be preserved and improved through selection. However, limited information is available on the factors affecting the productive and reproductive performances of these breeds, especially in the unimproved feral goat and their crossbreds. A study with the aim of evaluating the effect of litter size on gestation length, birth weight and mortality rate was evaluated in South African indigenous goats artificially inseminated (AI) during the natural breeding season.

Ninety (42 Boer goat and 48 unimproved indigenous) does were synchronized for 16 days with medroxyprogesterone acetate (MAP) intravaginal sponges, followed by an intra-muscular injection of 300IU eCG at sponge withdrawal. AI was performed at fixed times (48 and 60 h) following sponge withdrawal with fresh diluted Boer goat semen.

The mean conception rate for the groups was 52.5%, the mean litter size 2.0 ± 0.9 and the gestation period 149.1 ± 4.1d. Breed had no significant effect on gestation length, litter size, birth weight and neonatal loss and the data of the two breeds was combined. Does with quadruplets had a significantly ($P < 0.05$) shorter gestation length (143.8 ± 1.1d) than those with singletons, twins or triplets (149.9 ± 0.9d; 149.3 ± 1.0d and 150.0 ± 0.6d, respectively). No significant differences were recorded between the later three groups. The mean kid birth weight was 2.7 ± 0.5kg with male kids (2.9 ± 0.1kg) being significantly ($P < 0.05$) heavier than female kids (2.6 ± 0.1kg). Kid birth weight decreased with an increase in litter size, resulting in mean birth weights for singles and twins being significantly ($P < 0.05$) heavier than triplets and quadruplets (3.5 ± 0.1kg and 3.0 ± 0.1kg vs 2.3 ± 0.1kg and 2.0 ± 0.3kg, respectively). However, no significant differences were recorded between singletons and twins as well as between triplets and quadruplet kids at birth. The overall neonatal loss within a 48h postpartum period was 22.2%, which increased significantly with an increase in litter size (0.0%, 17.9%, 18.2% and 83.3% for singles, twins, triplets and quadruplets, respectively). Quadruplets recorded a significant ($P < 0.05$) higher and the singletons a significant ($P < 0.05$) lower mortality rate than any other group. No differences were recorded between twins and triplets in this parameter. Following this critical 48 h period, the mortalities up to weaning were negligible. From the results it can be concluded that an increase in litter size reduces gestation length and most importantly the weight at birth of the South African goat kids. Selection for increased litter size has to be done with some care in these breeds as twins are desirable, but quadruplets are undesirable as most kids die soon after birth.

79. Postpartum ovarian activity in high and average producing Holstein cows under a heat-stressed condition

A Mirzaei , M Kafi & M Ghavami

Dept of Clinical Science, School of Veterinary Medicine, Shiraz University, Shiraz, 71345, Iran

The onset of postpartum ovarian activity is an essential event for high-producing dairy cows to regain maximum reproductive potential. Information on the resumption of postpartum ovarian activity in high producing Holstein cows under heat-stressed conditions is limited. Therefore, the following study was designed to study ovarian activity of high and average producing postpartum dairy cows under heat-stressed conditions. The study was carried out in a dairy herd with a history of intensive genetic selection for increased milk production. Cows were fed to meet the production requirement according to NRC (1988). The study was performed in the northeast of Iran where the peak summer temperature reaches 40 °C. Only healthy cows free of detectable reproductive disorders and free of clinical disease were included in the study. Ovarian activity of 40 high producing (HP, FCM = 8707.3 kg) and 30 average producing (AP, FCM = 6115.1 kg) multiparous Holstein cows was monitored from day 7 to at least 65 days postpartum using a transrectal ultrasound scanner (5MHz, Ami Co., Canada) twice weekly. During the study up to 90 days postpartum, cows were observed visually four times a day to determine the length of the oestrous cycle and the duration of standing oestrus. Blood samples were collected twice weekly from all cows through the coccygeal vein for measuring progesterone concentration. Serum progesterone concentration was determined using a radio-immunoassay kit (Immunotech, France). The mean peak milk yield (within 65 days postpartum) of high and average yielding dairy cows were 41.14 ± 4.9 and 34.6 ± 4.2 kg/day, respectively. Based on the serum progesterone RIAs, the reproductive profiles of each cow were classified into normal and abnormal patterns of ovarian activity[1]. Eleven cows (7 HP and 4 AP cows) were excluded from the analysis due to the occurrence of mastitis, metritis, lameness and abomasal displacement during the study. Data were statistically analysed using an independent t-test and a chi-squired test where appropriate.

Normal ovarian activity, as defined by the occurrence of ovulation in ≤ 45 days postpartum followed by regular ovarian cycles, was diagnosed in 36.4% (12/33) of HP and 53.8% (14/26) of AP cows. Abnormal ovarian activities including delayed first ovulation, prolonged luteal phase, short luteal phase and cessation of ovarian cyclicity were further diagnosed in 63.6 % (21/33) of HP and 46.2 % (12/26) of AP cows ($p > 0.05$). The most frequent type of abnormal ovarian activity was delayed first postpartum ovulation as defined by the occurrence of the first ovulation after 45 days postpartum (36.4% (12/33) in HP and 38.5% (10/26) in AP cows ($p > 0.05$). The least frequent type of abnormal ovarian activity was the cessation of ovarian cyclicity in HP and AP cows (3.0% (1/33) vs 3.8% (1/26)), respectively. The mean ($\pm$SD) duration of standing oestrus was 6.1 ± 2.7 hours in HP cows compared to that of 10.5 ± 2.9 hours in AP cows ($p < 0.05$). The mean ($\pm$SD) calving to first detected oestrus interval in HP and AP cows was 35.3 ± 16 and 33.2 ± 12 days, respectively ($p > 0.05$).

In conclusion, a high percentage of abnormal ovarian activity occurred in postpartum dairy cows under heat-stressed conditions with the occurrence of a non-significantly higher incidence of abnormal ovarian activity in high producing compared to that of the average producing dairy cows. Further, delayed first ovulation was the most important abnormalities of ovarian activity in either high or average producing dairy cows under a heat-stressed condition.

[1]Taylor *et al.* 2000. Occasional Publication (No. 26) of Br Soc Anim Sci 495-498.

80. The effects of administration of ketoprofen on ovarian function in dairy cows

M Kafi, S Nazifi, R Bagheri-Nejad & M Rahmani

Department of Clinical Studies, School of Veterinary Medicine, Shiraz University, Shiraz, 71345, Iran

Ketoprofen is a propionic acid NSAIA with a strong anti-inflammatory property, reducing the biosynthesis of prostaglandin F2α through the inhibition of cyclooxygenase enzymes in the arachidonic acid cascade. No information is available regarding the effect of ketoprofen on ovarian function in dairy cows. The objective of the present experiment was to investigate the effects of ketoprofen on the growth of the dominant ovulatory follicle and ovulation in dairy cows. Five non-milking dairy cows were administered a 0.9% saline solution daily from day 8 (day -3) of ensuing synchronised oestrous cycle at 24 hours intervals over 4 days (control observation). After a complete oestrous cycle had elapsed, the cows were daily given the recommended therapeutic dose (3 mg/kg, im) of ketoprofen (Ketofen 100; 10%, Merieux/Webster, Australia) initiating from day 8 (day -3) of the synchronised oestrous cycle at 24 h intervals over 4 days. All cows received an im administration of $PGF_2\alpha$ (30mg, Lutalyse) either 6 h prior to the first treatment of saline solution or 6 h prior to the first treatment of ketoprofen. Ultrasonography of the ovaries was performed daily from the day before commencing (day -4) experimental treatments until 2 d after induced oestrus to monitor the growth of the dominant ovulatory follicle and ovulation. Coccygeal vein blood samples were collected from all cows once a day commencing 1 day prior to initiation of either saline or ketoprofen administration through to 24 h after administration of $PGF_2\alpha$, then every 4 h for 48 h, then every 6 h for a further 24 h and then every second day until day 9 after oestrus. The preovulatory oestradiol-17ß peak and progesterone concentrations on days 0, 3, and 6 after induced oestrus were determined. Data were statistically analysed using a Wilcoxon signed rank test and repeated measures ANOVA where appropriate. The mean ($\pm$SEM) diameter of the dominant ovulatory follicle on day of oestrus was significantly ($p < 0.05$) higher in saline administered oestrous cycles compared to that of ketoprofen administered cycles (11.3 ± 0.3 vs 8.9 ± 1.1 mm). Ultrasonography revealed that ovulation had taken place in all saline administered oestrous cycles within 58 ± 12.2 h after the oestradiol-17ß peak while the dominant ovulatory follicles remained unovulated by 72 ± 9.8 h after the oestradiol-17ß peak in four ketoprofen-administered cycles. A significantly ($p < 0.05$) higher mean ($\pm$SD) pre-ovulatory oestradiol-17ß peak (62.3 ± 27.9 pg/ml) was observed in ketoprofen administered oestrous cycles at oestrus compared to that of the control oestrous cycles (23.7 ± 5.9 pg/ml). Repeated measures ANOVA demonstrated that the increase in plasma progesterone concentration in the ketoprofen administered oestrous cycles was significantly ($p < 0.05$) greater than that observed in the control cycles between days 0 to 6 after oestrus. The results of the present study demonstrate that administration of ketoprofen during the pre-and periovulatory period of the oestrous cycle in dairy cows may cause a disturbance in ovarian oestradiol secretion, impairment of the final growth of the ovulatory follicle and ovulation.

81. Effects of semen, cow and embryo breeds on non-return rate to artificial inseminations in dairy cows in New Zealand

ZZ Xu

Livestock Improvement Corporation Ltd, Private Bag 3016, Hamilton, New Zealand

Previous studies have identified heterosis in fertility in the cow population in New Zealand[1]. The objective of the present study was to further investigate the effect of cow breed, semen bull breed and their interactions on the 2-24-day non-return rate (NRR) to artificial inseminations (AI).

AI records for the spring breeding season of 2005 were extracted from the national database. The data set was edited to exclude records for AI in primiparous heifers, AI in synchronised cows, AI with semen from unproven sires, AI not performed by AI technicians, or AI within 24 days of the end of the AI breeding period. The breed composition of cows was expressed in portions of 16th and grouped into 5 classes according to their predominant breeds of Holstein-Friesian (HF) or Jersey (J): HF containing ≥13/16 HF; FX containing between 8/16 and 12/16 HF; FJ containing 8/16 of HF and 8/16 of J; JX containing between 8/16 and 12/16 J; and J containing ≥13/16 J. There were 3 breed classes for AI sires: HF (N = 16), J (N = 8) or Crossbred (X, N = 9). A total of 779,715 qualifying records were used in the analyses. The data were analysed using the Genmod procedure of SAS. Results presented are least squares estimates transformed back to the original scale.

Jersey bulls had higher (P < 0.0001) NRR than HF bulls (J–HF = 1.02%), which in turn had higher (P=0.007) NRR than crossbred bulls (HF-X = 0.75%). Compared with the NRR for HF cows, NRR were similar for FX (+0.16%, P = 0.41) and FJ (+0.04%, P=0.78) cows, but were lower for JX (–0.59%, P = 0.004) and J (–3.15%, P < 0.0001) cows. There were also significant interactions between bull and cows breed classes. With HF semen, NRR decreased with decreasing proportions of HF blood in cows; deviations in NRR from HF cows being –0.33% (P=0.07), –0.55% (P=0.01), –1.77% (P<0.0001), and –4.16% (P<0.0001) for FX, FJ, JX and J cows, respectively. With J semen, NRR was highest for FX cows (HF–FX = 0.75%, P=0.02) and lowest in J cows (HF–J=1.30%, P<0.0001). For crossbred semen, NRR was about 4% lower in J cows compared with cows of other breeds, which did not differ significantly.

For J cows, the NRR for matings using HF semen was 2.54% lower than that using J semen, a difference that was greater than the semen breed effect of 1.02%, indicating that crossbred embryos had a negative impact on NRR. For HF cows, the NRR for matings using J semen was only 0.3% higher than that using HF semen, which was less than the effect due to semen breed, again indicating that crossbred embryos had a negative impact on NRR. To further investigate the impact of embryo breed on NRR, embryo breed was included in the model after sire and dam breed. Embryo breed had a significant (P < 0.05) effect on NRR. NRR was lowest for FJ embryos, with NRR increased as embryos becoming more HF or J; NRR deviations from FJ embryos being 1.43%, 0.64%, 0.40%, and 1.32% for embryos of HF, FX, JX and J breeds.

In conclusion, results from this study showed that the breed of sire and dam interact to affect NRR of artificial inseminations. While semen from J sires had better NRR than semen from HF sires, J dams had lower NRR compared with dams of predominantly HF breed. The finding that matings resulting in pure-breed embryos had better NRR than those resulting in crossbred embryos is interesting and requires further study to understand the underlying mechanisms.

[1]Harris, *et al.* 2001. In: Diskin, MG, Editor. Fertility in the high-producing dairy cow, Occasional publication No 26, British Society of Animal Science. pp. 491-493.

82. Use of a GnRH agonist (leuprolide) to suppress rut-associated events in farmed male red deer (*Cervus elaphus*)

GK Barrell[a] & LA Miller[b]

[a]*Agriculture and Life Sciences Division, P.O. Box 84, Lincoln University, Canterbury 8150, New Zealand and* [b]*Product Development Research Program, USDA APHIS WS National Wildlife Research Center, 4101 LaPorte Avenue, Fort Collins, Colorado 80521, USA*

This study examined the effectiveness of leuprolide, a GnRH agonist, for suppressing some of the rut-associated events in farmed male red deer that cause problems for management of these animals. About 6 weeks prior to commencement of the rut period adult red deer stags in 3 groups (n = 10) received leuprolide, administered subcutaneously in a 90-day release formulation, at zero (0 mg, control), low (22.5 mg) or high (45 mg) doses. Treatment with leuprolide caused a suppression of mean plasma LH concentration that was significant ($P < 0.05$) at 9 weeks. Mean plasma testosterone concentration of all three groups rose initially, then declined prematurely in the leuprolide-treated groups, so that it was significantly ($P < 0.05$) suppressed (0.66 ± 0.29 and 2.0 ± 0.88 ng/ml, low and high dose respectively) in the mid rut period when the peak value (9.0 ± 1.94 ng/ml) was recorded from control stags. A reduction in mean live weight occurred in all 3 groups throughout the 3 months of rut period but this did not differ between treatments. However a corresponding reduction in mean body condition score was greater in the control stags ($P < 0.05$). There was some evidence that leuprolide treatment stimulated aggressive behaviour initially, but it suppressed roaring behaviour later in the rut.

Although the results show a dose-related suppressive effect of this GnRH agonist on LH and testosterone secretion in male red deer, there was only a minimal effect of the treatment on aspects of major concern to farmers such as weight loss and aggressive behaviour.

We acknowledge M Ridgway, S Schaafsma and M Wellby for technical support, Dr AF Parlow for LH assay reagents and Dr GD Niswender for testosterone assay reagents. Atrix Laboratories, Fort Collins, Colorado, USA are thanked for the generous donation of the leuprolide acetate formulation.

83. The delay of puberty in tropical beef cattle is not due to a limitation in ovarian follicular growth during the prepubertal period

F Samadi & MJ D'Occhio

School of Animal Studies, The University of Queensland, Gatton Campus, Queensland, Australia 4343

Beef cattle that are adapted to tropical environments (e.g. Brahman, *Bos indicus*) typically attain puberty at older ages compared with temperate beef cattle (*Bos taurus*). The aims in the present study were to (1) ascertain whether there are features of ovarian follicular growth during the prepubertal period in tropical cattle that may help explain the delay in onset of puberty and (2) examine whether prepubertal follicular dynamics can be influenced by nutrition and growth rate. Brahman heifers (50 $\pm$ 1 wk of age and 200 $\pm$ 3 kg live weight (LW); mean $\pm$ SEM) were randomly assigned on LW to relatively good (GN, n = 11) and moderate (MN, n = 11) nutrition, which was controlled by pasture quality and quantity and protein and energy supplementation. Ovarian follicular status, LW and body condition score (BCS, scale 1-5) were recorded at 2-wk intervals from 50-70 wk of age. Daily LW gain was greater ($P < 0.01$) for GN heifers (0.65 $\pm$ 0.02 kg) than MN heifers (0.52 $\pm$ 0.02 kg) and at 70 wk GN heifers (292 $\pm$ 7 kg) were heavier ($P < 0.05$) than MN heifers (273 $\pm$ 5 kg). GN heifers also had a higher ($P < 0.05$) BCS than MN heifers at 70 wk (3.4 $\pm$ 0.1 and 3.0 $\pm$ 0.1, respectively). Follicular characteristics did not differ between GN and MN heifers and the data were pooled. The largest follicle (LF) observed for each heifer increased ($P < 0.01$) in size from 50 wk (7.5 $\pm$ 0.3 mm) to around 58 wk (11.0 $\pm$ 0.4 mm) and then remained constant to 70 wk. From 58-70 wk, maximum LF size ranged from 14-16 mm. Heifers with 2 or more LF ($\geq$10 mm) were observed in 9.9% (24/242) of the ovarian scans. The number of follicles $\geq$4 mm for both ovaries combined, at each ovarian scan, was greater ($P < 0.01$) from 50-60 wk (8.0 $\pm$ 0.4) compared with 62-70 wk (5.9 $\pm$ 0.3). Heifers tended to have a relatively large number of ß-follicles (2-3 mm) and in 23.5% (57/242) of the ovarian scans at least one ovary had > 20 ß-follicles. The differences in LW gain and absolute LW at 70 wk between GN and MN heifers were not reflected in major differences in ovarian follicular status. From 54 wk, heifers were attaining a LF size ($\geq$13 mm) equivalent to an ovulatory follicle in cyclic Brahman heifers, yet they had not ovulated by 70 wk.[1] Temperate beef cattle can show puberty from around 50 wk of age whereas puberty in Brahman typically occurs after 65 wk, but this can vary depending on LW and environmental conditions.[2,3] Therefore, in the present study, Brahman heifers had ovulatory size follicles at around the same age as temperate cattle achieve puberty. This finding was interpreted to suggest that the delay in puberty in tropical cattle is not due to the failure to achieve an ovulatory size follicle. It is possible that large follicles in prepubertal tropical cattle do not synthesise sufficient oestradiol to exert positive feedback on gonadotrophin secretion until older ages than temperate cattle. It is also possible that the sensitivity of the brain to oestradiol negative feedback is maintained for longer during the prepubertal period in tropical cattle compared with temperate cattle. These potential features of brain-ovarian maturation in prepubertal tropical cattle are likely to be related to adaptation to environmental cycles in tropical regions.

We gratefully acknowledge the provision of Brahman heifers by Tartrus Station, Marlborough, Queensland and management of heifers by Mr Tom Connolly.

[1]Murray *et al.* 1998. Anim Reprod Sci 50:11-26.
[2]Rawlings *et al.* 2003. Anim Reprod Sci 78:259-270.
[3]Rodrigues *et al.* 1999. Anim Reprod Sci 56:1-10.

84. Proteome profile and chromatin stability in sperm of beef bulls

K Hengstberger[a], S Stansfield[b], R Smith[b] & MJ D'Occhio[a]

[a]School of Animal Studies, The University of Queensland, Gatton Campus, Queensland, 4343 and
[b]School of Molecular and Microbial Sciences, The University of Queensland, St Lucia Campus, Queensland, Australia 4072

Sperm chromatin stability can be measured using the sperm chromatin structure assay which yields a DNA fragmentation index (DFI).[1] In men, DFI categories of < 15%, 15-27% and > 27-30% correspond to stable, moderately stale and unstable sperm chromatin, respectively, and individuals with a DFI > 27-30% achieve fertilisation but there is typically a high incidence of embryonic mortality.[2] The relationship of DFI to other features of sperm are poorly understood. The aim in this study was to undertake a preliminary investigation of the relationship between chromatin stability and the proteome profile in bulls. Sperm samples from sexually mature Brahman (*Bos indicus*) beef bulls (4.6 ± 0.7 years; body condition score 6.7 ± 0.2, scale 1-10) that had undergone a breeding soundness evaluation were allocated to 2 groups based on divergent (P < 0.01) DFI values of < 15% (2.5 ± 0.7 DFI, n = 4) and > 27% (53.7 ± 12.3 DFI, n = 4). Sperm were washed free of seminal plasma and sperm protein lysates were prepared using a cell lysis buffer. Equal amounts of protein from each sperm cell lysate were pooled within DFI category. For each pooled protein sample, 4 replicate separations were resolved by 2-dimensional polyacrylamide gel electrophoresis (2D PAGE) and protein abundances quantitatively assessed using the fluorescent dye Sypro Ruby®. Image analysis was performed by Delta-2D software and statistical differences determined for 1069 protein spots matched across all separations. It was found that 16% (172/1069) of the protein spots were altered in abundance (P < 0.01; Students *t*-test) between pooled protein samples from sperm with < 15 % DFI and > 27 % DFI. There was a significant relationship therefore between the relative abundance of specific proteins and the DFI. As whole sperm protein lysates were prepared in the present study the results are for both nucleoproteins and cytoplasmic proteins. There were no differences (P > 0.05) between bulls with < 15% DFI and > 27% DFI in scrotal circumference (35.3 ± 1.8 and 35.7 ± 1.3 mm, respectively) and testicular tone (3.2 ± 0.2 and 3.0 ± 0.4, respectively, scale 1-5). However, bulls with < 15% DFI had a higher (P = 0.04) percentage of normal sperm than bulls with > 27% DFI (78.5 ± 5.7% and 36.0 ± 17.5%, respectively). It is possible, therefore, that differences in specific protein abundance between bulls with < 15% DFI and > 27% DFI could be related to other features of sperm morphology and function. Because pooled protein lysates were used the possibility cannot be excluded that the differences in proteome profiles between bulls with < 15% DFI and > 27% DFI may have been contributed, at least in part, by an individual bull. In this regard, bulls with < 15% DFI had a relatively small range (26%) in percent normal sperm compared with the range (81%) for bulls with > 27% DFI, and one bull in the latter category had only 5% normal sperm. Notwithstanding, further investigation of the sperm proteome is likely to lead to novel associations between protein abundance and sperm function and, for example, tubulin differed in abundance between sperm with < 15% and > 27% DFI.

The study was supported, in part, by Meat and Livestock Australia.

[1]Evenson *et al.* 1980. Chromosoma 78:225-238.
[2]Larson-Cook *et al.* 2003. Fertil Steril 80:895-902.

85. Detection of tissue dependant differentially methylated regions in bovine fetal brain and placental membrane

<u>K Kizaki</u>[a], K Koshi[a], M Kremenskoy[b], K Ushizawa[c], K Imai[d], T Takahashi[c], K Shiota[b] & K Hashizume[a]

[a]Laboratory of Veterinary Physiology, Iwate University, Morioka, Japan, and [b]Laboratory of Cellular Biochemistry, Graduate School of Agricultural and Life Sciences, The University of Tokyo, Tokyo, Japan, [c]National Institute of Agrobiological Sciences, Tsukuba, Japan and [d]National Livestock Breeding Center, Fukushima, Japan

DNA methylation is the major epigenetic modification and is involved in development and differentiation by tissue or cell type-specific gene expression. In mammalian cells, 60-80% of CpG nucleotides are methylated. CpG-rich DNA fragments, called CpG islands, are preferentially located at the transcription start site of housekeeping genes and are also associated with tissue specific genes. Tissue-dependent differentially methylated regions (T-DMRs) in the CpG islands have been identified in human and rodent genomes, and methylation pattern of T-DMRs was specific but varied according to cell lineage and tissue type. In this study, we examined the methylation status of T-DMRs in bovine fetal and placental tissues using restriction landmark genomic scanning for methylation (RLGS-M) and bisulfite sequencing. Genomic DNA was extracted from placental tissue (intercotyledon, ICOT), fetal brain and skin of Japanese black cattle (around Days 50, 100, 150 and 250 of gestation). RLGS-M analysis was performed using fetal brain and skin on Day 50. The GC content and CpG frequency of DNA fragments were analyzed with EMBOSS CpG Plot program (http://www.ebi.ac.jp/emboss/). Further, methylation status was directly analysed with bisulfite sequencing. We obtained 12 DNA fragments (#1-12) of different methylation status in ICOT and fetal brain on Day 50. We identified the location of 9 DNA fragments (#1-7, 9 and 10) in the bovine genome following a GeneBank database search and screening of a bovine genome library. Analysis of the GC content and CpG frequency in these genomic sequences revealed that 6 (#2-5, 7 and 10) possess the CpG island region. Three loci (#4, 7 and 9) were analyzed more extensively by bisulfite sequencing to confirm the RLGS-M results and determine directly the density of methylation in the CpG island region. Bisulfite genomic sequencing indicated that in #4 and 7, the CpG island regions were densely methylated in fetal brain and skin, but hypo-methylated in ICOT. These results indicate that # 4 and 7 are T-DMRs in the bovine genome. We further investigated the methylation status of #7 in the ICOT during gestation. Although a hypo-methylated status in the CpG island region was detected in ICOT on Day 50, methylation increased with the progress of gestation to Days 100, 150 and 250. In this study, we identified differences in methylation status between fetal brain and ICOT, and these differences may be involved in bovine trophoblast differentiation.

Supported by a grant of Research Project for Utilizing Advanced Technologies (05-1770).

86. Comparative mapping of sheep chromosome X

MC French[a], GH Davis[b] & SM Galloway[a]

[a]AgResearch Molecular Biology Unit, University of Otago, Dunedin, New Zealand and [b]AgResearch Invermay Agricultural Centre, Mosgiel, New Zealand

The X chromosome in mammals contains several genes or genetic loci which play a role in reproduction. In sheep models, the X-linked gene BMP15 has been shown to be crucial for folliculogenesis. The human X chromosome also contains regions that are critical for normal reproductive function, although not all the specific genes have been identified. Premature Ovarian Failure (POF) and Turners syndrome (XO) map to the human X chromosome and X-linked genes such as BMP15, FMR1, FMR2, DACH2 and DIAPH2 have been implicated in reproductive function in humans. Comparisons of X chromosomes have shown that the same genes on the X chromosome of one mammalian species are also on the X chromosome of others, although the orders of these genes are often rearranged. With the human genome having been fully sequenced, it can provide information about potential candidate genes for reproductive traits in ruminants. However, the sheep X chromosome consists mainly of anonymous microsatellite markers, which don't correspond to human locations. To provide better links between the sheep and human X chromosomes we are mapping known genes from humans on to the sheep X chromosome to identify syntenic regions and rearrangements between sheep and human.

With the increased availability of cattle genomic sequence and the similarity between ovine and bovine sequences we have used two main methods for mapping genes on the sheep X chromosome. These techniques involve aligning either ovine or bovine sequences to known human genes, enabling us to take advantage of microsatellite repeats within these genes and to sequence across introns (300 – 1000bp) to determine single nucleotide polymorphisms (SNPs). These microsatellites and SNPs are then mapped using the sheep International Mapping Flock (IMF). We have sequenced 23 genes from across the human X chromosome, identified 48 polymorphisms and mapped 9 new genes. We have also identified 30 new bovine microsatellites, 21 of which amplified in sheep, but only 7 were polymorphic. Linkage mapping of these genes has identified one major breakpoint and inversion, and other potential minor rearrangements between sheep and human X chromosomes. Further genes are being mapped to better define these rearrangements.

87. Genotype frequency analysis and screening for the FecG[H] mutation of growth differentiation factor 9 gene in the Iranian Ghezel sheep population

M Akbarpour[a,c], H Hayatgheybi[b] & SA Ghorashi[a]

[a]Animal Biotechnology Dept., National Institute for Genetic Engineering and Biotechnology (NIGEB), Tehran-IRAN, [b]Physiology Dept., Veterinary Faculty, Urmia University, Urmia- IRAN and [c]Veterinary Faculty, Urmia Islamic Azad University, Young Researchers Club, Po.Box: 969Urmia-IRAN

Ghezel sheep are highly prolific and are one of the local sheep breeds in Iran and Turkey. Growth differentiation factor 9 (GDF9) gene has been found to be essential for growth and differentiation of early ovarian follicles. Novel mutations in GDF9 have been associated with increased ovulation rates and high litter sizes in heterozygous carriers. Therefore, mutations in GDF9 were considered as a possible candidate for the increased litter size observed in Ghezel ewes. The genetic polymorphism of the GDF9 gene exon I was detected in 33 ewes of Ghezel sheep by PCR-SSCP. Sequencing and PCR-RFLP.SSCP analysis identified five fragments that contained conformational differences. Combined results from sequence data revealed one single nucleotide polymorphism across the coding region. The G to A nucleotide changes give rise to an amino acid (R87H) change and disrupts a HhaI restriction enzyme cleavage site. RFLP of the 462bp PCR product including this SNP indicates wild type and heterozygous individuals carry this polymorphism. The frequencies analyzed by PopGene32 software and results showed Hardy-Weinberg equilibration in the population.

Also, the GDF9 gene exon II was investigated by these techniques to screen whether they are FecG[H] (S395F) carriers or not. SSCP analysis identified four fragments that contained conformational differences but the combined results with sequencing analysis data did not reveal the FecG[H] mutation (C to T) in GDF9 gene in Iranian Ghizel ewes.

88. Expression, purification and antibody generation of bovine bone morphogenetic protein (BMP) family proteins

T Takahashi[a], M Hosoe[a], K Ushizawa[a] & K Hashizume[b]

[a]National Institute of Agrobiological Sciences, Tsukuba, Ibaraki 305-8602, Japan and [b]Iwate University, Morioka, Iwate 020-8550, Japan

Bone morphogenetic proteins (BMPs) were initially identified as proteins to induce bone formation. Of these BMP members, BMP-4, -15 and growth differentiation factor (GDF) -9 have been reported to play critical roles on reproduction in mammals. To obtain better knowledge about the biology of BMPs in bovine, we have purified recombinant bovine BMP-4, -15 and GDF-9 generated by mammalian cell culture and cell-free bacterial lysate systems.

Bovine cDNAs encoding the mature protein regions of BMP-4, 15 and GDF-9 were cloned by RT-PCR with high fidelity DNA polymerase (Pfu, Stratagene). Amino termini of mature protein regions were predicted from a consensus motif R-X-X-R. For mammalian expression, cloned sequences were subcloned to pFLAG-CMV expression vector (Sigma) and transfected into human embryonic kidney cell line (HEK-293) as hosts for recombinant expression. Expression of recombinant protein was analyzed by Western blotting with anti-FLAG M2 antibody (Sigma). For cell-free expression, cDNAs encoding mature protein regions were subcloned to pIVEX 2.4a expression vector (Roche) and utilized for Rapid Translation System (RTS, Roche). Expression of recombinant protein was analyzed by Western blotting with anti-6xHis antibody (Sigma).

None of these three BMP members were secreted into conditioned media of HEK-293 cells, although a little recombinant protein was detected in transfected cell lysate. In contrast, BMP-4, 15 and GDF-9 were successfully expressed by RTS strategy. Both monomeric and dimeric forms of BMP-4, 15 and GDF-9 were purified from the precipitate of RTS mixtures after solubilization with 6M urea. Antisera against BMP-4, 15 and GDF-9 were generated in rabbits and found to react with immunized antigens in Western blotting. Immunohistochemical examination with generated antisera showed that BMP-15 and GDF-9 were expressed in the ovary, and BMP-4 was expressed in the placenta. In summary, we generated recombinant proteins and immunologic probes of bovine BMP-4, 15 and GDF-9 for studying the biology of BMP family proteins in the cow.

This study was supported by the grant-in-aid of Research Project for Utilizing Advanced Technologies (05-1770) from the Ministry of Agriculture, Forestry and Fisheries of Japan.

89. The expression of mRNA to the luteinizing hormone receptor in ovine cervical tissue is stimulated by the prostaglandin E analogue, Misoprostol

S Leethongdee[a,b], CM Kershaw[a], M Khalid[a] & RJ Scaramuzzi[a]

[a]Department of Veterinary Basic Sciences, Royal Veterinary College, Hatfield AL9 7TA, United Kingdom and [b]Department of Animal Technology, Mahasarakham University, Amphur Muang Mahasarakham, 44000, Thailand

Functional luteinizing hormone receptor (LH-R) is present in the cervix of the non-pregnant cow[1]. The pattern of expression varies over the oestrous cycle and is highest during luteal phase.[1] Luteinizing hormone was able to stimulate Cyclooxygenase-2 in incubated bovine cervical tissue[1] leading to increased prostaglandin E_2 production by the cervix[1] The local application of the prostaglandin E_1 analogue, Misoprostol induced cervical ripening at parturition[2]. This study determined the effect of intra-cervical Misoprostol on LH-R mRNA expression in cervical tissue from ewes in oestrus.

Fifteen Welsh Mountain ewes were randomly assigned to 3 groups of 5. Oestrus was synchronised using intravaginal progestogen sponges for 12 days and 48h after sponge removal, the ewes were given intracervical Misoprostol as follows: controls (vehicle), 200 μg or 400 μg per ewe. Ewes were euthanised 6h after treatment and cervical tissue collected immediately and divided into 6 transverse sections from the uterine (section A) to the vaginal end (section F). Alternate sections (A, C and E) were fixed in buffered formalin, wax embedded and sectioned at 9 μm. The expression of LHR mRNA was determined by ISH using digoxigenin-11-uridine-5'-triphosphate labelled riboprobes for ovine LH-R. Expression was quantified using an index, in 6 layers (luminal epithelium, sub-epithelial stroma, irregular, longitudinal and transverse smooth muscle and serosa) in each region. Data were analysed using a mixed model ANOVA.

There were significant effects of cervical region (P< 0.001) and tissue layer (P< 0.001) on the expression of LHR-mRNA. Expression in the middle region was higher than the vaginal (P< 0.001) and uterine regions (P< 0.001) and expression was higher in the vaginal compared to the uterine region (P= 0.007). Expression was highest in the luminal epithelium and lowest in the serosa (P< 0.001). Other layers had intermediate levels of expression with the sub-epithelial stroma > irregular muscle > transverse muscle > longitudinal muscle (P= 0.001). Misoprostol increased expression in the cervix, but the effect was only significant in the interactions with region (P< 0.001) and layer (P< 0.001). Ewes that received 400 μg Misoprostol had higher expression in epithelium and stroma at the vaginal (P< 0.001) and mid regions (P< 0.001) but not the uterine region. There was no significant effect of the 200 μg dose on LH-R mRNA expression.

The minimum dose of Misoprostol that stimulated LH-R mRNA expression in the ovine cervix was 400 μg and the effect was confined to the middle and vaginal regions of the cervix. Expression was highest in luminal epithelium and may be associated with secretion of cervical mucus at oestrus. However there was expression in stroma and muscle layers, the probable sites of prostaglandin synthesis, and expression was increased by Misoprostol suggesting a paracrine interaction between LH and prostaglandin in cervical function that may be related to remodelling the extracellular matrix of the cervix at oestrus.

[1]Mizrachi & Shemesh, 1999. Biology of Reproduction 66:776-784.
[2]Fletcher et al. 1994. Obstetrics and Gynaecology 83:244-247.

90. Effect of preovulatory estradiol concentrations on conceptus and uterine development

GA Bridges[a], ML Mussard[a], LA Helser[a], DE Grum[a], DM Lantz[a], TL Ott[b], JL Pate[a] & ML Day[a]

[a]Department of Animal Sciences, The Ohio State University, Columbus, OH, USA and [b]Department of Animal and Veterinary Sciences, University of Idaho, Moscow, ID, USA

In previous research from our lab, decreasing the length of proestrus (interval from luteolysis to the LH surge) before inducing ovulation of similar sized follicles decreased peripheral progesterone (P4) concentrations and reduced fertility. Subsequent research using this animal model demonstrated that a primary effect of manipulating length of proestrus was to alter peripheral, preovulatory estradiol concentrations (E2). The present experiment was conducted to determine the effect of altering preovulatory E2, through manipulation of length of proestrus, on peripheral P4 concentrations, conceptus development and interferon-tau (IFN) production, and steady-state levels of mRNA for nuclear progesterone receptor (NPr), membrane-bound progesterone receptor (MPr) and estradiol receptor alpha (Era).

On d -6.75 of the experiment, all follicles > 4 mm were removed from the ovaries of cyclic heifers via transvaginal aspiration, thus initiating a new wave of follicular development ~ 1.5 d later. All heifers received an ovulatory dose of GnRH on d 0, preceded by PGF_{2alpha} on either d -2.5 (2.5 days of proestrus; HiE; n = 5) or d -1.5 (1.5 days of proestrus; LoE; n = 5). Ovarian ultrasonography was performed on d -6.75, -3.5, -2.5, -1.5, 0, 2, and 7 to evaluate follicular dynamics and confirm ovulation and CL development. Blood samples, collected on d -2.5, -2.0, -1.5, -1.0, -0.5, -0.25, and 0 were analyzed for E2 concentrations and samples collected daily from d 2 to 7 and on d 9, 11, 13, and 15.5 were analyzed for P4 concentrations. On d 7 embryos were implanted into all heifers using standard ET procedures. On d 15.5 heifers were slaughtered and the reproductive tract was flushed to collect the conceptus. A segment of uterine endometrium in the uterine horn ipsilateral to the CL was snap frozen in liquid nitrogen for subsequent mRNA analyses and flush media was collected for analysis of luminal IFN content (anti-viral assay). Steady-state levels of mRNA for NPr, MPr and ERa in the uterine endometrium were determined by quantitative real-time PCR and standardized against G3PDH mRNA. Statistical analyses were conducted using Mixed procedures of SAS using repeated measures when appropriate.

Mean, and peak concentrations of E2, during the preovulatory period were greater (P < 0.05) in the HiE than LoE treatment (peak E2 concentrations; 8.9 ± 0.4 and 6.7 ± 0.8 pg/mL, respectively). Ovulatory follicle diameter did not differ between the HiE (12.6 ± 0.2 mm) and LoE (12.4 ± 0.5 mm) treatments and P4 concentrations were similar between treatments. For heifers from which intact conceptus tissue was recovered (HiE, 3/5; LoE, 3/5), total uterine IFN content did not differ between treatments. In the remaining heifers, embryos appeared to be degenerating and IFN was not detectable in the uterine flush media, except in one heifer in the HiE treatment that had very low but detectable IFN concentrations. Analyses of mRNA were completed in 4/5 heifers in the LoE and 5/5 heifers in the HiE treatment. Levels of NPr mRNA in the uterine endometrium tended (P < 0.10) to be greater in the HiE than LoE treatment. Levels of mRNA for MPr and Era did not differ between treatments. Treatments that reduced preovulatory E2 concentrations tended to reduce expression of nuclear P4 receptors in the endometrium but did not influence the aspects of conceptus development evaluated. The mechanisms by which reduced preovulatory estradiol concentrations may be responsible for lower fertility in cattle warrant further investigation.

91. Early embryo survival following repeated embryo transfer in primiparous dairy cows

S Meier[a], PJS Gore[a], M Berg[b], L McGowan[b] & AJ Peterson[b]

[a]Dexcel Limited, Hamilton, New Zealand and [b]AgResearch Limited, Hamilton, New Zealand

One of the key factors that contribute to poor reproductive performance in lactating dairy cows is poor conception rates and early embryonic loss. Conception rates to first insemination in New Zealand are 53% with a range of 42% to 62%[1]. Up to 40% of pregnancies are lost between fertilisation and day 42 of gestation[2]. There are many components that contribute to early embryonic survival, including the health of the oocyte and early embryonic development, which may be influenced by the metabolic status of the lactating cow[3].

This project examined early embryonic survival in primiparous dairy cows (age 3 years old; n = 36). Twin or single IVP embryos were transferred on 5 occasions at 72 ± 1, 125 ± 3, 200 ± 1, 271 ± 2 and 326 ± 2 days post-calving (mean $\pm$SEM), following a prostaglandin-based synchrony programme. The last 2 embryo transfers, 271 and 326 days post-calving, were undertaken when the animals had been dried off. The final embryo transfer round consisted of a single embryo being transferred with the embryo recovered at slaughter (Day 17 post-oestrus) and classified as viable when intact. During the earlier embryo transfer rounds, embryo survival was confirmed when a heartbeat was detected at the time of ultrasonography, 28 to 36 days post-oestrus (embryo transfer having taken place on day 6-7 following observed oestrus).

Embryo survival rates for each round of embryo transfer was 33%, 57%, 57%, 58% and 81%, for embryos transferred 72, 125, 200, 271, and 326 days post-calving (rounds 1 to 5). Embryo survival increased from round 1 to 2 ($P < 0.07$) and again from round 4 to 5 ($P < 0.05$). An embryo survival score (ESS) was calculated for individual animals over the first 3 embryo transfer rounds (ESS = proportion of embryos survived/total embryos transferred). The ESS was used to rank individual animals as having high, medium or low ESS, with a mean ($\pm$SEM) ESS of $80 \pm 4\%$, $55 \pm 2\%$, $27 \pm 2\%$. Trophoblast size was measured in the embryos recovered after the final round of embryo transfer and was not different between the high, medium or low ESS groups (18.7 ± 2.6, 14.2 ± 2.5 and 17.1 ± 2.9 cm, respectively).

The increase in embryo survival with increasing days postpartum may have been associated with the recovery of animals from the metabolic stresses associated with early lactation. Thus, the increase in embryo survival observed at 326 days post-calving may be attributed to animals having recovered from lactation. Previous work has shown that non-lactating cows have higher early embryonic survival than lactating cows[2]. Alternatively, this difference in embryo survival may reflect the rate of embryonic loss between days 17 and 35 post-oestrus, as early embryonic loss, between days 28 up to 50 post-oestrus, have been reported to be as high as 13%[2]. Work in this area continues as we focus on improving our understanding of the interactions between in early embryonic loss and lactation.

This project was made possible with the help of technical and farm staff from Dexcel No 5 Dairy. Funding for this project was made available through Dairy InSight.

[1]Xu & Burton, 2003. Monitoring Fertility Industry Report, pp5.
[2]Santos et al. 2004. Anim Reprod Sci 82-83:513-535.
[3]Sreenan & Diskin, 1983. Vet Record 112:517-521.

92. Use of intrauterine catheters to study the uterine microenvironment during early pregnancy in cattle

M Binelli[a], AF Freire[a], MC Lima[a], E Zimberknopf[b] & MA Miglino[a]

[a]College of Veterinary Medicine and Animal Science, University of São Paulo, São Paulo, Brazil,
[b]Laboratory of Physiology and Pharmacology, UNIFEOB, São João da Boa Vista, Brazil

In cattle, embryonic mortality associated with failure in the process of maternal recognition of pregnancy reaches 30 to 40%. Successful pregnancies depend on appropriate biochemical interactions between the maternal endometrium and conceptus. The overall objective of this study was to develop a surgical technique to probe the uterine microenvironment of cyclic and pregnant cows from days 14 to 20 post-estrus. Specific objectives were to verify whether the presence and operation of uterine catheters would affect (1) maintenance of pregnancy, (2) luteal function and (3) follicular growth. Non-lactating, cyclic, Holstein cows were fitted with a medical grade silicone catheter (2.5mm outer diameter, 1.2mm inner diameter; HpBio, Brazil) in each uterine horn on day 2 after estrus. On day 15 they received an injection of D-cloprostenol and ovulations were confirmed by transrectal ultrasonography (US; experimental day 1). On experimental day 7, cows either did (n = 6) or did not (n = 3) receive embryos by trans-cervical transfer to the uterine horn ipsilateral to the ovary containing the corpus luteum (CL). On experimental days 14, 16, 18 and 20, ovarian structures were observed by US and each uterine horn was washed through the catheter (three sessions of 6 ml each). Blood samples were collected from experimental days 1 to 20 and progesterone (P_4) concentrations were measured by radioimmunoassay. Cows were slaughtered on experimental day 20 and pregnancies were diagnosed by macroscopic visualization of a conceptus after dissection of uterus. Conception rate at day 20 was 0%. On experimental day 7, both a CL and a large follicle (9 to 17mm) were present in ovaries of all cows. The rate of increase of P_4 concentrations from days 1 to 5 was 0ng/ml/day in 7/9 cows. Luteolysis occurred before day 15 in 3/9 cows, between days 16 and 20 in 3/9 cows and after day 20 on 2/9 cows. No luteal phase rise in P_4 was noticed for one cow. Ovulation before day 20 was verified in 3/9 cows. Rate of growth of the last dominant follicle was 1.3mm/day in 2/9 cows and less in the remaining. Follicular cysts, poorly luteinized CL and endometrites were diagnosed in 2/9, 2/9 and 1/9 cows respectively. In summary, alterations in ovarian and uterine functions were caused by presence and operation of uterine catheters and such alterations were incompatible with maintenance of pregnancy. In conclusion, the surgical approach tested was not adequate for studying the uterine microenvironment of pregnant cows.

Authors want acknowledge FAPESP and CNPq for financial support, INNOVARE for donation of drugs, HPBio for donation of catheters, Bellman for donation of mineral supplements, and Drs. RP Arruda, JA Visintin, MEOD Assumpção, students from the Department of Animal Reproduction and employees of the Pirassununga Campus for technical assistance.

93. Somatic cell nuclear transfer and gene expression in bovine blastocysts

C Smith, J Somers, D Berg, S Beaumont, DN Wells & PL Pfeffer

AgResearch Crown Research Institute, Hamilton, New Zealand

Nuclear transfer (NT) involves the replacement of a matured egg's DNA with that of a donor cell. The resulting 'zygote' is artificially activated, thus bypassing fertilisation. It is remarkable that this process can lead to viable offspring considering that the donor DNA has to be reprogrammed to a high degree by the egg cytoplasm. We have observed that NT blastocysts showing implantation rates equal to in vitro fertilised embryos exhibited high mortality rates during later development. This suggested that these embryos, though morphologically normal, were defective. We reasoned that such defects should be manifested in the gene expression profile of the reconstructed embryos.

In a candidate approach, we optimised a RNA isolation and quantitative real-time RT-PCR protocol and used this to examine the levels of developmentally important genes in individual blastocysts. Whole mount in situ hybridisation was performed to detect spatial changes in gene expression. We noted that *Oct4* expression was normal but *Ifitm3/Fragilis* was expressed precociously in NT reconstructs.

In a global approach, we constructed a bovine blastocyst cDNA library and randomly picked 5000 clones to prepare a microarray. This microarray was screened with probes prepared from twice-linearly amplified cDNA derived from single NT and control blastocyst embryos. We found 141 genes consistently differentially expressed. Of these, 70% were expressed at lower levels in NT embryos, suggesting that NT embryos fail to correctly regulate blastocyst specific genes. A subset of these genes were further analysed by real-time RT-PCR on 32 individual embryos. We noted the gene expression changes between IVP and NT embryos were generally less than two fold and that misexpression of one gene in an embryo was not predictive of that of a second gene suggesting that reprogramming errors are stochastic.

94. *IGF2* and *IGFBP-2* expression in bovine somatic cell nuclear transfer (SCNT) conceptuses compared with AI and in vitro produced controls

N Li, DN Wells & RSF Lee

Reproductive Technologies, AgResearch, Ruakura Research Centre, East St, Private Bag 3123, Hamilton, New Zealand

The cloning of ruminants by somatic cell nuclear transfer (SCNT) is associated with a high incidence of pregnancy failure and fetal overgrowth. The insulin-like growth factors (IGFs) are important regulators of fetal size as they influence both proliferative and metabolic activity. Placental-specific *IGF2* has been shown to regulate nutrient delivery to the fetus in the mouse[1]. Excessive IGF-II is associated with increased birth weight, disproportionate fetal organ overgrowth, a phenotype common in SCNT fetuses. IGF-II actions are regulated in an autocrine and paracrine manner by tissue-specific expression of the IGF receptors and the IGF-binding proteins (IGFBPs)[2]. The IGFBPs can have both inhibitory and stimulatory activity with regards to IGFs. IGFBP2 levels were reported to be elevated in the plasma and liver of fetal sheep exhibiting the large offspring syndrome[3].

IGF2 and *IGFBP2* expression was investigated in SCNT conceptuses and fetuses and compared with controls generated by artificial insemination (AI) or *in vitro* produced (IVP) embryos. The donor cell line (EFC) for SCNT was derived from a Friesian dairy cow. The expression of these two genes was examined in Day 20, 26 and 30 conceptuses and in placental and organ tissues from Day 100 and 150 fetuses.

Mean *IGF2* expression levels tended to be higher in the SCNT and IVP trophoblast at Days 20 and 26 compared with AI but was not significantly different at Day 30. *IGF2* mRNA levels were not significantly different between the groups in the placentomes, liver, kidney and heart tissues from Day 100 and 150 fetuses.

IGFBP2 was expressed in the trophoblast of all conceptuses at Day 20 but by Day 26, the expression was silenced in trophoblast from all AI conceptuses. Day 26 IVP and SCNT conceptuses continued to express relatively high levels of *IGFBP2* in the trophoblast and even at Day 30, 1/6 and 1/7 IVP and SCNT trophoblasts respectively, were still expressing *IGFBP2*. Protein levels in trophoblastic tissue, as determined by Western blots, reflected the relative levels of mRNA in each sample.

In the bovine fetus, *IGFBP2* was expressed in the liver and kidney at relatively high levels but only low levels of mRNA were present in the heart and muscle. The protein was detectable by Western blots in Day 150 fetal plasma. Mean mRNA levels were not significantly different among the groups in the heart, muscle and kidney tissues but tended to be higher in the livers of SCNT fetuses at Day 150. In the placentomes, *IGFBP2* expression was observed predominantly in the maternal caruncular tissue; mRNA levels were lower in the SCNT and IVP placentomes at Day 150 but not at Day 100, when compared with AI samples.

Thus, *in vitro* culture, independent of SCNT, can have an effect in the onset of *IGF2* and the persistence of *IGFBP2* expression during embryonic development. The outcome of pregnancies resulting from these two types of embryos is however, very different. In our hands, IVP fetuses do not show the fetal overgrowth and pathology commonly encountered in SCNT. Thus, the contribution of dysregulated *IGF2* and *IGFBP2* expression to the SCNT phenotype is unclear.

[1]Constancia, *et al.* 2002. Nature 417:913-914.
[2]Wathes, *et al.* 1998. J Dairy Sci 81:1778-1789.
[3]Young *et al.* 1999. Theriogenology 51:196 (Abs)

95. Targeting the differential transcriptome of bovine NT, IVF and AI preimplantation embryos: global and candidate gene approaches

RJ Moser[a], N Bower[a], T Reverter[a], J Hill[b] & SA Lehnert[a]

[a]CSIRO Livestock Industries, 1Queensland Bioscience Precinct, 306 Carmody Rd., St. Lucia, QLD 4105, Australia, and [b]FD McMaster Laboratory, Chiswick, New England Highway, Armidale NSW 2350, Australia

The fate of embryos generated by assisted reproductive technologies (ART) is compromised when compared to superovulated and artificially inseminated (AI) counterparts. Somatic cell nuclear transfer (NT) derived embryos show the lowest efficiency with regards to their potential to give rise to healthy offspring followed by embryos produced by in-vitro fertilisation (IVF). Aberrant gene expression, due to reprogramming errors, is thought to impact on the embryo and fetal development.

We conducted two microarray experiments with 6 bovine blastocysts in each group comparing NT vs. IVF and NT vs. AI using a bovine cDNA array[1]. 439 differentially expressed (DE) array elements were identified in the first and 282 DE elements in the second study with 95 DE elements overlapping between both lists. NT embryos exhibited the greatest variance in gene expression compared to IVF and AI embryos.

In order to validate and investigate selected candidates further, we developed a universal reference gene approach (URG) to quantify gene expression in preimplantation embryos produced by different methods where no stable reference gene is available for quantitative reverse-transcribed PCR (QRT-PCR). The method uses a plant specific transcript (pt-RNA) containing a poly (A) tail and its corresponding PCR product (pt-DNA) as exogenous spikes to measure nucleic acid extraction efficiencies as well as quantifying mRNA transcript expression levels during RT-PCR.

This approach has been used to quantify the gene expression of candidate genes identified to be DE in the above microarray experiments, such as a member of the H2A histone family (H2A), myeloid cell leukemia 1 (MCL1), connective tissue growth factor (CTGF), and ferritin heavy polypeptide 1 (FTH1). We were able to confirm the microarray expression patterns of some genes, while others showed a different profile when measured by QRT-PCR. CTGF, identified as up-regulated in NT compared to IVF embryos in microarray experiments, showed a down-regulated profile in the NT group compared to AI. The expression of H2A, MCL1, and FTH1 was shown to be up-regulated in AI embryos when compared to NT embryos by QRT-PCR. These experiments show that a combination of microarray and QRT-PCR expression profiling is a powerful tool to investigate the differential transcriptome of bovine preimplantation embryos.

[1]Donaldson et al. 2005. BMC Genomics 6:135-156.

96. Mammalian target of rapamycin (FRAP1) in ovine conceptus development

FW Bazer, H Gao, TE Spencer, GWu & GA Johnson

Center for Animal Biotechnology and Genomics, Departments of Animal Science and Veterinary Integrative Biosciences[a], Texas A&M University, College Station, TX, USA

Improving reproductive efficiency in ruminants requires an understanding of hormonal, cellular and molecular mechanisms regulating conceptus development. Specific components of histotroph, i.e., glucose, leucine, arginine, glutamine and Secreted Phosphoprotein 1 (SPP1 or osteopontin) can activate the FRAP1 nutrient sensing cell signaling pathway to stimulate migration, hypertrophy and hyperplasia, and gene expression by trophectoderm cells. FRAP1 null mice experience post-implantation embryonic death[1]. Little is known of uterine factors that stimulate FRAP1 expression. In ovine allantoic fluid (ALF) collected between Days (D) 30 and 140 of gestation, alanine, citrulline, and glutamine account for 80% of total α-amino acids in early gestation, while serine accounts for 60% of total α-amino acids during late gestation[2]. We (unpublished data) found that total recoverable arginine, glutamine, leucine and glucose increase by 16-, 42-, 9-, and 12-fold, respectively between Days 10 and 15 of pregnancy, but not between Days 10 and 15 of the estrous cycle. The present study characterized changes in ovine fetal-placental development and changes in glucose, fructose and protein in ALF on D 25, 30, 35, 40, 50, 60, 70, 80, 90, 95, 100, 110, 120, 130 and 140 of gestation (N = 3 to 5 ewes/ D). Data were analyzed by ANOVA and are expressed as least square means. Placental weights increased (D, $P < 0.001$) from D25 (5.2 g) to D80 (435 g), and were similar on D 140 (348 g). Fetal weights increased (D, $P < 0.001$) from D25 (0.2 g) to D140 (2,956g). ALF volume (D, $P < 0.001$) increased from D25 (21 ml) to D 40 (91 ml), decreased to D60 (27 ml) and then increased to D140 (439 ml). ALF glucose (D, $P < 0.02$) was 0.10 to 0.30 mg/ml throughout gestation, while total glucose in ALF (D, $P < 0.001$) increased from D25 (4.5 mg) to D50 (16 mg), decreased to D70 (4.0 mg) and then increased to Day 140 (63 mg). Total fructose in ALF increased ($P < 0.03$) from D25 (46 mg) to D50 (217 mg), decreased to D70 (148 mg) and then increased to D120 (589 mg), while ALF fructose concentrations ranged ($P < 0.05$) from 0.92 mg/ml on D140 to 4.7, 9.6 and 5.7 mg/ml on D30, D 50 and D80, respectively. ALF protein concentration (D, $P < 0.001$) ranged from 3.2 mg/ml on D80 to 12.5 mg/ml on D70, while total protein in ALF (D, $P < 0.001$) was similar between D50 (305 mg) and D80 (289 mg) before increasing to D90 (604 mg), D120 (1,935 mg) and D140 (2,991 mg). SPP1mRNA and protein expression by uterine glandular epithelium increased from D13 and D18 of pregnancy with highly bioactive 45 kDa and >80 kDa multimeric SPP1 in ALF increasing dramatically between D40 and D120. These results indicate that selected amino acids, glucose, fructose and SPP1 in the uterine environment of ovine conceptuses may stimulate FRAP1 and, in turn, development and gene expression, e.g., orinithine decarboxylase, nitric oxide synthase and insulin-like growth factor-II, by ovine conceptuses.

[1] Murakami *et al.* 2004. Mol Cell Biol 24:6710-6718.
[2] Kwon *et al.* 2004. Biol Reprod 68:1813-1820.

97. Uterine environment differences between superior and inferior recipient cows

AM Ledgard[a], S Meier[b], PJS Gore[b], M Berg[a] & AJ Peterson[a]

[a]AgResearch Ruakura, Hamilton, New Zealand and [b]Dexcel Limited, Hamilton, New Zealand

Early embryo loss accounts for 22% of pregnancy failure in cattle and has been narrowed to a window between day 8 and 18 after insemination with one study in beef heifers suggesting losses had occurred by day 14.[1] Previous work identifying recipient cows that had high or low holding rates, selected using serial embryo transfers, revealed no overt differences between ovarian factors suggesting that the uterine environment influences early embryo mortality. [2] To further confirm the selection process and define the factors involved in embryo-maternal interactions, we looked in detail at the uterine proteins secreted during this stage of development.

After an oxytocin challenge thirty-five cows (3yrs old, 327 ± 2 days post calving) were selected as superior recipient (SR n = 17) or inferior recipient (IR n = 18) and were synchronised for estrus using a standard prostaglandin (EstroPlan, Parnell Laboratories, NZ) based synchrony. On day 7 of the oestrous cycle, a grade one *in vitro* produced blastocyst was transferred into the ipsilateral horn of each cow. Animals were slaughtered 11 days later and the horns of the uterine tracts were flushed separately with 20ml saline. The presence of a conceptus was determined and trophoblast length was measured prior to the conceptus being frozen. The uterine luminal fluid (ULF) was centrifuged to remove cellular debris and then lyophilised before being reconstituted, dialysed and the protein concentration adjusted to 1mg/ml. Endometrial tissue from each horn was frozen in liquid N or fixed in 10% formalin. Total conceptus length in SR cows was greater than IR cows: 1 SR and 4 IR were non-pregnant, 3 SR and 6 IR had 1-7 cm conceptuses and 13 SR and 8 IR cows had conceptuses $\geq$ 10cm. The ULF was examined by Western blotting for the amount of the pregnancy signalling protein interferon tau (IFN-τ) released by the conceptus and for maternal proteins that change in response to pregnancy. The mean of total IFN-τ was less in the IR compared to SR cows, which reflected conceptus size however there was no significant difference in mean IFN-τ concentration in the ULF from the pregnant horns of SR compared to IR cows. The difference was due to IFN-τ concentration in the empty (or contra) horns where amounts were less in IR ULF.

The difference in total IFN-τ resulted in lower mean concentrations of total ubiquitin cross-reactive protein (UCRP) in IR than SR uterine tracts. Tissue inhibitor metalloproteinase 2 (TIMP2) protein decreases in ULF in early pregnancy and although total mean TIMP2 was not different in SR compared to IR cows, comparing ULF from conceptuses that were > 24cm in length (n = 15), there was significantly less (P = 0.05) TIMP2 in the embryo containing than the empty horn ULF's of SR cows whereas IR cows had similar levels of TIMP2 in both horns.

Selection of cows with a high pregnancy holding ability that have early differences in their uterine environment in response to pregnancy has been confirmed by this trial. Further work to examine differential gene expression is to be carried out on the endometrial tissue collected.

This project was made possible with the help of technical and farm staff from AgResearch Ruakura and Dexcel No 5 Dairy. Funding for this project was made available through Dairy InSight.

[1]Dunne *et al.* 2000. Anim Reprod Sci 58:39-44.
[2]Peterson & Lee, 2003. Therio 59:687-697.

98. Evidence of early pregnancy failure: a comparison of conceptus fluid and fetometric parameters in the first trimester of naturally conceived, IVF and SCNT bovine pregnancies

GA Riding[ad], SA Lehnert[ad], AJ French[b] & JR Hill[c]

[a]CSIRO Livestock Industries, Queensland Bioscience Precinct, St Lucia, QLD, Australia [b]Stemagen Corporation, La Jolla, CA, USA, [c]CSIRO Livestock Industries, FD McMaster Laboratory, Armidale, NSW, Australia and [d]CRC for Innovative Dairy Products, Melbourne, Victoria, Australia

The potential of somatic cell nuclear transfer (SCNT) as an alternative reproductive technology for the dissemination of elite bovine genotypes has yet to be fully realised. With SCNT and to a lesser extent *in vitro* fertilisation (IVF) generated pregnancies, the overall inefficiency and significant costs related to excessive embryo wastage, particularly in the first trimester of pregnancy, remains a problem. Fetal developmental anomalies in early pregnancy can often be diagnosed by transrectal ultrasonography with the *in vivo* measurement of conceptus and fetometric parameters. To gain a more accurate and reliable measurement of conceptus fluid and fetometric parameters, we recently collected *ex utero* conceptus measurements from a sample (n = 108) of abattoir derived naturally conceived pregnancies (NCPs) from young beef breed females (Riding et al. in preparation). The aim of this current study is to compare *ex utero* conceptus measurements from Day 45 IVF/SCNT and Day 90 IVF derived pregnancies with those of NCPs.

IVF and SCNT pregnancies were established in suitably synchronised recipients. Four concepti from each group were harvested at Day 45 with the remaining four IVF pregnancies collected at Day 90 of gestation. Conceptus fluid and fetometric parameters were measured and compared with the normal ranges from NCPs. Developmental integrity of the fetus was determined using the expected crown-rump length (CRL) measurements[1,2,3]. All Day 45 IVF (4/4) and SCNT fetuses (4/4) and three out of four Day 90 IVF pregnancies appeared clinically normal by ultrasound and *ex utero* inspection. One Day 90 IVF fetus displayed retarded growth and placental anomalies which included large placentomes, placentomal necrosis and chorioallantoic membrane fragility. The allantoic fluid appeared discoloured compared to normal fluid observed at the same time-point.

In this experiment, the four SCNT Day 46-47 concepti appeared normal, with the majority of conceptus parameters observed being within normal range. In contrast, both Day 45 and Day 90 IVF concepti exhibited some deviation from the normal range (NCPs) conceptus parameters and 3 out of 4 Day 90 fetuses had CRLs greater than those predicted [145 mm[2] and 159mm[3] compared with day 90 IVF; 170, 165, 154 and 170mm]. The results suggest that in assisted reproductive technologies derived pregnancies, early evidence of fetal overgrowth and placental anomalies can be readily observed and diagnosed in the first trimester of gestation.

[1]Pace *et al.* 2002. Biol Reprod 67(1):334-339.
[2]Evans & Sack, 1973. Anat. Histol Embryol 2:11-45
[3]Rexroad *et al.* 1974. J Dairy Sc 57(3):346-347.

99. Diversity of bovine prolactin-related proteins: expression and localization of new family

K Ushizawa[a], T Takahashi[a], M Hosoe[a], K Kaneyama[b] & K Hashizume[c]

[a]Reproductive Biology and Technology Laboratory, Developmental Biology Department, National Institute of Agrobiological Sciences, Tsukuba, Ibaraki, Japan, [b]National Livestock Breeding Center, Odakura, Fukushima, Japan and [c]Department of Veterinary medicine, Iwate University, Morioka, Iwate, Japan

In cows, trophoblast cells produce various peptide hormones, like interferon-tau, placental lactogens (PLs), prolactin-related proteins (PRPs), and pregnancy-associated proteins (PAGs). However, their roles in the implantation process are still unclear. Recently we identified some new PRP family molecules in bovine placenta. The function of these molecules is also unknown. The present study investigated the expression profiles of PRP family members during gestation and the functional relationships among them.

Previously, PRP-I was the only identified protein in the bovine placenta in this family and its immunoreactivity was detected on Day 18 of gestation in binucleate cells. Microarray analysis suggested the diversity of PRP family in bovine placenta and we have identified three homologs, PRP-VII, -VIII and -IX. The cDNA library screening revealed a novel PRP, temporally named PRP-X. The structural and amino acid sequence analysis suggested that most of the PRP family members have no prolactin like activity because the important amino acid for functional activity of prolactin, which is Serine (S) at aa position 94, was phosphorylated and there were mutations of arginine (R) at position 93 and asparatic acid (D) at position 97. (The beginning of the processed PRP is referred as position 1.) From the alignment and comparison of deduced PRPs and prolactin aa sequence, amino acid residues 91 to 95 in PRPs are highly conserved and may not affect lactogenic or somatotrophic functions. Most of the PRPs have a substition at aa position 97 of aspargine (N) or R (D97N/D97R) and this change may be related to a reduction in biological functions of the PRPs when compared to bovine prolactin.[1] The novel PRP was mainly expressed in placentomes similar to other members of the PRP family and found mainly in binucleate cells. Alkaryphosphatase-conjugated PRP-I bound to the extracellular matrix just under epithelia in endometrium during the implantation period. These data suggest that bPRPs have at least two different biological functions, corresponding to trophoblast cell attachment to the endometrium, and to fetal and placental growth.

Supported by Kiban-kenkyu C(17580284) from JSPS, a grant of Animal remodeling Project (05-201) from NIAS, the grant-in-aid of Research Project for Utilizing Advanced Technologies (05-1770) from MAFF, and a grant from The Ito Foundation (Tokyo, Japan).

[1]Schenck et al. 2003. Mol Cell Endocrinol. 204:117-125.

100. Expression of the imprinted gene *IPL/PHLDA2* in the bovine placenta

M Guillomot, G Taghouti, S Degrelle, E Campion, B Tycko[a] & I Hue

Unité Biologie du Développement et Reproduction, National Institute of Agronomical Research, 78352 Jouy-en-Josas, France and [a]Institute for Cancer Genetics, Columbia University Medical Center, New York, NY, 10032, USA

Placentomegaly is one of the major complications during pregnancy of cloned embryos in cattle. Most of the well-studied imprinted genes control tissue growth and many of them affect the placenta. Loss of imprinting of the *IPL/PHLDA2* gene (*I*mprinted in *P*lacenta and *L*iver; *PH*-like *D*omain *A2*) causes placental growth retardation in the mouse[1] and this gene appears as a growth regulator of the human placenta[2]. To investigate a putative role of *IPL* in bovine placental development, we have undertaken studies on its pattern of expression during pregnancy. Placental tissues were collected from day 35 of pregnancy to term from artificially inseminated cows. The kinetics of expression was followed by Northern blots of total RNA from placental extracts and cellular localisation of mRNA was analyzed by in situ hybridisation (ISH). 32P-ds DNA (Northern blots) and digoxygenin-RNA (ISH) probes were derived from the bovine *IPL* cDNA. *IPL* was specifically expressed in the fetal compartment of the placenta throughout pregnancy from day 35 to term. By Northern blot a single band (0.85 kb) was observed in all placental extracts. No expression could be detected in liver extracts. Cellular localisation by ISH indicated a specific expression in the mononucleated trophoblast cells of the chorionic villi with a lack of expression in the binucleated cells or in the uterine tissues. This pattern of expression correlates with the previous observations in mouse and human placentae. These findings suggest homology between the murine Type II labyrinthine trophoblast, the human villous cytotrophoblast and the bovine villous trophoblast, all of which are highly proliferative tissues. Studies are in progress to determine the level of expression of IPL in placental overgrowth observed after somatic cloning in cattle.

[1]Salas *et al.* 2004. Mech Dev 121:1199-1210.
[2]Saxena *et al.* 2003. Placenta 24:835-842.

101. Bovine placentomal gene expression profiles during gestation

K Hashizume[a], K Ushizawa[b], K Kizaki[a], K Kaneyama[c], Y Yoshikawa[d], M Hosoe[b], T Takahashi[b], & K Imai[c]

[a]Department of Veterinary Medicine, Iwate University, Morioka, Japan, [b]Department of Developmental Biology, National Institute of Agrobiological Sciences, Tsukuba, Japan, [c]National Livestock Breeding Center, Odakura, Fukushima, Japan and [d]Hokkaido University Graduate School of Medicine, Sapporo, Japan

The mechanisms for implantation and maintenance of gestation are still unclear in cattle. The purpose of this study was to examine the genes involved in implantation, development and function of the placenta during gestation in cattle using an utero-placental cDNA microarray. Placentomes were collected and separated into cotyledon (COT) and caruncle (CAR) tissues in early, middle and late gestation. Gene expression profiles were determined by a combination of a custom-made cDNA microarray, real-time PCR for quantitative analysis and in situ hybridization for cellular localization. About 480 genes displayed more than a 2-fold difference in expression levels between Day 250 COT samples compared to Day 25 fetal membrane comprising 247 annotated genes and 232 EST genes. The k-means clustering analysis classified these genes into 5 clusters. Bovine trophoblastic binucleate cell specific molecules, such as pregnancy–associated glycoproteins (PAGs), placental lactogen (PL) and prolactin-related proteins (PRPs) were classified in clusters 2 and 3. Hierarchical clustering was only applied on annotated genes and indicated various period-specific features. Expression of PRPs rose earlier than that of PL in COT. Some genes related to adhesion such as LECAM-1 and alpha E-catenin were expressed in the early period of gestation in CAR. Estrogen sulfotransferase expression increased in mid-gestation and was maintained thereafter. We were able to determine the temporal patterns of gene expression by these clustering analyses, and simultaneously determined the tissue and/or cell specific profiles in caruncular and cotyledonary placental tissues. The real-time PCR data supported the reliability of expression by these cluster analysis. In situ hybridization analyses of PAGs, PRPs, PL and BCL suggested binucleate cells expressed these molecules simultaneously. Future studies will further investigate the role of these specific molecules in binucleate cells on placental development.

Supported by Hoga-kenkyu (16658105), Kiban-kenkyu B (17380172), Kiban-kenkyu C (17580284) from JSPS, a grant of Animal remodeling from NIAS, and BRAIN.

102. Effects of agonists on bovine luteal or caruncular endometrial secretion of prostaglandins E_2 and $F_{2\alpha}$ (PGE_2; $PGF_{2\alpha}$) or progesterone (P_4)

CW Weems[a], YS Weems[a], AW Lewis[b], DA Neuendorff[b] & RD Randel[b]

[a]Deparment of Human Nutrition, Food & Animal Sciences, University of Hawaii, Honolulu, HI, USA and [b]Texas Agricultural Experiment Station, Texas A&M University, Overton, TX, USA

Rodent blastocysts remain spherical after hatching; implantation is invasive and embryos erode the endometrium; implantation occurs rapidly on day-4; and requires progesterone (P_4), estradiol-17ß, LH, PGE_2, nitric oxide (NO), leukemia inhibitory factor (LIF), and possibly other agonists, which may increase PGE_2 secretion[1,2]. After ruminant embryos hatch, the embryo elongates; the endometrium is not invaded; and implantation to form the many placentomes is asynchronous with a loose association of the embryo trophoblast and uterine luminal epithelium, which occurs over weeks[3]. Chronic intraovarian or intrauterine PGE_1, PGE_2, or NO infusion delays luteolysis up to 48 days in ewes[3]. The objective of this experiment was to evaluate various agonists that may play a role in cow luteal P_4, PGE_2, or $PGF_{2\alpha}$ secretion or cow caruncular endometrial secretion of PGE_2 or $PGF_{2\alpha}$ in vitro. Day-14-15 Brahman endometrial diced slices were incubated in M-199 at pH 7.2, 95:5 % air and CO_2, respectively, at 39C for 1 h without and for 4&8 h with treatments. Treatments for luteal tissue were: Vehicle, BAY 11-7-82, Angipoietin-2 (ANG-2), Bradykinin (BK), Bradyzide (BZ), Ciglitazone (CIG), HOE-140, Leptin (LEP), RU-486, or PGE_2 and for endometrium were: Vehicle, Angiogenin (AG), Angiopoietin-1 (ANG-1), ANG-2, dbcAMP, Atrial Natriuretic peptide (ANP), BAY 11-7-82, Bone Morphogenic Protein-4 (BMP-4), BK, BZ, CIG, DETA, DETA-NONOate (NO donor), Endothelin-1 (ET-1), Staurosporine (STA), Epidermal Growth Factor (EGF), Forskolin, HOE-140, Interleukin-1$_\alpha$ (IL-1$_\alpha$), IL-1$_\beta$, LEP, LIF, bLH, PGE_1, PGE_2, bLH+ PGE_1, bLH+ PGE_2, RU-486, Osteopontin (OP), Phorbol-12-Mystriate-13-acetate (PMA), Sphingosphine-1-PO_4 (SP-1P), Thrombospondin (TSP), Thyroclacitonin (TCT), Transforming Growth Factor$_\alpha$ (TGF$_\alpha$), TGF$_{\beta1}$, Tumor Necrosis Factora (TNFa), Methotrexate (MEX), or Vascular Endothelial Growth Factor (VEGF). Hormones in media at 4&8 h were quantified for $PGF_{2\alpha}$, PGE_2, or P_4 secretion by RIA. Luteal hormone secretion data were analyzed by a 2X10 Factorial Design for ANOVA and endometrial hormones secreted were analyzed by a 2X38 Factorial Design for ANOVA. Luteal $PGF_{2\alpha}$ and PGE_2, secretion increased ($P \leq 0.05$) with time in culture, but not ($P \geq 0.05$) luteal P_4 secretion. Only PGE_2 increased ($P \leq 0.05$) luteal P_4 secretion and no treatment affected luteal ($P \leq 0.05$) $PGF_{2\alpha}$ or PGE_2 secretion. Endometrial PGE_2 and $PGF_{2\alpha}$ secretion increased ($P \leq 0.05$) with time in culture. PGE_2 secretion was increased ($P \leq 0.05$) at 4 h by Forskolin; at 4&8 h by by ANG, BK, IL-1$_\alpha$, PMA, STA, and EGF; and by ET-1, IL-1$_\beta$, TGF$_\alpha$, TNF$_\alpha$, VEGF, LIF, SP-1P, BZ, and DETA-NONOate at 8 h. $PGF_{2\alpha}$ secretion was increased ($P \leq 0.05$) at 4 h by Forskolin; MEX; and TGF$_\alpha$; at 8 h by IL-1$_\alpha$; at 4&8 h by STA, TSP, TGF$_{\beta1}$, and TNFa. It is concluded that luteal P_4 does not affect its own secretion, since RU-486 did not affect luteal P_4, PGE_2, or $PGF_{2\alpha}$ secretion. It is also concluded that agonists involved in rodent implantation may also be involved in the asynchronous implantation in cows.

[1]Lim et al. 2002. Vit Horm 64:43-76.
[2]Salamonsen et al. 2001. Reprod Fertil Dev 13:41-49.
[3]Weems et al. 2006. The Veterinary Journal 171/172:206-228.

103. Successful germ cell depletion and germ cell transplant in irradiated testes of pubertal Merino sheep

JA Olejnik[a], M Jackson[b], N Suchowerska[c], G Hinch[d] & J Hill[a]

[a]Reproductive Technology Centre, CSIRO Livestock Industries, Armidale, NSW, Australia, [b]Faculty of Medicine and [c]Department of Physics, University of Sydney, Sydney, NSW, Australia and [d]School of Rural Science and Agriculture, University of New England, Armidale, NSW, Australia

The germ cell transplant technique has many potential applications but has yet to be demonstrated with high success in livestock species. To ensure the success of this technique, the recipient testes must be depleted of its endogenous germ cells, without affecting the somatic cells. Irradiation is one method of depletion that was investigated.

The two objectives studied were:

1. To examine the effects of irradiation on the testis histology.
2. To test if the irradiated testes could support transplanted donor cells.

To examine the first objective, 4 groups (n = 3) of pubertal Merino ram lambs received 0, 9, 12, or 15 Gy of irradiation (6MV photon beam) directly to the testis. Testis samples were collected at 3 weeks and 3 months after irradiation and tubule morphology and germ cell identification were compared by Chi Square.

To examine the second objective, donor germ cells stained with the fluorescent dye PKH26 were transplanted into the left testis of 6 animals (2 controls, 4 irradiated), 10 weeks post irradiation. The animals were castrated 3 weeks later and whole mount tubules and frozen sections were examined for all animals.

The histology results from 3 weeks showed that at all doses, irradiation successfully (>98%) cleared the tubules of germ cells, leaving only the cells on the basement membrane (0 Gy: 19% vs 9 Gy: 100%, 12 Gy: 98%, 15 Gy: 99%: $p < 0.05$). The samples taken at 3 months showed that the majority of tubules had recovered and showed spermatogenesis, with only a few cleared tubules. The 15 Gy treatment had the highest percentage of cleared tubules (0 Gy: 0%, 9 Gy: 6%, 12Gy: 4% vs 15 Gy: 15%: $p < 0.05$). Although spermatogenesis was observed in the majority of tubules, the 12 and 15 Gy groups were slightly more delayed in the spermatogenic cycles, as reflected by a lower percentage of tubules containing round and elongated spermatids and half the number of tubules containing spermatozoa.

The results from the germ cell transplantation showed that the transplanted cells had migrated to the basement membrane of the seminiferous tubules. In 5 out of 6 animals, fluorescent cells were observed in both whole mount tubules and in frozen sections.

It appears that irradiation doses of 9 Gy and above are able to deplete the endogenous germ cells in pubertal ram testes. The seminiferous tubules are also able to recover after three months, even at doses up to 15 Gy, demonstrating that the testis environment was still able to support spermatogenesis.

This study has shown that irradiation can clear the seminiferous tubules and these tubules can support transplanted germ cells.

The contribution of M Herrid, R Davey, K Hutton, J Flack and M Hillard is greatly appreciated.

104. Trends in Sertoli cell proliferation in the developing bovine testis during prepubertal ages

SMLC Mendis-Handagama[a], G Michel[b], T Krychowski[b] & B Jégou[c]

[a]Department of Comparative Medicine, The University of Tennessee College of Veterinary Medicine, Knoxville, TN 37996, U.S.A, [b]URCEO, 69 rue de la Motte Brûlon, BP.80225, 35702 Rennes cedex 7 France and [c]Inserm, U625,GERHM; IFR 140; Univ Rennes I, Campus de Beaulieu, Rennes, F-35042 France

Testis size in bulls determines the sperm producing capacity and fertility. Sperm producing capacity of a testis depends on the amount of Sertoli cells in a testis. Proliferation of the Sertoli cells in the neonatal-prepubertal testis ceases at some point in every mammalian species so that the number of the Sertoli cells/testis in the adult is constant. The present study was designed to understand at what prepubertal age in bull calves it is desirable to intervene in order to lengthen the normal duration of the period of Sertoli cell proliferation. This is to achieve an increased number of Sertoli cells/testis at adulthood and thereby increase their sperm producing capacity. Testicles of bulls (n = 5-13 bulls/group) of 14, 17, 20, 25, 30, 35 and 40 weeks of age were obtained after slaughter. One testicle of each bull was weighed fresh to obtain the fresh testis/volume and the other testicle was subjected to perfusion fixation with 5% glutaraldehyde in 0.1M cacodylate buffer by canulating the testicular artery. Fixed testes were cut into 2-3 mm cubes, which were postfixed in osmium tetroxide-potassium ferrocyanide mixture, dehydrated in a series of graded ethanols and embedded in epon-aradite to be used for stereology. Using two $1\mu m$ sections three sections apart and stained with Methylene Blue Azure II stain, the Sertoli cell numerical density (Nv, number/unit volume of testis) and number per testis was determined by the disector method as described previously (1).

The average weight of a testis in bulls increased with age, from 16.1g at 14 weeks to 98.4g at 40 weeks. The Nv of Sertoli cells was unchanged from 14-20 weeks, but showed a concomitant decline thereafter, and reached a value which is 25% of what is observed at 14 weeks of age. However, the Sertoli cell number per testis gradually increased from 6287×10^6/testis to reach the highest value of 9341×10^6/testis at 40 weeks.

These findings show that the Sertoli cell number per testis in bulls increase with age along with the increase in the testis volume. The results of Nv values suggest that Sertoli cell proliferation rate in the bovine testis is highest at very early in life and gradually declines with the advancement of age. Based on the above findings,14-20 postnatal weeks are likely to correspond to a developmental age at which it could be envisaged to intervene to prolong the proliferative phase of Sertoli cells in bulls.

Supported by URCEO, Rennes, Région Bretagne, France and The University of Tennessee Center of Excellence and Professional Development Award Programs.

[1]Mendis-Handagama & Ewing, 1990. J Microscop (Oxford) 159:73-82.

105. Expression patterns of alpha-smooth muscle actin and vimentin in bovine testes alter by spontaneous cryptorchidism, scrotal insulation and mixed atrophy

B Devkota[a], M Sasaki[b], M Matsui[a], K-I Takahashi[c], S Matsuzaki[c], C Amaya Montoya[a] & Y-I Miyake[a]

[a]Depart of Clinical Vet. Sci., [b]Department of Basic Vet. Sci., Obihiro University of Agric. and Vet. Med., Obihiro and [c]Genetics Hokkaido, Shimizu, Japan

Cytoskeletal proteins provide structural and functional support to cells. In the testis, alpha-smooth muscle actin (SMA) and vimentin are localized in seminiferous peritubular cells and Sertoli cells respectively, and their expression patterns in bovine testes indicate the maturation status[1]. Thermal stress in the testis is lethal to germ cells, which also impairs the structures and functions of the peritubular and Sertoli cells[2]. In the human testis, different types of spermatogenic arrest found in adjacent tubules are termed as mixed atrophy. It might be due to functional impairment of the associated Sertoli cells that maintain or regain an undifferentiated state[3]. Some studies have shown increased Sertoli cell vimentin in cryptorchidism, and in atrophied tubules. The aim of the present study was to investigate the immunohistochemical expression of alpha-SMA and vimentin in bovine testes under cryptorchidism, scrotal insulation and mixed atrophy condition to evaluate the changes in expression patterns for cellular impairment.

Testis samples were collected from postnatally growing spontaneous unilateral cryptorchid bulls (n = 10), postpubertal bulls (n = 2) after unilateral castration and contralateral scrotal wrapping for one week, and from sub-fertile mature bulls (n = 4). Normal testes from the cryptorchid (n = 8) and pre-scrotal-insulation (n = 2) bulls were used as control against cryptorchid and insulated testes. Samples were fixed in Bouin's solution. The sections were stained with hematoxylin-eosin and immunohistochemical staining techniques. Different types of spermatogenic lesions were found in adjacent tubules of mature sub-fertile bull testes, and diagnosed as testicular mixed atrophy.

Seminiferous peritubular alpha-SMA did not appear in the cryptorchid testis until 8 months of age, except for very weak intermittent expressions in relatively larger seminiferous tubules. Distortion of peritubular alpha-SMA was common in testes after scrotal wrapping and its severity was related to the degree of spermatogenic cell loss. Such distortions were not observed in the tubules showing spermatogenic lesions in mixed atrophy. The Sertoli cell vimentin in the normal testis revealed a transforming pattern from 5 to 8 months of age and a mature pattern then onwards[1]. The cryptorchid testis, however, revealed a highly immature pattern at 5 months of age and a transforming pattern at 8 months and 18 months of age, with considerable weakening of the vimentin filaments in 18-month-testis. No considerable alteration of the vimentin expression was observed in insulated testis. A remarkable increase in vimentin intensity, which almost resembled with a transforming pattern, was common in the Sertoli-cell-only tubules and tubules with spermatogenic cell arrest under mixed atrophy. However, no such alterations were observed in the relatively normal tubules of the same testis.

In the bovine, cryptorchidism may cause considerable delay in testicular myoid cell differentiation and in attaining transforming pattern of the Sertoli cell vimentin. The vimentin expression may fail to attain a mature pattern by prolonged cryptorchidism, and it may increase and maintain or regain a transforming pattern under mixed atrophy. We suggest that the alpha-SMA and vimentin expressions change under thermal stress, and alpha-SMA may be more heat sensitive as compared to vimentin.

[1]Devkota et al. 2006. J Reprod Dev 52:in press.
[2]Kerr et al. 1979. Biol Reprod 21:823-838.
[3]Bergmann & Kliesch, 1994. Anat Embryol 190:515-520.

106. The role of the recently discovered sperm protein SPRASA in bovine fertilization

A Wagner, A Shelling & LW Chamley

Department of Obstetrics & Gynaecology, University of Auckland, Auckland, New Zealand

SPRASA is an acrosomal protein that was discovered recently based on its reactivity with antisperm antibodies from some infertile men[1]. SPRASA is highly conserved and has been identified, by immunohistochemistry, to be present in the acrosome of sperm from several species, including red and fallow deer, elk, cattle, horses, mice and humans. SPRASA shows high homology to the alpha-lactalbumin/C-type lysozyme family of proteins and it has previously been shown that alpha-lactalbumin, which is not naturally present in semen, can inhibit the binding of sperm to oocytes, potentially by modifying the substrate specificity of sperm/oocyte receptor galactosyl transferase (gal-T).

It has been previously shown that antibodies reactive with SPRASA inhibited sperm/oocyte binding in the zona-free hamster oocyte binding assay[2]. Based on existing data, we postulate that SPRASA plays an integral role in fertilization, and that binding of antibodies to SPRASA may inhibit its function leading to infertility. Using a cross-reactive antiserum, we investigated the effects of inhibiting SPRASA on bovine *in vitro* fertilisation.

Motile sperm were obtained from frozen-thawed bovine sperm by centrifugation through a discontinuous Percoll gradient. Bovine oocytes were obtained from the abattoir, and matured *in vitro*. Three different treatment groups were examined: (1) sperm was incubated for an hour with the SPRASA antiserum or a control irrelevant antiserum, washed and used to fertilize oocytes (n = 100) in IVF droplets; (2) matured oocytes (n = 100) were incubated in the presence of the SPRASA antiserum or a control irrelevant antiserum for an hour and fertilized with untreated motile sperm; and (3) sperm and oocytes (n = 100) were co-incubated with the SPRASA antiserum or a control irrelevant antiserum. Fertilisations conducted in plain IVF medium were used as baseline controls for each treatment. Fertilization was assessed by counting the numbers of embryos that reached cleavage and subsequent developmental stages.

Analysis of the results showed that there were no significant differences between the controls and the three treatment groups in fertilisation as measured by cleavage rates nor was there any difference in the development to morula or blastocyst in group 1 between the treated oocytes and the controls. However in groups 2 and 3, there was a significant inhibition of embryo development to later developmental stages in the SPRASA antiserum treated samples ($P < 0.05$), compared to the controls. This somewhat surprising set of results suggests that SPRASA maybe expressed by oocytes, as well as pre-implantation embryos, allowing the SPRASA antiserum to inhibit embryonic development without affecting fertilisation rates.

[1]Chiu *et al.* 2004. Human Reproduction 19(2):243-9.
[2]Mandal *et al.* 2003. Biology of Reproduction 68:1525-1537.

107. Expression of vascular endothelial growth factor (VEGF), basic fibroblast growth factor (FGF-2), endothelial nitric oxide synthase (eNOS), and their receptors in ovarian follicles during the periovulatory period in sheep

C Navanukraw[a], DA Redmer[b], ML Johnson[b], AT Grazul-Bilska[b], & LP Reynolds[b]

[a]Department of Animal Science, Faculty of Agriculture, Khon Kaen University, Khon Kaen, Thailand 40002, and [b]Department of Animal and Range Sciences, North Dakota State University, Fargo, ND, USA 58105

Knowledge of the expression pattern of angiogenic factors and their related receptors is valuable for a better understanding of angiogenesis and its regulation[1]. Significant remodelling in vascularization occurs during the transition of the preovulatory follicle into the corpus luteum, a critical event that involves structural and functional changes in the granulosa and theca layers[2]. Thus, the aim of this study was to evaluate the mRNA expression of three major angiogenic factors, VEGF, FGF-2, eNOS, and receptors for VEGF (Flt-1 and KDR), and eNOS (sGC) in granulosa and thecal layers during the periovulatory period. To induce multiple follicular growth and then ovulation, ewes were treated twice daily with intramuscular injections of follicle stimulating hormone (FSH) on d 13 and 14, and both FSH and 600 IU hCG on d 15 of the estrous cycle, respectively. At 0, 2, 4, 8, 12, 24, and 48 h (n = 5-7 ewes/group) after hCG treatment, follicular fluid (FF) was aspirated from follicles > 4 mm for determination of estradiol-17ß (E) and progesterone (P) concentrations by radioimmunoassay, and granulosa and theca layers were collected and pooled within a ewe and then snap-frozen for isolation of total cellular RNA. Angiogenic factor/receptor mRNA expression was quantified by real-time RT-PCR[3]. Expression of each factor was normalized to 18S rRNA expression. The pattern of steroid concentration and mRNA changes from 0 h to 48 h after hCG-treatment were evaluated by regression analysis. Follicular fluid E concentration and E:P ratio decreased (P < 0.002), but P concentration increased (P < 0.001) from 0 h to 48 h after hCG-treatment. In granulosa cells, Flt-1 and KDR mRNA expression increased (P < 0.03) from 0 h to 48 h after hCG-treatment. In thecal tissues, VEGF, Flt-1 and eNOS mRNA expression increased (P < 0.03) from 0 h to 48 h after hCG-treatment. No significant changes in mRNA expression of VEGF, FGF, eNOS and sGC in granulosa, and KDR, FGF-2, sGC in theca layer were observed. Progesterone concentration in FF was positively correlated (P < 0.03) with expression of VEGF ($r^2 = 0.498$) and eNOS ($r^2 = 0.354$) expression in thecal tissues. In addition, in thecal tissues, positive correlations (P < 0.0001) between expression of VEGF and eNOS ($r^2 = 0.862$), VEGF and Flt-1 ($r^2 = 0.781$), and eNOS and Flt-1 ($r^2 = 0.902$) were detected. These data demonstrate that ovine preovulatory follicles express mRNA for VEGF, FGF-2, eNOS and related receptors (Flt-1, KDR and sGC) in both granulosa and thecal layers. Moreover, during follicular development, the pattern of mRNA expression of the major angiogenic factors and related receptors changed in thecal and granulosa tissues which indicates that these factors play a role in maintaining health, growth and vascularization in the ovarian follicles during the periovulatory period. USDA NRI/CGP 2002-35203-12246 to DAR and LPR.

[1]Reynolds et al. 2002. Int J Exp Pathol 83:151-163.
[2]Redmer et al. 2001. Biol Reprod 65:879-889.
[3]Redmer et al. 2005. Biol Reprod 72:1004-1009.

108. Factors that affect the length of the first postpartum luteal phase

LR McNaughton, KS Sanders, GE Bracefield & RJ Spelman

Livestock Improvement Corporation Limited, Private Bag 3016, Hamilton, New Zealand

The first postpartum oestrous cycle is often shorter than subsequent oestrous cycles, due to a short luteal phase[1]. This study investigates factors that affect the length of the first postpartum luteal phase in Friesian x Jersey dairy cattle.

Daughters of six crossbred sires, born in 2000 and 2001 were enrolled in this study. Data from lactation one and two for all animals and lactation three of the 2000 born animals (1718 lactations from 793 animals) was used. Calving began in mid-July and mating began in mid-October each year. Progesterone concentrations were measured 2 x per week by enzyme immunosorbent assay (from 1 milk and 1 blood sample) from calving until the end of the artificial breeding period. Samples were defined as elevated when concentrations reached a threshold (>3ng/ml for milk samples and >1ng/ml for blood samples). The interval from calving to first luteal activity (CLA) was defined as the interval from calving until two consecutive elevated samples. Luteal phase length was calculated as the period from the first elevated sample to the final consecutive elevated sample, twice weekly sampling underestimates the actual luteal phase length by an average of 3.6 days. All figures in this abstract have been adjusted by this factor. A short luteal phase was defined as less than 9 days, and an extended luteal phase as greater than 23 days.

The first luteal phase was short, normal and long in 63, 34 and 3% of animals, respectively. The length of the luteal phase ($\pm$se) increased from 9.46 $\pm$ 0.16 days in the first to 14.47 $\pm$ 0.14 in the second and 15.54 $\pm$ 0.16 in the third oestrous cycles, respectively. The full model of year, calving date (relative to the start of calving), CLA (all significant, $p < 0.05$), age ($p = 0.06$) and sire (not significant) explained 17% of the variation in luteal phase length. A 1-day increase in CLA resulted in a reduction in first luteal phase length of 0.2 days. The interval from calving to luteal activity was significantly different between age groups (2yo 40.9 $\pm$ 0.7, 3yo 25.9 $\pm$ 0.5, 4yo 21.2 $\pm$ 0.6 days, $P < 0.001$), so the model was also fitted by age, giving a reduction in first luteal phase of 0.1, 0.2 and 0.4 days per day increase in CLA in 2, 3 and 4-year-olds, respectively. Animals with a CLA of <25 days had significantly longer first luteal phases than those with a CLA ≥25 days (12.2 $\pm$ 0.22 vs. 6.8 $\pm$ 0.24, $p < 0.001$). In agreement with the current findings the first oestrous cycle tended to be short when the first ovulatory follicle was detected after day 20 postpartum.[2] A possible explanation is that the high progesterone concentrations during pregnancy provide a progesterone priming effect for follicles ovulated early in the postpartum period[2]. This is able to prevent the premature release of $PGF_2\alpha$, which is the main cause of the short luteal phase commonly observed in the postpartum period.[3]

[1]Schams *et al.* 1978. Theriogenology. 10:453-468.
[2]Savio *et al.* 1990. J Reprod Fert 88:581-591
3Zollers *et al.* 1989. Biology of Reproduction. 41:262-267.

109. An extended period of low plasma progesterone levels by repeated follicle aspiration induces follicular cysts in cows

M Matsui[a], G Matsuda[a], K-G Hayashi[b], K Kida[c], A Miyamoto[b] & Y-I Miyake[a]

[a] Dept of Clinical Vet. Sci., [b]Graduate School of Anim. and Food Hygiene and [c]Field Center of Anim. Sci. and Agric., Obihiro University of Agric. and Vet. Med., Obihiro, Hokkaido, Japan

Ovarian cysts have long been recognized as an important cause of infertility in lactating dairy cows. Failure of a positive feedback response to estradiol has been previously reported for cows with follicular cysts[1]. It was reported that progesterone (P4) exposure of the hypothalamus was necessary to reinitiate the estradiol feedback mechanism[2]. Our recent study demonstrated that follicle aspiration before LH surge following corpus luteum (CL) regression prevented new CL formation and maintained low P4 levels[3]. In the present study, the effect of continuous low plasma P4 levels after CL regression, which was extended by repeated follicle aspiration, on follicular cyst formation was investigated.

In Experiment 1, ten cows received $PGF_{2\alpha}$ 10 days after spontaneous ovulation. Follicular fluid of all follicles with a diameter > 6 mm were aspirated using an ultrasound-guided technique 36 h after $PGF_{2\alpha}$ injection. Follicle aspirations were repeated at 4 and 8 days after first aspiration (d 0). Ovarian morphology was examined on every 4 days from d0 using ultrasonography. Blood samples for hormonal analysis were also collected at the same time as ovarian observations. CL did not form in 6 cows for 12 days from the first aspiration. Four cows showed no CL formation and maintained low plasma P4 levels for 24 days after first follicle aspiration. In these 4 cows, the largest follicle reached preovulatory size (19.0 ± 1.3 mm, mean $\pm$ SEM) at d16, and continuously grew to 26.9 ± 3.8 mm in diameter at d 24 without ovulating.

In Experiment 2, to examine the function of the hypothalamus in cows with an induced follicular cyst, the 4 cows with persistent follicles of Experiment 1 were treated with estradiol benzoate (EB, 4mg) at d 24, and blood samples were collected for LH analysis at 12, 15, 18 and 21 h after EB injection. Ovarian morphology was monitored by ultrasonography throughout the experiment. In one cow, a LH surge was observed in response to EB injection, and a CL was formed. In the other 3 cows, the LH surge and CL formation were not detected, and persistent follicles remained until d 34. To investigate the effect of P4/EB treatment on hypothalamus function in cows with induced follicular cyst, these 3 cows were treated with a P4-releasing intravaginal device (PRID) from d 34 to d 41 and then received an EB (4mg) injection at d41. Blood was collection for LH analysis as described previously. Two cows with an induced follicular cyst showed a LH surge in response to EB injection, and a CL was formed. In the third cow, the LH surge and CL formation were not observed, and the cystic follicle persisted.

Our results demonstrated that repeated follicle aspiration could prevent CL formation and maintain low plasma P4 levels. The data also showed that, under an extended period of low P4 levels, the hypothalamus became more refractory to estradiol and a persistent follicle developed. Exogenous P4 appeared to reinitiate estradiol responsiveness in the hypothalamus. These alterations in function of the hypothalamus observed in the present study also suggested that the repeated follicle aspiration after CL regression can be a useful model of follicular cyst formation in cows.

Supported by JSPS, the 21st century COE program (A - 1) and the Akiyama Foundation.

[1]Dobson & Alam, 1987. J Endocrinol 113:167-171.
[2]Gümen & Wiltbank, 2002. Biol Reprod 66:1689-1695.
[3]Hayashi et al. 2005. J Reprod Dev 52:129-135.

110. The response of ovine granulosa cells cultured in serum free conditions with different concentrations of glucose in the media

NR Kendall[a], RJ Scaramuzzi[b] & BK Campbell[a]

[a]Department of Obstetrics and Gynaecology, University of Nottingham, Nottingham, United Kingdom and [b]Department of Veterinary Basic Science, Royal Veterinary College, London, United Kingdom

Glucose is the major energy source in most physiological processes. Differences in glucose concentrations have been linked with differences in reproductive function, such as the nutritional stimulation of fecundity in sheep. The aim of this experiment was to culture ovine granulosa cells in a range of glucose concentrations and to examine the effects on oestradiol production and cell proliferation.

Sheep ovaries from a local abattoir were used to provide granulosa cells from 1-3mm follicles. The cells were cultured in 96 well plates using 75,000 cells per well, for 192h under optimum serum free culture conditions[1]. Commercial McCoy's 5a media was replaced with custom made McCoy's 5a (JRH Biosciences) containing no glucose. Glucose was added at 0, 0.19, 0.38, 0.75, 1.5 and 3.0 g/l. Media was changed every 48h and retained at -20°C for analysis. Oestradiol production was measured by radioimmunoassay, cell number at 192 h using neutral red and lactate production colourimetrically. Three replicate cultures were carried out. Data was analysed using (repeated measures and univariate) general linear model on SPSS 14.0.1 with appropriate data transformations where required to ensure normality.

Results are shown in order of increasing glucose concentrations from 0 to 3g/l with the highest representing that normally found in most culture media. The cell number at 192h (7.8, 40.9, 41.6, 45.8, 46.2, 43.1 kcells) and oestradiol production throughout the culture (143, 404, 399, 435, 392, 319 pg/48h) were significantly depressed for 0 glucose compared to all other glucose doses ($P < 0.001$), but there were no significant differences between any of the other individual glucose doses. There was no significant effect of glucose dose on oestradiol production per cell at 192h (10.8, 19.3, 20.4, 17.4, 14.2, 11.3 pg/kcells). However, there was a dose related decline ($P < 0.05$) in both the oestradiol production at 192 h (90, 642, 652, 690, 591, 444 pg) and oestradiol production per cell as the dose increased from 0.19 up to 3 g/l. Lactate production (0.7, 8.3, 10.5, 9.9, 8.6, 11.5 mg/dl) was significantly depressed by the 0 glucose dose with no difference between any of the other glucose doses.

Culturing cells in the absence of glucose resulted in a depression in cell number and oestradiol production. Surprisingly, the maximum production of oestradiol did not occur at the highest glucose dose (3 g/l) but at a level (0.38-0.75 g/l) closer to the physiological level in blood (40~60 mg/dl = 0.4-0.6 g/l). The high rates of inclusion of glucose (3 g/l or greater) in many culture media may be detrimental and may inhibit FSH-induced differentiation *in vitro*. Furthermore, this may be more pertinent for ruminants which generally have lower blood glucose concentrations than non ruminants, because of their use of volatile fatty acids as an energy source and other glucose sparing mechanisms.

It is concluded that ovine granulosa cells in culture have an absolute requirement for glucose for cell survival and steroidogenic activity but that higher doses of glucose can inhibit FSH-induced differentiation *in vitro*.

The authors wish to acknowledge the technical assistance of Mrs Catherine Pincott-Allen.

[1]Campbell *et al.* 1986. JRF 106:7-16.

111. Effects of active immunization against growth differentiation factor 9 (GDF9) and/or bone morphogenetic protein 15 (BMP15) on ovarian function in cattle

NL Hudson[a], M Berg[b], K Hamel[a], P Smith[a], SB Lawrence[a], L Whiting[a], JL Juengel[a], & KP McNatty[a]

[a]AgResearch, Wallaceville Animal Research Centre, Ward Street, PO Box 40063, Upper Hutt, New Zealand and [b]Reproductive Technologies, AgResearch, Hamilton, New Zealand

The oocyte-derived growth factors, growth differentiation factor 9 (GDF9) and bone morphogenetic protein 15 (BMP15) are essential for follicular growth and regulate ovulation rate in sheep; however, the role of these proteins differ among species. Thus, the aim of this experiment was to determine the effect of neutralizing GDF9 and/or BMP15 in vivo on ovarian follicular development in cattle.

Post-pubertal heifers (n = 10 heifers per antigen) were immunized 3 times at 4 week intervals using an oil based adjuvant with Keyhole Limpet haemocyanin (KLH; control), E. coli expressed ovine GDF9 mature region conjugated to KLH, E. coli expressed ovine BMP15 mature region conjugated to KLH, GDF9 peptide (amino acids 1-14 of the mature region) conjugated to KLH, BMP15 peptide (amino acids 1- 14 of the mature region) conjugated to KLH, or both GDF9 and BMP15 peptide conjugated to KLH. Blood samples were collected at the time of the first immunization and every 2 weeks until the end of the experiment for determination of antibody titres and progesterone concentrations. Ovaries were collected from heifers approximately 2 weeks after the final immunization at slaughter. At collection, the number and appearance of corpora lutea were noted and ovaries were fixed in Bouins fluid for histological examination.

Ovulation rate was assessed by counting the number of corpora lutea (CL) visible on the surface of the ovary. CL were considered to have regressed (and thus not counted in ovulation rate score) if they appeared non-functional by morphological criteria (small, lacking vascularization, pale colour) and endocrinological criteria (low levels of progesterone in blood 2 weeks prior and at time of ovarian collection). Follicular development was assessed on 5 μm histological sections stained with haematoxylin and eosin.

Immunization with the BMP15 peptide with/or without GDF9 peptide significantly altered (increased or decreased) ovulation rate. Immunization with GDF9 and/or BMP15 (either the E. coli derived mature regions or peptides) altered ovarian follicular development. For all groups, the percentage of the area of the cortex occupied by antral follicular space was reduced when compared to controls. This was particularly noticeable in the peptide immunized animals where antral space decreased from 27 % in controls to < 2% in peptide immunized animals. Similarly, the density of follicles decreased when compared to controls in all except those immunized against the GDF9 mature region. In addition, immunization against GDF9 and/or BMP15 peptide reduced follicular size to < 25 % of controls.

In conclusion, immunization against GDF9 and BMP15, alone or together, altered follicular development in cattle. Immunizations against peptides at the N-terminal of the mature region of either GDF9 or BMP15 appeared more efficacious in regulating ovarian follicular development than immunization against the whole mature region indicating that antibodies against the 14 mer peptide were effective in neutralizing the biological activity of GDF9 and BMP15. Thus, as has been observed in sheep, these finding suggest that both GDF9 and BMP15 are key regulators of normal follicular development in cattle.

Supported by the New Zealand Foundation for Research, Science and Technology and Ovita Limited, Dunedin New Zealand.

112. Growth and regression of benign tumours in streak ovaries of homozygous Inverdale sheep

PR Hurst[a] & BJ McLeod[b]

[a]Department of Anatomy and Structural Biology, School of Medical Sciences, University of Otago, New Zealand and [b]AgResearch, Invermay, Mosgiel New Zealand

Homozygous Inverdale ewes are characterised by non-functional streak ovaries. These contain numerous primordial and primary follicles, some of which initiate follicular growth but their oocytes eventually degenerate and the granulosa cells aggregate into nodular structures[1]. Repeated laparoscopic observations have also noted periodic growth and then regression of large cystic tumours. Samples of streak ovaries (n = 4) and ovaries with tumours (n = 6) were perfuse-fixed with glutaraldehyde fluid and systematic histological sections were prepared for light microscopic evaluation with a research grade light microscope.

Streak ovaries were noted for the presence of nodular structures in which the constituent cells were granulosa-like and enclosed by a basement membrane, as noted previously[1]. A small proportion of these nodules had lumina.

All protruding tumours were cystic and their tissue margins characterised by the presence of granulosa and/or thecal layers. Where the cysts contained clear fluid, the wall was typical of a large healthy antral follicle with regular granulosa and thecal tissue separated by a basement membrane. Other cysts had areas of wall tissue in which granulosa contained pycnotic cells and also areas where the granulosa and basement membrane was missing with the theca exposed directly to cyst lumina in which blood cells were also noted. Cysts with luteinized cells and dense capsular walls were also observed. Where tumours were green coloured at gross inspection, the cyst was dominated by haematous, green crystalline pigment and an unstructured wall in which macrophage-like cells contained aggregated pigment.

We hypothesise that cystic structures are derived from the spontaneous growth of nodular structures that attempt to differentiate into large preovulatory follicles (so called functional cysts). Subsequently the cyst wall either degenerates by atretic processes and becomes filled with blood, or alternatively becomes progressively luteinised. Finally, both of these phenotypic structures regress and become absorbed within the ovarian cortex as would normally occur to corpora lutea in functional ovaries.

Further sampling and histological correlation with the temporal development and gross form of the tumours are required to address this hypothesis. Interestingly, the structure with the morphologically normal granulosa/thecal wall observed in this study was known by sequential laparoscopic observations to have developed from a streak ovary within a period of less than 12 days, whereas those with either a luteinised or green appearance were recorded as such 12 days prior to tissue collection. This supports our suggestion that normal granulosa tissue is associated with early tumour development and granulosa degeneration and disruption of the thecal layer occurs as the tumours age.

Previous studies monitoring hormonal profiles showing that inhibin levels were raised periodically[2], indicates the probability of early cystic development analogous to that during growth of a large preovulatory follicle.

[1]Braw-Tal et al. 1993. Biol Reprod 49:895-907.
[2]McLeod et al. 1995. Proceedings NZ Society Animal Production 55:304-306.

113. Comparisons of ovarian follicular growth in early postpartum primiparous and multiparous dairy cows

S Guntaprom[a], C Navanukraw[a], A Boonsaom[a], S Promkasikorn[a], & J Yaeram[b]

[a]Department of Animal Science, Faculty of Agriculture, Khon Kaen University, Khon Kaen, Thailand 40002 and [b]Department of Animal Science, Rajamangala University of Technology ISAN Kalasin Campus, Thailand 46000

Understanding the pattern of follicular growth, in a specific physiological condition, of domestic ruminants is a prerequisite to obtain more precise control of the estrous cycle and ovulation. Thus, the objective of this study was to characterise differences in follicular dynamics between early postpartum primiparous and multiparous cows. Primiparous Holstein dairy cows (n = 4; 349.5 ± 12.1 kg BW; 10.7 ± 2.1 daily milk production) and multiparous Holstein dairy cows (n = 4; 386.0 ± 33.6 kg BW; 13.8 ± 1.8 daily milk production) were individually housed and fed ad libitum of roughage and concentrate to meet requirements for any lactating cow[1]. From days 5 to 90 postpartum, ovarian follicular growth was evaluated every other day using transrectal ultrasonography. The number of follicles on each ovary, the position and the size of each follicle as visualized on the ultrasound monitor were recorded and then classified into 5 classes as previously described[2]. A dominant follicle was defined as a follicle > 10 mm in diameter in the absence of other large growing follicles. Ovulation detected by ultrasound was confirmed by plasma progesterone radioimmunoassay. Body weight and body condition score (BCS; a quarter-point scale from 1 to 5) were assessed on a weekly basis during the study. The pattern of follicular growth, BW and BCS changes were evaluated and compared between primiparous and multiparous cows using regression analysis and Student t test. The follicular wave-like patterns were observed in both primiparous and multiparous cows. Most follicles of primiparous (50.2 %) and multiparous cows (47.8%) between days 5 to 90 postpartum fell into class 2 (3-5 mm in diameter). Although the numbers of follicles in class 1-4 were not statistically different between primiparous and multiparous cows, primiparous cows have fewer class 5 follicles (> 15 mm in a diameter) than did multiparous cows (0.5 ± 0.5 vs. 7.8 ± 6.1, P< 0.05). Follicular growth rates (mm/day) of dominant follicles were not different between primiparous and multiparous cows (0.8 ± 0.1 vs. 1.2 ± 0.2, P> 0.05). Follicular atretic rate (mm/day) of anovulatory dominant follicles were 1.0 ± 0.9 and 1.3 ± 1.1 (P< 0.05) in primiparous and multiparous cows respectively. Neither the interval between calving and the first detection of a dominant follicle nor the size and life span of a dominant follicle differed between primiparous and multiparous cows. However, the interval to until detection of a follicle ≥10 mm in diameter was longer in primiparous than multiparous cows (22.0 ± 0.9 days vs. 16.8 ± 2.4 days; P< 0.1). A relationship between BW during the postpartum period and maximal size of follicle was only observed in multiparous cows (r = 0.484, r[2] = 0.233, P< 0.01), but not in primiparous cows. These data described herein indicates that there are differences in follicular dynamics between early postpartum primiparous and multiparous cows.

This study was financially supported by Thailand Research Fund (TRF) and Commision on Higher Education, Ministry of Education, Thailand on MRG4880039 to CN.

[1]NRC. 1989. Nutrient requirements of dairy cattle. 6th edition.
[2] Beam & Butler, 1997. Biol Reprod 56:133-142.

114. FSH receptor expression in ovaries from Inverdale sheep using laser capture microdissection

JM Mora & PR Hurst

Department of Anatomy and Structural Biology, School of Medical Sciences, University of Otago, New Zealand

Oocyte-derived bone morphogenetic protein-15 (BMP-15) is essential for ovarian follicular growth. BMP-15 regulates granulosa cell proliferation and differentiation by several mechanisms including the suppression of follicle-stimulating hormone receptor (FSH-R) expression.

Inverdale sheep have a point mutation in the BMP-15 gene that disrupts its function resulting in abnormal follicular development and changes in ovulation number. In the homozygous state (I/I), ewes are infertile with "streak" ovaries and arrested follicular development. Interestingly, heterozygotes (I/+) have an increased ovulation quota compared to wildtypes (+/+) and a greater number of mature pre-ovulatory follicles that are smaller with fewer granulosa cells[1].

It is hypothesised that the increased ovulation rate observed in I/+ Inverdale ewes is due to increased FSH-R mRNA expression in granulosa and cumulus cells of antral follicle. Since BMP-15 is reported to suppress FSH-R mRNA expression[2], we would expect increased levels of this receptor in follicles of I/+ ewes, as only half the amount of functional BMP-15 is produced in these animals compared to wildtypes. An increase in FSH-R numbers and therefore an increase in responsiveness to FSH would lead to a greater number of developing healthy follicles with increased expression of LH receptors and overall an increased ovulation rate.

This preliminary study aimed to establish the techniques required to analyse the expression of FSH-R mRNA in different cellular compartments of healthy antral follicles from I/+ and +/+ Inverdale ewes. Ovaries from each genotype (n = 5) were frozen in OCT over cooled isopentane and stored at -80°C until sectioning. Frozen 8 μm sections were stained and laser capture microdissection (LCM) was used to isolate oocytes, mural granulosa, theca and cumulus cells. Extracted RNA was reverse transcribed and PCR was performed to determine the presence of expression of FSH-R mRNA. RNA extracted from whole tissue was used as a positive control.

The presence of FSH-R mRNA was detected in granulosa and cumulus cells from both genotypes and in the whole tissue control. No expression was seen in oocytes or thecal cells from antral follicles from these ewes.

Further research using LCM in combination with quantitative RT-PCR will address this hypothesis. Using frozen sections from I/+ and +/+ ewes we will carry out a similar LCM/RNA extraction process as mentioned above and quantify expression of mRNA encoding FSH receptor and the housekeeping gene ribosomal protein L19 (RPL19).

[1]Shackell *et al.* 1993. Biol Reprod 48:1150-1156.
[2]Otsuka *et al.* 2001. J Biol Chem 276:11387-11392.

115. Expression of mRNAs encoding factor in the germline, alpha (FIGα) within the developing ovary and in small growing follicles in sheep

JL Juengel, LJ Haydon, L Still, P Smith, NL Hudson & KP McNatty

AgResearch, Wallaceville Animal Research Centre, Upper Hutt, New Zealand

Factor in the germline, alpha (FIGα) is known to be expressed prior to follicle formation and throughout follicular growth in mice and humans. Directed deletion of FIGα revealed its essential role in follicular formation in mice[1]. However, whether FIGα could play a similar role in ruminants is unclear as the spatial-temporal pattern of expression of this gene is unknown in ruminants. Therefore, the objectives of these experiments were to determine the expression pattern of mRNA encoding FIGα in the developing ovary and during follicular growth in sheep.

A cDNA corresponding to a portion of the coding region of ovine FIGα was generated using primers based on the mouse and human sequences. The resulting cDNA was 99% identical to the predicted bovine FIGα sequence. Ovaries were collected from ovine fetuses (n = 4-5 per age group) on day 35, 55, 75, 90 and 135 of gestation. Ovaries were also collected from 4 week old lambs (n = 5) and adult ewes (n = 5) to obtain small growing follicles. Expression of FIGα mRNA was determined in ovarian sections following in situ hybridization with nonspecific hybridization monitored by examining hybridization of the sense RNA to at least 2 animals of each age.

Expression of FIGα mRNA was not observed in ovaries collected on days 35 or 55 of gestation. At day 75 of gestation, oocytes in the ovigerous cords as well as those in formed type 1/1a (primordial) follicles expressed FIGα. Within the ovigerous cords, expression was noted to be restricted to oocytes toward the centre of the ovary that contained nuclei with a well defined membrane and pale even-toned appearance suggesting that expression was restricted to oocytes that had undergone meiotic arrest. Thereafter, during fetal development, expression was restricted to oocytes of forming and formed follicles. In lamb and adult ovaries, expression of mRNA encoding FIGα was observed in the oocytes of all follicles observed. This included type 1/1a (primordial), type 2 (primary), type 3 (small pre-antral), type 4 (large pre-antral) and type 5 (antral) follicles. In all tissue sections examined, expression was limited to the oocytes.

In summary, expression of FIGα mRNA was first observed in oocytes after meiotic arrest. Thereafter, FIGα was expressed in oocytes during follicular formation and through all stages of follicular growth (types 1-5) in sheep. This expression pattern is similar to what has been observed in mice and humans and thus is consistent for an essential role for FIGα in follicular formation in sheep as has been observed in mice.

Supported by the New Zealand Foundation for Research, Science and Technology.

[1]Pangas & Rajkovic, 2006. Human Reprod Update 12:65-76.

116. Knockdown of protein kinase Cε (PKC ε) in bovine luteal steroidogenic cells mitigates the effectiveness of prostaglandin F 2α (PGF2α) in reducing LH-stimulated accumulation of progesterone (P$_4$)

JA Flores[a], A Sen[a], MP Goravanahally[a] & EK Inskeep[b]

Departments of [a]Biology and [b]Animal and Veterinary Sciences, West Virginia University Morgantown, WV

Increases in intracellular calcium ion concentration ($[Ca^{2+}]_i$), diacylglycerol and activation of PKC mediate PGF$_2$α-induced regression of bovine CL[1]. Expression of luteal PKC epsilon (PKCε) is greater by mid-cycle (d 10)[2]. Furthermore, inhibitors of PKC ε negate PGF$_2$a-stimulated increases in $[Ca^{2+}]_i$ and inhibit LH-stimulated accumulation of P$_4$ in d-10 cells[2]. Thus we hypothesized that PKCε might mediate the PGF$_2$α-stimulated rise in $[Ca^{2+}]_i$ and low PKC ε might be responsible for resistance of early CL to PGF$_2$α. Furthermore, the steroidogenic luteal cells targeted by PGF$_2$α express PKC ε[3], a necessary condition for our hypothesis to be feasible.

Here, siRNA was used to reduce experimentally the expression of PKC ε in bovine luteal steroidogenic cells in order to examine its role in facilitating the ability of PGF$_2$α to inhibit LH-stimulated accumulation of P$_4$. Enriched steroidogenic cells from d 6 bovine CL were isolated[2] and time-course siRNA experiments were performed to reduce PKC ε expression after 48, 72 after 96 h.

Expression of mRNA and protein corresponding to PKC ε was reduced 57 and 75% by 72 and 96 h of treatment, respectively. This reduction was specific because no similar changes were observed when cells were kept in experimental media, with only non-specific siRNA duplexes (Non-specific siRNA), or transfection reagents. Importantly, down-regulating expression of PKC e altered the effectiveness of PGF$_2$α in reducing LH-stimulated accumulation of P$_4$. Basal accumulation of P$_4$ was not affected by siRNA, by PGF$_2$α or their combined treatment. As previously reported, LH induced increased accumulation of P$_4$ ($P < 0.05$) This effect was not altered by the siRNA treatment during 4 h incubation. However, in cells in which PKC ε expression was ablated by 75%, the inhibitory effect of PGF$_2$α on LH-stimulated P$_4$ accumulation was mitigated; it was only 29% lower than in the LH-stimulated group. In contrast, it was reduced by 75% in the group where PKC ε had not been knocked down with siRNA ($P < 0.05$). These results support the hypothesis that differential expression of PKC ε is important for acquisition of luteolytic response to PGF$_2$α.

This study was supported in part by the Eberly College of Arts and Sciences, West Virginia University, by and award from the United States Israel, Binational Agriculture Research and Development fund and by Hatch Project 427(NE1007).

[1]Davis J et al. 1987. Proc Natl Acad Sci USA 84:3728-3732.
[2]Sen A et al. 2005. Biol Reprod 72:976-984.
[3]Sen A et al. 2005. Domest Anim Endocrinol Epub ahead of print PMID 16388928.

117. Expression of growth differentiation factor (GDF) 9, bone morphogenetic protein (BMP) 15, activin receptor like kinase 5 (ALK5) BMP receptor (BMPR) type IB and BMPRII mRNA in ewes carrying the Woodlands FecX2^W gene

ES Feary[a], JL Juengel[a], P Smith[a], AR O'Connell[b], L Morrison[a], GH Davis[b] & KP McNatty[a]

[a]AgResearch, Wallaceville Animal Research Centre, Upper Hutt, New Zealand and [b]AgResearch, Invermay Agricultural Centre, Mosgiel, New Zealand

Woodlands sheep have a major genetic mutation (FecX2^W) that increases ovulation rate. At present the identity of the Woodlands genetic mutation is unknown. While X-linked, this gene does not appear to be related directly to any of the known mutations in BMP15, GDF9 or BMPRIB that affect ovulation rate in sheep. Nevertheless, the FecX2^W mutation may be affecting ovulation rate by changing the expression patterns of the BMPRIB, BMP15 or GDF9 genes. In addition, since BMPRII is involved in signal transduction for both BMP15 and GDF9, and ALK5 is involved in signal transduction for GDF9, changes in expression pattern of these genes would also be likely to affect ovulation rate. Thus, the objective of this experiment was to examine the patterns of expression of mRNAs encoding GDF9, BMP15, BMPRIB, ALK 5 and BMPRII during follicular development in ewes heterozygous for the Woodlands mutation and their wild-type contemporaries.

Expression patterns of selected genes were determined following in situ hybridisation. Intensity of the signal was established by quantification of silver grains in the cell(s) of interest. As GDF9 and BMP15 are expressed exclusively in oocytes, the signal was only quantified in this cell-type. Silver grain counts for BMPRIB, ALK5 and BMPRII were made separately in the oocyte, granulosa and theca. Between 1-3 slides were analysed per animal and each follicle type (i.e. types 1-5) was observed in at least 3 animals of each assigned genotype.

Expression of GDF9 was observed to increase in oocytes from type 1 to 3 follicles and then decreased, whereas BMP15 increased in follicles from type 2 to 4 and then decreased. However, there were no differences between genotypes for expression of these two genes. No significant differences were found within theca cells among follicle types or genotype for any of the receptor genes. There was a significant effect of follicular type in the granulosa cells for all 3 genes (p < 0.01) and a significant genotype effect for BMPRIB, where overall, the expression in the wild-type was consistently higher (p < 0.01). In the oocyte, there was a significant effect of follicle type and genotype on expression of ALK 5 and BMPRIB, whereas for BMPRII no significant effect was noted on follicle type. Expression of ALK5 was increased in oocytes of Woodlands ewes whereas expression of BMPRIB was decreased.

Mutations in BMPRIB result in increased ovulation rates in sheep, and therefore the differences in expression level of BMPRIB in both the oocyte and granulosa cells may play a role in the increase in ovulation rate observed in Woodlands ewes.

Acknowledgements Marsden fund, Foundation for Research, Science and Technology and Ovita for an enterprise scholarship

118. Expression of fibroblast growth factor-10 and its receptor, fibroblast growth factor receptor-2b, during luteal development in cattle

IC Giometti[a], CA Price[c], ACS Castilho[b], PB Andrade[b], MF Machado[b], PC Papa[d] & J Buratini Jr[b]

[a]Department of Animal Reproduction, Faculty of Veterinary Medicine and [b]Department of Physiology, Institute of Biosciences, Sao Paulo State University, Brazil, [c]Centre de Recherche en Reproducion Animale, University of Montreal, Canada and [d]Department of Surgery, Faculty of Veterinary Medicine, University of Sao Paulo, Brazil

The fibroblast growth factor (FGF) family is involved in the regulation of angiogenesis and cellular proliferation and differentiation of the corpus luteum (CL)[1]. FGF-10 mediates paracrine interactions between mesenchymal and epithelial cells, regulating cellular proliferation and differentiation via activation of FGFR-2b[2]. The objective of this study was to investigate the expression of FGF-10 and FGFR-2b in the bovine CL.

Ovaries were obtained from a local abattoir and ten CL from each of four developmental stages (stage 1, corpus hemorrhagicum; 2, developing CL; 3, mature CL; and 4, regressing CL)[3] were selected for RNA extraction. FGF-10 and FGFR-2b gene expression was examined by semiquantitative RT-PCR using bovine-specific primers and GAPDH as the internal control. FGF-10 protein was assessed by immunohistochemistry using a commercial antibody against human FGF-10.

FGF-10 expression was detected at all stages of luteal development (mean $\pm$ sem relative values of 67.9 $\pm$ 17.4, 72.5 $\pm$ 15.9, 52.6 $\pm$ 16.4 and 95.4 $\pm$ 28.3 for stages 1, 2, 3 and 4, respectively), and there were no differences among stages. Immunohistochemistry demonstrated the presence of FGF10 protein in the CL. FGFR-2b gene expression was also detected at all stages of luteal development, but was highest in the stage 4 regressing CL ($P < 0.005$; relative values of 7.3 $\pm$ 2.8, 8.6 $\pm$ 1.9, 12.2 $\pm$ 2.5 and 28.0 $\pm$ 8.2 for stages 1, 2, 3 and 4, respectively).

The present data indicate that the FGF-10/FGFR-2b system is regulated at the receptor level. It is noteworthy that FGFR-2b expression is increased in regressing CL, implying a role for FGFR2b activation at this time, possibly mediated by FGF10. We propose that FGF-10 may have a role in the paracrine or autocrine control of luteal development and regression in ruminants.

Supported by FAPESP, Brazil.

[1]Salli et al. 1998. Biol Reprod 59:77-83.
[2]Igarashi et al. 1998. J Biol Chem 273:13230-13235.
[3]Ireland et al. 1980. J Dairy Sci 63:155-160.

119. Follicular development in response to exogenous FSH is affected within breeding season

A Veiga-Lopez[a], AS McNeilly[b] & A Gonzalez-Bulnes[a]

[a]Dpto Reproducción Animal, INIA. Avda P Hierro s/n, Madrid, Spain and[b]MRC, Human Reproductive Sciences Unit, Centre for Reproductive Biology, University of Edinburgh, Edinburgh, EH16 4TJ, UK

Embryo viability in superovulatory protocols has been found to be significantly affected by season in sheep from high latitudes[1] and, although to a lesser extent, in ewes from temperate areas[2]. We hypothesised the existence of season-related alterations in the function of follicles stimulated to grow by exogenous FSH. This is supported by previous studies reporting alterations in the follicle dynamics at the onset of the anovulatory season in ewes[3]. Thus, the aim of the present study was to evaluate possible differences in follicular development and functionality between sheep superovulated in the mid- or late breeding season.

Thirty four Manchega ewes were treated, in Madrid (Spain, 40°N), with an intravaginal progestagen for 14 days and superovulated with 8 decreasing oFSH doses (OVAGEN™, ICP, Auckland, NZ) twice daily, starting on Day 12 after sponge insertion. The first group (MS; $n = 14$) was treated in mid breeding season (end of November), while the second group (LS; $n = 20$) was treated in the late breeding season (end of February). The follicle population (≥ 2mm) was assessed by ultrasonography every 12h from first FSH dose (0h) to progestagen withdrawal (60h). Blood samples were collected every 12 from 0-60h for plasma inhibin A concentration.

The mean total number of follicles (≥ 2mm) was similar in both groups throughout the observation period (19.0 ± 0.6 vs 20.1 ± 0.6, n.s.). However, the size-distribution of these follicles during treatments varied between groups. The mean size of the two largest follicles (LF1 and LF2) and the mean size of remaining follicles (RF), were always higher in the MS group (LF1: 6.5 ± 0.1 vs 5.8 ± 0.1; LF2: 5.2 ± 0.1 vs 4.6 ± 0.1 and RF: 3.3 ± 0.1 vs 2.9 ± 0.1, $P < 0.0001$). The mean numbers of follicles categorized by size also differed between groups; large follicles (≥ 6mm) were greater in the MS group (3.8 ± 0.4 vs 1.7 ± 0.2), being the inverse of small follicles (2-3mm; 7.5 ± 0.4 vs 10.5 ± 0.5). However, the rates of follicular recruitment, growth and atresia did not differ between groups.

Inhibin A values also varied between groups. Concentrations in the LS group were higher from 24h of the FSH treatment ($P < 0.01$), but showed a higher individual variability throughout the treatment than in the MS group ($P < 0.0005$). At 60h, mean inhibin A values reached 2508.8 ± 409.2 pg/ml in MS and 3117.2 ± 359.1 pg/ml in the LS group.

In conclusion, superovulatory protocols carried out in temperate areas in the late breeding season showed a lower number and a smaller size of the follicles stimulated to grow with FSH. These findings suggest an altered follicular function close to the non-breeding season in some females which is supported by higher individual variability in inhibin A secretion.

[1]Mitchell et al. 2002. Anim Reprod Sci 74:163–174.
[2]Gonzalez-Bulnes et al. 2003. Theriogenology 60:281-288.
[3]Bartlewski et al. 1999. Anim Reprod Sci 57:51-66.

120. The BMPRIB Booroola mutation does not influence the effects of BMPs on suppressing FSH expression or secretion in the sheep pituitary *in vitro*

M-O Faure[a], M Laird[b], J McNeilly[b], L Nicol[b], J Fontaine[a], S Fabre[a], <u>AS McNeilly</u>[b] & C Taragnat[a]

[a]UMR 6175 INRA-CNRS-Université de TOURS -Haras Nationaux, Physiologie de la Reproduction et des Comportements, Nouzilly, France and [b]MRC Human Reproductive Sciences Unit, Queens Medical Research Institute, Edinburgh, Scotland

We have previously shown the complete repertoire of bone morphogenetic protein (BMP) receptors and ligands is present in the sheep pituitary. Furthermore in primary cultures of sheep pituitary cells BMPs suppress FSH production through effects on FSHß mRNA expression, inhibits the positive effects of activin, and enhances the negative effects of follistatin, and estradiol[1]. While the point mutation in BMPRIB in the Booroola leads to increased ovulation rates, there is disagreement over the role of FSH in this effect. Plasma concentrations of FSH may be elevated in some Booroola cross ewes, but not in others. To address this issue we have examined the effects of BMPs and activin on FSH secretion from primary pituitary cultures from gene (FF) and non-gene (+ +) carrying Merinos d'Arles ewes, and examined the cellular localisation of the BMPRIB in the pituitary. There was no difference between FF and + + genotype on the effects of BMP4 and activin in suppressing and stimulating FSH secretion respectively, and the effects of BMPs in negating the effects of activin, and amplifying the suppressive effects of follistatin in pituitary cultures. Immunolocalisation of the BMPRIB confirmed its absence in gonadotropes. Thus the absence of expression in gonadotropes would explain the lack of effect of the Booroola mutation in altering FSH secretion.

[1]Faure *et al.* 2005. Journal of Endocrinology 186,109-

121. Salsolinol stimulates prolactin release *in vivo* in ruminants

T Hashizume[a], R Shida[a], S Suzuki[a], S Nonaka[a], C Yonezawa[a], T Yamashita[a], E Kasuya[b], M Sutoh[c], M Oláh[d], D Szekacs[d] & GM Nagy[d]

[a]Faculty of Agriculture, Iwate University, Morioka, Japan [b]Laboratory of Animal Neurophysiology, National Institute of Agrobiological Science, Tsukuba, Japan [c]Laboratory of Growth Regulation, National Institute of Livestock and Grassland Science, Japan and [d]Neuroendocrine Research Laboratory, Department of Human Morphology, Hungarian Academy of Science and Semmelweis University, Budapest, Hungary

Prolactin (PRL) secretion is under a dominant and tonic inhibitory control of dopamine. However, recently, it has been reported that administration of salsonilol (SAL), a derivate of dopamine, specifically increases plasma levels of PRL in rats[1, 2]. This finding suggests that the secretion of PRL may be associated with the ratio of dopamine to SAL synthesis and release from the hypothalamic neuroendocrine dopaminergic system. Furthermore, in our preliminary study, SAL was found to be present in the supernatant of the bovine posterior pituitary (PP) extract using HPLC-EC. In our present study, the PRL-releasing activity of SAL in ruminants was examined following intravenous (i.v.) and intracerebroventricular (i.c.v.) injection of SAL in goats and in cattle, respectively.

Four Japanese native female goats (age, 12 months; mean live weight 15 kg) were used. The animals were injected with a single i.v. injection of SAL (5 or 10 mg/kg) dissolved in 5 ml of saline or 5 ml of the vehicle alone (as a control) via an indwelling catheter previously inserted into the external jugular veins. Six castrated Holstein calves (age, 5 months; mean live weight 172 kg) were used for the i.c.v. injection of SAL via a guide cannula stereotactically implanted into the third ventricle of each calf at least 3 weeks before the experiment[3]. On the day of the experiment, a microinjection cannula was inserted into the third ventricle through the implanted guide cannula, and SAL (1 or 5 mg/calf) dissolved in 200 μL of saline or 200 μL of saline vehicle alone was injected. In each experiment, blood samples were collected via the indwelling jugular catheter at 10-min (for 60 min) and 20-min (for 60 min before and for 60-180 min after injection) intervals. Plasma PRL concentrations were measured by a radioimmunoassay. Experimental and animal care protocols were approved by the Animal Care and Use Committee of Iwate University and the National Institute of Agrobiological Sciences.

A single i.v. injection of SAL (5 as well as 10 mg/kg) significantly, and dose-dependently stimulated PRL release in goats ($P < 0.05$). Plasma levels of PRL reached the peak value at 10 min after injection, then gradually declined and returned back to the basal values 60 to 80 min after injection. The i.c.v. injection of 1 mg of SAL had no significant effect on the release of PRL in calves; however, i.c.v. injection of 5 mg of SAL significantly stimulated the release of PRL ($P < 0.05$). Plasma PRL concentrations were significantly increased at 30 and 40 min after the injection, compared with the control values ($P < 0.05$).

This study is the first to evaluate the effect of SAL on the secretion of PRL in ruminants. Our results clearly show that SAL could not only be detected in the supernatant of the PP extract but it is a potent stimulator of PRL release. Therefore, SAL may be involved in the regulatory processes of the secretion of PRL in ruminants.

[1]Tòth *et al.* 2001. J Neuroendocrinology 13:1042-1050.
[2]Tòth *et al.* 2002. Neurotoxico Teratol 24:655-666.
[3]Kasuya *et al.* 2005. J Neurosci Methods 30:115-124.

122. The effects of intra-hypothalamic injections of leptin on the release of luteinizing hormone and growth hormone in male calves

S Nonaka[a], T Hashizume[a] & E Kasuya[b]

[a]Faculty of Agriculture, Iwate University, Morioka, Japan and [b]Laboratory of Animal Neurophysiology, National Institute of Agrobiological Science, Tsukuba, Japan

Leptin has been suggested to be important not only in the regulation of food intake, but also in functions as a neuroendocrine hormone. Leptin stimulates the secretion of several pituitary hormones *in vitro* and *in vivo* not only in rats but also in domestic animals such as pigs, sheep and cattle. However, contradictory results have been obtained regarding the secretion of luteinizing hormone (LH) and growth hormone (GH) in response to leptin *in vivo* in ruminants. The discrepancies in ruminants might be related to age, dose of leptin and length of treatment, method for administration, metabolic state and steroid milieu. The purpose of the present study was to clarify the hypothalamic action of leptin on the secretion of LH and GH in cattle. In the present study, leptin was infused into the medial basal hypothalamus in fully-fed male calves, and plasma LH and GH levels were examined.

Four male Holstein calves (age, 8 to 9 months; mean body weight, 196 kg) were used. The calves were fed hay and concentrate at 13:00 daily. Water was available continuously. On the experimental days, the calves were fed after the collection of blood samples. A unilateral guide cannula was stereotactically implanted into the medial basal hypothalamus in each calf at least 3 weeks before the experiment[1]. Catheters for the blood collections were inserted into a jugular vein at least 1 day before the experiment. On the day of the experiment, a microinfusion cannula was inserted into the hypothalamus through the implanted guide cannula, and 10 µg of human leptin dissolved in 200 µl of saline, or 200 µl of saline vehicle alone (as a control) was infused over a period of 30 sec. The blood samples were collected every 10 min from 50 min before to 5 h after infusion. At the termination of experiments, the infused site in the brain in each calf was histologically confirmed. Plasma LH and GH concentrations were measured by a radioimmunoassay. Experimental and animal care protocols were approved by the Animal Care and Use Committee of Iwate University and the National Institute of Agrobiological Science.

Plasma LH levels in control and leptin-injected calves varied similarly and leptin had no appreciable effect on the release of LH in four calves. Leptin had no appreciable effect on the release of GH in one of four calves; however, the administration of leptin stimulated the release of GH in three calves. Plasma GH concentrations began to rise after 220 min, and peaked at 230 to 250 min after the injection in two calves. Plasma GH concentrations in another one calf began to rise after 50 min, and peaked at 140 and 230 min each after the injection. From the histologic analysis, leptin was infused into or adjacent to the arcuate nucleus in four calves.

We have reported that a single intracerebroventricular (the third ventricle) injection of leptin stimulated the release of GH but not LH in calves[2]. The present results supported our previous report, and suggest that leptin may act partly on the medial basal hypothalamus to stimulate the release of GH in fully-fed male calves.

S. N. is supported by a grant from the Research Fellowship of the JSPS for Young Scientists.

[1]Kasuya *et al.* 2005. J Neurosci Methods 30:115-124.
[2]Nonaka *et al.* 2006. Anim Sci J 77:196-200.

123. GABA inhibition on LH secretion in prepubertal Nelore heifers

GP Nogueira[a] & DJC Oliveira[b]

[a]UNESP Veterinary Curse Araçatuba, Brazil and [b]APTA- Votuporanga, Brazil

Neurotransmitters play an important role in regulating gonadotropin secretion in prepubertal animals of several species. In order to test the hypothesis that GABA acts as an inhibitory neurotransmitter restricting the release of GnRH in prepubertal heifers, we examined the effect of picrotoxin (GABA antagonist) in the pulsatile release of LH during sexual development in prepubertal Nellore (*Bos indicus*) heifers.

After weaning (6 months) 5 Nellore heifers were ovariectomized (OVX) and used as their own control (24 hs before and immediately after picrotoxin injection 0.18 mg/kg, i.v.). Blood samples were collected every 15 min for 10 h at 8, 10, 14 and 17 months of age, plasma LH concentration was quantified by RIA. Data were analyzed by repeated measures ANOVA, using Duncan as a post test, $P < 0.05$ was considered significant.

Although average plasma LH concentration increased with age, there were no differences comparing the periods before (1.2 ± 0.1; 1.4 ± 0.05; 1.9 ± 0.2; 1.5 ± 0.07 ng/ml) and after picrotoxin injection (0.96 ± 0.3; 1.2 ± 0.3; 1.6 ± 0.2; 1.7 ± 0.2 ng/ml, respectively) for each month. Picrotoxin administration increased the number of LH peaks (0.4 ± 0.24 vs 2.0 ± 0.6) and total area of LH peak secretion (7 ± 5 vs 102 ± 25 ng/ml x min) in OVX Nellore heifers, when compared to the control period, but only at 10 months of age.

GABA inhibition on LH secretion seemed to be active in Nellore heifers at 10 months of age as reflected on LH pulsatility, but the GABA antagonist did not increase mean LH secretion. It is possible that GABA inhibition may be more important in an earlier age than close to puberty.

124. Correlation between secretion of leptin and pulses of luteinizing hormone in early postpartum Holstein cows

H Kadokawa[a], D Blache[b] & GB Martin[b]

[a]Animal Reproduction, Faculty of Agricultural Science, University of Yamaguchi, Yoshida 1677-1, Yamaguchi, 753-8515, Japan and [b]School of Animal Biology, Faculty of Natural & Agricultural Sciences, The University of Western Australia, Crawley 6009, Australia

Leptin concentrations decrease in early lactation, and the timing and duration of this decrease is correlated to the length of ovarian quiescence in the early post-partum period[1]. Ovulatory activity is largely controlled by the frequency of LH pulses so we tested whether this measure of hypothalamic-pituitary activity is related to energy status or to the circulating profiles of FFA, insulin, IGF-I, leptin, or GH[2] in dairy cattle during early lactation. On Day 14 postpartum, before the first ovulation and during the period of negative energy balance, blood plasma was sampled from 18 multiparous cows at 10-min intervals for 8 h. All samples were assayed for LH and GH and hourly samples were assayed for FFA, insulin, IGF-I and leptin. We measured milk yield and composition, body condition score and then calculated energy balance. The frequency of LH pulses was correlated positively with energy balance (r = 0.51) and plasma leptin concentrations (r = 0.73), and negatively with milk fat content (r = -0.52). The amplitude of the LH pulses was correlated only with leptin values (r = 0.53). The frequency of GH pulses was not correlated with any measure of LH secretion. The frequency of GH pulses was inversely related to the plasma concentrations of insulin (r = -0.62) and IGF-I (r = -0.61), suggesting the inability of GH to stimulate hepatic IGF-I production during periods of negative energy balance, labelled 'GH-resistance'[3]. These observations further support the existence of a link between pulsatile LH secretion and blood leptin concentrations during the early postpartum period in dairy cows under negative energy balance and may explain the delay in ovulation.

This study was supported in by a grant-in-aid (Biological Resource Management For Sustainable Agricultural Systems) from the OEC D.

[1]Kadokawa et al. 2000. Reprod Fertil Dev 12:405-411.
[2]Kadokawa et al. 2003. Domest Anim Endocrinol 24:219-229.
[3]Donaghy & Baxter, 1996. Baillieres Clin Endocrinol Metab 10:421-426.

125. The effect of visual and audio-visual stimuli from the ram on LH secretion in the anoestrous ewe

PAR Hawken, D Blache & GB Martin

School of Animal Biology, University of Western Australia, 35 Stirling Highway, Crawley 6009, Australia

Exposure to rams stimulates LH secretion in anoestrous ewes, a phenomenon termed the 'ram effect'. This communication between the male and female was thought to be predominantly mediated through olfactory cues similar to that observed in analogous effects in rodents[1]. However the importance of non-olfactory socio-sexual signals was highlighted by the observation that anosmic ewes can also respond to the ram stimulus[2]. Therefore, we hypothesised that exposure of anoestrous ewes to visual and audio-visual stimuli would stimulate an increase in LH secretion that was comparable to that observed in ewes given full contact with rams.

During seasonal anoestrus (November, Southern hemisphere), mature Merino ewes (6 years old) were allocated to one of four groups balanced to ensure the same average weight (54-55kg): "Ram" ewes were given full ram contact; "Picture" ewes (n = 8) were exposed to still images of rams; "Audio-visual" ewes (n = 8) were exposed to a video of ewes and rams mating; and "Control" ewes were completely isolated from rams. Blood was sampled from all ewes twice weekly for two weeks before the day of exposure to the stimulus (Day 0) to confirm their anovulatory status. On Day 0, ewes were exposed to their treatment midway through a frequent blood sampling regime that described their LH secretion before and after exposure to the stimulus. Data were logarithmically transformed and analysed using ANOVA for repeated measures.

All ewes had comparable LH concentrations (P > 0.1) during the period leading up to stimulus exposure (Control, 0.26 $\pm$ 0.07 ng/mL; Picture, 0.33 $\pm$ 0.07; Audio-Visual, 0.29 $\pm$ 0.08; Ram, 0.35 $\pm$ 0.05). Ewes exposed to the still images of the rams (Picture) had a significant increase in LH concentrations after exposure, from 0.33 $\pm$ 0.07 to 0.77 $\pm$ 0.29 ng/mL (P < 0.05). However, this increase was significantly less (P < 0.002) than that observed in ewes given full ram contact (Ram: from 0.35 $\pm$ 0.05 to 1.32 $\pm$ 0.10 ng/mL; P < 0.001). There was no change (P > 0.1) in LH concentrations in Audio-visual ewes (0.29 $\pm$ 0.08 versus 0.32 $\pm$ 0.09 n/mL) or in Control ewes isolated from ram contact (0.26 $\pm$ 0.07 versus 0.20 $\pm$ 0.12 ng/mL).

The observed increase in LH concentrations in the ewes exposed to still images of rams indicates an important role for visual cues in eliciting the gonadotrophic response of ewes to the introduction of rams. This concurs with the ability of ewes to distinguish between the "faces" of ewes and rams from a visual projected image[3]. The lack of a response to the video of ewes and rams mating may be due to an inability of ewes to process moving images and thus identify the rams in the video.

This work was supported by the Australian Research Council.

[1]Gelez & Fabre-Nys, 2004. Hormones and Behaviour 46:257-271.
[2]Cohen-Tannoudji *et al.* 1986. Physiology and Behaviour 36:921-924.
[3]Kendrick *et al.* 1995. Animal Behaviour 49:1665-1676.

126. A mathematical model of the GnRH-LH-testosterone system in rams – role of time-delays in the feedback control of GnRH pulses

TR Ferasyi[a], PHR Barrett[b], D Blache[a] & GB Martin[a]

[a]Faculty of Natural & Agricultural Sciences, and [b]Faculty of Medicine, Dentistry & Health Sciences, The University of Western Australia, Crawley 6009, Australia

In males, testosterone exerts negative feedback, mainly at the level of the preoptic-hypothalamic continuum, to control the frequency of GnRH and LH pulses. However, in addition to steroid feedback, GnRH pulse frequency is modulated by internal factors (eg, metabolic inputs) and external factors (eg, photoperiod, socio-sexual cues), all of which interact with genotype and can override the inhibitory action of testosterone[1]. Most experimental studies on this system use "steady-state" models such as the steroid-implanted castrated ram, but a steady state is unusual in homeostatic systems and does not account for delays in the processes that constitute the feedback loop. These delays might explain dynamic changes in GnRH pulse frequency[2] and need to be assessed for each critical control point in the GnRH-LH-testosterone negative feedback loop. This is difficult with animal models because the endocrine control system responds to complex interactions among internal and external factors. We have therefore taken the alternative approach of using a mathematical model.

We developed a three compartment model (GnRH, LH and testosterone) using SAAM II software. The values of the parameters were obtained by fitting the model to experimental data. The basic assumptions of the model are: the minimum and maximum pulse intervals for GnRH are 45 min and 300 min; pulsatile GnRH secretion controls pulsatile LH secretion in a simple positive feedforward and that, in turn, controls testosterone secretion; in contrast, testosterone concentration, rather than pulse parameters, inhibits GnRH secretion (principally pulse frequency) through a dose-response relationship. Importantly, at this point we have introduced a time-delay in the action of testosterone (32 min).

Simulations of the responses to changes in both testosterone concentration and the time-delay successfully emulated experimental observations with mature male sheep. An increase in testosterone concentration increased the interval between successive GnRH pulses and, when testosterone concentration declined, the effect was reversed. In addition, these two basic responses depended on time delay following a change in testosterone concentration – a longer delay allowed the frequency of GnRH pulses to increase briefly, as in the dynamic phenomenon of "mini-castration" observed in non-steady-state animal experiments[3]. A steady state in GnRH-LH-testosterone homeostasis could not be reached unless we reduced the parameter for testosterone delay by 32% of the fitted parameter value or set the minimum pulse interval between GnRH pulses at about 90 min, neither of which reflects real data. The delay to the increase in GnRH pulse frequency was 180 min when the model was switched to "castration state" (zero testosterone) and 130 min when we simulated the introduction of exogenous testosterone into the "castration state". Thus, it is clear that both testosterone concentration and a time delay in negative feedback play major roles in controlling GnRH pulse frequency. Future studies will expand the model to include environmental inputs, such as nutrition, photoperiod, and socio-sexual cues.

TR Ferasyi was supported by a UWA International Postgraduate Research Scholarship.

[1]Blache et al. 2003. Reprod Suppl 61:387-402.
[2]Blache et al. 1997. J Neuroendocrinol 9:887-892.
[3]Martin et al. 1987. Anim Reprod Sci 12:267-281.

Subject index